Progress in Mathematics
Volume 246

Sorin Dragomir
Giuseppe Tomassini

Differential Geometry and Analysis on CR Manifolds

Birkhäuser
Boston • Basel • Berlin

Sorin Dragomir
Università degli Studi della Basilicata
Dipartimento di Matematica
Contrada Macchia Romana
85100 Potenza
Italy

Giuseppe Tomassini
Scuola Normale Superiore
Classe di Scienze
Piazza dei Cavalieri 7
65126 Pisa
Italy

Mathematics Subject Classification (2000): 32T15, 32V05, 32V15, 32V20, 35B65, 35D05, 35D10, 53C07, 53C10, 53C12, 53C15, 53C17, 53C21, 53C25, 53C26, 53C40, 53C43, 53C50, 53D10

Library of Congress Control Number: 2006921475

ISBN-10 0-8176-4388-5 e-ISBN: 0-8176-4483-0
ISBN-13 978-0-8176-4388-1

Printed on acid-free paper.

Printed in the United States of America. (TXQ/EB)

9 8 7 6 5 4 3 2 1

www.birkhauser.com

Contents

Preface ix

1 CR Manifolds 1

1.1 CR manifolds 3

1.1.1 CR structures 3

1.1.2 The Levi form 5

1.1.3 Characteristic directions on nondegenerate CR manifolds 8

1.1.4 CR geometry and contact Riemannian geometry 10

1.1.5 The Heisenberg group 11

1.1.6 Embeddable CR manifolds 14

1.1.7 CR Lie algebras and CR Lie groups 18

1.1.8 Twistor CR manifolds 21

1.2 The Tanaka–Webster connection 25

1.3 Local computations 31

1.3.1 Christoffel symbols 32

1.3.2 The pseudo-Hermitian torsion 36

1.3.3 The volume form 43

1.4 The curvature theory 46

1.4.1 Pseudo-Hermitian Ricci and scalar curvature 50

1.4.2 The curvature forms Ω_α^β 51

1.4.3 Pseudo-Hermitian sectional curvature 58

1.5 The Chern tensor field 60

1.6 CR structures as G-structures 68

1.6.1 Integrability 70

1.6.2 Nondegeneracy 72

1.7 The tangential Cauchy–Riemann complex 73

1.7.1 The tangential Cauchy–Riemann complex 73

1.7.2 CR-holomorphic bundles 82

1.7.3 The Frölicher spectral sequence 84

1.7.4 A long exact sequence 94

1.7.5 Bott obstructions 96

1.7.6 The Kohn–Rossi Laplacian 99
1.8 The group of CR automorphisms 106

2 **The Fefferman Metric** 109
2.1 The sub-Laplacian 111
2.2 The canonical bundle 119
2.3 The Fefferman metric 122
2.4 A CR invariant 135
2.5 The wave operator 140
2.6 Curvature of Fefferman's metric 141
2.7 Pontryagin forms 142
2.8 The extrinsic approach 147
2.8.1 The Monge–Ampère equation 147
2.8.2 The Fefferman metric 150
2.8.3 Obstructions to global embeddability 151

3 **The CR Yamabe Problem** 157
3.1 The Cayley transform 161
3.2 Normal coordinates 165
3.3 A Sobolev-type lemma 176
3.4 Embedding results 193
3.5 Regularity results 196
3.6 Existence of extremals 200
3.7 Uniqueness and open problems 205
3.8 The weak maximum principle for Δ_b 207

4 **Pseudoharmonic Maps** 211
4.1 CR and pseudoharmonic maps 212
4.2 A geometric interpretation 215
4.3 The variational approach 218
4.4 Hörmander systems and harmonicity 231
4.4.1 Hörmander systems 233
4.4.2 Subelliptic harmonic morphisms 236
4.4.3 The relationship to hyperbolic PDEs 239
4.4.4 Weak harmonic maps from $C(\mathbf{H}_n)$ 242
4.5 Generalizations of pseudoharmonicity 246
4.5.1 The first variation formula 248
4.5.2 Pseudoharmonic morphisms 252
4.5.3 The geometric interpretation of F-pseudoharmonicity 255
4.5.4 Weak subelliptic F-harmonic maps 256

5 **Pseudo-Einsteinian Manifolds** 275
5.1 The local problem 275
5.2 The divergence formula 279
5.3 CR-pluriharmonic functions 280

5.4 More local theory 294
5.5 Topological obstructions 296
5.5.1 The first Chern class of $T_{1,0}(M)$ 296
5.5.2 The traceless Ricci tensor 299
5.5.3 The Lee class 301
5.6 The global problem 304
5.7 The Lee conjecture 312
5.7.1 Quotients of the Heisenberg group by properly discontinuous groups of CR automorphisms 313
5.7.2 Regular strictly pseudoconvex CR manifolds 318
5.7.3 The Bockstein sequence 320
5.7.4 The tangent sphere bundle 321
5.8 Pseudo-Hermitian holonomy 329
5.9 Quaternionic Sasakian manifolds 331
5.10 Homogeneous pseudo-Einsteinian manifolds 341

6 Pseudo-Hermitian Immersions 345
6.1 The theorem of H. Jacobowitz 347
6.2 The second fundamental form 351
6.3 CR immersions into $\mathbf{H}_{n+k}$ 356
6.4 Pseudo-Einsteinian structures 361
6.4.1 CR-pluriharmonic functions and the Lee class 361
6.4.2 Consequences of the embedding equations 364
6.4.3 The first Chern class of the normal bundle 369

7 Quasiconformal Mappings 377
7.1 The complex dilatation 377
7.2 K-quasiconformal maps 384
7.3 The tangential Beltrami equations 385
7.3.1 Contact transformations of $\mathbf{H}_n$ 388
7.3.2 The tangential Beltrami equation on $\mathbf{H}_1$ 394
7.4 Symplectomorphisms 398
7.4.1 Fefferman's formula and boundary behavior of symplectomorphisms 398
7.4.2 Dilatation of symplectomorphisms and the Beltrami equations 401
7.4.3 Boundary values of solutions to the Beltrami system 403
7.4.4 A theorem of P. Libermann 403
7.4.5 Extensions of contact deformations 405

8 Yang–Mills Fields on CR Manifolds 407
8.1 Canonical S-connections 407
8.2 Inhomogeneous Yang–Mills equations 410
8.3 Applications 412
8.3.1 Trivial line bundles 414
8.3.2 Locally trivial line bundles 414

8.3.3 Canonical bundles . . . 416
8.4 Various differential operators . . . 418
8.5 Curvature of S-connections . . . 421

9 Spectral Geometry . . . 423
9.1 Commutation formulas . . . 423
9.2 A lower bound for λ_1 . . . 427
9.2.1 A Bochner-type formula . . . 430
9.2.2 Two integral identities . . . 431
9.2.3 A. Greenleaf's theorem . . . 434
9.2.4 A lower bound on the first eigenvalue of a Folland–Stein operator . . . 439
9.2.5 Z. Jiaqing and Y. Hongcang's theorem on CR manifolds . . . 441

A Parametrix for $\Box_b$ . . . 445

References . . . 463

Index . . . 483

Preface

A *CR manifold* is a C^∞ differentiable manifold endowed with a complex subbundle $T_{1,0}(M)$ of the complexified tangent bundle $T(M)\otimes\mathbf{C}$ satisfying $T_{1,0}(M)\cap\overline{T_{1,0}(M)} = (0)$ and the Frobenius (formal) integrability property

$$\left[\Gamma^\infty(T_{1,0}(M)),\ \Gamma^\infty(T_{1,0}(M))\right] \subseteq \Gamma^\infty(T_{1,0}(M)).$$

The bundle $T_{1,0}(M)$ is the *CR structure* of M, and C^∞ maps $f : M \to N$ of CR manifolds preserving the CR structures (i.e., $f_* T_{1,0}(M) \subseteq T_{1,0}(N)$) are *CR maps*. CR manifolds and CR maps form a category containing that of complex manifolds and holomorphic maps. The most interesting examples of CR manifolds appear, however, as real submanifolds of some complex manifold. For instance, any real hypersurface M in $\mathbf{C}^n$ admits a CR structure, naturally induced by the complex structure of the ambient space

$$T_{1,0}(M) = T^{1,0}(\mathbf{C}^n) \cap [T(M) \otimes \mathbf{C}].$$

Let $(z^1, \ldots, z^n)$ be the natural complex coordinates on $\mathbf{C}^n$. Locally, in a neighborhood of each point of M, one may produce a frame $\{L_\alpha : 1 < \alpha \leq n-1\}$ of $T_{1,0}(M)$. Geometrically speaking, each L_α is a (complex) vector field tangent to M. From the point of view of the theory of PDEs, the L_α's are purely tangential first-order differential operators

$$L_\alpha = \sum_{j=1}^{n} a_\alpha^j(z) \frac{\partial}{\partial z^j}, \quad 1 \leq \alpha \leq n-1,$$

and $T_{1,0}(M)$ may be thought of as a bundle-theoretic recasting of the first-order PDE system with complex-valued C^∞ coefficients

$$\overline{L}_\alpha u(z) = 0, \quad 1 \leq \alpha \leq n-1,$$

called the *tangential Cauchy–Riemann equations*. These may be equally thought of as being induced on M by the Cauchy–Riemann equations in $\mathbf{C}^n$. *CR functions* are solutions $u(z)$ to the tangential Cauchy–Riemann equations, and any holomorphic function defined on a neighborhood of M will restrict to a CR function on M.

These introductory remarks lead to two fundamental problems in CR geometry and analysis. Given an (abstract) CR manifold, is it possible to realize it as a CR submanifold of $\mathbf{C}^n$ (of some complex manifold)? This is known as the *embeddability problem*, introduced to mathematical practice by J.J. Kohn [246]. The second problem is whether a given CR function $u : M \to \mathbf{C}$ extends to a holomorphic function defined on some neighborhood of M (the *CR extension problem*). Both these problems have local and global aspects, present many intricacies, and involve scientific knowledge from many mathematical fields. The solution to the local embeddability problem is due to A. Andreotti and C.D. Hill [13] in the real analytic category. Partial solutions in the C^∞ category are due to L. Boutet de Monvel [77], M. Kuranishi [263], and T. Akahori [2]. As to the CR extension problem, it is the object of intense investigation, cf. the monographs by A. Boggess [70] and M.S. Baouendi, P. Ebenfelt, and L.P. Rothschild [31] for an account of the present scientific achievements in this direction.

It should become clear from this discussion that CR manifolds and their study lie at the intersection of three main mathematical disciplines: the theory of partial differential equations, complex analysis in several variables, and differential geometry. While the analysis and PDE aspects seem to have captured most of the interest within the mathematical community, there has been, over the last ten or fifteen years, some effort to understand the differential-geometric side of the subject as well. It is true that A. Bejancu's discovery [55] of CR submanifolds signaled the start of a large number of investigations in differential geometry, best illustrated by the monographs by K. Yano and M. Kon [446], A. Bejancu [56], and S. Dragomir and L. Ornea [125]. Here by a *CR submanifold* we understand a real submanifold M of a Hermitian manifold (X, J, g), carrying a distribution $H(M)$ that is J-invariant (i.e., $JH(M) = H(M)$) and whose g-orthogonal complement is J-anti-invariant (i.e., $JH(M)^\perp \subseteq T(M)^\perp$, where $T(M)^\perp \to M$ is the normal bundle of M in X). The notion (of a CR submanifold of a Hermitian manifold) unifies concepts such as invariant, anti-invariant, totally real, semi-invariant, and generic submanifolds. Also, the observation (due to D.E. Blair and B.Y. Chen [64]) that proper CR submanifolds, in the sense of A. Bejancu, are actually CR manifolds shows that these investigations have the same central object, the CR category, as defined at the beginning of this preface, or by S. Greenfield [187]. The study of CR submanifolds in Hermitian manifolds, in the sense of A. Bejancu, has led to the discovery of many refined differential-geometric properties (e.g., K. Yano and M. Kon's classification of CR submanifolds of a complex projective space, with semiflat normal connection, parallel f-structure in the normal bundle, and the covariant derivative of the second fundamental form of constant length [445]) and will surely develop further within its own borders. It should be remarked nevertheless that as confined to Riemannian geometry (i.e., to the theory of submanifolds in Riemannian manifolds, cf., e.g., [91]), the above-mentioned study is perhaps insufficiently related to the (pseudo) convexity properties of submanifolds in complex manifolds, as understood in analysis in several complex variables. To be more precise, if M is a real hypersurface in $\mathbf{C}^n$ then the first and second fundamental forms of the given immersion describe the way M is shaped, both intrinsically (Riemannian curvature) and extrinsically, yet do not describe a priori the intrinsic properties of M as related to its Levi form. As an extreme case, M may be Levi flat yet will always exhibit, say, curvature properties arising from its first fundamental form. Or to give a nondegenerate example, the boundary of the Siegel domain $\Omega_n = \{z = (z', z_n) \in \mathbf{C}^{n-1} \times \mathbf{C} : \mathrm{Im}(z_n) > \|z'\|^2\}$ (i.e., the *Heisenberg group*) admits a contact form θ with a positive definite Levi form G_θ, and hence $g_\theta = \pi_H G_\theta + \theta \otimes \theta$ (the *Webster metric*) is a Riemannian metric, yet none of the metrics $g_{\lambda\theta}$, $\lambda \in C^\infty(M)$, $\lambda > 0$, coincides with the metric induced on M by the flat Kähler metric of $\mathbf{C}^n$.

Central to the present monograph is the discovery, around 1977–78, of a canonical linear connection ∇ on each nondegenerate CR manifold M of hypersurface type (the *Tanaka–Webster connection*) due to independent investigations by N. Tanaka [398] and S. Webster [422]. ∇ parallelizes both the Levi form and the complex structure of the Levi, or maximally complex, distribution of M, resembles both the Levi-Civita connection of a Riemannian manifold, and the Chern connection of a Hermitian manifold, and is a foundational tool for the *pseudo-Hermitian geometry* of a (nondegenerate) CR manifold, which is the main subject of this book. Now the curvature properties of ∇ are indeed tied to the CR structure: for instance, the Chern curvature tensor $C_\beta{}^\alpha{}_{\lambda\overline{\sigma}}$, a CR invariant of M, is computable in terms of the curvature of ∇ (and its contractions, such as the *pseudo-Hermitian Ricci tensor* and the *pseudo-Hermitian scalar curvature*) and $C_\beta{}^\alpha{}_{\lambda\overline{\sigma}} = 0$ if and only if M is locally CR equivalent to the standard sphere in $\mathbf{C}^{n+1}$, $n > 1$ (cf. S.S. Chern and J. Moser [99]). Variants of the Tanaka–Webster connection are known already in different contexts, e.g., on CR manifolds of higher CR codimension (R. Mizner [312]) or on contact Riemannian manifolds (S. Tanno [401]), whose almost CR structure is not integrable, in general.

After a detailed exposition of the basic facts of pseudo-Hermitian geometry of nondegenerate CR manifolds in Chapter 1, the present monograph introduces the main geometric object, the *Fefferman metric*, both a tool and object of investigation of the first magnitude. It is due to C. Fefferman [138], who first devised it as a (Lorentz) metric on $(\partial\Omega) \times S^1$, for a given strictly pseudoconvex domain $\Omega \subset \mathbf{C}^n$, in connection with the boundary behavior of the Bergman kernel of Ω and the solution to the Dirichlet problem for the (inhomogeneous) complex Monge–Ampére equation

$$\begin{cases} (-1)^{n+1} \det \begin{pmatrix} u & \partial u / \partial \overline{z}^k \\ \partial u / \partial z^j & \partial^2 u / \partial z^j \partial \overline{z}^k \end{pmatrix} = 1 & \text{in } \Omega, \\ u = 0 & \text{on } \partial\Omega \end{cases}$$

(the existence, uniqueness, and regularity of the solution are due to S.Y. Cheng and S.T. Yau [97]). See Chapter 2 of this book. By the mathematical creation of F. Farris [137], and J.M. Lee [271], an intrinsic description of the Fefferman metric (as a Lorentz metric on

$$C(M) = \left(\Lambda^{n+1,0}(M) \setminus \{0\}\right) \big/ \mathbf{R}_+,$$

where n is the CR dimension) is available. Also, the work of G. Sparling [377], C.R. Graham [182], L. Koch [242]–[244], helped clarify a number of geometric facts (e.g., how the Fefferman metric may be singled out, in terms of curvature properties, from the set of all Lorentz metrics on $C(M)$ (cf. [182]), or providing a simple proof (cf. [242]) to H. Jacobowitz's theorem (cf. [220]) that nearby points on a strictly pseudoconvex CR manifold may be joined by a chain). Other properties are known, e.g., that certain Pontryagin forms of the Fefferman metric are obstructions to global CR equivalence to a sphere (and perhaps to global embeddability); cf. E. Barletta et al. [38]. The Fefferman metric remains however an insufficiently understood object and worth of further investigation.

One of the most spectacular results in this book is D. Jerison and J.M. Lee's solution (cf. [226]–[228] and our Chapter 3) to the *CR Yamabe problem*, which is the Yamabe problem for the Fefferman metric. As the Yamabe problem in Riemannian geometry (find a conformal transformation $\tilde{g} = fg,\ f > 0$, such that $\tilde{g}$ is of constant scalar curvature) the Yamabe problem for the Fefferman metric may be reformulated as a nonlinear PDE on $C(M)$ whose principal part is the Laplace–Beltrami operator of the metric; here the wave operator as the metric is Lorentzian, and hence nonelliptic. However, this equation may be shown (cf. [227]) to project on

$$b_n \Delta_b u + \rho u = \lambda u^{p-1},$$

the *CR Yamabe equation*, a nonlinear PDE on M, whose principal part is the *sub-Laplacian* Δ_b. The book presents the solution to the CR Yamabe problem only when $\lambda(M) \leq \lambda(S^{2n+1})$ (cf. Theorem 3.4 in Chapter 4), where $\lambda(M)$ is the CR analogue to the Yamabe invariant in Riemannian geometry; i.e.,

$$\lambda(M) = \inf\{\int_M \{b_n \|\pi_H \nabla u\|^2 + \rho u^2\}\theta \wedge (d\theta)^n : \int_M |u|^p \theta \wedge (d\theta)^n = 1\}.$$

The remaining case was dealt with by N. Gamara and R. Yacoub [164], who completed the solution to the CR Yamabe problem (see the comments at the end of Section 4.7). Δ_b is degenerate elliptic and subelliptic of order $1/2$ (and hence hypoelliptic). The authors of this book believe that subelliptic PDEs are bound to play within CR geometry the strong role played by elliptic theory in Riemannian geometry. A similar application is to use the Fefferman metric in the study of *pseudoharmonic maps*; cf. Chapter 4 (these are, locally, J. Jost and C.J. Xu's subelliptic harmonic maps; cf. [234]).

Another main theme of the book is represented by *pseudo-Einsteinian structures* (i.e., contact forms such that the pseudo-Hermitian Ricci tensor of their Tanaka–Webster connection is proportional to the Levi form) and the problem of local and global existence of pseudo-Einsteinian structures on CR manifolds. We present the achievements in the field, together with the *Lee conjecture* [that each compact strictly pseudoconvex CR manifold whose CR structure has a vanishing first Chern class ($c_1(T_{1,0}(M)) = 0$) must possess some global pseudo-Einsteinian structure]. The global problem turns out to be related to the theory of *CR immersions*, certain aspects of which are discussed in Chapter 6. The source mainly used for discussing pseudo-Einsteinian structures is, of course, the original paper [270]. However, our works [121] (solving the Lee conjecture on a compact strictly pseudoconvex CR manifold admitting a contact form whose corresponding characteristic direction is *regular* in the sense of R. Palais) [37] (demonstrating pseudo-Einsteinian contact forms on (total spaces of) tangent sphere bundles over real space forms $M^n(1)$) [68] (taking into account the relationship between the pseudo-Einsteinian condition and pseudo-Hermitian holonomy, i.e., the holonomy of the Tanaka–Webster connection), and the work by M.B. Stenzel [386] (producing pseudo-Einsteinian structures on boundaries of tubes $T^{*\epsilon}X$ over harmonic Riemannian manifolds (X, g)), extend the knowledge about pseudo-Einsteinian structures somewhat beyond the starting point of J.M. Lee [270]. As to the relationship between the global existence problem of pseudo-Einsteinian structures and the theory of CR, or rather pseudo-Hermitian, immersions (cf. [424] and [120]), let us mention that the Lee class may be interpreted as an obstruction to the existence of a pseudo-Hermitian immersion $f : M \to S^{2N+1}$, of a strictly pseudoconvex CR manifold M into an odd-dimensional sphere, such that f has flat normal Tanaka–Webster connection $\nabla^{\perp}$ (cf. [36] and the corollary to Theorem 6.1 in this book). The *Lee class* is a cohomology class $\gamma(M) \in H^1(M, \mathcal{P})$ with coefficients in the sheaf $\mathcal{P}$ of CR-pluriharmonic functions on M, as devised by J.M. Lee [270], such that $\gamma(M) = 0$ if and only if M admits a globally defined pseudo-Einsteinian contact form.

We deal with quasiconformal mappings of CR manifolds (a subject developed mainly by A. Korányi and H. Reimann [254]–[255]) in Chapter 7, with H. Urakawa's Yang–Mills connections (cf. [412]) on CR manifolds in Chapter 8, and with spectral geometry of CR manifolds (cf. A. Greenleaf [186]) in Chapter 9. A previous version of this text contained material devoted to the interplay between CR geometry and foliation theory, which in the meanwhile grew into an independent volume. While the presentation in Chapter 7 owes, as mentioned above, to A. Korányi and H. Reimann (cf. op. cit.), the observation that the ordinary Beltrami equations in several complex variables (cf. [419]) induce on $tial\Omega_n$ (the boundary of the Siegel domain Ω_n) the (tangential) Beltrami equations considered by A. Korányi and H. Reimann is new (cf. [41]). It is interesting to note that given a strictly pseudoconvex domain $\Omega \subset \mathbf{C}^n$, any biholomorphism F of Ω lifts to a C^∞ map

$$F^\sharp : \partial\Omega \times S^1 \to \partial\Omega \times S^1 \, , \quad F^\sharp(z, \gamma) := (F(z), \gamma - \arg(\det F'(z))),$$

preserving the "extrinsic" Fefferman metric (2.62) up to a conformal factor

$$\left(F^\sharp\right)^* g = \left|\det F'(z)\right|^{2/(n+1)} g \ ,$$

(cf. [138], p. 402, or by a simple calculation based on (2.62) in Chapter 2 of this book). When F is only a symplectomorphism of (Ω, ω), with $\omega := -i\partial\overline{\partial}\log K(z, z))$, extending smoothly to the boundary, a fundamental result of A. Korányi and H. Reimann, presented in Chapter 7, is that the boundary values f of such F constitute a contact transformation. Thus, in general, F is not a holomorphic map, nor are its boundary values f a CR map, both phenomena manifesting in the presence of a "dilatation" (dil(F) for F, and μ_f for f, themselves related in the limit as $z \to \partial\Omega$, cf. Theorem 7.7 in Chapter 7). Although $f^* G_\theta = \lambda_f G_\theta$ fails to hold (since f is not CR), one may "adjust" the complex structure J on $H(\partial\Omega)$ (cf. section 7.1) and get a new complex structure J_f such that $f^* G_\theta = \lambda_f G_f$, where $G_f(X, Y) := (d\theta)(X, J_f Y)$, $X, Y \in H(M)$. The problem of computing the Fefferman metric of (M, J_f, θ), or more generally of investigating the relationship (if any) between F_θ and the symplectomorphisms of (Ω, ω), remains unsolved.

As to Chapter 8, let us mention that while solving the inhomogeneous Yang–Mills equation

$$d_D^* R^D = 4ni\{d_M^c \rho - \rho\,\theta\} \otimes I \tag{0.1}$$

for a Hermitian connection $D \in \mathcal{C}(E, h)$, pseudo-Einsteinian structures come once again into the picture in a surprising way. The canonical line bundle $K(M) \to M$ over a pseudo-Einsteinian manifold M is a *quantum bundle* (in the sense of [259]): this is just the condition that the canonical S-connection of $K(M)$ has curvature of type (1, 1), and one may use Theorem 8.2 to *explicitly* solve (0.1) (demonstrating among other things the strength of the purely differential-geometric approach to the study of the inhomogeneous Yang–Mills equations on CR manifolds).

This book also aims to explain how certain results in classical analysis apply to CR geometry (part of the needed material is taken from the fundamental paper by G.B. Folland and E.M. Stein [150]). This task, together with the authors' choice to give detailed proofs to a number of geometric facts, is expected to add to the clarity of exposition. It surely added in volume and prevented us from including certain modern, and still growing, subjects. A notable example is the theory of *homogeneous CR manifolds* (cf. H. Azad, A. Huckleberry, and W. Richthofer [26], A. Krüger [262], R. Lehmann and D. Feldmueller [277], and D.V. Alekseevski and A. Spiro [9]–[10]). See however our notes at the end of Chapter 5. Another absent protagonist is the theory of *deformation of CR structures* (cf. T. Akahori [3]–[6], T. Akahori and K. Miyajima [7], R.O. Buchweitz and J.J. Millson [78], J.J. Millson [302], and K. Miyajima [306]–[311]). The same holds for more recent work, such as H. Baum's (cf. [49]) on spinor calculus in the presence of the Fefferman metric, and F. Loose's (cf. [288]) initiating a study of the CR moment map, perhaps related to that of CR orbifolds (cf. [128]).

We may conclude that such objects as the Tanaka–Webster connection, the Fefferman metric, and pseudo-Einsteinian structures constitute the leitmotif of this book. More precisely, this book is an attempt to understand certain aspects of the relationship between Lorentzian geometry (on $(C(M), F_\theta)$) and pseudo-Hermitian geometry (on (M, θ)), a spectacular part of which is the relationship between hyperbolic and subelliptic PDEs (as demonstrated in Sections 2.5 and 4.4.3 of this monograph). The authors found a powerful source of techniques and ideas in the scientific creation of S.M. Webster and J.M. Lee, to whose papers they returned again and again over the years, and to whom they wish to express their gratitude.

Sorin Dragomir
Giuseppe Tomassini
August 2005

Differential Geometry and Analysis on CR Manifolds

1

CR Manifolds

Let Ω be a smooth domain in $\mathbf{C}^{n+1}$, i.e., there is an open neighborhood $U \supset \overline{\Omega}$ and a real-valued function $\rho \in C^2(U)$ such that $\Omega = \{z \in U : \rho(z) > 0\}$, $\mathbf{C}^{n+1} \setminus \overline{\Omega} = \{z \in U : \rho(z) < 0\}$, the boundary of Ω is given by $\partial\Omega = \{z \in U : \rho(z) = 0\}$, and $D\rho(z) \neq 0$ for any $z \in \partial\Omega$. Here $D\rho$ is the gradient

$$D\rho = \left(\frac{\partial \rho}{\partial x^1}, \ldots, \frac{\partial \rho}{\partial x^{2n+2}}\right)$$

and $(x^1, \ldots, x^{2n+2})$ are the Cartesian coordinates on $\mathbf{R}^{2n+2} \simeq \mathbf{C}^{n+1}$.

The Cauchy–Riemann equations in $\mathbf{C}^{n+1}$ induce on $\partial\Omega$ an overdetermined system of PDEs with smooth complex-valued coefficients

$$\overline{L}_\alpha u(z) \equiv \sum_{j=1}^{n+1} a_\alpha^j(z) \frac{\partial u}{\partial \overline{z}^j} = 0, \quad 1 \leq \alpha \leq n \tag{1.1}$$

(the *tangential Cauchy–Riemann equations*), $z \in V$, with $V \subseteq (\partial\Omega) \cap U$ open. Here L_α are linearly independent (at each point of V) and

$$\sum_{j=1}^{n+1} \overline{a}_\alpha^j(z) \frac{\partial \rho}{\partial z^j} = 0, \quad 1 \leq \alpha \leq n, \tag{1.2}$$

for any $z \in V$, i.e., L_α are purely tangential first-order differential operators (tangent vector fields on $\partial\Omega$). It then follows that

$$\left[L_\alpha, L_\beta\right] = C_{\alpha\beta}^\gamma(z) L_\gamma \tag{1.3}$$

for some complex-valued smooth functions $C_{\alpha\beta}^\gamma$ on V.

At each point $z \in V$ the $L_{\alpha,z}$'s span a complex n-dimensional subspace $T_{1,0}(\partial\Omega)_z$ of the complexified tangent space $T_z(\partial\Omega) \otimes_{\mathbf{R}} \mathbf{C}$. The bundle $T_{1,0}(\partial\Omega) \to \partial\Omega$ is the *CR structure* of $\partial\Omega$, and a bundle-theoretic recast of (1.1)–(1.3) consists in observing that

$$T_{1,0}(\partial\Omega) = [T(\partial\Omega) \otimes \mathbf{C}] \cap T^{1,0}(\mathbf{C}^{n+1}), \tag{1.4}$$

where $T^{1,0}(\mathbf{C}^{n+1})$ is the holomorphic tangent bundle over $\mathbf{C}^{n+1}$, and that $M = \partial\Omega$ satisfies the axioms (1.5)–(1.6) below. A C^1 function $u : \partial\Omega \to \mathbf{C}$ is a *CR function* if $\overline{Z}(u) = 0$ for any $Z \in T_{1,0}(\partial\Omega)$. Locally, a CR function is a solution of (1.1).

The pullback (via $j : \partial\Omega \subset U$) of the complex 1-form $\frac{i}{2}(\overline{\partial} - \partial)\rho$ is a *pseudo-Hermitian structure* θ on $\partial\Omega$. When $\partial\Omega$ is *nondegenerate* θ is a contact form. Everything stated above holds should one replace the boundary $\partial\Omega$ by some (open piece of a) smooth real hypersurface in $\mathbf{C}^{n+1}$.

As observed by N. Tanaka [398], and S. Webster [422], when θ is a contact form M may be described in terms of *pseudo-Hermitian geometry* (a term coined as a *seguito* of the—fundamental to this book—paper [422]), which complements the (better-known) *contact Riemannian geometry* (cf. [62]) and is well suited for capturing the convexity properties of M (as familiar in the analysis in several complex variables). M carries a semi-Riemannian metric g_θ (Riemannian, if M is strictly pseudoconvex) coinciding with the Levi form along the maximal complex distribution of M. This is the *Webster metric* (cf. Section 1.1.3). Of course, M carries also the Riemannian metric induced from the (flat Kähler) metric of $\mathbf{C}^{n+1}$, and the pseudo-Hermitian and contact Riemannian geometries do interact. However, that the two are quite different in character should be emphasized: for instance, *none* of the Webster metrics $g_{\lambda\theta}$, for every smooth $\lambda : M \to (0, +\infty)$, of the boundary of the Siegel domain $\Omega_{n+1} = \{(z, w) \in \mathbf{C}^n \times \mathbf{C} : \mathrm{Im}(w) > \|z\|^2\}$ coincides with the metric induced from $\mathbf{C}^{n+1}$.

CR manifolds as in (1.3), or in (1.12) below, are *embedded*. The currently accepted concept of a CR manifold as a tool for studying the tangential Cauchy–Riemann equations by geometric methods is, however, more general. The manifold may be abstract, and not all CR manifolds embed, even locally (cf. Section 1.6). The CR codimension (cf. Section 1.1) may be > 1, and distinct from the codimension (when M is a CR submanifold). The Levi form may be vector, rather than scalar, valued (and then there is no natural notion of strict pseudoconvexity) or degenerate (and then the tools of pseudo-Hermitian geometry are not available).

According to our purposes in this book, that is, to describe (1.1) by means of pseudo-Hermitian geometry, we shall assume integrability, nondegeneracy, and CR codimension 1. The reader should nevertheless be aware of the existence of a large literature, with similar expectations, and not subject to our hypothesis.[1]

---

[1] For instance, H. Rossi and M. Vergne [356], deal with CR manifolds $\Sigma = \Sigma(V, N, E)$ of the form $\Sigma = \{(x + iy, u) : x, y \in \mathbf{R}^n,\ u \in E,\ y - N(u) \in V\}$, where E is a domain in $\mathbf{C}^m$, $N : E \to \mathbf{R}^n$ is a smooth function, and V is a submanifold in $\mathbf{R}^n$. These are in general degenerate (the Levi form has a nontrivial null space); yet this is not relevant to the purpose of analysis on Σ. Indeed, it may be shown (by a partial Fourier transform technique) that the CR functions on Σ satisfy a Paley–Wiener (type) theorem (cf. [356], p. 306). An application of their result (to constant-coefficient PDEs on the Heisenberg group) is given by D.E. Blair et al. [67]. To give one more example, we may quote the long series of papers by C.D. Hill and M. Nacinovich [202], in which the CR codimension is always > 1, and eventually only a small amount of pseudoconcavity is prescribed (cf. [203]).

Chapter 1 is organized as follows. In Section 1.1 we discuss the fundamentals (CR structures, the Levi form, characteristic directions, etc.) and examples (e.g., CR Lie groups). Sections 1.2 to 1.5 are devoted to the construction and principal properties of what appears to be the main geometric tool through this book, the *Tanaka–Webster connection*. Ample space is dedicated to curvature properties, details of which appear nowhere else in the mathematical literature, and to applications (due to S. Webster [422]) of the Chern–Moser theorem (CR manifolds with a vanishing Chern tensor are locally CR isomorphic to spheres) to several *pseudo-Hermitian space forms*. In Section 1.6 we discuss CR structures as G-structures and hint at some open problems.

1.1 CR manifolds

1.1.1 CR structures

Let M be a real m-dimensional C^∞ differentiable manifold. Let $n \in \mathbf{N}$ be an integer such that[2] $1 \leq n \leq [m/2]$. Let $T(M) \otimes \mathbf{C}$ be the complexified tangent bundle over M. Elements of $T(M) \otimes \mathbf{C}$ are of the form $u \otimes 1 + v \otimes i$, where $u, v \in T(M)$ are real tangent vectors (and $i = \sqrt{-1}$). For simplicity, we drop the tensor products and write merely $u + iv$ (a *complex* tangent vector on M). The following definition is central to this book.

Definition 1.1. Let us consider a complex subbundle $T_{1,0}(M)$ of the complexified tangent bundle $T(M) \otimes \mathbf{C}$, of complex rank n. If

$$T_{1,0}(M) \cap T_{0,1}(M) = (0) \tag{1.5}$$

then $T_{1,0}(M)$ is called an *almost CR structure* on M. Here $T_{0,1}(M) = \overline{T_{1,0}(M)}$ and throughout an overbar denotes complex conjugation. The integers n and $k = m - 2n$ are respectively the *CR dimension* and *CR codimension* of the almost CR structure and (n, k) is its *type*. A pair $(M, T_{1,0}(M))$ consisting of an almost CR structure of type (n, k) is an *almost CR manifold* (of type (n, k)). □

It is easy to see that an almost CR manifold of type $(n, 0)$ is an almost complex manifold (cf., e.g., [241], vol. II, p. 121).

Given a vector bundle $E \to M$ we denote by $\Gamma^\infty(U, E)$ the space of all C^∞ cross-sections in E defined on the open subset $U \subseteq M$. We often write $\Gamma^\infty(E)$ for $\Gamma^\infty(M, E)$ (the space of globally defined smooth sections). Also E_x is the fiber in E over $x \in M$.

Definition 1.2. An almost CR structure $T_{1,0}(M)$ on M is (*formally*) *integrable* if for any open set $U \subseteq M$,

$$\left[\Gamma^\infty(U, T_{1,0}(M)), \Gamma^\infty(U, T_{1,0}(M))\right] \subseteq \Gamma^\infty(U, T_{1,0}(M)). \tag{1.6}$$

[2] If $a \in \mathbf{R}$ then $[a] \in \mathbf{Z}$ denotes the integer part of a.

That is, for any two complex vector fields Z, W (defined on $U \subseteq M$) belonging to $T_{1,0}(M)$, their Lie bracket $[Z, W]$ belongs to $T_{1,0}(M)$, i.e., $[Z, W]_x \in T_{1,0}(M)_x$ for any $x \in U$. An integrable almost CR structure (of type (n, k)) is referred to as a *CR structure* (of type (n, k)), and a pair $(M, T_{1,0}(M))$ consisting of a C^∞ manifold and a CR structure (of type (n, k)) is a *CR manifold* (of type (n, k)). □

CR manifolds are the objects of a category whose arrows are smooth maps preserving CR structures. Precisely we have the following definition:

Definition 1.3. Let $(M, T_{1,0}(M))$ and $(N, T_{1,0}(N))$ be two CR manifolds (of arbitrary, but fixed type). A C^∞ map $f : M \to N$ is a *CR map* if

$$(d_x f)T_{1,0}(M)_x \subseteq T_{1,0}(N)_{f(x)}, \tag{1.7}$$

for any $x \in M$, where $d_x f$ is the (**C**-linear extension to $T_x(M) \otimes_{\mathbf{R}} \mathbf{C}$ of the) differential of f at x. □

It is easy to see that the complex manifolds and holomorphic maps form a subcategory of the category of CR manifolds and CR maps.

Let $(M, T_{1,0}(M))$ be a CR manifold of type (n, k). Its *maximal complex*, or *Levi, distribution* is the real rank $2n$ subbundle $H(M) \subset T(M)$ given by

$$H(M) = \mathrm{Re}\{T_{1,0}(M) \oplus T_{0,1}(M)\}.$$

It carries the complex structure $J_b : H(M) \to H(M)$ given by

$$J_b(V + \overline{V}) = i(V - \overline{V}),$$

for any $V \in T_{1,0}(M)$. Here $i = \sqrt{-1}$. The (formal) integrability requirement (1.6) is equivalent to

$$[J_b X, Y] + [X, J_b Y] \in \Gamma^\infty(U, H(M)), \tag{1.8}$$

$$[J_b X, J_b Y] - [X, Y] = J_b\{[J_b X, Y] + [X, J_b Y]\}, \tag{1.9}$$

for any $X, Y \in \Gamma^\infty(U, H(M))$; cf. S. Greenfield [187]. This is formally similar to the notion of integrability of an almost complex structure (cf., e.g., [241], vol. II, p. 124). It should be noted, however, that in contrast to the case of an almost complex manifold, J_b is not defined on the whole of $T(M)$, and $[J_b X, Y] + [X, J_b Y]$ may not lie in $H(M)$ (thus requiring the axiom (1.8)). Proving the equivalence of (1.6) and (1.8)–(1.9) is an easy exercise and therefore left to the reader.

Let $f : M \to N$ be a CR mapping. Then (1.7) is equivalent to the prescriptions

$$(d_x f)H(M)_x \subseteq H(N)_{f(x)} \tag{1.10}$$

and

$$(d_x f) \circ J_{b,x} = J^N_{b,f(x)} \circ (d_x f), \tag{1.11}$$

for any $x \in M$. Here $J^N_b : H(N) \to H(N)$ denotes the complex structure in the Levi distribution $H(N)$ of N.

Definition 1.4. $f : M \to N$ is a *CR isomorphism* (or a *CR equivalence*) if f is a C^∞ diffeomorphism and a CR map. □

CR manifolds arise mainly as real submanifolds of complex manifolds. Let V be a complex manifold, of complex dimension N, and let $M \subset V$ be a real m-dimensional submanifold. Let us set

$$T_{1,0}(M) = T^{1,0}(V) \cap [T(M) \otimes \mathbf{C}], \tag{1.12}$$

where $T^{1,0}(V)$ is the holomorphic tangent bundle over V, i.e., locally the span of $\{\partial/\partial z^j : 1 \leq j \leq N\}$, where $(z^1, \dots, z^N)$ are local complex coordinates on V. The following result is immediate:

Proposition 1.1. *If M is a real hypersurface ($m = 2N - 1$) then $T_{1,0}(M)$ is a CR structure of type $(N - 1, 1)$.*

In general the complex dimension of $T_{1,0}(M)_x$ may depend on $x \in M$. Nevertheless, if

$$\dim_{\mathbf{C}} T_{1,0}(M)_x = n \quad (= \text{ const})$$

then $(M, T_{1,0}(M))$ is a CR manifold (of type (n, k)). The reader will meet no difficulty in checking both statements as a consequence of the properties $T^{1,0}(V) \cap T^{0,1}(V) = (0)$ (where $T^{0,1}(V) = \overline{T^{1,0}(V)}$) and $Z, W \in T^{1,0}(V) \Longrightarrow [Z, W] \in T^{1,0}(V)$.

Definition 1.5. If $k = 2N - m$ (i.e., the CR codimension of $(M, T_{1,0}(M))$ and the codimension of M as a real submanifold of V coincide) then $(M, T_{1,0}(M))$ is termed *generic*. □

1.1.2 The Levi form

Significant portions of the present text will be devoted to the study of CR manifolds of CR codimension $k = 1$ (referred to as well as CR manifolds of *hypersurface type*). Let M be a connected CR manifold of type $(n, 1)$. Assume M to be orientable. Let

$$E_x = \{\omega \in T^*_x(M) : \operatorname{Ker}(\omega) \supseteq H(M)_x\},$$

for any $x \in M$. Then $E \to M$ is a real line subbundle of the cotangent bundle $T^*(M) \to M$ and $E \simeq T(M)/H(M)$ (a vector bundle isomorphism). Since M is orientable and $H(M)$ is oriented by its complex structure J_b, it follows that E is orientable. Any orientable real line bundle over a connected manifold is trivial, so there exist globally defined nowhere vanishing sections $\theta \in \Gamma^\infty(E)$.

Definition 1.6. Any such section θ is referred to as a *pseudo-Hermitian structure* on M. Given a pseudo-Hermitian structure θ on M the *Levi form* L_θ is defined by

$$L_\theta(Z, \overline{W}) = -i(d\theta)(Z, \overline{W}), \tag{1.13}$$

for any $Z, W \in T_{1,0}(M)$. □

Since $E \to M$ is a real line bundle, any two pseudo-Hermitian structures $\theta, \hat{\theta} \in \Gamma^\infty(E)$ are related by

$$\hat{\theta} = \lambda\, \theta, \tag{1.14}$$

for some nowhere-zero C^∞ function $\lambda : M \to \mathbf{R}$. Let us apply the exterior differentiation operator d to (1.14). We get

$$d\hat{\theta} = d\lambda \wedge \theta + \lambda d\theta.$$

Since $\operatorname{Ker}(\theta) = H(M)$ (the $\mathbf{C}$-linear extension of) θ vanishes on $T_{1,0}(M)$ and $T_{0,1}(M)$ as well. Consequently, the Levi form changes according to

$$L_{\hat{\theta}} = \lambda L_\theta \tag{1.15}$$

under any transformation (1.14) of the pseudo-Hermitian structure. This leads to a largely exploited analogy between CR and conformal geometry (cf., e.g., J.M. Lee [270, 271], D. Jerison and J.M. Lee [227, 228], C.R. Graham [182], etc.), a matter we will treat in detail in the subsequent chapters of this text.

Let $(M, T_{1,0}(M))$ be an orientable CR manifold of type $(n, 1)$ (of hypersurface type) and θ a fixed pseudo-Hermitian structure on M. Define the bilinear form G_θ by setting

$$G_\theta(X, Y) = (d\theta)(X, J_b Y), \tag{1.16}$$

for any $X, Y \in H(M)$. Note that L_θ and (the $\mathbf{C}$-bilinear extension to $H(M) \otimes \mathbf{C}$ of) G_θ coincide on $T_{1,0}(M) \otimes T_{0,1}(M)$. Then

$$G_\theta(J_b X, J_b Y) = G_\theta(X, Y), \tag{1.17}$$

for any $X, Y \in H(M)$. Indeed, we may (by (1.8)–(1.9)) perform the following calculation:

$$\begin{aligned} G_\theta(J_b X, J_b Y) - G_\theta(X, Y) &= -(d\theta)(J_b X, Y) - (d\theta)(X, J_b Y) \\ &= \frac{1}{2}\{\theta([J_b X, Y]) + \theta([X, J_b Y])\} \\ &= \frac{1}{2}\theta(J_b\{[X, Y] - [J_b X, J_b Y]\}) = 0, \end{aligned}$$

since $\theta \circ J_b = 0$. In particular, G_θ is symmetric.

Definition 1.7. We say that $(M, T_{1,0}(M))$ is *nondegenerate* if the Levi form L_θ is nondegenerate (i.e., if $Z \in T_{1,0}(M)$ and $L_\theta(Z, \overline{W}) = 0$ for any $W \in T_{1,0}(M)$ then $Z = 0$) for some choice of pseudo-Hermitian structure θ on M. If L_θ is positive definite (i.e., $L_\theta(Z, \overline{Z}) > 0$ for any $Z \in T_{1,0}(M)$, $Z \neq 0$) for some θ, then $(M, T_{1,0}(M))$ is said to be *strictly pseudoconvex*. □

If θ and $\hat{\theta} = \lambda\theta$ are two pseudo-Hermitian structures, then as a consequence of (1.15), L_θ is nondegenerate if and only if $L_{\hat{\theta}}$ is nondegenerate. Hence nondegeneracy is a *CR-invariant* property, i.e., it is invariant under a transformation (1.14). Of course, strict pseuoconvexity is not a CR-invariant property. Indeed, if L_θ is positive definite then $L_{-\theta}$ is negative definite.

Definition 1.8. Let M be a nondegenerate CR manifold and θ a fixed pseudo-Hermitian structure on M. The pair (M, θ) is referred to as a *pseudo-Hermitian manifold.* □

Let $f : M \to N$ be a CR map and θ, θ_N pseudo-Hermitian structures on M and N, respectively. Then $f^*\theta_N = \lambda\, \theta$, for some $\lambda \in C^\infty(M)$.

Definition 1.9. Let M and N be two CR manifolds and θ, θ_N pseudo-Hermitian structures on M and N, respectively. We say that a CR map $f : M \to N$ is a *pseudo-Hermitian map* if $f^*\theta_N = c\,\theta$, for some $c \in \mathbf{R}$. If $c = 1$ then f is referred to as an *isopseudo-Hermitian map.* □

Let us go back for a moment to the case of CR manifolds of arbitrary type. If $(M, T_{1,0}(M))$ is a CR manifold of type (n, k) then its *Levi form* is defined as follows. Let $x \in M$ and $v, w \in T_{1,0}(M)_x$. We set

$$L_x(v, w) = i\, \pi_x \left[V, \overline{W}\right]_x ,$$

where $\pi : T(M) \otimes \mathbf{C} \to (T(M) \otimes \mathbf{C})/(H(M) \otimes \mathbf{C})$ is the natural bundle map and $V, W \in \Gamma^\infty(T_{1,0}(M))$ are arbitrary C^∞ extensions of v, w (i.e., $V_x = v$, $W_x = w$). The definition of $L_x(v, w)$ does not depend on the choice of extensions of v, w because of

$$\pi_x \left[V, \overline{W}\right]_x = v^\alpha \overline{w^\beta} \pi_x \left[T_\alpha, T_{\overline{\beta}}\right]_x ,$$

where $v = v^\alpha T_{\alpha,x}$ and $w = w^\alpha T_{\alpha,x}$, for some local frame $\{T_1, \dots, T_n\}$ of $T_{1,0}(M)$ defined on an open neighborhood of x (and $T_{\overline{\alpha}} = \overline{T_\alpha}$). Then $(M, T_{1,0}(M))$ is said to be *nondegenerate* if L is nondegenerate. However, since for $k \geq 2$ the Levi form L is vector valued, there isn't any obvious way to generalize the notion of strict pseudo-convexity (to the arbitrary CR codimension case).

Let us check that the new and old concepts of Levi form coincide (up to an isomorphism) when $k = 1$. Let $(M, T_{1,0}(M))$ be an oriented CR manifold (of hypersurface type) and θ a pseudo-Hermitian structure on M. Consider the bundle isomorphism

$$\Phi_\theta : \frac{T(M) \otimes \mathbf{C}}{H(M) \otimes \mathbf{C}} \to E$$

given by

$$(\Phi_\theta)_x \left(v + H(M)_x \otimes_{\mathbf{R}} \mathbf{C}\right) = \theta_x(v)\, \theta_x,$$

for any $v \in T_x(M) \otimes \mathbf{C}$, $x \in M$. Then

$$(\Phi_\theta)_x\, L_x(v, w) = 2\, (L_\theta)_x\, (v, w)\, \theta_x,$$

for any $v, w \in T_{1,0}(M)_x$. □

1.1.3 Characteristic directions on nondegenerate CR manifolds

Let $(M, T_{1,0}(M))$ be a CR manifold and θ a fixed pseudo-Hermitian structure on M. As a consequence of the formal integrability property,

$$(d\theta)(Z, W) = 0, \quad Z, W \in T_{1,0}(M).$$

Of course, in the preceding identity $d\theta$ is thought of as extended by **C**-linearity (to $T(M) \otimes \mathbf{C}$). Moreover, we also have $(d\theta)(\overline{Z}, \overline{W}) = 0$ (by complex conjugation), for any $Z, W \in T_{1,0}(M)$. If $(M, T_{1,0}(M))$ is nondegenerate then $d\theta$ is nondegenerate on $H(M)$. Indeed, let us assume that $X = Z + \overline{Z} \in H(M)$ $(Z \in T_{1,0}(M))$ and

$$(d\theta)(X, Y) = 0, \tag{1.18}$$

for any $Y \in H(M)$. Let $X \rfloor$ denote the interior product with X. For instance, $(X \rfloor d\theta)(Y) = (d\theta)(X, Y)$ for any $Y \in \mathcal{X}(M)$. When $X \rfloor d\theta$ is extended by **C**-linearity, (1.18) continues to hold for any $Y \in H(M) \otimes \mathbf{C} = T_{1,0}(M) \oplus T_{0,1}(M)$; hence

$$0 = (d\theta)(X, \overline{W}) = i\, L_\theta(Z, \overline{W}),$$

for any $W \in T_{1,0}(M)$. It follows that $Z = 0$. □

Consequently we may establish the following proposition:

Proposition 1.2. *There is a unique globally defined nowhere zero tangent vector field T on M such that*

$$\theta(T) = 1, \quad T \rfloor d\theta = 0. \tag{1.19}$$

T is transverse to the Levi distribution $H(M)$.

To prove Proposition 1.2 one uses the following fact of linear algebra (together with the orientability assumption).

Proposition 1.3. *Let V be a real $(n+1)$-dimensional linear space and $H \subset V$ an n-dimensional subspace. Let ω be a skew-symmetric bilinear form on V. Assume that ω is nondegenerate on H. Then there is $v_0 \in V$, $v_0 \neq 0$, such that $\omega(v_0, v) = 0$ for any $v \in V$.*

Proof. Let us set

$$K = \{v \in V : \omega(v, u) = 0\,,\ \forall\, u \in H\}.$$

The proof of Proposition 1.3 is organized in two steps, as follows.

Step 1. *K is 1-dimensional.*

Indeed, given a linear basis $\{e_1, \ldots, e_{n+1}\}$ of V with $\{e_1, \ldots, e_n\} \subset H$, then for each $v = \sum_{j=1}^{n+1} \lambda_j e_j \in K$ we have $A\lambda = 0$ where $A = [a_{jk}]$, $a_{jk} = \omega(e_j, e_k)$, $1 \leq j \leq n+1$, $1 \leq k \leq n$, and $\lambda = (\lambda_1, \ldots, \lambda_{n+1})^T$. Thus K is the solution space to the homogeneous system $AX = 0$, so that $\dim_{\mathbf{R}} K = n + 1 - \text{rank}(A)$ and $\text{rank}(A) = n$ (by the nondegeneracy of ω on H). $\square$

Step 2. $H \cap K = (0)$.

If $v \in H \cap K$ then $v \in K$ yields $\omega(v, u) = 0$ for any $u \in H$ such that $v = 0$ (since $v \in H$ and ω is nondegenerate). $\square$

By **Steps 1** and **2**, $V = H \oplus K$. Let (by **Step 1**) $v_0 \in K$, $v_0 \neq 0$. This is the vector we are looking for. Indeed, given any $v \in V$ there are $w \in H$ and $\lambda \in \mathbf{R}$ with $v = w + \lambda v_0$ such that $\omega(v_0, v) = 0$ by the skew-symmetry of ω. $\square$

The tangent vector field T determined by (1.19) is referred to as the *characteristic direction* of (M, θ). Next, we may state the following result:

Proposition 1.4. *Let $(M, T_{1,0}(M))$ be a nondegenerate CR manifold, θ a pseudo-Hermitian structure on M, and T the corresponding characteristic direction. Then*

$$T(M) = H(M) \oplus \mathbf{R}T. \tag{1.20}$$

Indeed, let $X \in T(M)$ and set $Y = X - \theta(X)T$. Then $\theta(Y) = 0$, i.e., $Y \in \text{Ker}(\theta) = H(M)$. Proposition 1.4 is proved.

Using (1.20) we may extend G_θ to a semi-Riemannian metric g_θ on M, which will play a crucial role in the sequel.

Definition 1.10. Let $(M, T_{1,0}(M))$ be a nondegenerate CR manifold and θ a pseudo-Hermitian structure on M. Let g_θ be the semi-Riemannian metric given by

$$g_\theta(X, Y) = G_\theta(X, Y), \quad g_\theta(X, T) = 0, \quad g_\theta(T, T) = 1,$$

for any $X, Y \in H(M)$. This is called the *Webster metric* of (M, θ). $\square$

Assume that $(M, T_{1,0}(M))$ is nondegenerate. It is not difficult to check that the signature (r, s) of $L_{\theta,x}$ doesn't depend on $x \in M$. Also, (r, s) is a CR-invariant. Moreover, the signature of the Webster metric g_θ is $(2r + 1, 2s)$. If $(M, T_{1,0}(M))$ is strictly pseudoconvex and θ is chosen in such a way that L_θ is positive definite, then g_θ is a Riemannian metric on M. Let $\pi_H : T(M) \to H(M)$ be the projection associated with the direct sum decomposition (1.20). If $\pi_H G_\theta$ denotes the $(0, 2)$-tensor field on M given by $(\pi_H G_\theta)(X, Y) = G_\theta(\pi_H X, \pi_H Y)$, for any $X, Y \in T(M)$, then the Webster metric may be written as

$$g_\theta = \pi_H G_\theta + \theta \otimes \theta.$$

g_θ is not a CR-invariant. To write the transformation law for g_θ (under a transformation $\hat{\theta} = \lambda\theta$ of the pseudo-Hermitian structure) is a rather tedious exercise. Of course $G_{\hat{\theta}} = \lambda G_\theta$ and $\hat{\theta} \otimes \hat{\theta} = \lambda^2 \, \theta \otimes \theta$; yet π_H transforms as well. We will return to this matter in Chapter 2.

1.1.4 CR geometry and contact Riemannian geometry

We start be recalling a few notions of contact Riemannian geometry, following for instance D.E. Blair [62]. Let M be a $(2n+1)$-dimensional C^∞ manifold. Let (ϕ, ξ, η) be a synthetic object, consisting of a $(1,1)$-tensor field $\phi : T(M) \to T(M)$, a tangent vector field $\xi \in \mathcal{X}(M)$, and a differential 1-form η on M. (ϕ, ξ, η) is an *almost contact structure* if

$$\phi^2 = -I + \eta \otimes \xi, \quad \phi\xi = 0, \ \eta(\xi) = 1, \quad \eta \circ \phi = 0.$$

An almost contact structure (ϕ, ξ, η) is said to be *normal* if

$$[\phi, \phi] + 2(d\eta) \otimes \xi = 0,$$

where $[\phi, \phi]$ is the Nijenhuis torsion of ϕ. A Riemannian metric g on M is said to be *compatible* with the almost contact structure (ϕ, ξ, η) if

$$g(\phi X, \phi Y) = g(X, Y) - \eta(X)\eta(Y),$$

for any $X, Y \in T(M)$. A synthetic object (ϕ, ξ, η, g) consisting of an almost contact structure (ϕ, ξ, η) and a compatible Riemannian metric g is said to be an almost contact metric structure. Given an almost contact metric structure (ϕ, ξ, η, g) one defines a 2-form Ω by setting $\Omega(X, Y) = g(X, \phi Y)$. (ϕ, ξ, η, g) is said to satisfy the *contact condition* if $\Omega = d\eta$, and if this is the case, (ϕ, ξ, η, g) is called a *contact metric* structure on M. A contact metric structure (ϕ, ξ, η, g) that is also normal is called a *Sasakian structure* (and g is a *Sasakian metric*).

Let $(M, T_{1,0}(M))$ be a nondegenerate CR manifold and θ a pseudo-Hermitian structure on M. Let T be the characteristic direction of (M, θ). Let us extend J_b to a $(1,1)$-tensor field on M by requiring that

$$J_b T = 0. \tag{1.21}$$

Then (by summarizing properties, old and new)

$$\begin{aligned} J_b^2 &= -I + \theta \otimes T, \\ J_b T &= 0, \quad \theta \circ J_b = 0, \quad g_\theta(X, T) = \theta(X), \\ g_\theta(J_b X, J_b Y) &= g_\theta(X, Y) - \theta(X)\theta(Y), \end{aligned}$$

for any $X, Y \in T(M)$. Therefore, if $(M, T_{1,0}(M))$ is a strictly pseudoconvex CR manifold and θ is a contact form such that L_θ is positive definite, then $(J_b, T, \theta, g_\theta)$ is an almost contact metric structure on M. Also

$$\Omega = -d\theta, \tag{1.22}$$

where Ω is defined by

$$\Omega(X, Y) = g_\theta(X, J_b Y),$$

for any $X, Y \in T(M)$. That is, if we set $\phi = J_b$, $\xi = -T$, $\eta = -\theta$, and $g = g_\theta$, then (ϕ, ξ, η, g) is a *contact metric structure* on M, provided that $(M, T_{1,0}(M))$ is strictly pseudoconvex. By (1.20) it suffices to check (1.22) on $H(M) \otimes H(M)$, respectively on $H(M) \otimes \mathbf{R}T$ and $\mathbf{R}T \otimes \mathbf{R}T$. Let $X, Y \in H(M)$. Then (by (1.17))

$$\Omega(X, Y) = g_\theta(X, J_b Y) = (d\theta)(X, J_b^2 Y) = -(d\theta)(X, Y).$$

Finally (by (1.19))

$$\Omega(T, X) = 0 = -(d\theta)(T, X),$$

for any $X \in T(M)$. Strictly pseudoconvex CR manifolds are therefore contact Riemannian manifolds, in a natural way. However, they might fail to be normal. As we shall see in the sequel, the almost contact structure (J_b, T, θ) is normal if and only if the Tanaka–Webster connection of (M, θ) (to be introduced in Section 1.2) has a vanishing pseudo-Hermitian torsion ($\tau = 0$). The converse, that is, which almost contact manifolds are CR manifolds, was taken up by S. Ianuş [214]. Indeed, an almost contact manifold $(M, (\phi, \xi, \eta))$ possesses a natural almost CR structure $T_{1,0}(M)$ defined as the eigenbundle $\text{Eigen}(J_b^{\mathbf{C}}; i)$ of $J_b^{\mathbf{C}}$ corresponding to the eigenvalue i. Here J_b is the restriction of ϕ to $H(M) = \text{Ker}(\eta)$ and $J_b^{\mathbf{C}}$ is the $\mathbf{C}$-linear extension of J_b to $H(M) \otimes \mathbf{C}$. In general, $T_{1,0}(M)$ may fail to be integrable. By a result of S. Ianuş (cf. op. cit.), if (ϕ, ξ, η) is normal then $T_{1,0}(M)$ is a CR structure. The converse is not true, in general. The question (of characterizing the almost contact manifolds whose natural almost CR structure is integrable) has been settled by S. Tanno [401], who built a tensor field Q (in terms of (ϕ, ξ, η)) such that $Q = 0$ if and only if $T_{1,0}(M)$ is integrable. This matter (together with the implication $[\phi, \phi] + 2(d\eta) \otimes \xi = 0 \Longrightarrow Q = 0$) will be examined in the sequel.

1.1.5 The Heisenberg group

Let us set $\mathbf{H}_n = \mathbf{C}^n \times \mathbf{R}$, thought of as endowed with the natural coordinates $(z, t) = (z^1, \ldots, z^n, t)$. $\mathbf{H}_n$ may be organized as a group with the group law

$$(z, t) \cdot (w, s) = (z + w, t + s + 2\text{Im}\langle z, w\rangle),$$

where $\langle z, w\rangle = \delta_{jk} z^j \overline{w^k}$. This actually makes $\mathbf{H}_n$ into a Lie group, referred to as the *Heisenberg group*. A good bibliographical reference is the paper by G.B. Folland and E.M. Stein [150], pp. 434–437, yet the mathematical literature (dealing with both geometric and analysis aspects) on the Heisenberg group occupies a huge (and still growing) volume. Let us consider the complex vector fields on $\mathbf{H}_n$,

$$T_j = \frac{\partial}{\partial z^j} + i\overline{z}^j \frac{\partial}{\partial t}, \tag{1.23}$$

where

$$\frac{\partial}{\partial z^j} = \frac{1}{2}\left(\frac{\partial}{\partial x^j} - i\frac{\partial}{\partial y^j}\right)$$

and $z^j = x^j + iy^j$, $1 \leq j \leq n$. Let us define $T_{1,0}(\mathbf{H}_n)_{(z,t)}$ as the space spanned by the $T_{j,(z,t)}$'s, i.e.,

$$T_{1,0}(\mathbf{H}_n)_{(z,t)} = \sum_{j=1}^{n} \mathbf{C} T_{j,(z,t)}, \tag{1.24}$$

for any $(z, t) \in \mathbf{H}_n$. Since

$$[T_j, T_k] = 0, \quad 1 \leq j, k \leq n,$$

it follows that $(\mathbf{H}_n, T_{1,0}(\mathbf{H}_n))$ is a CR manifold of type $(n, 1)$ (a CR manifold of hypersurface type). Next, let us consider the real 1-form θ_0 on $\mathbf{H}_n$ defined by

$$\theta_0 = dt + i \sum_{j=1}^{n} \left(z^j d\bar{z}^j - \bar{z}^j dz^j \right). \tag{1.25}$$

Then θ_0 is a pseudo-Hermitian structure on $(\mathbf{H}_n, T_{1,0}(\mathbf{H}_n))$. By differentiating (1.25) we obtain

$$d\theta_0 = 2i \sum_{j=1}^{n} dz^j \wedge d\bar{z}^j.$$

Then, by taking into account (1.13), it follows that

$$L_{\theta_0}(T_j, T_{\bar{k}}) = \delta_{jk},$$

where $T_{\bar{j}} = \overline{T_j}$, $1 \leq j \leq n$.

Our choice of θ_0 shows that $(\mathbf{H}_n, T_{1,0}(\mathbf{H}_n))$ is a strictly pseudoconvex CR manifold. Its Levi distribution $H(\mathbf{H}_n)$ is spanned by the (left-invariant) tangent vector fields $\{X_1, \ldots, X_n, Y_1, \ldots, Y_n\}$, where

$$X_j = \frac{\partial}{\partial x^j} + 2y^j \frac{\partial}{\partial t}, \quad Y_j = \frac{\partial}{\partial y^j} - 2x^j \frac{\partial}{\partial t}, \quad 1 \leq j \leq n.$$

The reader may easily check that $T = \partial/\partial t$ is the characteristic direction of $(\mathbf{H}_n, \theta_0)$.

For $n = 1$, $T_{\bar{1}} = \partial/\partial\bar{z} - iz\partial/\partial t$ is the *Lewy operator* (discovered by H. Lewy [284], in connection with the boundary behavior of holomorphic functions on $\Omega_2 = \{(z, w) \in \mathbf{C}^2 : \mathrm{Im}(w) > |z|^2\}$). The Lewy operator exhibits interesting unsolvability features described, for instance, in [232], pp. 235–239. See also L. Ehrenpreis [134], for two new approaches to the Lewy unsolvability phenomenon (one based on the existence of peak points in the kernel of $T_{\bar{1}}$, and the second on a Hartogs-type extension property), both with ramifications in the area of topological algebra.

Definition 1.11. The map $D_\delta : \mathbf{H}_n \to \mathbf{H}_n$ given by $D_\delta(z, t) = (\delta z, \delta^2 t)$, for any $(z, t) \in \mathbf{H}_n$, is called the *dilation* by the factor $\delta > 0$. □

It is an easy exercise to prove the following result:

Proposition 1.5. *Each dilation is a group homomorphism and a CR isomorphism.*

The Euclidean norm of $x = (z, t) \in \underline{\mathrm{H}}_n$ is denoted by $\|x\|$ (i.e., $\|x\|^2 = \|z\|^2 + t^2$). The Euclidean norm is not homogeneous with respect to dilations. However, $\mathbf{H}_n$ carries another significant function, the Heisenberg norm, which enjoys the required homogeneity property.

Definition 1.12. The *Heisenberg norm* is

$$|x| = (\|z\|^4 + t^2)^{1/4},$$

for any $x \in \mathbf{H}_n$. □

The Heisenberg norm is homogeneous with respect to dilations, i.e.,

$$|D_\delta x| = \delta |x|, \quad x \in \mathbf{H}_n.$$

Let us consider the transformation $T^\delta = D_{1/\delta}$. Then

$$(d_x T^\delta) T_{j,x} = \delta^{-1} T_{j, T^\delta(x)},$$

i.e., the T_j are homogeneous of degree -1 with respect to dilations. As to the form θ_0 given by (1.25), it satisfies

$$(D_\delta^* \theta)_x = \delta^2 \theta_{0, D_\delta x}.$$

The following inequalities hold on the Heisenberg group:

Proposition 1.6.

$$\|x\| \leq |x| \leq \|x\|^{1/2}, \tag{1.26}$$

for any $x \in \mathbf{H}_n$ *such that* $|x| \leq 1$.

We shall occasionally need the following:

Proposition 1.7. *There is a constant* $\gamma \geq 1$ *such that*

$$|x + y| \leq \gamma(|x| + |y|), \tag{1.27}$$

$$|xy| \leq \gamma(|x| + |y|), \tag{1.28}$$

for any $x, y \in \mathbf{H}_n$.

Definition 1.13. The inequalities (1.27)–(1.28) are called the *triangle inequalities.* □

The proof of (1.27)–(1.28) is elementary. Indeed, by homogeneity we may assume that $|x| + |y| = 1$. The set of all $(x, y) \in \mathbf{H}_n \times \mathbf{H}_n$ satisfying this equation is compact; hence we may take γ to be the larger of the maximum values of $|x + y|$ and $|xy|$ on this set. □

The Heisenberg group may be identified with the boundary of a domain in $\mathbf{C}^{n+1}$. Indeed, let Ω_{n+1} be the *Siegel domain*, i.e.,

$$\Omega_{n+1} = \{(z, w) \in \mathbf{C}^n \times \mathbf{C} : v > \|z\|^2\},$$

where $z = (z^1, \dots, z^n)$ and $w = u + iv$ $(u, v \in \mathbf{R})$. Also $\|z\|^2 = \sum_{j=1}^n |z^j|^2$. Let then

$$f : \mathbf{H}_n \to \partial\Omega_{n+1}, \quad f(z, t) = (z, t + i\|z\|^2),$$

for any $(z, t) \in \mathbf{H}_n$. f is a CR isomorphism, where the boundary $\partial\Omega_{n+1} = \{(z, w) : \|z\|^2 = v\}$ of the Siegel domain is thought of as a CR manifold (of hypersurface type) with the CR structure induced from $\mathbf{C}^{n+1}$. Computing the differential df (on the generators T_j, $T_{\bar{j}}$, and T) is a tedious but useful exercise (left to the reader).

Another useful identification is that of the Heisenberg group and the sphere $S^{2n+1} \subset \mathbf{C}^{n+1}$ minus a point. Let $\Omega \subset \mathbf{C}^N$ be a domain with smooth boundary $\partial\Omega$, i.e., there is an open neighborhood U of the closure $\overline{\Omega}$ in $\mathbf{C}^N$ and a C^∞ function $\rho : U \to \mathbf{R}$ such that $\Omega = \{x \in U : \rho(x) > 0\}$ (and $\partial\Omega = \{x \in U : \rho(x) = 0\}$) and $(D\rho)_x \neq 0$ at any $x \in \partial\Omega$. Let $T_{1,0}(\partial\Omega)$ be the induced CR structure on $\partial\Omega$, as a real hypersurface in $\mathbf{C}^N$. Let θ be the pullback to $\partial\Omega$ of the real 1-form $i(\overline{\partial} - \partial)\rho$ on U. Then θ is a pseudo-Hermitian structure on $(\partial\Omega, T_{1,0}(\partial\Omega))$. As we just saw, the boundary of the Siegel domain is a strictly pseudoconvex CR manifold. Also, the sphere $S^{2n+1} \subset \mathbf{C}^{2n+1}$ is a strictly pseudoconvex manifold, since the boundary of the unit ball $B_{n+1} = \{z \in \mathbf{C}^{n+1} : |z| < 1\}$, and the (restriction to $S^{2n+1} \setminus \{e_1\}$ of the) Cayley transform

$$\Phi : \mathbf{C}^{n+1} \setminus \{z_1 = 1\} \to \mathbf{C}^{n+1},$$
$$\Phi(z) = i\,\frac{e_1 + z}{1 - z_1}, \quad z_1 \neq 1, \; e_1 = (1, 0, \dots, 0),$$

gives a CR isomorphism $S^{2n+1} \setminus \{e_1\} \simeq \partial\Omega_{n+1}$ (and thus a CR isomorphism $S^{2n+1} \setminus \{e_1\} \simeq \mathbf{H}_n$).

1.1.6 Embeddable CR manifolds

Let $M \subset \mathbf{C}^N$ be a real m-dimensional submanifold. If M is a CR manifold (of type (n, k)) whose CR structure is given by (1.12) with $V = \mathbf{C}^N$, then M is referred to as an *embedded* (or *realized*) *CR manifold*.

Let $(M, T_{1,0}(M))$ be a CR manifold. If in some neighborhood of each point $x \in M$, $(M, T_{1,0}(M))$ is CR isomorphic to an embedded CR manifold, then $(M, T_{1,0}(M))$ is termed *locally embeddable*. If a global isomorphism with an embedded CR manifold exists, then $(M, T_{1,0}(M))$ is called *embeddable* (or *realizable*).

Let $(M, T_{1,0}(M))$ be a CR manifold. We say that $(M, T_{1,0}(M))$ is *real analytic* if M is a real analytic manifold and $T_{1,0}(M)$ is a real analytic subbundle of $T(M) \otimes \mathbf{C}$, i.e., $T_{1,0}(M)$ is locally generated by real analytic vector fields. By a (classical) result of A. Andreotti and C.D. Hill [13], any real analytic CR manifold $(M, T_{1,0}(M))$

of type (n, k), $k \geq 1$, is locally embeddable. Precisely, for any $x \in M$, there is a neighborhood U of x in M such that $(U, T_{1,0}(M)_{|U})$ is CR isomorphic via a real analytic CR map to a real analytic generic embedded CR manifold in $\mathbf{C}^{n+k}$. Here $T_{1,0}(M)_{|U}$ is the pullback of $T_{1,0}(M)$ by $\iota : U \subseteq M$. A proof of the embeddability result of A. Andreotti and C.D. Hill (cf. op. cit.) is already available in book form, cf. A. Boggess [70], pp. 169–172, and will not be reproduced here. A discussion of *characteristic coordinates* (the relevant ingredient in the proof) is also available in the book by D.E. Blair [62], pp. 57–60, in the context of the geometric interpretation of normal almost contact structures ([62], pp. 61–63). By a result of A. Andreotti and G.A. Fredricks [15], real analytic CR manifolds are also *globally embeddable*, i.e., globally isomorphic to a generic CR submanifold of some complex manifold. This generalizes a result by H.B. Shutrick [373], on the existence of complexifications.

The *embedding problem* (i.e., decide whether a given abstract CR manifold is (locally) embeddable) was first posed by J.J. Kohn [246], and subsequently solved to a large extent by M. Kuranishi [263]. By M. Kuranishi's result, each strictly pseudoconvex CR manifold of real dimension $2n + 1 \geq 9$ is locally embeddable in $\mathbf{C}^{n+1}$. T. Akahori [2], settled the question in dimension 7. A much simpler proof was given by S. Webster [429]. The embedding problem is open in dimension 5, while L. Nirenberg [326], built a counterexample in dimension 3 (as a perturbation of the CR structure of $\mathbf{H}_1$). More generally, by a result of H. Jacobowitz and F. Trèves [224], analytically small perturbations of 3-dimensional embeddable strictly pseudoconvex CR manifolds are known to be nonembeddable. M.S. Baouendi, L.P. Rothschild, and F. Treves [30], showed that the existence of a local transverse CR action implies local embeddability. As a global version of this result, though confined to the 3-dimensional case, we may quote a result by L. Lempert [278]: Let M be a 3-dimensional ($n = 1$) CR manifold admitting a smooth CR action of $\mathbf{R}$ that is transverse. Then M is the boundary of a strictly pseudoconvex complex surface, i.e., it is embeddable. More recently, by a result of Z. Balogh and C. Leuenberger [29], if a CR manifold M of hypersurface type admits a local semi-extendable $\mathbf{R}$-action then M is locally realizable as the boundary of a complex manifold.

Embeddability is related to solvability of certain PDEs, as shown by the following example, due to C.D. Hill [200]. Let $t = (t_1, t_2, t_3)$ be the Cartesian coordinates on $\mathbf{R}^3$. The natural coordinates on $\mathbf{R}^5 = \mathbf{R}^3 \times \mathbf{C}$ are denoted by (t, z). Let

$$L = T_{\overline{1}} = \frac{1}{2}\left(\frac{\partial}{\partial t_1} + i\,\frac{\partial}{\partial t_2}\right) - i(t_1 + it_2)\frac{\partial}{\partial t_3}$$

be the Lewy operator on $\mathbf{H}_1 = \mathbf{R}^3$. Given a C^∞ function $\omega : \mathbf{H}_1 \to \mathbf{C}$ we consider the first-order PDE

$$L\chi = \omega. \tag{1.29}$$

Definition 1.14. We say that (1.29) is *solvable* at a point $t_0 \in \mathbf{R}^3$ if there is an open set $U \subseteq \mathbf{R}^3$ such that $t_0 \in U$ and there is a C^∞ function $\chi : U \to \mathbf{C}$ such that $L\chi = \omega$ on U. □

Moreover, let us consider the complex vector fields $P, Q \in T(\mathbf{H}_1 \times \mathbf{C}) \otimes \mathbf{C}$ given by

$$P = \frac{\partial}{\partial \overline{z}}, \quad Q = L + \omega(t)\frac{\partial}{\partial z}. \tag{1.30}$$

Clearly $\{P, Q, \overline{P}, \overline{Q}\}$ are linearly independent at each point of $\mathbf{H}_1 \times \mathbf{C}$ and $[P, Q] = 0$. Consequently

$$T_{0,1}(\mathbf{H}_1 \times \mathbf{C})_{(t,z)} = \mathbf{C}P_{(t,z)} + \mathbf{C}Q_{(t,z)}, \quad (t, z) \in \mathbf{H}_1 \times \mathbf{C},$$

gives a CR structure on $\mathbf{H}_1 \times \mathbf{C}$. We shall prove the following theorem:

Theorem 1.1. (C.D. Hill [200])
The CR structure (1.30) *is locally embeddable at* $(t_0, z_0) \in \mathbf{H}_1 \times \mathbf{C}$ *if and only if* (1.29) *is solvable at* t_0.

On the other hand, by a result of H. Lewy [284], there is $\omega \in C^\infty(\mathbf{R}^3)$ such that for any open set $U \subseteq \mathbf{R}^3$ the equation (1.29) has no solution $\chi \in C^1(U)$. Hence the CR structure (1.30) is not locally embeddable in general.

Proof of Theorem 1.1. As shown in Section 1.1.5, the map

$$\mathbf{R}^3 \to \mathbf{C}^2, \quad t \mapsto (v_1(t), v_2(t)),$$
$$v_1(t) = t_1 + it_2, \quad v_2(t) = t_3 + i(t_1^2 + t_2^2),$$

embeds $\mathbf{H}_1$ globally into $\mathbf{C}^2$. The functions $v_j : \mathbf{H}_1 \to \mathbf{C}$ form a maximal set of functionally independent *characteristic coordinates* (in the sense of A. Andreotti and C.D. Hill [13]), i.e., $Lv_j = 0$ and $dv_1 \wedge dv_2 \neq 0$. If we adopt the terminology in Chapter 6, the map $v = (v_1, v_2) : \mathbf{H}_1 \to \mathbf{C}^2$ is a *CR immersion*, i.e., an immersion and a CR map (and it determines a CR isomorphism $\mathbf{H}_1 \simeq \partial\Omega_2$).

Let us prove first the sufficiency. Assume that (1.29) is solvable at t_0, i.e., there is $\chi \in C^\infty(U)$ with $t_0 \in U \subseteq \mathbf{H}_1$ and $L\chi = \omega$ on U. Let us consider the function

$$v_3 : U \times \mathbf{C} \to \mathbf{C}, \quad v_3(t, z) = z - \chi(t).$$

A calculation shows that

$$Pv_j = 0, \quad Qv_j = 0, \quad j \in \{1, 2, 3\}, \quad dv_1 \wedge dv_2 \wedge dv_3 \neq 0;$$

that is,

$$\varphi : U \times \mathbf{C} \to \mathbf{C}^3, \quad \varphi(t, z) = (v(t), v_3(t, z)), \quad t \in U, \quad z \in \mathbf{C},$$

is a CR immersion (of a neighborhood of (t_0, z_0)).

The proof of necessity is more involved. Let us assume that there is a CR immersion $u = (u_1, u_2, u_3) : V \to \mathbf{C}^3$ of an open set $V \subseteq \mathbf{H}_1 \times \mathbf{C}$ with $(t_0, z_0) \in V$, that is, $Pu_j = 0$, $Qu_j = 0$, and $du_1 \wedge du_2 \wedge du_3 \neq 0$ on V. In particular, each u_j is holomorphic with respect to the z-variable. Consequently the Jacobian matrix of u has the form

$$\begin{pmatrix} \frac{\partial u}{\partial t} & \frac{\partial u}{\partial z} & 0 \\ \frac{\partial \overline{u}}{\partial t} & 0 & \frac{\partial \overline{u}}{\partial \overline{z}} \end{pmatrix}$$

with six rows and five linearly independent columns. It follows that

$$\frac{\partial u_j}{\partial z}(t_0, z_0) \neq 0$$

for some $j \in \{1, 2, 3\}$, say $(\partial u_3/\partial z)(t_0, z_0) \neq 0$. Thus $\partial u_3/\partial z \neq 0$ at each point of a neighborhood of (t_0, z_0), denoted by the same symbol V. It is easily seen that

$$Pv_1 = Pv_2 = Pu_3 = 0, \quad Qv_1 = Qv_2 = Qu_3 = 0.$$

On the other hand,

$$\begin{aligned} dv_1 \wedge dv_2 \wedge du_3 = {} & (2v_1 \frac{\partial u_3}{\partial t_3} - \frac{\partial u_3}{\partial t_2} + i \frac{\partial u_3}{\partial t_1}) dt_1 \wedge dt_2 \wedge dt_3 \\ & + \frac{\partial u_3}{\partial z}(2v_1\, dt_1 \wedge dt_2 + dt_1 \wedge dt_3 + i dt_2 \wedge dt_3) \wedge dz \neq 0; \end{aligned}$$

hence

$$\varphi : V \to \mathbf{C}^3, \quad \varphi(t, z) = (v_1(t), v_2(t), u_3(t, z)), \quad (t, z) \in V,$$

is another CR immersion of V into $\mathbf{C}^3$, so that $M = \varphi(V)$ is a real hypersurface of $\mathbf{C}^3$. In other words, (v_1, v_2, u_3) is another maximal set of functionally independent characteristic coordinates. By eventually restricting the open set V we may assume that φ is injective, so that $\varphi : V \to M$ is a CR isomorphism. Let us consider the functions

$$f : V \to \mathbf{C}, \quad f(t, z) = -\frac{1}{(\partial u_3/\partial z)(t, z)}, \quad (t, z) \in V,$$
$$F : M \to \mathbf{C}, \quad F(\zeta) = f(\varphi^{-1}(\zeta)), \quad \zeta \in M.$$

If A is an open subset of some CR manifold, let $\mathrm{CR}^\ell(A)$ denote the set of all CR functions on A of class C^ℓ. Since $u_3 \in \mathrm{CR}^\infty(V)$ it follows that $\partial u_3/\partial z \in \mathrm{CR}^\infty(V)$ and then $f \in \mathrm{CR}^\infty(V)$. As $\varphi : V \to M$ is a CR diffeomorphism we may conclude that $F \in \mathrm{CR}^\infty(M)$. Let $\varphi^{-1} = (\psi_1, \psi_2, \psi_3, z \circ \varphi^{-1})$ be the components of $\varphi^{-1} : M \to V \subset \mathbf{R}^3 \times \mathbf{C}$. Then

$$\begin{aligned} \frac{\partial F}{\partial \overline{\zeta}_3}(\zeta) = \frac{\partial (f \circ \varphi^{-1})}{\partial \overline{\zeta}_3}(\zeta) = {} & \sum_j \frac{\partial f}{\partial t_j}(t, z) \frac{\partial \psi_j}{\partial \overline{\zeta}_3}(\zeta) \\ & + \frac{\partial f}{\partial z}(t, z) \frac{\partial (z \circ \varphi^{-1})}{\partial \overline{\zeta}_3}(\zeta) + \frac{\partial f}{\partial \overline{z}}(t, z) \frac{\partial (\overline{z} \circ \varphi^{-1})}{\partial \overline{\zeta}_3}(\zeta), \end{aligned}$$

for any $\zeta = (\zeta_1, \zeta_3, \zeta_3) \in M$, $\zeta = \varphi(t, z)$. Now on the one hand, $(\partial f/\partial \overline{z})(t, z) = (Pf)_{(t,z)} = 0$. On the other hand, $\varphi^{-1} \circ (v_1, v_2, u_3) = 1_V$, so that

$$\psi_j(v_1(t), v_2(t), u_3(t, z)) = t_j,$$

and differentiation with respect to $\overline{z}$ gives

$$\frac{\partial \psi_j}{\partial \overline{\zeta}_3}(\varphi(t, z)) = 0.$$

Similarly (i.e., applying $\partial/\partial\overline{z}$ to $(z \circ \varphi^{-1})(v_1(t), v_2(t), u_3(t, z)) = z$), we obtain

$$\frac{\partial (z \circ \varphi^{-1})}{\partial \overline{\zeta}_3}(\varphi(t, z)) = 0.$$

We may conclude that $\partial F/\partial \overline{\zeta}_3 = 0$. Next let us set

$$G : M \to \mathbf{C}, \quad G(\zeta) = \int_O^{\zeta_3} F(\zeta_1, \zeta_2, \eta) d\eta, \quad \zeta \in M,$$

where integration is carried along a curve γ joining the origin O to ζ_3 in the complex ζ_3-plane. For a fixed $(t, z) \in V$ we set $V_t = \{w \in \mathbf{C} : (t, w) \in V\}$ and define $h_t : V_t \to \mathbf{C}$ by $h_t(w) = u_3(t, w)$. Then $(\partial h_t/\partial w)(w) \neq 0$ so that h_t is a diffeomorphism of a neighborhood of w onto a neighborhood of $u_3(t, w)$. Therefore γ may be covered with open sets that are images by h_t of open subsets of V_t, so that $(\zeta_1, \zeta_2, \eta) \in M$ for any $\zeta \in M$ and any point η on γ. It follows that $G(\zeta)$ is well defined and $G \in \mathrm{CR}^\infty(M)$. Finally, let us set

$$g(t, z) = G(\varphi(t, z)), \quad (t, z) \in V.$$

Then $g \in \mathrm{CR}^\infty(V)$, again because $\varphi : V \to M$ is a CR isomorphism. Now on the one hand, $\partial g/\partial \overline{z} = Pg = 0$, and on the other,

$$\frac{\partial g}{\partial z}(t, z) = \frac{\partial G}{\partial \zeta_3}(\varphi(t, z))\frac{\partial u_3}{\partial z}(t, z) = F(v_1(t), v_2(t), u_3(t, z))\frac{\partial u_3}{\partial z}(t, z) = -1;$$

hence $g(t, z) = \chi(t) - z$, where $\chi(t)$ is a "constant of integration." Then $Qg = 0$ implies $L\chi - \omega = 0$ and the proof of Theorem 1.1 is complete.

1.1.7 CR Lie algebras and CR Lie groups

Definition 1.15. Let $\mathcal{G}$ be a $(2n + k)$-dimensional real Lie algebra. A *CR structure* (of type (n, k)) on $\mathcal{G}$ is an n-dimensional complex subalgebra $\mathbf{a} \subset \mathcal{G} \otimes_{\mathbf{R}} \mathbf{C}$ such that $\mathbf{a} \cap \overline{\mathbf{a}} = (0)$. A pair $(\mathcal{G}, \mathbf{a})$ consisting of a real Lie algebra $\mathcal{G}$ and a CR structure $\mathbf{a}$ on $\mathcal{G}$ is a *CR Lie algebra* (cf. G. Gigante and G. Tomassini [176]). □

Let $(\mathcal{G}, \mathbf{a})$ be a CR Lie algebra and set $\mathbf{h} = \mathrm{Re}\{\mathbf{a} \oplus \overline{\mathbf{a}}\}$. Then $\mathbf{h} \subset \mathcal{G}$ is a $2n$-dimensional real subspace (though in general, not a subalgebra). Define $J : \mathbf{h} \to \mathbf{h}$ by setting $J(Z+\overline{Z}) = i(Z-\overline{Z}),\ Z \in \mathbf{a}$. Then $\mathbf{h}$ is an n-dimensional complex linear space (with the multiplication $iX = JX,\ X \in \mathbf{h}$) and the map $X \mapsto X\otimes 1-(JX)\otimes i,\ X \in \mathbf{h}$, gives $\mathbf{h} \simeq \mathbf{a}$ (a complex linear isomorphism).

Define $d_{\mathcal{G}} : \mathcal{G}^* \to \Lambda^2\mathcal{G}^*$ by setting

$$(d_{\mathcal{G}}\alpha)(X, Y) = -\alpha([X, Y])$$

for any $\alpha \in \mathcal{G}^*,\ X, Y \in \mathcal{G}$. Next, assume that $k = 1$ and let $\theta \in \mathcal{G}^*$ be some linear functional such that $\mathrm{Ker}(\theta) = \mathbf{h}$. Let us set

$$\langle Z, W\rangle_\theta = -i(d_{\mathcal{G}}\theta)(Z, \overline{W})$$

for any $Z, W \in \mathbf{a}$.

Definition 1.16. One says that $(\mathcal{G}, \mathbf{a})$ is *nondegenerate* if $\langle , \rangle_\theta$ is nondegenerate on $\mathbf{a}$ for some θ (and thus for all). □

Definition 1.17. Let G be a real $(2n + k)$-dimensional Lie group and $T_{1,0}(G)$ a CR structure of type (n, k) on G. The pair $(G, T_{1,0}(G))$ is a *CR Lie group* if $T_{1,0}(G)$ is left invariant, i.e.,

$$(d_g L_h)T_{1,0}(G)_g = T_{1,0}(G)_{hg}$$

for any $g, h \in G$, where $L_h : G \to G,\ L_h(g) = hg$. □

That is, each left translation L_h is a CR map. For instance, *the Heisenberg group* $\mathbf{H}_n$ *is a CR Lie group* (with the CR structure $T_{1,0}(\mathbf{H}_n)$ given by (1.24)). Indeed, the components L^j, L^0 of the left translation $L_{(z,t)} : \mathbf{H}_n \to \mathbf{H}_n$ are given by

$$L^j(w, s) = z^j + w^j\,, \quad L^0(w, s) = t + s + 2\,\mathrm{Im}\langle z, w\rangle\,,$$

so that

$$\frac{\partial L^j}{\partial w^k} = \delta^j_k, \qquad \frac{\partial L^j}{\partial \bar{w}^k} = 0, \qquad \frac{\partial L^j}{\partial s} = 0,$$
$$\frac{\partial L^0}{\partial w^k} = i\,\bar{z}^k, \qquad \frac{\partial L^0}{\partial \bar{w}^k} = -i\,z^k, \qquad \frac{\partial L^0}{\partial s} = 1,$$

and hence

$$\left(d_{(w,s)}L_{(z,t)}\right)\left.\frac{\partial}{\partial w^j}\right|_{(w,s)} = \left.\frac{\partial}{\partial z^j}\right|_{(z,t)\cdot(w,s)} + 2\bar{z}^j \left.\frac{\partial}{\partial t}\right|_{(z,t)\cdot(w,s)},$$

$$\left(d_{(w,s)}L_{(z,t)}\right)\left.\frac{\partial}{\partial s}\right|_{(w,s)} = \left.\frac{\partial}{\partial t}\right|_{(z,t)\cdot(w,s)},$$

whence

$$\left(d_{(w,s)}L_{(z,t)}\right)T_{j,(w,s)} = T_{j,(z,t)\cdot(w,s)}\,. \qquad \square$$

A family of examples (containing $\mathbf{H}_n \simeq \partial\Omega_{n+1}$) of CR Lie groups is furnished by the so-called *quadric submanifolds*. Let $q : \mathbf{C}^{n-d} \times \mathbf{C}^{n-d} \to \mathbf{C}^d$ be a quadratic form (i.e., q is $\mathbf{C}$-bilinear, symmetric, and $\overline{q(z,w)} = q(\overline{z},\overline{w})$, for any $z, w \in \mathbf{C}$). A submanifold $M \subset \mathbf{C}^n$ given by

$$M = \{(x+iy, w) \in \mathbf{C}^d \times \mathbf{C}^{n-d} : y = q(w,\overline{w})\}$$

is called a *quadric submanifold* of $\mathbf{C}^n$. Let us set

$$(z_1, w_1) \circ (z_2, w_2) = (z_1 + z_2 + 2iq(w_1, \overline{w}_2), w_1 + w_2).$$

Then (by Lemma 1 in [70], p. 112) the operation $\circ$ defines a group structure on $\mathbf{C}^d \times \mathbf{C}^{n-d}$ that restricts to a group structure on M. Also $\circ : M \times M \to M$ is C^∞, hence $(M, \circ)$ is a Lie group. Finally, the CR structure $T_{1,0}(M)$ is easily seen to be spanned by

$$T_j = \frac{\partial}{\partial w^j} + 2i\frac{\partial q^\ell}{\partial w^j}\frac{\partial}{\partial z^\ell},\ 1 \leq j \leq n-d,$$

and (according to Theorem 1 in [70], p. 113) each T_j is left invariant, and hence $(M, T_{1,0}(M))$ is a CR Lie group.

Let $(G, T_{1,0}(G))$ be a CR Lie group of type (n,k) and $\mathcal{G} = L(G)$ its Lie algebra (of left-invariant tangent vector fields on G). Then

$$\mathbf{a} = j^{-1}\left(T_{1,0}(G)_e\right)$$

is a CR structure on $\mathcal{G}$ (of type (n,k)), where j is the ($\mathbf{C}$-linear extension to $\mathcal{G} \otimes_{\mathbf{R}} \mathbf{C}$ of the) $\mathbf{R}$-linear isomorphism $j : \mathcal{G} \to T_e(G)$, $X \mapsto X_e$, $X \in \mathcal{G}$, and e is the identity element in G. Conversely, any CR structure $\mathbf{a}$ on $\mathcal{G} = L(G)$ determines a unique left-invariant CR structure on G (by simply setting $T_{1,0}(G)_g = (d_eL_g)j(\mathbf{a})$, $g \in G$). Classifications of left-invariant CR structures on classical groups are available in particular cases (cf., e.g., S. Donnini and G. Gigante [118], for a classification of left-invariant CR structures on $\mathrm{GL}^+(3,\mathbf{R})$).

Assume from now on that $k = 1$ and let $\theta_0 \in L(G)^*$ such that $\mathrm{Ker}(\theta_0) = \mathbf{h}$. Then θ_0 determines a left-invariant pseudo-Hermitian structure θ on G. Indeed, one may set

$$\theta_g = \theta_0 j^{-1}(d_g L_{g^{-1}}),\ g \in G,$$

so that $\mathrm{Ker}(\theta) = H(G)$ and $(L_h)^*\theta = \theta$. $\square$

In particular, $(G, T_{1,0}(G))$ is nondegenerate if and only if $(L(G), \mathbf{a})$ is nondegenerate.

Let $(G, T_{1,0}(G))$ be a nondegenerate CR Lie group and let $T \in \mathcal{G}$ be the unique vector such that

$$\theta_0(T) = 1, T \rfloor d_{\mathcal{G}}\theta_0 = 0.$$

Then T is the characteristic direction of (G, θ) (and, by its very definition, a left-invariant vector field on G).

CR structures of CR codimension 1 and of k-torsion zero on a reductive Lie algebra $\mathcal{G}$ of the first category are studied in [176]. Their main result is that any such CR structure is determined by a CR structure on a compact Cartan subalgebra $\mathbf{h} \subset \mathcal{G}$ and by a direct sum of root spaces.

1.1.8 Twistor CR manifolds

As another example of CR manifold we briefly discuss the twistorial construction due to C. LeBrun [268], itself generalizing work by R. Penrose [342]. Let M be an n-dimensional manifold and $\mathcal{X} := T^*(M) \otimes \mathbf{C}$. Let $\pi : \mathcal{X} \to M$ be the projection. Let $(U, \tilde{x}^j)$ be a local coordinate system on M and define local coordinates $x^j : \pi^{-1}(U) \to \mathbf{R}$ and $\zeta_j : \pi^{-1}(U) \to \mathbf{C}$ on $\mathcal{X}$ by setting $x^j(\chi) := \tilde{x}^j(\pi(\chi))$ and $\zeta_j(\chi) := \langle \chi, \partial/\partial\tilde{x}^j \rangle$, for any $\chi \in \mathcal{X}$. Let $A \subset T(\mathcal{X}) \otimes \mathbf{C}$ be the span of $\{\partial/\partial\overline{\zeta}^j : 1 \leq j \leq n\}$, i.e., the portion of A over a fiber $\pi^{-1}(x)$ (clearly a complex n-dimensional manifold) is the antiholomorphic tangent bundle over $\pi^{-1}(x)$, for any $x \in M$.

Definition 1.18. The *canonical* 1*-form* θ of $\mathcal{X}$ is given by

$$\theta_\chi(w) := \langle \chi, (d_\chi \pi) w \rangle, \quad w \in T_\chi(\mathcal{X}), \quad \chi \in \mathcal{X}. \qquad \square$$

In local coordinates one has the expression $\theta = \zeta_j dx^j$.

Definition 1.19. The *Hamiltonian form* of $\mathcal{X}$ is $\omega := d\theta = d\zeta_j \wedge dx^j$. A codimension-2 submanifold $\mathcal{Y} \subset \mathcal{X}$ is an *energy surface* if $\pi^{-1}(x) \cap \mathcal{Y}$ is a complex hypersurface in $\pi^{-1}(x)$, for any $x \in \mathcal{X}$. $\square$

Let

$$N_x^* := \{f \in T_\chi^*(\pi^{-1}(x)) : \mathrm{Ker}(f) \supseteq T_\chi(\pi^{-1}(x) \cap \mathcal{Y})\}, \; \chi \in \pi^{-1}(x) \cap \mathcal{Y},$$

be the conormal bundle of $\pi^{-1}(x) \cap \mathcal{Y}$ in $\pi^{-1}(x)$. Consider the map

$$f \in T_\chi^*(\pi^{-1}(x)) \simeq \pi^{-1}(x) = T_x^*(M) \otimes_{\mathbf{R}} \mathbf{C} \to T_x(M) \otimes_{\mathbf{R}} \mathbf{C} \ni z_f,$$

where the last arrow is the double-dual identification, i.e.,

$$f = f^j d\zeta_j\Big|_\chi + \overline{f}^j d\overline{\zeta}_j\Big|_\chi \longmapsto z_f := f^j \frac{\partial}{\partial \tilde{x}^j}\Big|_x . \qquad (1.31)$$

Definition 1.20. An energy surface $\mathcal{Y} \subset \mathcal{X}$ is *generic* if

$$\{f \in N_x^* : z_f = \overline{z_f}\} = (0),$$

that is, N_x^* contains no nonzero vector corresponding to a real vector in $T_x(M) \otimes_{\mathbf{R}} \mathbf{C}$, under the linear map (1.31). $\square$

If $\iota : \mathcal{Y} \to \mathcal{X}$ is the canonical inclusion we set $\tilde{\omega} := \iota^*\omega$ and consider the bundle morphism

$$\rfloor_{\tilde{\omega}} : T(\mathcal{Y}) \otimes \mathbf{C} \to T^*(\mathcal{Y}) \otimes \mathbf{C}, \quad W \mapsto \tilde{\omega}(W, \cdot), \quad W \in T(\mathcal{Y}) \otimes \mathbf{C}.$$

According to C. LeBrun [268], the following result holds:

Theorem 1.2. *Let $\mathcal{Y} \subset \mathcal{X}$ be an energy surface. Then $D := Ker(\rfloor_{\tilde{\omega}}) \subset T(\mathcal{Y}) \otimes \mathbf{C}$ is involutive. If additionally $\mathcal{Y}$ is generic then $D \cap \overline{D} = (0)$. In particular, if M is a 3-dimensional manifold then D is a CR structure on $\mathcal{Y}$, of CR codimension* 1.

C. LeBrun applied (cf. op. cit.) his finding (that any generic energy surface associated with a 3-dimensional manifold is a 7-dimensional CR manifold (of hypersurface type)) to the following situation. Let M be a 3-dimensional CR manifold endowed with a conformal structure $G = \{e^u g : u \in C^\infty(M)\}$, where g is a fixed Riemannian metric on M. Next, let us consider the 7-dimensional manifold $\hat{N} := \{\chi \in \mathcal{X} : g^*(\chi, \chi) = 0, \ \chi \neq 0\}$. Then $\mathbf{C}^* := \mathbf{C} \setminus \{0\}$ acts freely on $\hat{N}$ and we may consider the quotient space $N := \hat{N}/\mathbf{C}^*$ (a 5-dimensional manifold). Note that $\hat{N}$, and then N, depends only on the conformal structure G (rather than on the fixed metric g). According to [268], $\hat{N} \subset \mathcal{X}$ is a generic energy surface; hence (by Theorem 1.2) carries a CR structure D. Its projection $T_{1,0}(N) := \mathbf{P}_* D$ is a CR structure on N, and N is referred to as the *twistor CR manifold* of M. Here $\mathbf{P} : \hat{N} \to N = \hat{N}/\mathbf{C}^*$ is the natural projection. There is a natural projection $\pi : N \to M$, all of whose fibers are complex lines. Again together with [268], one concludes that $(N, T_{1,0}(N))$ is a nondegenerate CR manifold carrying a smooth foliation by $\mathbf{C}P^1$'s, and $(N, T_{1,0}(N))$ *is embeddable if and only if M admits a real analytic atlas with respect to which the conformal structure G contains a real analytic representative* (a generalization of the construction and embeddability theorem to n dimensions is due to H. Rossi [355]). When M is a totally umbilical real hypersurface of a real analytic 4-dimensional Lorentzian manifold, the twistor CR manifold N coincides with R. Penrose's CR manifold, cf. [342]. We end this section by giving a brief account of the original R. Penrose construction (cf. op. cit.). Let $M = (\mathbf{R}^4, S)$ be the *Minkowski space*, where

$$S(x, x) = x_0^2 - x_1^2 - x_2^2 - x_3^2, \quad x = (x_0, x_1, x_2, x_3) \in \mathbf{R}^4.$$

We identify $\mathbf{R}^4$ with the vector space $H(2)$ of 2×2 Hermitian matrices by

$$\phi(x) = \begin{pmatrix} x_0 + x_3 & x_1 + i x_2 \\ x_1 - i x_2 & x_0 - x_3 \end{pmatrix}, x \in \mathbf{R}^4.$$

Then $S(x, x) = \det \phi(x)$. Let

$$\eta = \begin{pmatrix} 0 & \sigma_0 \\ \sigma_0 & 0 \end{pmatrix} \in \mathcal{M}_4(\mathbf{R}),$$

where

$$\sigma_0 = \begin{pmatrix} 1 & 0 \\ 0 & 1 \end{pmatrix}.$$

Definition 1.21. Let $\mathbf{T} = (\mathbf{C}^4, \Sigma)$ be the *twistor space*, where $\Sigma(W) = \overline{W}\eta W^t$, $W \in \mathbf{C}^4$. According to R. Penrose [342], $W \in \mathbf{T}$ and $x \in M$ are *incident* if

$$(W_2, W_3) = \frac{1}{i\sqrt{2}}(W_0, W_1)\phi(x). \tag{1.32}$$

□

Let $W \in \mathbf{T}$. If there is $x \in M$ incident with W then

$$\Sigma(W) = 0. \tag{1.33}$$

Definition 1.22. Let $\mathbf{P}(\mathbf{T}) = (\mathbf{T} \setminus \{0\})/\tilde{\mathbf{C}}$ be the *projective twistor space*. Let $[W]$ denote the point of $\mathbf{P}(\mathbf{T})$ of homogeneous coordinates W. If $X \subset \mathbf{T}$ then $\mathbf{P}(X) = \{[W] : W \in X \setminus \{0\}\}$. Let $\mathbf{T}_+$ (respectively $\mathbf{T}_-$) consist of all W such that $\Sigma(W) > 0$ (respectively $\Sigma(W) < 0$). Also, let $\mathbf{T}_0$ be the hypersurface in $\mathbf{T}$ defined by (1.33). □

Then $\mathbf{P}(\mathbf{T}_\pm)$ are open subsets of $\mathbf{P}(\mathbf{T})$ and $\mathbf{P}(\mathbf{T}_0)$ is their common boundary. According to R. Penrose [342], each point of $\mathbf{P}(\mathbf{T}_+)$ describes a right-handed (helicity $s > 0$) "classical" spinning photon. Moreover,

$$\dim_{\mathbf{R}} \mathbf{P}(\mathbf{T}_+) = 6,$$

and, roughly speaking, two of these real dimensions correspond to the direction of motion, one to energy, and three to position. Also, points in $\mathbf{P}(\mathbf{T}_-)$ represent the left-handed (helicity $s < 0$) "classical" photons. For our purposes, the *helicity* $s : \mathbf{T} \to \mathbf{R}$ is given[3] by $s(W) = \frac{1}{2}\Sigma(W)$, $W \in \mathbf{T}$.

Let I be the projective line $I = \{[W] : W_0 = W_1\}$. Since (1.32) is homogeneous with respect to W, the incidence relation on $\mathbf{T} \times M$ naturally induces an incidence relation on $\mathbf{P}(\mathbf{T}) \times M$. Then $\mathbf{P}(\mathbf{T}_0) \setminus I$ *consists of all points* $[W]$ *of the projective twistor space that are incident with some* $x \in M$. Indeed, let $[W]$ be incident with some $x \in M$. Then, as observed before, the homogeneous coordinates W satisfy (1.33), i.e., $[W] \in \mathbf{P}(\mathbf{T}_0)$. On the other hand, $[W] \notin I$, because (by (1.32)) no element of the form $[(0, 0, W_2, W_3)]$ may be incident with some $x \in M$. To check the opposite inclusion, let $[W] \in \mathbf{P}(\mathbf{T}_0) \setminus I$ and denote by $N_{[W]}$ the set of all $x \in M$ that are incident with $[W]$. Note that (1.32) may be written as

$$W_1 x_1 + W_0 x_3 = i\sqrt{2}W_2 - W_0 x_0 + iW_1 x_2$$
$$\overline{W}_0 x_1 - \overline{W}_1 x_3 = -i\sqrt{2}\overline{W}_3 - \overline{W}_1 x_0 + i\overline{W}_0 x_2$$

and $\begin{vmatrix} W_1 & W_0 \\ \overline{W}_0 & -\overline{W}_1 \end{vmatrix} \neq 0$ because $[W] \notin I$. Hence $N_{[W]} \neq \emptyset$. □

We shall need the following lemma:

[3] It may be shown that $S_a = s\, p_a$, where p_a is the *momentum* (of a massless particle in flat space-time) and S_a the *Pauli–Lubanski spin vector*; cf., e.g., [342], p. 69.

Lemma 1.1. *Let* $(W_0, W_1) \in \mathbf{C}^2$. *There is a unique null vector* $v \in M$, *i.e.*, $S(v, v) = 0$, *such that*

$$\phi(v) = \begin{pmatrix} |W_1|^2 & -W_1\overline{W}_0 \\ -W_0\overline{W}_1 & |W_0|^2 \end{pmatrix}. \tag{1.34}$$

Let $f : \mathbf{C}^2 \to M$ be given by $f(W_0, W_1) = v$, where v is determined by (1.34). Then

$$f(\lambda(W_0, W_1)) = |\lambda|^2 f(W_0, W_1)$$

for any $\lambda \in \mathbf{C}$.

Proposition 1.8. *Let* $[W] \in \mathbf{P}(\mathbf{T}_0) \setminus I$ *and* $x_0 \in N_{[W]}$. *Then*

$$N_{[W]} = \{x_0 + tv : t \in \mathbf{R}\}, \tag{1.35}$$

where $v = f(W_0, W_1)$. *Hence* $N_{[W]}$ *is a null geodesic of* M.

Proof. Since $\phi : \mathbf{R}^4 \to H(2)$ is $\mathbf{R}$-linear and (by (1.34)) $(W_0, W_1)\phi(v) = 0$, the inclusion $\supseteq$ in (1.35) follows from $(W_0, W_1)\phi(x_0+tv) = i\sqrt{2}(W_2, W_3)$. The opposite inclusion is a bit trickier. For each $(W_0, W_1) \in \mathbf{C}^2 \setminus \{0\}$ set

$$X_{(W_0, W_1)} = \{\phi(y) : y \in M,\ (W_0, W_1)\phi(y) = 0\}.$$

Let $x \in N_{[W]}$. Then $\phi(x - x_0) \in X_{(W_0, W_1)}$. On the other hand, $X_{(W_0, W_1)}$ may be explicitly described as

$$X_{(W_0, W_1)} = \{t\phi(f(W_0, W_1)) : t \in \mathbf{R}\};$$

hence $x - x_0 = tf(W_0, W_1)$ for some $t \in \mathbf{R}$. □

Let Ω_0 be the space of all null geodesics in M, i.e., if $N \in \Omega_0$ then $N = \{x_0 + tv : t \in \mathbf{R}\}$ for some $x_0 \in M$, $v \in \mathbf{R}^4 \setminus \{0\}$, $S(v, v) = 0$. Then

$$\mathbf{P}(\mathbf{T}_0) \setminus I \simeq \Omega_0$$

(a bijection). Indeed, the map $\mathbf{P}(\mathbf{T}_0) \setminus I \to \Omega_0,\ [W] \mapsto N_{[W]}$, is bijective.

Let $x \in M$ and let us set

$$L_x = \{[W] \in \mathbf{P}(\mathbf{T}) : (W_0, W_1)\phi(x) = i\sqrt{2}(W_2, W_3)\}.$$

Then L_x is a projective line entirely contained in $\mathbf{P}(\mathbf{T}_0) \setminus I$. Actually, the map $x \mapsto L_x$ is a bijection of M onto the set of all projective lines entirely contained in $\mathbf{P}(\mathbf{T}_0) \setminus I$.

Definition 1.23. The projective lines in $\mathbf{P}(\mathbf{T}_0)$ that meet I are the *points at infinity* for M, and the totality of all projective lines in $\mathbf{P}(\mathbf{T}_0)$ forms the *conformal compactification* $M^\sharp$ of M. The family $(\Omega_0)_x$ of all light rays (null geodesics) through a point $x \in M$ may be thought of as the *field of vision* of an observer situated at x. □

Since $(\Omega_0)_x = \{N_{[W]} : [W] \in L_x\} \simeq L_x$, the field of vision $(\Omega_0)_x$ has a natural structure of a Riemann surface.

R. Penrose has proposed (cf. [341, 342]) a program aiming at a reconstruction of the foundations of relativistic physics involving a transformation of the latter into a part of complex geometry on the space of complex light lines. A development of Penrose's program is carried out in [197]. G.M. Henkin shows (cf. op. cit.) that the theory of classical Yang–Mills fields, and Higgs and Dirac fields on Minkowski spaces can be transformed (via the so-called *Radon–Penrose transform*) into the theory of tangential Cauchy–Riemann equations on a 1-concave submanifold of the twistor space.

1.2 The Tanaka–Webster connection

Let $(M, T_{1,0}(M))$ be a nondegenerate CR manifold and θ a fixed pseudo-Hermitian structure. Let T be the characteristic direction of (M, θ). If ∇ is a linear connection on M then let T_∇ denote its torsion tensor field.

Definition 1.24. We say that T_∇ is *pure* if

$$T_\nabla(Z, W) = 0, \tag{1.36}$$

$$T_\nabla(Z, \overline{W}) = 2i L_\theta(Z, \overline{W})T, \tag{1.37}$$

$$\tau \circ J + J \circ \tau = 0, \tag{1.38}$$

for any $Z, W \in T_{1,0}(M)$. Here $\tau : T(M) \to T(M)$ is defined by

$$\tau X = T_\nabla(T, X),$$

for any $X \in T(M)$. □

On each nondegenerate CR manifold (of hypersurface type) on which a pseudo-Hermitian structure has been fixed, there is a canonical linear connection compatible with both the complex structure of the Levi distribution and the Levi form. Precisely, we may state the following theorem:

Theorem 1.3. *Let $(M, T_{1,0}(M))$ be a nondegenerate CR manifold and θ a pseudo-Hermitian structure on M. Let T be the characteristic direction of (M, θ) and J the complex structure in $H(M)$ (extended to an endomorphism of $T(M)$ by requiring that $JT = 0$). Let g_θ be the Webster metric of (M, θ). There is a unique linear connection ∇ on M satisfying the following axioms:*

(i) $H(M)$ is parallel with respect to ∇, that is,

$$\nabla_X \Gamma^\infty(H(M)) \subseteq \Gamma^\infty(H(M))$$

for any $X \in \mathcal{X}(M)$.
(ii) $\nabla J = 0, \quad \nabla g_\theta = 0.$
(iii) The torsion T_∇ of ∇ is pure.

Definition 1.25. The connection ∇ given by Theorem 1.3 is the *Tanaka–Webster connection* of $(M, T_{1,0}(M), \theta)$. The vector-valued 1-form τ on M is the *pseudo-Hermitian torsion* of ∇. □

The Tanaka–Webster connection was first built by N. Tanaka in [398], a monograph that appears to have remained little known to Western scientists up to the early 1980s. Independently from the work of N. Tanaka, S. Webster gave (cf. [422]) another approach to the Tanaka–Webster connection (though only as a connection in $T_{1,0}(M)$). Clearly, since both the Levi distribution $H(M)$ and its complex structure J are parallel with respect to the Tanaka–Webster connection ∇, it parallelizes the eigenbundles $T_{1,0}(M)$ and $T_{0,1}(M)$ of J and therefore ∇ is reducible to a connection in $T_{1,0}(M)$. The proof of Theorem 1.3 in [422] is based on exterior differential calculus with respect to *admissible* coframes in $T_{1,0}(M)^*$ (arising from the integrability of the given CR structure). We adopt the proof in [398] and relegate all local considerations to further sections.

Let $\pi_+ : T(M) \otimes \mathbf{C} \to T_{1,0}(M)$, respectively $\pi_- : T(M) \otimes \mathbf{C} \to T_{0,1}(M)$, be the natural projections associated with the direct sum decomposition

$$T(M) \otimes \mathbf{C} = T_{1,0}(M) \oplus T_{0,1}(M) \oplus \mathbf{C}T\,.$$

Then $\pi_- Z = \overline{\pi_+ \overline{Z}}$ for any $Z \in T(M) \otimes \mathbf{C}$. We establish first the uniqueness of a linear connection ∇ on M obeying the axioms (i)–(iii). Let $Y, Z \in T_{1,0}(M)$. We may write (by (1.37)) the identity

$$[\overline{Y}, Z] = \nabla_{\overline{Y}} Z - \nabla_Z \overline{Y} + 2i L_\theta(Z, \overline{Y}) T,$$

whence (as $\nabla_{\overline{Y}} Z \in \Gamma^\infty(T_{1,0}(M))$ and $\nabla_Z \overline{Y} \in \Gamma^\infty(T_{0,1}(M))$) we obtain

$$\pi_+[\overline{Y}, Z] = \nabla_{\overline{Y}} Z. \tag{1.39}$$

Let $\Omega = -d\theta$. Then $T \rfloor \Omega = 0$. The axiom $\nabla g_\theta = 0$ may be written as

$$X(g_\theta(Y, Z)) = g_\theta(\nabla_X Y, Z) + g_\theta(Y, \nabla_X Z)$$

for any $X, Y, Z \in T(M)$. In particular, for $Y = T$ we obtain

$$X(\theta(Z)) = g_\theta(\nabla_X T, Z) + \theta(\nabla_X Z). \tag{1.40}$$

We distinguish two cases: (I) $Z \in H(M)$ and (II) $Z = T$. If $Z \in H(M)$ then (1.40) yields $g_\theta(\nabla_X T, Z) = 0$ or

$$\pi_H \nabla_X T = 0,$$

where $\pi_H : T(M) \to H(M)$ is the natural projection associated with the direct sum decomposition (1.20). Finally, set $Z = T$ in (1.40) to obtain

$$2\theta(\nabla_X T) = 0.$$

By taking into account (1.20) we may conclude that

$$\nabla T = 0. \tag{1.41}$$

Note that

$$\nabla \Omega = 0,$$

as a consequence of axiom (ii) in Theorem 1.3. Therefore

$$X(\Omega(Y, \overline{Z})) = \Omega(\nabla_X Y, \overline{Z}) + \Omega(Y, \nabla_X \overline{Z})$$

for any $X, Y, Z \in T_{1,0}(M)$. Using (1.39) we may rewrite this identity as

$$\Omega(\nabla_X Y, \overline{Z}) = X(\Omega(Y, \overline{Z})) - \Omega(Y, \pi_-[X, \overline{Z}]), \tag{1.42}$$

which, in view of the nondegeneracy of Ω on $H(M)$, determines $\nabla_X Y$ for any $X, Y \in T_{1,0}(M)$. We shall need the bundle endomorphism K_T given by

$$K_T = -\frac{1}{2} J \circ (\mathcal{L}_T J),$$

where $\mathcal{L}$ denotes the Lie derivative. On the other hand (by $\nabla T = 0$),

$$\nabla_T X = \tau X + \mathcal{L}_T X \tag{1.43}$$

for any $X \in T(M)$. Note that as a consequence of axiom (1.38), τ is $H(M)$-valued. We may use $\nabla J = 0$ and (1.43) to perform the following calculation:

$$\begin{aligned}
0 &= (\nabla_T J)X = \nabla_T J X - J \nabla_T X \\
&= \tau(JX) + \mathcal{L}_T(JX) - J(\tau X + \mathcal{L}_T X) \\
&= -J\tau X + \mathcal{L}_T(JX) - J(\tau X + \mathcal{L}_T X) = -2J\tau X + (\mathcal{L}_T J)X \, .
\end{aligned}$$

Let us apply J in both members of this identity and use the fact that τ is $H(M)$-valued to obtain

$$\tau = K_T \, .$$

Therefore (1.43) may be rewritten as

$$\nabla_T X = K_T X + \mathcal{L}_T X \tag{1.44}$$

for any $X \in T(M)$. At this point, the identities (1.39), (1.41)–(1.42), and (1.44) account for the uniqueness statement in Theorem 1.3. To prove existence, we must show that the same expressions may be taken as the definition of a linear connection obeying (i)–(iii) in Theorem 1.3. This is, however, somewhat trickier, and may be carried out as follows. Let $\nabla : \Gamma^\infty(T(M) \otimes \mathbf{C}) \times \Gamma^\infty(T(M) \otimes \mathbf{C}) \to \Gamma^\infty(T(M) \otimes \mathbf{C})$ be the differential operator defined by

$$\begin{aligned} &\nabla_{\overline{X}} Y = \pi_+[\overline{X}, Y], && \nabla_X \overline{Y} = \overline{\nabla_{\overline{X}} Y}, \\ &\nabla_X Y = U_{XY}, && \nabla_{\overline{X}} \overline{Y} = \overline{\nabla_X Y}, \\ &\nabla_T X = \mathcal{L}_T X + K_T X, && \nabla_T \overline{X} = \overline{\nabla_T X}, \\ &\nabla T = 0, \end{aligned}$$

for any $X, Y \in T_{1,0}(M)$. Here

$$U : \Gamma^\infty(T_{1,0}(M)) \times \Gamma^\infty(T_{1,0}(M)) \to \Gamma^\infty(T_{1,0}(M))$$

is defined by

$$\Omega(U_{XY}, \overline{Z}) = X(\Omega(Y, \overline{Z})) - \Omega(Y, \pi_-[X, \overline{Z}]),$$

for any $X, Y, Z \in T_{1,0}(M)$. Then ∇ is a linear connection on $T(M)$. The rest of the proof is devoted to showing that ∇ satisfies (i)–(iii) in Theorem 1.3. We need some preparation. Note that

$$\mathcal{L}_T \theta = 0.$$

Indeed,

$$\begin{aligned} (\mathcal{L}_T \theta) X &= T(\theta(X)) - \theta(\mathcal{L}_T X) = T(\theta(X)) - X(\theta(T)) - \theta([T, X]) \\ &= 2(d\theta)(T, X) = 2(T \rfloor d\theta) X = 0, \end{aligned}$$

for any $X \in T(M)$. Using $\mathcal{L}_T \theta = 0$ and $J^2 = -I + \theta \otimes T$ we may perform the following calculation:

$$\begin{aligned} &(J \circ (\mathcal{L}_T J) + (\mathcal{L}_T J) \circ J)\, X \\ &\qquad = J\,([T, JX] - J[T, X]) + [T, J^2 X] - J[T, JX] \\ &\qquad = (T(\theta(X)) - \theta([T, X]) - X(\theta(T)))\, T = 2(d\theta)(T, X) T = 0. \end{aligned}$$

We may conclude that

$$J \circ (\mathcal{L}_T J) + (\mathcal{L}_T J) \circ J = 0. \tag{1.45}$$

Note that

$$\mathcal{L}_T \Gamma^\infty(H(M)) \subseteq \Gamma^\infty(H(M)).$$

Indeed,

$$\theta(\mathcal{L}_T X) = -2(d\theta)(T, X) = 0$$

for any $X \in H(M)$. Note that

$$J \circ K_T + K_T \circ J = 0 \tag{1.46}$$

as a direct consequence of (1.45) and the definition (of K_T). Next, we need to notice that

$$(\mathcal{L}_T + K_T)\,\Gamma^\infty(T_{1,0}(M)) \subseteq \Gamma^\infty(T_{1,0}(M)). \tag{1.47}$$

To prove (1.47), let $X \in T_{1,0}(M)$ (so that $JX = iX$). Clearly $\mathcal{L}_T X + K_T X \in H(M)$. Moreover (by (1.46)),

$$\begin{aligned} J(\mathcal{L}_T X + K_T X) &= -(\mathcal{L}_T J)X + i\mathcal{L}_T X + JK_T X \\ &= (\mathcal{L}_T J)(J^2 X) + i\mathcal{L}_T X - K_T JX \\ &= -J \circ (\mathcal{L}_T J)(JX) + i(\mathcal{L}_T X - K_T X) \\ &= -iJ(\mathcal{L}_T J)X + i\mathcal{L}_T X + \frac{i}{2} J \circ (\mathcal{L}_T J)X = i(\mathcal{L}_T X + K_T X). \end{aligned}$$

Let $X, Y \in T_{1,0}(M)$. Then

$$\begin{aligned} 0 = (d^2\theta)(T, X, \overline{Y}) &= -(d\Omega)(T, X, \overline{Y}) \\ &= -\frac{1}{3}(T(\Omega(X, \overline{Y}) + \overline{Y}(\Omega(T, X)) + X(\Omega(\overline{Y}, T)) \\ &\quad - \Omega([T, X], \overline{Y}) - \Omega([\overline{Y}, T], X) - \Omega([X, \overline{Y}], T)) \\ &= -\frac{1}{3}(\mathcal{L}_T \Omega)(X, \overline{Y}). \end{aligned}$$

We have obtained the identity

$$(\mathcal{L}_T \Omega)(X, \overline{Y}) = 0 \tag{1.48}$$

for any $X, Y \in T_{1,0}(M)$. Note that

$$\Omega(JX, JY) = \Omega(X, Y)$$

for any $X, Y \in T(M)$ (as a consequence of the integrability of the CR structure). We wish to show that

$$\Omega(K_T X, Y) + \Omega(X, K_T Y) = 0 \tag{1.49}$$

for any $X, Y \in T(M)$. To this end we perform the following calculation:

$$\begin{aligned} &\Omega(K_T X, Y) + \Omega(X, K_T Y) \\ &\quad = \frac{1}{2}(\Omega(\mathcal{L}_T JX, JY) - \Omega(J\mathcal{L}_T X, JY) + \Omega(JX, \mathcal{L}_T JY) - \Omega(JX, J\mathcal{L}_T Y)) \\ &\quad = \frac{1}{2}\left((\mathcal{L}_T \Omega)(X, Y) - (\mathcal{L}_T \Omega)(JX, JY)\right) = 0 \end{aligned}$$

and (1.49) is proved. Next

$$X(\Omega(Y, \overline{Z})) + Y(\Omega(\overline{Z}, X)) + \Omega(\overline{Z}, [X, Y]) \\ + \Omega(Y, \pi_-[\overline{Z}, X]) + \Omega(X, \pi_-[Y, \overline{Z}]) = 0 \quad (1.50)$$

for any $X, Y, Z \in T_{1,0}(M)$, as a consequence of

$$(d\Omega)(X, Y, \overline{Z}) = 0.$$

Finally, we may use (1.46)–(1.47) and (1.49)–(1.50) to check the axioms. Note that

$$T_\nabla(X, Y) \in T_{1,0}(M)$$

for any $X, Y \in T_{1,0}(M)$ because of

$$T_\nabla(X, Y) = U_{XY} - U_{YX} - [X, Y] \in T_{1,0}(M).$$

Therefore

$$\begin{aligned}\Omega(T_\nabla(X, Y), \overline{Z}) &= \Omega(U_{XY}, \overline{Z}) - \Omega(U_{YX}, \overline{Z}) - \Omega([X, Y], \overline{Z}) \\ &= X(\Omega(Y, \overline{Z})) - \Omega(Y, \pi_-[X, \overline{Z}]) - Y(\Omega(X, \overline{Z})) \\ &\quad + \Omega(X, \pi_-[Y, \overline{Z}]) - \Omega([X, Y], \overline{Z}) = 0\end{aligned}$$

(by (1.50)) for any $Z \in T_{1,0}(M)$. By the nondegeneracy of Ω on $H(M)$ we get $T_\nabla(X, Y) = 0$ (i.e., (1.36) holds). At this point we may check (1.38). From the definitions,

$$\begin{aligned}\tau X &= T_\nabla(T, X) = \nabla_T X - \nabla_X T - [T, X] \\ &= \nabla_T X - [T, X] = \mathcal{L}_T X + K_T X - [T, X] = K_T X,\end{aligned}$$

and therefore (by (1.46))

$$(\tau \circ J + J \circ \tau)X = K_T J X + J K_T X = 0.$$

Next, let us prove that $\nabla g_\theta = 0$. This is equivalent to $\nabla \Omega = 0$ and $\nabla T = 0$. Note that $\nabla_X \Omega$ vanishes on complex tangent vectors of the same type (i.e., both in $T_{1,0}(M)$, or both in $T_{0,1}(M)$). Let $Y, Z \in T_{1,0}(M)$. We have

$$\begin{aligned}(\nabla_X \Omega)(Y, \overline{Z}) &= X(\Omega(Y, \overline{Z})) - \Omega(\nabla_X Y, \overline{Z}) - \Omega(Y, \nabla_X \overline{Z}) \\ &= X(\Omega(Y, \overline{Z})) - \Omega(U_{XY}, \overline{Z}) - \Omega(Y, \pi_-[X, \overline{Z}]) = 0,\end{aligned}$$

for any $X \in H(M)$, by definition. Moreover,

$$\begin{aligned}(\nabla_T \Omega)(X, \overline{Y}) &= T(\Omega(X, \overline{Y})) - \Omega(\nabla_T X, \overline{Y}) - \Omega(X, \nabla_T \overline{Y}) \\ &= T(\Omega(X, \overline{Y})) - \Omega(\mathcal{L}_T X + K_T X, \overline{Y}) - \Omega(X, \mathcal{L}_T \overline{Y} + K_T \overline{Y}) \\ &= T(\Omega(X, \overline{Y})) - \Omega(\mathcal{L}_T X, \overline{Y}) - \Omega(X, \mathcal{L}_T \overline{Y}) \\ &\quad - \Omega(K_T X, \overline{Y}) - \Omega(X, K_T \overline{Y}) \\ &= T(\Omega(X, \overline{Y})) - \Omega(\mathcal{L}_T X, \overline{Y}) - \Omega(X, \mathcal{L}_T \overline{Y}) = (\mathcal{L}_T \Omega)(X, \overline{Y}) = 0,\end{aligned}$$

(by (1.48)). Let us check that $H(M)$ is parallel with respect to ∇. By definition, $\nabla_{\overline{X}}Y$, $\nabla_X Y$, $\nabla_T X \in T_{1,0}(M)$ for any $X, Y \in T_{1,0}(M)$, so that ∇ parallelizes $T_{1,0}(M)$ (and therefore, by complex conjugation, since ∇ is a real operator, it parallelizes $T_{0,1}(M)$, as well). Finally, if $Y = V + \overline{V}$, $V \in T_{1,0}(M)$, then

$$\begin{aligned}\nabla_X Y &= \nabla_X V + \nabla_X \overline{V} \\ &= \nabla_X V + \overline{\nabla_X V} \in \mathrm{Re}\{T_{1,0}(M) \oplus T_{0,1}(M)\} = H(M),\end{aligned}$$

for any $X \in T(M)$. It remains to be checked that $\nabla J = 0$. Indeed,

$$(\nabla_X J)Y = \nabla_X JY - J\nabla_X Y = \nabla_X iY - i\nabla_X Y = 0,$$

for any $X \in T(M)$ and $Y \in T_{1,0}(M)$ (since $\nabla_X Y \in T_{1,0}(M)$), etc. Finally, let us check (1.37):

$$\begin{aligned}T_\nabla(X, \overline{Y}) &= \nabla_X \overline{Y} - \nabla_{\overline{Y}} X - [X, \overline{Y}] \\ &= \pi_-[X, \overline{Y}] + \pi_+[X, \overline{Y}] - [X, \overline{Y}] \\ &= -\theta([X, \overline{Y}])T = 2(d\theta)(X, \overline{Y})T = 2iL_\theta(X, \overline{Y})T,\end{aligned}$$

for any $X, Y \in T_{1,0}(M)$. Theorem 1.3 is completely proved. □

We end the section with the following remark. Let ∇ be the Tanaka–Webster connection of (M, θ). The purity properties (1.36)–(1.38) of T_∇ may be formulated compactly as

$$\pi_+ T_\nabla(Z, \overline{W}) = 0$$

for any $Z \in T_{1,0}(M)$, $W \in T(M) \otimes \mathbf{C}$. The proof is left as an exercise to the reader.

1.3 Local computations

Given a CR manifold M (of hypersurface type), as long as one is concerned with differential-geometric applications, e.g., calculations of characteristic classes of $T_{1,0}(M)$, one needs a connection adapted to the given CR structure. Assume M to be nondegenerate. Then its Levi distribution $H(M)$ is as far from being integrable as possible, so that any linear connection ∇ on M that is reducible to a connection in $H(M)$ must have a nonvanishing torsion tensor field (any distribution parallel with respect to a torsion-free linear connection is involutive). Also, since the Levi form L_θ is precisely the obstruction to the integrability of $H(M)$, one expects L_θ to be somehow encoded in T_∇. By Theorem 1.3 we see that this is indeed the case, and (1.37) gives the explicit relation between T_∇ and L_θ for the Tanaka–Webster connection. Also, T_∇ may have some additional nontrivial part, the pseudo-Hermitian torsion τ, the properties of which are soon to be investigated.

1.3.1 Christoffel symbols

For all local calculations, let $\{T_\alpha : 1 \le \alpha \le n\}$ be a local frame of $T_{1,0}(M)$ defined on the open set $U \subseteq M$. Since the Tanaka–Webster connection parallelizes the eigenbundles of J there exist uniquely defined complex 1-forms $\omega_\beta^\alpha \in \Gamma^\infty(T^*(M) \otimes \mathbf{C})$ (locally defined on U) such that

$$\nabla T_\beta = \omega_\beta^\alpha \otimes T_\alpha\,.$$

These are the *connection* 1*-forms*. Let us set $T_{\overline{\alpha}} = \overline{T_\alpha}$. Then

$$\{T_1, \ldots, T_n, T_{\overline{1}}, \ldots, T_{\overline{n}}, T\}$$

is a frame of $T(M) \otimes \mathbf{C}$ on U. Let us set $\omega_{\overline{\beta}}^{\overline{\alpha}} = \overline{\omega_\beta^\alpha}$. Then (since ∇ is a real operator) we have

$$\nabla T_{\overline{\beta}} = \omega_{\overline{\beta}}^{\overline{\alpha}} \otimes T_{\overline{\alpha}}.$$

We shall need the *Christoffel symbols* $\Gamma_{A\beta}^\alpha : U \to \mathbf{C}$ given by

$$\Gamma_{A\beta}^\alpha = \omega_\beta^\alpha(T_A).$$

Unless otherwise stated, we adopt the following conventions as to the range of indices. The Greek indices $\alpha, \beta, \gamma, \ldots$ vary from 1 to n, while the block Latin indices $A, B, C, \ldots$ vary in $\{0, 1, \ldots, n, \bar{1}, \ldots, \bar{n}\}$, with the convention $T_0 = T$. Therefore

$$\nabla_{T_\gamma} T_\beta = \Gamma_{\gamma\beta}^\alpha T_\alpha,\ \nabla_{T_{\overline{\gamma}}} T_\beta = \Gamma_{\overline{\gamma}\beta}^\alpha T_\alpha,\ \nabla_T T_\beta = \Gamma_{0\beta}^\alpha T_\alpha\,.$$

In applications, given a concrete nondegenerate CR manifold (of hypersurface type) we shall compute its Tanaka–Webster connection by computing its Christoffel symbols with respect to an arbitrary (local) frame of $T_{1,0}(M)$. We denote by

$$h_{\alpha\overline{\beta}} = L_\theta(T_\alpha, T_{\overline{\beta}})$$

the components of the Levi form. Recall that

$$\nabla_X Y = U_{XY}\,,$$

for any $X, Y \in T_{1,0}(M)$. Let us set for simplicity

$$U_{\alpha\beta} = U_{T_\alpha T_\beta}\,.$$

Then on the one hand,

$$U_{\alpha\beta} = \Gamma_{\alpha\beta}^\gamma T_\gamma\,.$$

On the other hand, taking into account that

$$\Omega(T_\alpha, T_{\overline{\beta}}) = -i h_{\alpha\overline{\beta}}$$

we have

$$
\begin{aligned}
-i\Gamma^{\alpha}_{\gamma\beta}h_{\alpha\overline{\sigma}} &= \Omega(\Gamma^{\alpha}_{\gamma\beta}T_{\alpha}, T_{\overline{\sigma}}) = \Omega(U_{\gamma\beta}, T_{\overline{\sigma}}) \\
&= T_{\gamma}(\Omega(T_{\beta}, T_{\overline{\sigma}})) - \Omega(T_{\beta}, \pi_{-}[T_{\gamma}, T_{\overline{\sigma}}]) \\
&= -iT_{\gamma}(h_{\beta\overline{\sigma}}) - \Omega(T_{\beta}, [T_{\gamma}, T_{\overline{\sigma}}])
\end{aligned}
$$

and contraction by $h^{\overline{\sigma}\mu}$ leads to

$$
\Gamma^{\alpha}_{\gamma\beta} = h^{\overline{\sigma}\alpha}\left(T_{\gamma}(h_{\beta\overline{\sigma}}) - g_{\theta}(T_{\beta}, [T_{\gamma}, T_{\overline{\sigma}}])\right). \tag{1.51}
$$

Throughout $[h^{\overline{\alpha}\beta}] = [h_{\alpha\overline{\beta}}]^{-1}$, so that

$$
h^{\overline{\alpha}\beta}h_{\beta\overline{\sigma}} = \delta^{\alpha}_{\sigma}.
$$

The next step is to compute the Christoffel symbols $\Gamma^{\alpha}_{\overline{\gamma}\beta}$. To this end, by recalling the definitions, we start from

$$
\pi_{+}[T_{\overline{\gamma}}, T_{\beta}] = \nabla_{T_{\overline{\gamma}}}T_{\beta} = \Gamma^{\alpha}_{\overline{\gamma}\beta}T_{\alpha}.
$$

Let us take the inner product (using the Webster metric g_{θ}) with $T_{\overline{\mu}}$ and contract the resulting identity with $h^{\overline{\mu}\beta}$. This procedure furnishes

$$
\Gamma^{\alpha}_{\overline{\gamma}\beta} = h^{\overline{\mu}\alpha}g_{\theta}([T_{\overline{\gamma}}, T_{\beta}], T_{\overline{\mu}}). \tag{1.52}
$$

Note that

$$
\forall X \in \Gamma^{\infty}(H(M)) : [T, X] \in \Gamma^{\infty}(H(M)).
$$

Using this observation we may compute the remaining Christoffel symbols $\Gamma^{\alpha}_{0\beta}$ of the Tanaka–Webster connection. Indeed, recalling the definitions, we have

$$
\begin{aligned}
\Gamma^{\alpha}_{0\beta}T_{\alpha} &= \nabla_{T}T_{\beta} = \mathcal{L}_{T}T_{\beta} + K_{T}T_{\beta} = [T, T_{\beta}] - \frac{1}{2}J \circ (\mathcal{L}_{T}J)T_{\beta} \\
&= [T, T_{\beta}] - \frac{1}{2}J\left([T, JT_{\beta}] - J[T, T_{\beta}]\right) = \frac{1}{2}\left([T, T_{\beta}] \quad iJ[T, T_{\beta}]\right).
\end{aligned}
$$

Taking the inner product with $T_{\overline{\mu}}$ and contracting with $h^{\overline{\mu}\rho}$ in the resulting identity we obtain

$$
\Gamma^{\alpha}_{0\beta} = h^{\overline{\mu}\alpha}g_{\theta}([T, T_{\beta}], T_{\overline{\mu}}). \tag{1.53}
$$

Summing up (by (1.51)–(1.53)) we have proved the following identities:

$$
\begin{aligned}
\Gamma^{\alpha}_{\gamma\beta} &= h^{\bar{\mu}\alpha}\left\{T_{\gamma}(h_{\beta\bar{\mu}}) - g_{\theta}(T_{\beta}, [T_{\gamma}, T_{\bar{\mu}}])\right\}, \\
\Gamma^{\alpha}_{\bar{\gamma}\beta} &= h^{\bar{\mu}\alpha}g_{\theta}([T_{\bar{\gamma}}, T_{\beta}], T_{\bar{\mu}}), \\
\Gamma^{\alpha}_{0\beta} &= h^{\bar{\mu}\alpha}g_{\theta}([T, T_{\beta}], T_{\bar{\mu}}).
\end{aligned} \tag{1.54}
$$

As an application of (1.54), we compute the (Christoffel symbols of the) Tanaka–Webster connection of the family of CR manifolds $\mathbf{H}_n^{n-\nu}$, $n/2 \leq \nu \leq n$ (which constitute the Heisenberg group).

Let us consider $\mathbf{C}^n \times \mathbf{R}$ with the coordinates $(z^1, \ldots, z^n, t)$. Let us assume that $n \geq 3$ and let $\nu \in \mathbf{Z}$ such that $n/2 \leq \nu \leq n$. Let us set

$$W_j = \frac{\partial}{\partial w^j} + \epsilon^j \, i \, \overline{w^j} \, \frac{\partial}{\partial t}, \tag{1.55}$$

where

$$w^j = \begin{cases} z^j & \text{if } 1 \leq j \leq \nu, \\ \overline{z}^j & \text{if } \nu + 1 \leq j \leq n, \end{cases}$$

and

$$\epsilon^j = \begin{cases} +1 & \text{if } 1 \leq j \leq \nu, \\ -1 & \text{if } \nu + 1 \leq j \leq n, \end{cases}$$

for any $1 \leq j \leq n$. Let us set

$$\mathcal{H}_x = \sum_{j=1}^{n} \mathbf{C} W_{j,x}$$

for any $x \in \mathbf{C}^n \times \mathbf{R}$. Since

$$[W_j, W_k] = 0 \tag{1.56}$$

it follows that $\mathcal{H}$ is a CR structure on $\mathbf{C}^n \times \mathbf{R}$. Let θ_0 be given by (1.25). Then θ_0 is a pseudo-Hermitian structure on $(\mathbf{C}^n \times \mathbf{R}, \mathcal{H})$. Then, the commutation formula

$$[W_j, W_{\overline{k}}] = -i\delta_{jk}(\epsilon_j + \epsilon_k)\frac{\partial}{\partial t} \tag{1.57}$$

yields

$$L_{\theta_0}(W_j, W_{\overline{k}}) = \frac{1}{2}\delta_{jk}(\epsilon_j + \epsilon_k),$$

so that $(\mathbf{C}^n \times \mathbf{R}, \mathcal{H})$ is a nondegenerate CR manifold (of hypersurface type) whose Levi form has signature $(\nu, n - \nu)$. Let $\mathbf{H}_n^{n-\nu}$ denote $\mathbf{C}^n \times \mathbf{R}$ together with the CR structure $\mathcal{H}$ spanned by the complex vector fields (1.55). Then $\mathbf{H}_n^0$ is the Heisenberg group $\mathbf{H}_n$ (carrying the standard strictly pseudoconvex CR structure). Using the commutation formulas (1.56)–(1.57), the identities (1.51)–(1.53) yield

$$\Gamma^{\beta}_{\mu\alpha} = 0, \; \Gamma^{\beta}_{\overline{\mu}\alpha} = 0, \; \Gamma^{\beta}_{0\alpha} = 0.$$

Also $\tau = 0$. Therefore the Tanaka–Webster connection of $\mathbf{H}_n^{n-\nu}$ is flat and has vanishing pseudo-Hermitian torsion. Finally, note that $\mathbf{H}_n^{n-\nu}$ is a Lie group with the group structure

$$(z,t)\cdot(w,s)=(z+w\,,\;t+s+2\,\mathrm{Im}\,Q_{n-\nu}(z,w))\,,$$

where $Q_{n-\nu}$ is the Hermitian form given by

$$Q_{n-\nu}(z,w)=\sum_{j=1}^{n}\epsilon^j z^j\overline{w^j}\,,$$

for any $z,w\in\mathbf{C}^n$. As with $\mathbf{H}_n$, we may provide the following geometric interpretation of $\mathbf{H}_n^{n-\nu}$. Let $M\subset\mathbf{C}^{n+1}$ be the real hypersurface defined by

$$M=\{(z,u+iv):v=Q_{n-\nu}(z,z)\}\,.$$

Then the map $f:\mathbf{H}_n^{n-\nu}\to M$ given by

$$f(z,t)=(z,t+i\;Q_{n-\nu}(z,z))$$

is a CR isomorphism, where M is thought of as carrying the CR structure induced from $\mathbf{C}^{n+1}$. See also S. Kaneyuki [238].

Next, we wish to compute the Tanaka–Webster connection of a CR Lie group. Let $(G,T_{1,0}(G))$ be a nondegenerate CR Lie group, $\mathcal{G}=L(G)$ its Lie algebra, and $\mathbf{a}=j^{-1}T_{1,0}(G)_e$ the CR structure on $\mathcal{G}$ induced by $T_{1,0}(G)$. With the notation and conventions in Section 1.1.7, let $\theta_0\in\mathcal{G}^*$ be an annihilator of $\mathbf{h}$, $\theta_0\neq 0$, and let θ be the corresponding left-invariant pseudo-Hermitian structure of $(G,T_{1,0}(G))$. Let us compute the Tanaka–Webster connection of (G,θ). To this end, let $\{T_\alpha\}$ be a linear basis of $\mathbf{a}$ over $\mathbf{C}$ (hence $\{T_A\}=\{T,T_\alpha,T_{\bar\alpha}\}$ spans $\mathcal{G}\otimes_{\mathbf{R}}\mathbf{C}$). Let us set

$$[T_B,T_C]=C^A_{BC}T_A\,.$$

Of course, the structure constants C^A_{BC} are not independent (because of the skew-symmetry of the Lie algebra product and the Jacobi identity). Also (since $\mathbf{a}\subset\mathcal{G}\otimes_{\mathbf{R}}\mathbf{C}$ is a subalgebra)

$$C^{\bar\gamma}_{\alpha\beta}=0,\quad C^0_{\alpha\beta}=0.$$

Other relations may be obtained from $\overline{[T_A,T_B]}=[T_{\bar A},T_{\bar B}]$ (with the convention $\bar 0=0$). The identity

$$[T_\alpha,T_{\bar\beta}]=\Gamma^{\bar\gamma}_{\alpha\bar\beta}T_{\bar\gamma}-\Gamma^{\gamma}_{\bar\beta\alpha}T_\gamma-2ih_{\alpha\bar\beta}T$$

leads to

$$\Gamma^{\bar\gamma}_{\alpha\bar\beta}=C^{\bar\gamma}_{\alpha\bar\beta}\,,\quad h_{\alpha\bar\beta}=\frac{i}{2}C^0_{\alpha\bar\beta}\,.$$

Similarly, the identity

$$[T,T_\beta]=\Gamma^{\gamma}_{0\beta}T_\gamma-A^{\bar\gamma}_{\beta}T_{\bar\gamma}$$

leads to

$$\Gamma^{\gamma}_{0\beta} = C^{\gamma}_{0\beta}\,, \quad A^{\bar\gamma}_{\beta} = -C^{\bar\gamma}_{0\beta}\,.$$

The coefficients $A^{\overline{\alpha}}_{\beta}$ (describing locally the pseudo-Hermitian torsion) are introduced shortly after Lemma 1.2. Finally (by (1.51)),

$$\Gamma^{\gamma}_{\alpha\beta} = -h^{\gamma\bar\mu} C^{\bar\lambda}_{\alpha\bar\mu} h_{\beta\bar\lambda}\,.$$

1.3.2 The pseudo-Hermitian torsion

We proceed by establishing a few elementary properties of the pseudo-Hermitian torsion τ of the Tanaka–Webster connection.

Lemma 1.2. $\tau(T_{1,0}(M)) \subseteq T_{0,1}(M)$.

The proof follows from (1.38). By Lemma 1.2 there exist uniquely defined C^∞ functions $A^{\overline{\alpha}}_{\beta} : U \to \mathbf{C}$ such that

$$\tau\, T_\beta = A^{\overline{\alpha}}_{\beta} T_{\overline{\alpha}}\,.$$

The calculation

$$\tau\, T_\alpha = T_\nabla(T, T_\alpha) = \nabla_T T_\alpha - [T, T_\alpha]$$

shows that

$$A^{\overline{\beta}}_{\alpha} T_{\overline{\beta}} = -\pi_-[T, T_\alpha]\,.$$

Let us take the inner product with T_μ and contract the resulting identity with $h^{\mu\overline{\sigma}}$. This procedure leads to

$$A^{\overline{\beta}}_{\alpha} = -h^{\mu\overline{\beta}} g_\theta([T, T_\alpha], T_\mu). \tag{1.58}$$

Let us set

$$A(X, Y) = g_\theta(\tau X, Y),$$

for any $X, Y \in T(M)$. Also, we set $A_{\alpha\beta} = A(T_\alpha, T_\beta)$. Then

$$A_{\alpha\beta} = A^{\overline{\gamma}}_{\alpha} h_{\overline{\gamma}\beta}\,.$$

The matrix of τ with respect to the local frame $\{T_\alpha, T_{\overline{\alpha}}, T\}$ is given by

$$\tau : \begin{pmatrix} 0 & A^{\overline{\beta}}_{\alpha} & 0 \\ A^{\beta}_{\overline{\alpha}} & 0 & 0 \\ 0 & 0 & 0 \end{pmatrix},$$

where $A^{\beta}_{\overline{\alpha}} = \overline{A^{\overline{\beta}}_{\alpha}}$. Therefore, since the trace of an endomorphism of a linear space coincides with the trace of its extension (by **C**-linearity) to the complexified linear space, it follows that

$$\operatorname{trace}(\tau) = 0. \tag{1.59}$$

To explain the geometric meaning of (1.59) we need to recall the notion of minimality of a C^∞ distribution in Riemannian geometry.

Definition 1.26. Let (N, g) be a Riemannian manifold and $\mathcal{D} : x \mapsto \mathcal{D}_x \subseteq T_x(N)$ a C^∞ distribution on N. Let $\mathcal{D}^\perp$ be the orthogonal complement (with respect to g) of $\mathcal{D}$ in $T(N)$ (so that $T(N) = \mathcal{D} \oplus \mathcal{D}^\perp$). Let ∇^N be the Levi-Civita connection of N. Define

$$B(X, Y) = \pi^\perp \nabla^N_X Y$$

for any $X, Y \in \mathcal{D}$. Here $\pi^\perp : T(N) \to \mathcal{D}^\perp$ is the natural projection. Finally, set $H = \operatorname{trace}(B)$. We say that $\mathcal{D}$ is *minimal* in (N, g) if $H = 0$. □

Returning to pseudo-Hermitian geometry, we may state the following result:

Theorem 1.4. *Let $(M, T_{1,0}(M))$ be a strictly pseudoconvex CR manifold and θ a pseudo-Hermitian structure on M such that the Levi form L_θ is positive definite. Let g_θ be the corresponding Webster metric. Then the Levi distribution $H(M)$ is minimal in (M, g_θ).*

Proof. Since L_θ is positive definite, g_θ is a Riemannian metric on M. The orthogonal complement (with respect to g_θ) of $H(M)$ in $T(M)$ is precisely $\mathbf{R}T$ (where T is the characteristic direction of $d\theta$). The projection of a tangent vector field $X \in T(M)$ on $\mathbf{R}T$ is $\theta(X)T$. Let ∇^θ be the Levi-Civita connection of (M, g_θ). Consider

$$B(X, Y) = \theta(\nabla^\theta_X Y)T,$$

for any $X, Y \in H(M)$. We shall show that $\operatorname{trace}(B) = 0$. We need the following lemma:

Lemma 1.3. *Let $(M, T_{1,0}(M))$ be a nondegenerate CR manifold and θ a fixed pseudo-Hermitian structure on M. Let ∇ be the Tanaka–Webster connection of (M, θ). Then the torsion tensor field T_∇ of ∇ is given by*

$$T_\nabla = 2(\theta \wedge \tau - \Omega \otimes T). \tag{1.60}$$

Moreover, the Levi-Civita connection ∇^θ of the semi Riemannian manifold (M, g_θ) is related to ∇ by

$$\nabla^\theta = \nabla + (\Omega - A) \otimes T + \tau \otimes \theta + 2\theta \odot J. \tag{1.61}$$

Here $\odot$ denotes the symmetric tensor product.[4] The proof of the identity (1.60) in Lemma 1.3 is a straightforward consequence of the purity conditions (1.36)–(1.38). As to (1.61), it follows from the Christoffel process applied to the metric connection ∇. Indeed, since $\nabla g_\theta = 0$ we may write

$$X(g_\theta(Y, Z)) = g_\theta(\nabla_X, Y, Z) + g_\theta(Y, \nabla_X Z)$$

for any $X, Y, Z \in T(M)$. The Christoffel process consists in the cyclic permutation of X, Y, Z in the above identity (thus producing two more identities of the kind), summing the first two, and subtracting the third. This leads (by the very definition of T_∇) to

$$\begin{aligned}X(g_\theta(Y, Z)) + Y(g_\theta(Z, X)) - Z(g_\theta(X, Y)) = \\ 2g_\theta(\nabla_X Y, Z) + g_\theta(T_\nabla(X, Z), Y) + g_\theta(T_\nabla(Y, Z), X) - g_\theta(T_\nabla(X, Y), Z) \\ + g_\theta([X, Z], Y) + g_\theta([Y, Z], X) - g_\theta([X, Y], Z)\end{aligned}$$

for any $X, Y, Z \in T(M)$. Next we use Proposition 2.3 in [241], vol. I, p. 160, to conclude that

$$\begin{aligned}2g_\theta(\nabla^\theta_X Y, Z) = \\ 2g_\theta(\nabla_X Y, Z) + g_\theta(T_\nabla(X, Z), Y) + g_\theta(T_\nabla(Y, Z), X) - g_\theta(T_\nabla(X, Y), Z).\end{aligned}$$

Lemma 1.4. *The pseudo-Hermitian torsion* $\tau : T(M) \to T(M)$ *is self-adjoint with respect to the Webster metric, that is,*

$$g_\theta(\tau X, Y) = g_\theta(X, \tau Y)$$

for any $X, Y \in T(M)$.

We shall prove Lemma 1.4 later on. Using Lemma 1.4 we may finish the proof of (1.61). Indeed, by (1.60) and Lemma 1.4 we may perform the calculation of the torsion terms,

$$\begin{aligned}g_\theta(T_\nabla(X, Z), Y) + g_\theta(T_\nabla(Y, Z), X) - g_\theta(T_\nabla(X, Y), Z) = \\ 2\theta(Y)g_\theta(\tau Z, X) - 2\theta(Z)g_\theta(\tau X, Y) \\ - 2\big(\Omega(X, Z)\theta(Y) + \Omega(Y, Z)\theta(X) - \Omega(X, Y)\theta(Z)\big)\end{aligned}$$

yielding

$$\nabla^\theta_X Y = \nabla_X Y - g_\theta(\tau X, Y)T + \Omega(X, Y)T + \theta(Y)\tau X + \theta(X)JY + \theta(Y)JX$$

for any $X, Y \in T(M)$. □

[4] For instance, $2(\theta \odot J)(X, Y) = \theta(X)JY + \theta(Y)JX$ for any $X, Y \in T(M)$.

Using (1.61) in Lemma 1.3 we may conclude the proof of Theorem 1.4 as follows. Let $X, Y \in H(M)$. Then (1.61) may be written

$$\nabla^\theta_X Y = \nabla_X Y + (\Omega - A)(X, Y)T$$

such that

$$B(X, Y) = \big(\Omega(X, Y) - A(X, Y)\big)T.$$

Let $\{E_1, \dots, E_{2n}\}$ be a local orthonormal frame of $H(M)$. Then (by the skew-symmetry of Ω) we have

$$\text{trace}(B) = \sum_{j=1}^{2n} B(E_j, E_j) = -\sum_{j=1}^{2n} A(E_j, E_j)T = -(\text{trace}(\tau))T = 0$$

(by (1.59)). Our Theorem 1.4 is completely proved. □

At this point we ought to prove Lemma 1.3. Let $\{T_1, \dots, T_n\}$ be a local frame of $T_{1,0}(M)$ and $\{\theta^1, \dots, \theta^n\}$ the dual coframe in $T_{1,0}(M)^*$, that is,

$$\theta^\alpha(T_\beta) = \delta^\alpha_\beta, \quad \theta^\alpha(T_{\overline{\beta}}) = 0, \quad \theta^\alpha(T) = 0.$$

Then $d\theta$ may be written in the following form:

$$d\theta = B_\alpha \theta \wedge \theta^\alpha + B_{\overline{\alpha}} \theta \wedge \theta^{\overline{\alpha}} + B_{\alpha\beta}\theta^\alpha \wedge \theta^\beta + B_{\overline{\alpha}\overline{\beta}}\theta^{\overline{\alpha}} \wedge \theta^{\overline{\beta}} + B_{\alpha\overline{\beta}}\theta^\alpha \wedge \theta^{\overline{\beta}}$$

where the coefficient functions satisfy (since $d\theta$ is a real 2-form) the identities

$$B_{\overline{\alpha}} = \overline{B_\alpha}, \quad B_{\overline{\alpha}\overline{\beta}} = \overline{B_{\alpha\beta}}, \quad B_{\alpha\overline{\beta}} = -\overline{B_{\beta\overline{\alpha}}}\,.$$

Next $T \rfloor d\theta = 0$ yields $B_\alpha = 0$. Also (by the integrability of the CR structure)

$$0 = -\theta([T_\sigma, T_\rho]) = 2(d\theta)(T_\sigma, T_\rho) = B_{\sigma\rho} - B_{\rho\sigma},$$

so that

$$B_{\alpha\beta}\theta^\alpha \wedge \theta^\beta = 0.$$

Finally,

$$i h_{\lambda\overline{\mu}} = (d\theta)(T_\lambda, T_{\overline{\mu}}) = \frac{1}{2} B_{\lambda\overline{\mu}}.$$

We conclude that

$$d\theta = 2i h_{\alpha\overline{\beta}} \theta^\alpha \wedge \theta^{\overline{\beta}}. \tag{1.62}$$

Let us define the local 1-forms $\tau^\alpha \in \Gamma^\infty(U, T^*(M) \otimes \mathbf{C})$ by setting

$$\tau^\alpha = A^\alpha_{\overline{\beta}} \theta^{\overline{\beta}} .$$

Then

$$\tau = \tau^\alpha \otimes T_\alpha + \tau^{\overline{\alpha}} \otimes T_{\overline{\alpha}}, \tag{1.63}$$

where $\tau^{\overline{\alpha}} = \overline{\tau^\alpha}$. The proof of (1.63) is straightforward. Next

$$d\theta^\alpha = \theta^\beta \wedge \omega^\alpha_\beta + \theta \wedge \tau^\alpha . \tag{1.64}$$

The proof of (1.64) follows from the identity

$$2(d\theta^\alpha)(X, Y) = (\nabla_X \theta^\alpha) Y - (\nabla_Y \theta^\alpha) X + \theta^\alpha (T_\nabla (X, Y))$$

for any $X, Y \in T(M)$. At this point we may prove Lemma 1.4. To this end, let us differentiate (1.62) to obtain

$$0 = 2i dh_{\alpha\overline{\beta}} \wedge \theta^\alpha \wedge \theta^{\overline{\beta}} + 2i h_{\alpha\overline{\beta}} d\theta^\alpha \wedge \theta^{\overline{\beta}} - 2i h_{\alpha\overline{\beta}} \theta^\alpha \wedge d\theta^{\overline{\beta}} .$$

We may substitute from (1.64) so that

$$\begin{aligned} 0 = dh_{\alpha\overline{\beta}} \wedge \theta^\alpha \wedge \theta^{\overline{\beta}} + h_{\alpha\overline{\beta}} (\theta^\gamma \wedge \omega^\alpha_\gamma + \theta \wedge \tau^\alpha) \wedge \theta^{\overline{\beta}} \\ - h_{\alpha\overline{\beta}} \theta^\alpha \wedge (\theta^{\overline{\gamma}} \wedge \omega^{\overline{\beta}}_{\overline{\gamma}} + \theta \wedge \theta^{\overline{\beta}}) . \end{aligned} \tag{1.65}$$

On the other hand, we may use the local coordinate representation of $\nabla g_\theta = 0$, that is,

$$dh_{\alpha\overline{\beta}} = h_{\alpha\overline{\gamma}}\, \omega^{\overline{\gamma}}_{\overline{\beta}} + \omega^\gamma_\alpha\, h_{\gamma\overline{\beta}},$$

to rewrite (1.65) as

$$\begin{aligned} 0 &= (\omega^\gamma_\alpha h_{\gamma\overline{\beta}} + h_{\alpha\overline{\gamma}} \omega^{\overline{\gamma}}_{\overline{\beta}}) \wedge \theta^\alpha \wedge \theta^{\overline{\beta}} - h_{\mu\overline{\beta}} \omega^\mu_\alpha \wedge \theta^\alpha \wedge \theta^{\overline{\beta}} + h_{\alpha\overline{\beta}} \theta \wedge \tau^\alpha \wedge \theta^{\overline{\beta}} \\ &\quad - h_{\alpha\overline{\mu}} \omega^{\overline{\mu}}_{\overline{\beta}} \wedge \theta^\alpha \wedge \theta^{\overline{\beta}} - h_{\alpha\overline{\beta}} \theta^\alpha \wedge \theta \wedge \tau^{\overline{\beta}} \\ &= h_{\alpha\overline{\beta}} A^\alpha_{\overline{\mu}} \theta \wedge \theta^{\overline{\mu}} \wedge \theta^{\overline{\beta}} - h_{\alpha\overline{\beta}} A^{\overline{\beta}}_\lambda \theta^\alpha \wedge \theta \wedge \theta^\lambda, \end{aligned}$$

or

$$A_{\overline{\mu}\overline{\beta}} \theta \wedge \theta^{\overline{\mu}} \wedge \theta^{\overline{\beta}} = A_{\lambda\alpha} \theta^\alpha \wedge \theta \wedge \theta^\lambda,$$

that is,

$$A_{\lambda\alpha} \theta^\alpha \wedge \theta \wedge \theta^\lambda = 0.$$

Finally, applying this identity to (T_μ, T, T_β) we get

$$A_{\beta\mu} = A_{\mu\beta},$$

and Lemma 1.4 is completely proved. □

The following remark (relating the theory of CR manifolds to the theory of CR submanifolds in Hermitian manifolds, in the sense of A. Bejancu [55]), is in order. The fact that $H(M)$ is minimal in (M, g_θ) (cf. Theorem 1.4) is an analogue of a result by B.Y. Chen [90], i.e., *the holomorphic distribution of a CR submanifold M of a Kählerian manifold is minimal in* (M, g), where g is the metric induced on M by the ambient Kähler metric. To make this statement precise we need to recollect a few definitions and conventions (cf., e.g., [56]).

Definition 1.27. Let $(N, J, \tilde{g})$ be a Hermitian manifold, where J denotes the complex structure and $\tilde{g}$ the Hermitian metric. Let M be a real submanifold of N and $g = j^*\tilde{g}$ the induced metric, where $j : M \to N$ is the inclusion. We say that M is a *CR submanifold* of N if M carries a distribution $H(M)$ such that (i) $H(M)$ is *holomorphic* (i.e., $J_x H(M)_x = H(M)_x$ for any $x \in M$) and (ii) the orthogonal complement (with respect to g) $H(M)^\perp$ of $H(M)$ in $T(M)$ is *anti-invariant* (i.e., $J_x H(M)_x^\perp \subseteq \nu(j)_x$ for any $x \in M$). Here $\nu(j) \to M$ is the normal bundle of j. □

For a CR submanifold $(M, H(M))$ of a Hermitian manifold $(N, J, \tilde{g})$ we set $\dim_{\mathbf{R}} H(M)_x = 2n$ and $\dim_{\mathbf{R}} H(M)_x^\perp = k$, for any $x \in M$. Clearly the complex structure J descends to a complex structure $J_b : H(M) \to H(M)$. Let us extend J_b to $H(M) \otimes \mathbf{C}$ by $\mathbf{C}$-linearity and let $T_{1,0}(M) = \text{Eigen}(i)$ be the eigenbundle of J_b corresponding to the eigenvalue i. By a result of D.E. Blair and B.Y. Chen [64], for any *proper* (i.e., $n \neq 0, k \neq 0$) CR submanifold $(M, H(M))$ of a Hermitian manifold, $(M, T_{1,0}(M))$ is a CR manifold of type (n, k) (and $H(M)$ is its Levi distribution). See also K. Yano and M. Kon [446], pp. 83–85. Moreover, if $(N, J, \tilde{g})$ is a Kähler manifold and $(M, H(M))$ a (proper) CR submanifold of N, then the anti-invariant distribution $H(M)^\perp$ is (cf. [90]) integrable (so that each CR submanifold comes equipped with a natural totally real foliation) and the holomorphic distribution $H(M)$ is minimal in (M, g). This is, however, distinct from the result in our Theorem 1.4, as we emphasize below.

The CR manifolds considered in [90] have arbitrary CR codimension k. Therefore, to draw a comparison line among strictly pseudoconvex CR manifolds and CR submanifolds, let us consider an orientable real hypersurface M of a Kahlerian manifold $(N, J, \tilde{g})$. M is a proper CR submanifold in a natural way.[5] Moreover, $(M, T_{1,0}(M))$ is a CR manifold (of hypersurface type) for which $\theta X = g(X, J\xi)$ (with $X \in T(M)$) is a pseudo-Hermitian structure. Yet in general, the induced metric g and the Webster metric g_θ do not coincide (hence our Theorem 1.4 doesn't follow from B.Y. Chen's result; cf. op. cit.). For instance, none of the Webster metrics of $\partial\Omega_2$ (the boundary of the Siegel domain in $\mathbf{C}^2$) coincides with the metric induced on $\partial\Omega_2$ from the standard (flat) Kähler metric of $\mathbf{C}^2$. Let us check this statement, in the following more general situation. Let

$$M^3 = \{(z, w) \in \mathbf{C}^2 : v = h(z)\}$$

[5] Indeed, let $\xi \in \Gamma^\infty(\nu(j))$ be a unit normal vector field on M. Then $J\xi$ is tangent to M. Let $H(M)$ be the orthogonal complement of $\mathbf{R}J\xi$ in $T(M)$. Then $H(M)$ is holomorphic and $\mathbf{R}J\xi$ is anti-invariant.

(a *rigid hypersurface* in $\mathbf{C}^2$) for some $\mathbf{R}$-valued C^2 function h, $z = x+iy$, $w = u+iv$. Let us set

$$g = (1 + h_x^2 + h_y^2)^{1/2}.$$

The Levi distribution $H(M^3)$ is generated by

$$X_1 = e_1 + h_y e_3, \quad X_2 = e_2 - h_x e_3,$$

where

$$e_1 = \frac{\partial}{\partial x} + h_x \frac{\partial}{\partial v}, \quad e_2 = \frac{\partial}{\partial y} + h_y \frac{\partial}{\partial v}, \quad e_3 = \frac{\partial}{\partial v}$$

(the generators of $T(M^3)$). Thus $T_{1,0}(M^3)$ is the span of $Z = \partial/\partial z + 2ih_z \partial/\partial w$. Next

$$\theta = 4(h_y dx - h_x dy - du)$$

is a contact form on M^3 and

$$d\theta = -4\Delta h \, dx \wedge dy,$$

so that the Levi form of M^3 is given by

$$L_\theta(Z, \overline{Z}) = \Delta h.$$

Thus M^3 is nondegenerate (respectively strictly pseudoconvex) if $\Delta h \neq 0$ everywhere (respectively h is strictly subharmonic). Let $j : M^3 \subset \mathbf{C}^2$ be the inclusion and g_{can} the flat Kähler metric of $\mathbf{C}^2$. Then $j^* g_{\text{can}} = \text{diag}(g^2, g^2, g^2)$ (with respect to $\{X_1, X_2, X = \partial/\partial u - h_y \partial/\partial x + h_x \partial/\partial y\}$). Also $g_{u\theta} = \text{diag}(2u\Delta h, 2u\Delta h, 16u^2 g^4)$, where $u \in C^\infty(M^3)$ is any smooth $(0, \infty)$-valued function. Thus the induced metric on M^3 coincides with the Webster metric $g_{u\theta}$ if and only if $4u = 1/g$ and $\Delta h = 2g^3$. □

The vanishing of the pseudo-Hermitian torsion ($\tau = 0$) admits an important geometric interpretation, due to S. Webster [422]. Precisely, we may state the following theorem:

Theorem 1.5. *Let M be a nondegenerate CR manifold, θ a contact form on M, and T the characteristic direction of $d\theta$. Then the Tanaka–Webster connection of (M, θ) has a vanishing pseudo-Hermitian torsion ($\tau = 0$) if and only if T is an infinitesimal CR automorphism.*

Definition 1.28. By an *infinitesimal CR automorphism* we mean a real tangent vector field whose (local) 1-parameter group of (local) transformations of M consists of (local) CR automorphisms. □

The following simple characterization of infinitesimal CR automorphisms is available:

Lemma 1.5. *Let X be a tangent vector field on a CR manifold (of hypersurface type). Let θ be a pseudo-Hermitian structure. Then X is an infinitesimal CR automorphism if and only if $\mathcal{L}_X\theta \equiv 0 \bmod \theta$ and $\mathcal{L}_X\theta^\alpha \equiv 0 \mod \theta, \theta^\beta$, for any (local) frame $\{\theta^\alpha\}$ in $T_{1,0}(M)^*$.*

Here $\mathcal{L}_X$ indicates the Lie derivative in the direction X. Also $\equiv$ denotes the ordinary congruence relation among forms (for instance, $\mathcal{L}_X\theta \equiv 0 \mod \theta$ if $\mathcal{L}_x\theta$ belongs to the ideal spanned by θ in the de Rham algebra of M). The proof of Lemma 1.5 is straightforward (and therefore left as an exercise to the reader). Recall that $\mathcal{L}_X = i_X d + d\, i_X$. Then

$$\begin{aligned}
\mathcal{L}_T\theta &= i_T d\theta = 0,\\
\mathcal{L}_T\theta^\alpha &= i_T d\theta^\alpha + d(i_T\theta^\alpha) = i_T\{\theta^\beta \wedge \omega_\beta^\alpha + \theta \wedge \tau^\alpha\}\\
&= -\frac{1}{2}\omega_\beta^\alpha(T)\theta^\beta - \frac{1}{2}\tau^\alpha(T)\theta + \frac{1}{2}\tau^\alpha,
\end{aligned}$$

for any admissible (local) coframe $\{\theta^\alpha\}$. Finally, to prove Theorem 1.5, note that $\tau = 0$ is equivalent to the prescriptions in Lemma 1.5 since $\tau^\alpha = A^\alpha_{\overline{\beta}}\theta^{\overline{\beta}}$. □

1.3.3 The volume form

Let M be a nondegenerate CR manifold and θ a pseudo-Hermitian structure on M.

Proposition 1.9. $\Psi = \theta \wedge (d\theta)^n$ *is a volume form on M.*

In other words, θ is a *contact form* on M. This is done by explicitly computing Ψ with respect to an (admissible) coframe $\{\theta^\alpha : 1 \le \alpha \le n\}$. By taking the nth exterior power of the identity

$$d\theta = 2ih_{\alpha\overline{\beta}}\theta^\alpha \wedge \theta^{\overline{\beta}}$$

we obtain

$$(d\theta)^n = 2^n i^n h_{\alpha_1\overline{\beta}_1} \cdots h_{\alpha_n\overline{\beta}_n}\theta^{\alpha_1} \wedge \theta^{\overline{\beta}_1} \wedge \cdots \wedge \theta^{\alpha_n} \wedge \theta^{\overline{\beta}_n}.$$

A straightforward exercise of multilinear algebra shows that the wedge product may be rearranged as

$$\theta^{\alpha_1} \wedge \theta^{\overline{\beta}_1} \wedge \cdots \wedge \theta^{\alpha_n} \wedge \theta^{\overline{\beta}_n} = (-1)^{n(n-1)/2}\theta^{\alpha_1\cdots\alpha_n} \wedge \theta^{\overline{\beta}_1\cdots\overline{\beta}_n},$$

where

$$\theta^{\alpha_1\cdots\alpha_n} = \theta^{\alpha_1} \wedge \cdots \wedge \theta^{\alpha_n}, \quad \theta^{\overline{\alpha}_1\cdots\overline{\alpha}_n} = \overline{\theta^{\alpha_1\cdots\alpha_n}}.$$

The proof is simply elementary combinatorics. Indeed, we may set

$$\theta^{\alpha_1} \wedge \theta^{\overline{\beta}_1} \wedge \cdots \wedge \theta^{\alpha_n} \wedge \theta^{\overline{\beta}_n} = I(n)\theta^{\alpha_1\cdots\alpha_n} \wedge \theta^{\overline{\beta}_1\cdots\overline{\beta}_n}$$

and obtain the recurrence relation $I(n+1) = (-1)^n I(n)$, whence $I(n) = (-1)^{n(n-1)/2}$. Therefore

$$(d\theta)^n = 2^n i^n (-1)^{n(n-1)/2} h_{\alpha_1\overline{\beta}_1} \cdots h_{\alpha_n\overline{\beta}_n} \theta^{\alpha_1\cdots\alpha_n} \wedge \theta^{\overline{\beta}_1\cdots\overline{\beta}_n}.$$

Furthermore, note that

$$h_{\alpha_1\overline{\beta}_1} \cdots h_{\alpha_n\overline{\beta}_n} \theta^{\alpha_1\cdots\alpha_n} \wedge \theta^{\overline{\beta}_1\cdots\overline{\beta}_n} = n!\det(h_{\alpha\overline{\beta}})\theta^{1\cdots n\cdot\overline{1}\cdots\overline{n}},$$

where

$$\theta^{1\cdots n\cdot\overline{1}\cdots\overline{n}} = \theta^1 \wedge \cdots \wedge \theta^n \wedge \theta^{\overline{1}} \wedge \cdots \wedge \theta^{\overline{n}}.$$

Once again, the proof is an exercise in multilinear algebra. Precisely, we may carry out the calculation

$$\begin{aligned} h_{\alpha_1\overline{\beta}_1} \cdots h_{\alpha_n\overline{\beta}_n} \theta^{\alpha_1\cdots\alpha_n} \wedge \theta^{\overline{\beta}_1\cdots\overline{\beta}_n} &= \sum_{f,g\in\sigma_n} h_{f(1)\overline{g(1)}} \cdots h_{f(n)\overline{g(n)}} \epsilon(f)\epsilon(g)\theta^{1\cdots n\cdot\overline{1}\cdots\cdots\overline{n}} \\ &= \sum_{f,g\in\sigma_n} h_{1\,\overline{g(f^{-1}(1))}} \cdots h_{n\,\overline{g(f^{-1}(n))}} \epsilon(gf^{-1})\theta^{1\cdots n\cdot\overline{1}\cdots\cdots\overline{n}} \\ &= n!\ \det(h_{\alpha\overline{\beta}})\theta^{1\cdots n\cdot\overline{1}\cdots\overline{n}}. \end{aligned}$$ □

We conclude that

$$(d\theta)^n = 2^n\, i^{n^2}\, n!\ \det(h_{\alpha\overline{\beta}})\theta^{1\cdots n\cdot\overline{1}\cdots\overline{n}},$$

so that Ψ is given by

$$\Psi = 2^n\, i^{n^2}\, n!\ \det(h_{\alpha\overline{\beta}})\theta \wedge \theta^1 \wedge \cdots \theta^n \wedge \theta^{\overline{1}} \wedge \cdots \wedge \theta^{\overline{n}}. \tag{1.66}$$

In particular, let $x \in M$ and let $\{\theta^\alpha : 1 \leq \alpha \leq n\}$ be a local admissible coframe defined on an open neighborhood U of x. By (1.66) it follows that $\Psi_x \neq 0$, i.e., Ψ is a volume form. □

Now let us discuss the divergence of a vector field on a nondegenerate CR manifold. The divergence $\mathrm{div}(X)$ of a vector field $X \in \mathcal{X}(M)$ is defined by

$$\mathcal{L}_X \Psi = \mathrm{div}(X)\, \Psi,$$

where $\mathcal{L}_X$ denotes the Lie derivative. As usual, we may extend div to complex vector fields (by **C**-linearity). Then, since div is a real operator,

$$\mathrm{div}(\overline{Z}) = \overline{\mathrm{div}(Z)},$$

for any $Z \in \Gamma^\infty(T_{1,0}(M))$. The divergence $\mathrm{div}(X)$ may be computed in yet another way, i.e.,

$$\mathrm{div}(X) = \mathrm{trace}\{Y \in T(M) \mapsto \nabla_Y X\}, \tag{1.67}$$

where ∇ is the Tanaka–Webster connection of (M, θ). A remark is in order: Note that the traces of the endomorphisms

$$T_{1,0}(M) \to T_{1,0}(M), \quad W \mapsto \nabla_W Z,$$

and

$$T(M) \otimes \mathbf{C} \to T(M) \otimes \mathbf{C}, \quad W \mapsto \nabla_W Z,$$

coincide for any $Z \in \Gamma^\infty(T_{1,0}(M))$, so that (1.67) yields

$$\operatorname{div}(Z) = \operatorname{trace}\{W \in T_{1,0}(M) \mapsto \nabla_W Z\}, \tag{1.68}$$

for any $Z \in T_{1,0}(M)$. Consequently, if $Z = Z^\alpha T_\alpha$ then

$$\operatorname{div}(Z) = T_\alpha(Z^\alpha) + Z^\beta \Gamma^\alpha_{\alpha\beta}.$$

We end this section by proving (1.68). The reader may easily complete this to a full proof of (1.67). Let us set

$$\theta \wedge \theta^{1\cdots n \cdot \overline{1} \cdots \overline{n}} = \theta^{01\cdots n \cdot \overline{1} \cdots \overline{n}}, \quad \theta = \theta^0, \quad f = \det(h_{\alpha\overline{\beta}}),$$

for simplicity. Note that

$$\theta^{01\cdots n \cdot \overline{1} \cdots \overline{n}}(T, T_1, \ldots, T_n, T_{\overline{1}}, \ldots, T_{\overline{n}}) = \frac{1}{(2n+1)!}.$$

Then (by applying $\mathcal{L}_Z \Psi = (\operatorname{div} Z)\Psi$ at $(T, T_1, \ldots, T_n, T_{\overline{1}}, \ldots, T_{\overline{n}})$) we obtain

$$f \operatorname{div}(Z) = Z(f) - f \sum_{A=0}^{2n} \theta^A(\mathcal{L}_Z T_A), \tag{1.69}$$

where we agree to relabel for a moment $\{T, T_\alpha, T_{\overline{\alpha}}\}$ as $\{T_A : 0 \le A \le 2n\}$. Taking into account the identities

$$\begin{aligned}
[Z, T] &= -\left(T(Z^\alpha) + Z^\beta \Gamma^\alpha_{0\beta}\right) T_\alpha + Z^\beta A^{\overline{\alpha}}_\beta T_{\overline{\alpha}}, \\
[Z, T_\mu] &= -\left(T_\mu(Z^\alpha) + Z^\beta \Gamma^\alpha_{\mu\beta}\right) T_\alpha, \\
[Z, T_{\overline{\mu}}] &= -\left(T_{\overline{\mu}}(Z^\alpha) + Z^\beta \Gamma^\alpha_{\overline{\mu}\beta}\right) T_\alpha + Z^\beta \Gamma^{\overline{\alpha}}_{\beta\overline{\mu}} T_{\overline{\alpha}} - 2i Z^\beta h_{\beta\overline{\mu}} T,
\end{aligned}$$

we may write (1.69) as

$$f \operatorname{div}(Z) = Z(f) + f\left(T_\alpha(Z^\alpha) + Z^\beta \Gamma^\alpha_{\alpha\beta} - Z^\beta \Gamma^\alpha_{\beta\alpha} - Z^\beta \Gamma^{\overline{\alpha}}_{\beta\overline{\alpha}}\right). \tag{1.70}$$

Note that

$$h^{\lambda\overline{\mu}} = \frac{1}{f} \frac{\partial f}{\partial h_{\lambda\overline{\mu}}}.$$

Then

$$Z(f) = f\, h^{\lambda\overline{\mu}} Z(h_{\lambda\overline{\mu}}),$$

so that (1.70) becomes

$$\operatorname{div}(Z) = \operatorname{trace}\{W \mapsto \nabla_W Z\} + h^{\lambda\overline{\mu}} Z(h_{\lambda\overline{\mu}}) - \Gamma^{\alpha}_{\beta\alpha} Z^{\beta} - \Gamma^{\overline{\alpha}}_{\beta\overline{\alpha}} Z^{\beta},$$

which yields (1.68) because of

$$dh_{\alpha\overline{\beta}} = h_{\alpha\overline{\gamma}} \omega^{\overline{\gamma}}_{\overline{\beta}} + \omega^{\gamma}_{\alpha} h_{\gamma\overline{\beta}}.$$

1.4 The curvature theory

We start by relating semi-Riemannian and pseudo-Hermitian curvature. Let $(M, T_{1,0}(M))$ be a nondegenerate CR manifold (of hypersurface type) of CR dimension n. Let θ be a fixed pseudo-Hermitian structure on M, T the characteristic direction of $d\theta$, and g_θ the corresponding Webster metric. Let ∇ and ∇^θ be the Tanaka–Webster connection of (M, θ), respectively the Levi-Civita connection of the semi-Riemannian manifold (M, g_θ). Let R and R^θ be the curvature tensor fields of ∇ and ∇^θ, respectively. We recall that (cf. (1.61))

$$\nabla^\theta = \nabla + (\Omega - A) \otimes T + \tau \otimes \theta + 2\theta \odot J,$$

so that we may derive an identity relating the curvature tensor fields R^θ and R. This will indeed prove to be useful in circumventing the difficulties arising from the failure of $R_{ijk\ell}$ to satisfy the identity $R_{ijk\ell} = R_{k\ell ij}$. Note, however, that

$$R_{ijk\ell} + R_{ij\ell k} = 0$$

(because R is a 2-form) and

$$R_{ijk\ell} + R_{jik\ell} = 0$$

(because of $\nabla g_\theta = 0$). Here, given a local coordinate system (U, x^i) on M, we set

$$R_{ijk\ell} = g_\theta(R(\partial_k, \partial_\ell)\partial_j, \partial_i)$$

and $\partial_i = \partial/\partial x^i$. As is well known (cf., e.g., [241], vol. I, pp. 198–199), the missing property $R_{ijk\ell} = R_{k\ell ij}$ would hold if

$$R^i_{jk\ell} + R^i_{k\ell j} + R^i_{\ell jk} = 0.$$

Yet this no longer holds for the curvature of the Webster connection (due to the fact that the Webster connection has nontrivial torsion). Instead, we have the following first Bianchi identity

$$\sum_{XYZ} R(X,Y)Z = \sum_{XYZ} ((\nabla_X T_\nabla)(Y,Z) + T_\nabla(T_\nabla(X,Y),Z))$$

(cf. Theorem 5.3 in [241], vol. I, p. 135) for any $X, Y, Z \in T(M)$. Here $\sum_{XYZ}$ denotes the cyclic sum over X, Y, Z.

Let us use (1.61) to perform the following calculations:

$$\begin{aligned}
\nabla^\theta_X \nabla^\theta_Y Z &= \nabla_X \nabla_Y Z + \Omega(X, \nabla_Y Z)T - A(X, \nabla_Y Z)T \\
&\quad + (\tau X)\theta(\nabla_Y Z) + \theta(X)J\nabla_Y Z + \theta(\nabla_Y Z)JX + X(\Omega(Y,Z))T \\
&\quad + \Omega(Y,Z)LX - X(A(Y,Z))T - A(Y,Z)LX + X(\theta(Z))\tau Y \\
&\quad + \theta(Z)\nabla^\theta_X \tau Y + X(\theta(Y))JZ + \theta(Y)\nabla^\theta_X JZ + X(\theta(Z))JY + \theta(Z)\nabla^\theta_X JY
\end{aligned}$$

because of

$$\nabla^\theta_X T = LX,$$

where the (1, 1)-tensor field L on M is given by

$$L = \tau + J.$$

Furthermore,

$$\begin{aligned}
\nabla^\theta_X \nabla^\theta_Y Z &= \nabla_X \nabla_Y Z + \Omega(X, \nabla_Y Z)T - A(X, \nabla_Y Z)T \\
&\quad + \theta(\nabla_Y Z)\tau X + \theta(X)J\nabla_Y Z + \theta(\nabla_Y Z)JX + X(\Omega(Y,Z))T \\
&\quad + \Omega(Y,Z)LX - X(A(Y,Z))T - A(Y,Z)LX + X(\theta(Z))\tau Y \\
&\quad + \theta(Z)\{\nabla_X \tau Y + \Omega(X, \tau Y)T - A(X, \tau Y)T + \theta(X)J\tau Y\} \\
&\quad + X(\theta(Y))JZ + \theta(Y)\{\nabla_X JZ + \Omega(X, JZ)T - A(X, JZ)T + \theta(X)J^2 Z\} \\
&\quad + X(\theta(Z))JY + \theta(Z)\{\nabla_X JY + \Omega(X, JY)T - A(X, JY)T + \theta(X)J^2 Y\}
\end{aligned}$$

due to $\theta \circ J = 0$ and $\theta \circ \tau = 0$. Similarly, one computes $\nabla^\theta_Y \nabla^\theta_X Z$. Furthermore,

$$\begin{aligned}
\nabla^\theta_{[X,Y]} Z &= \nabla_{[X,Y]} Z + \Omega([X,Y], Z)T - A([X,Y], Z)T \\
&\quad + (\tau[X,Y])\theta(Z) + \theta([X,Y])JZ + \theta(Z)J[X,Y].
\end{aligned}$$

We may replace the Lie bracket from

$$[X,Y] = \nabla_X Y - \nabla_Y X - T_\nabla(X,Y).$$

Taking into account the identity (1.60) in Lemma 1.3 we may actually express $[X, Y]$ as

$$[X,Y] = \nabla_X Y - \nabla_Y X - \theta(X)\tau Y + \theta(Y)\tau X + 2\Omega(X,Y)T$$

Therefore

$$
\begin{aligned}
\nabla^{\theta}_{[X,Y]}Z &= \nabla_{[X,Y]}Z + \theta([X,Y])JZ \\
&\quad + \{\Omega(\nabla_X Y, Z) - \Omega(\nabla_Y X, Z) - \theta(X)\Omega(\tau Y, Z) + \theta(Y)\Omega(\tau X, Z)\}T \\
&\quad - \{A(\nabla_X Y, Z) - A(\nabla_Y X, Z) - \theta(X)A(\tau Y, Z) + \theta(Y)A(\tau X, Z)\}T \\
&\quad + \theta(Z)\{\tau\nabla_X Y - \tau\nabla_Y X - \theta(X)\tau^2 Y + \theta(Y)\tau^2 X\} \\
&\quad + \theta(Z)\{J\nabla_X Y - J\nabla_Y X - \theta(X)J\tau Y + \theta(Y)J\tau X\},
\end{aligned}
$$

due to $JT = 0$ and $\tau T = 0$. At this point we may perform the following calculations:

$$
\begin{aligned}
R^{\theta}(X,Y)Z &= R(X,Y)Z + \Omega(X, \nabla_Y Z)T - \Omega(Y, \nabla_X Z)T \\
&\quad - A(X, \nabla_Y Z)T + A(Y, \nabla_X Z)T + \theta(\nabla_Y Z)\tau X - \theta(\nabla_X Z)\tau Y \\
&\quad + \theta(X)J\nabla_Y Z - \theta(Y)J\nabla_X Z + \theta(\nabla_Y Z)JX - \theta(\nabla_X Z)JY \\
&\quad + X(\Omega(Y,Z))T - Y(\Omega(X,Z))T + \Omega(Y,Z)LX - \Omega(X,Z)LY \\
&\quad - X(A(Y,Z))T + Y(A(X,Z))T - A(Y,Z)LX + A(X,Z)LY \\
&\quad + X(\theta(Z))\tau Y - Y(\theta(Z))\tau X \\
&\quad + \theta(Z)\{\nabla_X \tau Y - \nabla_Y \tau X + \Omega(X, \tau Y)T - \Omega(Y, \tau X)T \\
&\quad - A(X, \tau Y)T + A(Y, \tau X)T + \theta(X)J\tau Y - \theta(Y)J\tau X\} \\
&\quad + X(\theta(Y))JZ - Y(\theta(X))JZ + \theta(Y)\nabla_X JZ - \theta(X)\nabla_Y JZ \\
&\quad + \theta(Y)\Omega(X, JZ)T - \theta(X)\Omega(Y, JZ)T \\
&\quad - \theta(Y)A(X, JZ)T + \theta(X)A(Y, JZ)T \\
&\quad + X(\theta(Z))JY - Y(\theta(Z))JX + \theta(Z)\nabla_X JY - \theta(Z)\nabla_Y JX \\
&\quad + \theta(Z)\Omega(X, JY)T - \theta(Z)\Omega(Y, JX)T \\
&\quad - \theta(Z)A(X, JY)T + \theta(Z)A(Y, JX)T \\
&\quad + \theta(Z)\theta(X)J^2 Y - \theta(Z)\theta(Y)J^2 X - \theta([X,Y])JZ \\
&\quad - \Omega(\nabla_X Y, Z)T + \Omega(\nabla_Y X, Z)T \\
&\quad + \theta(X)\Omega(\tau Y, Z)T - \theta(Y)\Omega(\tau X, Z)T \\
&\quad + A(\nabla_X Y, Z)T - A(\nabla_Y X, Z)T \\
&\quad - \theta(X)A(\tau Y, Z)T + \theta(Y)A(\tau X, Z)T \\
&\quad - \theta(Z)\tau\nabla_X Y + \theta(Z)\tau\nabla_Y X + \theta(X)\theta(Z)\tau^2 Y - \theta(Y)\theta(Z)\tau^2 X \\
&\quad - \theta(Z)J\nabla_X Y + \theta(Z)J\nabla_Y X + \theta(X)\theta(Z)J\tau Y - \theta(Y)\theta(Z)J\tau X.
\end{aligned}
$$

At this point we may use the identities

$$
\nabla\theta = 0, \; \nabla J = 0, \; \nabla\Omega = 0,
$$
$$
(\nabla_X A)(Y,Z) = g_\theta((\nabla_X \tau)Y, Z).
$$

Also, due to

$$
-\Omega(X,Y) + A(X,Y) = g_\theta(LX, Y)
$$

we have

$$
\begin{aligned}
(-\Omega(X,Z) + A(X,Z))LY - (-\Omega(Y,Z) + A(Y,Z))LX \\
&= g_\theta(LX,Z)LY - g_\theta(LY,Z)LX \\
&= (LX \wedge LY)Z.
\end{aligned}
$$

The wedge product of two tangent vector fields X, Y on (M, g_θ) is defined as usual by $(X \wedge Y)Z = g_\theta(X,Z)Y - g_\theta(Y,Z)X$ for any $Z \in T(M)$. Then

$$
\begin{aligned}
R^\theta(X,Y)Z = R(X,Y)Z &+ \big\{(\nabla_Y A)(X,Z) - (\nabla_X A)(Y,Z)\big\}T \\
&+ (LX \wedge LY)Z + \theta(Z)\{(\nabla_X \tau)Y - (\nabla_Y \tau)X\} \\
&\quad + 2(d\theta)(X,Y)JZ \\
&+ \big\{ - \theta(Y)g_\theta(JX,JZ) + \theta(X)g_\theta(JY,JZ) \\
&\quad - \theta(Y)g_\theta(\tau X, JZ) + \theta(X)g_\theta(\tau Y, JZ) \\
&\quad - \theta(X)g_\theta(J\tau Y, Z) + \theta(Y)g_\theta(J\tau X, Z) \\
&\quad - \theta(X)g_\theta(\tau^2 Y, Z) + \theta(Y)g_\theta(\tau^2 X, Z)\big\}T \\
&+ \theta(Z)\big\{\theta(X)J\tau Y - \theta(Y)J\tau X + \theta(X)J^2 Y - \theta(Y)J^2 X \\
&\quad + \theta(X)\tau^2 Y - \theta(Y)\tau^2 X + \theta(X)J\tau Y - \theta(Y)J\tau X\big\}.
\end{aligned}
$$

On the other hand, one may observe that

$$
\begin{aligned}
&- \theta(Y)g_\theta(JX,JZ) + \theta(X)g_\theta(JY,JZ) - \theta(Y)g_\theta(\tau X, JZ) + \theta(X)g_\theta(\tau Y, JZ) \\
&- \theta(X)g_\theta(J\tau Y, Z) + \theta(Y)g_\theta(J\tau X, Z) - \theta(X)g_\theta(\tau^2 Y, Z) + \theta(Y)g_\theta(\tau^2 X, Z) \\
&\qquad = \theta(Y)g_\theta(\mathcal{O}X, Z) - \theta(X)g_\theta(\mathcal{O}Y, Z) = -2g_\theta((\theta \wedge \mathcal{O})(X,Y), Z)
\end{aligned}
$$

where

$$
\mathcal{O} = \tau^2 + 2J\tau - I\,.
$$

These calculations yield the following theorem:

Theorem 1.6. *Let M be a nondegenerate CR manifold and θ a pseudo-Hermitian structure on M. The curvature tensor fields R and R^θ (of the Tanaka–Webster connection ∇ of (M,θ) and the Levi-Civita connection ∇^θ of (M, g_θ)) are related by*

$$
\begin{aligned}
R^\theta(X,Y)Z = R(X,Y)Z &+ (LX \wedge LY)Z - 2\Omega(X,Y)JZ \\
&- g_\theta(S(X,Y),Z)T + \theta(Z)S(X,Y) \\
&- 2g_\theta((\theta \wedge \mathcal{O})(X,Y),Z)T + 2\theta(Z)(\theta \wedge \mathcal{O})(X,Y) \qquad (1.71)
\end{aligned}
$$

for any $X, Y, Z \in \mathcal{X}(M)$, where S is given by

$$
S(X,Y) = (\nabla_X \tau)Y - (\nabla_Y \tau)X\,.
$$

1.4.1 Pseudo-Hermitian Ricci and scalar curvature

We shall examine the consequences of (1.71) later on. Let us look at the local manifestation of R with respect to a local frame $\{T_1, \dots, T_n\}$ of $T_{1,0}(M)$ on U. We adopt the unifying notation

$$\{T_A\} = \{T, T_\alpha, T_{\overline{\alpha}}\},$$

where

$$T_0 = T, \quad A \in \{0, 1, \dots, n, \overline{1}, \dots, \overline{n}\}.$$

We define the functions $R_A{}^D{}_{BC} : U \to \mathbf{C}$ by setting

$$R(T_B, T_C)T_A = R_A{}^D{}_{BC} T_D = R_A{}^\sigma{}_{BC} T_\sigma + R_A{}^{\overline{\sigma}}{}_{BC} T_{\overline{\sigma}} + R_A{}^0{}_{BC} T .$$

If $A = \alpha \in \{1, \dots, n\}$ then $R(T_B, T_C)T_\alpha \in T_{1,0}(M)$, so that

$$R_\alpha{}^{\overline{\sigma}}{}_{BC} = 0, \; R_\alpha{}^0{}_{BC} = 0 .$$

Similarly, if $A = \overline{\alpha}$ then

$$R_{\overline{\alpha}}{}^\sigma{}_{BC} = 0, \; R_{\overline{\alpha}}{}^0{}_{BC} = 0.$$

Also, $\nabla T = 0$ yields

$$R_0{}^D{}_{BC} = 0.$$

Definition 1.29. The *Ricci tensor* of the Webster connection is defined by

$$\mathrm{Ric}(Y, Z) = \mathrm{trace}\{X \longmapsto R(X, Z)Y\},$$

for any $Y, Z \in T(M)$. The *pseudo-Hermitian Ricci tensor* is then given by $R_{\lambda\overline{\mu}} = \mathrm{Ric}(T_\lambda, T_{\overline{\mu}})$. □

Note that

$$R_{\lambda\overline{\mu}} = R_\lambda{}^\alpha{}_{\alpha\overline{\mu}} .$$

The pseudo-Hermitian Ricci tensor was introduced by S. Webster [422]. It is a natural question whether $R_{\alpha\overline{\mu}}$ determines Ric. As we shall demonstrate later on, unless $\tau = 0$ the answer is negative (there are other nontrivial components of Ric that may be computed as certain (contractions of) covariant derivatives of the pseudo-Hermitian torsion).

Definition 1.30. Let us set $\rho = h^{\lambda\overline{\mu}} R_{\lambda\overline{\mu}}$. This is the *pseudo-Hermitian scalar curvature.* □

It will be shortly shown (at the end of Section 1.4.2) that $\rho = \frac{1}{2}\mathrm{trace}\,(\mathrm{Ric})$.

1.4.2 The curvature forms Ω_α^β

We may perform the following calculation:

$$\begin{aligned} R(X,Y)T_\alpha &= \nabla_X \nabla_Y T_\alpha - \nabla_Y \nabla_X T_\alpha - \nabla_{[X,Y]} T_\alpha \\ &= \nabla_X(\omega_\alpha^\beta(Y)T_\beta) - \nabla_Y(\omega_\alpha^\beta(X)T_\beta) - \omega_\alpha^\beta([X,Y])T_\beta \\ &= 2(d\omega_\alpha^\beta)(X,Y)T_\beta + (\omega_\alpha^\beta(Y)\omega_\beta^\gamma(X) - \omega_\alpha^\beta(X)\omega_\beta^\gamma(Y))T_\gamma \,. \end{aligned}$$

Consequently,

$$R(X,Y)T_\alpha = 2(d\omega_\alpha^\beta - \omega_\alpha^\gamma \wedge \omega_\gamma^\beta)(X,Y)T_\beta \,. \tag{1.72}$$

Let us take the inner product of (1.71) with W to obtain

$$\begin{aligned} & g_\theta(R^\theta(X,Y)Z,W) = \\ & \quad g_\theta(R(X,Y)Z,W) + g_\theta((LX \wedge LY)Z,W) + 2(d\theta)(X,Y)g_\theta(JZ,W) \\ & \quad - g_\theta(S(X,Y),Z)\theta(W) + g_\theta(S(X,Y),W)\theta(Z) \\ & \quad - 2g_\theta((\theta \wedge \mathcal{O})(X,Y),Z)\theta(W) + 2g_\theta((\theta \wedge \mathcal{O})(X,Y),W)\theta(Z) \,. \end{aligned} \tag{1.73}$$

Using (1.73) and the symmetry of the Riemann–Christoffel tensor field of (M, g_θ),

$$g_\theta(R^\theta(X,Y)Z,W) = g_\theta(R^\theta(W,Z)Y,X),$$

we obtain

$$\begin{aligned} & g_\theta(R(X,Y)Z,W) = \\ & \quad g_\theta(R(W,Z)Y,X) - g_\theta((LX \wedge LY)Z,W) + g_\theta((LW \wedge LZ)Y,X) \\ & \quad + g_\theta(S(X,Y),Z)\theta(W) - g_\theta(S(W,Z),Y)\theta(X) \\ & \quad - \theta(Z)g_\theta(S(X,Y),W) + \theta(Y)g_\theta(S(W,Z),X) \\ & \quad + 2g_\theta((\theta \wedge \mathcal{O})(X,Y),Z)\theta(W) - 2g_\theta((\theta \wedge \mathcal{O})(W,Z),Y)\theta(X) \\ & \quad - 2\theta(Z)g_\theta((\theta \wedge \mathcal{O})(X,Y),W) + 2\theta(Y)g_\theta((\theta \wedge \mathcal{O})(W,Z),X) \end{aligned} \tag{1.74}$$

for any $X, Y, Z, W \in T(M)$. In particular, for $X, Y, Z, W \in H(M)$ the identity (1.74) becomes

$$\begin{aligned} & g_\theta(R(X,Y)Z,W) = \\ & \quad g_\theta(R(W,Z)Y,X) - g_\theta((LX \wedge LY)Z,W) + g_\theta((LW \wedge LZ)Y,X). \end{aligned} \tag{1.75}$$

Furthermore, let us set $X = T$ in (1.74) to obtain (since $LT = 0$)

$$\begin{aligned} & g_\theta(R(T,Y)Z,W) = \\ & \quad \theta(R(W,Z)Y) + \theta((LW \wedge LZ)Y) \\ & \quad + g_\theta(S(T,Y),Z)\theta(W) - g_\theta(S(W,Z),Y) - \theta(Z)g_\theta(S(T,Y),W) \\ & \quad + 2g_\theta((\theta \wedge \mathcal{O})(T,Y),Z)\theta(W) - 2g_\theta((\theta \wedge \mathcal{O})(W,Z),Y) \\ & \quad - 2\theta(Z)g_\theta((\theta \wedge \mathcal{O})(T,Y),W) + 2\theta(Y)\theta((\theta \wedge \mathcal{O})(W,Z)) \end{aligned}$$

for any $Y, Z, W \in T(M)$. We may use

$$\begin{gathered}\tau H(M) \subseteq H(M), \\ T \rfloor S = \nabla_T \tau, \quad \theta(S(X,Y)) = 0, \\ T \rfloor (\theta \wedge \mathcal{O}) = \frac{1}{2}(\mathcal{O} + \theta \otimes T), \\ \theta \circ (\theta \wedge \mathcal{O}) = 0, \quad \theta \circ L = 0, \\ \theta((LW \wedge LZ)Y) = 0,\end{gathered}$$

to obtain

$$\begin{aligned} g_\theta(R(T,Y)Z,W) = \\ \theta(R(W,Z)Y) + g_\theta((\nabla_T\tau)Y, Z)\theta(W) - g_\theta((\nabla_T\tau)Y, W)\theta(Z) \\ - g_\theta(S(W,Z),Y) - 2g_\theta((\theta \wedge \mathcal{O})(W,Z),Y) \\ + \theta(W)g_\theta(\mathcal{O}Y,Z) - \theta(Z)g_\theta(\mathcal{O}Y,W). \quad (1.76)\end{aligned}$$

In particular, for $Y, Z, W \in H(M)$ we obtain

$$g_\theta(R(T,Y)Z,W) = g_\theta(S(Z,W),Y). \quad (1.77)$$

Note that with respect to a local frame $\{T_\alpha\}$ in $T_{1,0}(M)$ we have

$$LT_\alpha = iT_\alpha + A_\alpha^{\overline{\beta}} T_{\overline{\beta}}, \quad LT_{\overline{\alpha}} = -iT_{\overline{\alpha}} + A_{\overline{\alpha}}^{\beta} T_\beta .$$

Next, we wish to express $d\omega_\alpha^\beta - \omega_\alpha^\gamma \wedge \omega_\gamma^\beta$ with respect to the components $R_A{}^D{}_{BC}$ of the curvature tensor R of the Tanaka–Webster connection. To this end we write

$$\begin{aligned} 2(d\omega_\alpha^\beta - \omega_\alpha^\gamma \wedge \omega_\gamma^\beta) = B^\beta_{\alpha\lambda\mu}\theta^\lambda \wedge \theta^\mu + B^\beta_{\alpha\overline{\lambda}\overline{\mu}}\theta^{\overline{\lambda}} \wedge \theta^{\overline{\mu}} \\ + B^\beta_{\alpha\lambda\overline{\mu}}\theta^\lambda \wedge \theta^{\overline{\mu}} + B^\beta_{\alpha 0\overline{\mu}}\theta \wedge \theta^{\overline{\mu}} + B^\beta_{\alpha\lambda 0}\theta^\lambda \wedge \theta . \quad (1.78)\end{aligned}$$

Hence, using (1.72) we obtain

$$\begin{aligned} R(\cdot,\cdot)T_\alpha = (\theta^\lambda \wedge \theta^\mu) \otimes (B^\beta_{\alpha\lambda\mu}T_\beta) + (\theta^{\overline{\lambda}} \wedge \theta^{\overline{\mu}}) \otimes (B^\beta_{\alpha\overline{\lambda}\overline{\mu}}T_\beta) \\ + (\theta^\lambda \wedge \theta^{\overline{\mu}}) \otimes (B^\beta_{\alpha\lambda\overline{\mu}}T_\beta) + (\theta \wedge \theta^{\overline{\mu}}) \otimes (B^\beta_{\alpha 0\overline{\mu}}T_\beta) + (\theta^\lambda \wedge \theta) \otimes B^\beta_{\alpha\lambda 0}T_\beta) .\end{aligned}$$

Therefore

$$R_\alpha{}^\beta{}_{\sigma\rho} = \frac{1}{2}(B^\beta_{\alpha\sigma\rho} - B^\beta_{\alpha\rho\sigma}), \quad (1.79)$$

$$R_\alpha{}^\beta{}_{\bar\sigma\bar\rho} = \frac{1}{2}(B^\beta_{\alpha\bar\sigma\bar\rho} - B^\beta_{\bar\rho\bar\sigma}), \quad (1.80)$$

$$R_\alpha{}^\beta{}_{\sigma\overline{\rho}} = \frac{1}{2}B^\beta_{\alpha\sigma\overline{\rho}}, \quad (1.81)$$

$$R_\alpha{}^\beta{}_{0\overline{\rho}} = \frac{1}{2}B^\beta_{\alpha 0\overline{\rho}}, \quad (1.82)$$

$$R_\alpha{}^\beta{}_{\sigma 0} = \frac{1}{2}B^\beta_{\alpha\sigma 0}. \quad (1.83)$$

Using (1.79)–(1.83) to substitute in (1.78) we obtain

$$
\begin{aligned}
2(d\omega_\alpha^\beta - \omega_\alpha^\gamma \wedge \omega_\gamma^\beta) &= R_\alpha{}^\beta{}_{\lambda\mu}\theta^\lambda \wedge \theta^\mu + R_\alpha{}^\beta{}_{\bar\lambda\bar\mu}\theta^{\overline{\lambda}} \wedge \theta^{\overline{\mu}} \\
&\quad + 2(R_\alpha{}^\beta{}_{\lambda\bar\mu}\theta^\lambda \wedge \theta^{\overline{\mu}} + R_\alpha{}^\beta{}_{0\bar\mu}\theta \wedge \theta^{\overline{\mu}} + R_\alpha{}^\beta{}_{\lambda 0}\theta^\lambda \wedge \theta). \qquad (1.84)
\end{aligned}
$$

Let us set $X = T_\lambda$, $Y = T_\mu$, $Z = T_\alpha$, and $W = T_{\overline{\sigma}}$ in (1.75) and note that

$$g_\theta(R(T_{\overline{\sigma}}, T_\alpha)T_\mu, T_\lambda) = 0$$

to obtain

$$R_\alpha{}^\beta{}_{\lambda\mu}h_{\beta\overline{\sigma}} = -g_\theta((LT_\lambda \wedge LT_\mu)T_\alpha, T_{\overline{\sigma}}) + g_\theta((LT_{\overline{\sigma}} \wedge LT_\alpha)T_\mu, T_\lambda).$$

On the other hand,

$$
\begin{aligned}
(LT_\lambda \wedge LT_\mu)T_\alpha &= A_{\lambda\alpha}LT_\mu - A_{\mu\alpha}LT_\lambda\,, \\
(LT_{\overline{\sigma}} \wedge LT_\alpha)T_\mu &= -ih_{\overline{\sigma}\mu}LT_\alpha - A_{\alpha\mu}LT_{\overline{\sigma}}\,.
\end{aligned}
$$

Consequently

$$R_\alpha{}^\rho{}_{\lambda\mu} = 2i(A_{\mu\alpha}\delta_\lambda^\rho - A_{\lambda\alpha}\delta_\mu^\rho). \qquad (1.85)$$

Similarly, set $X = T_{\overline{\lambda}}$, $Y = T_{\overline{\mu}}$, $Z = T_\alpha$, and $W = T_{\overline{\sigma}}$ in (1.75) to obtain

$$R_\alpha{}^\beta{}_{\bar\lambda\bar\mu}h_{\beta\overline{\sigma}} = -g_\theta((LT_{\overline{\lambda}} \wedge LT_{\overline{\mu}})T_\alpha, T_{\overline{\sigma}}) + g_\theta((LT_{\overline{\sigma}} \wedge LT_\alpha)T_{\overline{\mu}}, T_{\overline{\lambda}})$$

because of

$$g_\theta(R(T_{\overline{\sigma}}, T_\alpha)T_{\overline{\mu}}, T_{\overline{\lambda}}) = 0.$$

On the other hand, due to

$$
\begin{aligned}
(LT_{\overline{\lambda}} \wedge LT_{\overline{\mu}})T_\alpha &= -ih_{\overline{\lambda}\alpha}LT_{\overline{\mu}} + ih_{\overline{\mu}\alpha}LT_{\overline{\lambda}}, \\
(LT_{\overline{\sigma}} \wedge LT_\alpha)T_{\overline{\mu}} &= A_{\bar\sigma\bar\mu}LT_\alpha - ih_{\alpha\overline{\mu}}LT_{\overline{\sigma}},
\end{aligned}
$$

we may conclude that

$$R_\alpha{}^\rho{}_{\bar\lambda\bar\mu} = 2i(h_{\alpha\overline{\lambda}}A_{\overline{\mu}}^\rho - h_{\alpha\overline{\mu}}A_{\overline{\lambda}}^\rho)\,. \qquad (1.86)$$

Next we apply (1.77) for $Y = T_{\overline{\mu}}$, $Z = T_\alpha$, and $W = T_{\overline{\sigma}}$ to obtain

$$R_\alpha{}^\rho{}_{0\overline{\mu}} = h_{\overline{\mu}\lambda}h^{\overline{\sigma}\rho}S_{\alpha\overline{\sigma}}^\lambda, \qquad (1.87)$$

where

$$S(T_\alpha, T_{\overline{\sigma}}) = S_{\alpha\overline{\sigma}}^\lambda T_\lambda + S_{\alpha\overline{\sigma}}^{\overline{\lambda}} T_{\overline{\lambda}}\,.$$

Similarly, again from (1.77) for $Y = T_\lambda$, $Z = T_\alpha$, and $W = T_{\overline{\sigma}}$ we get

$$R_\alpha{}^\rho{}_{\lambda 0} = -h^{\overline{\sigma}\rho} h_{\overline{\mu}\lambda} S^{\overline{\mu}}_{\alpha\overline{\sigma}} \,. \tag{1.88}$$

At this point we may substitute from (1.85)–(1.88) into (1.84) to derive the identity

$$\begin{aligned} d\omega_\alpha^\beta - \omega_\alpha^\gamma \wedge \omega_\gamma^\beta &= i(A_{\mu\alpha}\delta_\lambda^\beta - A_{\lambda\alpha}\delta_\mu^\beta)\theta^\lambda \wedge \theta^\mu + i(h_{\alpha\overline{\lambda}} A^\beta_{\overline{\mu}} - h_{\alpha\overline{\mu}} A^\beta_{\overline{\lambda}})\theta^{\overline{\lambda}} \wedge \theta^{\overline{\mu}} \\ &+ R_\alpha{}^\beta{}_{\lambda\overline{\mu}}\theta^\lambda \wedge \theta^{\overline{\mu}} + h_{\overline{\mu}\lambda} h^{\overline{\sigma}\beta} S^\lambda_{\alpha\overline{\sigma}}\theta \wedge \theta^{\overline{\mu}} - h^{\overline{\sigma}\beta} h_{\overline{\mu}\lambda} S^{\overline{\mu}}_{\alpha\overline{\sigma}}\theta^\lambda \wedge \theta \,. \end{aligned}$$

We define the covariant derivative $A_{CD,B}$ of $A(X,Y) = g_\theta(X, \tau Y)$ (with respect to the Tanaka–Webster connection) by setting

$$A_{CD,B} = (\nabla_{T_B} A)(T_C, T_D)$$

and note that

$$h_{\overline{\mu}\lambda} S^\lambda_{\alpha\overline{\sigma}} = A_{\bar\mu\bar\sigma,\alpha} \,, \quad h_{\overline{\mu}\lambda} S^\lambda_{\alpha\overline{\sigma}} = -A_{\alpha\lambda,\overline{\sigma}} \,,$$

because of

$$(\nabla_{T_{\overline{\sigma}}}\tau)T_\alpha \in T_{0,1}(M).$$

Finally, we obtain

$$\begin{aligned} d\omega_\alpha^\beta - \omega_\alpha^\gamma \wedge \omega_\gamma^\beta &= R_\alpha{}^\beta{}_{\lambda\overline{\mu}}\theta^\lambda \wedge \theta^{\overline{\mu}} \\ &+ W^\beta_{\alpha\lambda}\theta^\lambda \wedge \theta - W^\beta_{\alpha\overline{\mu}}\theta^{\overline{\mu}} \wedge \theta + 2i\theta^\beta \wedge \tau_\alpha + 2i\theta_\alpha \wedge \tau^\beta, \end{aligned} \tag{1.89}$$

where

$$W^\beta_{\alpha\overline{\mu}} = h^{\overline{\sigma}\beta} A_{\bar\mu\bar\sigma,\alpha}, \quad W^\beta_{\alpha\lambda} = h^{\overline{\sigma}\beta} A_{\alpha\lambda,\overline{\sigma}},$$

and

$$\tau_\alpha = h_{\alpha\overline{\beta}}\tau^{\overline{\beta}}, \quad \theta_\alpha = h_{\alpha\overline{\beta}}\theta^{\overline{\beta}} \,.$$

Indeed,

$$(A_{\alpha\mu}\delta^\beta_\lambda - A_{\alpha\lambda}\delta^\beta_\mu)\theta^\lambda \wedge \theta^\mu = 2A_{\alpha\mu}\theta^\beta \wedge \theta^\mu = 2\theta^\beta \wedge \tau_\alpha$$

and

$$(A^\beta_{\overline{\mu}} h_{\alpha\overline{\lambda}} - A^\beta_{\overline{\lambda}} h_{\alpha\overline{\mu}})\theta^{\overline{\lambda}} \wedge \theta^{\overline{\mu}} = 2A^\beta_{\overline{\mu}} h_{\alpha\overline{\lambda}} = 2\theta_\alpha \wedge \tau^\beta \,.$$

Let us define the local 2-forms $\Pi_\alpha^\beta, \Omega_\alpha^\beta \in \Gamma^\infty(U, \Lambda^2 T^*(M) \otimes \mathbf{C})$ by setting

$$\begin{aligned} \Pi_\alpha^\beta &= d\omega_\alpha^\beta - \omega_\alpha^\gamma \wedge \omega_\gamma^\beta \,, \\ \Omega_\alpha^\beta &= \Pi_\alpha^\beta - 2i\theta_\alpha \wedge \tau^\beta + 2i\tau_\alpha \wedge \theta^\beta \,. \end{aligned}$$

With this notation (1.89) may be written as follows:

Theorem 1.7. (S.M. Webster [422])
Let M be a nondegenerate CR manifold and θ a pseudo-Hermitian structure on M. Then

$$\Omega_\alpha^\beta = R_\alpha{}^\beta{}_{\lambda\overline{\mu}}\theta^\lambda \wedge \theta^{\overline{\mu}} + W^\beta_{\alpha\lambda}\theta^\lambda \wedge \theta - W^\beta_{\alpha\overline{\lambda}}\theta^{\overline{\lambda}} \wedge \theta \,. \tag{1.90}$$

Next, we set by definition

$$R_{\alpha\overline{\beta}\lambda\overline{\mu}} = g_\theta(R(T_\lambda, T_{\overline{\mu}})T_\alpha, T_{\overline{\beta}}) = h_{\gamma\overline{\beta}}R_\alpha{}^\gamma{}_{\lambda\overline{\mu}} \,. \tag{1.91}$$

Let us look at the various symmetry properties satisfied by $R_{\alpha\overline{\beta}\lambda\overline{\mu}}$. First,

$$R_{\alpha\overline{\beta}\lambda\overline{\mu}} + R_{\alpha\bar{\beta}\bar{\mu}\lambda} = 0 \tag{1.92}$$

because R is a 2-form. Moreover,

$$R_{\alpha\overline{\beta}\lambda\overline{\mu}} + R_{\overline{\beta}\alpha\lambda\overline{\mu}} = 0 \tag{1.93}$$

because $\nabla g_\theta = 0$. Other symmetries may be obtained from (1.90) as follows. Let us set

$$\Omega_{\alpha\overline{\gamma}} = \Omega_\alpha^\beta h_{\beta\overline{\gamma}} \,.$$

Let us contract with $h_{\beta\overline{\gamma}}$ in (1.90) to obtain

$$\Omega_{\alpha\overline{\gamma}} = R_{\alpha\overline{\gamma}\lambda\overline{\mu}}\theta^\lambda \wedge \theta^{\overline{\mu}} + \lambda_{\alpha\overline{\gamma}} \wedge \theta, \tag{1.94}$$

where

$$\lambda_{\alpha\overline{\gamma}} = W^\beta_{\alpha\lambda}h_{\beta\overline{\gamma}}\theta^\lambda - W^\beta_{\alpha\overline{\lambda}}h_{\beta\overline{\gamma}}\theta^{\overline{\lambda}} \,.$$

We recall the identity (1.64):

$$d\theta^\alpha = \theta^\beta \wedge \omega_\beta^\alpha + \theta \wedge \tau^\alpha \,.$$

Exterior differentiation leads to

$$0 = d\theta^\beta \wedge \omega_\beta^\alpha - \theta^\beta \wedge d\omega_\beta^\alpha + d\theta \wedge \tau^\alpha - \theta \wedge d\tau^\alpha \,.$$

Next, we replace from

$$d\theta = 2ih_{\alpha\overline{\beta}}\theta^\alpha \wedge \theta^{\overline{\beta}} \,,$$

and rearrange terms to obtain

$$\theta^\gamma \wedge (\omega_\gamma^\beta \wedge \omega_\beta^\alpha - d\omega_\gamma^\alpha + 2i\theta_\gamma \wedge \tau^\alpha) + \theta \wedge (\tau^\beta \wedge \omega_\beta^\alpha - d\tau^\alpha) = 0,$$

and taking into account that

$$\theta^\gamma \wedge \tau_\gamma = 0.$$

we obtain

$$0 = \theta^\gamma \wedge (-\Pi_\gamma^\alpha + 2i\theta_\gamma \wedge \tau^\alpha - 2i\tau_\gamma \wedge \theta^\alpha) + \theta \wedge (\tau^\beta \wedge \omega_\beta^\alpha - d\tau^\alpha),$$

or

$$\theta^\gamma \wedge \Omega_\gamma^\alpha + \theta \wedge (d\tau^\alpha - \tau^\beta \wedge \omega_\beta^\alpha) = 0. \tag{1.95}$$

Let us set by definition

$$\Omega^\alpha = d\tau^\alpha - \tau^\beta \wedge \Omega_\beta^\alpha, \quad \Omega_{\overline{\beta}} = \Omega^\alpha h_{\alpha\overline{\beta}}.$$

With this notation, let us contract (1.95) by $h_{\alpha\overline{\beta}}$. We obtain

$$\theta^\gamma \wedge \Omega_{\gamma\overline{\beta}} + \theta \wedge \Omega_{\overline{\beta}} = 0. \tag{1.96}$$

Using (1.94) we have

$$\theta^\gamma \wedge (R_{\gamma\overline{\beta}\lambda\overline{\mu}}\theta^\lambda \wedge \theta^{\overline{\mu}} + \lambda_{\gamma\overline{\beta}} \wedge \theta) + \theta \wedge \Omega_{\overline{\beta}} = 0,$$

that is,

$$R_{\gamma\overline{\beta}\lambda\overline{\mu}}\theta^\gamma \wedge \theta^\lambda \wedge \theta^{\overline{\mu}} + \theta \wedge (\Omega_{\overline{\beta}} - \lambda_{\gamma\overline{\beta}} \wedge \theta^\gamma) = 0. \tag{1.97}$$

Applying (1.97) to $(T_\rho, T_\nu, T_{\overline{\alpha}})$ we obtain the following theorem:

Theorem 1.8. (S.M. Webster [422])
Let M be a nondegenerate CR manifold, with a fixed pseudo-Hermitian structure θ. Then the curvature tensor $R_{\alpha\overline{\beta}\lambda\overline{\mu}}$ of the Tanaka–Webster connection of (M, θ) satisfies

$$R_{\rho\overline{\beta}\nu\overline{\alpha}} = R_{\nu\overline{\beta}\rho\overline{\alpha}}. \tag{1.98}$$

Let us recall that the pseudo-Hermitian Ricci curvature $R_{\lambda\overline{\mu}}$ is given by

$$R_{\lambda\overline{\mu}} = R_\lambda{}^\alpha{}_{\alpha\overline{\mu}}.$$

Contraction with $h^{\overline{\beta}\mu}$ in (1.98) gives

$$R_\rho{}^\mu{}_{\nu\overline{\alpha}} = R_\nu{}^\mu{}_{\rho\overline{\alpha}},$$

whence

$$R_{\lambda\overline{\mu}} = R_\alpha{}^\alpha{}_{\lambda\overline{\mu}}, \tag{1.99}$$

which agrees with the definition (2.16) in [422], p. 33. □

We end this section by discussing the Ricci operator and by proving that (up to a factor of $1/2$) the pseudo-Hermitian scalar curvature is the trace of Ric.

Let $(M, T_{1,0}(M))$ be a nondegenerate CR manifold (of hypersurface type) of CR dimension n. Let F be a complex vector bundle over M and φ a global C^∞ section of $T^*(M) \otimes T^*(M) \otimes F$ (i.e., an F-valued (0, 2)-tensor field on M). Let $\Lambda_\theta \varphi$ be the trace of φ with respect to the Levi form L_θ. Precisely, let $x \in M$ and $\{E_1, \ldots, E_n\}$ an orthonormal frame of $T_{1,0}(M)$ defined on some open neighborhood $U \subseteq M$ of x, i.e., $L_\theta(E_\alpha, E_{\bar\beta}) = \epsilon_\alpha \delta_{\alpha\beta}$, where $\epsilon_1 = \cdots = \epsilon_r = -\epsilon_{r+1} = \cdots = -\epsilon_{r+s} = 1$, $r + s = n$. Here (r, s) is the signature of L_θ. Then

$$i\,(\Lambda_\theta \varphi)_x = \sum_{\alpha=1}^{n} \epsilon_\alpha \varphi(E_\alpha, E_{\bar\alpha})_x.$$

Let $U(r, s)$ consist of all $A \in \mathrm{GL}(n, \mathbf{C})$ such that $A^{-1} = D\overline{A}^{\,t} D$, where t denotes the transpose and $D = \mathrm{diag}(\epsilon_1, \ldots, \epsilon_n)$. If $\{\hat{E}_\alpha\}$ is another orthonormal frame of $T_{1,0}(M)$, defined on the open set $V \subseteq M$, $x \in V$, then $\hat{E}_\alpha = a_\alpha^\beta E_\beta$ for some $A = \left[a_\alpha^\beta\right] : U \cap V \to U(r, s)$, i.e., $\sum_{\alpha=1}^n \epsilon_\alpha a_\alpha^\lambda a_{\bar\alpha}^{\bar\mu} = \epsilon^\lambda \delta^{\lambda\mu}$, where $\epsilon^\alpha = \epsilon_\alpha$. Thus the definition of $(\Lambda_\theta \varphi)_x$ does not depend on the choice of (local) orthonormal frame at x and $\Lambda_\theta \varphi$ is a global C^∞ section of F.

Definition 1.31. The *Ricci operator* R_* is given by

$$R_* X = \Lambda_\theta R(\cdot, \cdot) J X$$

for any $X \in H(M)$. □

On a strictly pseudoconvex CR manifold (the case in [398], p. 34)

$$R_* X = -i \sum_{\alpha=1}^{n} R(E_\alpha, E_{\bar\alpha}) J X$$

for some (local) orthonormal frame $\{E_\alpha\}$ of $T_{1,0}(M)$. Hence, with respect to an arbitrary (local) frame $\{T_\alpha\}$ of $T_{1,0}(M)$,

$$R_* X = -i h^{\alpha\bar\beta} R(T_\alpha, T_{\bar\beta}) J X,$$

and a calculation shows that

$$R_* T_\alpha = R_\alpha{}^\beta T_\beta,$$

where $R_\alpha{}^\beta = h^{\beta\bar\gamma} R_{\alpha\bar\gamma}$.

Proposition 1.10. *The following identities hold:*

$$R_* T_{1,0}(M) \subseteq T_{1,0}(M), \quad g_\theta(R_* Z, \overline{W}) = g_\theta(Z, \overline{R_* W}),$$

for any $Z, W \in T_{1,0}(M)$.

As previously remarked, $R_{\alpha\bar\beta}$ is only a fragment of Ric and (as a consequence of (1.71)) we have

$$
\begin{aligned}
R^{\theta}_{\alpha\bar\beta} &= h_{\alpha\bar\beta} - \frac{1}{2} R_{\alpha\bar\beta} , \\
R_{\alpha\beta} &= i(n-1) A_{\alpha\beta} , \\
R_{0\beta} &= S^{\bar\alpha}_{\bar\alpha\beta}, \quad R_{\alpha 0} = R_{00} = 0.
\end{aligned}
\tag{1.100}
$$

Here $S^{\bar\alpha}_{\bar\beta\gamma}$ are (among) the complex components of S. Also we set

$$
R^{\theta}_{\alpha\bar\beta} = \operatorname{trace}\{X \mapsto R^{\theta}(X, T_\alpha) T_{\bar\beta}\} .
$$

In particular (by the first of the identities (1.100)), $R_{\alpha\bar\beta} = R_{\bar\beta\alpha}$ (and the second of the formulas in Proposition 1.10 may be easily proved).

Finally, we wish to prove that $\operatorname{trace}(\operatorname{Ric}) = 2\rho$. Again by (1.71), the following formulas hold:

$$
\begin{aligned}
\operatorname{Ric}(X_\alpha, X_\beta) &= i(n-1)(A_{\alpha\beta} - A_{\bar\alpha\bar\beta}) + R_{\alpha\bar\beta} + R_{\bar\alpha\beta} , \\
\operatorname{Ric}(J X_\alpha, X_\beta) &= -(n-1)(A_{\alpha\beta} + A_{\bar\alpha\bar\beta}) + i(R_{\alpha\bar\beta} - R_{\bar\alpha\beta}) , \\
\operatorname{Ric}(X_\alpha, J X_\beta) &= -(n-1)(A_{\alpha\beta} + A_{\bar\alpha\bar\beta}) + i(R_{\bar\alpha\beta} - R_{\alpha\bar\beta}) , \\
\operatorname{Ric}(J X_\alpha, J X_\beta) &= -i(n-1)(A_{\alpha\beta} - A_{\bar\alpha\bar\beta}) + R_{\alpha\bar\beta} + R_{\bar\alpha\beta} ,
\end{aligned}
$$

where $X_\alpha = T_\alpha + T_{\bar\alpha}$. Then we may compute $\operatorname{trace}(\operatorname{Ric})$ as

$$
\operatorname{trace}(\operatorname{Ric}) = g^{ij} \operatorname{Ric}(X_i, X_j),
$$

where $\{X_j\} = \{X_\alpha, J X_\alpha\}$ and

$$
\begin{aligned}
g^{\alpha+n,\beta+n} &= g^{\alpha\beta}, & g^{\alpha,\beta+n} &= -g^{\alpha+n,\beta} , \\
g^{\alpha 0} &= g^{0\alpha} = 0, & g^{00} &= 1 , \\
g^{\alpha\beta} &= \frac{1}{4}(h^{\alpha\bar\beta} + h^{\bar\alpha\beta}), & g^{\alpha,\beta+n} &= \frac{i}{4}(h^{\alpha\bar\beta} - h^{\bar\alpha\beta}) .
\end{aligned}
$$

1.4.3 Pseudo-Hermitian sectional curvature

We devote this section to a pseudo-Hermitian analogue (cf. S. Webster [422]) of the notion of holomorphic sectional curvature in Hermitian geometry (cf., e.g., [241], vol. II, p. 168). First, we consider the following pseudo-Hermitian analogue of the Riemann–Christoffel tensor field (of a Riemannian manifold):

$$
R(Z, W, X, Y) = g_\theta(R(X, Y)Z, W),
$$

for any $X, Y, Z, W \in T(M)$. This agrees with our preceding conventions (cf. (1.91)), that is,

$$
R(T_\alpha, T_{\overline\beta}, T_\lambda, T_{\overline\mu}) = R_{\alpha\overline\beta\lambda\overline\mu} .
$$

Let $x \in M$. Let $G_1(H(M))_x$ consist of all 2-planes $\sigma \subset T_x(M)$ such that (i) $\sigma \subset H(M)_x$ and (ii) $J_x(\sigma) = \sigma$. Then $G_1(H(M))$ (the disjoint union of all $G_1(H(M))_x$) is a fiber bundle over M with standard fiber $\mathbf{C}P^{n-1}$. Define a function

$$k_\theta : G_1(H(M)) \to \mathbf{R}$$

by setting

$$k_\theta(\sigma) = -\frac{1}{4} R_x(X, J_x X, X, J_x X) \tag{1.101}$$

for any $\sigma \in G_1(H(M))$ and any linear basis $\{X, J_x X\}$ in σ satisfying $G_\theta(X, X) = 1$. It is a simple matter that the definition of $k_\theta(\sigma)$ does not depend on the choice of orthonormal basis $\{X, J_x X\}$, as a consequence of the following properties:

$$R(Z, W, X, Y) + R(Z, W, Y, X) = 0,$$
$$R(Z, W, X, Y) + R(W, Z, X, Y) = 0.$$

Definition 1.32. k_θ is referred to as the *(pseudo-Hermitian) sectional curvature* of (M, θ). □

With respect to an arbitrary (not necessarily orthonormal) basis $\{X, J_x X\}$ of the 2-plane σ, the sectional curvature $k_\theta(\sigma)$ is also expressed by

$$k_\theta(\sigma) = -\frac{1}{4} \frac{R_x(X, J_x X, X, J_x X)}{G_\theta(X, X)^2}$$

(to see this, one merely applies the definition (1.101) for the orthonormal basis $\{U, J_x U\}$, with $U = G_\theta(X, X)^{-1/2} X$). Since $X \in H(M)_x$, there is $Z \in T_{1,0}(M)_x$ such that $X = Z + \overline{Z}$. Thus

$$k_\theta(\sigma) = \frac{1}{4} \frac{R_x(Z, \overline{Z}, Z, \overline{Z})}{g_\theta(Z, \overline{Z})^2}.$$

Therefore, we have the following:

Proposition 1.11. (S.M. Webster [422])
If $Z = \xi^\alpha T_\alpha$ with respect to some (local) frame $\{T_\alpha\}$ in $T_{1,0}(M)$, then $k_\theta(\sigma)$ may be expressed as

$$k_\theta(\sigma) = \frac{1}{4} \frac{R_{\alpha\overline{\beta}\lambda\overline{\mu}} \xi^\alpha \xi^{\overline{\beta}} \xi^\lambda \xi^{\overline{\mu}}}{\left(h_{\alpha\overline{\beta}} \xi^\alpha \xi^{\overline{\beta}}\right)^2},$$

where $\xi^{\overline{\alpha}} = \overline{\xi^\alpha}$.

The coefficient $1/4$ makes the sphere $S^{2n+1} \subset \mathbf{C}^{n+1}$ have constant curvature $+1$. Indeed, if $M = S^{2n+1}$ then

$$R^\theta(X,Y)Z = g_\theta(Y,Z)X - g_\theta(X,Z)Y.$$

Also $\tau = 0$ (hence $S = 0$, $\mathcal{O} = I$, $L = J$) and (1.72) shows that

$$R(X,Y)Z = g_\theta(Y,Z)X - g_\theta(X,Z)Y \\ + g_\theta(JY,Z)JX - g_\theta(JX,Z)JY - 2g_\theta(JX,Y)JZ,$$

for any $X, Y, Z \in H(S^{2n+1})$. We have thus proved the following result:

Proposition 1.12. *Let $\theta = \iota^* \frac{i}{2}(\partial - \overline{\partial})|z|^2$ be the natural pseudo-Hermitian structure on S^{2n+1}. Then the pseudo-Hermitian sectional curvature of (S^{2n+1}, θ) is $k_\theta(\sigma) = 1$, for any $\sigma \in G_1(H(S^{2n+1}))$.*

Let $\mathrm{Psh}(M, \theta)$ be the group of CR transformations $f : M \to M$ such that $f^*\theta = \theta$. The following result is due to S. Webster [422].

Theorem 1.9. *If (M, θ) is a $(2n+1)$-dimensional pseudo-Hermitian manifold then* $\mathrm{Psh}(M, \theta)$ *is a Lie group of dimension $\leq (n+1)^2$, with isotropy groups of dimension $\leq n^2$. If M is strictly pseudoconvex then the isotropy groups are compact, and if M is compact then* $\mathrm{Psh}(M, \theta)$ *is compact.*

The Riemannian counterpart of Theorem 1.9 is Theorem 3.4 in [241], Vol. I, p. 239. The proof is beyond the scope of this book (cf. also Theorem 1.2 in [422], p. 31). By a well-known result in Riemannian geometry (cf., e.g., [241], Vol. I, p. 238), if N is a connected n-dimensional Riemannian manifold then its Lie algebra $\mathbf{i}(N)$ of infinitesimal isometries has dimension $\leq n(n+1)/2$. Moreover, if $\mathbf{i}(N)$ has maximal dimension, i.e., $\dim_{\mathbf{R}} \mathbf{i}(N) = n(n+1)/2$, then N has constant sectional curvature. The pseudo-Hermitian analogue of this result has been analyzed by E. Musso [319]. Let $\mathbf{H}^n$ be the standard hyperbolic complex space form.

Theorem 1.10. (E. Musso [319])
Let (M, θ) be a connected $(2n+1)$-dimensional pseudo-Hermitian manifold such that L_θ is positive definite. If $\dim \mathrm{Psh}(M, \theta) = (n+1)^2$ *then M is contact homothetic to one of the following spaces:*

(i) a canonical pseudo-Hermitian manifold $B^{(k)}$ over $\mathbf{C}P^n$,
(ii) $\mathbf{H}^n \times S^1$ or $\mathbf{H}^n \times \mathbf{R}$ equipped with their canonical pseudo-Hermitian structures,
(iii) $\mathbf{C}^n \times S^1$ or $\mathbf{C}^n \times \mathbf{R}$, equipped with their canonical pseudo-Hermitian structures.

For the definitions of the objects in Theorem 1.10 the reader may see Section 5.10 in this book.

1.5 The Chern tensor field

Let $(M, T_{1,0}(M))$ be a nondegenerate CR manifold (of hypersurface type) of CR dimension n. Let θ be a fixed pseudo-Hermitian structure on M and ∇ the Tanaka–Webster connection of (M, θ).

Definition 1.33. The *Chern tensor field* C is defined by

$$\begin{aligned} C(X,Y)Z = R(X,Y)Z \\ - \frac{1}{n+2}\{g_\theta(X,Y)\,\mathrm{Ric}^\sharp(Z) + g_\theta(Z,Y)\,\mathrm{Ric}^\sharp(X) \\ + \mathrm{Ric}(X,Y)\,Z + \mathrm{Ric}(Z,Y)\,X\} \\ + \frac{\rho}{(n+1)(n+2)}\{g_\theta(X,Y)Z + g_\theta(Z,Y)X\} \end{aligned} \tag{1.102}$$

for any $X, Y, Z \in T(M)$. Here R, Ric, and ρ are respectively the curvature tensor, the Ricci tensor, and the (pseudo-Hermitian) scalar curvature of ∇. Also $\mathrm{Ric}^\sharp : T(M) \to T(M)$ is defined by $g_\theta(\mathrm{Ric}^\sharp X, Y) = \mathrm{Ric}(X, Y)$ for any $X, Y \in T(M)$. □

Definition 1.34. Let $\{T_\alpha\}$ be a local frame of $T_{1,0}(M)$ and set

$$C(T_\lambda, T_{\bar\sigma})T_\beta = C_\beta{}^A{}_{\lambda\bar\sigma} T_A.$$

Then $C_\beta{}^\alpha{}_{\lambda\bar\sigma}$ is Chern's *pseudoconformal curvature tensor*. □

Explicitly (by (1.102)),

$$\begin{aligned} C_\beta{}^\alpha{}_{\lambda\bar\sigma} = R_\beta{}^\alpha{}_{\lambda\bar\sigma} - \frac{1}{n+2}\left\{R_\beta{}^\alpha h_{\lambda\bar\sigma} + R_\lambda{}^\alpha h_{\beta\bar\sigma} + \delta^\alpha_\beta R_{\lambda\bar\sigma} + \delta^\alpha_\lambda R_{\beta\bar\sigma}\right\} \\ + \frac{\rho}{(n+1)(n+2)}\left\{\delta^\alpha_\beta h_{\lambda\bar\sigma} + \delta^\alpha_\lambda h_{\beta\bar\sigma}\right\}. \end{aligned}$$

This is similar to H. Weyl's conformal curvature tensor of a Riemannian manifold (cf., e.g., L.P. Eisenhart [132]). Note that $C_\alpha{}^\alpha{}_{\lambda\bar\sigma} = 0$ hence Chern's pseudoconformal curvature tensor vanishes identically when $n = 1$.

Theorem 1.11. (S.S. Chern and J. Moser [99])
When $n > 1$, $C_\beta{}^\alpha{}_{\lambda\bar\sigma} = 0$ if and only if M is locally CR isomorphic to the unit sphere $S^{2n+1} \subset \mathbf{C}^{n+1}$.

A CR manifold is called *spherical* if it is locally CR isomorphic to S^{2n+1}. Therefore, nondegenerate CR manifolds with a vanishing Chern pseudoconformal curvature tensor are spherical. Compact spherical CR manifolds M with amenable[6] holonomy may be classified.[7] By a result of R.R. Miner [303], such M is finitely covered by S^{2n+1}, $S^1 \times S^{2n}$, or a compact quotient of the Heisenberg group $\mathbf{H}_n/\Gamma$, where Γ is a lattice in $\mathbf{H}_n$.

While the proof of the *Chern–Moser theorem* is beyond the scope of this book, we wish to look at few instances in which it may be applied. Let $[g_{\alpha\bar\beta}] \in \mathrm{GL}(n, \mathbf{C})$

[6] A topological group G is *amenable* if for every continuous G-action on a compact metrizable space X, there exists a G-invariant probability measure on X.

[7] W. Goldman [181], obtained the same classification under the assumption that the holonomy of M is nilpotent instead of amenable.

be a Hermitian matrix, of signature (p, q), $p + q = n$, and let $c \in (0, +\infty)$. Let $(z, w) = (z^1, \dots, z^n, w)$ be the natural complex coordinates in $\mathbf{C}^{n+1}$ and consider

$$Q_0 : r_0(z, w) = g_{\alpha\bar{\beta}} z^\alpha \bar{z}^\beta + \frac{i}{2}(w - \bar{w}) = 0,$$
$$Q_+(c) : r_+(z, w) = g_{\alpha\bar{\beta}} z^\alpha \bar{z}^\beta + w\bar{w} - c = 0,$$
$$Q_-(c) : r_-(z, w) = g_{\alpha\bar{\beta}} z^\alpha \bar{z}^\beta - w\bar{w} + c = 0.$$

Each $M \in \{Q_0, Q_\pm(c)\}$ is a real hypersurface in $\mathbf{C}^{n+1}$.

Definition 1.35. Let M be a nondegenerate CR manifold (of hypersurface type). Let θ be a pseudo-Hermitian structure on M. We call (M, θ) a *pseudo-Hermitian space form* if $k_\theta = \text{const}$ (where k_θ is defined by (1.101)). □

Let r be one of the defining functions $\{r_0, r_\pm\}$ and set

$$\theta = j^* \left[i(\overline{\partial} - \partial) r \right]$$

(where $j : M \subset \mathbf{C}^{n+1}$). We shall show that

Theorem 1.12. (S.M. Webster [422])
(M, θ) is a pseudo-Hermitian space form, for each $M \in \{Q_0, Q_\pm(c)\}$.

Let us consider the biholomorphism

$$F : \mathbf{C}^{n+1} \setminus \{w = 0\} \to \mathbf{C}^{n+1} \setminus \{w = 0\}, \quad F(z, w) = \left(\frac{\sqrt{c}}{w} z, \frac{c}{w} \right).$$

Note that F maps $Q_-(c) \setminus \{w = 0\}$ onto $Q_+(c) \setminus \{w = 0\}$. Hence $Q_-(c) \setminus \{w = 0\}$ and $Q_+(c) \setminus \{w = 0\}$ are CR equivalent. However, the induced CR transformation f is not isopseudo-Hermitian. Let $j_\pm : Q_\pm(c) \subset \mathbf{C}^{n+1}$ and set $\theta_\pm = j_\pm^* \left[i(\overline{\partial} - \partial) r_\pm \right]$. The identities

$$F^* dz^\alpha = \frac{\sqrt{c}}{w} dz^\alpha - \frac{\sqrt{c}}{w^2} z^\alpha dw, \quad F^* dw = -\frac{c}{w^2} dw,$$

yield

$$f^* \theta_+ = \frac{c}{|w|^2} \theta_- .$$

A transformation mapping Q_0 onto $Q_+(c)$ minus a point was devised in [99]. Let G_0 denote the group of all matrices

$$a = \begin{pmatrix} 1 & b^\beta & b \\ 0 & B_\alpha^\beta & b_\alpha \\ 0 & 0 & 1 \end{pmatrix} \in \mathcal{M}_n(\mathbf{C})$$

satisfying

$$B^{\gamma}_{\alpha} g_{\gamma\bar{\rho}} B^{\bar{\rho}}_{\bar{\beta}} = g_{\alpha\bar{\beta}}, \quad b_{\alpha} = 2i B^{\rho}_{\alpha} g_{\rho\bar{\gamma}} b^{\bar{\gamma}}, \quad 0 = \frac{i}{2}(b - \bar{b}) + g_{\alpha\bar{\beta}} b^{\alpha} b^{\bar{\beta}},$$

where $B^{\bar{\beta}}_{\bar{\alpha}} = \overline{B^{\beta}_{\alpha}}$ and $b^{\bar{\alpha}} = \overline{b^{\alpha}}$. There is a natural action of G_0 on $\mathbf{C}^{n+1}$ given by

$$\begin{aligned} G_0 \times \mathbf{C}^{n+1} \longrightarrow \mathbf{C}^{n+1}, (a, (z, w)) \mapsto a \cdot (z, w) &= (z', w'), \\ z'^{\alpha} &= B^{\alpha}_{\beta} z^{\beta} + b^{\alpha}, \\ w' &= w + b_{\beta} z^{\beta} + b, \end{aligned}$$

for any

$$a = \begin{pmatrix} 1 & b^{\beta} & b \\ 0 & B^{\beta}_{\alpha} & b_{\alpha} \\ 0 & 0 & 1 \end{pmatrix} \in G_0, \qquad (z, w) \in \mathbf{C}^{n+1}.$$

A short calculation shows that $r_0(a \cdot (z, w)) = r_0(z, w)$, for any $a \in G_0$, $(z, w) \in \mathbf{C}^{n+1}$, i.e., the action of G_0 on $\mathbf{C}^{n+1}$ preserves r_0. Hence it descends to an action of G_0 on Q_0. Each right action $R_a : Q_0 \to Q_0$ is a CR map (because $(z, w) \mapsto a \cdot (z, w)$ is holomorphic). Moreover, the identities

$$R_a^* dz^{\alpha} = B^{\alpha}_{\beta} dz^{\beta}, \quad R_a^* dw = b_{\beta} dz^{\beta} + dw$$

yield $R_a^* \theta_0 = \theta_0$, i.e., R_a is isopseudo-Hermitian. Here

$$\theta_0 = j_0^* \left[i(\bar{\partial} - \partial) r_0 \right]$$

and $j_0 : Q_0 \subset \mathbf{C}^{n+1}$. The isotropy group at $(0, 0) \in Q_0$ is given by

$$\{a \in G_0 : a \cdot (0, 0) = (0, 0)\} = \left\{ \begin{pmatrix} 1 & 0 & 0 \\ 0 & B^{\beta}_{\alpha} & 0 \\ 0 & 0 & 1 \end{pmatrix} : [B^{\beta}_{\alpha}] \in U(p, q) \right\} \simeq U(p, q),$$

where by $U(p, q)$ one denotes the unitary group of the Hermitian form $g_{\alpha\bar{\beta}}$. Hence $Q_0 \simeq G_0 / U(p, q)$. Let us set $\theta^{\alpha} = j_0^* dz^{\alpha}$ and $\theta^{\bar{\alpha}} = \overline{\theta^{\alpha}}$. Then

$$d\theta_0 = 2i g_{\alpha\bar{\beta}} \theta^{\alpha} \wedge \theta^{\bar{\beta}};$$

hence $\{\theta^{\alpha}\}$ is admissible. Next $d\theta^{\alpha} = 0$ yields $\Omega^{\alpha}_{\beta} = 0$ and $\tau^{\alpha} = 0$. Thus (Q_0, θ_0) is Tanaka–Webster flat. In particular, $C = 0$; hence Q_0 is locally CR equivalent to S^{2n+1}. Let $\langle\ ,\ \rangle_+$ be the Hermitian form given by

$$\langle (z, w), (\zeta, \Omega) \rangle_+ = g_{\alpha\bar{\beta}} z^{\alpha} \overline{\zeta}^{\beta} + w \overline{\Omega}$$

for any $(z, w), (\zeta, \Omega) \in \mathbf{C}^{n+1}$. Let $U(p + 1, q)$ be the unitary group of $\langle\ ,\ \rangle_+$. Then $U(p + 1, q)$ acts transitively on $Q_+(c)$ and preserves θ_+. On the other hand, the isotropy group at $(0, \sqrt{c})$ is $U(p, q)$; hence

$$Q_+(c) \simeq U(p+1,q)/U(p,q).$$

Similarly, if $U(p, q+1)$ is the unitary group of the Hermitian form

$$\langle (z,w),(\zeta,\Omega)\rangle_- = g_{\alpha\bar\beta} z^\alpha \overline{\zeta}^\beta - w\overline{\Omega},$$

then it may be shown that

$$Q_-(c) \simeq U(p,q+1)/U(p,q).$$

We wish to compute the curvature $R_{\alpha\bar\beta\lambda\bar\mu}$ of $Q_\pm(c)$. We do this in a slightly more general situation, as follows. Note that both defining functions $r_\pm$ are of the form $r(z,w) = p(z) + q(w)$. Let us compute the torsion and curvature of $M = \{r = 0\}$ on the open subset $U = M \cap \{q_w \neq 0\}$. We adopt the notation

$$p_\alpha = \frac{\partial p}{\partial z^\alpha}, \quad p_{\bar\alpha} = \frac{\partial p}{\partial \overline{z}^\alpha}, \quad p_{\alpha\bar\beta} = \frac{\partial^2 p}{\partial z^\alpha \partial \overline{z}^\beta}, \quad q_w = \frac{\partial q}{\partial w}, \quad q_{\bar w} = \frac{\partial q}{\partial \bar w}.$$

A pseudo-Hermitian structure on M is given by the (pullback to M via $j : M \subset \mathbf{C}^{n+1}$ of the following) form

$$\theta = i(p_{\bar\alpha} d\bar z^\alpha - p_\alpha dz^\alpha + q_{\bar w} d\bar w - q_w dw).$$

Hence

$$d\theta = 2i(p_{\alpha\bar\beta} dz^\alpha \wedge d\bar z^\beta + q_{w\bar w} dw \wedge d\bar w). \tag{1.103}$$

We shall make use of the following lemma:

Lemma 1.6. *Let $M = \{r = 0\}$ and let $j : M \to \mathbf{C}^{n+1}$ be the inclusion. Then (i) $j^*(\partial r \wedge \overline{\partial} r) = 0$ and (ii) $j^* \overline{\partial} r = -j^* \partial r$.*

Using the identities

$$\partial r = p_\alpha dz^\alpha + q_w dw, \quad \overline{\partial} r = p_{\bar\alpha} d\bar z^\alpha + q_{\bar w} d\bar w$$

and (i) of Lemma 1.6 we may compute the (pullback via $j : M \to \mathbf{C}^{n+1}$ of the) form $dw \wedge d\bar w$ as

$$dw \wedge d\bar w = \frac{1}{q_w q_{\bar w}} \left(p_\alpha p_{\bar\beta} dz^\alpha \wedge d\bar z^\beta - \partial r \wedge p_{\bar\beta} d\bar z^\beta - p_\alpha dz^\alpha \wedge \overline{\partial} r \right).$$

Next (by (ii) of Lemma 1.6) θ is expressed both by $\theta = 2ij^*\overline{\partial} r$ and by $\theta = -2ij^*\partial r$; hence

$$dw \wedge d\overline{w} = \frac{1}{q_w q_{\bar w}} \left(p_\alpha p_{\bar\beta} dz^\alpha \wedge d\bar z^\beta + \frac{1}{2i} \theta \wedge p_{\bar\beta} d\bar z^\beta - \frac{1}{2i} p_\alpha dz^\alpha \wedge \theta \right). \tag{1.104}$$

Let us substitute from (1.104) into (1.103) to obtain

$$d\theta = 2ih_{\alpha\bar{\beta}}dz^\alpha \wedge d\bar{z}^\beta + \eta_\alpha dz^\alpha \wedge \theta + \eta_{\bar{\alpha}} d\bar{z}^\alpha \wedge \theta,$$

where

$$Q = \frac{q_{w\bar{w}}}{q_w q_{\bar{w}}}, \quad h_{\alpha\bar{\beta}} = p_{\alpha\bar{\beta}} + Q p_\alpha p_{\bar{\alpha}}, \quad \eta_\alpha = -Q p_\alpha, \quad \eta_{\bar{\alpha}} = \overline{\eta_\alpha}.$$

Finally, set

$$\theta^\alpha = dz^\alpha + \frac{i}{2}\eta^\alpha\theta, \quad \eta^\alpha = h^{\alpha\bar{\beta}}\eta_{\bar{\beta}}.$$

Then

$$d\theta = 2ih_{\alpha\bar{\beta}}\theta^\alpha \wedge \theta^{\bar{\beta}}. \tag{1.105}$$

The characteristic direction T of $d\theta$ is given by

$$T = -\frac{i}{2}\eta^\alpha \frac{\partial}{\partial z^\alpha} + \frac{i}{2}\eta^{\bar{\alpha}} \frac{\partial}{\partial \bar{z}^\alpha} + \frac{i}{q_w}\left(1 + \eta^\alpha p^\alpha\right)\frac{\partial}{\partial w} - \frac{i}{q_{\bar{w}}}\left(1 + \eta^{\bar{\alpha}} p_{\bar{\alpha}}\right)\frac{\partial}{\partial \bar{w}}.$$

A local frame of $T_{1,0}(M)$ (on $U = M \cap \{q_w \neq 0\}$) is given by

$$T_\alpha = \frac{\partial}{\partial z^\alpha} - \frac{1}{q_w} p_\alpha \frac{\partial}{\partial w}.$$

At this point, we may compute the connection and torsion components (of the Tanaka–Webster connection of (M, θ)). To this end, we differentiate $\theta^\alpha = dz^\alpha + (i/2)\eta^\alpha\theta$ to get

$$d\theta^\alpha = \theta^\beta \wedge \omega'^\alpha_\beta + \theta \wedge \left(-\frac{i}{2}T_{\bar{\beta}}(\eta^\alpha)\theta^{\bar{\beta}}\right), \tag{1.106}$$

where

$$\omega'^\alpha_\beta = -\eta^\alpha h_{\beta\bar{\gamma}}\theta^{\bar{\gamma}} + \frac{i}{2}T_\beta(\eta^\alpha)\theta.$$

Thus, on one hand, the pseudo-Hermitian torsion $\tau^\alpha = A^\alpha_{\bar{\beta}}\theta^{\bar{\beta}}$ is given by

$$A^\alpha_{\bar{\beta}} = -\frac{i}{2}T_{\bar{\beta}}(\eta^\alpha). \tag{1.107}$$

On the other, a comparison between (1.106) and (1.65) yields

$$\theta^\beta \wedge (\omega'^\alpha_\beta - \omega^\alpha_\beta) = 0;$$

hence (by successively applying this identity to $(T_\lambda, T_{\bar{\mu}})$, respectively to (T_λ, T))

$$\Gamma^\alpha_{\bar{\mu}\lambda} = -\eta^\alpha h_{\lambda\bar{\mu}}, \quad \Gamma^\alpha_{0\lambda} = \frac{i}{2}T_\lambda(\eta^\alpha).$$

The same identity applied to the pair (T_λ, T_μ) gives only as much as $\gamma^\alpha_{\lambda\mu} = \Gamma^\alpha_{\mu\lambda}$. Therefore, to compute $\Gamma^\alpha_{\lambda\mu}$, we recall the identity (1.51), that is,

$$\Gamma^\alpha_{\gamma\beta} = h^{\bar\sigma\alpha}\{T_\gamma(h_{\beta\bar\sigma}) - g_\theta(T_\beta, [T_\gamma, T_{\bar\sigma}])\}.$$

The bracket $[T_\gamma, T_{\bar\sigma}]$ may be computed directly from

$$T_\alpha = \frac{\partial}{\partial z^\alpha} - \frac{1}{q_w} p_\alpha \frac{\partial}{\partial w}.$$

However, to avoid lengthy computations, we may exploit the identity (1.37) (since the $\Gamma^\alpha_{\bar\mu\lambda}$ are already determined). This procedure yields

$$[T_\gamma, T_{\bar\sigma}] = h_{\gamma\bar\sigma}\left(\eta^\alpha T_\alpha - \eta^{\bar\alpha} T_{\bar\alpha} - 2iT\right)$$

and therefore

$$\Gamma^\alpha_{\gamma\beta} = \delta^\alpha_\gamma \eta_\beta + h^{\bar\sigma\alpha} T_\gamma(h_{\beta\bar\sigma}).$$

Finally, by collecting the identities expressing the Christoffel symbols, we derive

$$\omega^\alpha_\beta = \left(\delta^\alpha_\mu \eta_\beta + h^{\bar\sigma\alpha} T_\mu(h_{\beta\bar\sigma})\right)\theta^\mu - \eta^\alpha h_{\beta\bar\mu}\theta^{\bar\mu} + \frac{i}{2} T_\beta(\eta^\alpha)\theta. \tag{1.108}$$

For the rest of this section, let $\equiv$ denote the congruence relation mod $\theta^\alpha \wedge \theta^\beta, \theta^{\bar\alpha} \wedge \theta^{\bar\beta}$, $\theta^\alpha \wedge \theta, \theta^{\bar\alpha} \wedge \theta$. Differentiation of (1.108) furnishes

$$\begin{aligned} d\omega^\alpha_\beta \equiv -\Big\{ & [\delta^\alpha_\rho \eta_\beta + h^{\bar\sigma\alpha} T_\rho(h_{\beta\bar\sigma})]\eta^\rho h_{\lambda\bar\mu} \\ & + \delta^\alpha_\lambda T_{\bar\mu}(\eta_\beta) + h^{\alpha\bar\sigma} T_{\bar\mu} T_\lambda h_{\beta\bar\sigma} + T_{\bar\mu}(h^{\bar\sigma\alpha}) T_\lambda(h_{\beta\bar\sigma}) \\ & + T_\lambda(\eta^\alpha) h_{\beta\bar\beta} + \eta^\alpha T_\lambda(h_{\beta\bar\mu}) + \eta^\alpha h_{\beta\bar\rho}\eta^{\bar\rho} h_{\lambda\bar\mu} + T_\beta(\eta^\alpha) h_{\lambda\bar\mu}\Big\}\theta^\lambda \wedge \theta^{\bar\mu}. \end{aligned}$$

Also (again by (1.108))

$$\begin{aligned} & \omega^\gamma_\beta \wedge \omega^\alpha_\gamma \equiv \\ & \Big\{-\left(\delta^\gamma_\lambda \eta_\beta + h^{\bar\sigma\gamma} T_\lambda(h_{\beta\bar\sigma})\right)\eta^\alpha h_{\gamma\bar\mu} + \left(\delta^\alpha_\lambda \eta_\gamma + h^{\bar\sigma\alpha} T_\lambda(h_{\gamma\bar\sigma})\right)\eta^\gamma h_{\beta\bar\mu}\Big\}\theta^\lambda \wedge \theta^{\bar\mu}. \end{aligned}$$

Next (by (1.107))

$$\theta_\alpha \wedge \tau^\beta \equiv 0, \quad \tau_\alpha \wedge \theta^\beta \equiv 0.$$

Recall (cf. (1.90)) that

$$\begin{aligned} \Pi^\beta_\alpha &= d\omega^\beta_\alpha - \omega^\gamma_\alpha \wedge \omega^\beta_\gamma, \\ \Omega^\beta_\alpha &= \Pi^\beta_\alpha - 2i\theta_\alpha \wedge \tau^\beta + 2i\tau_\alpha \wedge \theta^\beta, \\ \Omega^\beta_\alpha &= R_\alpha{}^\beta{}_{\lambda\bar\mu}\theta^\lambda \wedge \theta^{\bar\mu} + W^\beta_{\alpha\lambda}\theta^\lambda \wedge \theta - W^\beta_{\alpha\bar\lambda}\theta^{\bar\lambda} \wedge \theta. \end{aligned}$$

Hence, by collecting the various identities established above, we have

$$\begin{aligned} R_\beta{}^\alpha{}_{\lambda\bar\mu} = &-[\delta^\alpha_\rho \eta_\beta + h^{\bar\sigma\alpha} T_\rho(h_{\beta\bar\sigma})]\eta^\rho h_{\lambda\bar\mu} \\ &- \delta^\alpha_\lambda T_{\bar\mu}(\eta_\beta) - h^{\alpha\bar\sigma} T_{\bar\mu} T_\lambda h_{\beta\bar\sigma} - T_{\bar\mu}(h^{\bar\sigma\alpha}) T_\lambda(h_{\beta\bar\sigma}) \\ &- T_\lambda(\eta^\alpha) h_{\beta\bar\mu} - \eta^\alpha T_\lambda(h_{\beta\bar\mu}) - \eta^\alpha h_{\beta\bar\rho}\eta^{\bar\rho} h_{\lambda\bar\rho} - T_\beta(\eta^\alpha) h_{\lambda\bar\mu} \\ &+ \left(\delta^\gamma_\lambda \eta_\beta + h^{\bar\sigma\gamma} T_\lambda(h_{\beta\bar\sigma})\right)\eta^\alpha h_{\gamma\bar\mu} - \eta^\gamma h_{\beta\bar\mu}\left(\delta^\alpha_\lambda \eta_\gamma + h^{\bar\sigma\alpha} T_\lambda(h_{\gamma\bar\sigma})\right), \end{aligned}$$

or (after some simplification)

$$\begin{aligned} R_{\beta\bar\epsilon\lambda\bar\mu} = &-T_{\bar\mu} T_\lambda h_{\beta\bar\epsilon} + h^{\bar\sigma\alpha} T_{\bar\mu}(h_{\alpha\bar\epsilon}) T_\lambda(h_{\beta\bar\epsilon}) - h_{\lambda\bar\mu}\eta^\rho T_\rho(h_{\beta\bar\epsilon}) + h_{\lambda\bar\mu}\eta^\alpha T_\beta(h_{\alpha\bar\epsilon}) \\ &- h_{\lambda\bar\epsilon} T_{\bar\mu}(\eta_\beta) - h_{\beta\bar\mu} T_\lambda(\eta_{\bar\epsilon}) - h_{\lambda\bar\mu} T_\beta(\eta_{\bar\epsilon}) - \eta_{\bar\epsilon}\eta_\beta h_{\lambda\bar\mu} - \eta_\gamma \eta^\gamma h_{\beta\bar\mu} h_{\lambda\bar\epsilon}. \end{aligned} \quad (1.109)$$

See also S. Webster [422], p. 38. Let us return now to $M \in \{Q_\pm(c)\}$. Because $Q_\pm(c)$ are homogeneous, it suffices to compute their torsion and curvature at points where $z = 0$. To fix the ideas, let $M = Q_+(c)$. Then

$$h_{\alpha\bar\beta} = g_{\alpha\bar\beta} + |w|^{-2} h_{\alpha\bar\mu}\bar z^\mu h_{\lambda\bar\beta} z^\lambda, \; q\eta_\alpha = -|w|^{-2} g_{\alpha\bar\mu}\bar z^\mu,$$
$$T_\alpha = \frac{\partial}{\partial z^\alpha} - \frac{1}{w} g_{\alpha\bar\mu}\bar z^\mu \frac{\partial}{\partial w}.$$

Then (1.107) shows that $A^\alpha_{\bar\beta} = 0$ at $z = 0$. Also, substitution into (1.109) shows that

$$R_{\beta\bar\epsilon\lambda\bar\mu} = |w|^{-2}\{h_{\beta\bar\mu} h_{\lambda\bar\epsilon} + h_{\lambda\bar\mu} h_{\beta\bar\epsilon}\}$$

at $z = 0$. Finally $(0, w) \in Q_+(c)$ yields $|w|^2 = c$. Similar computations may be performed for $M = Q_-(c)$. We may conclude that

$$R_{\alpha\bar\beta\lambda\bar\mu} = \frac{\epsilon}{c}\{h_{\alpha\bar\mu} h_{\lambda\bar\beta} + h_{\lambda\bar\mu} h_{\alpha\bar\beta}\}, \quad (1.110)$$

at any point $(0, w) \in M$, $M \in \{Q_\pm(c)\}$, where

$$\epsilon = \begin{cases} 1 & \text{if } M = Q_+(c), \\ -1 & \text{if } M = Q_-(c). \end{cases}$$

Let $\sigma \in G_1(H(M))$ and $X \in H(M)_x$ such that $\{X, J_x X\}$ is a basis of σ. Let us set $X = Z + \overline{Z}$, $Z = \xi^\alpha T_\alpha$. Then

$$k_\theta(\sigma) = \frac{1}{4}\left(h_{\alpha\bar\beta}\xi^\alpha \bar\xi^\beta\right)^{-2} \frac{\epsilon}{c}\left(h_{\alpha\bar\mu} h_{\lambda\bar\beta} + h_{\lambda\bar\mu} h_{\alpha\bar\beta}\right)\xi^\alpha \bar\xi^\beta \xi^\lambda \bar\xi^\mu,$$

that is, $k_{\theta_\pm}(\sigma) = \pm 1/(2c)$. □

As another consequence of (1.110), we have the following:

Corollary 1.1. *The Chern tensor vanishes for each* $M \in \{Q_\pm(c)\}$.

Properties of the pseudo-Hermitian sectional curvature of CR submanifolds (e.g., CR analogues of Proposition 9.2 and 9.3 in [241], vol. II, p. 176, regarding the holomorphic sectional curvature of Kähler submanifolds) are not known. A theory of pseudo-Hermitian immersions is presented in Chapter 6, yet no applications to CR submanifolds of pseudo-Hermitian space forms are known so far (while CR submanifolds of complex space forms are known to be rich in geometric properties, cf., e.g., [446], pp. 76–136).

1.6 CR structures as G-structures

Consider the group G_0 consisting of all $(2n+1) \times (2n+1)$ matrices of the form

$$\begin{pmatrix} v & 0 & 0 \\ a^\alpha & a^\alpha_\beta & b^\alpha_\beta \\ b^\alpha & -b^\alpha_\beta & a^\alpha_\beta \end{pmatrix}, \qquad (1.111)$$

where $v \in \mathbf{R} \setminus \{0\}$, $a^\alpha, b^\beta \in \mathbf{R}$, $1 \leq \alpha \leq n$, and $[a^\alpha_\beta + i b^\alpha_\beta] \in \mathrm{GL}(n, \mathbf{C})$. Then G_0 is a Lie subgroup of $\mathrm{GL}(2n+1, \mathbf{R})$. At this point, we need some notation and terminology. If $E \to M$ is a real vector bundle of rank r over a C^∞ manifold M, then we denote by $L(E) \to M$ the principal $\mathrm{GL}(r, \mathbf{R})$-bundle of all frames in the fibers of E, i.e., if $x \in M$ then $u \in L(E)_x$ is an $\mathbf{R}$-linear isomorphism $u : \mathbf{R}^r \to E_x$. We also adopt the notation $F(M) = L(T(M))$ (the principal $\mathrm{GL}(m, \mathbf{R})$-bundle of linear frames tangent to M, $m = \dim_{\mathbf{R}} M$).

Definition 1.36. Given a Lie subgroup $G \subset \mathrm{GL}(m, \mathbf{R})$, a *$G$-structure* on M is a principal G-subbundle of $F(M) \to M$. □

The theory of G-structures is a classical, yet central chapter of differential geometry. See S. Sternberg [388], P. Mollino [314], for a treatment of the main themes of this theory.

Let M be a real $(2n+1)$-dimensional C^∞ manifold and $B_{G_0}(M) \to M$ a G_0-structure on M, where G_0 is the Lie group of all matrices of the form (1.111), as before.

Proposition 1.13. *Any G_0-structure determines an almost CR structure on M and conversely.*

Indeed, let $x \in M$ and set

$$H(M)_x = p(j(\mathbf{R}^{2n}))$$

for some $p \in B_{G_0}(M)_x$, where $j : \mathbf{R}^{2n} \to \mathbf{R}^{2n+1}$, $j(\xi) = (0, \xi) \in \mathbf{R} \times \mathbf{R}^{2n}$, $\xi \in \mathbf{R}^{2n}$. Any $g \in G_0$ preserves $j(\mathbf{R}^{2n})$; hence $H(M)_x$ is well defined, i.e., its definition doesn't depend on the choice of linear frame p at x, adapted to the G_0-structure. Then $H(M)$ is a rank-$2n$ subbundle of $T(M)$. If $\{e_0, e_1, \ldots, e_{2n}\}$ is the canonical basis in $\mathbf{R}^{2n+1}$, we set

$$X_\alpha = p(e_\alpha), \quad Y_\alpha = p(e_{\alpha+n})$$

for $p \in B_{G_0}(M)_x$ fixed. Next, define $J_x : H(M)_x \to H(M)_x$ by setting

$$J_x X_\alpha = Y_\alpha, \quad J_x Y_\alpha = -X_\alpha.$$

Then J_x is a (well-defined) complex structure in $H(M)_x$. Extend J to $H(M) \otimes \mathbf{C}$ by $\mathbf{C}$-linearity and note that $\mathrm{Spec}(J) = \{\pm i\}$. Then $T_{1,0}(M) = \mathrm{Eigen}(i)$ is an almost CR structure on M. In a moment we shall investigate under what conditions this almost CR structure is integrable, i.e., a CR structure. Conversely, any almost CR structure $T_{1,0}(M)$ determines a G_0-structure on M, for we may set

$$B_x = \{u : \mathbf{R}^{2n+1} \to T_x(M) : u(e_\alpha) \in H(M)_x,\ u(e_{\alpha+n}) = J_x u(e_\alpha),\ 1 \le \alpha \le n\},$$

for any $x \in M$. To see that B (the disjoint union of the B_x, $x \in M$) is a G_0-structure on M we may use the following standard criterion (cf., e.g., M. Crampin [107]):

Theorem 1.13. *Let $G \subset \mathrm{GL}(2n+1, \mathbf{R})$ be a Lie subgroup and $B \subset F(M)$ a submanifold. If*

(i) the projection $\pi : F(M) \to M$ maps B onto M,
(ii) given $p \in B$ and $q \in F(M)$ such that $q = pg$ for some $g \in \mathrm{GL}(2n+1, \mathbf{R})$ then $q \in B \Longleftrightarrow g \in G$,
(iii) for any $x \in M$ there is an open neighborhood U and a C^∞ section $\sigma : U \to F(M)$ such that $\sigma(U) \subseteq B$,

then B is a G-structure on M.

Clearly (i) holds. Let $p, q \in B_x$ such that $q = pg$, $g \in \mathrm{GL}(2n+1, \mathbf{R})$. We wish to show that $g \in G_0$. Indeed

$$q(e_\alpha) = (pg)(e_\alpha) = p(e_j)g^j_\alpha = p(e_0)g^0_\alpha + p(e_\beta)g^\beta_\alpha + p(e_{\beta+n})g^{\beta+n}_\alpha .$$

Yet $q(e_\alpha) \in H(M)_x =$ span of $p(e_\alpha)$, $p(e_{\alpha+n})$, $1 \le \alpha \le n$; hence we may conclude that

$$g^0_\alpha = 0, \quad 1 \le \alpha \le n. \tag{1.112}$$

Similarly, from

$$q(e_{\alpha+n}) = p(e_0)g^0_{\alpha+n} + p(e_\beta)g^\beta_{\alpha+n} + p(e_{\beta+n})g^{\beta+n}_{\alpha+n}$$

we get

$$g^0_{\alpha+n} = 0. \tag{1.113}$$

Moreover, from $Jq(e_\beta) = q(e_{\beta+n})$ we deduce

$$g^\beta_\alpha = g^{\beta+n}_{\alpha+n}, \quad g^{\beta+n}_\alpha = -g^\beta_{\alpha+n}. \tag{1.114}$$

The identities (1.112)–(1.114) show that $g \in G_0$. Finally, we need to check (iii). Let $\{X_\alpha, JX_\alpha\}$ be a local frame of $H(M)$ on U. Let θ be a nowhere-zero real 1-form on U such that $\mathrm{Ker}(\theta) = H(M)\,|_U$ (as we showed previously, θ may be chosen to be global if M is oriented). Let T be a tangent vector field on U such that $\theta(T) = 1$ (for instance, fix a Riemannian metric g_{ij} on U and set $T^i = \|\theta\|^{-2} g^{ij}\theta_j$). Then $\{X_\alpha, JX_\alpha, T\}$ is a (local) frame of $T(M)$ on U (giving rise to a C^∞ section $\sigma : U \to F(M)$ with $\sigma(U) \subseteq B$). □

1.6.1 Integrability

Originally, S.S. Chern and J.K. Moser, in their paper [99], regarded almost CR structures as principal subbundles of the principal $\mathrm{GL}(2n+1, \mathbf{C})$-bundle $L(T^*(M) \otimes \mathbf{C})$ (rather than G_0-structures on M). Let us see how the two points of view match. Consider the group monomorphism $h : G_0 \to \mathrm{GL}(2n+1, \mathbf{C})$ given by

$$h : \begin{pmatrix} v & 0 & 0 \\ a^\alpha & a^\alpha_\beta & b^\alpha_\beta \\ b^\alpha & -b^\alpha_\beta & a^\alpha_\beta \end{pmatrix} \mapsto \begin{pmatrix} v & 0 & 0 \\ a^\alpha + ib^\alpha & a^\alpha_\beta + ib^\alpha_\beta & 0 \\ a^\alpha - ib^\alpha & 0 & a^\alpha_\beta - ib^\alpha_\beta \end{pmatrix}.$$

Also set $G_0^{\mathbf{C}} = h(G_0)$ (so that $G_0^{\mathbf{C}}$ is a Lie subgroup of $\mathrm{GL}(2n+1, \mathbf{C})$). Let $B_{G_0}(M) \to M$ be a G_0-structure on M and let $p : \mathbf{R}^{2n+1} \to T_x(M)$ be a linear frame adapted to $B_{G_0}(M)$. If $\{e_j\} = \{e_0, e_\alpha, e_{\alpha+n}\}$ is the canonical basis in $\mathbf{R}^{2n+1}$, set $X_j = p(e_j)$, $0 \le j \le 2n$, and let $\{\Omega^j\}$ be the dual basis in $T_x^*(M)$, i.e., $\Omega^j(X_i) = \delta_i^j$. Think of $\{e_j\}$ as a linear basis (over $\mathbf{C}$) in $\mathbf{C}^{2n+1}$ and let $\{e^j\}$ be the dual basis in $(\mathbf{C}^{n+1})^*$. Define

$$p^{\mathbf{C}} : (\mathbf{C}^{n+1})^* \to T_x^*(M) \otimes_{\mathbf{R}} \otimes \mathbf{C}, \;\; p^{\mathbf{C}}(e^j) = \Omega^j \otimes 1.$$

Next, let us consider $B_{G_0}(M)_x^{\mathbf{C}} = \{p^{\mathbf{C}} : p \in B_{G_0}(M)_x\}$.

Proposition 1.14. *$B_{G_0}(M)^{\mathbf{C}}$ is a principal G_0-subbundle of $L(T^*(M) \otimes \mathbf{C})$ isomorphic to $B_{G_0}(M)$.*

Let us address now the question of integrability and formal (Frobenius) integrability.

Definition 1.37. A G-structure $B_G(M) \to M$ is *integrable* if for any $x \in M$ there is a coordinate neighborhood (U, x^i) such that $\sigma_0(y) \in B_G(M)$ for any $y \in U$, where

$$\sigma_0(y) : \mathbf{R}^{2n+1} \to T_y(M), \;\; \sigma_0(y)(e_j) = \left.\frac{\partial}{\partial x^j}\right|_y . \qquad \square$$

Let $T_{1,0}(M)$ be an (almost) CR structure on M and $B \to M$ the G_0-structure on M deduced from $T_{1,0}(M)$, as before.

Definition 1.38. We say $(M, T_{1,0}(M))$ is *Levi flat* if the Levi form of M vanishes, i.e., $L = 0$ (equivalently, if $H(M)$ is integrable). □

Let us look at the following example. Let $z_j = x_j + iy_j$, $j = 1, 2$, be coordinates in $\mathbf{C}^2$ and let

$$M = \{z \in \mathbf{C}^2 : y_2 = 0\} \tag{1.115}$$

be a real hyperplane in $\mathbf{C}^2$. In his well-known 1907 paper, H. Poincaré pointed out (cf. [350]) a natural exterior differential system $H(M)$ on a real hypersurface $M \subset \mathbf{C}^2$ (generated by the tangential holomorphic cotangent vectors and their complex conjugates) possessing the property that $H(M)$ is completely integrable if and only if M is locally CR equivalent to the example (1.115). The obstruction to the integrability of $H(M)$ turns out to be the Levi form, and Poincaré's result was later reformulated by B. Segre [368], in the form that a real hypersurface in $\mathbf{C}^2$ is locally CR equivalent to a hyperplane if and only if the Levi form vanishes. This was shown to be true in $\mathbf{C}^n$ for arbitrary $n \geq 2$ by F. Sommer [376].

If B is integrable (as a G_0-structure) then $(M, T_{1,0}(M))$ is Levi flat (indeed, since $\sigma_0(x) \in B$ for any $x \in U$, it follows that $\{\partial/\partial x^\alpha, \partial/\partial x^{\alpha+n}\}$ is a (local) frame of $H(M)$; hence $H(M)$ is involutive), so that for a generic (almost) CR structure $T_{1,0}(M)$ the G_0-structure B is not integrable. Consequently, it would be useful to study (compute) the structure functions[8] (cf., e.g., S. Sternberg [388], pp. 317–318) of B (an open problem, as yet).

To understand for which G_0-structures the corresponding almost CR structure is integrable, let $\sigma : U \to B$ be a section and define $X_j : U \to T(M)$ by $X_j(x) = \sigma(x)(e_j)$, $x \in U$. Next, let $\omega^j : U \to T^*(M)$ be the dual 1-forms and define $\theta^\alpha : U \to T^*(M) \otimes \mathbf{C}$ by $\theta^\alpha = \omega^\alpha + i\omega^{\alpha+n}$, $1 \leq \alpha \leq n$ (hence $\sigma(x)^{\mathbf{C}}(e^\alpha + ie^{\alpha+n}) = \theta^\alpha(x)$, $x \in U$). Also set $T = X_0$ and $\theta = \omega^0$. We may prove the following theorem:

Theorem 1.14. *Let $(M, T_{1,0}(M))$ be an almost CR manifold. The following statements are equivalent:*

(1) The almost CR structure is formally integrable, i.e.,

$$\left[\Gamma^\infty(T_{1,0}(M)), \Gamma^\infty(T_{1,0}(M))\right] \subseteq \Gamma^\infty(T_{1,0}(M)).$$

(2) For any (local) section $\sigma : U \to B$ one has

$$d\theta \equiv 0 \bmod \theta,\ \theta^\beta,$$
$$d\theta^\alpha = 0 \bmod \theta,\ \theta^\beta.$$

Proof. For instance, let us prove the implication (1) $\Longrightarrow$ (2). We have

$$d\theta = B_{\alpha\beta}\theta^\alpha \wedge \theta^\beta + B_{\alpha\bar{\beta}}\theta^\alpha \wedge \theta^{\bar{\beta}} + B_{\bar{\alpha}\bar{\beta}}\theta^{\bar{\alpha}} \wedge \theta^{\bar{\beta}} + B_\alpha \theta \wedge \theta^\alpha + B_{\bar{\alpha}}\theta \wedge \theta^{\bar{\alpha}},$$

for some C^∞ functions B_α, $B_{\alpha\beta}$, $B_{\alpha\bar{\beta}} : U \to \mathbf{C}$, where $B_{\bar{\alpha}} = \overline{B_\alpha}$, $B_{\bar{\alpha}\bar{\beta}} = \overline{B_{\alpha\beta}}$ (since θ is real). Let us set

[8] For example, $c : B \to \mathrm{Hom}(\mathbf{R}^{2n+1} \wedge \mathbf{R}^{2n+1}, \mathbf{R}^{2n+1})/\partial \mathrm{Hom}(\mathbf{R}^{2n+1}, L(G_0))$ (the *first structure function* [388], p. 318), carrying information about the nonintegrability of B, where $\partial : \mathrm{Hom}(\mathbf{R}^{2n+1}, L(G_0)) \to \mathrm{Hom}(\mathbf{R}^{2n+1} \wedge \mathbf{R}^{2n+1}, \mathbf{R}^{2n+1})$ is the map $(\partial T)(\xi \wedge \eta) = T(\xi)\eta - T(\eta)\xi$.

$$T_\alpha = \frac{1}{2}(X_\alpha - iX_{\alpha+n}) \in \Gamma^\infty(U, T_{1,0}(M)).$$

Then $\theta(T_\alpha) = 0$. Also, since $T_{1,0}(M)$ is formally integrable, $(d\theta)(T_\alpha, T_\beta) = 0$ hence $B_{\alpha\beta} = B_{\beta\alpha}$ and $d\theta$ may be written as

$$d\theta = \theta \wedge \Phi + \theta^\alpha \wedge \Phi_\alpha,$$

where $\Phi = B_\alpha \theta^\alpha + B_{\bar\alpha}\theta^{\bar\alpha}$ and $\Phi_\alpha = B_{\alpha\bar\beta}\theta^{\bar\beta}$. The proof that $d\theta^\alpha = \theta \wedge \Phi^\alpha + \theta^\beta \wedge \Phi^\alpha_\beta$, for some 1-forms Φ^α, Φ^α_β, is similar and thus left as an exercise to the reader. □

Let $(M, T_{1,0}(M))$ be an almost CR manifold (of hypersurface type) and $B \to M$ the G_0-structure corresponding to $T_{1,0}(M)$. Let $\sigma : U \to B$ be a (local) section and θ, θ^α (respectively T, T_α) associated with σ, as before. Let $h_{\alpha\bar\beta}(\sigma) : U \to \mathbf{C}$ be defined by

$$h_{\alpha\bar\beta}(\sigma) = -2i\,d\theta(T_\alpha, T_{\bar\beta}),$$

so that $h_{\alpha\bar\beta}(\sigma) = \overline{h_{\beta\bar\alpha}(\sigma)}$.

Theorem 1.15. *If $T_{1,0}(M)$ is integrable then*

$$d\theta \equiv i h_{\alpha\bar\beta}(\sigma)\theta^\alpha \wedge \theta^{\bar\beta} \mod \theta.$$

Proof. As seen above, $d\theta = \theta \wedge \Phi + \theta^\alpha \wedge \Phi_\alpha$ for some real-valued 1-form Φ and some 1-form $\Phi_\alpha = B_{\alpha\bar\beta}\theta^{\bar\beta}$. Then

$$B_{\alpha\bar\beta} = 2d\theta(T_\alpha, T_{\bar\beta}) = i h_{\alpha\bar\beta}(\sigma).$$ □

1.6.2 Nondegeneracy

Consider the complex line bundle $F \to M$ constructed as follows. Let $x \in M$ and $\sigma : U \to B$ a (local) section in B with $x \in U$. Let θ be associated with σ, as before, and set $F_x = \mathbf{C}\theta(x) \subset T^*_x(M) \otimes_{\mathbf{R}} \mathbf{C}$. To see that F_x is well defined, let $\hat\sigma : \hat U \to B$ be another local section, $x \in \hat U$. Then there is $g : U \cap \hat U \to G$ such that $\hat\sigma = \sigma g$ on $U \cap \hat U$. In particular, $\theta = g^0_0\,\hat\theta$ on $U \cap \hat U$. □

Note that F is the annihilator of $H(M) \otimes \mathbf{C}$ in $T^*(M) \otimes \mathbf{C}$.

Theorem 1.16. *Let $(M, T_{1,0}(M))$ be a CR manifold (of hypersurface type). The following statements are equivalent*

(i) $(M, T_{1,0}(M))$ *is nondegenerate.*
(ii) $\det\left[h_{\alpha\bar\beta}(\sigma)\right] \neq 0$ *on U for any section $\sigma : U \to B$.*

To prove the theorem, one considers the bundle morphism

$$\Phi_\sigma : \frac{T(M)\otimes\mathbf{C}}{H(M)\otimes\mathbf{C}}\,|_U \to F\,|_U, \qquad (\Phi_\sigma)_x\,(v + H(M)_x \otimes \mathbf{C}) = \theta_x(v)\theta_x,$$

for any $x \in U$, and observes that

$$(\Phi_\sigma)_x\, L_x(v, w) = \left(h_{\alpha\bar\beta}(\sigma)(x) v^\alpha \overline{w^\beta}\right)\theta_x;$$

hence L_x is nondegenerate if and only if $\det\left[h_{\alpha\bar\beta}(\sigma)(x)\right] \neq 0$. □

1.7 The tangential Cauchy–Riemann complex

The tangential Cauchy–Riemann operator (on functions, i.e., $(\overline{\partial}_b f)\overline{Z} = \overline{Z}(f)$, $Z \in T_{1,0}(M)$) of an almost CR manifold admits a natural extension to forms of arbitrary type $(0, q)$, $q \geq 1$, thus giving rise to a pseudocomplex (in the sense of I. Vaisman [416]) satisfying the complex condition (1.119) below precisely when the almost CR structure is integrable. We give a detailed proof of (1.119), an elementary but rather involved matter (our discussion is based on the monograph by G. Taiani [394]).

When M is a nondegenerate CR manifold (of hypersurface type) the sections of $\Lambda^q T_{0,1}(M)^*$, $q \geq 1$, may be identified with ordinary differential forms and the $\overline{\partial}_b$ operator admits a useful reformulation (due to [271]) in terms of a fixed contact form. We only hint at the problem of holomorphic extension of CR objects (functions, forms, etc.), a subject that is not within the purposes of this book. The reader may find a precise account of the matter in the recent monographs by A. Boggess [70], and M.S. Baouendi, P. Ebenfelt, and L.P. Rothschild [31].

As another topic of this section we present CR-holomorphic bundles, and the discussion is based on the monograph by N. Tanaka [398]. Moreover, we show how a filtration of the de Rham complex leads to the *Frölicher spectral sequence* of a CR manifold, which in turn may be used to study the Kohn–Rossi cohomology. Our discussion relies again on [398] and is confined to the nondegenerate case (where an explicit relation among d, the exterior differentiation operator, and $\overline{\partial}_b$, the tangential Cauchy–Riemann operator, is available). We establish a CR analogue of a result by R. Bott and P. Baum [76], in the theory of holomorphic foliations. The proof of Theorem 1.18 relies on the formal analogy between a connection in $\hat{T}(M) := [T(M) \otimes \mathbf{C}]/T_{0,1}(M)$ extending the $\overline{\partial}_{\hat{T}(M)}$ operator and the Bott connection of a holomorphic foliation. The last section is technical and provides the explicit expression (1.146) of the Kohn–Rossi Laplacian (on forms of arbitrary type) of a nondegenerate CR manifold, in terms of covariant derivatives and curvature (of the Tanaka–Webster connection). Applications of (1.146) are given in Chapter 4.

1.7.1 The tangential Cauchy–Riemann complex

We start by defining the tangential Cauchy–Riemann operator on forms. Let $(M, T_{1,0}(M))$ be a CR manifold of type (n, k). We define a differential operator

$$\overline{\partial}_b^q : \Gamma^\infty(\Lambda^q T_{0,1}(M)^*) \to \Gamma^\infty(\Lambda^{q+1} T_{0,1}(M)^*), \quad q \geq 0.$$

Here $\Gamma^\infty(\Lambda^0 T_{0,1}(M)^*) = \Gamma^\infty(M \times \mathbf{C})$ consists of the C^∞ functions $f : M \to \mathbf{C}$. For $q = 0$ we set

$$(\overline{\partial}_b f)\overline{Z} = \overline{Z}(f), \tag{1.116}$$

for any C^∞ function $f : M \to \mathbf{C}$ and any $Z \in \Gamma^\infty(T_{1,0}(M))$. In general, if $q \geq 1$, we set

$$
\begin{aligned}
(\overline{\partial}_b^q \varphi)(\overline{Z}_1, \ldots, \overline{Z}_{q+1}) &= \frac{1}{q+1} \Big\{ \sum_{i=1}^{q+1} (-1)^{i+1} \overline{Z}_i (\varphi(\overline{Z}_1, \ldots, \hat{\overline{Z}}_i, \ldots, \overline{Z}_{q+1})) \\
&+ \sum_{1 \le i < j \le q+1} (-1)^{i+j} \varphi([\overline{Z}_i, \overline{Z}_j], \overline{Z}_1, \ldots, \hat{\overline{Z}}_i, \ldots, \hat{\overline{Z}}_j, \ldots, \overline{Z}_{q+1}) \Big\}
\end{aligned} \tag{1.117}
$$

for any $\varphi \in \Gamma^\infty(\Lambda^q T_{0,1}(M)^*)$ and any $Z_1, \ldots, Z_{q+1} \in \Gamma^\infty(T_{1,0}(M))$. As usual, a hat denotes the suppression of a term. By the integrability of the CR structure $[\overline{Z}_i, \overline{Z}_j] \in \Gamma^\infty(T_{0,1}(M))$, so that the term

$$\varphi([\overline{Z}_i, \overline{Z}_j], \ldots, \hat{\overline{Z}}_i, \ldots, \hat{\overline{Z}}_j, \ldots)$$

makes sense. It is an elementary matter that $\overline{\partial}_b^q \varphi$ is skew-symmetric and $C^\infty(M) \otimes \mathbf{C}$-multilinear. Then $\overline{\partial}_b^q$ is referred to as the *tangential Cauchy–Riemann operator*. We obtain a sequence of $C^\infty(M) \otimes \mathbf{C}$-modules and differential operators

$$\Gamma^\infty(M \times \mathbf{C}) \xrightarrow{\overline{\partial}_b} \Gamma^\infty(T_{0,1}(M)^*) \xrightarrow{\overline{\partial}_b^1} \Gamma^\infty(\Lambda^2 T_{0,1}(M)^*) \xrightarrow{\overline{\partial}_b^2} \cdots . \tag{1.118}$$

The central result of the present section is the following theorem:

Theorem 1.17. *The sequence* (1.118) *is a cochain complex, that is,*

$$\overline{\partial}_b^{q+1}(\overline{\partial}_b^q \varphi) = 0 \tag{1.119}$$

for any $\varphi \in \Gamma^\infty(\Lambda^q T_{0,1}(M)^*)$ *and any* $q \ge 0$.

Consequently, we may consider the cohomology

$$
\begin{aligned}
H^{0,q}(M) &= H^q(\Gamma^\infty(\Lambda^\cdot T_{0,1}(M)^*), \overline{\partial}_b^q) \\
&= \frac{\operatorname{Ker}\{\overline{\partial}_b^q : \Gamma^\infty(\Lambda^q T_{0,1}(M)^*) \to \Gamma^\infty(\Lambda^{q+1} T_{0,1}(M)^*)\}}{\overline{\partial}_b^{q-1} \Gamma^\infty(\Lambda^{q-1} T_{0,1}(M)^*)}
\end{aligned}
$$

of the cochain complex (1.118). This is the *Kohn–Rossi cohomology* of the CR manifold $(M, T_{1,0}(M))$. See J.J. Kohn and H. Rossi [252]. The proof of Theorem 1.17 is a rather lengthy exercise in multilinear algebra (cf. also G. Taiani [394], pp. 23–26).

Proof of Theorem 1.17. We start with the following lemma:

Lemma 1.7. *For any* $\varphi \in \Gamma^\infty(\Lambda^p T_{0,1}(M)^*)$ *and* $\psi \in \Gamma^\infty(\Lambda^q T_{0,1}(M)^*)$ *the following identity holds:*

$$\overline{\partial}_b^{p+q}(\varphi \wedge \psi) = (\overline{\partial}_b^p \varphi) \wedge \psi + (-1)^p \varphi \wedge \overline{\partial}_b^q \psi. \tag{1.120}$$

As to multilinear algebra, we adopt the notation and conventions in [241], vol. I, pp. 26–38. It is an elementary matter that

$$(\varphi \wedge \psi)(\overline{Z}_1, \dots, \overline{Z}_{q+1}) = \frac{1}{q+1} \sum_{i=1}^{q+1} (-1)^{i+1} \varphi(\overline{Z}_i) \psi(\overline{Z}_1, \dots, \hat{\overline{Z}}_i, \dots, \overline{Z}_{q+1}), \quad (1.121)$$

for any

$$\varphi \in \Gamma^\infty(T_{0,1}(M)^*), \quad \psi \in \Gamma^\infty(\Lambda^q T_{0,1}(M)^*),$$

and any $Z_1, \dots, Z_{q+1} \in \Gamma^\infty(T_{0,1}(M))$. To prove Lemma 1.7 we look first at the case $p = 0$, that is, we need to show that

$$\overline{\partial}_b^q (f\psi) = (\overline{\partial}_b f) \wedge \psi + f \overline{\partial}_b^q \psi.$$

Using (1.121) for $\varphi = \overline{\partial}_b f$ we may perform the following calculation:

$$\begin{aligned}
\overline{\partial}_b^q (f\psi)(\overline{Z}_1, \dots, \overline{Z}_{q+1}) &= \frac{1}{q+1} \Big\{ \sum_{i=1}^{q+1} \overline{Z}_i (f\psi(\dots, \hat{\overline{Z}}_i, \dots)) \\
&\quad + \sum_{1 \le i < j \le q+1} (-1)^{i+j} f\psi([\overline{Z}_i, \overline{Z}_j], \dots, \hat{\overline{Z}}_i, \dots, \hat{\overline{Z}}_j, \dots) \Big\} \\
&= \frac{1}{q+1} \Big\{ \sum_{i=1}^{q+1} (-1)^{i+1} \{ \overline{Z}_i(f) \psi(\dots, \hat{\overline{Z}}_i, \dots) + f \overline{Z}_i (\psi(\dots, \hat{\overline{Z}}_i, \dots)) \} \\
&\quad + \sum_{i<j} (-1)^{i+j} f\psi([\overline{Z}_i, \overline{Z}_j] \dots, \hat{\overline{Z}}_i, \dots, \hat{\overline{Z}}_j, \dots) \Big\} \\
&= ((\overline{\partial}_b f) \wedge \psi)(\overline{Z}_1, \dots, \overline{Z}_{q+1}) + f (\overline{\partial}_b^q \psi)(\overline{Z}_1, \dots, \overline{Z}_{q+1}).
\end{aligned}$$

Next we examine the case $p = 1$. Given $\varphi \in \Gamma^\infty(T_{0,1}(M)^*)$ note that

$$(q+1)(q+2)((\overline{\partial}_b^1 \varphi) \wedge \psi)(\overline{Z}_1, \dots, \overline{Z}_{q+2}) = S_1 + S_2 + S_5, \quad (1.122)$$

where

$$S_1 = \sum_{i=1}^{q+2} (-1)^{i+k} \sum_{1 \le k \le i-1} \overline{Z}_i (\varphi(\overline{Z}_k)) \psi(\dots, \hat{\overline{Z}}_k, \dots, \hat{\overline{Z}}_i, \dots),$$

$$S_2 = \sum_{i=1}^{q+2} \sum_{i \le k \le q+1} (-1)^{i+k} \overline{Z}_i (\varphi(\overline{Z}_{k+1})) \psi(\dots, \hat{\overline{Z}}_i, \dots, \hat{\overline{Z}}_{k+1}, \dots),$$

$$S_5 = \sum_{1 \le i < j \le q+2} (-1)^{i+j} \varphi([\overline{Z}_i, \overline{Z}_j]) \psi(\dots, \hat{\overline{Z}}_i, \dots, \hat{\overline{Z}}_j, \dots).$$

We may perform the following calculation (by (1.121)):

$$
\begin{aligned}
&(q+2)\overline{\partial}_b^{q+1}(\varphi\wedge\psi)(\overline{Z}_1,\dots,\overline{Z}_{q+2})\\
&=\sum_{i=1}^{q+2}(-1)^{i+1}\overline{Z}_i((\varphi\wedge\psi)(\overline{Z}_1,\dots,\hat{\overline{Z}}_i,\dots,\overline{Z}_{q+2}))\\
&\qquad+\sum_{1\le i<j\le q+2}(-1)^{i+j}(\varphi\wedge\psi)([\overline{Z}_i,\overline{Z}_j],\dots,\hat{\overline{Z}}_i,\dots,\hat{\overline{Z}}_j,\dots)\\
&=\sum_{i=1}^{q+2}(-1)^{i+1}\overline{Z}_i\Big(\frac{1}{q+1}\sum_{j=1}^{q+1}(-1)^{j+1}\varphi(\overline{W}_{ij})\psi(\overline{W}_{i1},\dots,\hat{\overline{W}}_{ij},\dots,\overline{W}_{i,q+1})\Big)\\
&+\sum_{1\le i<j\le q+2}(-1)^{i+j}\frac{1}{q+1}\sum_{k=1}^{q+1}(-1)^{k+1}\varphi(\overline{V}_{ijk})\psi(\overline{V}_{ij1},\dots,\hat{\overline{V}}_{ijk},\dots,\overline{V}_{ij,q+1}),
\end{aligned}
$$

where

$$
W_{ij}=\begin{cases} Z_j & \text{if } 1\le j\le i-1,\\ Z_{j+1} & \text{if } i\le j\le q+1,\end{cases}
$$

for any $i\in I_{q+2}$, $j\in I_{q+1}$, and

$$
V_{ijk}=\begin{cases} [Z_i,Z_j] & \text{if } k=1,\\ Z_{k-1} & \text{if } 2\le k\le i,\\ Z_k & \text{if } i+1\le k\le j-1,\\ Z_{k+1} & \text{if } j\le k\le q+1,\end{cases}
$$

for any $1\le i<j\le q+2$ and any $k\in I_{q+1}$. Then

$$
\begin{aligned}
&(q+1)(q+2)\overline{\partial}_b^{q+1}(\varphi\wedge\psi)(\overline{Z}_1,\dots,\overline{Z}_{q+2})\\
&=\sum_{i=1}^{q+2}\sum_{1\le j<i}(-1)^{i+j}\overline{Z}_i(\varphi(\overline{Z}_j)\psi(\dots,\hat{\overline{Z}}_j,\dots,\hat{\overline{Z}}_i,\dots))\\
&+\sum_{i=1}^{q+2}\sum_{i\le j\le q+1}(-1)^{i+j}\overline{Z}_i(\varphi(\overline{Z}_{j+1})\psi(\dots,\hat{\overline{Z}}_i,\dots,\hat{\overline{Z}}_{j+1},\dots))\\
&\qquad+\sum_{1\le i<j\le q+2}(-1)^{i+j}\varphi([\overline{Z}_i,\overline{Z}_j])\psi(\dots,\hat{\overline{Z}}_i,\dots,\hat{\overline{Z}}_j,\dots,\overline{Z}_{q+2})\\
&+\sum_{1\le i<j\le q+2}(-1)^{i+j+k+1}\sum_{2\le k\le i}\varphi(\overline{Z}_{k-1})\psi([\overline{Z}_i,\overline{Z}_j],\overline{Z}_1,\dots\\
&\qquad\qquad\dots,\hat{\overline{Z}}_{k-1},\dots,\hat{\overline{Z}}_i,\dots,\hat{\overline{Z}}_j,\dots,\overline{Z}_{q+2})\\
&+\sum_{1\le i<j\le q+2}(-1)^{i+j+k+1}\sum_{j\le k\le q+1}\varphi(\overline{Z}_{k+1})\psi([\overline{Z}_i,\overline{Z}_j],\overline{Z}_1,\dots\\
&\qquad\qquad\dots,\hat{\overline{Z}}_i,\dots,\hat{\overline{Z}}_j,\dots,\hat{\overline{Z}}_{k+1},\dots,\overline{Z}_{q+2}),
\end{aligned}
$$

that is,

$$(q+1)(q+2)\overline{\partial}_b^{q+1}(\varphi \wedge \psi)(\overline{Z}_1, \dots, \overline{Z}_{q+2}) = \sum_{i=1}^{8} S_i, \tag{1.123}$$

where

$$\begin{aligned}
S_3 &= \sum_{i=1}^{q+2} \sum_{1 \le j < i} (-1)^{i+j} \varphi(\overline{Z}_j) \overline{Z}_i(\psi(\dots, \hat{\overline{Z}}_j, \dots, \hat{\overline{Z}}_i, \dots)), \\
S_4 &= \sum_{i=1}^{q+2} \sum_{i < j \le q+2} (-1)^{i+j+1} \varphi(\overline{Z}_j) \overline{Z}_i(\psi(\dots, \hat{\overline{Z}}_i, \dots, \hat{\overline{Z}}_j, \dots)), \\
S_6 &= \sum_{1 \le i < j \le q+2} (-1)^{i+j+k} \sum_{1 \le k < i} \varphi(\overline{Z}_k) \\
&\qquad \times \psi([\overline{Z}_i, \overline{Z}_j], \overline{Z}_1, \dots, \hat{\overline{Z}}_k, \dots, \hat{\overline{Z}}_i, \dots, \hat{\overline{Z}}_j, \dots, \overline{Z}_{q+2}), \\
S_7 &= \sum_{1 \le i < j \le q+2} (-1)^{i+j+k+1} \sum_{i < k < j} \varphi(\overline{Z}_k) \\
&\qquad \times \psi([\overline{Z}_i, \overline{Z}_j], \overline{Z}_1, \dots, \hat{\overline{Z}}_i, \dots, \hat{\overline{Z}}_k, \dots, \hat{\overline{Z}}_j, \dots, \overline{Z}_{q+2}), \\
S_8 &= \sum_{1 \le i < j \le q+2} (-1)^{i+j+k} \sum_{j < k \le q+2} \varphi(\overline{Z}_k) \\
&\qquad \times \psi([\overline{Z}_i, \overline{Z}_j], \overline{Z}_1, \dots, \hat{\overline{Z}}_i, \dots, \hat{\overline{Z}}_j, \dots, \hat{\overline{Z}}_k, \dots, \overline{Z}_{q+2}).
\end{aligned}$$

Next

$$\begin{aligned}
&(q+2)(\varphi \wedge \overline{\partial}_b^q \psi)(\overline{Z}_1, \dots, \overline{Z}_{q+2}) \\
&= \sum_{i=1}^{q+2} (-1)^{i+1} \varphi(\overline{Z}_i)(\overline{\partial}_b^q \psi)(\cdots, \hat{\overline{Z}}_i, \dots) \\
&- \sum_{i=1}^{q+2} (-1)^{i+1} \varphi(\overline{Z}_i) \left\{ \frac{1}{q+1} \sum_{j=1}^{q+1} (-1)^{j+1} \overline{W}_{ij} \big(\psi(\overline{W}_{i1}, \dots, \hat{\overline{W}}_{ij}, \dots, \overline{W}_{i,q+1})\big) \right. \\
&\left. + \frac{1}{q+1} \sum_{1 \le j < k \le q+1} (-1)^{j+k} \psi\big([\overline{W}_{ij}, \overline{W}_{ik}], \overline{W}_{i1}, \dots, \hat{\overline{W}}_{ij}, \dots, \hat{\overline{W}}_{ik}, \dots, \overline{W}_{i,q+1}\big) \right\},
\end{aligned}$$

so that

$$(q+1)(q+2)(\varphi \wedge \overline{\partial}_b^q \psi)(\overline{Z}_1, \dots, \overline{Z}_{q+2}) = S_3 + S_4 + S_6 + S_7 + S_8. \tag{1.124}$$

Comparing (1.123) with (1.122) and (1.124) we may conclude that (1.120) holds for any

$$\varphi \in \Gamma^\infty(T_{0,1}(M)^*), \quad \psi \in \Gamma^\infty(\Lambda^q T_{0,1}(M)^*).$$

The proof of (1.120) for arbitrary p is by induction over p. Let us assume that (1.120) holds for any $\varphi \in \Gamma^\infty(\Lambda^p T_{0,1}(M)^*))$. Any $(0, p+1)$-form φ may be written as a sum of terms of the form $\lambda \wedge \mu$ with $\lambda \in \Gamma^\infty(T_{0,1}(M)^*)$ and $\mu \in \Gamma^\infty(\Lambda^p T_{0,1}(M)^*)$. Since both sides of (1.120) are **C**-linear in φ it suffices to prove (1.120) for $\varphi = \lambda \wedge \mu$. We may perform (by (1.120) with $\varphi = \lambda$, respectively with $\varphi = \mu$) the following calculation:

$$\begin{aligned}
\overline{\partial}_b^{p+q+1}((\lambda \wedge \mu) \wedge \psi) &= \overline{\partial}_b^{p+q+1}(\lambda \wedge (\mu \wedge \psi)) \\
&= (\overline{\partial}_b^1 \lambda) \wedge (\mu \wedge \psi) - \lambda \wedge \overline{\partial}_b^{p+q}(\mu \wedge \psi) \\
&= (\overline{\partial}_b^1 \lambda) \wedge (\mu \wedge \psi) - \lambda \wedge ((\overline{\partial}_b^p \mu) \wedge \psi + (-1)^p \mu \wedge \overline{\partial}_b^q \psi) \\
&= \left[(\overline{\partial}_b^1 \lambda) \wedge \mu - \lambda \wedge \overline{\partial}_b^p \mu \right] \wedge \psi + (-1)^{p+1} (\lambda \wedge \mu) \wedge \overline{\partial}_b^q \psi.
\end{aligned}$$

Finally (by (1.120) for $\varphi = \lambda$ and $\psi = \mu$)),

$$\overline{\partial}_b^{p+q+1}((\lambda \wedge \mu) \wedge \psi) = \overline{\partial}_b^{p+1}(\lambda \wedge \mu) \wedge \psi + (-1)^{p+1} (\lambda \wedge \mu) \wedge \overline{\partial}_b^q \psi,$$

and our Lemma 1.7 is completely proved. At this point, we may prove Theorem 1.17. To this end, we consider the operators

$$\Gamma^\infty(\Lambda^q T_{0,1}(M)^*) \xrightarrow{\overline{\partial}_b^q} \Gamma^\infty(\Lambda^{q+1} T_{0,1}(M)^*) \xrightarrow{\overline{\partial}_b^{q+1}} \Gamma^\infty(\Lambda^{q+2} T_{0,1}(M)^*).$$

We shall prove that $\overline{\partial}_b^{q+1} \circ \overline{\partial}_b^q = 0$ by induction over q. First, let $q = 0$, i.e., $\varphi = f \in \Gamma^\infty(M \times \mathbf{C})$. Then

$$(\overline{\partial}_b^1 \overline{\partial}_b f)(\overline{Z}_1, \overline{Z}_2) = \frac{1}{2} \left(\overline{Z}_1(\overline{Z}_2 f) - \overline{Z}_2(\overline{Z}_1 f) - [\overline{Z}_1, \overline{Z}_2] f \right) = 0,$$

for any $Z_1, Z_2 \in \Gamma^\infty(T_{1,0}(M))$. Next, let $\varphi \in \Gamma^\infty(T_{0,1}(M)^*)$. Then

$$\begin{aligned}
(\overline{\partial}_b^2 \overline{\partial}_b^1 \varphi)(\overline{Z}_1, \overline{Z}_2, \overline{Z}_3) = \frac{1}{3} \Big\{ & \overline{Z}_1 \left((\overline{\partial}_b^1 \varphi)(\overline{Z}_2, \overline{Z}_3) \right) - \overline{Z}_2 \left((\overline{\partial}_b^1 \varphi)(\overline{Z}_1, \overline{Z}_3) \right) \\
& + \overline{Z}_3 \left((\overline{\partial}_b^1 \varphi)(\overline{Z}_1, \overline{Z}_2) \right) \\
& - (\overline{\partial}_b^1 \varphi)([\overline{Z}_1, \overline{Z}_2], \overline{Z}_3) + (\overline{\partial}_b^1 \varphi)([\overline{Z}_1, \overline{Z}_3], \overline{Z}_2) \\
& - (\overline{\partial}_b^1 \varphi)([\overline{Z}_2, \overline{Z}_3], \overline{Z}_1) \Big\}
\end{aligned}$$

$$
\begin{aligned}
&= \frac{1}{6}\Big\{ \overline{Z}_1 \left(\overline{Z}_2(\varphi \overline{Z}_3) - \overline{Z}_3(\varphi \overline{Z}_2) - \varphi([\overline{Z}_2, \overline{Z}_3])\right) \\
&\quad - \overline{Z}_2 \left(\overline{Z}_1(\varphi \overline{Z}_3) - \overline{Z}_3(\varphi \overline{Z}_1) - \varphi([\overline{Z}_1, \overline{Z}_3)\right) \\
&\quad + \overline{Z}_3 \left(\overline{Z}_1(\varphi \overline{Z}_2) - \overline{Z}_2(\varphi \overline{Z}_1) - \varphi([\overline{Z}_1, \overline{Z}_2])\right) \\
&\quad - [\overline{Z}_1, \overline{Z}_2](\varphi \overline{Z}_3) + \overline{Z}_3(\varphi[\overline{Z}_1, \overline{Z}_2]) + \varphi([[\overline{Z}_1, \overline{Z}_2], \overline{Z}_3]) \\
&\quad + [\overline{Z}_1, \overline{Z}_3](\varphi \overline{Z}_3) - \overline{Z}_2(\varphi[\overline{Z}_1, \overline{Z}_3]) - \varphi([[\overline{Z}_1, \overline{Z}_3], \overline{Z}_2]) \\
&\quad - [\overline{Z}_2, \overline{Z}_3](\varphi \overline{Z}_1) + \overline{Z}_1(\varphi[\overline{Z}_2, \overline{Z}_3]) + \varphi([[\overline{Z}_2, \overline{Z}_3], \overline{Z}_1)\Big\} \\
&= 0,
\end{aligned}
$$

by Jacobi's identity. Assume now that the identity (1.119) holds for any $\varphi \in \Gamma^\infty(\Lambda^q T_{0,1}(M)^*)$. As noticed above, it suffices to check (1.119) for $\varphi = \lambda \wedge \mu$, where $\lambda \in \Gamma^\infty(T_{0,1}(M)^*)$ and $\mu \in \Gamma^\infty(\Lambda^q T_{0,1}(M)^*)$. We have (by Lemma 1.7 and by (1.119) for $\varphi = \lambda$, respectively for $\varphi = \mu$)

$$
\begin{aligned}
\overline{\partial}_b^{q+2}\overline{\partial}_b^{q+1}(\lambda \wedge \mu) &= \overline{\partial}_b^{q+2}\left((\overline{\partial}_b^1 \lambda) \wedge \mu - \lambda \wedge \overline{\partial}_b^q \mu\right) \\
&= (\overline{\partial}_b^2 \overline{\partial}_b^1 \lambda) \wedge \mu + (\overline{\partial}_b^1 \lambda) \wedge \overline{\partial}_b^q \mu - (\overline{\partial}_b^1 \lambda) \wedge \overline{\partial}_b^q \mu + \lambda \wedge \overline{\partial}_b^{q+1} \overline{\partial}_b^q \mu \\
&= 0
\end{aligned}
$$

and (1.119) is completely proved. □

We proceed by discussing the reformulation of the $\overline{\partial}_b$ operator on a nondegenerate CR manifold. Let $(M, T_{1,0}(M))$ be a nondegenerate CR manifold, of CR dimension n, and θ a pseudo-Hermitian structure on M. Let T be the characteristic direction of $d\theta$.

Definition 1.39. A complex-valued q-form φ on M is said to be a *$(0, q)$-form* if $T_{1,0}(M) \rfloor \varphi = 0$ and $T \rfloor \varphi = 0$. □

Let $\Lambda^{0,q}(M)$ be the bundle of all $(0, q)$-forms on M. Clearly $\Lambda^{0,0}(M) = M \times \mathbf{C}$, the trivial line bundle over M. Let $\{T_\alpha\}$ be a local frame of $T_{1,0}(M)$ and $\{\theta^\alpha\}$ the corresponding *admissible* local coframe, i.e.,

$$
T_\beta \rfloor \theta^\alpha = \delta^\alpha_\beta , \; T_{\bar\beta} \rfloor \theta^\alpha = 0 , \; T \rfloor \theta^\alpha = 0.
$$

We define a differential operator

$$
\overline{\partial}_b : \Gamma^\infty(\Lambda^{0,q}(M)) \to \Gamma^\infty(\Lambda^{0,q+1}(M))
$$

as follows.

Definition 1.40. Let φ be a $(0, q)$-form on M. Then, by definition, $\overline{\partial}_b \varphi$ is the unique $(0, q+1)$-form on M that agrees with $d\varphi$ when both are restricted to $T_{0,1}(M) \otimes \cdots \otimes T_{0,1}(M)$ ($q + 1$ terms). □

For instance, given a C^∞ function $f : M \to \mathbf{C}$, then (locally)

$$\overline{\partial}_b f = T_{\bar\alpha}(f)\theta^{\bar\alpha}.$$

Also, let φ be a (0, 1)-form (locally) given by

$$\varphi = \varphi_{\bar\alpha}\theta^{\bar\alpha}.$$

Let ∇ be the Tanaka–Webster connection of (M, θ) and let us set

$$\nabla_{\bar\alpha}\varphi_{\bar\beta} = (\nabla_{\bar\alpha}\varphi)T_{\bar\beta}$$

(so that $\nabla_{\bar\alpha}\varphi_{\bar\beta} = T_{\bar\alpha}(\varphi_{\bar\beta}) - \Gamma^{\bar\rho}_{\bar\alpha\bar\beta}\varphi_{\bar\rho}$). Then

$$\overline{\partial}_b\varphi = (\nabla_{\bar\alpha}\varphi_{\bar\beta})\,\theta^{\bar\alpha} \wedge \theta^{\bar\beta}.$$

Moreover,

$$\begin{aligned}(q+1)(\overline{\partial}_b\varphi)(\overline{Z}_1, \dots, \overline{Z}_{q+1}) = \\ \sum_{i=1}^{q+1}(-1)^{i+1}\overline{Z}_i\left(\varphi(\overline{Z}_1, \dots, \overline{Z}_{i-1}, \overline{Z}_{i+1}, \dots, \overline{Z}_{q+1})\right) \\ + \sum_{1\le i<j\le q+1}(-1)^{i+j}\varphi([\overline{Z}_i, \overline{Z}_j], \dots, \hat{\overline{Z}}_i, \dots, \hat{\overline{Z}}_j, \dots)\end{aligned}$$

for any $Z_1, \dots, Z_{q+1} \in \Gamma^\infty(T_{1,0}(M))$. In other words, modulo the identification (in the presence of an admissible local frame $\{\theta^\alpha\}$) of $T_{0,1}(M)^*$ with a subbundle of $T^*(M) \otimes \mathbf{C}$, the differential operator $\overline{\partial}_b$ is precisely the tangential Cauchy–Riemann operator of $(M, T_{1,0}(M))$ as previously introduced. We obtain a sequence of $\Gamma^\infty(M \times \mathbf{C})$-modules and differential operators

$$\begin{aligned}\Gamma^\infty(\Lambda^{0,0}(M)) \xrightarrow{\overline{\partial}_b} \Gamma^\infty(\Lambda^{0,1}(M)) \xrightarrow{\overline{\partial}_b} \Gamma^\infty(\Lambda^{0,2}(M)) \xrightarrow{\overline{\partial}_b} \cdots \\ \cdots \xrightarrow{\overline{\partial}_b} \Gamma^\infty(\Lambda^{0,n+1}(M)) \xrightarrow{\overline{\partial}_b} 0 \end{aligned} \tag{1.125}$$

satisfying

$$\overline{\partial}_b^2 = 0, \tag{1.126}$$

that is, (1.125) is a cochain complex. The alternative approach to $\overline{\partial}_b$ on a nondegenerate CR manifold (of hypersurface type) through the use of the given pseudo-Hermitian structure, rather than as a CR-invariant differential operator (cf. also J.M. Lee [270], p. 165) has the advantage that since elements in $\Gamma^\infty(\Lambda^{0,q}(M))$ are genuine differential forms on M, the properties of $\overline{\partial}_b$ may be easily deduced from the corresponding properties enjoyed by the exterior differentiation operator d. Indeed, for any $\varphi \in \Gamma^\infty(\Lambda^{0,q}(M))$ and any $Z_1, \dots, Z_{q+2} \in \Gamma^\infty(T_{1,0}(M))$ we have

$$(q+2)(\overline{\partial}_b^2\varphi)(\overline{Z}_1,\ldots,\overline{Z}_{q+2}) = \sum_{i=1}^{q+2}(-1)^{i+1}\overline{Z}_i((\overline{\partial}_b\varphi)(\cdots,\hat{\overline{Z}}_i,\ldots)$$
$$+ \sum_{1\le i<j\le q+2}(-1)^{i+j}(\overline{\partial}_b\varphi)([\overline{Z}_i,\overline{Z}_j],\ldots,\hat{\overline{Z}}_i,\ldots,\hat{\overline{Z}}_j,\ldots).$$

Since (by the integrability of the CR structure) $[\overline{Z}_i,\overline{Z}_j] \in \Gamma^\infty(T_{0,1}(M))$, it follows that

$$(\overline{\partial}_b\varphi)([\overline{Z}_i,\overline{Z}_j],\ldots,\hat{\overline{Z}}_i,\ldots,\hat{\overline{Z}}_j,\ldots) = (d\varphi)([\overline{Z}_i,\overline{Z}_j],\ldots,\hat{\overline{Z}}_i,\ldots,\hat{\overline{Z}}_j,\ldots)$$

(by the very definition of $\overline{\partial}_b$ on a nondegenerate CR manifold). Then

$$(q+2)(\overline{\partial}_b^2\varphi)(\overline{Z}_1,\ldots,\overline{Z}_{q+2}) = \sum_{i=1}^{q+2}(-1)^{i+1}\overline{Z}_i((d\varphi)(\overline{Z}_1,\ldots,\hat{\overline{Z}}_i,\ldots,\overline{Z}_{q+2})$$
$$+ \sum_{1\le i<j\le q+2}(-1)^{i+j}(d\varphi)([\overline{Z}_i,\overline{Z}_j],\ldots,\hat{\overline{Z}}_i,\ldots,\hat{\overline{Z}}_j,\ldots)$$
$$= (q+2)(d^2\varphi)(\overline{Z}_1,\ldots,\overline{Z}_{q+2}) = 0$$

and (1.126) is proved. Similarly, one may check that

$$\overline{\partial}_b(\varphi\wedge\psi) = (\overline{\partial}_b\varphi)\wedge\psi + (-1)^q\varphi\wedge\overline{\partial}_b\psi \tag{1.127}$$

for any $(0,q)$-form φ and any $(0,p)$-form ψ on M. □

We wish to make a few remarks on the tangential CR equations on functions. Let $(M, T_{1,0}(M))$ be a CR manifold (of arbitrary, but fixed, type).

Definition 1.41. A function $f \in \Gamma^\infty(M\times\mathbf{C})$ is *CR-holomorphic* (or simply *CR*) if

$$\overline{\partial}_b f = 0 \tag{1.128}$$

and (1.128) are the *tangential Cauchy–Riemann equations.* □

We do not discuss CR functions (and the holomorphic extension problem) in this book. Excellent monographs treating the argument are those of A. Boggess [70], M.S. Baouendi, P. Ebenfelt, and L.P. Rothschild [31]. Let us look, however, at a couple of simple examples.

(1) Let $U \subseteq \mathbf{C}^N$ be an open subset and $F : U \to \mathbf{C}$ a holomorphic function. Let $M \subset U$ be an embedded CR manifold. Then $f = F\circ\iota$ is a CR-holomorphic function on M (where $\iota : M \to U$ is the inclusion). The converse is not true in general. That is, given a CR-holomorphic function $f : M \to \mathbf{C}$ and a point $x \in M$, then in general, f may not extend to a function that is holomorphic in a neighborhood of x in $\mathbf{C}^N$. For instance, let

$$M = \{(z^1, z^2) \in \mathbf{C}^2 : \mathrm{Im}(z^1) = 0\}.$$

Then $T_{0,1}(M)$ is spanned by $T_{\overline{1}} = \partial/\partial\overline{z}^2$ and the tangential Cauchy–Riemann equations may be written

$$\frac{\partial f}{\partial \overline{z}^2} = 0. \tag{1.129}$$

Let us set $z^1 = x + iy$. Then any C^∞ function $f(x)$ is a solution of (1.129); yet $f(x)$ may not extend to a holomorphic function in $\mathbf{C}^2$ unless it is real analytic in x to start with. See A. Boggess [70], p. 141. □

(2) Let $F : \mathbf{H}_n \to \mathbf{H}_n$ be a C^∞ map, where $\mathbf{H}_n$ is the Heisenberg group. Then F is a CR map if and only if F^j, $1 \leq j \leq n$, and $|F|^2 - if$ are CR-holomorphic. Here we have set $F = (F^1, \ldots, F^n, f)$ and $|F|^2 = \delta_{jk} F^j \overline{F^k}$. □

The problem whether (or under what assumptions) a given CR function on an embedded CR manifold extends (locally or globally) to a holomorphic function occupies a large amount of literature (cf. [70] and references therein). For instance (by a classical result of F. Severi [370], and G. Tomassini [403], who extended Severi's result to the case of CR manifolds of arbitrary CR codimension), if M is a real analytic embedded CR manifold then any real analytic CR-holomorphic function extends to a function that is holomorphic in some neighborhood of M.

A generalization of the Cauchy–Riemann equations (in one complex variable) has been proposed by I.N. Vekua (cf. [418]), who studied solutions to the PDE $\partial_{\overline{z}} w = aw + b\overline{w}$ (*generalized analytic functions*), where a, b are C^2 functions on a domain in $\mathbf{C}$. The ideas of I.N. Vekua were carried over to the case of several complex variables by A. Koohara ([253]) and L.G. Mikhailov and A.V. Abrosimov ([301]), who considered systems of the form

$$\partial_{\overline{z}_j} w = a_j w + b_j \overline{w} + c_j, \quad 1 \leq j \leq n \tag{1.130}$$

(and Y. Hayashi ([195]) extended their work to nonlinear equations of the form $\partial_{\overline{z}_j} w = a_j f_j(\overline{w})$). The problem of studying traces of solutions to (1.130) on a smooth real hypersurface in $\mathbf{C}^n$ (*generalized CR functions*) is open.

1.7.2 CR-holomorphic bundles

Let $(M, T_{1,0}(M))$ be a CR manifold and $E \to M$ a C^∞ complex vector bundle over M. Let

$$\overline{\partial}_E : \Gamma^\infty(E) \to \Gamma^\infty(T_{0,1}(M)^* \otimes E)$$

be a differential operator satisfying the following requirements:

$$\overline{\partial}_E(fu) = f\overline{\partial}_E u + (\overline{\partial}_b f) \otimes u, \tag{1.131}$$

$$[\overline{Z}, \overline{W}]u = \overline{Z}\,\overline{W}\,u - \overline{W}\,\overline{Z}\,u, \tag{1.132}$$

for any $f \in C^\infty(M) \otimes \mathbf{C}$, $u \in \Gamma^\infty(E)$, $Z, W \in \Gamma^\infty(T_{1,0}(M))$. As to the notation in (1.132), we have set $\overline{Z}u = (\overline{\partial}_E u)\overline{Z}$. Following N. Tanaka [398], we make the following definition:

Definition 1.42. A pair $(E, \overline{\partial}_E)$ consisting of a complex vector bundle (over a CR manifold) and a differential operator satisfying (1.131)–(1.132) is said to be a *CR-holomorphic vector bundle.* □

Let us look at a few examples of CR-holomorphic vector bundles.

(i) Let V be a complex manifold. Let $M \subset V$ be a CR submanifold (carrying the CR structure (1.12) induced on M by the complex structure of V). Assume that M is regularly embedded in V as a closed subset. Let $\pi : F \to V$ be a holomorphic vector bundle over V. Then the portion $E = \pi^{-1}(M)$ of F over M is CR-holomorphic. Indeed, since F is holomorphic, there is a natural differential operator

$$\overline{\partial}_F : \Gamma^\infty(F) \to \Gamma^\infty(T^{0,1}(V)^* \otimes F),$$

where $T^{0,1}(V)$ is the antiholomorphic tangent bundle over V. Given $u \in \Gamma^\infty(E)$ let $\tilde{u} \in \Gamma^\infty(F)$ be a C^∞ extension of u as a cross-section in F and set $\left(\overline{\partial}_E u\right)_x = \left(\overline{\partial}_F \tilde{u}\right)_x$ for any $x \in M$. The definition of $\left(\overline{\partial}_E u\right)_x$ does not depend on the choice of extension $\tilde{u}$ of u because $(\overline{\partial} f)_{|T_{0,1}(M)} = \overline{\partial}_b(f_{|M})$ for any C^∞ function $f : V \to \mathbf{C}$. Let $\{\Phi_\alpha : \pi^{-1}(\Omega_\alpha) \to \Omega_\alpha \times \mathbf{C}^m : \alpha \in I\}$ be a trivialization atlas for F and $G_{\beta\alpha} : \Omega_\beta \cap \Omega_\alpha \to \mathrm{GL}(m, \mathbf{C})$ the corresponding transition functions. Let us set $U_\alpha = \Omega_\alpha \cap M$ and $g_{\beta\alpha} = G_{\beta\alpha |U_\alpha \cap U_\beta}$. Since the $G_{\beta\alpha}$ are holomorphic, it follows that $E \to M$ is a peculiar type of CR-holomorphic vector bundle (called *locally trivial* by C. Le Brun [268]) in that its transition functions are matrix-valued CR functions on M. □

(ii) Let $(M, T_{1,0}(M))$ be a CR manifold and consider the quotient bundle

$$\hat{T}(M) = \frac{T(M) \otimes \mathbf{C}}{T_{0,1}(M)}$$

and the canonical bundle map $\pi : T(M) \otimes \mathbf{C} \to \hat{T}(M)$. We define a differential operator

$$\overline{\partial}_{\hat{T}(M)} : \Gamma^\infty(\hat{T}(M)) \to \Gamma^\infty(T_{0,1}(M)^* \otimes \hat{T}(M))$$

as follows. For any $u \in \Gamma^\infty(\hat{T}(M))$ there is $W \in \Gamma^\infty(T(M) \otimes \mathbf{C})$ such that $\pi(W) = u$. Then we set

$$(\overline{\partial}_{\hat{T}(M)} u)\overline{Z} = \pi[\overline{Z}, W],$$

for any $Z \in \Gamma^\infty(T_{1,0}(M))$. The definition of $\overline{\partial}_{\hat{T}(M)} u$ does not depend on the choice of W with $\pi(W) = u$ because of the formal integrability property (1.6) of the CR structure $T_{1,0}(M)$. Then $(\hat{T}(M), \overline{\partial}_{\hat{T}(M)})$ is a CR-holomorphic vector bundle. It is interesting to note that the construction of $\overline{\partial}_{\hat{T}(M)}$ is very similar to that of the complex Bott connection of a holomorphic foliation (compare with R. Bott and P. Baum [76]). This analogy is further exploited in the next section. □

(iii) The CR structure $T_{1,0}(M)$ itself may be organized, when M is nondegenerate, as a CR-holomorphic vector bundle over M. Indeed, if θ is a contact structure on

M and ∇ the corresponding Tanaka–Webster connection, then we may consider the first-order differential operator

$$\overline{\partial}_{T_{1,0}(M)} : \Gamma^\infty(T_{1,0}(M)) \to \Gamma^\infty(T_{0,1}(M)^* \otimes T_{1,0}(M)),$$

$$(\overline{\partial}_{T_{1,0}(M)} Z)\overline{W} := \nabla_{\overline{W}} Z, \ \ Z, W \in T_{1,0}(M).$$

Then the pair $(T_{1,0}(M), \overline{\partial}_{T_{1,0}(M)})$ is a CR-holomorphic vector bundle because the curvature R_∇ of ∇ satisfies $R_\nabla(\overline{Z}, \overline{W})V = 0$, for any $Z, V, W \in T_{1,0}(M)$ (cf. H. Urakawa [412], p. 569). □

(iv) The *canonical bundle* $K(M)$ is organized as a CR-holomorphic vector bundle in Chapter 8 of this book (in connection with the solutions to the inhomogeneous Yang–Mills equation). □

Let $(E, \overline{\partial}_E)$ be a CR-holomorphic vector bundle. We define a differential operator

$$\overline{\partial}_E^q : \Gamma^\infty(\Lambda^q T_{0,1}(M)^* \otimes E) \to \Gamma^\infty(\Lambda^{q+1} T_{0,1}(M)^* \otimes E)$$

by setting

$$(\overline{\partial}_E^q \varphi)(\overline{Z}_1, \ldots, \overline{Z}_{q+1}) =$$
$$\frac{1}{q+1}\left\{ \sum_{i=1}^{q+1} (-1)^{i+1} \left(\overline{\partial}_E \varphi(\overline{Z}_1, \ldots, \hat{\overline{Z}}_i, \ldots, \overline{Z}_{q+1})\right) \overline{Z}_i \right.$$
$$\left. + \sum_{i<j} (-1)^{i+j} \varphi([\overline{Z}_i, \overline{Z}_j], \overline{Z}_1, \ldots, \hat{\overline{Z}}_i, \ldots, \hat{\overline{Z}}_j, \ldots, \overline{Z}_{q+1}) \right\}$$

for any $\varphi \in \Gamma^\infty(\Lambda^q T_{0,1}(M)^* \otimes E)$ and any $Z_j \in \Gamma^\infty(T_{1,0}(M))$, $1 \le j \le q+1$. Just as in the case of the Cauchy–Riemann operator $\overline{\partial}_b$ one may show that $\overline{\partial}_E^q \varphi$ is a C^∞ section in $\Lambda^{q+1} T_{0,1}(M)^* \otimes E$ and that $\overline{\partial}_E^{q+1} \circ \overline{\partial}_E^q = 0$ for any $q \ge 0$.

Proposition 1.15. *$\{\Gamma^\infty(\Lambda^q T_{0,1}(M)^* \otimes E), \overline{\partial}_E^q\}_{q \ge 0}$ is a cochain complex.*

Let $H^q(M, E)$ be the corresponding cohomology groups.

1.7.3 The Frölicher spectral sequence

We start by introducing a certain filtration of the de Rham complex. Let

$$\{\Omega^k(M) := \Gamma^\infty(\Lambda^k T^*(M) \otimes \mathbf{C}), d\}_{k \ge 0}$$

be the de Rham complex of M with complex coefficients and $H^k(M, \mathbf{C})$ the corresponding de Rham cohomology groups. Let $x \in M$. Let $F^p\Omega^k(M)_x$ consist of all $\varphi \in \Lambda^k T^*(M)_x \otimes_{\mathbf{R}} \mathbf{C}$ such that

$$\varphi(W_1, \ldots, W_{p-1}, \overline{V}_1, \ldots, \overline{V}_{k-p+1}) = 0,$$

for any $W_i \in T_x(M) \otimes_{\mathbf{R}} \mathbf{C}$ and any $V_j \in T_{1,0}(M)_x$, $1 \le i \le p-1$, $1 \le j \le k-p+1$. Then $F^p\Omega^k(M)$ is a subbundle of $\Lambda^k T^*(M) \otimes \mathbf{C}$ satisfying the following properties:

$$F^p\Omega^k(M) \supset F^{p+1}\Omega^k(M), \quad F^{p+1}\Omega^p(M) = 0.$$

We also set $F^0\Omega^k(M) = \Lambda^k T^*(M) \otimes \mathbf{C}$. Note that

$$d\,\Gamma^\infty(F^p\Omega^k(M)) \subset \Gamma^\infty(F^p\Omega^{k+1}(M)).$$

Hence

Proposition 1.16. $\{\Gamma^\infty(F^p\Omega^k(M))\}_{p,k\ge 0}$ *is a (decreasing) filtration of the de Rham complex*

$$\begin{array}{ccccccc}
\vdots & & \vdots & & \vdots & & \\
\downarrow & & \downarrow & & \downarrow & & \\
\Omega^k(M) & \supset & \Gamma^\infty(F^1\Omega^k(M)) & \supset & \Gamma^\infty F^2\Omega^k(M)) & \supset & \cdots \\
\downarrow d & & \downarrow d & & \downarrow d & & \\
\Omega^{k+1}(M) & \supset & \Gamma^\infty(F^1\Omega^{k+1}(M)) & \supset & \Gamma^\infty(F^2\Omega^{k+1}(M)) & \supset & \cdots \\
\downarrow & & \downarrow & & \downarrow & & \\
\vdots & & \vdots & & \vdots & &
\end{array}$$

Let $\{E_r(M)\}_{r\ge 1}$ be the spectral sequence associated with this filtration. In analogy with [158] we make the following definition:

Definition 1.43. $\{E_r(M)\}_{r\ge 1}$ is referred to as the *Frölicher spectral sequence* of the CR manifold M. □

Precisely, we set

$$C_r^{p,q}(M) = \Gamma^\infty(F^p\Omega^{p+q}(M)) \cap d^{-1}\Gamma^\infty(F^{p+r}\Omega^{p+q+1}(M)),$$

and

$$D_r^{p,q}(M) = \Gamma^\infty(F^p\Omega^{p+q}(M)) \cap d\,\Gamma^\infty(F^{p-r}\Omega^{p+q-1}(M)).$$

Note that

$$D_r^{p,q}(M) = dC_r^{p-r,q+r-1}(M).$$

Finally, we set

$$E_r(M) = \{E_r^{p,q}(M)\}_{p,q\ge 0},$$
$$E_r^{p,q}(M) = \frac{C_r^{p,q}(M)}{C_{r-1}^{p+1,q-1}(M) + D_{r-1}^{p,q}(M)}.$$

We shall be particularly interested in $E_1^{p,q}(M)$ and $E_2^{k,0}(M)$, which are denoted by $H^{p,q}(M)$ and $H_0^k(M)$. This is consistent with our previous notation, since, as will be shortly shown, $E_1^{0,q}(M)$ are precisely the Kohn–Rossi cohomology groups. CR analogues of Theorems 1 and 3 of L. Cordero, M. Fernandez, L. Ugarte, and A. Gray [106] (describing the Frölicher spectral sequence of a complex manifold) are not known, so far.

Next, we express $\overline{\partial}_b$ in terms of covariant derivatives. For the rest of this section, we assume $(M, T_{1,0}(M))$ to be a nondegenerate CR manifold (of hypersurface type) of CR dimension n. Let θ be a fixed pseudo-Hermitian structure on M. Let $\{T_1, \dots, T_n\}$ be a (local) frame of $T_{1,0}(M)$, defined on some open set $U \subseteq M$. Let $\{\theta^1, \dots, \theta^n\}$ be the corresponding dual admissible coframe. As a useful consequence of our nondegeneracy assumption, we may express $\overline{\partial}_b$ in terms of covariant derivatives, with respect to the Tanaka–Webster connection ∇ of (M, θ). Any $(0, q)$-form φ on M may be represented locally as

$$\varphi = \frac{1}{q!} \varphi_{\bar{\alpha}_1 \cdots \bar{\alpha}_q} \theta^{\bar{\alpha}_1} \wedge \cdots \wedge \theta^{\bar{\alpha}_q}.$$

The covariant derivatives $\nabla_A \varphi_{\bar{\alpha}_1 \cdots \bar{\alpha}_q}$ of φ with respect to ∇ are defined by

$$\nabla_A \varphi_{\bar{\alpha}_1 \cdots \bar{\alpha}_q} = (\nabla_{T_A} \varphi)(T_{\bar{\alpha}_1}, \dots, T_{\bar{\alpha}_q}),$$

where $A \in \{0, 1, \dots, n, \bar{1}, \dots, \bar{n}\}$ and $T_0 = T$. Explicitly,

$$\nabla_A \varphi_{\bar{\alpha}_1 \cdots \bar{\alpha}_q} = \frac{1}{q!} \left\{ T_A(\varphi_{\bar{\alpha}_1 \cdots \bar{\alpha}_q}) - \sum_{j=1}^{q} \varphi_{\bar{\alpha}_1 \cdots \bar{\alpha}_{j-1} \bar{\mu} \bar{\alpha}_{j+1} \cdots \bar{\alpha}_q} \Gamma^{\bar{\mu}}_{A \bar{\alpha}_j} \right\}. \tag{1.133}$$

If T_∇ is the torsion tensor field of ∇, we recall that $T_\nabla(T_{\bar{\alpha}}, T_{\bar{\beta}}) = 0$, that is,

$$[T_{\bar{\alpha}}, T_{\bar{\beta}}] = \left(\Gamma^{\bar{\mu}}_{\bar{\alpha}\bar{\beta}} - \Gamma^{\bar{\mu}}_{\bar{\beta}\bar{\alpha}} \right) T_{\bar{\mu}} . \tag{1.134}$$

We may perform the following calculation:

$$\begin{aligned}
\frac{1}{(q+1)!} (\overline{\partial}_b \varphi)_{\bar{\alpha}_1 \cdots \bar{\alpha}_{q+1}} &= (\overline{\partial}_b \varphi)(T_{\bar{\alpha}_1}, \dots, T_{\bar{\alpha}_{q+1}}) = (d\varphi)(T_{\bar{\alpha}_1}, \dots, T_{\bar{\alpha}_{q+1}}) \\
&= \frac{1}{q+1} \left\{ \sum_{i=1}^{q+1} (-1)^{i-1} T_{\bar{\alpha}_i} (\varphi(\cdots \hat{T}_{\bar{\alpha}_i} \cdots)) \right. \\
&\qquad \left. + \sum_{i<j} (-1)^{i+j} \varphi([T_{\bar{\alpha}_i}, T_{\bar{\alpha}_j}], \dots, \hat{T}_{\bar{\alpha}_i}, \dots, \hat{T}_{\bar{\alpha}_j}, \dots) \right\};
\end{aligned}$$

hence (by (1.134))

$$\begin{aligned}
(\overline{\partial}_b \varphi)_{\bar{\alpha}_1 \cdots \bar{\alpha}_{q+1}} &= \sum_{i=1}^{q+1} (-1)^{i-1} T_{\bar{\alpha}_i} (\varphi_{\bar{\alpha}_1 \cdots \hat{\bar{\alpha}}_i \cdots \bar{\alpha}_{q+1}}) \\
&\quad + \sum_{i=1}^{q+1} \sum_{i<j} (-1)^{i+j} \Gamma^{\bar{\mu}}_{\bar{\alpha}_i \bar{\alpha}_j} \varphi_{\bar{\mu} \bar{\alpha}_1 \cdots \hat{\bar{\alpha}}_i \cdots \hat{\bar{\alpha}}_j \cdots \bar{\alpha}_{q+1}} \\
&\quad - \sum_{j=1}^{q+1} \sum_{i<j} (-1)^{i+j} \Gamma^{\bar{\mu}}_{\bar{\alpha}_j \bar{\alpha}_i} \varphi_{\bar{\mu} \cdots \hat{\bar{\alpha}}_i \cdots \hat{\bar{\alpha}}_j \cdots \bar{\alpha}_{q+1}} .
\end{aligned}$$

Finally (interchanging i and j in the last sum), by the skew-symmetry of $\varphi_{\bar{\alpha}_1 \cdots \bar{\alpha}_q}$ we obtain the following result:

Proposition 1.17. *Let M be a nondegenerate CR manifold. For any $(0,q)$-form φ on M,*

$$\overline{\partial}_b\varphi = \frac{1}{(q+1)!}(\overline{\partial}_b\varphi)_{\bar{\alpha}_1\cdots\bar{\alpha}_{q+1}}\theta^{\bar{\alpha}_1}\wedge\cdots\wedge\theta^{\bar{\alpha}_{q+1}},$$

$$(\overline{\partial}_b\varphi)_{\bar{\alpha}_1\cdots\bar{\alpha}_{q+1}} = q!\sum_{i=1}^{q+1}(-1)^i\nabla_{\bar{\alpha}_i}\varphi_{\bar{\alpha}_1\cdots\hat{\bar{\alpha}}_i\cdots\bar{\alpha}_{q+1}}. \tag{1.135}$$

Next, we wish to establish an explicit identity relating the exterior differentiation operator d and the tangential Cauchy–Riemann operator $\overline{\partial}_b$. Precisely, we relate $d\varphi$ to $\overline{\partial}_b\varphi$, for any $(0,q)$-form φ on M. Let $\pi_q : \Omega^q(M) \to \Omega^{0,q}(M)$ be the natural projection. Then

$$\overline{\partial}_b\varphi = \pi_{q+1}d\varphi.$$

Another way to put it is that

$$d\varphi = \overline{\partial}_b\varphi + \psi,$$

for some uniquely defined $(q+1)$-form ψ such that $\psi(\overline{Z}_1,\ldots,\overline{Z}_{q+1}) = 0$ for any $Z_j \in \Gamma^\infty(T_{1,0}(M))$, $1 \le j \le q+1$. The $(q+1)$-form ψ may be explicitly computed in terms of covariant derivatives of φ. Indeed, using the identities

$$\begin{aligned}
df &= T_\alpha(f)\theta^\alpha + T_{\bar{\alpha}}(f)\theta^{\bar{\alpha}} + T(f)\theta,\\
d\theta^\alpha &= \theta^\beta\wedge\omega_\beta^\alpha + \theta\wedge\tau^\alpha,\\
\omega_\alpha^\beta &= \Gamma_{\mu\alpha}^\beta\theta^\mu + \Gamma_{\bar{\mu}\alpha}^\beta\theta^{\bar{\mu}} + \Gamma_{0\alpha}^\beta\theta,\\
\tau^\alpha &= A_{\bar{\beta}}^\alpha\theta^{\bar{\beta}},
\end{aligned}$$

we may perform the calculation

$$\begin{aligned}
q!d\varphi &= (d\varphi_{\bar{\alpha}_1\cdots\bar{\alpha}_q})\wedge\theta^{\bar{\alpha}_1}\wedge\cdots\wedge\theta^{\bar{\alpha}_q}\\
&\quad + \varphi_{\bar{\alpha}_1\cdots\bar{\alpha}_q}\sum_{i=1}^{q}(-1)^{i-1}\theta^{\alpha_1}\wedge\cdots\wedge\theta^{\alpha_{i-1}}\wedge d\theta^{\alpha_i}\wedge\theta^{\alpha_{i+1}}\wedge\cdots\wedge\theta^{\alpha_q}\\
&= T_A(\varphi_{\bar{\alpha}_1\cdots\bar{\alpha}_q})\theta^A\wedge\theta^{\bar{\alpha}_1}\wedge\cdots\wedge\theta^{\bar{\alpha}_q}\\
&\quad + \varphi_{\bar{\alpha}_1\cdots\bar{\alpha}_q}\sum_{i=1}^{q}(-1)^{i-1}\theta^{\bar{\alpha}_1}\wedge\cdots\wedge\theta^{\bar{\alpha}_{i-1}}\wedge\\
&\Big(\Gamma_{\mu\bar{\beta}}^{\bar{\alpha}_i}\theta^{\bar{\beta}}\wedge\theta^\mu + \Gamma_{\bar{\mu}\bar{\beta}}^{\bar{\alpha}_i}\theta^{\bar{\beta}}\wedge\theta^{\bar{\mu}} + \Gamma_{0\bar{\beta}}^{\bar{\alpha}_i}\theta^{\bar{\beta}}\wedge\theta + A_\mu^{\bar{\alpha}_i}\theta\wedge\theta^\mu\Big)\wedge\theta^{\bar{\alpha}_{i+1}}\wedge\cdots\wedge\theta^{\bar{\alpha}_q};
\end{aligned}$$

hence (by interchanging β and α_i)

$$\begin{aligned}
q!d\varphi = q!\overline{\partial}_b\varphi + T_\mu(\varphi_{\bar\alpha_1\cdots\bar\alpha_q})\theta^\mu \wedge \theta^{\bar\alpha_1}\wedge\cdots\wedge\theta^{\bar\alpha_q} \\
- \sum_{i=1}^q \Gamma^{\bar\beta}_{\mu\bar\alpha_i}\varphi_{\bar\alpha_1\cdots\bar\alpha_{i-1}\bar\beta\bar\alpha_{i+1}\cdots\bar\alpha_q}\theta^\mu\wedge\theta^{\bar\alpha_1}\wedge\cdots\wedge\theta^{\bar\alpha_q} \\
+ T(\varphi_{\bar\alpha_1\cdots\bar\alpha_q})\theta\wedge\theta^{\bar\alpha_1}\wedge\cdots\wedge\theta^{\bar\alpha_q} \\
- \sum_{i=1}^q \Gamma^{\bar\beta}_{0\bar\alpha_i}\varphi_{\bar\alpha_1\cdots\bar\alpha_{i-1}\bar\beta\bar\alpha_{i+1}\cdots\bar\alpha_q}\theta\wedge\theta^{\bar\alpha_1}\wedge\cdots\wedge\theta^{\bar\alpha_q} \\
+\sum_{i=1}^q(-1)^{i-1}A^{\bar\beta}_\mu\varphi_{\bar\alpha_1\cdots\bar\alpha_{i-1}\bar\beta\bar\alpha_{i+1}\cdots\bar\alpha_q}\theta^{\bar\alpha_1}\wedge\cdots\wedge\theta^{\bar\alpha_{i-1}}\wedge(\theta\wedge\theta^\mu)\wedge\theta^{\bar\alpha_{i+1}}\wedge\cdots\wedge\theta^{\bar\alpha_q}.
\end{aligned}$$

Finally (by (1.133)),

$$\begin{aligned}
d\varphi = \overline{\partial}_b\varphi + \Big\{\left(\nabla_\mu\varphi_{\bar\alpha_1\cdots\bar\alpha_q}\right)\theta^\mu + \left(\nabla_0\varphi_{\bar\alpha_1\cdots\bar\alpha_q}\right)\theta\Big\}\wedge\theta^{\bar\alpha_1}\wedge\cdots\wedge\theta^{\bar\alpha_q} \\
+\theta\wedge\frac{1}{q!}\sum_{i=1}^q\varphi_{\bar\alpha_1\cdots\bar\alpha_q}\theta^{\bar\alpha_1}\wedge\cdots\theta^{\bar\alpha_{i-1}}\wedge\tau^{\bar\alpha_i}\wedge\theta^{\bar\alpha_{i+1}}\wedge\cdots\wedge\theta^{\bar\alpha_q}. \quad (1.136)
\end{aligned}$$

Next, we look at the Kohn–Rossi cohomology groups.

Proposition 1.18. (N. Tanaka [398])

$$E_1^{0,q}(M)\simeq H^{0,q}(M)$$

(*an isomorphism*).

First, note that

$$\begin{aligned}
E_1^{0,q}(M) &= \frac{C_1^{0,q}(M)}{C_0^{1,q-1}(M)+D_0^{0,q}(M)}, \\
C_1^{0,q}(M) &= \Big\{\varphi\in\Omega^q(M) : (d\varphi)(\overline{Z}_1,\ldots,\overline{Z}_{q+1}) = 0, \\
&\qquad \forall Z_j\in T_{1,0}(M),\ 1\le j\le q+1\Big\}, \\
C_0^{1,q-1}(M)+D_0^{0,q}(M) &= \Big\{\psi + d\varphi : \varphi\in\Gamma^\infty(\omega^{q-1}(M)\otimes\mathbf{C}),\ \psi\in\Omega^q(M), \\
&\qquad \psi(\overline{Z}_1,\ldots,\overline{Z}_q) = 0,\ \forall Z_j\in T_{1,0}(M),\ 1\le j\le q\Big\}.
\end{aligned}$$

We build an isomorphism $\Phi : H^{0,q}(M)\to E_1^{0,q}(M)$ as follows. Let ω be a $\overline{\partial}_b$-closed $(0,q)$-form on M and $[\omega]_{H^{0,q}(M)}$ its Kohn–Rossi cohomology class. Then ω is a complex q-form and $(d\omega)(\overline{Z}_1,\ldots,\overline{Z}_{q+1}) = 0$ by the definition of $\overline{\partial}_b$, so that the class $[\omega]_{E_1^{0,q}(M)}$ of ω (modulo $C_0^{1,q-1}(M)+D_0^{0,q}(M)$) is well defined. We set

$$\Phi([\omega]_{H^{0,q}(M)}) = [\omega]_{E_1^{0,q}(M)}.$$

The definition of $\Phi([\omega]_{H^{0,q}(M)})$ doesn't depend on the choice of representative because

$$\overline{\partial}_b \Omega^{0,q-1}(M) \subset C_0^{1,q-1}(M) + D_0^{0,q}(M).$$

Indeed, if φ is a $(0, q-1)$-form on M then (by (1.136)) $d\varphi = \overline{\partial}_b \varphi - \psi$ for some uniquely determined q-form ψ such that $\psi(\overline{Z}_1, \ldots, \overline{Z}_q) = 0$; hence $\overline{\partial}_b \varphi \in C_0^{1,q-1}(M) + D_0^{0,q}(M)$. To prove that Φ is one-to-one, assume that $\Phi([\omega]_{H^{0,q}(M)}) = 0$. Then $\omega = \psi + d\varphi$ for some $\psi \in \Omega^q(M)$, $\varphi \in \Omega^{q-1}(M)$ with $\psi(\overline{Z}_1, \ldots, \overline{Z}_q) = 0$. Yet we may write $\varphi = \pi_{q-1}\varphi + \eta$ for some uniquely determined $(q-1)$-form η such that $\eta(\overline{Z}_1, \ldots, \overline{Z}_{q-1}) = 0$. Thus $(d\eta)(\overline{Z}_1, \ldots, \overline{Z}_q) = 0$, as a consequence of the formal integrability property of $T_{1,0}(M)$. Finally, $\omega = \psi + d\pi_{q-1}\varphi + d\eta$ yields $\omega = \overline{\partial}_b \pi_{q-1}\varphi$, that is, $[\omega]_{H^{0,q}(M)} = 0$. To check the surjectivity of Φ, let $[\eta]_{E_1^{0,q}(M)} \in E_1^{0,q}(M)$. It suffices to observe that $\eta - \pi_q \eta \in C_0^{1,q-1}(M) + D_0^{0,q}(M)$ (indeed, set $\psi = \eta - \pi_q \eta$ and $\varphi = 0$). □

Our next task is to look at the cohomology groups $H^{p,q}(M)$. Let $(M, T_{1,0}(M))$ be a CR manifold and set

$$\Lambda^{p,q}(M) = \frac{F^p \Omega^{p+q}(M)}{F^{p+1}\Omega^{p+q}(M)}.$$

In particular,

$$\Lambda^{0,q}(M) = \frac{\Lambda^q T^*(M) \otimes \mathbf{C}}{\{\varphi \in \Lambda^q T^*(M) \otimes \mathbf{C} : \varphi(\overline{Z}_1, \ldots, \overline{Z}_q) = 0, \forall Z_j \in T_{1,0}(M), 1 \le j \le q\}};$$

hence, if M is a nondegenerate CR manifold (of hypersurface type) then $\Lambda^{0,q}(M)$ is the bundle of all $(0, q)$-forms on M. Let us set

$$\Omega^{p,q}(M) := \Gamma^\infty(\Lambda^{p,q}(M)).$$

Since the filtration $\{\Gamma^\infty\left(F^p \Omega^k(M)\right)\}$ is stable under d (the exterior differentiation operator), there is a naturally induced differential operator

$$\overline{\partial}_b^{p,q} : \Omega^{p,q}(M) \to \Omega^{p,q+1}(M)$$

such that $\overline{\partial}_b^{p,q+1} \circ \overline{\partial}_b^{p,q} = 0$, i.e.,

$$\left\{\Omega^{p,\cdot}(M), \overline{\partial}_b^{p,\cdot}\right\}$$

is a cochain complex. Clearly, for $p = 0$ this is the tangential Cauchy–Riemann complex (1.118).

Proposition 1.19. (N. Tanaka [398])

$$H^q\left(\Omega^{p,\cdot}(M), \overline{\partial}_M^{p,\cdot}\right) \simeq H^{p,q}(M)$$

(an isomorphism), i.e., $H^{p,q}(M)$ are the cohomology groups of the complex $\{\Omega^{p,\cdot}(M), \overline{\partial}_b^{p,\cdot}\}$.

Proof. We have

$$H^{p,q}(M) = E_1^{p,q}(M) = \frac{C_1^{p,q}}{C_0^{p+1,q-1} + D_0^{p,q}}$$
$$= \frac{\Gamma^\infty(F^p\Omega^{p+q}(M)) \cap d^{-1}\Gamma^\infty(F^{p+1}\Omega^{p+q+1}(M))}{\Gamma^\infty(F^{p+1}\Omega^{p+q}(M)) + d\Gamma^\infty(F^p\Omega^{p+q-1}(M))}.$$

We define a homomorphism

$$\Psi : H^{p,q}(M) \to H^q(\Omega^{p,\cdot}(M), \overline{\partial}_b^{p,\cdot})$$

by setting

$$\Psi([\varphi]_{E_1^{p,q}}) = [\varphi + F^{p+1}\Omega^{p+q}]$$

for any $\varphi \in \Gamma^\infty(F^p\Omega^{p+q}(M))$ such that $d\varphi \in \Gamma^\infty(F^{p+1}\Omega^{p+q+1}(M))$. Here $[\varphi]_{E_1^{p,q}}$ is the class of φ modulo

$$\Gamma^\infty(F^{p+1}\Omega^{p+q}(M)) + d\Gamma^\infty(F^p\Omega^{p+q-1}(M)).$$

Also

$$\varphi + F^{p+1}\Omega^{p+q} \in \Gamma^\infty(\Omega^{p,q})$$

and $[\varphi + F^{p+1}\Omega^{p+q}]$ is the class of $\varphi + F^{p+1}\Omega^{p+q}$ modulo

$$\overline{\partial}_b^{p,q-1}\Gamma^\infty(\Omega^{p,q-1}(M)).$$

Since $d\varphi \in \Gamma^\infty(F^{p+1}\Omega^{p+q+1}(M))$ it follows that

$$\overline{\partial}_b^{p,q}(\varphi + F^{p+1}\Omega^{p+q}) = 0;$$

hence $\varphi + F^{p+1}\Omega^{p+q}$ does define a cohomology class in

$$H^q(\Omega^{p,\cdot}(M), \overline{\partial}_b^{p,\cdot}).$$

To see that the definition of $\Psi([\varphi]_{E_1^{p,q}})$ doesn't depend on the choice of representative, it suffices to check that given $\varphi \in C_0^{p+1,q-1} + D_0^{p,q}$ one has $\varphi + F^{p+1}\Omega^{p+q} \in \overline{\partial}_b^{p,q-1}\Omega^{p,q-1}(M)$. Indeed $\varphi = \alpha + d\beta$ for some $\alpha \in \Gamma^\infty(F^{p+1}\Omega^{p+q}(M))$ and $\beta \in \Gamma^\infty(F^p\Omega^{p+q-1}(M))$; therefore

$$\varphi + F^{p+1}\Omega^{p+q} = d\beta + F^{p+1}\Omega^{p+q} = \overline{\partial}_b^{p,q-1}(\beta + F^{p+1}\Omega^{p+q-1}).$$

The reader may check that Ψ is an isomorphism. □

We shall need the complex vector bundle

$$E^p = \Lambda^p \hat{T}(M)^*.$$

We organize $E^p \to M$ as a CR-holomorphic bundle by considering the differential operator

$$\overline{\partial}_{E^p} : \Gamma^\infty(E^p) \to \Gamma^\infty(T_{0,1}(M)^* \otimes E^p)$$

defined by

$$\left((\overline{\partial}_{E^p}\varphi)\overline{Z}\right)(u_1, \ldots, u_p) = \overline{Z}(\varphi(u_1, \ldots, u_p)) - \sum_{i=1}^{p} \varphi(u_1, \ldots, u_{i-1}, \overline{Z}u_i, u_{i+1}, \ldots, u_p),$$

for any $\varphi \in \Gamma^\infty(E^p)$ and any $Z \in \Gamma^\infty(T_{1,0}(M))$, $u_i \in \Gamma^\infty(\hat{T}(M))$, $1 \leq i \leq p$. Here $\overline{Z}u_i = (\overline{\partial}_{\hat{T}(M)}u_i)\overline{Z}$.

Proposition 1.20. (N. Tanaka [398])

$$\Lambda^{p,q}(M) \simeq \Lambda^q T_{0,1}(M)^* \otimes E^p$$

(a complex vector bundle isomorphism).

Proof. Define

$$f^{p,q} : F^p\Omega^{p+q}(M) \to \Lambda^q T_{0,1}(M)^* \otimes E^p$$

by setting

$$\left((f^{p,q}\varphi)(\overline{Z}_1, \ldots, \overline{z}_q)\right)(u_1, \ldots, u_p) = \varphi(W_1, \ldots, W_p, \overline{Z}_1, \ldots, \overline{Z}_q),$$

for any $\varphi \in F^p\Omega^{p+q}(M)$, $Z_j \in T_{1,0}(M)$, $1 \leq j \leq q$, and $u_i \in \hat{T}(M)$, $1 \leq i \leq p$. Here $W_i \in T(M) \otimes \mathbf{C}$ is a lift of u_i, i.e., $\pi(W_i) = u_i$. The definition doesn't depend on the choice of lifts W_i of u_i because $\varphi \in F^p\Omega^{p+q}(M)$. Note that

$$0 \to F^{p+1}\Omega^{p+q}(M) \hookrightarrow F^p\Omega^{p+q}(M) \xrightarrow{f^{p,q}} \Lambda^q T_{0,1}(M)^* \otimes E^p \to 0 \qquad (1.137)$$

is a short exact sequence of complex vector bundles and complex vector bundle homomorphisms. It is not difficult to check that $\mathrm{Ker}(f^{p,q}) = F^{p+1}\Omega^{p+q}(M)$. However, the surjectivity of $f^{p,q}$ is somewhat trickier. Let $\psi \in \Lambda^q T_{0,1}(M)^* \otimes E^p$. To build a $\varphi \in F^p\Omega^{p+q}(M)$ such that $f^{p,q}(\varphi) = \psi$ we choose just any complement F of $T_{0,1}(M)$ in $T(M) \otimes \mathbf{C}$ (e.g., if M is a nondegenerate CR manifold of hypersurface type and T the characteristic direction of (M, θ), then we set $F = T_{1,0}(M) \oplus \mathbf{C}T$) and let $\pi_{0,1} : T(M) \otimes \mathbf{C} \to T_{0,1}(M)$ be the natural projection (associated with the direct sum decomposition $T(M) \otimes \mathbf{C} = T_{0,1}(M) \oplus F$). Next, we define $\varphi \in \Omega^{p+q}(M)$ by setting

$$\varphi(W_1, \dots, W_{p+q}) =$$
$$\frac{(-1)^{pq}}{p!q!} \sum_{\sigma \in \sigma_{p+q}} \epsilon(\sigma)\psi(\pi_{0,1} W_{\sigma(1)}, \dots, \pi_{0,1} W_{\sigma(q)})(\pi W_{\sigma(q+1)}, \dots, \pi W_{\sigma(q+p)}),$$

for any $W_j \in T(M) \otimes \mathbf{C}$, $1 \leq j \leq p+q$. Clearly φ so defined is an element of $\Gamma^\infty(F^p \Omega^{p+q}(M))$. For the convenience of the reader we compute $f^{p,q}(\varphi)$ explicitly. Given $f \in \sigma_p$ and $g \in \sigma_q$ let $\sigma(g, f) \in \sigma_{p+q}$ be defined by

$$\sigma(g, f) = \begin{pmatrix} 1 & \cdots & q & q+1 \cdots & q+p \\ p+g(1) & \cdots & p+g(q) & f(1) \ \cdots & f(p) \end{pmatrix}.$$

Then $\epsilon(\sigma(g, f)) = (-1)^{pq}\epsilon(f)\epsilon(g)$. Finally, one may perform the calculation

$$\varphi(W_1, \dots, W_p, \overline{Z}_1, \dots, \overline{Z}_q) =$$
$$\frac{(-1)^{pq}}{p!q!} \sum_{f \in \sigma_p,\, g \in \sigma_q} \epsilon(\sigma(g, f))\psi(\pi_{0,1} W_{p+g(1)}, \dots, \pi_{0,1} W_{p+g(q)})(\pi W_{f(1)}, \dots, \pi W_{f(p)})$$
$$= \psi(\overline{Z}_1, \dots, \overline{Z}_q)(\pi W_1, \dots, \pi W_p),$$

where $W_{p+j} = \overline{Z}_j$, $1 \leq j \leq q$. □

Finally, from (1.137) we have

$$\Lambda^q T_{0,1}(M)^* \otimes E^p = \mathrm{Im}(f^{p,q}) \simeq \frac{F^p \Omega^{p+q}(M)}{\mathrm{Ker}(f^{p,q})} = \Lambda^{p,q}(M). \qquad \square$$

Let $g^{p,q} : \Lambda^{p,q}(M) \to \Lambda^q T_{0,1}(M)^* \otimes E^p$ be the isomorphism furnished by the preceding proposition. If $f : E \to F$ is a complex vector bundle homomorphism then we denote by $f_\sharp : \Gamma^\infty(E) \to \Gamma^\infty(F)$ the induced map of sections. Consider the diagram

$$\begin{array}{ccc} \Omega^{p,q}(M) & \xrightarrow{g_\sharp^{p,q}} & \Gamma^\infty(\Lambda^q T_{0,1}(M)^* \otimes E^p) \\ \downarrow \overline{\partial}_b^{p,q} & & \downarrow \overline{\partial}_{E^p}^q \\ \Omega^{p,q+1}(M) & \xrightarrow{g_\sharp^{p,q+1}} & \Gamma^\infty(\Lambda^{q+1} T_{0,1}(M)^* \otimes E^p) \end{array}$$

Then

$$\overline{\partial}_{E^p}^q \circ g_\sharp^{p,q} = (-1)^p g_\sharp^{p,q+1} \circ \overline{\partial}_b^{p,q}. \tag{1.138}$$

The proof of (1.138) is left as an exercise to the reader.

Corollary 1.2. (N. Tanaka [398])

$$H^{p,q}(M) \simeq H^q(M, E^p)$$

(an isomorphism).

Proof. By (1.138) the map $g_\sharp^{p,q}$ induces a map $(g^{p,q})_*$ on cohomology. Finally, we inspect the diagram

$$\begin{array}{ccc} H^{p,q}(M) & \simeq & H^q\left(\Omega^{p,\cdot}(M), \overline{\partial}_b^{p,\cdot}\right) \\ & & \downarrow (g^{p,q})_* \\ H^q(M, E^p) = & H^q & \left(\Lambda^\cdot T_{0,1}(M)^* \otimes E^p, \overline{\partial}_{E^p}^\cdot\right). \end{array}$$

□

Let $(E, \overline{\partial}_E)$ be a CR-holomorphic vector bundle (over the CR manifold $(M, T_{1,0}(M))$).

Definition 1.44. *A cross-section $u \in \Gamma^\infty(E)$ is* CR-holomorphic *if $\overline{\partial}_E u = 0$.* □

The purpose of the present section is to establish the following result:

Proposition 1.21.

$$E_1^{k,0}(M) \simeq \operatorname{Ker}(\overline{\partial}_{E^k}), \quad H_0^k(M) \simeq H^k(E_1^{\cdot,0}(M), d_1^{\cdot,0}).$$

That is, $E_1^{k,0}(M)$ is isomorphic to the space of CR-holomorphic sections in $E^p = \Lambda^p \hat{T}(M)$ and $H_0^k(M)$ are the cohomology groups of the complex $\{E_1^{k,0}(M), d_1^{k,0}\}_{k\geq 0}$ (where $d_1^{k,0}$ is naturally induced by d).

Proof. Since $F^k\Omega^{k-1}(M) = 0$ it follows that

$$E_1^{k,0}(M) = \Gamma^\infty\left(F^k\Omega^k(M)\right) \cap d^{-1}\Gamma^\infty\left(F^{k+1}\Omega^{k+1}(M)\right).$$

Hence

$$d : \Gamma^\infty(F^k\Omega^k(M)) \to \Gamma^\infty(F^k\Omega^{k+1}(M))$$

descends to a differential operator

$$d_1^{k,0} : E_1^{k,0}(M) \to E_1^{k+1,0}(M)$$

such that

$$d_1^{k+1,0} \circ d_1^{k,0} = 0.$$

Note that $E_1^{k,0}(M)$ consists of all $\varphi \in \Gamma^\infty(\Omega^k(M) \otimes \mathbf{C})$ with

$$\varphi(W_1, \ldots, W_{k-1}, \overline{Z}) = 0$$

and

$$(d\varphi)(W_1, \ldots, W_k, \overline{Z}) = 0$$

for any $W_i \in T(M) \otimes \mathbf{C}$, $1 \leq i \leq k$, and any $Z \in T_{1,0}(M)$. Let us define

$$\Phi : \operatorname{Ker}(\overline{\partial}_{E^k}) \to E_1^{k,0}(M)$$

by setting

$$(\Phi\psi)(W_1, \dots, W_k) = \psi(\pi W_1, \dots, \pi W_k).$$

Since

$$(\Phi\psi)(W_1, \dots, W_{k-1}, \overline{Z}) = 0,$$
$$(\Phi\psi)(W_1, \dots, W_k, \overline{Z}) = (-1)^k \left((\overline{\partial}_{E^k}\psi)\overline{Z}\right)(\pi W_1, \dots, \pi W_k) = 0,$$

it follows that $\Phi\psi$ is well defined. The reader may check that Φ is an isomorphism with the (obvious) inverse

$$(\Phi^{-1}\varphi)(u_1, \dots, u_k) = \varphi(W_1, \dots, W_k),$$

where $W_i \in T(M) \otimes \mathbf{C}$ is some lift of $u_i \in \hat{T}(M)$, $1 \le i \le k$.

To prove the second statement in the proposition, note that

$$H_0^k(M) = E_1^{k,0}(M) = \frac{C_2^{k,0}(M)}{C_1^{k+1,-1}(M) + D_1^{k,0}(M)},$$
$$C_2^{k,0}(M) = \{\varphi \in \Gamma^\infty(F^k\Omega^k(M)) : d\varphi = 0\},$$
$$C_1^{k+1,-1}(M) + D_1^{k,0}(M) = \left(\Gamma^\infty(F^k\Omega^k(M))\right) \cap d\Gamma^\infty(F^{k-1}\Omega^{k-1}(M)).$$

Also

$$H^k(E_1^{\cdot,0}(M), d_1^{\cdot,0}) = \frac{\operatorname{Ker}(d_1^{k,0})}{d_1^{k-1,0}E_1^{k-1,0}(M)}.$$

Let $\varphi \in \Gamma^\infty(F^k\Omega^k(M))$ be such that $d\varphi = 0$. The reader may check that

$$[\varphi]_{H_0^k(M)} \longmapsto \varphi + d_1^{k-1,0}E_1^{k-1,0}(M)$$

is a (well-defined) isomorphism. □

1.7.4 A long exact sequence

Let $d^{p,q}$ be the restriction of d to $\Gamma^\infty(F^p\Omega^{p+q}(M))$. Since

$$d\,\Gamma^\infty(F^p\Omega^{p+q}(M)) \subset \Gamma^\infty(F^p\Omega^{p+q+1}(M))$$

it follows that $\{\Gamma^\infty(F^p\Omega^{p+\cdot}(M)), d^{p,\cdot}\}$ is a cochain complex. Let us set

$$H_*^{p,q}(M) = H^q\left(\Gamma^\infty(F^p\Omega^{p+\cdot}(M)), d^{p,\cdot}\right).$$

Proposition 1.22. (N. Tanaka [398])
There is a natural exact sequence of cohomology groups

$$
\begin{aligned}
0 \to H_0^k(M) \to H_*^{k-1,1}(M) \to H^{k-1,1}(M) \to H_*^{k,1}(M) \to \\
\to H_*^{k-1,2}(M) \to \cdots \to H_*^{k,p}(M) \to \\
\to H_*^{k-1,q+1}(M) \to H^{k-1,q+1}(M) \to H_*^{k,q+1}(M) \to \cdots
\end{aligned}
$$

Proof. Consider the short exact sequence

$$0 \to F^k\Omega^{k+p}(M) \to F^{k-1}\Omega^{k+q}(M) \to \Omega^{k-1,q+1}(M) \to 0,$$

for any $q \geq 0$. An inspection of the diagram

$$
\begin{array}{ccccc}
 & 0 & & 0 & \\
 & \downarrow & & \downarrow & \\
\cdots \to & \Gamma^\infty\left(F^k\Omega^{k+q}(M)\right) & \overset{d^{k,q}}{\longrightarrow} & \Gamma^\infty\left(F^k\Omega^{k+q+1}(M)\right) & \to \cdots \\
 & \downarrow & & \downarrow & \\
\cdots \to & \Gamma^\infty\left(F^{k-1}\Omega^{k+q}(M)\right) & \overset{d^{k-1,q+1}}{\longrightarrow} & \Gamma^\infty\left(F^{k-1}\Omega^{k+q+1}(M)\right) & \to \cdots \\
 & \downarrow & & \downarrow & \\
\cdots \to & \Gamma^\infty\left(\Omega^{k-1,q+1}(M)\right) & \overset{\overline{\partial}_b^{k-1,q+1}}{\longrightarrow} & \Gamma^\infty\left(\Omega^{k-1,q+2}(M)\right) & \to \cdots \\
 & \downarrow & & \downarrow & \\
 & 0 & & 0 &
\end{array}
$$

furnishes the cohomology sequence

$$H_*^{k,q}(M) \to H_*^{k-1,q+1}(M) \to H^{k-1,q+1}(M) \overset{\delta^{k-1,q+1}}{\longrightarrow} H_*^{k,q+1}(M),$$

where all but the connection homomorphism $\delta^{k-1,q+1}$ are naturally induced maps. As to $\delta^{k-1,q+1}$, it is defined by the following considerations. Let $\varphi + F^k\Omega^{k+q} \in \text{Ker}(\overline{\partial}_b^{k-1,q+1})$. Then

$$0 = \overline{\partial}_b^{k-1,q+1}\left(\varphi + F^k\Omega^{k+q}\right) = d\varphi + F^k\Omega^{k+q+1};$$

hence $d\varphi \in \Gamma^\infty\left(F^k\Omega^{k+q+1}(M)\right)$. Therefore the class $[d\varphi]_{H_*^{k,q+1}(M)}$ of $d\varphi$ (modulo $d^{k,q}\Gamma^\infty\left(F^k\Omega^{k+q}(M)\right)$) is well defined. Note that in general $d\varphi$ is not $d^{k,q}$-exact (because one only has $\varphi \in \Gamma^\infty(F^{k-1}\Omega^{k+q}(M))$, rather than $\varphi \in \Gamma^\infty(F^k\Omega^{k+q}(M))$). Finally, we set

$$\delta^{k-1,q+1} : [\varphi + F^k\Omega^{k+q}]_{H^{k-1,q+1}(M)} \longmapsto [d\varphi]_{H_*^{k,q+1}(M)}.$$

It is a standard exercise in homological algebra that the definition doesn't depend on the choice of representative. For $q = 0$ one obtains the sequence

$$H_*^{k,0}(M) \to H_*^{k-1,1}(M) \to H^{k-1,1}(M) \overset{\delta^{k-1,1}}{\longrightarrow} H_*^{k,1}(M) \to \cdots.$$

This is exact at all terms, except at $H_*^{k,0}(M)$. Finally, the sequence in the statement of the proposition may be obtained from the identities

$$H_*^{k,0}(M) = \{\varphi \in \Gamma^\infty\left(F^k\Omega^k(M)\right) : d\varphi = 0\},$$
$$\operatorname{Ker}\left(H_*^{k,0}(M) \to H_*^{k-1,1}(M)\right) = D_1^{k,0}(M),$$
$$H_0^k(M) \simeq H^k\left(E_1^{\cdot,0}(M), d_1^{\cdot,0}\right) = \frac{H_*^{k,0}(M)}{D_1^{k,0}(M)}. \qquad \square$$

1.7.5 Bott obstructions

Let $(M, T_{1,0}M))$ be a CR manifold of type (n, k). Recall the (CR holomorphic) vector bundle $\hat{T}(M) = (T(M) \otimes \mathbf{C})/T_{0,1}(M)$.

Definition 1.45. A connection D in $\hat{T}(M)$ is said to be *basic* if it extends the $\overline{\partial}_{\hat{T}(M)}$ operator, that is,

$$\overline{Z} \rfloor D\pi W = \pi[\overline{Z}, W]$$

for any $Z \in T_{1,0}(M)$ and any $W \in T(M) \otimes \mathbf{C}$. $\square$

For instance, if $(M, T_{1,0}(M))$ is a nondegenerate CR manifold of hypersurface type and $\theta \in \Omega^1(M)$ is a fixed pseudo-Hermitian structure on M then let T be the characteristic direction of $d\theta$ and $\sigma : \hat{T}(M) \to T_{1,0}(M) \oplus \mathbf{C}T$ the natural isomorphism (associated with the decomposition $T(M) \otimes \mathbf{C} = T_{1,0}(M) \oplus T_{0,1}(M) \oplus \mathbf{C}T$). Let ∇ be the Tanaka–Webster connection of (M, θ). Then the connection D in $\hat{T}(M)$ given by

$$D_V u = \begin{cases} \pi[V, \sigma(u)] & \text{if } V \in T_{0,1}(M), \\ \pi\nabla_V\sigma(u) & \text{if } V \in T_{1,0}(M) \oplus \mathbf{C}T, \end{cases}$$

is clearly basic. The reader may note the analogy with the transverse Levi-Civita connection of a Riemannian foliation (cf. (5.3) in [408], p. 48).

Let D be a connection in $\hat{T}(M)$ and

$$K(D) : \hat{T}(M) \to [\Lambda^2 T^*(M) \otimes \mathbf{C}] \otimes \hat{T}(M)$$

its curvature 2-form. The aim of the present section is to establish the following result:

Theorem 1.18. *Let M be a nondegenerate CR manifold (of hypersurface type) of CR dimension n and θ an arbitrary pseudo-Hermitian structure on M. Let D be a basic connection in $\hat{T}(M)$. Then*

$$K(D) \equiv 0 \bmod \theta^\alpha,\ \theta$$

for any admissible frame $\{\theta^1, \ldots, \theta^n\}$ of $T_{1,0}(M)$ on U.

Proof. As we shall see shortly, this is similar to the P. Baum and R. Bott vanishing theorem (cf. (0.51) in [76], p. 287). To prove Theorem 1.18, let $\{\theta^\alpha\}$ be an admissible, i.e., $\theta^\alpha(T_\beta) = \delta^\alpha_\beta$, $\theta^\alpha(T_{\bar\beta}) = \theta^\alpha(T) = 0$, frame of $T_{1,0}(M)^*$ on U. Next, let $I(U)$ be the ideal of $\Omega^*(U)$ (the de Rham algebra of U) spanned by $\{\theta^1, \dots, \theta^n, \theta\}$. We organize the proof in several steps, as follows.

Step 1 $dI(U) \subseteq I(U)$. To prove Step 1, let $\eta = \sum_{j=0}^n \theta^j \wedge \eta_j$, where $\theta^0 = \theta$. Then (by (1.64))

$$d\eta = \sum_{j=0}^{n} \left(d\theta^j \wedge \eta_j - \theta^j \wedge d\eta_j \right) \equiv$$

$$\equiv 2i h_{\alpha\bar\beta}\theta^\alpha \wedge \theta^{\bar\beta} \wedge \eta_0 + (\theta^\beta \wedge \omega^\alpha_\beta + \theta \wedge \tau^\alpha) \wedge \eta_\alpha \equiv 0.$$

All congruence relations are mod $I(U)$. Let D be a basic connection in $\hat T(M)$. Note that $\{\pi T_1, \dots, \pi T_n, \pi T\}$ is a (local) frame of $\hat T(M)$ on U. The corresponding connection 1-forms γ^i_j of D are given by

$$D\pi T_j = \gamma^i_j \otimes \pi T_i .$$

Here $0 \le i, j \le n$ and $T_0 = T$.

Step 2 $\gamma^0_0 \equiv 0, \ \ \gamma^0_\alpha \equiv 2i h_{\alpha\bar\beta}\theta^{\bar\beta}, \ \ \gamma^\alpha_0 \equiv A^\alpha_{\bar\beta}\theta^{\bar\beta}, \ \ \gamma^\alpha_\mu \equiv \Gamma^\alpha_{\bar\beta\mu}\theta^{\bar\beta}.$

Since D extends the $\overline\partial_{\hat T(M)}$ operator we have

$$\gamma^i_j(\overline Z)\pi T_i = \pi[\overline Z, T_j],$$

for any $Z \in T_{1,0}(M)$. Then (by (1.37))

$$\gamma^0_\alpha(\overline Z) = 2i L_\theta(T_\alpha, \overline Z),$$
$$\gamma^\beta_\alpha(\overline Z) = \nabla_{\overline Z} T_\alpha,$$
$$\gamma^0_0(\overline Z) = 0, \quad \gamma^\beta_0(\overline Z) T_\beta = \iota(\overline Z),$$

and Step 2 is proved. Let K^i_j be the curvature 2-forms of D with respect to $\{\pi T_j\}$, that is,

$$K(D)\pi T_j = K^i_j \otimes \pi T_i .$$

Then

$$K^i_j = d\gamma^i_j + \gamma^i_k \wedge \gamma^k_j .$$

It remains to prove that $K^i_j \equiv 0$. Using Steps 1 and 2 and the symmetry of $A_{\alpha\beta}$ (cf. the proof of Lemma 1.4) we may perform the calculation

$$K_0^0 = d\gamma_0^0 + \gamma_k^0 \wedge \gamma_0^k \equiv \gamma_\alpha^0 \wedge \gamma_0^\alpha \equiv 2ih_{\alpha\bar{\beta}}\theta^{\bar{\beta}} \wedge A_{\bar{\mu}}^\alpha \theta^{\bar{\mu}} = 2iA_{\bar{\beta}\bar{\mu}}\theta^{\bar{\beta}} \wedge \theta^{\bar{\mu}} = 0,$$

hence $K_0^0 \equiv 0$. Next, by $\nabla g_\theta = 0$ and (1.64), and again by Steps 1 and 2, we have

$$\begin{aligned}
K_\alpha^0 = d\gamma_\alpha^0 + \gamma_k^0 \wedge \gamma_\alpha^k &\equiv 2id(h_{\alpha\bar{\beta}}\theta^{\bar{\beta}}) + \gamma_\beta^0 \wedge \gamma_\alpha^\beta \\
&\equiv 2i\{dh_{\alpha\bar{\beta}} \wedge \theta^{\bar{\beta}} + h_{\alpha\bar{\beta}} d\theta^{\bar{\beta}}\} + 2ih_{\beta\bar{\lambda}}\theta^{\bar{\lambda}} \wedge \Gamma_{\bar{\mu}\alpha}^\beta \theta^{\bar{\mu}} \\
= 2i\{\left(h_{\alpha\bar{\mu}}\omega_{\bar{\beta}}^{\bar{\mu}} + \omega_\alpha^\mu h_{\mu\bar{\beta}}\right) \wedge \theta^{\bar{\beta}} &+ h_{\alpha\bar{\beta}}(\theta^{\bar{\mu}} \wedge \omega_{\bar{\mu}}^{\bar{\beta}} + \theta \wedge \tau^{\bar{\beta}})\} + 2ih_{\beta\bar{\lambda}}\Gamma_{\bar{\mu}\alpha}^\beta \theta^{\bar{\lambda}} \wedge \theta^{\bar{\mu}} \\
\equiv 2i\{(h_{\alpha\bar{\mu}}\Gamma_{\bar{\rho}\bar{\beta}}^{\bar{\mu}}\theta^{\bar{\rho}} + h_{\mu\bar{\beta}}\Gamma_{\bar{\rho}\alpha}^{\mu}\theta^{\bar{\rho}}) \wedge \theta^{\bar{\beta}} &+ h_{\alpha\bar{\beta}}\theta^{\bar{\mu}} \wedge \Gamma_{\bar{\rho}\bar{\mu}}^{\bar{\beta}}\theta^{\bar{\rho}} + h_{\beta\bar{\lambda}}\Gamma_{\bar{\mu}\alpha}^\beta \theta^{\bar{\lambda}} \wedge \theta^{\bar{\mu}}\} = 0;
\end{aligned}$$

hence $K_\alpha^0 \equiv 0$. Similarly

$$\begin{aligned}
K_0^\alpha = d\gamma_0^\alpha + \gamma_k^0 \wedge \gamma_0^k &\equiv d(A_{\bar{\beta}}^\alpha \theta^{\bar{\beta}}) + \gamma_\beta^\alpha \wedge \gamma_0^\beta \\
&\equiv (dA_{\bar{\beta}}^\alpha) \wedge \theta^{\bar{\beta}} + A_{\bar{\beta}}^\alpha d\theta^{\bar{\beta}} + \Gamma_{\bar{\lambda}\beta}^\alpha \theta^{\bar{\lambda}} \wedge A_{\bar{\mu}}^\beta \theta^{\bar{\mu}} \\
&\equiv T_{\bar{\rho}}(A_{\bar{\beta}}^\alpha)\theta^{\bar{\rho}} \wedge \theta^{\bar{\beta}} + A_{\bar{\beta}}^\alpha(\theta^{\bar{\rho}} \wedge \omega_{\bar{\rho}}^{\bar{\beta}} + \theta \wedge \tau^{\bar{\beta}}) + \Gamma_{\bar{\lambda}\beta}^\alpha A_{\bar{\mu}}^\beta \theta^{\bar{\lambda}} \wedge \theta^{\bar{\mu}} \\
&\equiv \{T_{\bar{\rho}}(A_{\bar{\beta}}^\alpha) + +\Gamma_{\bar{\rho}\lambda}^\alpha A_{\bar{\beta}}^\lambda - A_{\bar{\lambda}}^\alpha \Gamma_{\bar{\rho}\bar{\beta}}^{\bar{\lambda}}\}\theta^{\bar{\rho}} \wedge \theta^{\bar{\beta}} = h^{\alpha\bar{\gamma}}\left(\nabla_{\bar{\rho}} A_{\bar{\beta}\bar{\gamma}}\right)\theta^{\bar{\rho}} \wedge \theta^{\bar{\beta}}.
\end{aligned}$$

Yet

$$\nabla_\alpha A_{\beta\gamma} = \nabla_\beta A_{\alpha\gamma}; \tag{1.139}$$

hence $K_0^\alpha \equiv 0$. The proof of (1.139) is left as an exercise to the reader. See also (2.6) in [270], p. 163. Finally (by (1.89)),

$$\begin{aligned}
K_\beta^\alpha = d\gamma_\beta^\alpha + \gamma_k^\alpha \wedge \gamma_\beta^k &\equiv d(\Gamma_{\bar{\mu}\beta}^\alpha \theta^{\bar{\mu}}) + \gamma_0^\alpha \wedge \gamma_\beta^0 + \gamma_\mu^\alpha \wedge \gamma_\beta^\mu \\
&\equiv d\omega_\beta^\alpha + A_{\bar{\mu}}^\alpha \theta^{\bar{\mu}} \wedge 2ih_{\beta\bar{\lambda}}\theta^{\bar{\lambda}} + \Gamma_{\bar{\rho}\mu}^\alpha \theta^{\bar{\rho}} \wedge \Gamma_{\bar{\lambda}\beta}^\mu \theta^{\bar{\lambda}} \\
&\equiv d\omega_\beta^\alpha + \omega_\mu^\alpha \wedge \omega_\beta^\mu + 2ih_{\beta\bar{\lambda}}A_{\bar{\mu}}^\alpha \theta^{\bar{\mu}} \wedge \theta^{\bar{\lambda}} \\
&= \Pi_\beta^\alpha + 2i\tau^\alpha \wedge \theta_\beta \equiv \Omega_\beta^\alpha == R_\beta{}^\alpha{}_{\lambda\bar{\rho}} + W_{\beta\lambda}^\alpha \theta^\lambda \wedge \theta - W_{\beta\bar{\lambda}}^\alpha \theta^{\bar{\lambda}} \wedge \theta;
\end{aligned}$$

hence $K_\beta^\alpha \equiv 0$. Theorem 1.18 is completely proved. □

We close the section with the following remark. Let V be a complex manifold, of complex dimension n, and let F be an involutive, that is,

$$[\Gamma^\infty(F), \Gamma^\infty(F)] \subseteq \Gamma^\infty(F),$$

complex subbundle of the holomorphic tangent bundle $T^{1,0}(V)$ of V. Assume $\dim_{\mathbf{C}} F_x = k$, $x \in V$. Moreover, let $\varphi \in \mathbf{C}[X_1, \dots, X_n]$ be a symmetric homogeneous polynomial of degree $\ell \leq n$. Let $\sigma_1, \dots, \sigma_n$ be the elementary symmetric functions of $X_1, \dots, X_n$. Clearly φ may be written as a function $\tilde{\varphi}(\sigma_1, \dots, \sigma_\ell)$ of the first ℓ elementary symmetric functions. Define $\varphi(T^{1,0}(V)/F) \in H^*(V, \mathbf{C})$ by setting

$$\varphi(T^{1,0}(V)/F) = \tilde{\varphi}(c_1(T^{1,0}(V)/F), \ldots, c_\ell(T^{1,0}(V)/F)),$$

where $c_j(T^{1,0}(V)/F)$ are the Chern classes of $T^{1,0}(V)/F$. Then the P. Baum and R. Bott vanishing theorem is that

$$\varphi(T^{1,0}(V)/F) = 0, \tag{1.140}$$

for any symmetric homogeneous polynomial $\varphi \in \mathbf{C}[X_1, \ldots, X_n]$ of degree $n - k < \ell \leq n$. Let (U, z^α) be local complex coordinates on V such that $F_{|U}$ is the span of $\partial/\partial z^1, \ldots, \partial/\partial z^k$. Let $I(U, F)$ be the ideal of $\Omega^*(U)\otimes\mathbf{C}$ spanned by $dz^{k+1}, \ldots, dz^n$. Let $\pi : T^{1,0}(V) \to T^{1,0}(V)/F$ be the projection. Let K^α_β be the curvature forms of a basic (in the sense of [76], p. 295) connection in $T^{1,0}(V)/F$, with respect to the local frame $\{\pi\,\partial/\partial z^{k+1}, \ldots, \pi\,\partial/\partial z^n\}$. The proof of (1.140) is to show that $K^\alpha_\beta \in I(U, F)$ (hence, if the degree of φ is large enough (i.e., $\ell > n - k$) the characteristic form[9] $\tilde{\varphi}(\sigma_1(K), \ldots, \sigma_\ell(K))$ vanishes because of $\omega_1, \ldots, \omega_{n-k+1} \in I(U, F) \Longrightarrow \omega_1 \wedge \cdots \wedge \omega_{n-k+1} = 0$). In this respect, the P. Baum and R. Bott vanishing theorem is similar to our Theorem 1.18. However, Theorem 1.18 is the best one can get, for one may not interpret Theorem 1.18 as a vanishing of $\varphi(\hat{T}(M)) = 0$ for some φ. Indeed, if $\varphi \in \mathbf{C}[X_1, \ldots, X_n]$ is a symmetric homogeneous polynomial with $\deg(\varphi) \leq n$, then one has only

$$\omega_1, \ldots, \omega_{n+2} \in I(U) \Longrightarrow \omega_1 \wedge \cdots \wedge \omega_{n+2} = 0; \tag{1.141}$$

hence the degree of φ should be at least $n + 2$ (to allow the use of (1.141)). Yet, if this is the case, the characteristic form vanishes in its own right:

$$\tilde{\varphi}(\sigma_1(K), \ldots, \sigma_n(K)) = 0$$

(because it has degree $\geq 2n + 2$ and $\dim_{\mathbf{R}} M = 2n + 1$). □

1.7.6 The Kohn–Rossi Laplacian

Let $(M, T_{1,0}(M))$ be a strictly pseudoconvex CR manifold, of CR dimension n. Let θ be a choice of pseudo-Hermitian structure on M such that the Levi form L_θ is positive definite. Let $\{T_\alpha\}$ be a local frame of $T_{1,0}(M)$ and $\{\theta^\alpha\}$ the corresponding (admissible) dual frame. Given $\varphi, \psi \in \Omega^{0,q}(M)$ we define a pointwise inner product $\langle\varphi, \psi\rangle_\theta$ by setting

$$\langle\varphi, \psi\rangle_\theta = \frac{1}{q!}\varphi_{\overline{\alpha}_1\cdots\overline{\alpha}_q}\psi^{\overline{\alpha}_1\cdots\overline{\alpha}_q},$$

[9] Let D be a connection in $T^{1,0}(V)/F$ and $K(D)$ its curvature form. Define $\sigma_j(K)$, $1 \leq j \leq n$, by $\det(I_{n-k} + tK) = 1 + t\sigma_1(K) + \cdots + t^{n-k}\sigma_{n-k}(K)$ and $\sigma_a(K) = 0$, $n - k < a \leq n$. Patching together the local forms $(i/(2\pi))^\ell\tilde{\varphi}(\sigma_1(K), \ldots, \sigma_\ell(K))$ gives rise to a globally defined form $(i/(2\pi))^\ell\tilde{\varphi}(\sigma_1(K(D)), \ldots, \sigma_\ell(K(D)))$ representing the cohomology class $\varphi(T^{1,0}(V)/F)$.

where

$$\varphi = \frac{1}{q!}\varphi_{\overline{\alpha}_1\cdots\overline{\alpha}_q}\theta^{\overline{\alpha}_1}\wedge\cdots\wedge\theta^{\overline{\alpha}_q},$$

$$\psi = \frac{1}{q!}\psi_{\overline{\alpha}_1\cdots\overline{\alpha}_q}\theta^{\overline{\alpha}_1}\wedge\cdots\wedge\theta^{\overline{\alpha}_q},$$

$$\psi^{\overline{\alpha}_1\cdots\overline{\alpha}_q} = \overline{\psi^{\alpha_1\cdots\alpha_q}}, \quad \psi^{\alpha_1\cdots\alpha_q} = h^{\alpha_1\overline{\beta}_1}\cdots h^{\alpha_q\overline{\beta}_q}\psi_{\overline{\beta}_1\cdots\overline{\beta}_q}.$$

As usual, we define an L^2 inner product by setting

$$(\varphi, \psi)_\theta = \int_M \langle\varphi, \psi\rangle_\theta\ \theta\wedge(d\theta)^n$$

for any $(0, q)$-forms φ, ψ on M (at least one of compact support). The formal adjoint $\overline{\partial}_b^*$ of the tangential Cauchy–Riemann operator $\overline{\partial}_b$ is given by

$$\left(\overline{\partial}_b^*\psi, \varphi\right)_\theta = \left(\psi, \overline{\partial}_b\varphi\right)_\theta$$

for any $(0, q)$-form φ, respectively any $(0, q+1)$-form ψ on M. We shall need the following definition:

Definition 1.46. The *Kohn–Rossi Laplacian* $\Box_b$ given by

$$\Box_b = \overline{\partial}_b^*\overline{\partial}_b + \overline{\partial}_b\overline{\partial}_b^*. \qquad \Box$$

The main purpose of the present section is to express $\Box_b\varphi$ locally in terms of the covariant derivatives of φ (with respect to the Tanaka–Webster connection ∇ of (M, θ)). To compute $\overline{\partial}_b^*\psi$ we recall the local expression (1.135) of $\overline{\partial}_b\varphi$. Therefore, we may perform the calculation

$$\begin{aligned}\left(\overline{\partial}_b^*\psi, \varphi\right)_\theta &= \left(\psi, \overline{\partial}_b\varphi\right)_\theta = \int_M \langle\psi, \overline{\partial}_b\varphi\rangle_\theta\ \theta\wedge(d\theta)^n \\ &= \frac{1}{(q+1)!}\int_M \psi^{\alpha_1\cdots\alpha_{q+1}}(\overline{\partial}_b\varphi)_{\alpha_1\cdots\alpha_{q+1}}\ \theta\wedge(d\theta)^n \\ &= \frac{1}{q+1}\sum_{i=1}^{q+1}(-1)^{i-1}\int_M \psi^{\alpha_1\cdots\alpha_{q+1}}\nabla_{\alpha_i}\varphi_{\alpha_1\cdots\hat{\alpha}_i\cdots\alpha_{q+1}}\ \theta\wedge(d\theta)^n;\end{aligned}$$

hence (by (1.133)),

$$\begin{aligned}\left(\overline{\partial}_b^*\psi, \varphi\right)_\theta = \frac{1}{(q+1)!}\sum_{i=1}^{q+1}(-1)^{i-1}\int_M \psi^{\alpha_1\cdots\alpha_{q+1}}\{T_{\alpha_i}\left(\varphi_{\alpha_1\cdots\hat{\alpha}_i\cdots\alpha_{q+1}}\right) \\ - \sum_{j<i}\varphi_{\alpha_1\cdots\alpha_{j-1}\mu\alpha_{j+1}\cdots\hat{\alpha}_i\cdots\alpha_{q+1}}\Gamma^\mu_{\alpha_i\alpha_j} \\ - \sum_{i<j}\varphi_{\alpha_1\cdots\hat{\alpha}_i\cdots\alpha_{j-1}\mu\alpha_{j+1}\cdots\alpha_{q+1}}\Gamma^\mu_{\alpha_i\alpha_j}\}\ \theta\wedge(d\theta)^n.\end{aligned}$$

Let us define $Z_i \in T_{1,0}(M)$, $1 \leq i, j \leq q+1$, by setting

$$Z_i = \psi^{\alpha_1 \cdots \alpha_{q+1}} \varphi_{\alpha_1 \cdots \hat{\alpha}_i \cdots \alpha_{q+1}} T_{\alpha_i}.$$

As we shall see

$$\operatorname{div}(Z_i) = T_\alpha(Z_i^\alpha) + Z_i^\beta \Gamma^\alpha_{\alpha\beta}.$$

By Green's lemma

$$\int_M \operatorname{div}(Z_i)\, \theta \wedge (d\theta)^n = 0$$

hence

$$\begin{aligned}\left(\overline{\partial}_b^* \psi, \varphi\right)_\theta = \frac{1}{(q+1)!} \sum_{i=1}^{q+1} (-1)^i \int_M \Big\{ & \psi^{\alpha_1 \cdots \alpha_{q+1}} \varphi_{\alpha_1 \cdots \hat{\alpha}_i \cdots \alpha_{q+1}} \Gamma^\mu_{\mu\alpha_i} \\ & + \sum_{j<i} \psi^{\alpha_1 \cdots \alpha_{q+1}} \varphi_{\alpha_1 \cdots \alpha_{j-1} \mu \alpha_{j+1} \cdots \hat{\alpha}_i \cdots \alpha_{q+1}} \Gamma^\mu_{\alpha_i \alpha_j} \\ + \sum_{i<j} \psi^{\alpha_1 \cdots \alpha_{q+1}} \varphi_{\alpha_1 \cdots \hat{\alpha}_i \cdots \alpha_{j-1} \mu \alpha_{j+1} \cdots \alpha_{q+1}} \Gamma^\mu_{\alpha_i \alpha_j} & + \varphi_{\alpha_1 \cdots \hat{\alpha}_i \cdots \alpha_{q+1}} T_{\alpha_i}\left(\psi^{\alpha_1 \cdots \alpha_{q+1}}\right) \Big\} \\ = \frac{1}{(q+1)!} \sum_{i=1}^{q+1} (-1)^i \int_M \Big\{ & T_{\alpha_i}\left(\psi^{\alpha_1 \cdots \alpha_{q+1}}\right) \\ & + \sum_{j<i} \psi^{\alpha_1 \cdots \alpha_{j-1} \mu \alpha_{j+1} \cdots \alpha_{q+1}} \Gamma^{\alpha_j}_{\alpha_i \mu} + \psi^{\alpha_1 \cdots \alpha_{q+1}} \Gamma^\mu_{\mu \alpha_i} \\ & + \sum_{i<j} \psi^{\alpha_1 \cdots \alpha_{j-1} \mu \alpha_{j+1} \cdots \alpha_{q+1}} \Gamma^{\alpha_j}_{\alpha_i \mu} \Big\} \varphi_{\alpha_1 \cdots \hat{\alpha}_i \cdots \alpha_{q+1}} \theta \wedge (d\theta)^n.\end{aligned}$$

At this point we define the $(0, q)$-forms A_i, $1 \leq i \leq q+1$, by setting

$$\begin{aligned} A_i &= \frac{1}{q!} (A_i)_{\overline{\alpha}_1 \cdots \overline{\alpha}_q}\, \theta^{\overline{\alpha}_1} \wedge \cdots \wedge \theta^{\overline{\alpha}_q}, \\ (A_i)_{\overline{\alpha}_1 \cdots \overline{\alpha}_q} &- h_{\overline{\alpha}_1 \beta_1} \cdots h_{\overline{\alpha}_q \beta_q} A_i^{\beta_1 \cdots \beta_q}, \\ A_i^{\alpha_1 \cdots \hat{\alpha}_i \cdots \alpha_{q+1}} &= T_{\alpha_i}\left(\psi^{\alpha_1 \cdots \alpha_{q+1}}\right) + \sum_{j=1}^{q+1} \psi^{\alpha_1 \cdots \alpha_{j-1} \mu \alpha_{j+1} \cdots \alpha_{q+1}} \Gamma^{\alpha_j}_{\alpha_i \mu}. \end{aligned}$$

Therefore

$$\begin{aligned} \left(\overline{\partial}_b^* \psi, \varphi\right)_\theta &= \frac{1}{(q+1)!} \sum_{i=1}^{q+1} (-1)^i \int_M A_i^{\alpha_1 \cdots \hat{\alpha}_i \cdots \alpha_{q+1}} \varphi_{\alpha_1 \cdots \hat{\alpha}_i \cdots \alpha_{q+1}}\, \theta \wedge (d\theta)^n \\ &= \frac{1}{q+1} \sum_{i=1}^{q+1} (-1)^i \int_M \langle A_i, \varphi \rangle_\theta\, \theta \wedge (d\theta)^n, \end{aligned}$$

so that

$$\overline{\partial}_b^* \psi = \frac{1}{q+1} \sum_{i=1}^{q+1} (-1)^i A_i .$$

On the other hand, using the identities

$$dh_{\alpha\overline{\beta}} = h_{\alpha\overline{\gamma}} \omega_{\overline{\beta}}^{\overline{\gamma}} + \omega_\alpha^\gamma h_{\gamma\overline{\beta}}, \quad T_\rho(h^{\overline{\gamma}\mu}) = -h^{\overline{\beta}\mu} h^{\alpha\overline{\gamma}} T_\rho(h_{\alpha\overline{\beta}}),$$

and the expression (1.133) of the covariant derivative of a $(0, q)$-form, one may show that

$$A_i^{\alpha_1 \cdots \hat{\alpha}_i \cdots \alpha_{q+1}} = (q+1)!\, h^{\alpha_1 \overline{\beta}_1} \cdots h^{\alpha_{q+1} \overline{\beta}_{q+1}} (\nabla_{\alpha_i} \psi_{\overline{\beta}_1 \cdots \overline{\beta}_{q+1}});$$

hence

$$\overline{\partial}_b^* \psi = \sum_{i=1}^{q+1} (-1)^i h^{\alpha\overline{\beta}} (\nabla_\alpha \psi_{\overline{\mu}_1 \cdots \overline{\mu}_{i-1} \overline{\beta} \overline{\mu}_i \cdots \overline{\mu}_q}) \theta^{\overline{\mu}_1} \wedge \cdots \wedge \theta^{\overline{\mu}_q},$$

or (by the skew-symmetry of $\psi_{\overline{\alpha}_1 \cdots \overline{\alpha}_{q+1}}$)

$$\overline{\partial}_b^* \psi = (-1)^{q+1} (q+1)\, h^{\lambda\overline{\mu}} (\nabla_\lambda \psi_{\overline{\alpha}_1 \cdots \overline{\alpha}_q \overline{\mu}}) \theta^{\overline{\alpha}_1} \wedge \cdots \wedge \theta^{\overline{\alpha}_q} \tag{1.142}$$

for any $(0, q+1)$-form ψ on M. For simplicity, set $\varphi = \overline{\partial}_b^* \psi$. We wish to compute $\overline{\partial}_b \varphi$. Again by (1.135) we have

$$\overline{\partial}_b \varphi = \frac{1}{q+1} \sum_{i=1}^{q+1} (-1)^{i-1} (\nabla_{\overline{\alpha}_i} \varphi_{\overline{\alpha}_1 \cdots \hat{\overline{\alpha}}_i \cdots \overline{\alpha}_{q+1}}) \theta^{\overline{\alpha}_1} \wedge \cdots \wedge \theta^{\overline{\alpha}_{q+1}},$$

where (by (1.142))

$$\varphi_{\overline{\alpha}_1 \cdots \overline{\alpha}_q} = (-1)^{q+1} (q+1)!\, h^{\lambda\overline{\mu}} \nabla_\lambda \psi_{\overline{\alpha}_1 \cdots \overline{\alpha}_q \overline{\mu}}.$$

Hence we need to compute the covariant derivatives of φ. To this end we define

$$\nabla_{\overline{\mu}} \nabla_\lambda \psi_{\overline{\alpha}_1 \cdots \overline{\alpha}_{q+1}} = \left(\nabla_{T_{\overline{\mu}}} (\nabla \psi)\right) (T_\lambda, T_{\overline{\alpha}_1}, \ldots, T_{\overline{\alpha}_{q+1}})$$

for any $(0, q+1)$-form ψ on M. Explicitly

$$\nabla_{\overline{\mu}} \nabla_\lambda \psi_{\overline{\alpha}_1 \cdots \overline{\alpha}_{q+1}} =$$

$$T_{\overline{\mu}} (\nabla_\lambda \psi_{\overline{\alpha}_1 \cdots \overline{\alpha}_{q+1}}) - \Gamma_{\overline{\mu}\lambda}^\rho \nabla_\rho \psi_{\overline{\alpha}_1 \cdots \overline{\alpha}_{q+1}} - \sum_{j=1}^{q+1} \Gamma_{\overline{\mu}\overline{\alpha}_j}^{\overline{\rho}} \nabla_\lambda \psi_{\overline{\alpha}_1 \cdots \overline{\alpha}_{j-1} \overline{\rho} \overline{\alpha}_{j+1} \cdots \overline{\alpha}_{q+1}} .$$

Using this identity, we obtain

$$\nabla_{\overline{\alpha}_i} \varphi_{\overline{\alpha}_1 \cdots \hat{\overline{\alpha}}_i \cdots \overline{\alpha}_{q+1}} = (-1)^{q+1} (q+1)\, h^{\lambda\overline{\mu}} \nabla_{\overline{\alpha}_i} \nabla_\lambda \psi_{\overline{\alpha}_1 \cdots \hat{\overline{\alpha}}_i \cdots \overline{\alpha}_{q+1} \overline{\mu}}$$

and therefore

$$\overline{\partial}_b\overline{\partial}_b^*\psi = -\sum_{i=1}^{q+1} h^{\lambda\overline{\mu}}\nabla_{\overline{\alpha}_i}\nabla_\lambda\psi_{\overline{\alpha}_1\cdots\bar{\alpha}_{i-1}\bar{\mu}\bar{\alpha}_{i+1}\cdots\overline{\alpha}_{q+1}}\theta^{\overline{\alpha}_1}\wedge\cdots\wedge\theta^{\overline{\alpha}_{q+1}}. \qquad (1.143)$$

To compute $\overline{\partial}_b^*\overline{\partial}_b\psi$ we start from

$$\overline{\partial}_b^*\overline{\partial}_b\psi = (-1)^{q+2}(q+2)\,h^{\lambda\overline{\mu}}(\nabla_\lambda(\overline{\partial}_b\psi)_{\overline{\alpha}_1\cdots\overline{\alpha}_{q+1}\overline{\mu}})\theta^{\overline{\alpha}_1}\wedge\cdots\wedge\theta^{\overline{\alpha}_{q+1}},$$

$$(\overline{\partial}_b\psi)_{\overline{\alpha}_1\cdots\overline{\alpha}_{q+2}} = (q+1)!\sum_{i=1}^{q+2}(-1)^{i-1}\nabla_{\overline{\alpha}_i}\psi_{\overline{\alpha}_1\cdots\hat{\overline{\alpha}}_i\cdots\overline{\alpha}_{q+2}}.$$

First, we obtain

$$\nabla_\lambda(\overline{\partial}_b\psi)_{\overline{\alpha}_1\cdots\overline{\alpha}_{q+2}} = \frac{1}{q+2}\sum_{i=1}^{q+2}(-1)^{i-1}\nabla_\lambda\nabla_{\overline{\alpha}_i}\psi_{\overline{\alpha}_1\cdots\hat{\overline{\alpha}}_i\cdots\overline{\alpha}_{q+2}}. \qquad (1.144)$$

The proof of (1.144) is a rather lengthy computation in multilinear algebra. For the more pedantic reader, we give some of the details below. Let us start with

$$\nabla_\lambda(\overline{\partial}_b\psi)_{\overline{\alpha}_1\cdots\overline{\alpha}_{q+2}} = \frac{1}{(q+2)!}\left\{T_\lambda\big((\overline{\partial}_b\psi)_{\overline{\alpha}_1\cdots\overline{\alpha}_{q+2}}\big) - \sum_{j=1}^{q+2}(\overline{\partial}_b\psi)_{\overline{\alpha}_1\cdots\bar{\alpha}_{j-1}\bar{\mu}\bar{\alpha}_{j+1}\cdots\overline{\alpha}_{q+2}}\Gamma^{\overline{\mu}}_{\lambda\overline{\alpha}_j}\right\}$$

and define the multi-index

$$(\overline{\beta}_{j1},\ldots,\overline{\beta}_{j,q+2}) = (\overline{\alpha}_1,\ldots,\overline{\alpha}_{j-1},\overline{\mu},\overline{\alpha}_{j+1},\ldots,\overline{\alpha}_{q+2}).$$

Then

$$\begin{aligned}
&\nabla_\lambda(\overline{\partial}_b\psi)_{\overline{\alpha}_1\cdots\overline{\alpha}_{q+2}}\\
&= \frac{1}{q+2}\sum_{i=1}^{q+2}(-1)^{i-1}\left\{T_\lambda\big(\nabla_{\overline{\alpha}_i}\psi_{\overline{\alpha}_1\cdots\hat{\overline{\alpha}}_i\cdots\overline{\alpha}_{q+2}}\big) - \sum_{j=1}^{q+2}\big(\nabla_{\overline{\beta}_{ji}}\psi_{\overline{\beta}_{j1}\cdots\hat{\overline{\beta}}_{ji}\cdots\overline{\beta}_{j,q+2}}\big)\Gamma^{\mu}_{\lambda\overline{\alpha}_j}\right\}\\
&= \frac{1}{q+2}\sum_{i=1}^{q+2}(-1)^{i-1}\Big\{T_\lambda\big(\nabla_{\overline{\alpha}_i}\psi_{\overline{\alpha}_1\cdots\hat{\overline{\alpha}}_i\cdots\overline{\alpha}_{q+2}}\big)\\
&\quad - \sum_{j<i}\big(\nabla_{\overline{\alpha}_i}\psi_{\overline{\alpha}_1\cdots\bar{\alpha}_{j-1}\bar{\mu}\bar{\alpha}_{j+1}\cdots\hat{\overline{\alpha}}_i\cdots\overline{\alpha}_{q+1}}\big)\Gamma^{\overline{\mu}}_{\lambda\overline{\alpha}_j} - \big(\nabla_{\overline{\mu}}\psi_{\overline{\alpha}_1\cdots\hat{\overline{\alpha}}_i\cdots\overline{\alpha}_{q+1}}\big)\Gamma^{\overline{\mu}}_{\lambda\overline{\alpha}_i}\\
&\quad - \sum_{i<j}\big(\nabla_{\overline{\alpha}_i}\psi_{\overline{\alpha}_1\cdots\hat{\overline{\alpha}}_i\cdots\bar{\alpha}_{j-1}\bar{\mu}\bar{\alpha}_{j+1}\cdots\overline{\alpha}_{q+1}}\big)\Gamma^{\overline{\mu}}_{\lambda\overline{\alpha}_j}\Big\}\\
&= \frac{1}{q+2}\sum_{i=1}^{q+2}(-1)^{i-1}\nabla_\lambda\nabla_{\overline{\alpha}_i}\psi_{\overline{\alpha}_1\cdots\hat{\overline{\alpha}}_i\cdots\overline{\alpha}_{q+2}}
\end{aligned}$$

and (1.144) is proved. □

We have obtained

$$\overline{\partial}_b^* \overline{\partial}_b \psi = h^{\lambda\overline{\mu}} \left\{ \sum_{i=1}^{q+1} \nabla_\lambda \nabla_{\overline{\alpha}_i} \psi_{\overline{\alpha}_1 \cdots \overline{\alpha}_{i-1} \overline{\mu} \overline{\alpha}_{i+1} \cdots \overline{\alpha}_{q+1}} - \nabla_\lambda \nabla_{\overline{\mu}} \psi_{\overline{\alpha}_1 \cdots \overline{\alpha}_{q+1}} \right\} \theta^{\overline{\alpha}_1} \wedge \cdots \wedge \theta^{\overline{\alpha}_{q+1}}. \tag{1.145}$$

At this point, using (1.143) and (1.145) we get

$$\Box_b \psi = \frac{1}{(q+1)!} (\Box_b \psi)_{\overline{\alpha}_1 \cdots \overline{\alpha}_{q+1}} \theta^{\overline{\alpha}_1} \wedge \cdots \theta^{\overline{\alpha}_{q+1}},$$

where

$$\frac{1}{(q+1)!} (\Box_b \psi)_{\overline{\alpha}_1 \cdots \overline{\alpha}_{q+1}} + h^{\lambda\overline{\mu}} \nabla_\lambda \nabla_{\overline{\mu}} \psi_{\overline{\alpha}_1 \cdots \overline{\alpha}_{q+1}}$$
$$= \sum_{i=1}^{q+1} h^{\lambda\overline{\mu}} \left\{ \nabla_\lambda \nabla_{\overline{\alpha}_i} \psi_{\overline{\alpha}_1 \cdots \overline{\alpha}_{i-1} \overline{\mu} \overline{\alpha}_{i+1} \cdots \overline{\alpha}_{q+1}} - \nabla_{\overline{\alpha}_i} \nabla_\lambda \psi_{\overline{\alpha}_1 \cdots \overline{\alpha}_{i-1} \overline{\mu} \overline{\alpha}_{i+1} \cdots \overline{\alpha}_{q+1}} \right\}.$$

Of course, as in Riemannian geometry (cf., e.g., [178], p. 78) the presence of the term $\nabla_\lambda \nabla_{\overline{\alpha}_i} \psi_{\overline{\alpha}_1 \cdots \overline{\alpha}_{i-1} \overline{\mu} \overline{\alpha}_{i+1} \cdots \overline{\alpha}_{q+1}} - \nabla_{\overline{\alpha}_i} \nabla_\lambda \psi_{\overline{\alpha}_1 \cdots \overline{\alpha}_{i-1} \overline{\mu} \overline{\alpha}_{i+1} \cdots \overline{\alpha}_{q+1}}$ in the above identity is an indication that we may express $\Box_b \psi$ in terms of the curvature of the Tanaka–Webster connection of (M, θ). Using the identities

$$[T_\lambda, T_{\overline{\mu}}] = \Gamma^{\overline{\rho}}_{\lambda\overline{\mu}} T_{\overline{\rho}} - \Gamma^{\rho}_{\overline{\mu}\lambda} T_\rho - 2i h_{\lambda\overline{\mu}} T,$$
$$R_{\overline{\alpha}}{}^{\overline{\gamma}}{}_{\lambda\overline{\mu}} = T_\lambda(\Gamma^{\overline{\gamma}}_{\overline{\mu}\overline{\alpha}}) - T_{\overline{\mu}}(\Gamma^{\overline{\gamma}}_{\lambda\overline{\alpha}}) + \Gamma^{\overline{\rho}}_{\overline{\mu}\overline{\alpha}} \Gamma^{\overline{\gamma}}_{\lambda\overline{\rho}} - \Gamma^{\overline{\rho}}_{\lambda\overline{\alpha}} \Gamma^{\overline{\gamma}}_{\overline{\mu}\overline{\rho}}$$
$$- \Gamma^{\overline{\rho}}_{\lambda\overline{\mu}} \Gamma^{\overline{\gamma}}_{\overline{\rho}\overline{\alpha}} + \Gamma^{\rho}_{\overline{\mu}\lambda} \Gamma^{\overline{\gamma}}_{\rho\overline{\alpha}} + 2i h_{\lambda\overline{\mu}} \Gamma^{\overline{\gamma}}_{0\overline{\alpha}},$$

after some unenlightening calculations, which we omit, we obtain

$$q! \left(\nabla_\lambda \nabla_{\overline{\mu}} \varphi_{\overline{\alpha}_1 \cdots \overline{\alpha}_q} - \nabla_{\overline{\mu}} \nabla_\lambda \varphi_{\overline{\alpha}_1 \cdots \overline{\alpha}_q} \right)$$
$$= -2i\, q!\, h_{\lambda\overline{\mu}} \nabla_0 \varphi_{\overline{\alpha}_1 \cdots \overline{\alpha}_q} - \sum_{j=1}^{q} \varphi_{\overline{\alpha}_1 \cdots \overline{\alpha}_{j-1} \overline{\gamma} \overline{\alpha}_{j+1} \cdots \overline{\alpha}_q} R_{\overline{\alpha}_j}{}^{\overline{\gamma}}{}_{\lambda\overline{\mu}}$$

for any $(0, q)$-form φ on M. Finally, one may observe that

$$h^{\lambda\overline{\beta}} R_{\overline{\beta}}{}^{\overline{\gamma}}{}_{\lambda\overline{\mu}} = -R^{\overline{\gamma}}{}_{\overline{\mu}},$$

where

$$R^{\overline{\gamma}}{}_{\overline{\mu}} = h^{\alpha\overline{\gamma}} R_{\alpha\overline{\mu}},$$

to obtain the sought-after expression of the Kohn–Rossi Laplacian of (M, θ).

Theorem 1.19. *Let M be a strictly pseudoconvex CR manifold. Let θ be such that L_θ is positive definite. Then*

$$\Box_b \psi = \Big\{ - h^{\lambda\overline{\mu}} \nabla_\lambda \nabla_{\overline{\mu}} \psi_{\overline{\alpha}_1 \cdots \overline{\alpha}_{q+1}} - 2i(q+1) \nabla_0 \psi_{\overline{\alpha}_1 \cdots \overline{\alpha}_{q+1}}$$

$$+ \frac{1}{(q+1)!} \Big[\sum_{i=1}^{q+1} \psi_{\overline{\alpha}_1 \cdots \overline{\alpha}_{i-1} \overline{\gamma} \overline{\alpha}_{i+1} \cdots \overline{\alpha}_{q+1}} R^{\overline{\gamma}}{}_{\overline{\alpha}_i}$$

$$- \sum_{i=1}^{q+1} \Big(\sum_{j<i} R_{\overline{\alpha}_j}{}^{\overline{\gamma}\overline{\mu}}{}_{\overline{\alpha}_i} \psi_{\overline{\alpha}_1 \cdots \overline{\alpha}_{j-1} \overline{\gamma} \overline{\alpha}_{j+1} \cdots \overline{\alpha}_{i-1} \overline{\mu} \overline{\alpha}_{i+1} \cdots \overline{\alpha}_{q+1}}$$

$$+ \sum_{j>i} R_{\overline{\alpha}_j}{}^{\overline{\gamma}\overline{\mu}}{}_{\overline{\alpha}_i} \psi_{\overline{\alpha}_1 \cdots \overline{\alpha}_{i-1} \overline{\mu} \overline{\alpha}_{i+1} \cdots \overline{\alpha}_{j-1} \overline{\gamma} \overline{\alpha}_{j+1} \cdots \overline{\alpha}_{q+1}} \Big) \Big] \Big\} \theta^{\overline{\alpha}_1} \wedge \cdots \wedge \theta^{\overline{\alpha}_{q+1}} \tag{1.146}$$

for any $(0, q+1)$-form ψ on M.

The reader should compare our (1.146) with its Riemannian counterpart (2.12.4) in [178], p. 78. A combination of (2.12.4) in [178], Green's lemma, and the fundamental theorem[10] in [178], p. 76, is known to lead to the classical relationship[11] among curvature and homology (cf. Theorem 3.2.4 in [178], p. 88). It should be mentioned that although a CR analogue (in terms of Kohn–Rossi cohomology groups and $\overline{\partial}_b$-harmonic forms) of the fundamental theorem above has been known for quite some time (cf. J.J. Kohn [246]) similar applications (based on our (1.146)) seem to be unknown as yet. The very explicit identity (1.146) doesn't seem to appear anywhere in the literature (apparently only approximate formulas such as (A.5) in Appendix A, modulo error terms, have been derived).

In particular, for $q = 0$ the identity (1.146) becomes

$$\Box_b \psi = \left\{ -h^{\lambda\overline{\mu}} \nabla_\lambda \nabla_{\overline{\mu}} \psi_{\overline{\alpha}} - 2i \nabla_0 \psi_{\overline{\alpha}} + \psi_{\overline{\gamma}} R^{\overline{\gamma}}{}_{\overline{\alpha}} \right\} \theta^{\overline{\alpha}}$$

for any $\psi \in \Omega^{0,1}(M)$. Also, if $f : M \to \mathbf{C}$ is some C^∞ function then

$$\Box_b f = \overline{\partial}_b^* \overline{\partial}_b f = h^{\lambda\overline{\mu}} \nabla_\lambda f_{\overline{\mu}},$$

where $f_{\overline{\mu}} = T_{\overline{\mu}}(f)$. Note the presence of the second-order differential operator $h^{\lambda\overline{\mu}} \nabla_\lambda \nabla_{\overline{\mu}}$ in the expression of $\Box_b$. Since we wish $h^{\lambda\overline{\mu}}$ to be positive definite, it is only natural to employ a *Riemannian* background metric to define our L^2 inner product on $(0, q)$-forms and the corresponding formal adjoint $\overline{\partial}_b^*$ of the tangential Cauchy–Riemann operator. Hence, as shown above, when $(M, T_{1,0}(M))$ is strictly pseudoconvex one may use the Webster metric g_θ (for some fixed choice of pseudo-Hermitian

[10] Let M be a compact orientable Riemannian manifold. Then the number of linearly independent real harmonic forms of degree p is equal to the pth Betti number of M.

[11] If on a compact and orientable Riemannian n-dimensional manifold M the quadratic form $F(\alpha) = R_{ij} \alpha^{i i_2 \cdots i_p} \alpha^j{}_{i_2 \cdots i_p} + \frac{p-1}{2} R_{ijk\ell} \alpha^{i j i_3 \cdots i_p} \alpha^{k\ell}{}_{i_3 \cdots i_p}$ is positive definite then $b_p(M) = 0$, $0 < p < n$.

structure θ such that the Levi form L_θ is positive definite). When $(M, T_{1,0}(M))$ is only nondegenerate of signature $(k, n-k)$ then (by Lemma 13.2 in [150], p. 469) there is a Hermitian form $\langle \ , \ \rangle$ on $T_{1,0}(M)$ such that for any $x \in M$ there is an open neighborhood U of x and a frame $\{T_1, \dots, T_n\}$ of $T_{1,0}(M)$ on U such that $\langle T_\alpha, T_\beta \rangle = \delta_{\alpha\beta}$ and $\langle T_\alpha, T_\beta \rangle_\theta = \epsilon_\alpha \delta_{\alpha\beta}$ (i.e., a Hermitian, hence positive definite, form $\langle \ , \ \rangle$ such that $\langle \ , \ \rangle$ and $\langle \ , \ \rangle_\theta$ may be simultaneously diagonalized). Here $\langle Z, W \rangle_\theta = L_\theta(Z, \overline{W})$ for any $Z, W \in T_{1,0}(M)$. Also $\epsilon_\alpha = 1$ if $1 \leq \alpha \leq k$ and $\epsilon_\alpha = -1$ if $k+1 \leq \alpha \leq n$. One then uses $\langle \ , \ \rangle$ to build $\overline{\partial}_b^*$.

Let $(M, T_{1,0}(M))$ be a CR manifold (of hypersurface type). A *Levi metric* is a semi-Riemannian metric g on M such that (1) $g(JX, JY) = g(X, Y)$ for any $X, Y \in H(M)$, and (2) there is a pseudo-Hermitian structure θ on M such that $g^*(\theta, \theta) = 1$ and $g^{\mathbf{C}} = L_\theta$ on $T_{1,0}(M) \otimes T_{0,1}(M)$. Here g^* is the naturally induced cometric on $T^*(M)$ and $g^{\mathbf{C}}$ is the $\mathbf{C}$-linear extension of g to $T(M) \otimes \mathbf{C}$. For any pseudo-Hermitian structure θ on a nondegenerate CR manifold M, the Webster metric g_θ is a Levi metric.

A Riemannian metric on a CR manifold M is *compatible with the CR structure* of M if $T_{1,0}(M)$ and $T_{0,1}(M)$ are orthogonal with respect to the Hermitian form in $T(M) \otimes \mathbf{C}$ that is induced by this metric (and J.J. Kohn [248], employs such a compatible metric to build the L^2 closure of $\overline{\partial}_b$ and its L^2 adjoint).

If g is a Levi metric and $h(A, B) = g^{\mathbf{C}}(A, \overline{B})$, for any $A, B \in T(M) \otimes \mathbf{C}$, then $T_{1,0}(M) \perp T_{0,1}(M)$ with respect to h. Hence any Levi metric is compatible with the CR structure of M. However, when M is only nondegenerate the converse is not true (for $k \neq n$). Hence it is unclear whether $\overline{\partial}_b^*$ and $\Box_b$ may be expressed in terms of covariant derivatives (with respect to the Tanaka–Webster connection ∇) in this more general setting (i.e., when a compatible Riemannian metric on a nondegenerate CR manifold (with $k \neq n$) replaces the use of the Webster metric g_θ). Indeed, the proof of (1.142) (and hence that of (1.146)) uses the fact that ∇ parallelizes g_θ (rather than the additional compatible Riemannian metric). On the other hand, at least a Levi metric is needed in order to build intrinsic canonical connections (similar to the Tanaka–Webster connection) on a CR manifold; cf. C.M. Stanton [378].

1.8 The group of CR automorphisms

We close Chapter 1 of this book by an informal discussion of the group $\mathrm{Aut}_{CR}(M)$ of all CR automorphisms of a CR manifold. A fundamental result is the following:

Theorem 1.20. (J.M. Lee [273])
Let M be a compact, connected, strictly pseudoconvex CR manifold of dimension $2n+1 \geq 3$. Then the identity component of $\mathrm{Aut}_{CR}(M)$ is compact in the compact-open topology unless M is globally CR equivalent to S^{2n+1} with its standard CR structure.

When $2n+1 \geq 5$ the result above follows from known results on biholomorphism groups of complex manifolds with boundary (actually the full CR automorphism group is compact unless M is globally CR equivalent to the sphere). Indeed any such M can be realized as the boundary of an analytic variety whose biholomorphism group is isomorphic to $\mathrm{Aut}_{CR}(M)$. Then, by results of B. Wong [434], J.P. Rosay [353], and

D. Burns and S. Shnider [79], it follows that any such variety with a noncompact biholomorphism group is biholomorphic to the unit ball, so M is CR equivalent to the sphere. However, abstract 3-dimensional CR manifolds aren't in general realizable as boundaries, so the arguments above fail to apply. Previous to Theorem 1.20 it was known only that a compact, connected, strictly pseudoconvex CR manifold M with $\text{Aut}_{CR}(M)$ noncompact is locally CR equivalent to a sphere (cf. S. Webster [423]). By taking into account the analogy between conformal and CR geometry, J.M. Lee conjectured (cf. op. cit.) that for any connected, strictly pseudoconvex CR manifold M the full automorphism group $\text{Aut}_{CR}(M)$ acts properly[12] unless M is CR equivalent to either S^{2n+1} or $\mathbf{H}_n$. When M is compact the properness of the action implies the compactness of $\text{Aut}_{CR}(M)$. When M is noncompact it may be shown that $\text{Aut}_{CR}(M)$ acts properly if and only if it preserves some pseudo-Hermitian structure. The conjecture of J.M. Lee was later[13] proved by R. Schoen [366]. The final result is our next theorem:

Theorem 1.21. (R. Schoen [366])
The CR automorphism group of a strictly pseudoconvex CR manifold M acts properly unless M is globally CR equivalent to the sphere or the Heisenberg group with the standard CR structure.

The proof of Theorem 1.20 is imitative of that of results in conformal geometry by M. Obata [330], and J. Lafontaine [265]. It was actually shown by S.M. Webster [423], that except for one step the arguments in conformal geometry carry over easily to the CR case. S.M. Webster's result is (cf. op. cit.) that if M is a compact, connected, strictly pseudoconvex CR manifold that is locally CR isomorphic to the sphere, and for which there is a closed noncompact 1-parameter subgroup $G_1 \subset \text{Aut}_{CR}(M)$ with a fixed point, then M must be globally CR isomorphic to the sphere. The contribution of J.M. Lee (cf. op. cit.) is precisely to show that a closed noncompact 1-parameter group of $\text{Aut}_{CR}(M)$ has a fixed point.

The paper by R. Schoen (cf. op. cit.) establishes first a result in conformal geometry: the conformal automorphism group of a Riemannian manifold M acts properly unless M is conformally diffeomorphic to the sphere or the Euclidean space with the standard metric. R. Schoen provides a PDE-theoretic proof[14] that can be adapted easily to give Theorem 1.21.

Bochner-type formulas for operators related to CR automorphisms and spherical CR structures were obtained by J-H. Cheng [95]. Such formulas allow conclusions about rigidity of 3-dimensional CR manifolds. Let M be a 3-dimensional strictly pseudoconvex CR manifold and let $\text{Aut}_0(M)$ be the identity component of $\text{Aut}_{CR}(M)$. One of the main results in [95] is as follows:

---

12 A Lie group G acts *properly* on the manifold X if the map $G \times X \to X \times X$ given by $(g, x) \mapsto (g \cot x, x)$ is proper.

13 J.M. Lee's paper [273] was published in 1996 yet was available in preprint form since 1994.

14 Elliptic estimates for the (elliptic) equation satisfied by the *dilation factor* of a conformal diffeomorphism f (the factor by which f stretches lengths).

Theorem 1.22. (J-H. Cheng [95])
Let θ be a contact form on M such that L_θ is positive definite. If $\rho < 0$ and $\sqrt{3}T(\rho) - 2\mathrm{Im}(A_{11,\bar{1}\bar{1}}) > 0$ then $\mathrm{Aut}_0(M)$ *consists only of the identity automorphism. If $\rho < 0$ and $\sqrt{3}T(\rho) - 2\mathrm{Im}(A_{11,\bar{1}\bar{1}}) = 0$ then* $\mathrm{Aut}_0(M)$ *has dimension ≤ 1.*

A spherical CR structure J_b is *rigid* if for any smooth 1-parameter family $\{J_b^{(t)}\}$ of spherical CR structures on M with $J_b^{(0)} = J_b$ one has

$$\frac{d}{dt}\{J_b^{(t)}\}_{t=0} = \mathcal{L}_X J_b$$

for some tangent vector field X on M. Another result in [95] is the following:

Theorem 1.23. (J-H. Cheng [95])
Let M be a closed spherical 3-dimensional CR manifold. Assume that there is a contact form θ on M such that $\rho > 0$ and

$$\frac{3}{8}\rho^2 - 2\rho|A_{11,\bar{1}\bar{1}}|^{2/3} - 25|A_{11}|^2 > 0,$$

$$\begin{aligned}\Big(\frac{3}{8}\rho^2 - 2\rho\big|A_{11,\bar{1}\bar{1}}\big|^{2/3} - 25|A_{11}|^2\Big)\Big\{&\frac{83}{3456}\rho^2 + \frac{55}{1152}\Delta_b\rho \\ &+ \frac{5}{4}|A_{11}|^2 - \frac{5}{36}|A_{11,1}|^{4/3} + \frac{5i}{9}(A_{11,\bar{1}\bar{1}} - A_{\bar{1}\bar{1},11})\Big\} \\ &- \frac{15}{8}\Big(\frac{1}{8}\rho - \frac{2}{3}|A_{11,1}|^{2/3}\Big)\Big|\frac{5}{48}T_{\bar{1}}(\rho) - 2iA_{\bar{1}\bar{1},1}\Big|^2 > 0.\end{aligned}$$

Then the CR structure is rigid.

2

The Fefferman Metric

Let $\Omega \subset \mathbf{C}^{n+1}$ be a smoothly bounded strictly pseudoconvex domain. The *Fefferman metric* is a Lorentz metric on $(\partial\Omega) \times S^1$, originally discovered by C. Fefferman [138], in connection with the boundary behavior of the Bergman kernel $K(z, w)$ of Ω. This is the metric g induced on $(\partial\Omega) \times S^1$, as $z \to \partial\Omega$, by the nondegenerate complex $(0, 2)$-tensor field on $\Omega \times (\mathbf{C} \setminus \{0\})$:

$$\begin{gathered} G = \frac{\partial^2 H}{\partial z^A \partial \bar{z}^B} dz^A \odot d\bar{z}^B, \\ H(z, \zeta) = |\zeta|^{2/(n+1)} u(z), \quad u(z) = K(z, z)^{-1/(n+1)}. \end{gathered} \tag{2.1}$$

Except from certain particular domains, such as $\Omega = \{(z, w) \in \mathbf{C}^n \times \mathbf{C}^m : \|z\|^2 + \|w\|^{2p} < 1\}$ (for $p = 1$ this is the ball), cf. J.P. D'Angelo [111], or the complex ovals of G. Francsics and N. Hanges [156], the Bergman kernel cannot be computed explicitly, a fact that results in a lack of computability for g. As a remedy to this difficulty, C. Fefferman replaced (cf. op. cit.) $u(z)$ in (2.1) by a solution to the Dirichlet problem for the complex Monge–Ampére equation

$$\begin{cases} J(u) \equiv (-1)^{n+1} \det \begin{pmatrix} u & \partial u / \partial \bar{z}^k \\ \partial u / \partial z^j & \partial^2 u / \partial z^j \partial \bar{z}^k \end{pmatrix} = 1 & \text{in } \Omega, \\ u = 0 & \text{on } \partial\Omega, \end{cases} \tag{2.2}$$

and showed that (2.1) furnishes, in the limit $z \to \partial\Omega$, the same Lorentz metric g. The understanding of the fact that a solution $u(z)$ to (2.2) does the same job as $K(z, z)^{-1/(n+1)}$ was based on heuristic arguments, e.g., C. Fefferman's observation that a solution to $J(u) = 1$ transforms, under a biholomorphism of Ω, as a negative power of the Bergman kernel (cf. Proposition 2.10 below). Also, for the case of the ball, the Bergman kernel on the diagonal is $n!\pi^{-n}(1 - \|z\|^2)^{-(n+1)}$ and $1 - \|z\|^2$ is a solution to (2.2).

Of course, the interrelation between the Bergman kernel, the Dirichlet problem (2.2), and the geometric aspects associated with the metric g is interesting in itself. By a theorem of S.Y. Cheng and S.T. Yau [97], the solution $u(z)$ to (2.2) exists and is unique ($u \in C^\infty(\Omega) \cap C^{n+(3/2)-\epsilon}(\overline{\Omega})$). No explicit solution to (2.2) is available; yet, as shown by C. Fefferman, the 2-jet, along the boundary, of a solution is actually sufficient to produce g. The function $u(z) = u^{(2)}(z)$,

$$
\begin{aligned}
u^{(2)} &= \left\{ 1 + \frac{1 - J(u^{(1)})}{n+1} \right\} u^{(1)}, \\
u^{(1)} &= \frac{\psi}{J(\psi)^{1/(n+1)}}, \quad \Omega = \{\psi > 0\},
\end{aligned}
\tag{2.3}
$$

in his approximation scheme, leads to g and bypasses the original lack of computability of the solution to (2.2). However, even for explicit domains such as $\Omega_0 = \{\psi > 0\} \subset \mathbf{C}^2$, where

$$
\psi(z, w) = z + \overline{z} - w\overline{w} + i(z - \overline{z})w^2\overline{w}^2, \tag{2.4}
$$

the computation of g, and subsequently of its null geodesics, based on the approximate solution (2.3) to the solution of (2.2), turns out to be highly complicated (cf. [138], pp. 410–415). On the other hand, $\partial\Omega$ is a strictly pseudoconvex CR manifold and it is a natural question whether g may be computed in terms of pseudo-Hermitian invariants (of $\partial\Omega$). This is indeed the case, as demonstrated by F. Farris [137], and J.M. Lee [271], on whose work we report in this chapter.

The Fefferman metric F_θ, as rebuilt by J.M. Lee (cf. op. cit.), is well defined for each (abstract) strictly pseudoconvex CR manifold M, lives on the total space of the principal S^1-bundle

$$
C(M) := [\Lambda^{n+1,0}(M) \setminus 0]/\mathrm{GL}^+(1, \mathbf{R}) \xrightarrow{\pi} M,
$$

has a particularly simple expression (cf. also (2.28)–(2.29))

$$
\pi^*\tilde{G}_\theta + \frac{2}{n+2}(\pi^*\theta) \odot \left\{ d\gamma + \pi^*\left(i\omega_\alpha^\alpha - \frac{i}{2}h^{\alpha\overline{\beta}}dh_{\alpha\overline{\beta}} - \frac{\rho}{4(n+1)}\theta \right) \right\}, \tag{2.5}
$$

and its *restricted conformal class* $[F_\theta] = \{e^{u\circ\pi} F_\theta : u \in C^\infty(M)\}$ is a CR invariant (cf. Theorem 2.3). Also, it may be characterized (in terms of curvature restrictions) among all Lorentz metrics on $C(M)$ (cf. C.R. Graham [182], and also G. Sparling [377]) and its null geodesics project on S.S. Chern and J.K. Moser's *chains*; cf. [99] (a biholomorphically invariant system of curves in M) when $M = \partial\Omega$ (of course, $C(\partial\Omega) \simeq (\partial\Omega) \times S^1$ and F_θ given by (2.5) is the metric g originally defined by C. Fefferman, up to a conformal diffeomorphism). The fact that chains can be obtained as projections of null geodesics of g enabled C. Fefferman to show that the boundary $\partial\Omega_0$ of the domain (2.4) admits an infinite family of chains, which spiral in toward the origin (cf. [138]), and more recently L. Koch to give a simple proof (cf. [243]) of H. Jacobowitz's result that nearby points on a strictly pseudoconvex CR manifold can be joined by a chain (cf. [220]).

Among the most recent results on the Fefferman metric, it is known that the Pontryagin forms of g are CR invariants of M (cf. Theorem 2.6). Also, if $M \simeq S^{2n+1}$, a CR isomorphism, then the first Pontryagin form of M vanishes ($P_1(\Omega^2) = 0$) and the de Rham cohomology class of the corresponding transgression form has integer coefficients (cf. Theorem 2.7).

The Fefferman metric also leads to interesting new problems and geometric notions, such as the *CR Yamabe problem* (the Yamabe problem for the Fefferman metric), or the *pseudoharmonic maps* (coinciding locally with J. Jost and C.J. Xu's subelliptic harmonic maps [234]). Finally, we should mention the work by H. Baum [49], developing the spinor calculus in the presence of the Fefferman metric.

2.1 The sub-Laplacian

The scope of this section is to introduce the *sub-Laplacian* of a strictly pseudoconvex CR manifold, a second-order differential operator Δ_b in many ways similar to the Laplace–Beltrami operator of a Riemannian manifold. Δ_b is a degenerate elliptic (in the sense of J.M. Bony [73]) operator, which is subelliptic of order $1/2$ (in the sense of G.B. Folland [146]) and hence (by a result of L. Hörmander [213]) hypoelliptic. This is the main common feature enjoyed by Δ_b and the Laplacian of a Riemannian manifold. It led to a considerable development of the theory of second-order subelliptic equations on domains in $\mathbf{R}^n$, often by analogy with the elliptic theory; cf., e.g., G. Citti, N. Garofalo, and E. Lanconelli [104], N. Garofalo and E. Lanconelli [165], N. Garofalo, and D.M. Nhieu [166], D. Jerison and A. Sánchez-Calle [230], A. Parmeggiani [339], A. Parmeggiani and C.J. Xu [340], A. Sánchez-Calle [362, 363], C.J. Xu [438, 441], C.J. Xu and C. Zuily [442, 443]. The authors believe that subelliptic theory is bound to play within CR and pseudo-Hermitian geometry the strong role played by elliptic theory in Riemannian geometry.

Let $(M, T_{1,0}(M))$ be a strictly pseudoconvex CR manifold, of real dimension $2n+1$ and CR dimension n. Let θ be a pseudo-Hermitian structure and $\Psi = \theta \wedge (d\theta)^n$ the corresponding volume form. As in Chapter 1, if X is a vector field of class C^1 the divergence of X is meant with respect to Ψ, i.e.,

$$\mathcal{L}_X \Psi = \operatorname{div}(X)\, \Psi,$$

where $\mathcal{L}$ denotes the Lie derivative.

Definition 2.1. The *sub-Laplacian* is the differential operator Δ_b defined by

$$\Delta_b u = \operatorname{div}(\nabla^H u),$$

for any $u \in C^2(M)$. Here $\nabla^H u$ is the *horizontal gradient*, i.e., $\nabla^H u = \pi_H \nabla u$, where $\pi_H : T(M) \to H(M)$ is the natural projection associated with the direct sum decomposition $T(M) = H(M) \oplus \mathbf{R}T$ (T is the characteristic direction of (M, θ)). Also ∇u is the ordinary gradient of u with respect to the Webster metric, i.e., $g_\theta(\nabla u, X) = X(u)$, for any $X \in \mathcal{X}(M)$. □

Let ∇ be the Tanaka–Webster connection of (M, θ). Since $\nabla\theta = 0$ it follows that $\nabla\omega = 0$; hence we may compute the divergence of a vector field as in

$$\operatorname{div}(X) = \operatorname{trace}\{Y \in T(M) \mapsto \nabla_Y X \in T(M)\}.$$

This was explicitly proved in Chapter 1, though only for complex vector fields of type $(1, 0)$. Let $\{X_a : 1 \leq a \leq 2n\}$ be a local G_θ-orthonormal frame of $H(M)$ (i.e., $G_\theta(X_a, X_b) = \delta_{ab}$) defined on the open set $U \subseteq M$. Then

$$\Delta_b u = \operatorname{trace}\{Y \in T(M) \mapsto \nabla_Y(\nabla^H u)$$

(since $H(M)$ is parallel with respect to ∇)

$$= \operatorname{trace}\{X_a \mapsto \nabla_{X_a} \nabla^H u\} = \sum_{a=1}^{2n} G_\theta(\nabla_{X_a} \nabla^H u, X_a)$$

(since $\nabla g_\theta = 0$)

$$= \sum_a \{X_a(g_\theta(\nabla^H u, X_a)) - g_\theta(\nabla^H u, \nabla_{X_a} X_a)\},$$

or

$$\Delta_b u = \sum_{a=1}^{2n} \{X_a(X_a u) - (\nabla_{X_a} X_a)u\} \tag{2.6}$$

on U. Let $(U, x^1, \ldots, x^{2n+1})$ be a local coordinate system on M. We set $X_a = b_a^i \, \partial/\partial x^i$, where $b_a^i : U \to \mathbf{R}$ are C^∞ functions. Then

$$\sum_{a=1}^{2n} X_a X_a u = a^{ij} \frac{\partial^2 u}{\partial x^i \partial x^j} + \sum_{a=1}^{2n} b_a^i \frac{\partial b_a^j}{\partial x^i} \frac{\partial u}{\partial x^j},$$

$$\sum_{a=1}^{2n} (\nabla_{X_a} X_a) u = a^{ij} \Gamma_{ij}^k \frac{\partial u}{\partial x^k} + \sum_{a=1}^{2n} b_a^i \frac{\partial b_a^j}{\partial x^i} \frac{\partial u}{\partial x^j},$$

where we set by definition $a^{ij} = \sum_a b_a^i b_a^j$. Moreover, Γ^i_{jk} are the local coefficients of ∇ with respect to the local frame $\{\partial/\partial x^i : 1 \leq i \leq 2n+1\}$. Hence

$$\Delta_b u = a^{ij} \frac{\partial^2 u}{\partial x^i \partial x^j} - a^{ij} \Gamma_{ij}^k \frac{\partial u}{\partial x^k},$$

or

$$\Delta_b u = \frac{\partial}{\partial x^i} \left(a^{ij} \frac{\partial u}{\partial x^j} \right) + c^j \frac{\partial u}{\partial x^j}, \tag{2.7}$$

where we set

$$c^j = -\frac{\partial a^{ij}}{\partial x^i} - a^{ik} \Gamma_{ik}^j .$$

Let us observe that the matrix a^{ij} is symmetric and positive semidefinite; hence Δ_b is a *degenerate elliptic* operator (in the sense of M. Bony [73]). Let us consider the L^2 inner product

$$(u, v) = \int_M u\overline{v}\,\Psi,$$

where $u, v \in L^2(M)$. The formal adjoint X_a^* of X_a is given by

$$(X_a^* u, v) = (u, X_a v) = \int u\overline{X_a v}\Psi = \int u b_a^i \frac{\partial \overline{v}}{\partial x^i}\Psi$$

for any $u, v \in C_0^\infty(U)$. Then, by the well-known identity $\operatorname{div}(fX) = f\operatorname{div}(X) + X(f)$ it follows that

$$\begin{aligned}(X_a^* u, v) &= \int \left\{\frac{\partial}{\partial x^i}\left(u b_a^i \overline{v}\right) - \frac{\partial}{\partial x^i}\left(u b_a^i\right)\overline{v}\right\}\Psi \\ &= \int \left\{\operatorname{div}(u\overline{v}X_a) - u\overline{v}b_a^i \operatorname{div}\left(\frac{\partial}{\partial x^i}\right) - \frac{\partial}{\partial x^i}\left(u b_a^i\right)\overline{v}\right\}\Psi;\end{aligned}$$

hence by Green's lemma,

$$X_a^* u = -\frac{\partial}{\partial x^i}\left(b_a^i u\right) - b_a^i \operatorname{div}\left(\frac{\partial}{\partial x^i}\right)u$$

for any $u \in C_0^\infty(U)$. Let us observe that

$$\operatorname{div}(\partial_i) = \operatorname{trace}\{\partial_j \mapsto \nabla_{\partial_j}\partial_i\} = \Gamma^j_{ji},$$

where we set (for simplicity) $\partial_i = \partial/\partial x^i$. In the end

$$X_a^* u = -\frac{\partial}{\partial x^i}\left(b_a^i u\right) - b_a^j \Gamma^i_{ij} u. \tag{2.8}$$

Definition 2.2. The *Hörmander operator* is defined by

$$Hu = -\sum_{a=1}^{2n} X_a^* X_a u. \qquad \square$$

Let us substitute from (2.8). We obtain

$$Hu = \frac{\partial}{\partial x^i}\left(a^{ij}\frac{\partial u}{\partial x^j}\right) + a^{ij}\Gamma^k_{ki}\frac{\partial u}{\partial x^j}. \tag{2.9}$$

Let us write locally $T = b_0^i\,\partial/\partial x^i$. It is not difficult to check that $a^{ij} = g^{ij} - b_0^i b_0^j$, where $[g^{ij}] = [g_{ij}]^{-1}$ and $g_{ij} = g_\theta(\partial_i, \partial_j)$, and consequently $c^j = a^{ij}\Gamma^k_{ki}$. The verifications are left as an exercise to the reader. Therefore, if we compare (2.7) and (2.9) we come to

$$\Delta_b u = Hu \tag{2.10}$$

on U. Let us recall the following definition:

Definition 2.3. A formally self-adjoint second-order differential operator $L : C^\infty(M) \to C^\infty(M)$ is said to be *subelliptic* of order ϵ $(0 < \epsilon \leq 1)$ at a point $x \in M$ if there is an open neighborhood $U \subseteq M$ of x such that

$$\|u\|_\epsilon^2 \leq C\left(|(Lu, u)| + \|u\|^2\right),$$

for any $u \in C_0^\infty(U)$. Here $\|u\|_\epsilon$ is the Sobolev norm[1] of order ϵ. L is *subelliptic* (of order ϵ) if it is subelliptic (of order ϵ) at any $x \in M$. □

Definition 2.4. Let L be a differential operator with the formal adjoint L^*. If $T \in C_0^\infty(M)'$ is a distribution on M then the distribution LT is defined by $(LT)(\varphi) = T(L^*\varphi)$, for any $\varphi \in C_0^\infty(M)$. We say that L is *hypoelliptic* if $LT \in C^\infty(M)$ yields $T \in C^\infty(M)$. □

Theorem 2.1. *The sub-Laplacian Δ_b is subelliptic of order $1/2$. Consequently Δ_b is hypoelliptic, and more generally, for any $x \in M$ there is an open neighborhood U of x such that*

$$\|u\|_{s+1}^2 \leq C_s\left(\|\Delta_b u\|_s^2 + \|u\|^2\right), \quad s \geq 0,$$

for any $u \in C_0^\infty(U)$.

Proof. Let us start by checking that Δ_b is formally self-adjoint. For any $u, v \in C_0^\infty(M)$,

$$(\Delta_b^* u, v) = (u, \Delta_b v) = \int u \overline{\Delta_b v} \Psi = \int u \, \operatorname{div}(\nabla^H \overline{v}) \Psi$$
$$= \int \left\{ \operatorname{div}(u \nabla^H \overline{v}) - (\nabla^H \overline{v})(u) \right\} \Psi$$

(by Green's lemma)

$$= -\int g_\theta(\nabla u, \nabla^H \overline{v}) \Psi = -\int g_\theta(\nabla^H u, \nabla \overline{v}) \Psi = -\int (\nabla^H u)(\overline{v}) \Psi$$
$$= -\int \operatorname{div}(\overline{v} \nabla^H u) \Psi + \int \operatorname{div}(\nabla^H u) \, \overline{v} \Psi = (\Delta_b u, v).$$ □

We accept now (without proof) the following result:

Lemma 2.1. (E.V. Radkevic [351])
Let $K \subseteq M$ be a compact set and let $\{Z_1, \ldots, Z_N\}$ be a system of complex vector fields on M such that (i) the space spanned by $\{Z_1, \ldots, Z_N\}$ (over $\mathbf{C}$) is closed under complex conjugation and (ii) $\{Z_{1,x}, \ldots, Z_{N,x}\} \cup \{[Z_i, Z_j]_x : 1 \leq i, j \leq N\}$ span $T_x(M) \otimes_{\mathbf{R}} \mathbf{C}$, for any $x \in K$. Then there is a constant $C > 0$ such that

$$\|u\|_{1/2}^2 \leq C\Big(\sum_{j=1}^N \|Z_j u\|^2 + \|u\|^2\Big),$$

for any $u \in C_0^\infty(K)$.

[1] If $\hat{u}$ is the Fourier transform of $u \in C_0^\infty(\mathbf{R}^{2n+1})$ then $\|u\|_\epsilon^2 = \int (1 + |\xi|^2)^\epsilon |\hat{u}(\xi)|^2 \, d\xi$.

Let now $x \in M$ and let $\{T_1, \dots, T_n\}$ be a local orthonormal ($L_\theta(T_\alpha, T_{\overline{\beta}}) = \delta_{\alpha\beta}$) frame of $T_{1,0}(M)$, defined on an open neighborhood U of x. We set

$$X_\alpha = \frac{1}{\sqrt{2}}(T_\alpha + T_{\overline{\alpha}}), \quad X_{\alpha+n} = \frac{i}{\sqrt{2}}(T_\alpha - T_{\overline{\alpha}}).$$

If Z is a complex vector field on M then for $u, v \in C_0^\infty(M)$ one has

$$\begin{aligned}(Z^* u, v) = (u, Zv) &= \int u\overline{Zv} = \int \{\overline{Z}(u\overline{v}) - \overline{Z}(u)\overline{v}\} \\ &= \int \{\operatorname{div}(u\overline{v}\overline{Z}) - u\overline{v}\operatorname{div}(\overline{Z}) - \overline{Z}(u)\overline{v}\};\end{aligned}$$

hence

$$Z^* u = -\overline{Z}u - u \operatorname{div}(\overline{Z}).$$

Therefore

$$X_\alpha^* u = \frac{1}{\sqrt{2}}\left(T_\alpha^* u + T_{\overline{\alpha}}^* u\right), \quad X_{\alpha+n}^* u = -\frac{i}{\sqrt{2}}\left(T_\alpha^* u - T_{\overline{\alpha}}^* u\right).$$

Then (2.10) implies (on U)

$$\Delta_b u = -\sum_{\alpha=1}^{n}(T_\alpha^* T_\alpha + T_{\overline{\alpha}}^* T_{\overline{\alpha}})u. \tag{2.11}$$

The fields $\{T_\alpha, T_{\overline{\alpha}}\}$ satisfy the hypothesis of Lemma 2.1 because by the purity property of T_∇,

$$T_\nabla(T_\alpha, T_{\overline{\beta}}) = 2i\delta_{\alpha\beta}T,$$

it follows that

$$[T_\alpha, T_{\overline{\beta}}] = \Gamma_{\alpha\overline{\beta}}^{\overline{\gamma}} T_{\overline{\gamma}} - \Gamma_{\overline{\beta}\alpha}^{\gamma} T_\gamma - 2i\delta_{\alpha\beta}T$$

and $\{T_\alpha, T_{\overline{\alpha}}, T\}$ is a local frame of $T(M) \otimes \mathbf{C}$. Thus there is a constant $C > 0$ such that for any $u \in C_0^\infty(U)$ one has

$$\|u\|_{1/2}^2 \le C\left(\sum_{\alpha=1}^{n}(\|T_\alpha u\|^2 + \|T_{\overline{\alpha}} u\|^2) + \|u\|^2\right).$$

On the other hand,

$$\begin{aligned}(\Delta_b u, u) = \int (\Delta_b u)\overline{u} &= -\sum_\alpha \int (T_\alpha^* T_\alpha u + T_{\overline{\alpha}}^* T_{\overline{\alpha}} u)\overline{u} \\ &= -\sum_\alpha \{\|T_\alpha u\|^2 + \|T_{\overline{\alpha}} u\|^2\};\end{aligned}$$

hence

$$\|u\|_{1/2}^2 \leq C\left(|(\Delta_b u, u)| + \|u\|^2\right),$$

for any $u \in C_0^\infty(U)$. To justify the second statement in Theorem 2.1 we recall that by a result of J.J. Kohn and L. Nirenberg [250], any subelliptic operator L (of order ϵ) is hypoelliptic and satisfies the a priori estimates

$$\|u\|_{s+2\epsilon}^2 \leq C_s\left(\|Lu\|_s^2 + \|u\|^2\right), \quad s \geq 0,$$

for any $u \in C_0^\infty(U)$. Hence the second statement in Theorem 2.1 follows from the first. □

Another approach to the sub-Laplacian (on a strictly pseudoconvex CR manifold) on C^2 functions is as the trace of the pseudo-Hermitian Hessian. Let M be a nondegenerate CR manifold and θ a pseudo-Hermitian structure on M. Let ∇ be the Tanaka–Webster connection of (M, θ). Let $f \in C^2(M)$.

Definition 2.5. The *pseudo-Hermitian Hessian* of f is defined by

$$(\nabla^2 f)(X, Y) = (\nabla_X df)Y = X(Y(f)) - (\nabla_X Y)(f), \quad X, Y \in \mathcal{X}(M). \qquad \square$$

Let B be a bilinear form on $T(M)$. We denote by $\pi_H B$ the restriction of B to $H(M) \otimes H(M)$. Moreover, if φ is a bilinear form on $H(M)$, we make the following definition:

Definition 2.6. The *trace* of φ with respect to G_θ is the C^∞ function $\mathrm{trace}_{G_\theta}(\varphi) : M \to \mathbf{R}$ given by

$$\mathrm{trace}_{G_\theta}(\varphi)_x = \sum_{j=1}^{2n} \epsilon_j \, \varphi(X_j, X_j),$$

for some (local) orthonormal frame $\{X_j : 1 \leq j \leq 2n\}$ of $H(M)$ (with respect to G_θ, i.e., $G_\theta(X_i, X_j) = \epsilon_i \, \delta_{ij}$, $\epsilon_i^2 = 1$), defined on an open neighborhood U of x, $x \in M$. □

The definition of $\mathrm{trace}_{G_\theta}(\varphi)_x$ does not depend on the choice of (local) orthonormal frame at x.

Proposition 2.1. *Let M be a strictly pseudoconvex CR manifold and θ a pseudo-Hermitian structure on M. Then*

$$\Delta_b u = \mathrm{trace}_{G_\theta}\{\pi_H \nabla^2 u\} \tag{2.12}$$

for any C^2 function $u : M \to$ **R**.

Proof. Let $\{X_a : 1 \leq a \leq 2n\}$ be a local orthonormal frame of $H(M)$, i.e., $G_\theta(X_a, X_b) = \delta_{ab}$ ($\epsilon_\alpha^2 = 1$). Then (by (2.6))

$$\mathrm{trace}_{G_\theta}(\pi_H \nabla^2 u) = \sum_{a=1}^{2n} (\nabla^2 u)(X_a, X_a) = \sum_a \{X_a(X_a u) - (\nabla_{X_a} X_a)(u)\} = \Delta_b u.$$

Proposition 2.1 is proved. □

To obtain yet another useful expression of the sub-Laplacian on functions we need a few remarks from linear algebra. Let us consider a local orthonormal frame $\{X_\alpha, JX_\alpha : 1 \leq \alpha \leq n\}$ $(G_\theta(X_\alpha, X_\beta) = \delta_{\alpha\beta})$ of $H(M)$, defined on $U \subseteq M$. Next, let us set $Z_\alpha = \frac{1}{\sqrt{2}}(X_\alpha - iJX_\alpha)$. Then $L_\theta(Z_\alpha, Z_{\overline{\beta}}) = \delta_{\alpha\beta}$. If φ is a bilinear form on $H(M)$ then

$$\text{trace}_{G_\theta}(\varphi) = \sum_{\alpha=1}^{n} \{\varphi(Z_\alpha, Z_{\bar{\alpha}}) + \varphi(Z_{\bar{\alpha}}, Z_\alpha)\} \,. \tag{2.13}$$

If, in turn, $\{T_\alpha\}$ is just any local frame of $T_{1,0}(M)$ defined on U then $T_\alpha = U_\alpha^\beta Z_\beta$ for some C^∞ function $[U_\alpha^\beta] : U \to \text{GL}(n, \mathbf{C})$ satisfying

$$\sum_{\mu=1}^{n} U_\alpha^\mu U_{\bar{\beta}}^{\bar{\mu}} = h_{\alpha\bar{\beta}}$$

(where $h_{\alpha\overline{\beta}} = L_\theta(T_\alpha, T_{\overline{\beta}})$). Let $[V_\beta^\alpha] = [U_\beta^\alpha]^{-1}$. Then $\sum_{\lambda=1}^{n} V_\lambda^\alpha V_{\bar{\lambda}}^{\bar{\beta}} = h^{\alpha\bar{\beta}}$; hence (by $Z_\alpha = V_\alpha^\beta T_\beta$ and (2.13))

$$\text{trace}_{G_\theta}(\varphi) = h^{\alpha\bar{\beta}}\varphi(T_\alpha, T_{\bar{\beta}}) + h^{\alpha\bar{\beta}}\varphi(T_{\bar{\alpha}}, T_\beta) \tag{2.14}$$

or, if φ is real, i.e., $\overline{\varphi(Z, W)} = \varphi(\overline{Z}, \overline{W})$ for any $Z, W \in H(M) \otimes \mathbf{C}$, then

$$\text{trace}_{G_\theta}(\varphi) = 2\,\text{Re}\{h^{\alpha\bar{\beta}}\varphi(T_\alpha, T_{\bar{\beta}})\}.$$

At this point, by Proposition 2.1

$$\Delta_b u = \sum_\alpha \{h^{\alpha\overline{\beta}}(\nabla^2 u)(T_\alpha, T_{\overline{\beta}}) + h^{\overline{\alpha}\beta}(\nabla^2 u)(T_{\overline{\alpha}}, T_\beta)\}$$

hence we have proved the following result:

Proposition 2.2. *Let M be a strictly pseudoconvex CR manifold and θ a contact form on M. The sub-Laplacian Δ_b of (M, θ) is locally given by*

$$\Delta_b u = u^\alpha{}_\alpha + u^{\bar{\alpha}}{}_{\bar{\alpha}},$$

for any $u \in C^2(M)$.

Here we have adopted the notation $u_{AB} = (\nabla^2 u)(T_A, T_B)$. In particular, if $\{T_\alpha\}$ is chosen so that $h_{\alpha\bar{\beta}} = \delta_{\alpha\beta}$ then

$$\Delta_b u = \sum_{\alpha=1}^{n} (u_{\alpha\bar{\alpha}} + u_{\bar{\alpha}\alpha}). \tag{2.15}$$

The sub-Laplacian was first introduced by G.B. Folland [146], in the special context of the Heisenberg group. Precisely, G.B. Folland established (cf. op. cit.) the following

Theorem 2.2. *The differential operator $\mathcal{L}_0$ on the Heisenberg group $\mathbf{H}_n$ given by*

$$\mathcal{L}_0 = \sum_{\alpha=1}^{n} \left\{ -\frac{\partial^2}{\partial z^\alpha \partial \overline{z}^\alpha} - |z^\alpha|^2 \frac{\partial^2}{\partial t^2} + i \frac{\partial}{\partial t} \left(z^\alpha \frac{\partial}{\partial z^\alpha} - \overline{z}^\alpha \frac{\partial}{\partial \overline{z}^\alpha} \right) \right\}$$

is left invariant and is subelliptic of order $1/2$ at each $x \in \mathbf{H}_n$.

Proof. Let us observe that

$$\mathcal{L}_0 = -\frac{1}{2} \sum_{\alpha=1}^{n} (T_\alpha T_{\overline{\alpha}} + T_{\overline{\alpha}} T_\alpha) ,$$

where $T_\alpha = \partial/\partial z^\alpha + i\overline{z}^\alpha \partial/\partial t$, i.e., $\mathcal{L}_0$ is the Folland–Stein operator obtained for $\alpha = 0$; cf. Chapter 1 of this book. Thus the left invariance of $\mathcal{L}_0$ follows from that of T_α. On the other hand (by (2.11)), $\mathcal{L}_0 = \frac{1}{2}\Delta_b$ (where Δ_b is the sub-Laplacian on $\mathbf{H}_n$ with respect to the canonical contact form $\theta_0 = dt + i \sum_{j=1}^{n} \left(z^j d\overline{z}^j - \overline{z}^j dz^j \right)$), whence the last statement in Theorem 2.2. □

The Laplace–Beltrami operator (on functions) on a Riemannian manifold is often viewed as the differential operator $-d^*d$, where d denotes the exterior differential operator and d^* its formal adjoint (with respect to the given Riemannian structure). To derive a similar expression for Δ_b we need to introduce the operator d_b. Let

$$r : T^*(M) \to H(M)^*$$

be the natural restriction map $\omega \mapsto \omega|_{H(M)}$.

Definition 2.7. For any C^∞ function $u : M \to \mathbf{C}$ we define a section $d_b u \in \Gamma^\infty(H(M)^*)$ by setting

$$d_b u = r \circ du = du\big|_{H(M)}. \qquad \square$$

Let $\{T_\alpha : 1 \leq \alpha \leq n\}$ be a local frame of $H(M)$ on $U \subseteq M$ and let $\{\theta^\alpha : 1 \leq \alpha \leq n\}$ be the corresponding admissible frame. We extend the Levi form L_θ to a (pointwise) inner product L_θ^* on $(H(M) \otimes \mathbf{C})^*$ by merely requiring that

$$L_\theta^*(\theta^\alpha, \theta^{\overline{\beta}}) = h^{\alpha\overline{\beta}}, \quad L_\theta^*(\theta^\alpha, \theta^\beta) = 0, \quad L_\theta^*(\theta^{\overline{\alpha}}, \theta^{\overline{\beta}}) = 0.$$

Then we may consider the L^2 inner product

$$(\omega, \eta) = \int_M L_\theta^*(\omega, \overline{\eta})\, \Psi,$$

for any C^∞ sections ω, η in $(H(M) \otimes \mathbf{C})^*$ (i.e., for any 1-forms ω, η on M such that $T \rfloor \omega = 0$ and $T \rfloor \eta = 0$).

Proposition 2.3.

$$(\Delta_b u, v) = -(d_b u, d_b v),$$

for any C^∞ functions $u, v : M \to \mathbf{R}$, at least one of compact support.

Proof. Locally, we may write

$$d_b u = u_\alpha \theta^\alpha + u_{\overline{\alpha}} \theta^{\overline{\alpha}},$$

where $u_\alpha = T_\alpha(u)$, $u_{\overline{\alpha}} = T_{\overline{\alpha}}(u)$. Thus

$$L_\theta^*(d_b u, d_b v) = u_\alpha v_{\overline{\beta}} h^{\alpha\overline{\beta}} + u_{\overline{\alpha}} v_\beta h^{\overline{\alpha}\beta}.$$

Consequently

$$\begin{aligned}(d_b u, d_b v) &= \int_M \left(u_\alpha v_{\overline{\beta}} h^{\alpha\overline{\beta}} + u_{\overline{\alpha}} v_\beta h^{\overline{\alpha}\beta} \right) \Psi \\ &= \int_M \left(T_{\overline{\beta}}(u^{\overline{\beta}} v) - v T_{\overline{\beta}}(u^{\overline{\beta}}) + T_\beta(u^\beta v) - v T_\beta(u_\beta) \right) \Psi.\end{aligned}$$

Thus, we may perform the following calculation:

$$\begin{aligned}(d_b u, d_b v) = \int_M \Big(&\operatorname{div}(u^\beta v T_\beta) - u^\beta v \Gamma^\alpha_{\alpha\beta} - v T_\beta(u^\beta) \\ &+ \operatorname{div}(u^{\overline{\beta}} v T_{\overline{\beta}}) - u^{\overline{\beta}} v \Gamma^{\overline{\alpha}}_{\overline{\alpha}\overline{\beta}} - v T_{\overline{\beta}}(u^{\overline{\beta}}) \Big) \Psi\end{aligned}$$

(by Green's lemma)

$$\begin{aligned}&= -\int_M \left(u^\beta \Gamma^\alpha_{\alpha\beta} + u^{\overline{\beta}} \Gamma^{\overline{\alpha}}_{\overline{\alpha}\overline{\beta}} + T_\beta(u^\beta) + T_{\overline{\beta}}(u^{\overline{\beta}}) \right) v \, \Psi \\ &= -\int_M (u^\alpha{}_\alpha + u^{\overline{\alpha}}{}_{\overline{\alpha}}) v \, \Psi = -\int_M (\Delta_b u) v \, \Psi.\end{aligned}$$

Proposition 2.3 is proved. □

2.2 The canonical bundle

Let $(M, T_{1,0}(M))$ be a strictly pseudoconvex CR manifold (of CR dimension n) and θ a pseudo-Hermitian structure on M such that L_θ is positive definite. Let $K(M) = \Lambda^{n+1,0}(M)$ and $\pi_0 : K(M) \to M$ the natural projection.

Definition 2.8. $\mathbf{C} \to K(M) \to M$ is called the *canonical bundle* over M. □

Let $K^0(M) = K(M) \setminus \{zero \;\; section\}$. Let $\mathbf{R}_+ \times K^0(M) \to K^0(M)$ be the natural action of $\mathbf{R}_+ = (0, +\infty)$ on $K^0(M)$ and

$$C(M) = K^0(M)/\mathbf{R}_+ \stackrel{\pi}{\longrightarrow} M$$

the quotient bundle. Note that $C(M) \to M$ is a principal S^1-bundle.

Definition 2.9. $S^1 \to C(M) \to M$ is called the *canonical circle bundle* or the *Fefferman bundle* over M. □

A local frame $\{\theta^\alpha\}$ of $T_{1,0}(M)^*$ on $U \subseteq M$ induces the trivialization chart

$$\pi^{-1}(U) \to U \times S^1, \quad [\omega] \mapsto \left(x, \frac{\lambda}{|\lambda|}\right),$$

where $\omega \in K^0(M)$, $\pi_0(\omega) = x \in U$, and

$$\omega = \lambda \left(\theta \wedge \theta^{1\cdots n}\right)_x ,$$

with $\lambda \in \mathbf{C}$, $\lambda \neq 0$. We shall need the following definition:

Definition 2.10. The *tautologous form* $\xi \in \Omega^{n+1}(K(M))$ is defined by

$$\xi_\omega(Z_1, \ldots, Z_{n+1}) = \omega((d_\omega \pi_0)Z_1, \ldots, (d_\omega \pi_0)Z_{n+1}),$$

for any $Z_1, \ldots, Z_{n+1} \in T_\omega(K(M))$ and any $\omega \in K(M)$. □

Lemma 2.2. *For any $\omega \in \Gamma^\infty(K^0(M))$ there is a unique C^∞ function $\lambda : M \to (0, \infty)$ such that*

$$2^n i^{n^2} n! \theta \wedge (T \rfloor \omega) \wedge (T \rfloor \overline{\omega}) = \lambda \theta \wedge (d\theta)^n.$$

Proof. Let $\{\theta^\alpha\}$ be a (local) frame of $T_{1,0}(M)^*$ on $U \subseteq M$. Then $\omega \in \Gamma^\infty(K^0(M))$ may be represented as

$$\omega_{|U} = f\, \theta \wedge \theta^{1\cdots n}$$

for some C^∞ function $f : U \to \mathbf{C}$, $f \neq 0$ everywhere on U. Then

$$T \rfloor \omega = \frac{f}{n+1} \theta^{1\cdots n}, \quad T \rfloor \overline{\omega} = \frac{\overline{f}}{n+1} \theta^{\bar{1}\cdots\bar{n}}.$$

Let us set, by definition,

$$\lambda = \frac{1}{(n+1)^2} \frac{|f|^2}{\det(h_{\alpha\overline{\beta}})} > 0.$$

Then (cf. the local expression of the volume form Ψ in Chapter 1)

$$2^n\, i^{n^2}\, n! \,\theta \wedge (T \rfloor \omega) \wedge (T \rfloor \overline{\omega}) = 2^n\, i^{n^2}\, n! \,\frac{|f|^2}{(n+1)^2} \theta \wedge \theta^{1\cdots n, \bar{1}\cdots\bar{n}} = \lambda\, \theta \wedge (d\theta)^n.$$

Finally, it is easy to see that λ is invariant under any transformation

$$\theta'^\alpha = U^\alpha_\beta \theta^\beta, \quad \det[U^\alpha_\beta] \neq 0, \quad \text{on } U \cap U'.$$

Lemma 2.2 is proved. □

Proposition 2.4. *There is a natural embedding $i_\theta : C(M) \to K(M)$.*

Indeed, let $[\omega] \in C(M)$ with $\pi_0(\omega) = x$. By Lemma 2.2 there is a unique $\lambda \in (0, +\infty)$ such that

$$2^n \, i^{n^2} \, n! \, \theta_x \wedge (T_x \rfloor \omega) \wedge (T_x \rfloor \overline{\omega}) = \lambda \Psi_x .$$

Then we set

$$i_\theta([\omega]) = \frac{1}{\sqrt{\lambda}} \, \omega .$$

If ω' is another representative of $[\omega]$ then $\omega' = a\omega$ for some $a \in (0, +\infty)$; hence $\lambda' = a^2 \lambda$, such that $i_\theta([\omega])$ is well defined. □

Another way to put it is that (by Lemma 2.2) within each class $c = [\omega] \in C(M)$ there is a unique representative η such that

$$2^n \, i^{n^2} \, n! \, \theta_x \wedge (T_x \rfloor \eta) \wedge (T_x \rfloor \overline{\eta}) = \Psi_x$$

(where $x = \pi(c)$) and then $i_\theta(c) = \eta$ by definition.

Using the embedding i_θ we may define the form $\zeta \in \Omega^{n+1}(C(M))$ as the pullback of the tautologous $(n+1)$-form ξ on $K(M)$:

Definition 2.11.

$$\zeta = \frac{1}{n+1} \, i_\theta^* \, \xi . \qquad \square$$

Let $\{\theta^1, \ldots, \theta^n\}$ be a frame of $T_{1,0}(M)^*$ on U. Let us define the (local) form $\xi_0 \in \Gamma^\infty(U, K(M))$ as follows.

Definition 2.12. Let M be a strictly pseudoconvex CR manifold, $\{T_\alpha : 1 \le \alpha \le n\}$ a local frame of $T_{1,0}(M)$ on $U \subseteq M$, and $\{\theta^\alpha : 1 \le \alpha \le n\}$ the corresponding admissible local coframe. We set

$$\xi_0 = \det(h_{\alpha\overline{\beta}})^{1/2} \theta \wedge \theta^{1 \cdots n}$$

on U. Also, let us consider $\zeta_0 \in \Omega^{n+1}(C(U))$ given by $\zeta_0 = \pi^* \xi_0$ on U. □

If $c \in C(M)$ is fixed, we define the C^∞ curve $a_c : \mathbf{R} \to C(M)$ by setting

$$a_c(\theta) = e^{i\theta} c , \quad \theta \in \mathbf{R}.$$

Next, let S be the tangent vector field on $C(M)$ given by

$$S_c = \frac{da_c}{d\theta}(0).$$

S is a vertical vector field for the principal S^1-bundle $C(M) \to M$, in the sense that $S \in \mathrm{Ker}(d\pi)$. Indeed, a_c lies in the S^1-orbit of c so that $\pi \circ a_c = \text{const.}$ and consequently

$$(d_c \pi) S_c = d_0(\pi \circ a_c) \frac{d}{d\theta}\Big|_0 = 0.$$

Let $\{\theta^\alpha : 1 \le \alpha \le n\}$ be a local frame of $T_{1,0}(M)^*$ defined on U, as above.

Definition 2.13. Let us define $\gamma : \pi^{-1}(U) \to \mathbf{R}$ by setting

$$\gamma([\omega]) = \arg\left(\frac{f}{|f|}\right),$$

where

$$\omega = f\ (\theta \wedge \theta^{1\cdots n})_x\,, \quad \pi_0(\omega) = x, \quad f \in \mathbf{C}, \quad f \neq 0.$$

We call γ a local *fiber coordinate* on $C(M)$. □

Here $\arg : S^1 \to [0, 2\pi)$, so that if $w = e^{i\theta} \in S^1$ then $\theta \in \mathrm{Arg}(w) = \{\arg(w) + 2k\pi : k \in \mathbf{Z}\}$.

Lemma 2.3. *Let M be a strictly pseudoconvex CR manifold and $S^1 \to C(M) \xrightarrow{\pi} M$ the canonical circle bundle over M. Let γ be a local fiber coordinate on $C(M)$ and S the tangent to the S^1-action. Then (1) $\zeta = e^{i\gamma}\zeta_0$ and (2) $(d\gamma)S = 1$.*

Proof. Both ζ and S are thought of as restricted to $\pi^{-1}(U)$. To prove Lemma 2.3, let $\omega = f(\theta \wedge \theta^{1\cdots n})_x \in K^0(M)_x$. Since $\pi_0 \circ i_\theta = \pi$ we may perform the following calculation:

$$\begin{aligned}
\zeta_{[\omega]} &= \frac{1}{n+1}\left(i_\theta^* \xi\right)_{[\omega]} = \frac{1}{n+1}\xi_{i_\theta([\omega])}\left(d_{[\omega]} i_\theta\right) \\
&= \frac{1}{n+1} i_\theta([\omega]) \circ (d_{i_\theta([\omega])}\pi_0) \circ (d_{[\omega]} i_\theta) = \frac{f}{|f|}\xi_{0,x}(d_{[\omega]}\pi) \\
&= e^{i\arg(f/|f|)}(\pi^*\xi_0)_{[\omega]} = e^{i\gamma([\omega])}\zeta_{0,[\omega]}
\end{aligned}$$

and (1) is proved. To prove (2) note that

$$e^{i\gamma(a_{[\omega]}(\theta))} = e^{i(\theta+\gamma([\omega]))}.$$

Differentiation with respect to θ then gives

$$\frac{d}{d\theta}\left(\gamma \circ a_{[\omega]}\right) = 1.$$

Lemma 2.3 is proved. □

2.3 The Fefferman metric

Let $(M, T_{1,0}(M))$ be a strictly pseudoconvex CR manifold and θ a pseudo-Hermitian structure on M such that L_θ is positive definite. Let T be the characteristic direction of (M, θ).

Lemma 2.4. (J.M. Lee [271])
There is a unique complex n-form η on $C(M)$ such that

$$V \rfloor \eta = 0, \quad \zeta = (\pi^*\theta) \wedge \eta,$$

for any lift V of T to $C(M)$, i.e., for any $V \in \mathcal{X}(C(M))$ such that $\pi_* V = T$.

Proof. Let V be a lift of T to C and let us set

$$\eta = (n+1) V \rfloor \zeta.$$

First, we check that the definition of η doesn't depend on the choice of the lift. Indeed, if V' is another vector field on $C(M)$ with $\pi_* V' = T$ then $V' - V \in \text{Ker}(\pi_*)$, that is,

$$V' = V + f S$$

for some $f \in C^\infty(C(M))$. On the other hand (by Lemma 2.3),

$$S \rfloor \zeta = S \rfloor (e^{i\gamma} \zeta_0) = e^{i\gamma} S \rfloor (\pi^* \xi_0) = e^{i\gamma} (\pi_* S) \rfloor \xi_0 = 0.$$

Clearly $V \rfloor \eta = 0$ (because ζ is skew). To establish the second requirement we perform the calculation

$$\begin{aligned}(\pi^*\theta) \wedge \eta &= (n+1)(\pi^*\theta) \wedge (V \rfloor \zeta) \\ &= (n+1)e^{i\gamma} (\pi^*\theta) \wedge (V \rfloor \zeta_0) = (n+1)e^{i\gamma} (\pi^*\theta) \wedge (V \rfloor \pi^*\xi_0) \\ &= (n+1)e^{i\gamma} \pi^* \left(\theta \wedge (T \rfloor \xi_0)\right).\end{aligned}$$

On the other hand,

$$T \rfloor \xi_0 = \frac{1}{n+1} \det(h_{\alpha\overline{\beta}})^{1/2} \theta^1 \wedge \cdots \wedge \theta^n$$

and the proof is complete. □

Next, we need to establish the following proposition:

Proposition 2.5. (J.M. Lee [271])
There is a unique real 1*-form* $\sigma \in \Omega^1(C(M))$ *such that*

$$\begin{aligned} d\zeta &= i(n+2)\sigma \wedge \zeta, \\ \sigma \wedge d\eta \wedge \overline{\eta} &= \text{trace}(d\sigma)\, i\, \sigma \wedge (\pi^*\theta) \wedge \eta \wedge \overline{\eta}, \end{aligned}$$

where the 1*-form* η *is given by* Lemma 2.4.

Proof. First, we make use of the identities

$$d\theta = 2ih_{\alpha\overline{\beta}}\theta^\alpha \wedge \theta^{\overline{\beta}}, \quad d\theta^\alpha = \theta^\beta \wedge \omega_\beta^\alpha + \theta \wedge \tau^\alpha,$$

to compute the exterior derivative of $\theta^{01\cdots n}$, that is,

$$\begin{aligned} d\theta^{01\cdots n} &= d\theta \wedge \theta^{1\cdots n} - \theta \wedge \sum_{\alpha=1}^{n} (-1)^{\alpha-1} \theta^1 \wedge \cdots \wedge d\theta^\alpha \wedge \cdots \wedge \theta^n \\ &= 2ih_{\alpha\overline{\beta}}\theta^\alpha \wedge \theta^{\overline{\beta}} \wedge \theta^{1\cdots n} \\ &\quad + \theta \wedge \sum_{\alpha=1}^{n} (-1)^\alpha \theta^1 \wedge \cdots \wedge (\theta^\beta \wedge \omega_\beta^\alpha + \theta \wedge \tau^\alpha) \wedge \cdots \wedge \theta^n \\ &= \theta \wedge \sum_{\alpha=1}^{n} (-1)^\alpha \theta^1 \wedge \cdots \wedge (\theta^\alpha \wedge \omega_\alpha^\alpha) \wedge \cdots \wedge \theta^n. \end{aligned}$$

We obtain

$$d\theta^{01\cdots n} = -\omega_\alpha^\alpha \wedge \theta^{01\cdots n}. \tag{2.16}$$

Let us set $H = \det(h_{\alpha\overline{\beta}})^{1/2}$, for the sake of simplicity. Then (by (2.16))

$$d\xi_0 = dH \wedge \theta^{01\cdots n} + Hd\theta^{01\cdots n} = (dH - H\omega_\alpha^\alpha) \wedge \theta^{01\cdots n};$$

hence

$$d\xi_0 = (d\log H - \omega_\alpha^\alpha) \wedge \xi_0. \tag{2.17}$$

Let us set $h = \log H \in C^\infty(M)$. Define the local form $\omega \in \Omega^{1,0}(M)$ by setting

$$\omega = (h_\beta - \Gamma_{\beta\overline{\alpha}}^{\overline{\alpha}})\theta^\beta, \quad h_\beta = T_\beta(h).$$

Then (by (2.17))

$$d\xi_0 = \overline{\omega} \wedge \xi_0 = (\overline{\omega} - \omega) \wedge \xi_0$$

(since $\omega \wedge \xi_0 = 0$). Since $\overline{\omega} - \omega$ is purely imaginary we may define a real 1-form $\sigma_0 \in \Omega^1(M)$ by setting

$$\overline{\omega} - \omega = i(n+2)\sigma_0;$$

hence

$$d\xi_0 = i(n+2)\sigma_0. \tag{2.18}$$

At this point, we may differentiate $\zeta = e^{i\gamma}\zeta_0$ and use (2.18) to obtain

$$d\zeta = i(n+2)\sigma \wedge \zeta, \tag{2.19}$$

where $\sigma \in \Omega^1(C(M))$ is the real 1-form defined by

$$\sigma = \frac{1}{n+2}d\gamma + \pi^*\sigma_0. \tag{2.20}$$

We proceed by introducing a notion of *trace* of a 2-form. As before, let $(M, T_{1,0}(M))$ be a strictly pseudoconvex CR manifold, of CR dimension n, and let θ be a pseudo-Hermitian structure on M such that the corresponding Levi form L_θ is positive definite. For any differential 2-form $\omega \in \Omega^2(M)$, perhaps real, there is a natural concept of trace, defined as follows.

Definition 2.14. Let $\tilde{\omega} : T_{1,0}(M) \to T_{1,0}(M)$ be the bundle endomorphism naturally induced by the $(1,1)$-component of ω, that is,

$$(d\theta)(\tilde{\omega}Z, \overline{W}) = \omega(Z, \overline{W}),$$

for any $Z, W \in T_{1,0}(M)$. Locally, with respect to some (local) frame $\{T_\alpha\}$ of $T_{1,0}(M)$, we may write

$$\omega = \omega_{\alpha\beta}\theta^\alpha \wedge \theta^\beta + \omega_{\bar{\alpha}\bar{\beta}}\theta^{\overline{\alpha}} \wedge \theta^{\overline{\beta}} + \omega_{\alpha 0}\theta^\alpha \wedge \theta + \omega_{\overline{\alpha}0}\theta^{\overline{\alpha}} \wedge \theta + i\,\omega_{\alpha\overline{\beta}}\theta^\alpha \wedge \theta^{\overline{\beta}}.$$

It is also customary to write

$$\omega \equiv i\,\omega_{\alpha\overline{\beta}}\theta^\alpha \wedge \theta^{\overline{\beta}} \ (\text{mod } \theta^\alpha \wedge \theta^\beta, \theta^\alpha \wedge \theta^{\overline{\beta}}, \theta^\alpha \wedge \theta, \theta^{\overline{\alpha}} \wedge \theta).$$

Therefore, we have $\tilde{\omega} T_\alpha = \omega_\alpha^\beta T_\beta$, where $\omega_\alpha^\beta = \frac{1}{2} h^{\beta\overline{\gamma}} \omega_{\alpha\overline{\gamma}}$. Finally, by definition the *trace* of ω is the trace of the endomorphism $\tilde{\omega}$, i.e.,

$$\text{trace}(\omega) = \text{trace}(\tilde{\omega}) = \frac{1}{2} h^{\beta\overline{\gamma}} \omega_{\beta\overline{\gamma}}. \qquad \square$$

Let Ω be a 2-form on M. Then by definition,

$$\text{trace}(\pi^*\Omega) = \text{trace}(\Omega) \circ \pi$$

where trace(Ω) is taken in the sense of Definition 2.14. Since (by (2.20)) $d\sigma = \pi^* d\sigma_0$ and $d\sigma_0 \in \Omega^2(M)$ the above definition may be applied to make sense of trace$(d\sigma)$.

Let $f \in \Omega^0(M)$ be a real-valued C^∞ function and define $\sigma_f \in \Omega^1(C(M))$ by setting

$$\sigma_f = \sigma + \pi^*(f\theta).$$

Let η be the n-form on $C(M)$ furnished by Lemma 2.4. Let us take the exterior derivative of

$$\zeta = (\pi^*\theta) \wedge \eta$$

to get

$$d\zeta = (\pi^* d\theta) \wedge \eta - (\pi^*\theta) \wedge d\eta;$$

hence (by (2.19))

$$(\pi^*\theta) \wedge d\eta = (\pi^* d\theta) \wedge \eta - i(n+2)\sigma \wedge \zeta.$$

Consequently

$$\sigma_f \wedge d\eta \wedge \overline{\eta} = \sigma \wedge d\eta \wedge \overline{\eta} + f\left((\pi^* d\theta) \wedge \eta - i(n+2)\sigma \wedge \zeta\right) \wedge \overline{\eta};$$

hence

$$\sigma_f \wedge d\eta \wedge \overline{\eta} = \sigma \wedge d\eta \wedge \overline{\eta} - i(n+2) f \sigma \wedge (\pi^*\theta) \wedge \eta \wedge \overline{\eta} \qquad (2.21)$$

because of

$$(\pi^* d\theta) \wedge \eta \wedge \overline{\eta} = 0.$$

Indeed, to see that this is true it suffices to look at the explicit expression of η, that is (cf. the proof of Lemma 2.4),

$$\eta = e^{i\gamma} \pi^*(\det(h_{\alpha\overline{\beta}})^{1/2} \theta^{1\cdots n}).$$

Let us take the exterior derivative of

$$\sigma_f = \sigma + \pi^*(f\theta)$$

to get

$$d\sigma_f = d\sigma + \pi^*(df \wedge \theta + f d\theta).$$

We wish to compute trace$(d\sigma_f)$. Since

$$df \wedge \theta + f d\theta = (f_\alpha \theta^\alpha + f_{\overline{\alpha}} \theta^{\overline{\alpha}}) \wedge \theta + 2 f i h_{\alpha\overline{\beta}} \theta^\alpha \wedge \theta^{\overline{\beta}}$$

(where $f_\alpha = T_\alpha(f)$ and $f_{\overline{\alpha}} = T_{\overline{\alpha}}(f)$) it follows that

$$\text{trace}(df \wedge \theta + f d\theta) = nf;$$

hence

$$\text{trace}(d\sigma_f) = \text{trace}(d\sigma) + nf \circ \pi. \tag{2.22}$$

During the following calculations, for the sake of simplicity, we do not distinguish notationally between f and $f \circ \pi$, respectively θ and $\pi^*\theta$. Using (2.22) we obtain

$$\text{trace}(d\sigma_f) i \sigma_f \wedge \theta \wedge \eta \wedge \overline{\eta} = \text{trace}(d\sigma)\, i\, \sigma \wedge \theta \wedge \eta \wedge \overline{\eta} + nfi\sigma \wedge \theta \wedge \eta \wedge \overline{\eta}. \tag{2.23}$$

We wish to determine $f \in \Omega^0(M)$ such that

$$\sigma_f \wedge d\eta \wedge \overline{\eta} = \text{trace}(d\sigma_f) i \sigma_f \wedge \theta \wedge \eta \wedge \overline{\eta} \tag{2.24}$$

(then σ_f would be the real 1-form on $C(M)$ we are looking for, because of $(f\theta) \wedge \zeta = 0$). To solve (2.24) for f we substitute from (2.21) and (2.23). We obtain

$$\sigma \wedge d\eta \wedge \overline{\eta} - \text{trace}(d\sigma)\, i\, \sigma \wedge \theta \wedge \eta \wedge \overline{\eta} = 2(n+1) i f \sigma \wedge \theta \wedge \eta \wedge \overline{\eta}, \tag{2.25}$$

which uniquely determines f because $\sigma \wedge \theta \wedge \eta \wedge \overline{\eta}$ is a volume form on $C(M)$. However, we need to check that f (determined by (2.25)) is real-valued. Since $\sigma \wedge d\eta \wedge \overline{\eta}$ is a $(2n+2)$-form on $C(M)$, there is a C^∞ function $u : C(M) \to \mathbf{C}$ such that

$$\sigma \wedge d\eta \wedge \overline{\eta} = u i \sigma \wedge \theta \wedge \eta \wedge \overline{\eta}. \tag{2.26}$$

Lemma 2.5. *u is real-valued.*

Proof. Note that

$$\eta \wedge \overline{\eta} = \det(h_{\alpha\overline{\beta}})\theta^{1\cdots n,\overline{1}\cdots\overline{n}},$$

or (by taking into account the explicit expression of $(d\theta)^n$)

$$(d\theta)^n = 2^n \, i^{n^2} \, n! \, \eta \wedge \overline{\eta};$$

hence (by differentiating both sides)

$$0 = d\eta \wedge \overline{\eta} + (-1)^n \eta \wedge d\overline{\eta},$$

which may be written as

$$d\eta \wedge \overline{\eta} = (-1)^{n^2+1} d\overline{\eta} \wedge \eta. \tag{2.27}$$

Finally, it is easily seen that (2.26)–(2.27) yield $u = \overline{u}$. Lemma 2.5 is proved. □

Let us substitute from (2.26) into (2.25). Since $\sigma \wedge \theta \wedge \eta \wedge \overline{\eta}$ is a volume form, we obtain

$$u - \mathrm{trace}(d\sigma) = 2(n+1) f;$$

hence f is real-valued. The proof of Proposition 2.5 is complete. □

The following remark is in order. In the proof of Proposition 2.5 we made use several times of the fact that $\sigma \wedge (\pi^*\theta) \wedge \eta \wedge \overline{\eta}$ is a volume form. This follows by observing that

$$2^n \, i^{n^2} \, n! \, (n+2) \, \sigma \wedge (\pi^*\theta) \wedge \eta \wedge \overline{\eta} = d\gamma \wedge (\pi^*\Psi). \quad \square$$

We extend G_θ to a degenerate $(0,2)$-tensor field $\tilde{G}_\theta$ on M by setting (by definition)

$$\tilde{G}_\theta(X, Y) = G_\theta(X, Y), \quad X, Y \in H(M), \tag{2.28}$$

$$\tilde{G}_\theta(T, W) = 0, \quad W \in T(M) \tag{2.29}$$

(in particular T is null, i.e., $\tilde{G}_\theta(T, T) = 0$). At this point, we may define the semi-Riemannian metric F_θ on $C(M)$ by setting

$$F_\theta = \pi^* \tilde{G}_\theta + 2(\pi^*\theta) \odot \sigma, \tag{2.30}$$

where $\sigma \in \Omega^1(C(M))$ is the real-valued 1-form furnished by Proposition 2.5. It is easy to see that F_θ is a Lorentz metric on $C(M)$. Set

$$h(Z, W) = F_\theta(Z, \overline{W})$$

for any $Z, W \in T(C(M)) \otimes \mathbf{C}$. Since $\{\pi^*\theta, \pi^*\theta^\alpha, \pi^*\theta^{\overline{\alpha}}, \sigma\}$ are pointwise independent, we may consider the dual frame $\{V, V_\alpha, V_{\overline{\alpha}}, \Sigma\}$. Set $Z_\alpha = \pi_* V_\alpha$. Then $Z_\alpha \in T_{1,0}(M)$. Hence h is represented (with respect to the chosen frame $\{V, V_\alpha, V_{\overline{\alpha}}, \Sigma\}$) as

$$h : \begin{bmatrix} 0 & 0 & 0 & 1 \\ 0 & g_{\alpha\overline{\beta}} & 0 & 0 \\ 0 & 0 & g_{\overline{\alpha}\beta} & 0 \\ 1 & 0 & 0 & 0 \end{bmatrix},$$

where $g_{\alpha\overline{\beta}} = L_\theta(Z_\alpha, Z_\beta)$. The characteristic polynomial is

$$(1 - t^2) \left|\det(g_{\alpha\overline{\beta}} - t\delta_{\alpha\beta})\right|^2 = 0;$$

hence F_θ has signature $(+\cdots+-)$. The proof is an exercise in elementary linear algebra. Indeed, let $t \in \mathbf{R}$ and let us set

$$p(t) = \begin{vmatrix} -t & 0 & 0 & 1 \\ 0 & g_{\alpha\overline{\beta}} - t\delta_{\alpha\beta} & 0 & 0 \\ 0 & 0 & g_{\overline{\alpha}\beta} - t\delta_{\alpha\beta} & 0 \\ 1 & 0 & 0 & -t \end{vmatrix}.$$

Clearly $p(0) \neq 0$, so that F_θ is nondegenerate. Let C_j, $1 \leq j \leq 2n+2$, be the columns of $p(t)$. To compute $p(t)$ one performs the elementary transformation $C_1 \mapsto C_1 + tC_{2n+2}$ and applies (twice) the Laplace theorem to the resulting determinant.

Definition 2.15. The Lorentz metric F_θ (given by (2.30)) is the *Fefferman metric* of the (strictly pseudoconvex) CR manifold M. □

The main result in [271] is the following:

Theorem 2.3. (J.M. Lee [271])
Let M be a strictly pseudoconvex CR manifold and θ a contact form on M such that L_θ is positive definite. If $\hat{\theta} = e^{2u}\theta$ is another contact form on M, and $F_{\hat{\theta}}$ is the associated Lorentz metric, then $F_{\hat{\theta}} = e^{2u\circ\pi} F_\theta$. Thus the set $\{e^{2u\circ\pi} F_\theta : u \in C^\infty(M)\}$ is a CR invariant of M.

We relegate the proof of Theorem 2.3 to Section 2.4. The metric (2.30) was firstly discovered by C. Fefferman [138], for the case of a strictly pseudoconvex hypersurface M in $\mathbf{C}^{n+1}$. If this is the case, then $M \times S^1$ carries a Lorentz metric whose conformal class is invariant by biholomorphisms (cf. op. cit.). Moreover, the null geodesics of this metric project on the biholomorphically invariant system of curves on M called *chains* (cf. S.S. Chern and J.K. Moser [99]). Throughout this section we have followed the line of J.M. Lee [271], in order to find a description of the Fefferman metric in terms of the (intrinsic) CR structure of M, thus making the Fefferman metric available on an abstract (i.e., not necessarily embedded) CR manifold.

Other attempts at building an abstract version of the Fefferman metric belong to D. Burns, K. Diederich, and S. Schneider [81], who showed how the Fefferman metric may be obtained from the Chern connection of the Chern CR structure bundle of M, and to F. Farris [137]. The construction of the Fefferman metric by F. Farris (cf. op. cit.) makes use of a closed $(n+1, 0)$-form on M. As we shall see in Chapter 5 of this book, closed $(n+1, 0)$-forms do exist when M is embedded, yet they may fail to exist for an abstract, nonembeddable CR manifold; cf. H. Jacobowitz [222].

Our next goal is to express the Fefferman metric (2.30) in terms of the Tanaka–Webster connection. We prove

Theorem 2.4. (J.M. Lee [271])
The real 1*-form* $\sigma \in \Omega^1(C(M))$ *in* (2.30) *may be expressed as*

$$\sigma = \frac{1}{n+2}\left\{d\gamma + \pi^*\left(i\,\omega_\alpha{}^\alpha - \frac{i}{2}\,h^{\alpha\overline{\beta}}dh_{\alpha\overline{\beta}} - \frac{1}{4(n+1)}\rho\,\theta\right)\right\}. \tag{2.31}$$

Proof. Here $\rho = h^{\alpha\overline{\beta}}R_{\alpha\overline{\beta}}$ is the pseudo-Hermitian scalar curvature. Note that (due to the identity $\omega_{\alpha\overline{\beta}} + \omega_{\overline{\beta}\alpha} = dh_{\alpha\overline{\beta}}$) the 1-form

$$i\,\omega_\alpha{}^\alpha - \frac{i}{2}\,h^{\alpha\overline{\beta}}dh_{\alpha\overline{\beta}}$$

is real. Hence σ given by (2.31) is a real 1-form and (by the uniqueness statement in Proposition 2.5) to prove Theorem 2.4 we need to check only that σ satisfies the relations

$$d\zeta = i(n+2)\sigma\wedge\zeta, \tag{2.32}$$

$$\sigma\wedge d\eta\wedge\overline{\eta} = \text{trace}(d\sigma)\,i\,\sigma\wedge(\pi^*\theta)\wedge\eta\wedge\overline{\eta}, \tag{2.33}$$

where η is the complex n-form furnished by Lemma 2.4. Let us differentiate in $\zeta = e^{i\gamma}\zeta_0$ to get

$$d\zeta = i d\gamma\wedge\zeta + e^{i\gamma}d\zeta_0.$$

Recall that

$$\zeta_0 = \pi^*\xi_0,\quad \xi_0 = H\theta\wedge\theta^1\wedge\cdots\wedge\theta^n,\quad H = \det(h_{\alpha\overline{\beta}})^{1/2}.$$

Moreover

$$d\xi_0 = dH\wedge\theta^{01\cdots n} + H\,d\theta^{01\cdots n}$$

and (by (2.16))

$$d\theta^{01\cdots n} = -\omega_\alpha{}^\alpha\wedge\theta^{01\cdots n}.$$

Hence

$$d\xi_0 = (dH - H\omega_\alpha{}^\alpha)\wedge\theta^{01\cdots n}.$$

On the other hand, note that

$$dH = \frac{1}{2}\,Hh^{\lambda\overline{\mu}}dh_{\lambda\overline{\mu}}. \tag{2.34}$$

Indeed, let $\Delta_{\alpha\overline{\beta}}$ be the algebraic complement of $h_{\alpha\overline{\beta}}$. Then (2.34) follows from the identities

$$\det(h_{\alpha\overline{\beta}}) = h_{\alpha\overline{1}}\Delta_{\alpha\overline{1}} + \cdots + h_{\alpha\overline{n}}\Delta_{\alpha\overline{n}}\,,$$

$$\frac{\partial}{\partial h_{\lambda\overline{\mu}}}\left(\det(h_{\alpha\overline{\beta}})\right) = \Delta_{\lambda\overline{\mu}}, \quad h^{\lambda\overline{\mu}} = \frac{\Delta_{\lambda\overline{\mu}}}{det(h_{\alpha\overline{\beta}})}\,,$$

$$d\left(\det(h_{\alpha\overline{\beta}})\right) = \frac{\partial}{\partial h_{\lambda\overline{\mu}}}\left(\det(h_{\alpha\overline{\beta}})\right) dh_{\lambda\overline{\mu}}\,.$$

Finally, we obtain

$$d\zeta = i\, e^{i\gamma}(H \circ \pi)\{d\gamma + \pi^*(i\,\omega_\alpha{}^\alpha - \frac{i}{2}\, h^{\alpha\overline{\beta}} dh_{\alpha\overline{\beta}})\} \wedge \pi^*(\theta \wedge \theta^1 \wedge \cdots \wedge \theta^n),$$

so that

$$i\,(n+2)\sigma \wedge \zeta =$$
$$i\, e^{i\gamma}\{d\gamma + \pi^*(i\,\omega_\alpha{}^\alpha - \frac{i}{2}\, h_{\alpha\overline{\beta}} dh^{\alpha\overline{\beta}} - \frac{1}{4(n+1)}\rho\theta)\} \wedge \pi^*(H\,\theta^{01\cdots n}) = d\zeta,$$

that is, σ (given by (2.31)) satisfies (2.32). Next, we recall (cf. the proof of Lemma 2.4) that

$$\eta = e^{i\gamma}\pi^*(H\,\theta^{1\cdots n}).$$

Then

$$d\eta = i\, e^{i\gamma} d\gamma \wedge \pi^*(H\,\theta^{1\cdots n}) + e^{i\gamma}\pi^* d(H\,\theta^{1\cdots n}),$$

$$d\theta^{1\cdots n} = -\omega_\alpha{}^\alpha \theta^{1\cdots n} - \theta \wedge \sum_{\alpha=1}^{n} \theta^1 \wedge \cdots \wedge \theta^{\alpha-1} \wedge \tau^\alpha \wedge \theta^{\alpha+1} \wedge \cdots \wedge \theta^n$$

so that

$$d\eta = i\left\{d\gamma + \pi^*(i\,\omega_\alpha{}^\alpha - \frac{i}{2}\, h^{\alpha\overline{\beta}} dh_{\alpha\overline{\beta}})\right\}$$
$$\wedge\, \eta - e^{i\gamma}(H \circ \pi)\pi^*\left(\theta \wedge \sum_{\alpha=1}^{n} \theta^1 \wedge \cdots \wedge \theta^{\alpha-1} \wedge \tau^\alpha \wedge \theta^{\alpha+1} \wedge \cdots \wedge \theta^n\right).$$

Note that $\tau^\alpha \wedge \overline{\eta} = 0$. Thus

$$\sigma \wedge d\eta \wedge \overline{\eta} = \frac{1}{n+2}\left\{d\gamma + \pi^*(i\,\omega_\alpha{}^\alpha - \frac{i}{2}\, h^{\alpha\overline{\beta}} dh_{\alpha\overline{\beta}} - \frac{1}{4(n+1)}\,\rho\theta)\right\}$$
$$\wedge\, i\left\{d\gamma + \pi^*(i\,\omega_\alpha{}^\alpha - \frac{i}{2}\, h^{\alpha\overline{\beta}} dh_{\alpha\overline{\beta}})\right\} \wedge \eta \wedge \overline{\eta}$$
$$= \frac{i(\rho \circ \pi)}{4(n+1)}\,\sigma \wedge (\pi^*\theta) \wedge \eta \wedge \overline{\eta}.$$

Finally, to see that σ (given by (2.31)) satisfies (2.33), it remains to check that

$$\mathrm{trace}(d\sigma) = \frac{\rho \circ \pi}{4(n+1)}.$$

Note that

$$d(h^{\alpha\overline{\beta}} dh_{\alpha\overline{\beta}}) = 2\, d^2 \log H = 0;$$

hence

$$d\sigma = \frac{1}{n+2}\, \pi^* \left\{ i\, d\omega_\alpha{}^\alpha - \frac{1}{4(n+1)}\, \rho\, d\theta - \frac{1}{4(n+1)}\, d\rho \wedge \theta \right\}.$$

Taking into account (1.62) we get $\mathrm{trace}(d\theta) = n$. Also $\mathrm{trace}(d\rho \wedge \theta) = 0$. Let us contract α and β in (1.89) and note that $\omega_\alpha{}^\gamma \wedge \omega_\gamma{}^\alpha = 0$. Also, using the symmetry $A_{\alpha\beta} = A_{\beta\alpha}$, we obtain

$$d\omega_\alpha{}^\alpha = R_{\lambda\overline{\mu}} \theta^\lambda \wedge \theta^{\overline{\mu}} + W^\alpha_{\alpha\lambda} \theta^\lambda \wedge \theta - W^\alpha_{\alpha\overline{\mu}} \theta^{\overline{\mu}} \wedge \theta;$$

hence

$$\mathrm{trace}(i\, d\omega_\alpha{}^\alpha) = \frac{1}{2}\, h^{\lambda\overline{\mu}} R_{\lambda\overline{\mu}} = \frac{1}{2} \rho.$$

Therefore

$$\mathrm{trace}(d\sigma) = \frac{1}{n+2} \left\{ \mathrm{trace}(i\, d\omega_\alpha{}^\alpha) \circ \pi - \frac{n}{4(n+1)} (\rho \circ \pi) \right\} = \frac{\rho \circ \pi}{4(n+1)}.$$

Our Theorem 2.4 is completely proved. □

As an example, let us look at the boundary of a *pseudo-Siegel domain*

$$\mathcal{S}_{\mathbf{p}} = \left\{ z \in \mathbf{C}^n : \sum_{\alpha=1}^{n-1} |z_\alpha|^{2p_\alpha} + \mathrm{Im}(z_n^{p_n}) - 1 < 0 \right\},$$
$$\mathbf{p} = (p_1, \dots, p_n) \in \mathbf{Z}^n, \quad p_j \geq 1, \quad 1 \leq j \leq n.$$

Let us set

$$\psi(z) = -1 + \frac{1}{2i} (z_n^{p_n} - \overline{z}_n^{p_n}) + \sum_{\alpha=1}^{n-1} |z_\alpha|^{2p_\alpha},$$

so that $\mathcal{S}_{\mathbf{p}} = \{\psi(z) < 0\}$. Then

$$\frac{\partial \psi}{\partial z_\alpha} = p_\alpha |z_\alpha|^{2(p_\alpha - 1)} \overline{z}_\alpha, \qquad \frac{\partial \psi}{\partial z_n} = \frac{p_n}{2i} z_n^{p_n - 1},$$
$$\partial \overline{\partial} \psi = p_\alpha^2 |z_\alpha|^{2(p_\alpha - 1)} dz^\alpha \wedge d\overline{z}^\alpha$$

where $z^j = z_j$. The CR structure $T_{1,0}(\partial\mathcal{S}_{\mathbf{p}})$ is locally spanned by the complex vector fields $Z = z^j \partial/\partial z^j$ satisfying $Z(\psi) = 0$, i.e., $T_{1,0}(\partial\mathcal{S}_{\mathbf{p}})$ admits the local frame

$$T_\alpha = \frac{\partial}{\partial z^\alpha} - 2i \frac{p_\alpha |z_\alpha|^{2(p_\alpha - 1)} \overline{z}_\alpha}{p_n z_n^{p_n - 1}} \frac{\partial}{\partial z^n}, \quad 1 \leq \alpha \leq n-1,$$

defined on the open set $U = \partial\mathcal{S}_{\mathbf{p}} \setminus \{z_n = 0\}$. Therefore, the Levi form of $\partial\mathcal{S}_{\mathbf{p}}$ is

$$h_{\alpha\overline{\beta}} = \frac{1}{2} p_\alpha^2 |z_\alpha|^{2(p_\alpha - 1)} \delta_{\alpha\beta};$$

hence

$$\det(h_{\alpha\overline{\beta}}) = \frac{1}{2^{n-1}} \prod_{\alpha=1}^{n-1} p_\alpha^2 |z_\alpha|^{2(p_\alpha - 1)}.$$

Let us consider the set $A = \{\alpha \in \{1, \ldots, n-1\} : p_\alpha \geq 2\}$. Then $\det(h_{\alpha\overline{\beta}}) = 0$ precisely on

$$w(\partial\mathcal{S}_{\mathbf{p}}) = \bigcup_{\alpha \in A} (\partial\mathcal{S}_{\mathbf{p}}) \cap \{z_\alpha = 0\}$$

(the *weak pseudoconvexity locus* of $\partial\mathcal{S}_{\mathbf{p}}$), while $\det(h_{\alpha\overline{\beta}}) > 0$ on

$$M_{\mathbf{p}} = (\partial\mathcal{S}_{\mathbf{p}}) \setminus w(\partial\mathcal{S}_{\mathbf{p}})$$

(an open subset of $\partial\mathcal{S}_{\mathbf{p}}$), i.e., $M_{\mathbf{p}}$ is a strictly pseudoconvex CR manifold. Note also that

$$w(\partial\mathcal{S}_{\mathbf{p}}) = \bigcup_{\alpha \in A} \partial\mathcal{S}_{(p_1, \ldots, \hat{p}_\alpha, \ldots, p_n)}.$$

Let us set

$$f_\alpha(z) = \frac{p_\alpha |z_\alpha|^{2(p_\alpha - 1)} \overline{z}_\alpha}{p_n z_n^{p_n - 1}}, \quad z \in U,$$

for simplicity, so that

$$T_\alpha = \frac{\partial}{\partial z^\alpha} - 2i f_\alpha(z) \frac{\partial}{\partial z^n}.$$

The real 1-form

$$\theta = p_n z_n^{p_n - 1} dz^n + p_n \overline{z}_n^{p_n - 1} d\overline{z}^n + 2i p_\alpha |z_\alpha|^{2(p_\alpha - 1)} \overline{z}_\alpha dz^\alpha - 2i p_\alpha |z_\alpha|^{2(p_\alpha - 1)} z_\alpha d\overline{z}^\alpha$$

is a contact form on $M_{\mathbf{p}}$ and

$$d\theta = -4i p_\alpha^2 |z_\alpha|^{2(p_\alpha - 1)} dz^\alpha \wedge d\overline{z}^\alpha.$$

Then

$$T = \frac{1}{2p_n|z_n|^{2(p_n-1)}}\left\{\overline{z}_n^{p_n-1}\frac{\partial}{\partial z^n} + z_n^{p_n-1}\frac{\partial}{\partial \overline{z}^n}\right\}$$

is the characteristic direction of $d\theta$. We have the following commutation relations:

$$[T_\alpha, T] = 0, \quad [T_\alpha, T_\beta] = 0,$$
$$[T_\alpha, T_{\overline{\beta}}] = 2i\delta_{\alpha\beta}\frac{p_\alpha^2|z_\alpha|^{2(p_\alpha-1)}}{p_n|z_n|^{2(p_n-1)}}\left\{z_n^{p_n-1}\frac{\partial}{\partial \overline{z}^n} + \overline{z}_n^{p_n-1}\frac{\partial}{\partial z^n}\right\},$$

which may be also written

$$[T_\alpha, T_{\overline{\beta}}] = 4i\delta_{\alpha\beta}p_\alpha^2|z_\alpha|^{2(p_\alpha-1)}T.$$

The commutation formulas and the identity

$$\begin{aligned} 2g_\theta(\nabla_X Y, Z) &= X(g_\theta(Y, Z)) + Y(g_\theta(X, Z)) - Z(g_\theta(X, Y)) \\ &\quad + g_\theta([X, Y], Z) + g_\theta([Z, X], Y) - g_\theta([Y, Z], X) \\ &\qquad + g_\theta(T_\nabla(X, Y), Z) + g_\theta(T_\nabla(Z, X), Y) - g_\theta(T_\nabla(Y, Z), X) \end{aligned}$$

give the expressions of the Christoffel symbols (of the Tanaka–Webster connection) of $(M_{\mathbf{p}}, \theta)$:

$$\Gamma^\gamma_{\alpha\beta} = \frac{p_\alpha - 1}{z_\alpha}\delta_{\alpha\beta}\delta^\gamma_\alpha, \quad \Gamma^{\overline{\gamma}}_{\alpha\overline{\beta}} = 0, \quad \Gamma^\gamma_{0\beta} = 0.$$

Hence the connection 1-forms

$$\omega^\alpha_\beta = \Gamma^\alpha_{\mu\beta}\theta^\mu + \Gamma^\alpha_{\overline{\mu}\beta}\theta^{\overline{\mu}} + \Gamma^\alpha_{0\beta}\theta$$

are given by

$$\omega^\alpha_\beta = \frac{p_\alpha - 1}{z_\alpha}\delta_{\alpha\beta}\theta^\alpha,$$

where $\theta^\alpha = dz^\alpha$. Then

$$R(X, Y)T_\gamma = 2\{d\omega^\alpha_\gamma + \omega^\alpha_\beta \wedge \omega^\beta_\gamma\}(X, Y)T_\alpha$$

shows that $R = 0$, i.e., the Tanaka–Webster connection of $(M_{\mathbf{p}}, \theta)$ is flat. Finally, we wish to compute the 1-form

$$\sigma = \frac{1}{n+1}\left\{d\gamma + \pi^*\left(i\omega^\alpha_\alpha - \frac{i}{2}h^{\alpha\overline{\beta}}dh_{\alpha\overline{\beta}} - \frac{\rho}{4n}\theta\right)\right\}$$

on $C(M_{\mathbf{p}})$. To this end, note that

$$dh_{\alpha\overline{\beta}} = \frac{1}{2}p_\alpha^2(p_\alpha - 1)|z_\alpha|^{2(p_\alpha-2)}\delta_{\alpha\beta}\{\overline{z}_\alpha dz^\alpha + z_\alpha d\overline{z}^\alpha\}.$$

Then

$$\omega_\alpha^\alpha = \frac{p_\alpha - 1}{z_\alpha}\theta^\alpha, \quad h^{\alpha\overline{\beta}} dh_{\alpha\overline{\beta}} = \frac{p_\alpha - 1}{|z_\alpha|^2}\{\overline{z}_\alpha \theta^\alpha + z_\alpha \theta^{\overline{\alpha}}\},$$

lead to the following corollary:

Corollary 2.1. *The Fefferman metric F_θ of $(M_{\mathbf{p}}, \theta)$, the strictly pseudoconvex part of $\partial \mathcal{S}_{\mathbf{p}}$, is given by* (2.30), *where the* 1*-form σ is given by*

$$\sigma = \frac{1}{n+1}\left\{d\gamma + \frac{i}{2}(p_\alpha - 1)\pi^*\left(\frac{1}{z_\alpha}\theta^\alpha - \frac{1}{\overline{z}_\alpha}\theta^{\overline{\alpha}}\right)\right\}.$$

As previously mentioned, S.S. Chern and J.K. Moser have constructed (cf. [99]) a family of CR-invariant curves on each nondegenerate CR manifold M embedded in $\mathbf{C}^{n+1}$ as a real hypersurface, the chains of M. For any $X \in T_x(M)$ transverse to $H(M)_x$, there is a unique chain of initial data (x, X). S.S. Chern and J.K. Moser also showed (cf. op. cit.) that along a chain, M may be represented in a *normal form* that osculates M to maximal order by the holomorphic image of a quadric (an excellent exposition of these ideas is given by H. Jacobowitz [221]). Chains, thought of by S.S. Chern and J.K. Moser as several-complex-variables analogues of geodesics in Riemannian geometry, carry a significant amount of information about the CR structure. For instance, by a result of Jih-Hsin Cheng [94], a chain-preserving diffeomorphism of two nondegenerate CR manifolds is either a CR or a conjugate CR isomorphism. While the precise definition of chains is not needed through this text, let us mention that by a result of D. Burns, K. Diederich, and S. Shnider [81], on a strictly pseudoconvex CR manifold—the case of our main concern in this book—a chain is precisely the projection of a nonvertical[2] null geodesic of the Fefferman metric F_θ. Therefore, when the Levi form L_θ is positive definite, this may be taken as a definition for chains. However, when M is only nondegenerate of signature (r, s), with $r \neq 0$, $s \neq 0$, and $r+s = n$, the Fefferman metric makes sense (it is a semi-Riemannian metric on $C(M)$, of signature $(2r+1, 2s+1)$; cf. C.R. Graham [182]) and there exist nonvertical null geodesics whose projections on M are not chains (cf. L. Koch [242]). The characteristic direction T of a nondegenerate CR manifold M, on which a contact form θ has been fixed, and thus the flow obtained by locally integrating T, are transverse to the Levi distribution of M, as well as the chains of M. An interesting question, raised by M.B. Stenzel [386], is whether the two are related. M.B. Stenzel's finding (cf. op. cit.) is that given a compact connected real analytic Riemannian manifold (N, g) and the tube $T^{*\epsilon}N := \{\alpha \in T^*(N) : g^*(\alpha, \alpha)^{1/2} < \epsilon\}$, which for $\epsilon > 0$ sufficiently small carries a canonical complex structure (by a result of M.B. Stenzel and V. Guilemin [387], or L. Lempert and R. Szöke [280]), if the integral curves of T on the boundary $M^\epsilon := \partial T^{*\epsilon}N$ of the tube are chains of M^ϵ then (N, g) is an Einstein manifold. A partial converse, due to the same author, is that the integral curves of T are indeed chains when (N, g) is harmonic (in the sense of A. Besse [60]). The fact that M^ϵ is indeed nondegenerate (actually strictly pseudoconvex) was proved in [387] (and independently in [280]). When $\Omega = \{z \in \mathbf{C}^{n+1} : \varphi(z) < 0\}$ is a strictly pseudoconvex

[2] That is, not tangent to $\mathrm{Ker}(d\pi)$ at any of its points.

domain, C.R. Graham and J.M. Lee considered (cf. [185]) the foliation $\mathcal{F}$ by level sets of φ. The foliation $\mathcal{F}$ is defined on a one-sided neighborhood U of $M := \partial\Omega$. Then they built a connection ∇ on U such that the induced connection on each leaf of $\mathcal{F}$ is precisely the Tanaka–Webster connection of the leaf. Here the notion of an *induced connection* is similar to that appearing in the Gauss formula within the theory of submanifolds in Riemannian manifolds. ∇ is the *Graham–Lee connection*; cf. [129] for a new axiomatic description. It is interesting to note that the C.R. Graham and J.M. Lee construction carries over to the case of the boundaries $M^\epsilon = \partial T^{*\epsilon} N$ ($\epsilon > 0$ small) and that an important technical ingredient in M.B. Stenzel's proof of his result is the possibility of expressing the Fefferman metrics of the leaves M^ϵ in terms of ∇.

In view of the H. Jacobowitz theorem (cf. [220]) and, in general, in view of the philosophy, mentioned above, that chains should play in several complex variables the role of geodesics in Riemannian geometry, it is a natural question whether a variational theory of chains may be developed. Yet chains are projections of light rays of the Fefferman metric, hence one may ask for a variational theory of null geodesics. This indeed exists for certain classes of Lorentzian manifolds, as developed by F. Giannoni and A. Masiello [170, 172]. For instance, F. Giannoni and A. Masiello have studied (cf. op. cit.) the relation between the set of light-like geodesics joining a point p to a smooth time-like curve γ on a Lorentzian manifold and the topology of the space of light-like curves joining p and γ. The result by F. Giannoni and A. Masiello states the existence of a formal series $Q(r) = \sum_{n=0}^{\infty} a_n r^n$ with positive cardinal integer coefficients such that

$$\sum_{z \in Z^+} r^{\mu(z)} = \mathcal{P}(\mathcal{L}^+_{p,\gamma}\,,\ K)(r) + (1+r)Q(r) \tag{2.35}$$

holds. Here $\mathcal{L}^+_{p,\gamma}$ is the set of smooth light-like future pointing curves joining p and γ, while Z^+ is the set of light-like *geodesics* joining p and γ in the future of p. Also, given $z \in Z^+$, one denotes by $\mu(z)$ the index of z, i.e., the number of points $z(s)$ conjugate to $z(0)$, counted with their multiplicities. Finally,

$$\mathcal{P}(X, K)(r) = \sum_{q \in \mathbf{N}} \beta_q(X, K) r^q\ , \quad \beta_q(X, K) = \dim H_q(X, K),$$

is the Poincaré polynomial of $X = \mathcal{L}^+_{p,\gamma}$ and K is an arbitrary field. The beautiful finding (2.35) by F. Giannoni and A. Masiello (cf. op. cit.) cannot be applied directly to the case of $C(M)$ with the Fefferman metric. Indeed, in view of Theorem 1.6 of [172], p. 858, one should investigate whether $(C(M), F_\theta)$ admits some time function, whether p and γ may be chosen such that $\mathcal{L}^+_{p,\gamma} \neq \emptyset$ and such that $\mathcal{L}^+_{p,\gamma}$ is c-compact, for any $c \in \mathbf{R}$, etc. We leave this as an open problem.

2.4 A CR invariant

We wish to prove Theorem 2.3, i.e., show that the *restricted* conformal class $[F_\theta] = \{e^{2u\circ\pi} F_\theta : u \in \Omega^0(M)\}$ of the Fefferman metric F_θ is invariant under a transformation

$$\hat{\theta} = e^{2u}\theta$$

with $u \in C^\infty(M)$. If $F_{\hat{\theta}}$ is the Fefferman metric of $(M, \hat{\theta})$ then we are going to show that

$$F_{\hat{\theta}} = e^{2u\circ\pi} F_\theta \tag{2.36}$$

(hence $[F_\theta] = [F_{\hat{\theta}}]$, i.e., $[F_\theta]$ is a CR invariant). At first, we need to derive the transformation law of the connection 1-forms $\omega_\beta{}^\alpha$ under a transformation $\hat{\theta} = e^{2u}\theta$. We establish the following lemma:

Lemma 2.6. *Let $\hat{\omega}_\beta{}^\alpha$ be the connection 1-forms of the Tanaka–Webster connection of $(M, \hat{\theta})$. Then*

$$\begin{aligned}\hat{\omega}_\beta{}^\alpha = \omega_\beta{}^\alpha + \delta^\alpha_\beta du + 2(u_\beta\theta^\alpha - u^\alpha\theta_\beta) + (u_\mu\theta^\mu - u^\mu\theta_\mu)\delta^\alpha_\beta \\ + \frac{i}{2}(\nabla^\alpha u_\beta + \nabla_\beta u^\alpha + 4u_\beta u^\alpha + 4u_\mu u^\mu\delta^\alpha_\beta)\theta.\end{aligned} \tag{2.37}$$

Proof. Define the tangent vector field $\hat{T}$ by setting

$$\hat{T} = e^{-2u}(T + i\, u^{\overline{\gamma}} T_{\overline{\gamma}} - i\, u^\gamma T_\gamma).$$

A calculation then shows that

$$\hat{\theta}(\hat{T}) = 1, \quad \hat{T} \rfloor d\hat{\theta} = 0;$$

hence $\hat{T}$ is the characteristic direction of $(M, \hat{\theta})$. Let $\pi_H : T(M) \to H(M)$ (respectively $\hat{\pi}_H : T(M) \to H(M)$) be the natural projection associated with the direct sum decomposition $T(M) = H(M) \oplus \mathbf{R}T$ (respectively with $T(M) = H(M) \oplus \mathbf{R}\hat{T}$). Then

$$\hat{\pi}_H = \pi_H + i\,\theta \otimes (u^\alpha T_\alpha - u^{\overline{\alpha}} T_{\overline{\alpha}}). \tag{2.38}$$

Let ∇u be the gradient of u with respect to g_θ, i.e., $g_\theta(\nabla u, X) = X(u)$ for any $X \in T(M)$. Then (2.38) may be written also in the (index-free) form

$$\hat{\pi}_H = \pi_H + \theta \otimes J\pi_H \nabla u,$$

where J is the complex structure in $H(M)$. Let us recall that (by (1.51))

$$\Gamma^\alpha_{\gamma\beta} = h^{\overline{\sigma}\alpha}\left\{T_\gamma(h_{\beta\overline{\sigma}}) g_\theta(T_\beta, [T_\gamma, T_{\overline{\sigma}}])\right\}.$$

On the other hand,

$$g_{\hat{\theta}}(T_\beta, W) = e^{2u}\left\{g_\theta(T_\beta, W) - i\, u_\beta \theta(W)\right\}, \tag{2.39}$$

for any $W \in T(M) \otimes \mathbf{C}$. Using (2.39) and

$$\theta([T_\gamma, T_{\overline{\sigma}}]) = -2i\, h_{\gamma\overline{\sigma}}$$

one easily shows that

$$\hat{\Gamma}^\alpha_{\gamma\beta} = \Gamma^\alpha_{\gamma\beta} + 2(u_\gamma \delta^\alpha_\beta + u_\beta \delta^\alpha_\gamma). \tag{2.40}$$

Next, let us recall that (by (1.52))

$$\Gamma^\alpha_{\overline{\gamma}\beta} = h^{\overline{\mu}\alpha} g_\theta([T_{\overline{\gamma}}, T_\beta], T_{\overline{\mu}}).$$

Again by (2.39), a calculation leads to

$$\hat{\Gamma}^\alpha_{\overline{\gamma}\beta} = \Gamma^\alpha_{\overline{\gamma}\beta} - 2u^\alpha h_{\beta\overline{\gamma}}\,. \tag{2.41}$$

Moreover, we have

$$\Gamma^\gamma_{0\alpha} = h^{\gamma\overline{\beta}} \left\{ T(h_{\alpha\overline{\beta}}) + g_\theta(T_\alpha, [T_{\overline{\beta}}, T]) \right\}. \tag{2.42}$$

This may be obtained from the identities

$$[T(h_{\alpha\overline{\beta}}) = h_{\alpha\overline{\gamma}} \Gamma^{\overline{\gamma}}_{0\overline{\beta}} + h_{\gamma\overline{\beta}} \Gamma^\gamma_{0\alpha}\,, \quad [T, T_{\overline{\beta}}] = \Gamma^{\overline{\gamma}}_{0\overline{\beta}} T_{\overline{\gamma}} - A^\mu_{\overline{\beta}} T_\mu$$

(and the reader may check that the expressions (2.42) and (1.53) are equivalent). Using (2.39) and

$$\begin{aligned}
[T_\mu, T_{\overline{\beta}}] &= \Gamma^{\overline{\rho}}_{\mu\overline{\beta}} T_{\overline{\rho}} - \Gamma^\rho_{\overline{\beta}\mu} T_\rho - 2i\, h_{\mu\overline{\beta}} T, \\
[T_{\overline{\beta}}, \hat{T}] &= e^{-2u}\{-2u_{\overline{\beta}}(T + i\, u^{\overline{\mu}} T_{\overline{\mu}} - i\, u^\mu T_\mu) \\
&\quad + [T_{\overline{\beta}}, T] + i\, u^{\overline{\mu}}[T_{\overline{\beta}}, T_{\overline{\mu}}] + i\, T_{\overline{\beta}}(u^{\overline{\mu}}) T_{\overline{\mu}} \\
&\quad - i\, u^\mu [T_{\overline{\beta}}, T_\mu] - i\, T_{\overline{\beta}}(u^\mu) T_\mu\},
\end{aligned}$$

a calculation yields

$$e^{2u} \hat{\Gamma}^\gamma_{0\alpha} = \Gamma^\gamma_{0\alpha} + 2u_0 \delta^\gamma_\alpha + i\{\nabla^\gamma u_\alpha - 2u_\alpha u^\gamma + u^{\overline{\rho}} \Gamma^\gamma_{\overline{\rho}\alpha} - u^\rho \Gamma^\gamma_{\rho\alpha}\}. \tag{2.43}$$

Here we adopt the following notation for the second-order covariant derivatives

$$\nabla^\gamma u_\alpha = h^{\gamma\overline{\beta}} \nabla_{\overline{\beta}} u_\alpha\,, \quad \nabla_{\overline{\beta}} u_\alpha = T_{\overline{\beta}}(u_\alpha) - \Gamma^\mu_{\overline{\beta}\alpha} u_\mu\,.$$

Consider the 1-forms $\hat{\theta}^\alpha$ on M given by

$$\hat{\theta}^\alpha = \theta^\alpha + i\, u^\alpha \theta.$$

Then $\hat{\theta}^\alpha(T_\beta) = \delta^\alpha_\beta$, $\hat{\theta}^\alpha(T_{\overline{\beta}}) = 0$ and $\hat{\theta}^\alpha(\hat{T}) = 0$, that is, $\{\hat{\theta}^\alpha\}$ is an admissible coframe. At this point we may use the transformation laws (2.40)–(2.41) and (2.43) of the connection coefficients to prove Lemma 2.6. Indeed

$$\hat{\omega}_\beta{}^\alpha = \hat{\Gamma}^\alpha_{\mu\beta}\hat{\theta}^\mu + \hat{\Gamma}^\alpha_{\overline{\mu}\beta}\hat{\theta}^{\overline{\mu}} + \hat{\Gamma}^\alpha_{\hat{0}\beta}\hat{\theta}$$

may be written as

$$\begin{aligned}\hat{\omega}_\beta{}^\alpha &= \omega_\beta{}^\alpha + 2(u_\beta\theta^\alpha - u^\alpha\theta_\beta) \\ &\quad + 2u_\mu\theta^\mu\delta^\alpha_\beta + 2u_0\delta^\alpha_\beta\theta + i(\nabla^\alpha u_\beta + 2u_\beta u^\alpha + 2u_\mu u^\mu\delta^\alpha_\beta)\theta.\end{aligned}$$

Finally, to see that this is equivalent to (2.37) one may use

$$\nabla_{\overline{\mu}} u_\lambda = \nabla_\lambda u_{\overline{\mu}} + 2i\, h_{\lambda\overline{\mu}} u_0 \, .$$

Lemma 2.7. *If $\hat{A}_{\alpha\beta}$ is the pseudo-Hermitian torsion of the Tanaka–Webster connection of $(M, \hat{\theta})$ then*

$$\hat{A}_{\alpha\beta} = A_{\alpha\beta} + i\nabla_\alpha u_\beta - 2i\, u_\alpha u_\beta \, . \tag{2.44}$$

Proof. We recall (cf. (1.58)) that

$$A_\alpha^{\overline{\lambda}} = -h^{\mu\overline{\lambda}} g_\theta(T_\mu, [T, T_\alpha]),$$

and use once again (2.39). □

Let us contract the indices α and β in (2.37) and use the expression of the sub-Laplacian in Proposition 2.2. We obtain

$$\hat{\omega}_\alpha{}^\alpha = \omega_\alpha{}^\alpha + (n+2)(u_\alpha\theta^\alpha - u^\alpha\theta_\alpha) + \frac{i}{2}\,\{4(n+1)u_\alpha u^\alpha + \Delta_b u\}\theta + n\, du. \tag{2.45}$$

Differentiation of (2.45) gives

$$d\hat{\omega}_\alpha{}^\alpha \equiv d\omega_\alpha{}^\alpha + (n+2)d(u_\alpha\theta^\alpha - u^\alpha\theta_\alpha) + \frac{i}{2}\,\{4(n+1)u_\alpha u^\alpha + \Delta_b u\}\, d\theta \bmod \theta.$$

Here we use the standard relation of *congruence* of differential forms:

Definition 2.16. Two q-forms φ, ψ on M are *congruent* modulo θ (and we write $\varphi \equiv \psi \bmod \theta$) if $\varphi = \psi + \theta \wedge \eta$ for some $(q-1)$-form η on M. □

On the other hand, a calculation shows that

$$d(u_\alpha\theta^\alpha - u^\alpha\theta_\alpha) \equiv -(\nabla_\alpha u_{\overline{\beta}} + \nabla_{\overline{\beta}} u_\alpha)\theta^\alpha \wedge \theta^{\overline{\beta}} \bmod \theta^\alpha \wedge \theta^\beta\,,\ \theta^{\overline{\alpha}} \wedge \theta^{\overline{\beta}}\,,\ \theta;$$

hence

$$\begin{aligned}d\hat{\omega}_\alpha{}^\alpha &\equiv d\omega_\alpha{}^\alpha - (n+2)(\nabla_\alpha u_{\overline{\beta}} + \nabla_{\overline{\beta}} u_\alpha)\theta^\alpha \wedge \theta^{\overline{\beta}} \\ &\quad + \{2(n+1)u_\alpha u^\alpha + \Delta_b u\}d\theta \bmod \theta^\alpha \wedge \theta^\beta\,,\ \theta^{\overline{\alpha}} \wedge \theta^{\overline{\beta}}\,,\ \theta.\end{aligned} \tag{2.46}$$

The trace function of a differential 2-form obviously depends on the choice of the pseudo-Hermitian structure θ on M. Since we deal with the effect of $\hat{\theta} = e^{2u}\theta$ on various tensor fields (depending on θ) we refine our notation trace(ω) to $\mathrm{trace}_\theta(\omega)$ (for the remainder of this section only). Let us recall that

$$\mathrm{trace}_\theta(i\, d\omega_\alpha{}^\alpha) = \frac{1}{2}\rho.$$

Hence

$$\mathrm{trace}_\theta(i\, d\hat{\omega}_\alpha{}^\alpha) = \frac{1}{2}h^{\lambda\overline{\mu}}(i\, d\hat{\omega}_\alpha{}^\alpha)_{\lambda\overline{\mu}} = e^{2u}\,\mathrm{trace}_{\hat{\theta}}(i\, d\hat{\omega}_\alpha{}^\alpha) = \frac{1}{2}e^{2u}\hat{\rho}.$$

Let us apply the operator trace_θ to both sides of (2.46) to get

$$e^{2u}\hat{\rho} = \rho - 4(n+1)\Delta_b u - 4n(n+1)u_\alpha u^\alpha \tag{2.47}$$

(since $\mathrm{trace}_\theta(d\theta) = n$). At this point we may derive the transformation law for σ. Recall (cf. (2.31)) that

$$\sigma = \frac{1}{n+2}\left\{d\gamma + \pi^*(i\,\omega_\alpha{}^\alpha - \frac{i}{2}h^{\alpha\overline{\beta}}dh_{\alpha\overline{\beta}} - \frac{1}{4(n+1)}\rho\theta)\right\}.$$

Using (2.37) one first obtains

$$\begin{aligned} i\,\hat{\omega}_\alpha{}^\alpha &- \frac{i}{2}\hat{h}^{\alpha\overline{\beta}}d\hat{h}_{\alpha\overline{\beta}} - \frac{\hat{\rho}}{4(n+1)}\hat{\theta} \\ &= i\,\omega_\alpha{}^\alpha - \frac{i}{2}h^{\alpha\overline{\beta}}dh_{\alpha\overline{\beta}} - \frac{\rho}{4(n+1)}\theta + (n+2)\{i\,(u_\alpha\theta^\alpha - u^\alpha\theta_\alpha) - u_\alpha u^\alpha\theta\}. \end{aligned}$$

Let us see how the fiber coordinate $\gamma : \pi^{-1}(U) \to \mathbf{R}$ depends on the choice of θ. If $c \in [\omega] \in C(M)$ then $\omega = f\ (\theta^{01\cdots n})_x$ and $\omega = \hat{f}\ (\hat{\theta}^{01\cdots n})_x$ lead to $\hat{f} = e^{-2u(x)}f$ hence $f/|f| = \hat{f}/|\hat{f}|$ and $\gamma = \hat{\gamma}$. We may then conclude that

$$\hat{\sigma} = \sigma + \pi^*\{i\,(u_\alpha\theta^\alpha - u^\alpha\theta_\alpha) - u_\alpha u^\alpha\theta\}. \tag{2.48}$$

At this point we are able to prove (2.36). To this end, let us note that

$$\tilde{L}_\theta = 2\theta^\alpha \odot \theta_\alpha$$

on $T(M)\otimes\mathbf{C}$. Here $\tilde{L}_\theta$ is the degenerate extension to the whole of $T(M)\otimes\mathbf{C}$ of L_θ, obtained by setting $\tilde{L}_\theta = L_\theta$ on $T_{1,0}(M)\otimes T_{0,1}(M)$ and $L_\theta(T, W) = 0$ for any $W \in T(M)\otimes\mathbf{C}$. Clearly, the extension of $\tilde{G}_\theta$ (by $\mathbf{C}$-linearity) and L_θ coincide. Then

$$\tilde{L}_{\hat{\theta}} = e^{2u}\tilde{L}_\theta + 2e^{2u}\{i\,(u^\alpha\theta_\alpha - u_\alpha\theta^\alpha)\odot\theta + u_\alpha u^\alpha\theta^2\}$$

and

$$(\pi^*\hat{\theta})\odot\hat{\sigma} = e^{2u}(\pi^*\theta)\odot\sigma + e^{2u\circ\pi}\pi^*\{i(u_\alpha\theta^\alpha - u^\alpha\theta_\alpha)\odot\theta - u_\alpha u^\alpha\theta^2\}$$

lead to $\hat{h} = e^{2u\circ\pi}h$ and Theorem 2.3 is completely proved. □

2.5 The wave operator

Let $(M, T_{1,0}(M))$ be a strictly pseudoconvex CR manifold of CR dimension n and θ a pseudo-Hermitian structure on M such that L_θ is positive definite. Let F_θ be the Fefferman metric of (M, θ) and $\Box$ the Laplace–Beltrami operator, or the wave operator, of $(C(M), F_\theta)$. Then $\Box$ is invariant under any isometry of $(C(M), F_\theta)$:

Proposition 2.6. *Let $\phi \in \mathrm{Isom}(C(M), F_\theta)$. Let us set $v^\phi = v \circ \phi^{-1}$ and $\Box^\phi v = \left(\Box v^{\phi^{-1}}\right)^\phi$, for any $v \in \Omega^0(C(M))$. Then $\Box^\phi = \Box$.*

Cf., e.g., S. Helgason [196]. On the other hand,

$$S^1 \subset \mathrm{Isom}(C(M), F_\theta).$$

Indeed, if $z \in S^1$ then

$$e^{i\gamma \circ R_z} = z e^{i\gamma},$$

where $R_z : C(M) \to C(M)$ is the right translation by z. Hence $d(\gamma \circ R_z) = d\gamma$. Then (by (2.30)) $R_z^* g = g$.

Proposition 2.7. *$\Box$ is S^1-invariant.*

Given $u \in \Omega^0(M)$, by $\pi \circ R_z = \pi$ and by the S^1-invariance of $\Box$ we obtain

$$(\Box(u \circ \pi)) \circ R_z = \Box(u \circ \pi);$$

hence $\Box(u \circ \pi)$ descends to a function on M, denoted by the same symbol $\Box(u \circ \pi)$. That is to say, $\Box$ pushes forward to a differential operator

$$\pi_* \Box : \Omega^0(M) \to \Omega^0(M)$$

given by

$$(\pi_* \Box) u = \Box(u \circ \pi),$$

for any $u \in \Omega^0(M)$. We shall need the following result:

Proposition 2.8. (J.M. Lee [271])
The wave operator $\Box$ (i.e., the Laplacian of $(C(M), F_\theta)$) and the sub-Laplacian Δ_b of (M, θ) are related by

$$\pi_* \Box = \Delta_b .$$

Proof. The wave operator $\Box$ (on functions) of $(C(M), F_\theta)$ is given by

$$\Box = -d^* d,$$

where d^* is the formal adjoint of d with respect to the L^2 inner product

$$(v, w)_g = \int_{C(M)} v\overline{w} * 1$$

for any C^∞ functions $v, w : C(M) \to \mathbf{C}$ (at least one of compact support). Here $*$ is the Hodge operator (with respect to the Fefferman metric F_θ). To prove Proposition 2.8 we observe first that given a C^∞ function f on M with $\text{supp}(f) \subset U$, where U is the domain of a chart $\Phi : \pi^{-1}(U) \to U \times S^1$ of $C(M)$, we have

$$\int_{C(M)} (f \circ \pi) * 1 = 2\pi \int_M f\Psi. \tag{2.49}$$

Indeed, let $\{T_\alpha\}$ be a local frame of $T_{1,0}(M)$ on U, so that $h_{\alpha\overline{\beta}} = \delta_{\alpha\beta}$. Then (by (1.66))

$$\Psi = 2^n i^{n^2} n!\, \theta \wedge \theta^{1\cdots n} \wedge \theta^{\overline{1}\cdots\overline{n}}.$$

Let $x^i : U \to \mathbf{R}$ be local coordinates on U. Then

$$\Psi = 2^n i^{n^2} n! \det(\theta^A_B) dx^1 \wedge \cdots , dx^{2n+1}$$

where

$$\theta^j = \theta^j_A dx^A , \quad 0 \le j \le n, \quad \theta^{\alpha+n}_A = \overline{\theta^\alpha_A}.$$

Moreover,

$$*1 = c_n d\gamma \wedge \pi^*\Psi$$

(for some constant $c_n > 0$ depending only on n and on the orientation of M, cf., e.g., E. Barletta et al. [44]) and $(x^i \circ \pi, \gamma)$ are local coordinates on $C(M)$; hence $\int_{C(M)} (f \circ \pi) * 1$ may be calculated by the theorem of Fubini. Finally (using (2.49) twice),

$$\begin{aligned}
\int_M ((\pi_* \Box) u)\, \overline{v} \Psi &= -\frac{1}{2\pi} \int_{C(M)} (\Box(u \circ \pi))\, (\overline{v} \circ \pi) * 1 \\
&- \frac{1}{2\pi} \int_{C(M)} F^*_\theta \left(d(u \circ \pi), d(\overline{v} \circ \pi)\right) * 1 \\
&= -\frac{1}{2\pi} \int_{C(M)} L^*_\theta (d_b u, d_b \overline{v}) \circ \pi \ * 1 \\
&= - \int_M L^*_\theta (d_b u, d_b \overline{v}) \Psi = \int_M (\Delta_b u) \overline{v} \Psi.
\end{aligned}$$

□

2.6 Curvature of Fefferman's metric

Let K be the scalar curvature of the Fefferman metric F_θ on $C(M)$. By the S^1 invariance of F_θ it follows that K is constant on the fibers of π and thus it descends to a function on M, which will be denoted by $\pi_* K$. Define $D_\theta \in C^\infty(M)$ by setting

$$D_\theta = \pi_* K - \frac{2n+1}{n+1}\,\rho.$$

Let $F_{\hat\theta}$ be the Fefferman metric of $\hat\theta = e^{2u}\theta$. Then $F_{\hat\theta} = e^{2u\circ\pi} F_\theta$; hence a standard calculation (cf., e.g., T. Aubin [23]) shows that the scalar curvature $\hat K$ of $F_{\hat\theta}$ is related to K by

$$\hat K = e^{-2u\circ\pi}\{K + 2(2n+1)\Box(u\circ\pi) - 2n(2n+1)F^*_\theta(\pi^* du, \pi^* du)\}. \tag{2.50}$$

Let us observe that

$$F^*_\theta(\pi^* du, \pi^* du) = L^*_\theta(d_b u, d_b u) = 2u_\alpha u^\alpha.$$

At this point, using (2.50) and (2.47) we get

$$D_{\hat\theta} = e^{-2u} D_\theta.$$

This may be shown (cf. J.M. Lee [271], p. 426) to yield $D_\theta = 0$:

Proposition 2.9. (J.M. Lee [271])

$$\pi_* K = \frac{2n+1}{n+1}\,\rho.$$

The proof of Proposition 2.9 relies on the Chern–Moser *normal form* (cf. [99]) and falls beyond the purposes of this book. The reader may find an excellent account of the Chern–Moser normal form in the monograph by H. Jacobowitz [221].

This relationship between the pseudo-Hermitian scalar curvature ρ of (M, θ) and the scalar curvature K of the Fefferman metric F_θ suggests several geometric problems on the given CR manifold M. An important example, to be discussed in detail in Chapter 3 of this book, is the Yamabe problem for the Fefferman metric F_θ on $C(M)$, i.e., find a representative $\hat F_\theta$ in the restricted conformal class of F_θ such that its scalar curvature $\hat K$ is a constant λ. This may be reformulated as a nonelliptic problem, because the principal part of the relevant (nonlinear) equation is the wave operator, yet it may be reduced to a *subelliptic* problem on M, known as the *CR Yamabe problem*; cf. D. Jerison and J.M. Lee [226] [227]. Precisely, the Yamabe equation on $C(M)$ projects on M to an equation of the form

$$c_n \Delta_b u + \rho u = \lambda u^{p-1}. \tag{2.51}$$

As previously mentioned, the analysis of the solutions to (2.51) lies within the scope of Chapter 3 of this book.

2.7 Pontryagin forms

By classical work of S.S. Chern and J. Simons [100], the Pontryagin forms of a Riemannian manifold are conformal invariants. On the other hand, the restricted conformal class of the Fefferman metric was shown (cf. Theorem 2.3) to be a CR invariant.

This is evidence enough to ask whether the result by S.S. Chern and J. Simons (cf. op. cit.) may carry over to Lorentz geometry. One finds (cf. Theorem 2.6 below) that the Pontryagin forms $P(\Omega^\ell)$ of the Fefferman metric are CR invariants of M. Also, whenever $P(\Omega^\ell) = 0$, the de Rham cohomology class of the corresponding transgression form is a CR invariant as well. As an application, we shall show that a necessary condition for M to be globally CR equivalent to the sphere S^{2n+1} is that $P_1(\Omega^2) = 0$ (i.e., the first Pontryagin form of $(C(M), F_\theta)$ must vanish) and the corresponding transgression form gives an integral cohomology class.

We start with a brief review of Chern–Weil theory (cf. also S. Kobayashi and K. Nomizu [241], vol. II, pp. 293–320). Let G be a Lie group with finitely many components and let $\mathcal{G} = L(G)$ be its Lie algebra. Let $\{E_G, B_G\}$ be a universal bundle and classifying space for G. Its key property is that each principal G-bundle E over M admits a bundle map into $\{E_G, B_G\}$, and any two such maps of the same G-bundle $\{E, M\}$ are homotopic. Since G has finitely many components,

$$H^{2\ell-1}(B_G, \mathbf{R}) = 0,$$

for all ℓ. Also, E_G is contractible. Next, if Λ is any coefficient ring and $u \in H^k(B_G, \Lambda)$, then with any principal G-bundle $\alpha = \{E, M\}$ one may associate the characteristic class $u(\alpha) \in H^k(M, \Lambda)$, built by pulling back u under any bundle map.

Let us set $\mathcal{G}^\ell = \mathcal{G} \otimes \cdots \otimes \mathcal{G}$ (ℓ terms). A symmetric multilinear map $P : \mathcal{G}^\ell \to \mathbf{R}$ is a *polynomial* of degree ℓ. Let $I^\ell(G)$ be the space of *invariant* polynomials of degree ℓ, i.e., if $P \in I^\ell(G)$ then

$$P((\mathrm{ad}(g)A_1) \otimes \cdots \otimes (\mathrm{ad}(g)A_\ell)) = P(A_1 \otimes \cdots \otimes A_\ell),$$

for any $g \in G$ and any $A_j \in \mathcal{G}$, $1 \leq j \leq \ell$. Invariant polynomials multiply in a natural way; hence $I(G) = \sum_{\ell \geq 0} I^\ell(G)$ turns out to be a graded ring. These polynomials give information about the real cohomology of the classifying space B_G. We recall the *universal Weil homomorphism*

$$W : I^\ell(G) \to H^{2\ell}(B_G, \mathbf{R}).$$

The Chern–Weil theorem is that $P(\Omega^\ell)_M \in W(P)(\alpha)$ for any $P \in I^\ell(G)$ and any principal G-bundle $\alpha = \{E, M\}$ with a connection 1-form $\omega \in \Gamma^\infty(T^*(E) \otimes \mathcal{G})$. Here $\Omega = D\omega$ is the curvature 2-form of ω and $\Omega^\ell = \Omega \wedge \cdots \wedge \Omega$ (ℓ terms). Also $P(\Omega^\ell) \in \Gamma^\infty(\Lambda^{2\ell} T^* E)$ is given by $P(\Omega^\ell) = P \circ \Omega^\ell$. The 2ℓ-form $P(\Omega^\ell)$ is closed, invariant, and horizontal; hence it projects to a (closed) 2ℓ-form on M denoted by $P(\Omega^\ell)_M$.

Let $\epsilon(G)$ be the category whose objects are triples $\{E, M, \omega\}$, where $\{E, M\}$ is a principal G-bundle and ω a connection 1-form on E, and whose morphisms are connection-preserving bundle maps. An object $A \in \epsilon(G)$ is *n-classifying* if (i) for any $\alpha = \{E, M, \omega\} \in \epsilon(G)$ with $\dim(M) \leq n$ there is a morphism $\alpha \to A$, and (ii) any two such morphisms are homotopic through bundle maps (not necessarily through $\epsilon(G)$-morphisms). Then the M.S. Narasimhan and S. Ramanan theorem is that for each n there is an n-classifying object $A \in \epsilon(G)$; cf. [325].

Let $\alpha = \{E, M, \omega\} \in \epsilon(G)$ and set

$$TP(\omega) = \ell \int_0^1 P(\omega \wedge \Omega_t^{\ell-1}) dt,$$

where

$$\Omega_t = t\Omega + \frac{t(t-1)}{2}[\omega, \omega]$$

for any $P \in I^\ell(G)$. Then $TP(\omega)$ is an invariant $(2\ell - 1)$-form on E (the *transgression* form) and $d\,TP(\omega) = P(\Omega^\ell)$. Let us set

$$I_0^\ell(G) = \{P \in I^\ell(G) : W(P) \in H^{2\ell}(B_G, \mathbf{Z})\}.$$

Building on the M.S. Narasimhan and S. Ramanan theorem, S.S. Chern and J. Simons have shown (cf. Theorem 3.16 in [100], p. 56) that given $\alpha = \{E, M, \omega\} \in \epsilon(G)$ and $P \in I_0^\ell(G)$ with $P(\Omega^\ell) = 0$ there is $U \in H^{2\ell-1}(M, \mathbf{R}/\mathbf{Z})$ such that $q[TP(\omega)] = \pi^* U$. Here q is the natural homomorphism $H^{2\ell-1}(E, \mathbf{R}) \to H^{2\ell-1}(E, \mathbf{R}/\mathbf{Z})$ and $\pi : E \to M$ the projection.

Let $Q_\ell \in I^\ell(\mathrm{GL}(2n))$, $1 \le \ell \le 2n$, be the natural generators of the ring of invariant polynomials on $\mathbf{gl}(2n) = L(\mathrm{GL}(2n))$, where $\mathrm{GL}(2n)$ is short for $\mathrm{GL}(2n, \mathbf{R})$ (cf. [100], p. 57, for the explicit expression of the Q_ℓ). Let $(M, T_{1,0}(M))$ be a strictly pseudoconvex CR manifold of CR dimension $n-1$ and θ a pseudo-Hermitian structure on M such that L_θ is positive definite. Let F_θ be the Fefferman metric of (M, θ). Let $F(C(M)) \to C(M)$ be the principal $\mathrm{GL}(2n)$-bundle of all linear frames tangent to $C(M)$ and $\omega \in \Gamma^\infty(T^*(F(C(M))) \otimes \mathbf{gl}(2n))$ the connection 1-form (of the Levi-Civita connection) of the Lorentz manifold $(C(M), F_\theta)$.

Theorem 2.5. *The characteristic forms* $Q_{2\ell+1}(\Omega^{2\ell+1})$ *vanish for* $0 \le \ell \le n-1$.

Proof. Let $\mathcal{L}(C(M)) \to C(M)$ be the principal $\mathrm{O}(2n-1, 1)$-bundle of all Lorentz frames, i.e., $u = (c, \{X_i\}) \in \mathcal{L}(C(M))$ if $F_{\theta,c}(X_i, X_j) = \epsilon_i \delta_{ij}$, where $\epsilon_\alpha = 1$, $1 \le \alpha \le 2n-1$; and $\epsilon_{2n} = -1$ and $c \in C(M)$. Here $\mathrm{O}(2n-1, 1)$ is the Lorentz group. Let $\mathbf{o}(2n-1, 1)$ be its Lie algebra. By hypothesis

$$\omega_u(T_u(\mathcal{L}(C(M)))) \subseteq \mathbf{o}(2n-1, 1),$$

i.e., $\epsilon\, \omega_u(X) + \omega_u(X)^t \epsilon = 0$ for any $X \in T_u(\mathcal{L}(C(M)))$, $u \in \mathcal{L}(C(M))$. Here $\epsilon = \mathrm{diag}(\epsilon_1, \ldots, \epsilon_{2n})$. Let $\{E_j^i\}$ be the canonical basis of $\mathbf{gl}(2n)$ and set $\omega = \omega_j^i \otimes E_i^j$. We claim that

$$\epsilon^i \Omega_j^i + \epsilon^j \Omega_i^j = 0, \tag{2.52}$$

at all points of $\mathcal{L}(C(M))$, as a form on $F(C(M))$. Here $\epsilon^i = \epsilon_i$. Since Ω is horizontal, it suffices to check (2.52) on horizontal vectors (hence tangent to $\mathcal{L}(C(M))$). We have

$$\epsilon^i \Omega_j^i = \epsilon^i \left(d\omega_j^i + \omega_k^i \wedge \omega_j^k \right) = d(-\epsilon^j \omega_i^j) + \sum_k (-\epsilon^k \omega_i^k) \wedge \omega_j^k = -\epsilon^j \Omega_i^j$$

on $T_u(\mathcal{L}(C(M)))$ for any $u \in \mathcal{L}(C(M))$, etc. Next, note that for any $A \in \mathbf{o}(2n-1,1)$ one has (i) $\text{trace}(A) = 0$, (ii) $\text{trace}(AB) = 0$ for any $B \in \mathcal{M}_{2n}(\mathbf{R})$ satisfying $B = \epsilon B^t \epsilon$, and (iii) $\text{trace}(A^{2\ell+1}) = 0$. Then

$$\text{trace}(A_1 \cdots A_{2\ell+1}) = 0, \tag{2.53}$$

for any $A_1, \ldots, A_{2\ell+1} \in \mathbf{o}(2n-1,1)$ (the proof is by induction over ℓ). Since $Q_{2\ell+1}(\Omega^{2\ell+1})$ is invariant, we need only show that it vanishes at the points of $\mathcal{L}(C(M))$. But at those points the range of $\Omega^{2\ell+1}$ lies (by (2.52)–(2.53)) in the kernel of $Q_{2\ell+1}$. Our Theorem 2.5 is proved. □

By Theorem 2.5, the transgression forms $TQ_{2\ell+1}(\omega)$ are closed; hence we get the cohomology classes

$$[TQ_{2\ell+1}(\omega)] \in H^{4\ell+1}(F(C(M)), \mathbf{R}).$$

Let us observe that

$$[TQ_{2\ell+1}(\omega)] \in \text{Ker}(j^*), \tag{2.54}$$

where the homomorphism

$$j^* : H^{4\ell+1}(F(C(M)), \mathbf{R}) \to H^{4\ell+1}(\mathcal{L}(C(M)), \mathbf{R})$$

is induced by $j : \mathcal{L}(C(M)) \subset F(C(M))$. Indeed $TQ_{2\ell+1}(\omega)$ may be written as

$$TQ_{2\ell+1}(\omega) = \sum_{i=0}^{2\ell} B_i Q_{2\ell+1}(\omega \wedge [\omega, \omega]^i \wedge \Omega^{2\ell-i}),$$

for some constants $B_i > 0$. Since $j^*\omega$ is $\mathbf{o}(2n-1,1)$-valued, the same argument as in the proof of Theorem 2.5 shows that $j^* TQ_{2\ell+1}(\omega) = 0$.

A remark is in order. One has to work with $j^*\omega$ (rather than ω at points of $\mathcal{L}(C(M))$) because ω (unlike its curvature form) is not horizontal.

If g_0 is a Riemannian metric on $C(M)$ with connection 1-form ω_0 and $O(C(M)) \to C(M)$ is the principal $O(2n)$-bundle of orthonormal (with respect to g_0) frames tangent to $C(M)$, then orthonormalization of frames gives a deformation retract $F(C(M)) \to O(C(M))$ and hence (cf. Proposition 4.3 in [100], p. 58) the transgression forms $TQ_{2\ell+1}(\omega_0)$ are exact. As to the Lorentz case, in general (2.54) need not imply exactness of $TQ_{2\ell+1}(\omega)$. For instance $\mathbf{R}^2_1$ is a Lorentz manifold for which the homomorphism

$$j^* : H^1(F(\mathbf{R}^2_1), \mathbf{R}) \to H^1(\mathcal{L}(\mathbf{R}^2_1), \mathbf{R})$$

(induced by $j : \mathcal{L}(\mathbf{R}^2_1) \subset F(\mathbf{R}^2_1)$) has a nontrivial kernel. Here we have set

$$\mathbf{R}^N_\nu = \left(\mathbf{R}^N, \langle\, , \rangle_{N-\nu,\nu}\right), \quad \langle x, y\rangle_{N-\nu,\nu} = \sum_{i=1}^{N-\nu} x_i y_i - \sum_{i=N-\nu+1}^{N} x_i y_i \,.$$

Indeed, since both $F(\mathbf{R}^2_1)$ and $\mathcal{L}(\mathbf{R}^2_1)$ are trivial bundles, j^* may be identified with the homomorphism

$$j^* : H^1(\mathrm{GL}(2), \mathbf{R}) \to H^1(\mathrm{O}(1,1), \mathbf{R})$$

induced by $j : \mathrm{O}(1,1) \subset \mathrm{GL}(2)$. Now, the Lorentz group $\mathrm{O}(1,1)$ has four components, each diffeomorphic to $\mathbf{R}$. Hence $H^1(\mathrm{O}(1,1), \mathbf{R}) = 0$. Moreover $\mathrm{O}(2) \subset \mathrm{GL}(2)$ is a homotopy equivalence; hence $\mathrm{Ker}(j^*) = H^1(\mathrm{GL}(2), \mathbf{R}) = H^1(\mathrm{O}(2), \mathbf{R}) = \mathbf{R} \oplus \mathbf{R}$ (since $\mathrm{O}(2)$ has two components, each diffeomorphic to S^1).

Theorem 2.6. *Let M be a strictly pseudoconvex CR manifold of CR dimension $n-1$ and $P \in I^\ell(\mathrm{GL}(2n))$. Then $P(\Omega^\ell)$ is a CR invariant of M. Moreover, if $P(\Omega^\ell) = 0$ then the cohomology class $[TP(\omega)] \in H^{2\ell-1}(F(C(M)), \mathbf{R})$ is a CR invariant of M. In particular $[TQ_{2\ell+1}(\omega)] \in H^{4\ell+1}(F(C(M)), \mathbf{R})$ is a CR invariant.*

Proof. Let $\varphi \in \Gamma^\infty(T^*(F(C(M))) \otimes \mathbf{R}^{2n})$ be the canonical 1-form and set $\varphi = \varphi^i \otimes e_i$, where $\{e_i\}$ is the canonical basis in $\mathbf{R}^{2n}$. Moreover, let $E_i = B(e_i)$ be the corresponding standard horizontal vector fields (cf., e.g. [241], vol. I, p. 119). Let $u : M \to \mathbf{R}$ be a C^∞ function and let $\hat{F}_\theta$ be the Fefferman metric of $(M, e^{2u}\theta)$. Let $\hat{\omega}$ be the corresponding connection 1-form. Then

$$\hat{\omega}^i_j = \omega^i_j + d(u \circ \rho)\delta^i_j + E_j(u \circ \rho)\varphi^i - \epsilon_i E_i(u \circ \rho)\epsilon_j \varphi^i \tag{2.55}$$

at all points of $\mathcal{L}(C(M))$, as forms on $F(C(M))$. Here $\rho = \pi \circ p_F$ and $\pi : C(M) \to M$ respectively $p_F : F(C(M)) \to C(M)$ are projections. The proof of (2.55) is to relate the Levi-Civita connections of the conformally equivalent Fefferman metrics F_θ and $\hat{F}_\theta$, followed by a translation of the result in principal bundle terminology. The reader may supply the details. Consider the 1-parameter family of Lorentz metrics $g(s) = e^{2s(u \circ \pi)} F_\theta$, $0 \le s \le 1$, on $C(M)$. Let $\omega(s)$ be the corresponding connection 1-form and let us set

$$\omega' = \frac{d}{ds}\{\omega(s)\}_{s=0}\,.$$

By (2.55) (applied to su instead of u) we obtain

$$\omega'^i_j = d(u \circ \rho)\delta^i_j + E_i(u \circ \rho)\varphi^i - \epsilon_i E_i(u \circ \rho)\epsilon_j \varphi^j \tag{2.56}$$

at all points of $\mathcal{L}(C(M))$, as forms on $F(C(M))$. Let $P \in I^\ell(\mathrm{GL}(2n))$. We wish to show that $P(\Omega^\ell)$ is invariant under any transformation $\hat{\theta} = e^{2u}\theta$. Note that a relation of the form

$$TP(\hat{\omega}) = TP(\omega) + \text{exact} \tag{2.57}$$

yields $P(\hat{\Omega}^\ell) = P(\Omega^\ell)$; hence we need only prove (2.57). Since the Q_ℓ generate $I(\mathrm{GL}(2n))$ we may assume that P is a monomial in the Q_ℓ. Using Proposition 3.7 in [100], p. 53, an inductive argument shows that it is sufficient to prove (2.57) for $P = Q_\ell$. It is enough to prove that

$$\frac{d}{ds}\{T Q_\ell(\omega(s))\} = \text{exact}. \tag{2.58}$$

Since each point on the curve $s \mapsto g(s)$ is the initial point of another such curve, it suffices to prove (2.58) at $s = 0$. By Proposition 3.8 in [100], p. 53, we know that

$$\frac{d}{ds}\{T Q_\ell(\omega(s))\}_{s=0} = \ell Q_\ell(\omega' \wedge \Omega^{\ell-1}) + \text{exact};$$

hence it is enough to show that $Q_\ell(\omega' \wedge \Omega^{\ell-1}) = \text{exact}$. Using (2.56) and the identity

$$Q_\ell(\psi \wedge \Omega^{\ell-1}) = \sum_{i_1,\ldots,i_\ell} \psi^{i_1}_{i_2} \wedge \Omega^{i_2}_{i_3} \wedge \cdots \wedge \Omega^{i_\ell}_{i_1}$$

(cf. (4.2) in [100], p. 57) for any $\mathbf{gl}(2n)$-valued form ψ on $F(C(M))$, we may perform the following calculation:

$$\begin{aligned} Q_\ell(\omega' \wedge \Omega^{\ell-1}) &= \sum \omega'^{i_1}_{i_2} \wedge \Omega^{i_2}_{i_3} \wedge \cdots \wedge \Omega^{i_\ell}_{i_1} \\ &= \sum d(u \circ \rho) \wedge \Omega^{i_2}_{i_3} \wedge \cdots \wedge \Omega^{i_\ell}_{i_2} \\ &\quad + \sum \left(E_{i_2}(u \circ \rho)\varphi^{i_1} - \epsilon_{i_1} E_{i_1}(u \circ \rho)\epsilon_{i_2}\varphi^{i_2} \right) \wedge \Omega^{i_2}_{i_3} \wedge \cdots \wedge \Omega^{i_\ell}_{i_1}. \end{aligned}$$

Let us recall the structure equations, cf., e.g. [241], vol. I, p. 121. Since F_θ is Lorentz, ω is torsion-free. Hence $\varphi^{i_1} \wedge \Omega^{i_\ell}_{i_1} = 0$. This and (2.52) also yield $\epsilon_{i_2}\varphi^{i_2} \wedge \Omega^{i_2}_{i_3} = 0$. Hence

$$Q_\ell(\omega' \wedge \Omega^{\ell-1}) = d(u \circ \rho) \wedge Q_{\ell-1}(\Omega^{\ell-1}) = \text{exact}$$

(because $d Q_{\ell-1}(\Omega^{\ell-1}) = 0$) at each point of $\mathcal{L}(C(M))$, as a form on $F(C(M))$. This suffices because both

$$Q_\ell(\omega' \wedge \Omega^{\ell-1}) \text{ and } (u \circ \rho) Q_{\ell-1}(\Omega^{\ell-1})$$

are invariant forms. □

2.8 The extrinsic approach

2.8.1 The Monge–Ampère equation

Let u be a function on a strictly pseudoconvex domain $\Omega \subset \mathbf{C}^n$. We set

$$J(u) = (-1)^n \det \begin{pmatrix} u & u_{\overline{k}} \\ u_j & u_{j\overline{k}} \end{pmatrix},$$

where $u_j = \partial u / \partial z^j$ and $u_{\overline{k}} = \partial u / \partial \overline{z}^k$, etc. Next, we consider the following Dirichlet problem for the complex Monge–Ampère equation

$$
\begin{aligned}
J(u) &= 1 \quad \text{in } \Omega, \\
u &= 0 \quad \text{on } \partial\Omega.
\end{aligned}
\tag{2.59}
$$

The Bergman kernel of Ω is closely related to the solutions of (2.59). For instance, if Ω is the unit ball in $\mathbf{C}^n$ then its Bergman kernel on the diagonal is given by

$$
K(z, z) = \frac{C_n}{\left(1 - |z|^2\right)^{n+1}},
$$

where $C_n = n!/\pi^n$ (cf., e.g., S.G. Krantz [261], p. 50) while $u(z) = 1 - |z|^2$ satisfies (2.59). Therefore, a natural question is, how closely is $K(z, z)$ (the Bergman kernel of an arbitrary pseudoconvex domain Ω) approximated by $C_n / (u(z))^{n+1}$, where u is a solution of (2.59)? Equivalently, is $J(C_n^{1/(n+1)} K^{-1/(n+1)})$ nearly 1? By a result of L. Hörmander [212], $J(C_n^{1/(n+1)} K^{-1/(n+1)}) = 1$ on $\partial\Omega$. On the other hand, if for instance $\Omega = \{z \in \mathbf{C}^n : \phi(|z^1|, \dots, |z^n|) < 1\}$ (where $\phi \mathbf{R}^n \to \mathbf{R}$ is a smooth function) then

$$
J(C_n^{1/(n+1)} K^{-1/(n+1)})_{z=0} \neq 1;
$$

hence the relationship of (2.59) to the Bergman kernel is only asymptotic. To further investigate this phenomenon, we first show that the solutions of (2.59) transform like a negative power of the Bergman kernel (cf., e.g., S.G. Krantz [261], p. 44)

Proposition 2.10. (C. Fefferman [138])
Let $F : \Omega \to \tilde{\Omega}$ be a biholomorphic map and $\tilde{u}$ a real-valued function on $\tilde{\Omega}$. Let us set

$$
u(z) = \left|\det F'(z)\right|^{-2/(n+1)} \tilde{u}(F(z)),
$$

for any $z \in \Omega$. Then

$$
J(u) = J(\tilde{u}) \circ F.
$$

In particular, if $J(\tilde{u}) = 1$ then $J(u) = 1$.

Proof. Here $F'(z) = \left(\partial F^j / \partial z^k\right)$. To prove Proposition 2.10, let us lift F to the biholomorphic map

$$
\mathcal{F} : \mathbf{C}_* \times \Omega \to \mathbf{C}_* \times \tilde{\Omega}
$$

given by

$$
\mathcal{F}(\zeta, z) = \left(\frac{\zeta}{\det F'(z)}, F(z)\right),
$$

for any $\zeta \in \mathbf{C}_*$ and $z \in \Omega$. Here $\mathbf{C}_* = \mathbf{C} \setminus \{0\}$. On the other hand, let us define $U : \mathbf{C}_* \times \Omega \to \mathbf{R}$ (respectively $\tilde{U} : \mathbf{C}_* \times \tilde{\Omega} \to \mathbf{R}$) by setting

$$U(\zeta, z) = |\zeta|^{2/(n+1)} u(z)$$

for any $\zeta \in \mathbf{C}_*$, $z \in \Omega$ (respectively $\tilde{U}(\zeta, z) = |\zeta|^{2/(n+1)} \tilde{u}(z)$ for any $\zeta \in \mathbf{C}_*$, $z \in \tilde{\Omega}$). One may check that

$$U = \tilde{U} \circ \mathcal{F}, \quad \det\left(\mathcal{F}'(\zeta, z)\right) = 1,$$

for any $\zeta \in \mathbf{C}_*$, $z \in \Omega$. Consequently

$$\det(U_{A\overline{B}}) = [\det(\tilde{U}_{A\overline{B}})] \circ \mathcal{F}, \tag{2.60}$$

where $(U_{A\overline{B}})$ is the complex Hessian of U. Finally,

$$\begin{aligned}
\frac{\partial^2 U}{\partial \zeta \partial \overline{\zeta}} &= \frac{1}{(n+1)^2} |\zeta|^{2/(n+1)-2} u(z), \\
\frac{\partial^2 U}{\partial \zeta \partial \overline{z}^k} &= \frac{1}{n+1} \overline{\zeta} |\zeta|^{2/(n+1)-2} u_{\overline{k}}(z), \\
\frac{\partial^2 U}{\partial z^j \partial \overline{z}^k} &= |\zeta|^{2/(n+1)} u_{j\overline{k}}(z),
\end{aligned}$$

and hence

$$\det(U_{A\overline{B}}) = \frac{1}{(n+1)^2} J(u),$$

and (2.60) yields the statement in Proposition 2.10. □

C. Fefferman has developed (cf. [138], pp. 399–400) a formal technique for finding approximate solutions of (2.59). His goal was to produce smooth functions u on some neighborhood of $\overline{\Omega}$, vanishing at $\partial\Omega$, for which $J(u) - 1$ vanishes to high order at $\partial\Omega$. Precisely, let $\Omega \subset \mathbf{C}^n$ be a smoothly bounded domain, i.e., there is U open, $\overline{\Omega} \subset U$, and a smooth function $\psi : U \to \mathbf{R}$ such that $\Omega = \{z \in U : \psi(z) > 0\}$ and $D\psi(z) \neq 0$ for any $z \in \partial\Omega$. Let f be a function on a neighborhood of $\overline{\Omega}$. We say that $f = O(\psi^s)$ if $|f| < C\psi^s$ on $\overline{\Omega}$, for some constant $C > 0$. Next, let us assume that $\Omega = \{\psi > 0\}$ is strictly pseudoconvex and define recursively

$$\begin{aligned}
u^{(1)} &= \frac{\psi}{J(\psi)^{1/(n+1)}}, \\
u^{(s)} &= \left\{ 1 + \frac{1 - J(u^{(s-1)})}{(n+2-s)s} \right\} u^{(s-1)},
\end{aligned}$$

for $2 \leq s \leq n+1$. Then Fefferman's result is that each $u^{(s)}$ satisfies

$$J(u^{(s)}) = 1 + O(\psi^s).$$

For instance, let us look at the first approximation $u^{(1)}$. Clearly

$$J(u^{(1)}) = 1$$

on $\partial\Omega$ because

$$J(\eta\psi) = \eta^{n+1} J(\psi)$$

whenever $\psi = 0$, for any smooth function η. To see that

$$J(u^{(1)}) = 1 + O(\psi)$$

it suffices to use

$$\begin{aligned} u^{(1)}_j &= J(\psi)^{-1/(n+1)} \psi_j + O(\psi), \\ u^{(1)}_{j\bar{k}} &= J(\psi)^{-1/(n+1)} \psi_{j\bar{k}} \\ &\quad - \frac{1}{n+1} J(\psi)^{-1/(n+1)-1} \{\psi_{\bar{k}} J(\psi)_j + \psi_j J(\psi)_{\bar{k}}\} + O(\psi), \end{aligned}$$

as a determinant is a multilinear function of its rows (columns).

To end this section, let us mention that by a result of S.Y. Cheng and S.T. Yau [97], the solution u of (2.59) exists and is unique. Moreover, u is C^∞ in the interior of Ω and belongs to $C^{n+(3/2)-\epsilon}(\overline{\Omega})$.

2.8.2 The Fefferman metric

In this section we present Fefferman's original approach (cf. [138]) to the metric g on $C(M)$, as an induced metric. Let $\Omega \subset \mathbf{C}^n$ be a strictly pseudoconvex domain and let u be the solution of (2.59). This exists and is unique by the S.Y. Cheng and S.T. Yau theorem quoted above (cf. [97]). However, as we shall see shortly, the second approximation $u^{(2)}$ to the solution u of (2.59) is actually sufficient for our purposes. Define $H : U \times \mathbf{C}_* \to \mathbf{R}$ by setting

$$H(z, \zeta) = |\zeta|^{2/(n+1)} u(z),$$

and consider the $(0, 2)$-tensor field G on $U \times \mathbf{C}_*$ given by

$$G = \frac{\partial^2 H}{\partial z^A \partial \bar{z}^B} dz^A \odot d\bar{z}^B .$$

Here $0 \le A, B, \ldots \le n$ and $z^0 = \zeta$. The restriction of G to $\Omega \times \mathbf{C}_*$ is a biholomorphic invariant of Ω, in the following sense. Let $F : \Omega \to \Omega$ be a biholomorphic map. Then, by the proof of Proposition 2.10, F lifts to a biholomorphism $\mathcal{F}$ of $\Omega \times \mathbf{C}_*$ in itself and $H \circ \mathcal{F} = H$; hence $\mathcal{F}^* G = G$. Since $\det(H_{A\bar{B}}) = J(u)/(n+1)^2$ and Ω is strictly pseudoconvex, it follows that G is nondegenerate. Let $j : \partial\Omega \times S^1 \to U \times \mathbf{C}_*$ be the inclusion. We wish to compute the pullback $j^* G$ of G to $\partial\Omega \times S^1$. The metric G may be written explicitly as

$$
\begin{aligned}
G = {} & \frac{u(z)}{(n+1)^2}|\zeta|^{2/(n+1)-2} d\zeta \odot d\overline{\zeta} \\
& + \frac{|\zeta|^{2/(n+1)}}{n+1}(\partial u) \odot (\frac{1}{\overline{\zeta}} d\overline{\zeta}) + \frac{|\zeta|^{2/(n+1)}}{n+1}(\frac{1}{\zeta} d\zeta) \odot (\overline{\partial} u) \\
& + |\zeta|^{2/(n+1)} u_{j\bar{k}} dz^j \odot d\bar{z}^k . \qquad (2.61)
\end{aligned}
$$

Let $\gamma : S^1 \setminus \{(1,0)\} \to (0, 2\pi)$, $\gamma(w) = \arg(w)$, be a local coordinate on S^1. Then

$$
j^* d\zeta = i e^{i\gamma} d\gamma;
$$

hence

$$
j^* G = -\frac{i}{n+1} j^*(\partial u - \overline{\partial} u) \odot d\gamma + j^*(u_{j\bar{k}} dz^j \odot d\bar{z}^k). \qquad (2.62)
$$

Let us set $M = \partial\Omega$ and let $\iota : M \to \mathbf{C}^n$ be the inclusion. Then $\zeta_0 = \iota^* dz^1 \wedge \cdots \wedge dz^n$ is a global C^∞ section in $K^0(M)$; hence $C(M)$ is trivial. Then (2.30) and (2.62) agree (via the isomorphism $C(M) \simeq M \times S^1$). A calculation shows that ζ_0 satisfies the *volume normalization*

$$
2^n i^{n^2} n! \theta \wedge (T \rfloor \zeta_0) \wedge (T \rfloor \overline{\zeta}_0) = \theta \wedge (d\theta)^n
$$

if and only if u satisfies $J(u) = 1$ along M (cf. Proposition 5.2 in [137], p. 43). Let $\alpha_{\zeta_0} \in \Omega^0(M)$ be defined by

$$
d(T \rfloor \zeta_0) \wedge (T \rfloor \overline{\zeta}_0) = i\alpha\zeta_0\theta \wedge (T \rfloor \zeta_0) \wedge (T \rfloor \overline{\zeta}_0).
$$

Again by a result of F. Farris (cf. [137], p. 45) the Lorentz metric

$$
\pi^* L_\theta + \frac{2}{n+1}\theta \odot d\gamma + \frac{1}{n}\alpha_{\zeta_0}\theta \odot \theta,
$$

where $\theta = (i/2)\iota^*(\overline{\partial} - \partial)u$, agrees with (2.62). Hence, to prove that $g = j^* G$ it suffices to show that

$$
\sigma = \frac{1}{n+1} d\gamma + \frac{1}{2n}\alpha_{\zeta_0}\theta.
$$

See J.M. Lee [271], p. 424, for further details.

2.8.3 Obstructions to global embeddability

Let M be a strictly pseudoconvex CR manifold. Assume that M is realizable as the boundary of a smooth domain Ω in $\mathbf{C}^n$. If $\varphi : M \to \mathbf{C}^n$ is the given immersion then $\eta = \varphi^* dz^1 \wedge \cdots \wedge dz^n$ is a nowhere-zero global $(n, 0)$-form on M; hence $C(M)$ is a trivial bundle. There is a smooth defining function ψ of M satisfying the complex Monge–Ampère equation $J(\psi) = 1$ to second order along M, so that $F^* h$ is the

Fefferman metric of $(M, \hat{\theta})$, $\hat{\theta} = (i/2)\varphi^*(\overline{\partial} - \partial)\psi$, where h is the Lorentz metric given by

$$h = -\frac{i}{n+1} j^*\{(\partial - \overline{\partial})\psi\} \odot d\gamma + j^*\left\{\frac{\partial^2 \psi}{\partial z^j \partial \overline{z}^k} dz^j \odot d\overline{z}^k\right\}$$

and $F : C(M) \simeq M \times S^1$ is the diffeomorphism induced by η. Also γ is a local coordinate on S^1 and $j : M \times S^1 \subset \mathbf{C}^{n+1}$. Let θ be any pseudo-Hermitian structure on M (such that L_θ is positive definite). Then $\hat{\theta} = e^{2u}\theta$ for some smooth function u on M and F^*h and $g = F_\theta$ (given by (2.30)) turn out to be conformally equivalent Lorentz metrics. On the other hand, $h = j^*G$, where G is the semi-Riemannian metric given by (2.61) with $u = \psi$. Summing up, we have the following result:

Proposition 2.11. *If M is realizable as $\partial\Omega$ then $(C(M), g)$ admits a global conformal immersion in $(U \times \mathbf{C}_*, G)$, for some open neighborhood U of $\overline{\Omega}$, where $\mathbf{C}_* = \mathbf{C} \setminus \{0\}$.*

On the other hand, let us recall that given an n-dimensional Riemannian manifold (N, g), a necessary condition for the existence of a global conformal immersion $N \to \mathbf{R}^{n+k}$ is that $P_i^\perp(\Omega^{2i}) = 0$ and $[\frac{1}{2} T P_i^\perp(\omega)] \in H^{4i-1}(F(N), \mathbf{Z})$ for any $i > [k/2]$, where $P_i^\perp \in I_0^{2i}(\mathrm{GL}(n))$ are the inverse Pontryagin polynomials; cf. Theorem 5.14 of S.S. Chern and J. Simons [100], p. 64. Here ω is the connection 1-form of (the Levi-Civita connection of) (N, g) and Ω its curvature 2-form. In view of this result, it is reasonable to expect that some of the CR invariants furnished by Theorem 2.5 are obstructions to the global embeddability of a given abstract CR manifold M. While we leave this as an open problem, we address the following simpler situation. Assume M to be globally CR equivalent to S^{2n-1} (that is to say M globally embeds in $\mathbf{C}^n$ as the boundary of the unit ball). Then $C(M)$ is diffeomorphic to the Hopf manifold $H^n = S^{2n-1} \times S^1$.

Lemma 2.8. $I_{n+1} = \{\zeta \in \mathbf{C} : \zeta^{n+1} = 1\}$ *acts freely on* $\mathbf{C}^n \times \mathbf{C}_*$ *as a properly discontinuous group of complex analytic transformations.*

Proof. We organize the proof in two steps, as follows.

Step 1. *Let $z, z' \in \mathbf{C}^n \times \mathbf{C}_*$ such that $R_a(z) \neq z'$ for any $a \in I_{n+1}$. Then there exist neighborhoods U and U', of z respectively z', such that $R_a(U) \cap U' = \emptyset$ for any $a \in I_{n+1}$.*

Let $a_k = e^{2\pi i k/(n+1)}$ with $k \in \{0, 1, \ldots, n\}$. Set $z_k = a_k z$ and $s = \min\{|z' - z_k| : 0 \le k \le n\}$. Then $z' \neq z_k$ yields $s > 0$. Next, set $r = \min\{s/2, \sin\frac{\pi}{n+1}\}$ and $U = B(z, r)$. Then for any $a \in I_{n+1}$ one has $R_a(U) \subseteq B(az, r)$.

Step 2. *For any $z \in \mathbf{C}^n \times \mathbf{C}_*$ there is a neighborhood U of z such that $R_a(U) \cap U = \emptyset$ for any $a \in I_{n+1}$.*

To prove Step 2 let $z_k = a_k z$ and set $r = \min\{|z - z_k| : 1 \le k \le n\} = 2|z| \sin\frac{\pi}{n+1} > 0$. Finally, set $U = B(z, r/2)$. Then $R_a(U) \subseteq B(az, r/2)$; hence $R_a(U) \cap U = \emptyset$. □

By Lemma 2.8 the quotient space $V_{n+1} = (\mathbf{C}^n \times \mathbf{C}_*)/I_{n+1}$ is a complex $(n+1)$-dimensional manifold. Consider the biholomorphism $p : V_{n+1} \to \mathbf{C}^n \times \mathbf{C}_*$ given by

$$p([z,\zeta]) = \left(\frac{z}{\zeta}, \zeta^{n+1}\right),$$

for any $[z,\zeta] \in V_{n+1}$, and let us set

$$\phi_0 = p^{-1} \circ j \circ F.$$

Next

$$G_0 = \sum_{j=1}^{n} dz^j \odot d\bar{z}^j - d\zeta \odot d\bar{\zeta} \tag{2.63}$$

is I_{n+1}-invariant, hence gives rise to a globally defined semi-Riemannian metric of index 2 on V_{n+1}. Note that (V_{n+1}, G_0) is locally isometric to $\mathbf{R}_2^{2n+2}$.

Lemma 2.9. *$\phi_0 : (C(M), g) \to (V_{n+1}, G_0)$ is a conformal immersion.*

Proof. A calculation shows that $G_0 = p^* G$. Indeed,

$$p^* dz^j = \frac{1}{\zeta} dz^j - \frac{z^j}{\zeta^2} d\zeta, \quad p^* d\zeta = (n+1)\zeta^n d\zeta;$$

hence

$$\begin{aligned} p^* G &= \sum_j \left(dz^j - \frac{z^j}{\zeta} d\zeta\right) \odot \left(d\bar{z}^j - \frac{\bar{z}^j}{\bar{\zeta}} d\bar{\zeta}\right) + \sum_j \bar{z}^j \left(dz^j - \frac{z^j}{\zeta} d\zeta\right) \odot \left(\frac{1}{\bar{\zeta}} d\bar{\zeta}\right) \\ &\quad + \left(\frac{1}{\zeta} d\zeta\right) \odot \sum_k z^k \left(d\bar{z}^k - \frac{\bar{z}^k}{\bar{\zeta}} d\bar{\zeta}\right) + (|z|^2 - |\zeta|^2)\left(\frac{1}{\zeta} d\zeta\right) \odot \left(\frac{1}{\bar{\zeta}} d\bar{\zeta}\right) \\ &= \sum_{j=1}^{n} dz^j \odot d\bar{z}^j - d\zeta \odot d\bar{\zeta}. \end{aligned}$$

Finally, if we set $\psi(z) = |z|^2 - 1$ in the expression of h then it may be seen that $F : (C(M), g) \to (H^n, h)$ is a conformal diffeomorphism. Lemma 2.9 is proved. □

Let $P_i \in I^{2i}(\mathrm{GL}(2n))$ be the invariant polynomials given by

$$\det\left(\lambda I_{2n} - \frac{1}{2\pi} A\right) = \sum_{i=0}^{n} P_i(A \otimes \cdots \otimes A)\lambda^{2n-2i} + Q(\lambda^{2n-\mathrm{odd}}),$$

i.e., the P_i are the invariant polynomials obtained by ignoring the powers $\lambda^{2n-\mathrm{odd}}$ in the expression of $\det(\lambda I_{2n} - A/(2\pi))$. We obtain the following result:

Theorem 2.7. *Let M be a strictly pseudoconvex CR manifold of CR dimension $n-1$ and θ a pseudo-Hermitian structure on M such that L_θ is positive definite. Let $g = F_\theta$ be the Fefferman metric of (M, θ). Let ω be the connection 1-form of g and Ω its curvature 2-form. If M is globally CR equivalent to S^{2n-1}, then $P_1(\Omega^2) = 0$ and $[T P_1(\omega)] \in H^3(F(C(M)), \mathbf{Z})$ provided $n \geq 3$.*

Proof. To prove Theorem 2.7, we study the geometry of the second fundamental form of the immersion $\phi = p^{-1} \circ j : H^n \to (\mathbf{C}^n \times \mathbf{C}_*, G)$. Set $C_n = \sqrt{n+1}/\sqrt{2(n+1)}$. The tangent vector fields ξ_a given by

$$\xi_1 = C_n \Big(z^j \frac{\partial}{\partial z^j} + \bar{z}^j \frac{\partial}{\partial \bar{z}^j} + \zeta \frac{\partial}{\partial \zeta} + \bar{\zeta} \frac{\partial}{\partial \bar{\zeta}} \Big),$$
$$\xi_2 = C_n \Big(z^j \frac{\partial}{\partial z^j} + \bar{z}^j \frac{\partial}{\partial \bar{z}^j} - (n+2) \Big(\zeta \frac{\partial}{\partial \zeta} + \bar{\zeta} \frac{\partial}{\partial \bar{\zeta}} \Big) \Big),$$

are such that $G(\xi_1, \xi_2) = 0$, $G(\xi_1, \xi_1) = 1$, and $G(\xi_2, \xi_2) = -1$, and they form a frame of the normal bundle of ϕ. Since p is a biholomorphism with the inverse

$$p^{-1}(z, \zeta) = [z\zeta^{1/(n+1)}, \zeta^{1/(n+1)}]$$

we have

$$p_* \frac{\partial}{\partial z^j} = \zeta^{-1/(n+1)} \frac{\partial}{\partial z^j},$$
$$p_* \frac{\partial}{\partial \zeta} = \zeta^{-1/(n+1)} \Big(- z^j \frac{\partial}{\partial z^j} + (n+1)\zeta \frac{\partial}{\partial \zeta} \Big).$$

Due to the identity (2.63), the Christoffel symbols of the Levi-Civita connection ∇^0 of (V_{n+1}, G_0) vanish. The Levi-Civita connection ∇ of $(\mathbf{C}^n \times \mathbf{C}_*, G)$ is related to ∇^0 by

$$p_*(\nabla_X Y) = \nabla_{p_* X} p_* Y$$

for any $X, Y \in T(V_{n+1})$. A calculation shows that

$$\nabla_{\partial/\partial z^j} \frac{\partial}{\partial z^k} = 0, \quad \nabla_{\partial/\partial \zeta} \frac{\partial}{\partial \zeta} = - \frac{n}{n+1} \frac{1}{\zeta} \frac{\partial}{\partial \zeta}, \quad \nabla_{\partial/\partial \zeta} \frac{\partial}{\partial z^j} = \frac{1}{n+1} \frac{1}{\zeta} \frac{\partial}{\partial z^j},$$

Tangent vector fields on H^n are of the form $X + Y$ with $X = A^j \partial/\partial z^j + \overline{A^j} \partial/\partial \bar{z}^j$ and $Y = B\partial/\partial\zeta + \overline{B}\partial/\partial\bar{\zeta}$ satisfying $A^j \bar{z}_j + \overline{A^j} z_j = 0$, respectively $B\bar{\zeta} + \overline{B}\zeta = 0$. Here $z_j = z^j$. It follows that

$$\nabla_X \xi_1 = C_n \frac{n+2}{n+1} X, \quad \nabla_X \xi_2 = - \frac{C_n}{n+1} X, \tag{2.64}$$
$$\nabla_Y \xi_1 = \frac{C_n}{n+1} \Big\{ Y + B\bar{\zeta} z^j \frac{\partial}{\partial z^j} + \overline{B}\zeta \bar{z}^j \frac{\partial}{\partial \bar{z}^j} \Big\}, \tag{2.65}$$
$$\nabla_Y \xi_2 = \frac{C_n}{n+1} \Big\{ - (n+2)Y + B\bar{\zeta} z^j \frac{\partial}{\partial z^j} + \overline{B}\zeta \bar{z}^j \frac{\partial}{\partial \bar{z}^j} \Big\}. \tag{2.66}$$

Let $A_a = A_{\xi_a}$ be the Weingarten operator corresponding to the normal section ξ_a. We shall need the following lemma:

Lemma 2.10. *The first Pontryagin form of* (H^n, h) *is*

$$\frac{1}{4\pi^2} \Psi_{12} \wedge \Psi_{12},$$

where (with respect to a local coordinate system (x^i) on H^n)

$$\Psi_{12} = h\Big(\frac{\partial}{\partial x^i}, A_1 A_2 \frac{\partial}{\partial x^j}\Big) dx^i \wedge dx^j .$$

Proof. Let us recall (cf., e.g. [241], vol. II, p. 313) that

$$P_\ell(\Omega^{2\ell}) = c_\ell \sum \delta^{j_1 \cdots j_{2\ell}}_{i_1 \cdots i_{2\ell}} \Omega^{i_1}_{j_1} \wedge \cdots \wedge \Omega^{i_{2\ell}}_{i_{2\ell}} ,$$

where $c_\ell = 1/[(2\pi)^{2\ell}(2\ell)!]$ and the summation runs over all ordered subsets $(i_1, \ldots, i_{2\ell})$ of $\{1, \ldots, 2n\}$ and all permutations $(j_1, \ldots, j_{2\ell})$ of $(i_1, \ldots, i_{2\ell})$, and $\delta^{j_1 \cdots j_{2\ell}}_{i_1 \cdots i_{2\ell}}$ is the sign of the permutation. We need the Gauss equation (cf., e.g., (2.4) in [446], p. 21)

$$R^\ell_{kij} = B^a_{jk} A^\ell_{ai} - B^a_{ik} A^\ell_{aj} ,$$

where R^ℓ_{kij} , B^a_{jk} are respectively the curvature tensor field of (H^n, h) and the second fundamental form of ϕ (with respect to a local coordinate system (U, x^i) on H^n). Also $A_a \partial_i = A^j_{ai} \partial_j$, where ∂_i is short for $\partial/\partial x^i$. The Gauss equation and the identity

$$R(X, Y)Z = u(2\Omega(X^*, Y^*)_u (u^{-1} Z))$$

(cf. [241], vol. I, p. 133) for any $X, Y, Z \in T_x(H^n)$ and some $u \in F(H^n)_x$ furnish

$$2\Omega^r_s = Y^r_p X^k_s \left(B^a_{jk} A^p_{ai} - B^a_{ik} A^p_{aj} \right) dx^i \wedge dx^j$$

(where $X^i_j : p_F^{-1}(U) \to \mathbf{R}$ are fiber coordinates on $F(H^n)$ and $(Y^i_j) = (X^i_j)^{-1}$). Using

$$B^a_{jk} = A^r_{aj} h_{rk}$$

a calculation leads to

$$2 P_1(\Omega^2) = -c_1 \left(B^{a_1} j_1 k_1 A^{k_2}_{a_1 p_1} B^{a_2}_{j_2 k_2} A^{k_1}_{a_2 p_2} - B^{a_1}_{p_1 k_1} A^{k_2}_{a_1 j_1} B^{a_2}_{j_2 k_2} A^{k_1}_{a_2 p_2} \right) dx^{p_1} \wedge dx^{j_1} \wedge dx^{p_2} \wedge dx^{j_2};$$

hence

$$P_1(\Omega^2) = c_1 \sum_{a,b} \Psi_{ab} \wedge \Psi_{ab},$$

where Ψ_{ab} is the 2-form on $F(H^n)$ given by

$$\Psi_{ab} = h(A_a \partial_i, A_b \partial_j) dx^i \wedge dx^j .$$

Finally, let us observe that $\Psi_{11} = \Psi_{22} = 0$ and $\Psi_{21} = -\Psi_{12}$. Lemma 2.10 is proved. □

Note that the proof works for any codimension-two submanifold of a flat Riemannian manifold. At this point we may finish the proof of Theorem 2.7. Let us recall the Ricci equation (of the given immersion ϕ, cf., e.g., (2.7) in [446], p. 22)

$$G(R(X,Y)\xi,\xi') = G(R^{\perp}(X,Y)\xi,\xi') + h([A_{\xi}, A_{\xi'}]X, Y),$$

where R, $R^{\perp}$ are respectively the curvature tensor fields of $(\mathbf{C}^n \times \mathbf{C}_*, G)$ and those of the normal connection. As a consequence of (2.64)–(2.66), ξ_a are parallel in the normal bundle; hence the immersion ϕ has a flat normal connection ($R^{\perp} = 0$). On the other hand, $R = 0$ (because (V_{n+1}, G_0) is flat) and the Ricci equation shows that the Weingarten operators A_a commute. Then $\Psi_{12} = 0$ and our Lemmas 2.9 and 2.10 together with Theorem 2.5 yield $P_1(\Omega^2) = 0$. □

Let $q : H^3(F(C(M)), \mathbf{R}) \to H^3(F(C(M)), \mathbf{R}/\mathbf{Z})$ be the natural homomorphism. By Theorem 3.16 in [100], p. 56, since $P_1(\Omega^2) = 0$ there is a cohomology class $\alpha \in H^3(C(M), \mathbf{R}/\mathbf{Z})$ such that $p_F^*\alpha = q([TP_1(\omega)])$, where $p_F : F(C(M)) \to C(M)$ is the projection. Yet for the Hopf manifold, $H^3(H^n, \mathbf{R}/\mathbf{Z}) = 0$ provided that $n \geq 3$; hence $[TP_1(\omega)] \in \mathrm{Ker}(q)$ and then by the exactness of the Bockstein sequence

$$\cdots \to H^3(F(C(M)), \mathbf{Z}) \to H^3(F(C(M)), \mathbf{R}) \to \\ \to H^3(F(C(M)), \mathbf{R}/\mathbf{Z}) \to H^4(F(C(M)), \mathbf{Z}) \to \cdots$$

it follows that $[TP_1(\omega)]$ is an integral class. □

3

The CR Yamabe Problem

The scope of Chapter 3 is to present D. Jerison and J.M. Lee's solution (cf. [226, 228]) to the *CR Yamabe problem* (cf. Theorem 3.4 below). We start with a brief presentation of the Riemannian Yamabe problem. The reader less familiar with the Riemannian counterpart of Chapter 3 may consult the expository paper by J.M. Lee and T. Parker [275]. The high interest shown in the (Riemannian) Yamabe problem, over more than twenty years,[1] by an ample portion of the mathematical community, is perhaps the best motivation for introducing a CR version of the problem (in terms of the Fefferman metric). A parallel between the Riemannian and the CR Yamabe problems also offers to the reader a lively comparison between the elliptic and subelliptic theories, as applied to nonlinear problems arising from differential geometry.

Let (M, g) be an m-dimensional Riemannian manifold, $m \geq 3$. Let $\tilde{K}$ be the scalar curvature of the Riemannian metric

$$\tilde{g} = \phi^{q-2} g, \quad q := \frac{2m}{m-2}.$$

By a standard calculation (cf., e.g., T. Aubin [24])

$$\tilde{K} = \phi^{1-q} \left(a_m \Delta \phi + K \phi \right), \quad a_m := \frac{4(m-1)}{m-2},$$

where Δ is the Laplace–Beltrami operator of g and K the scalar curvature of g. Then the *Yamabe problem* is to find a metric $\tilde{g}$, conformally related to g, such that $\tilde{K}$ is a constant. If $\tilde{g} = \phi^{q-2} g$ then the equation $\tilde{K} = \mu =$ constant is

$$a_m \Delta \phi + K \phi = \mu \phi^{q-1}.$$

This is the *Yamabe equation*. The Yamabe equation is the Euler–Lagrange equation of the constrained variational principle

$$\mu(M) = \inf \left\{ \int_M \left(a_m \|d\phi\|^2 + K \phi^2 \right) dv_g : \int_M |\phi|^q dv_g = 1 \right\} \tag{3.1}$$

[1] H. Yamabe's work [444] was published in 1960. R. Schoen's completion [365] of the solution to the Yamabe problem appeared in 1984.

provided that $\phi \geq 0$. Exploiting the variational description of the Yamabe equation, the following fundamental result may be established:

Theorem 3.1. (H. Yamabe [444], N. Trudinger [409], T. Aubin [24])
Let (M, g) be a compact Riemannian manifold, $m \geq 3$. Then

(a) $\mu(M)$ is a conformal invariant.
(b) $\mu(M) \leq \mu(S^m)$.
(c) If $\mu(M) < \mu(S^m)$ then $\mu(M)$ is attained for some positive C^∞ solution ϕ of the Yamabe equation $a_m \Delta\phi + K\phi = \mu\phi^{q-1}$.

Therefore $\tilde{g} = \phi^{q-2} g$ has constant scalar curvature $\mu = \mu(M)$.

Theorem 3.2. (T. Aubin [24])
If $m \geq 6$ and M is not locally conformally flat then $\mu(M) < \mu(S^m)$.

Theorem 3.3. (R. Schoen [365])
$\mu(M) < \mu(S^m)$ unless M is the sphere.

The theorem by R. Schoen completes the solution to the Yamabe problem. The proof of (a) in the Yamabe–Trudinger–Aubin theorem relies on the observation that $\tilde{g} = t^{q-2} g$ implies that

$$\left(a_m \tilde{\Delta} + \tilde{K}\right) \tilde{\phi} = t^{1-q} \left(a_m \Delta + K\right) \phi, \quad \tilde{\phi} := t^{-1}\phi,$$

where $\tilde{\Delta}$, $\tilde{K}$ are respectively the Laplace–Beltrami operator and the scalar curvature of $\tilde{g} = t^{q-2} g$. As a consequence, it may be shown that the integrals $\int_M \left(a_m \|d\phi\|^2 + K\phi^2\right) dv_g$ and $\int_M |\phi|^q dv_g$ are unchanged if g, ϕ, and K are replaced by $\tilde{g}$, $\tilde{\phi}$, and $\tilde{K}$, respectively; hence $\mu(M)$ is a conformal invariant.

To prove (b) one starts with the special case of S^m. By a conformal change of variables one may convert the variational problem on S^m into a problem on $\mathbf{R}^m$,

$$\mu(S^m) = \inf\left\{a_m \int_{\mathbf{R}^m} \|df\|^2 dx : \int_{\mathbf{R}^m} |f|^q dx = 1\right\}, \tag{3.2}$$

which is just the problem of finding the best constant and extremal functions for the Sobolev inequality on $\mathbf{R}^m$

$$\mu(S^m) \left(\int_{\mathbf{R}^m} |f|^q dx\right)^{2/q} \leq a_m \int_{\mathbf{R}^m} \|df\|^2 dx.$$

T. Aubin (cf. [24]) showed that extremals exist and are of the form

$$\left(a + b\|x - x_0\|^2\right)^{-(m-2)/2}. \tag{3.3}$$

See also G. Talenti [396]. Now, given a compact Riemannian manifold M, one may use normal coordinates on M and the dilation invariance of the problem (3.2) to transplant a suitable extremal (3.3) from $\mathbf{R}^m$ to a sufficiently small neighborhood on M and conclude that $\mu(M) \le \mu(S^m)$.

To prove (c) one employs the *Sobolev lemma* for compact Riemannian manifolds. Let $W^{1,2}(M)$ be the Sobolev space with the norm

$$\|f\|_{W^{1,2}(M)} = \left(\int_M \left(\|df\|^2 + |f|^2 \right) dv_g \right)^{1/2}.$$

By the Sobolev lemma, there is an embedding

$$W^{1,2}(M) \to L^s(M), \quad \frac{1}{s} \ge \frac{1}{2} - \frac{1}{m},$$

and this embedding is compact if $\frac{1}{s} > \frac{1}{2} - \frac{1}{m}$. If $\phi_i \in W^{1,2}(M)$ is a minimizing sequence for the problem (3.1), then by the Sobolev lemma, $\{\phi_i\}$ is uniformly bounded in $W^{1,2}(M)$; hence there is a subsequence weakly convergent to some $\phi \in W^{1,2}(M)$. Unfortunately, the inclusion $W^{1,2}(M) \to L^q(M)$ is not compact; hence the constraint $\int_M |\phi|^q dv_g = 1$ might fail to be preserved in the limit. One is led to consider the perturbed equation

$$a_m \Delta\phi + K\phi = \mu_s \phi^{s-1}, \quad 2 \le s < q.$$

Now, given a minimizing sequence (for the corresponding variational principle), due to the compactness of $W^{1,2}(M) \to L^s(M)$, there will exist a sequence converging strongly in the L^s-norm to some $\phi_{(s)} \in W^{1,2}(M)$; hence $\phi_{(s)}$ satisfies the constraint. At this point, an iteration procedure of the standard L^p estimates for the Laplace–Beltrami operator shows that $\phi_{(s)}$ is smooth, while the strong maximum principle yields $\phi_{(s)} > 0$.

To complete the proof, one should show that $\phi_{(s)} \to \phi$ as $s \to q$, where ϕ is smooth and > 0. This has been shown by T. Aubin (cf. op. cit.) as follows. Let $\mu = \mu(S^m)$. Then for any compact manifold M and any $\epsilon > 0$ there is $C_{M,\epsilon} > 0$ such that

$$(\mu - \epsilon)\left(\int_M |f|^q dv_g \right)^{2/q} \le a_m \int_M \|df\|^2 dv_g + C_{M,\epsilon} \int_M |f|^2 dv_g \tag{3.4}$$

for any $f \in W^{1,2}(M)$. The inequality (3.4) is obtained by transplanting the inequality from $\mathbf{R}^m$ to a normal coordinate neighborhood in M, and then to the whole of M by a partition of unity argument. Let $\epsilon > 0$ be such that $\mu - \epsilon > \mu_s$, for s sufficiently close to q. Then (3.4) for $f = \phi_{(s)}$ shows that $\|\phi_{(s)}\|_{L^2(M)}$ is bounded away from zero as $s \to q$, and the proof is complete.

Next, we wish to discuss the following CR analogue of the Yamabe problem. Let M be a strictly pseudoconvex CR manifold of CR dimension n. Let θ be a pseudo-Hermitian structure on M such that the Levi form L_θ is positive definite. Let F_θ be the Fefferman metric of (M, θ). By a result in Chapter 2 of this book, F_θ is a Lorentz metric on the total space $C(M)$ of the canonical circle bundle $S^1 \to C(M) \xrightarrow{\pi} M$. By Theorem 2.3 of the same chapter, if θ is replaced by $\hat{\theta} = u^{p-2}\theta$, with $p = 2 + 2/n$, then F_θ goes over to

$$F_{\hat{\theta}} = (u \circ \pi)^{p-2} F_\theta,$$

so that the (restricted) conformal class of the Fefferman metric is a CR invariant of M. The reason for representing the conformal factor as u^{p-2}, $u > 0$, is that it simplifies the transformation law (2.47). Indeed, due to the identity

$$\Delta_b(\log u) = \frac{1}{u}\Delta_b u + \frac{1}{u^2} u_\alpha u^\alpha$$

the transformation law (2.47) (of the pseudo-Hermitian scalar curvature ρ, under a transformation $\hat{\theta} = u^{p-2}\theta$) may be written

$$\hat{\rho} = u^{1-p}\left(b_n \Delta_b u + \rho u\right),$$

where $b_n = 2 + 2/n$. Hence a necessary and sufficient condition for the contact form $\hat{\theta} = u^{p-2}\theta$ to have constant pseudo-Hermitian scalar curvature $\hat{\rho} = \lambda$ is that u satisfy

$$b_n \Delta_b u + \rho u = \lambda u^{p-1}. \tag{3.5}$$

This is the *CR Yamabe equation*. It is the Euler–Lagrange equation for the constrained variational problem

$$\lambda(M) = \inf\{A_\theta(u) : B_\theta(u) = 1\}, \tag{3.6}$$

where

$$A_\theta(u) = \int_M \{b_n \|\pi_H \nabla u\|^2 + \rho u^2\}\theta \wedge (d\theta)^n,$$

$$B_\theta(u) = \int_M |u|^p \theta \wedge (d\theta)^n.$$

Here ∇u is the gradient of u (given by $g_\theta(\nabla u, X) = X(u)$ for any $X \in T(M)$) and $\pi_H : T(M) \to H(M)$ is the projection (associated with the direct sum decomposition $T(M) = H(M) \oplus \mathbf{R}T$). The purpose of Chapter 3 is to describe the following result by D. Jerison and J.M. Lee (cf. [227]):

Theorem 3.4. *Let M be a compact strictly pseudoconvex CR manifold of CR dimension n. Let θ be a contact form on M for which L_θ is positive definite. Then*

(i) $\lambda(M)$ is a CR invariant of M.
(ii) $\lambda(M) \leq \lambda(S^{2n+1})$ (where the sphere $S^{2n+1} \subset \mathbf{C}^{n+1}$ carries the standard CR structure).
(iii) If $\lambda(M) < \lambda(S^{2n+1})$ then the infimum (3.6) is achieved by a positive solution u of (3.5). Hence the contact form $\hat{\theta} = u^{p-2}\theta$ has constant pseudo-Hermitian scalar curvature $\hat{\rho} = \lambda(M)$.

We refer to Theorem 3.4 as the *Jerison–Lee theorem*. S.S. Chern and R. Hamilton [88], studying contact structures on 3-manifolds, proved independently a result equivalent to (iii) in Theorem 3.4 in the case $\lambda(M) \leq 0$ and $n = 1$.

3.1 The Cayley transform

We start by discussing the bundle of densities and the CR invariant Laplacian Δ_b^c. Let M be a strictly pseudoconvex CR manifold and θ such that L_θ is positive definite. Let us set

$$E_x^+ = \{\lambda\theta_x : \lambda \in \mathbf{R}_+\},$$

for any $x \in M$, where $\mathbf{R}_+ = (0, \infty)$. This furnishes an $\mathbf{R}_+$-bundle $E^+ \to M$. Let $\alpha \in \mathbf{R}$ and let us set

$$E_x^\alpha = \{\mu : E_x^+ \to \mathbf{R} : \mu(\lambda\theta_x) = \lambda^{-\alpha}\mu(\theta_x)\,,\ \forall\lambda > 0\},$$

for any $x \in M$.

Definition 3.1. The resulting bundle $E^\alpha \to M$ is referred to as the *bundle of densities of CR weight* α on M. □

Let us define $\mu_\theta : M \to E^1$ by setting $\mu_\theta(x) = (\mu_\theta)_x$, where $(\mu_\theta)_x : E_x^+ \to \mathbf{R}$ is given by $(\mu_\theta)_x(\lambda\theta_x) = \lambda^{-1}$. Then $\mu_\theta \in \Gamma^\infty(E^1)$. Moreover, any C^∞ section in E^α is of the form $u\mu_\theta^\alpha$ for some $u \in C^\infty(M)$.

We consider the differential operator

$$\Delta_b^c : \Gamma^\infty(E^{n/2}) \to \Gamma^\infty(E^{n/2+1})$$

given by

$$\Delta_b^c(u\mu_\theta^{n/2}) = (b_n\Delta_b u + \rho u)\mu_\theta^{n/2+1}.$$

Note that Δ_b^c is invariant under a transformation $\hat\theta = u^{p-2}\theta$.

Definition 3.2. Δ_b^c is called the *CR invariant Laplacian* of M. □

The CR invariance of $\lambda(M)$ (i.e., the statement (*i*) in Theorem 3.4) follows from

$$\lambda(M) = \inf\left\{\int_M (\Delta_b^c\phi)\otimes\phi : \phi \in C^\infty(E^{n/2}),\ \ \phi > 0,\ \ \int_M \phi^p = 1\right\}$$

Next, we need to recall a few facts regarding the Cayley transform. Let $\mathbf{B}^{n+1} = \{z \in \mathbf{C}^{n+1} : |z| < 1\}$ be the unit ball in $\mathbf{C}^{n+1}$.

Definition 3.3. The *Cayley transform* is

$$\mathcal{C}(\zeta) = \left(\frac{\zeta'}{1+\zeta^{n+1}}\,,\ i\frac{1-\zeta^{n+1}}{1+\zeta^{n+1}}\right),\quad \zeta = (\zeta', \zeta^{n+1}),\ 1+\zeta^{n+1} \neq 0. \qquad \square$$

The Cayley transform gives a biholomorphism of $\mathbf{B}^{n+1}$ onto the Siegel domain Ω_{n+1}. Moreover, when restricted to the sphere minus a point, $\mathcal{C}$ gives a CR diffeomorphism (onto the boundary of the Siegel domain)

$$\mathcal{C} : S^{2n+1} \setminus \{(0, \ldots, 0, -1)\} \to \partial\Omega_{n+1}.$$

Let us recall (cf. Chapter 1 of this book) the CR diffeomorphism

$$f : \mathbf{H}_n \to \partial\Omega_{n+1}, \quad f(z,t) = (z, t + i|z|^2),$$

with the obvious inverse $f^{-1}(z, w) = (z, \mathrm{Re}(w))$, $z \in \mathbf{C}^n$, $w \in \mathbf{C}$. We obtain the CR equivalence

$$F : S^{2n+1} \setminus \{(0, \ldots, 0, -1)\} \to \mathbf{H}_n, \quad F := f^{-1} \circ \mathcal{C}.$$

We shall need the following lemma:

Lemma 3.1. *Let* $\theta_1 = j^* \left[i(\overline{\partial} - \partial)|\zeta|^2 \right]$ *be the standard contact form on the sphere, where* $j : S^{2n+1} \hookrightarrow \mathbf{C}^{n+1}$ *is the inclusion. Let* θ_0 *be the canonical contact form on* $\mathbf{H}_n$*. Then*

$$\left(F^*\theta_0\right)_\zeta = \frac{1}{|1 + \zeta^{n+1}|^2} \theta_{1,\zeta} \,,$$

for any $\zeta \in S^{2n+1} \setminus \{(0, \ldots, 0, -1)\}$.

To prove Lemma 3.1 we first carry out some local calculations for the sphere. The standard CR structure $T_{1,0}(S^{2n+1})$ admits the (local) frame

$$\left\{ T_\alpha := \frac{\partial}{\partial \zeta^\alpha} - \frac{\overline{\zeta}^\alpha}{\overline{\zeta}^{n+1}} \frac{\partial}{\partial \zeta^{n+1}} : 1 \le \alpha \le n \right\}$$

defined on the open set $S^{2n+1} \cap \{\zeta : \zeta^{n+1} \neq 0\}$. Moreover, the characteristic direction T of $d\theta_1$ is

$$T = \frac{i}{2} \sum_{j=1}^{n+1} \left(\zeta^j \frac{\partial}{\partial \zeta^j} - \overline{\zeta}^j \frac{\partial}{\partial \overline{\zeta}^j} \right).$$

Next, let us note that

$$(d_\zeta \mathcal{C}) \left. \frac{\partial}{\partial \zeta^j} \right|_\zeta = \begin{cases} \dfrac{1}{1 + \zeta^{n+1}} \left. \dfrac{\partial}{\partial z^\beta} \right|_{\mathcal{C}(\zeta)}, & j = \beta, \\ -\dfrac{1}{(1 + \zeta^{n+1})^2} \left[\zeta^\alpha \dfrac{\partial}{\partial z^\alpha} + 2i \dfrac{\partial}{\partial w} \right]_{\mathcal{C}(\zeta)}, & j = n+1; \end{cases}$$

hence

$$(d_\zeta \mathcal{C}) T_{\alpha,\zeta} = \frac{1}{1 + \zeta^{n+1}} \left[\delta^\beta_\alpha + \frac{\overline{\zeta}_\alpha \zeta^\beta}{\overline{\zeta}^{n+1}(1 + \zeta^{n+1})} \right] \left. \frac{\partial}{\partial z^\beta} \right|_{\mathcal{C}(\zeta)} + \frac{2i\overline{\zeta}_\alpha}{\overline{\zeta}^{n+1}(1 + \zeta^{n+1})^2} \left. \frac{\partial}{\partial w} \right|_{\mathcal{C}(\zeta)},$$

where $\zeta_j = \zeta^j$, and

$$(d_\gamma C)T_\zeta = \frac{i}{2}\left(\frac{\zeta^\alpha}{(1+\zeta^{n+1})^2}\,\frac{\partial}{\partial z^\alpha}\bigg|_{C(\zeta)} - \frac{\overline{\zeta}^\alpha}{(1+\overline{\zeta}^{n+1})^2}\,\frac{\partial}{\partial \overline{z}^\alpha}\bigg|_{C(\zeta)}\right)$$
$$+ \frac{\zeta^{n+1}}{(1+\zeta^{n+1})^2}\,\frac{\partial}{\partial w}\bigg|_{C(\zeta)} + \frac{\overline{\zeta}^{n+1}}{(1+\overline{\zeta}^{n+1})^2}\,\frac{\partial}{\partial \overline{w}}\bigg|_{C(\zeta)}.$$

Let us set $g := f^{-1}$. Taking into account

$$(d_{(z,w)}g)\,\frac{\partial}{\partial z^\alpha}\bigg|_{(z,w)} = \frac{\partial}{\partial z^\alpha}\bigg|_{(z,\mathrm{Re}(w))}, \quad (d_{(z,w)}g)\,\frac{\partial}{\partial w}\bigg|_{(z,w)} = \frac{1}{2}\,\frac{\partial}{\partial t}\bigg|_{(z,\mathrm{Re}(w))}$$

it follows that

$$(d_\zeta F)T_{\alpha,\zeta} = \frac{1}{1+\zeta^{n+1}}\left[\delta_\alpha^\beta + \frac{\overline{\zeta}_\alpha \zeta^\beta}{\overline{\zeta}^{n+1}(1+\zeta^{n+1})}\right] Z_{\beta,F(\zeta)}, \tag{3.7}$$
$$(d_\zeta F)T_\zeta = \frac{i\zeta^\alpha}{2(1+\zeta^{n+1})^2} Z_{\alpha,F(\zeta)} - \frac{i\overline{\zeta}^\alpha}{2(1+\overline{\zeta}^{n+1})^2} Z_{\overline{\alpha},F(\zeta)} + \frac{1}{|1+\zeta^{n+1}|^2}\,\frac{\partial}{\partial t}\bigg|_{F(\zeta)}$$

where $Z_\alpha = \partial/\partial z^\alpha + i\overline{z}^\alpha \partial/\partial t$. At this point we may prove Lemma 3.1. The pullback of $\theta_0 = dt + i\sum_{\alpha=1}^n (z^\alpha d\overline{z}^\alpha - \overline{z}^\alpha dz^\alpha)$ (the standard contact form of the Heisenberg group) is of the form

$$F^*\theta_0 = \lambda_j d\zeta^j + \overline{\lambda}_j d\overline{\zeta}^j.$$

Then, on the one hand,

$$\lambda_\alpha - \frac{\overline{\zeta}^\alpha}{\overline{\zeta}^{n+1}}\lambda_{n+1} = \theta_{0,F(\zeta)}(d_\zeta F)T_{\alpha,\zeta} = 0$$

(by (3.7)), and on the other,

$$\lambda_{n+1} = \theta_{0,F(\zeta)}(d_\zeta F)\,\frac{\partial}{\partial \zeta^{n+1}}\bigg|_\zeta = -\frac{i\overline{\zeta}^{n+1}}{|1+\zeta^{n+1}|^2},$$

i.e.,

$$F^*\theta_0 = \frac{i}{|1+\zeta^{n+1}|^2}\sum_{j=1}^{n+1}\left\{\zeta_j d\overline{\zeta}^j - \overline{\zeta}_j d\zeta^j\right\},$$

and Lemma 3.1 is proved. □

Let us consider the function

$$a(z,w) := \frac{4}{|i+w|^2}, \quad (z,w) \in U := \{(z,w) \in \mathbf{C}^{n+1} : w + i \neq 0\}.$$

Note that $\partial\Omega_{n+1} \subset U$ (hence we may restrict a to the boundary of the Siegel domain). Let us set

$$b(z,t) := a(f(z,t)) = \frac{4}{|t + i(1+|z|^2)|^2}, \quad (z,t) \in \mathbf{H}_n.$$

Then (by Lemma 3.1)

$$F^*(b\theta_0) = \theta_1 .$$

Let us differentiate and take into account that $b(F(\zeta)) = |1 + \zeta^{n+1}|^2$. We obtain

$$d\theta_1 = \Big(\frac{d\zeta^{n+1}}{1+\zeta^{n+1}} + \frac{d\overline{\zeta}^{n+1}}{1+\overline{\zeta}^{n+1}}\Big) \wedge \theta_1 + |1+\zeta^{n+1}|^2 F^* d\theta_0;$$

hence

$$\theta_1 \wedge (d\theta_1)^n = |1+\zeta^{n+1}|^{2(n+1)} F^* \left[\theta_0 \wedge (d\theta_0)^n\right].$$

Given a nondegenerate CR manifold M endowed with the contact form θ, let us extend the Hermitian form $\langle Z, W\rangle = L_\theta(Z, \overline{W}),\ \ Z, W \in T_{1,0}(M)$, to complex 1-forms by setting

$$\langle \theta^\alpha, \theta^\beta\rangle = h^{\alpha\overline{\beta}}, \quad \langle \theta^{\overline{\alpha}}, \theta^{\overline{\beta}}\rangle = h^{\overline{\alpha}\beta},$$

$$\langle \theta^\alpha, \theta^{\overline{\beta}}\rangle = 0 = \langle \theta^{\overline{\alpha}}, \theta^\beta\rangle, \quad \langle \theta^\alpha, \theta\rangle = 0 = \langle \theta^{\overline{\alpha}}, \theta\rangle, \quad \langle \theta, \theta\rangle = 0.$$

When we wish to emphasize on the choice of θ we also write $L^*_\theta(\lambda, \overline{\mu})$ instead of $\langle \lambda, \mu\rangle$, for any $\lambda,\ \mu \in \Omega^1(M)$. Also, we set $\|\lambda\|^2_\theta = \langle \lambda, \lambda\rangle$. Note that $\|\pi_H \nabla u\| = \|du\|_\theta$. Recall that (cf. Chapter 2)

$$\int_M (\Delta_b u) v\, \theta \wedge (d\theta)^n = -\int_M \langle du, dv\rangle \theta \wedge (d\theta)^n ,$$

for any $u, v \in C^\infty(M)$, at least one of compact support. It is also useful to note that for any $u \in C^1(\mathbf{H}_n)$,

$$\|du\|^2_{\theta_0} = \sum_{\alpha=1}^n |Z_\alpha u|^2.$$

The pseudo-Hermitian scalar curvature of $(\mathbf{H}_n, \theta_0)$ is zero; hence

$$\lambda(\mathbf{H}_n) = \inf \left\{ \int_{\mathbf{H}_n} b_n \sum_{\alpha=1}^n |Z_\alpha u|^2\, \theta_0 \wedge (d\theta_0)^n : \int_{\mathbf{H}_n} |u|^p \theta_0 \wedge (d\theta_0)^n = 1 \right\},$$

with $p = b_n = 2 + 2/n$. Consider the function

$$v(\zeta) = \frac{u(F(\zeta))}{|1+\zeta^{n+1}|^{n+1}}.$$

Then

$$\int_{S^{2n+1}} \left(b_n |dv|^2_{\theta_1} + \rho_n v^2\right) \theta_1 \wedge (d\theta_1)^n = \int_{\mathbf{H}_n} b_n \sum_{\alpha=1}^n |Z_\alpha u|^2 \theta_0 \wedge (d\theta_0)^n$$

$$\int_{S^{2n+1}} v^p \theta_1 \wedge (d\theta_1)^n = \int_{\mathbf{H}_n} u^p \theta_0 \wedge (d\theta_0)^n$$

(where $\rho_n = n(n+1)/2$ is the pseudo-Hermitian scalar curvature of the sphere (S^{2n+1}, θ_1)) for any $u \in C^1(\mathbf{H}_n),\ \ u \geq 0$, and hence the extremal problems (3.6) for $\mathbf{H}_n$ and S^{2n+1} are the same. In particular $\lambda(\mathbf{H}_n) = \lambda(S^{2n+1})$. □

3.2 Normal coordinates

Let $(M, T_{1,0}(M))$ be a strictly pseudoconvex[2] CR manifold of CR dimension n, on which one has fixed a contact 1-form θ such that the Levi form L_θ is positive definite. Let T be the characteristic direction of (M, θ).

Definition 3.4. Let $\{T_\alpha\}$ be a local orthonormal (i.e., $L_\theta(T_\alpha, T_{\bar\beta}) = \delta_{\alpha\beta}$) frame of $T_{1,0}(M)$ defined on the open subset $V \subseteq M$. Such a $\{T_\alpha\}$ is referred to as a *pseudo-Hermitian frame.* □

Following the ideas in [150], pp. 471–476, we are going to introduce, for any $x_0 \in M$, local coordinates $z^1, \dots, z^n, t$ at x_0, naturally associated with the given pseudo-Hermitian frame $\{T_\alpha\}$ in the sense that $T_\alpha = \partial/\partial z^\alpha + i\bar z^\alpha \partial/\partial t$ and $T = \partial/\partial t$ modulo suitably small error terms near x_0 (cf. Theorem 3.5 below).

Given a pseudo-Hermitian frame $\{T_\alpha\}$ let us set $X_\alpha = T_\alpha + T_{\bar\alpha}$ and $Y_\alpha = i(T_{\bar\alpha} - T_\alpha)$. When we think of $\{X_\alpha, Y_\alpha, T\}$ together we write $\{X_j\}$ (with $0 \le j \le 2n$ and $X_0 = T$, $X_{\alpha+n} = Y_\alpha$). Let also $\{\theta^j\}$ be the dual frame with respect to $\{X_j\}$ (here $\theta^0 = \theta$). We shall denote by $(\xi^\alpha, \eta^\alpha, \tau)$ the Cartesian coordinates in $\mathbf{R}^{2n+1}$ and we shall also write (ξ^j) (with $0 \le j \le 2n$ and $\xi^0 = \tau$, $\xi^{\alpha+n} = \eta^\alpha$) when $(\xi^\alpha, \eta^\alpha, \tau)$ will be thought of as together.

Let $x \in M$ be fixed. Given $\xi \in \mathbf{R}^{2n+1}$, let us consider the tangent vector field

$$X_\xi = \xi^j X_j \in \mathcal{X}(V).$$

For ξ sufficiently close to the origin in $\mathbf{R}^{2n+1}$, let $E_x(\xi)$ be the endpoint $C(1)$ of the integral curve $C : [0, 1] \to M$ of X_ξ issuing at x, i.e.,

$$\begin{aligned} \frac{dC}{dt}(t) &= X_\xi(C(t)), \\ C(0) &= x. \end{aligned} \tag{3.8}$$

By standard ODE theory E_x is a smooth map of a star-shaped neighborhood $\tilde U_x$ of $0 \in \mathbf{R}^{2n+1}$ into M. Also

$$(d_0 E_x) \left. \frac{\partial}{\partial \xi^j} \right|_0 = X_j(x);$$

hence E_x is a diffeomorphism of a perhaps smaller neighborhood $U_x \subseteq \tilde U_x$ of $0 \in \mathbf{R}^{2n+1}$ (which may be assumed to be star-shaped, too) onto a neighborhood V_x of x in M. Then $E_x^{-1} : V_x \to U_x$ is the local chart we were looking for.

Definition 3.5. The resulting local coordinates are referred to as the *Folland–Stein (normal) coordinates* at x. □

[2] The matters discussed in this section, i.e., existence of Folland–Stein normal coordinates and Heisenberg-type order may be easily generalized to the case of an arbitrary nondegenerate CR manifold (of hypersurface type).

Next, we discuss the notion of Heisenberg-type order. For $x \in M$ fixed let

$$\Theta_x = E_x^{-1} = (x^j) = (x^\alpha, y^\alpha, t) : V_x \to U_x \subseteq \mathbf{R}^{2n+1}$$

be Folland–Stein coordinates at x.

Definition 3.6. A function f on V_x is said to be O^1 (and we write $f = O^1$) if

$$f(y) = O\Big(\sum_{\alpha=1}^{n}(|x^\alpha(y)| + |y^\alpha(y)|) + |t(y)|^{1/2}\Big)$$

as $y \to x$ in V_x. For $k \in \mathbf{Z}$, $k \geq 2$, we define $f = O^k$ inductively, i.e., $f = O^k$ if and only if $f = O(O^1 \cdot O^{k-1})$. □

Note that $f = O^2$ if and only if

$$f(y) = O\Big(\sum_{\alpha=1}^{n}(|x^\alpha(y)|^2 + |y^\alpha(y)|^2) + |t(y)|\Big)$$

as $y \to x$ in V_x.

Theorem 3.5. (G.B. Folland and E.M. Stein [150])
With respect to the Folland–Stein normal coordinates

$$E_x^{-1} = (x^j) = (x^\alpha, y^\alpha, t)$$

on V_x one has

$$X_\alpha = \frac{\partial}{\partial x^\alpha} + 2y^\alpha \frac{\partial}{\partial t} + \sum_{\beta=1}^{n}\Big(O^1 \frac{\partial}{\partial x^\beta} + O^1 \frac{\partial}{\partial y^\beta}\Big) + O^2 \frac{\partial}{\partial t},$$

$$Y_\alpha = \frac{\partial}{\partial y^\alpha} - 2x^\alpha \frac{\partial}{\partial t} + \sum_{\beta=1}^{n}\Big(O^1 \frac{\partial}{\partial x^\beta} + O^1 \frac{\partial}{\partial y^\beta}\Big) + O^2 \frac{\partial}{\partial t},$$

$$T = \frac{\partial}{\partial t} + \sum_{\beta=1}^{n}\Big(O^1 \frac{\partial}{\partial x^\beta} + O^1 \frac{\partial}{\partial y^\beta}\Big) + O^2 \frac{\partial}{\partial t}.$$

Proof. Let us set

$$X_j = B_j^k \frac{\partial}{\partial x^k}$$

for some $B_k^j : V_x \to \mathbf{R}$. The coordinates $x^j : V_x \to \mathbf{R}$ are given by $x^j(y) = p^j(E_x^{-1}(y))$, where $p^j : \mathbf{R}^{2n+1} \to \mathbf{R}$ are the canonical projections. Let $y = E_x(\xi)$ for $\xi \in U_x$ be arbitrary. Then

$$x^j(E_x(\xi)) = \xi^j$$

and by differentiation one has

$$\frac{\partial E_x^j}{\partial \xi^k} = \delta_k^j, \tag{3.9}$$

or

$$(d_\xi E_x)\left.\frac{\partial}{\partial \xi^j}\right|_\xi = \left.\frac{\partial}{\partial x^j}\right|_{E_x(\xi)}.$$

It follows that

$$B_k^j(x) = \delta_k^j. \tag{3.10}$$

We shall need the following lemma:

Lemma 3.2. *Let $G_k^j : U_x \to \mathbf{R}$ be the local expression of B_k^j, i.e., $G_k^j = B_k^j \circ E_x$. Then*

$$G_k^j(s\xi)\xi^k = \xi^j \tag{3.11}$$

for any $\xi \in U_x$ and any $|s| \leq 1$.

Proof. Let $C(t)$ be the solution of the Cauchy problem (3.8) and set $C_s(t) = C(st)$. Hence

$$\frac{dC_s}{dt}(t) = s\frac{dC}{dt}(st) = sX_\xi(C(st)) = s\xi^j X_j(C(st)) = X_{s\xi}(C_s(t))$$

so that (since $C_s(0) = x$)

$$E_x(s\xi) = C_s(1) = C(s), \tag{3.12}$$

by the definition of E_x. Differentiation with respect to s gives (by (3.9))

$$\frac{dC}{ds}(s) = \frac{dC^j}{ds}(s)\left.\frac{\partial}{\partial x^j}\right|_{C(s)} = \frac{\partial E_x^j}{\partial \xi^k}(s\xi)\xi^k \left.\frac{\partial}{\partial x^j}\right|_{C(s)} = \xi^j \left.\frac{\partial}{\partial x^j}\right|_{C(s)},$$

or (by (3.8))

$$X_\xi(C(s)) = \xi^j \left.\frac{\partial}{\partial x^j}\right|_{C(s)},$$

i.e.,

$$\xi^j B_j^k(C(s)) = \xi^k,$$

which is equivalent (again by (3.12)) to (3.11). □

Next, let us consider the structure functions $C^i_{jk} : V \to \mathbf{R}$ defined by

$$[X_j, X_k] = C^i_{jk} X_i .$$

Let $A = B^{-1}$ and denote by F_k^j the local expression of each A_k^j, i.e., $F_k^j = A_k^j \circ E_x$. For arbitrary $\xi \in U_x$ and $|s| \le 1$ let us consider the matrices

$$\mathcal{A}(s,\xi) = \left[\mathcal{A}_k^j(s,\xi)\right]_{0\le j,k\le 2n}, \quad \mathcal{A}_k^j(s,\xi) = s F_k^j(s\xi),$$
$$\Gamma(s,\xi) = \left[\Gamma_k^j(s,\xi)\right]_{0\le j,k\le 2n}, \quad \Gamma_k^j(s,\xi) = C_{\ell k}^j(E_x(s\xi))\xi^\ell .$$

We shall need the following lemma:

Lemma 3.3.

$$\frac{\partial \mathcal{A}}{\partial s} = I - \Gamma \mathcal{A}.$$

Proof. Note that

$$C^i_{jk} B_i^\ell = B_j^m \frac{\partial B_k^\ell}{\partial x^m} - B_k^m \frac{\partial B_j^\ell}{\partial x^m} .$$

In particular, at the point $y = E_x(s\xi)$ one gets the identity

$$C^i_{jk}(E_x(s\xi)) = F_\ell^i(s\xi)\left\{G_j^m(s\xi)\frac{\partial G_k^\ell}{\partial \xi^m}(s\xi) - G_k^m(s\xi)\frac{\partial G_j^\ell}{\partial \xi^m}(s\xi)\right\}.$$

Let us contract with ξ^j and use Lemma 3.2. Then

$$\begin{aligned} C^i_{jk}(E_x(s\xi))\xi^j &= F_\ell^i(s\xi)\left\{\xi^m \frac{\partial G_k^\ell}{\partial \xi^m}(s\xi) - G_k^m(s\xi)\xi^j \frac{\partial G_j^\ell}{\partial \xi^m}(s\xi)\right\} \\ &= -\xi^m \frac{\partial F_\ell^i}{\partial \xi^m}(s\xi) G_k^\ell(s\xi) - F_\ell^i(s\xi) G_k^m(s\xi)\xi^j \frac{\partial G_j^\ell}{\partial \xi^m}(s\xi) \end{aligned}$$

(because of $F_\ell^i G_k^\ell = \delta_k^i$). We have derived the identity

$$sC^i_{jk}(E_x(s\xi))\xi^j = -F_\ell^i(s\xi)G_k^m(s\xi)s\xi^j \frac{\partial G_j^\ell}{\partial \xi^m}(s\xi) - s\frac{d}{ds}\left\{F_\ell^i(s\xi)\right\}G_k^\ell(s\xi). \quad (3.13)$$

Differentiating $G_j^\ell(s\xi)\xi^j = \xi^\ell$ with respect to ξ^m, one gets

$$s\xi^j \frac{\partial G_j^\ell}{\partial \xi^m}(s\xi) = \delta_m^\ell - G_m^\ell(s\xi),$$

which substituted into (3.13) leads to

$$sC^i_{jk}(E_x(s\xi))\xi^j = -\delta_k^i + G_k^i(s\xi) - s\frac{d}{ds}\left\{F_j^i(s\xi)\right\}G_k^j(s\xi),$$

or (by contraction with $F^k_\ell(s\xi)$)

$$\frac{\partial}{\partial s}\mathcal{A}^i_\ell(s,\xi) = \delta^i_\ell - \Gamma^i_k(s,\xi)\mathcal{A}^k_\ell(s,\xi)$$

and Lemma 3.3 is completely proved. □

Let us go back to the proof of Theorem 3.5. By Taylor's formula (with the remainder in Lagrange's form)

$$G^j_k(s\xi) = \delta^j_k + s\frac{\partial G^j_k}{\partial \xi^\ell}(0)\xi^\ell + \tfrac{1}{2}s^2\frac{\partial^2 G^j_k}{\partial \xi^\ell \partial \xi^m}(cs\xi)\xi^\ell\xi^m$$

for some $c \in (0,1)$ (depending on s). Let $0 < \delta < 1$ so that $\overline{B}(0,\delta) \subseteq U_x$ and set

$$a^j_{k\ell m} = \sup_{\overline{B}(0,\delta)}\left|\frac{\partial^2 G^j_k}{\partial \xi^\ell \partial \xi^m}\right|, \quad C^j_k = 4\max_{0\le \ell,m\le 2n} a^j_{k\ell m},$$

$$f^j_k(\xi) = \tfrac{1}{2}\frac{\partial^2 G^j_k}{\partial \xi^\ell \partial \xi^m}(c\xi)\xi^\ell\xi^m .$$

Then, for any $\xi \in B(0,\delta)$,

$$|f^j_k(\xi)| \le 2C^j_k\sum_{\ell,m}|\xi^\ell|\,|\xi^m| \le C^j_k\|\xi\|^2 \le C^j_k|\xi|^2 \le C^j_k\left(|w|^2 + |\tau|\right),$$

where $w^\alpha = \xi^\alpha + i\xi^{\alpha+n}$ and $\tau = \xi^0$ (because of (1.26) as $\|\xi\| < 1$); hence $f^j_k = O^2$ in U_x as $\xi \to 0$. We therefore have

$$G^j_k(\xi) = \delta^j_k + G^{(1)}(\xi)^j_k + O^2, \tag{3.14}$$

where

$$G^{(1)}(\xi)^j_k = \frac{\partial G^j_k}{\partial \xi^\ell}(0)\xi^\ell .$$

We wish to compute $G^{(1)}(\xi)$. The Taylor series expansions for $F(s\xi)$ and $G(s\xi)$ read

$$F(s\xi) = I + sF^{(1)}(\xi) + s^2F^{(2)}(\xi) + \cdots,$$
$$G(s\xi) = I + sG^{(1)}(\xi) + s^2G^{(2)}(\xi) + \cdots,$$

for some matrices $F^{(p)}(\xi)$, $G^{(p)}(\xi)$, $p \ge 1$. Then

$$G^{(1)}(\xi) = -F^{(1)}(\xi) \tag{3.15}$$

because of $FG = I$. Also, we may write

$$\mathcal{A}(s,\xi) = \mathcal{A}(0,\xi) + s\mathcal{A}^{(0)}(\xi) + s^2\mathcal{A}^{(1)}(\xi) + \cdots$$

for some matrices $\mathcal{A}^j(\xi)$. Note that $\mathcal{A}(0,\xi)=0$ (by the definition of $\mathcal{A}$) and

$$\mathcal{A}^{(0)}(\xi)=\left[\frac{d}{ds}\left\{sF_k^j(s\xi)\right\}_{s=0}\right]_{0\le j,k\le 2n}=\left[F_k^j(0)\right]_{0\le j,k\le 2n}=I;$$

hence

$$\mathcal{A}(s,\xi)=sI+s^2\mathcal{A}^{(1)}(\xi)+s^3\mathcal{A}^{(2)}(\xi)+\cdots.$$

Let us differentiate with respect to s and use Lemma 3.3. We get

$$I-\Gamma(s,\xi)\mathcal{A}(s,\xi)=I+2s\mathcal{A}^{(1)}(\xi)+3s^2\mathcal{A}^{(2)}(\xi)+\cdots.$$

Now the Taylor series expansion of $\Gamma(s,\xi)$ reads

$$\Gamma(s,\xi)=\left[C_{\ell k}^j(s)\xi^\ell\right]_{0\le j,k\le 2n}+s\Gamma^{(1)}(\xi)+s^2\Gamma^{(2)}(\xi)+\cdots,$$

for some matrices $\Gamma^{(p)}(\xi)$; hence

$$\mathcal{A}^{(1)}(\xi)=-\tfrac{1}{2}\left[C_{\ell k}^j(x)\xi^\ell\right].$$

Summing up, we have

$$\mathcal{A}(s,\xi)=sI-\tfrac{1}{2}s^2\left[C_{\ell k}^j(x)\xi^\ell\right]+O(s^3),$$

or

$$s\left(I+sF^{(1)}(\xi)+s^2F^{(2)}(\xi)+\cdots\right)=sI-\tfrac{1}{2}s^2\left[C_{\ell k}^j(x)\xi^\ell\right]+O(s^3);$$

hence (by (3.15))

$$G^{(1)}(\xi)=\tfrac{1}{2}\left[C_{\ell k}^j(x)\xi^\ell\right]. \tag{3.16}$$

Let us write $T_\alpha=\frac{1}{2}(X_\alpha+iX_{\alpha+n})$. Then

$$4[T_\alpha,T_\beta]=\left(C_{\alpha\beta}^j-C_{\alpha+n,\beta+n}^j+iC_{\alpha,\beta+n}^j-iC_{\alpha+n,\beta}^j\right)X_j.$$

Yet $[T_\alpha,T_\beta]\in\Gamma^\infty(T_{1,0}(M))$; hence the T-component must vanish:

$$\begin{aligned}C_{\alpha\beta}^0&=C_{\alpha+n,\beta+n}^0\\ C_{\alpha,\beta+n}^0&=-C_{\alpha+n,\beta}^0.\end{aligned} \tag{3.17}$$

Likewise

$$4[T_\alpha,T_{\bar\beta}]=\left(C_{\alpha\beta}^j+C_{\alpha+n,\beta+n}^j-iC_{\alpha,\beta+n}^j+iC_{\alpha+n,\beta}^j\right)X_j,$$

and the coefficient of T in this formula must be $-8i\delta_{\alpha\beta}$; hence

$$\begin{aligned} C^0_{\alpha\beta} &= -C^0_{\alpha+n,\beta+n} \\ C^0_{\alpha+n,\beta} &= C^0_{\alpha,\beta+n} - 8\delta_{\alpha\beta}. \end{aligned} \tag{3.18}$$

Finally, solving (3.17)–(3.18) yields

$$\begin{aligned} C^0_{\alpha\beta} &= C^0_{\alpha+n,\beta+n} = 0 \\ C^0_{\alpha,\beta+n} &= -C^0_{\alpha+n,\beta} = 4\delta_{\alpha\beta}. \end{aligned} \tag{3.19}$$

Using (3.19) one may write (3.14) as

$$G^j_k(\xi) = \delta^j_k + \tfrac{1}{2}C^j_{\ell k}(\xi)\xi^\ell + O^2.$$

Let $y = E_x(\xi)$ be an arbitrary point in V_x. Then

$$\begin{aligned} X_k(y) = G^j_k(\xi)\left.\frac{\partial}{\partial x^j}\right|_y &= \left.\frac{\partial}{\partial x^k}\right|_y + \tfrac{1}{2}C^\alpha_{\ell k}(x)\xi^\ell\left.\frac{\partial}{\partial x^j}\right|_y + \tfrac{1}{2}C^{\alpha+n}_{\ell k}(x)\xi^\ell\left.\frac{\partial}{\partial y^\alpha}\right|_y \\ &\quad + \tfrac{1}{2}C^0_{\ell k}(x)\xi^\ell\left.\frac{\partial}{\partial t}\right|_y + \sum_{j=0}^{2n} O^2\left.\frac{\partial}{\partial x^j}\right|_y. \end{aligned}$$

Now $\xi^\alpha = O^1$, $\xi^{\alpha+n} = \eta^\alpha = O^1$, and $\xi^0 = \tau = O^2$, and of course an O^2 is O^1 as well. Hence (by (3.19))

$$X_\alpha = \frac{\partial}{\partial x^\alpha} + \sum_{\beta=1}^n\left(O^1\frac{\partial}{\partial x^\beta} + O^1\frac{\partial}{\partial y^\beta}\right) - 2\xi^{\alpha+n}\frac{\partial}{\partial t} + O^2\frac{\partial}{\partial t}$$

and the first formula in Theorem 3.5 is proved. The proof of the remaining formulas is quite similar, hence omitted. □

As a corollary of Theorem 3.5 we have that the expression of the complex tangent vectors T_α in Folland–Stein local coordinates is

$$T_\alpha = \frac{\partial}{\partial z^\alpha} + i\bar{z}^\alpha\frac{\partial}{\partial t} + \sum_{\beta=1}^n\left(O^1\frac{\partial}{\partial z^\beta} + O^1\frac{\partial}{\partial \bar{z}^\beta}\right) + O^2\frac{\partial}{\partial t}. \tag{3.20}$$

Let us set

$$\Omega = \bigcup_{x\in M}\{x\}\times V_x$$

so that Ω is a neighborhood of the diagonal in $M \times M$. For any $(x, y) \in \Omega$ let us set

$$\Theta(x, y) = \Theta_x(y),$$

where $\Theta_x : V_x \to \mathbf{R}^{2n+1}$ are the Folland–Stein coordinates at x. The reader may well note that for $M = \mathbf{H}_n$ and $T_\alpha = \partial/\partial z^\alpha + i\bar{z}^\alpha\partial/\partial t$ one gets $\Theta(x, y) = x^{-1}y$. Finally, let us set

$$\rho(x, y) = |\Theta(x, y)|$$

(Heisenberg norm) for any $(x, y) \in \Omega$.

Theorem 3.6. (G.B. Folland and E.M. Stein [150])

(1) $\Theta(x, y) = -\Theta(y, x) = \Theta(x, y)^{-1} \in \mathbf{H}_n$ *(in particular* $\Theta(x, x) = 0$*).*
(2) $\Theta : \Omega \to \mathbf{H}_n$ *is* C^∞.
(3) Let $K \subset M$ *be a compact subset and* $x, y, z \in K$ *such that*

$$(x, y), (x, z), (y, z) \in \Omega,$$
$$\rho(x, y) \le 1, \rho(x, z) \le 1.$$

There exist constants $C_1 > 0$, $C_2 > 0$ *such that*

$$|\Theta(x, y) - \Theta(z, y)| \le C_1 \left(\rho(x, z) + \rho(x, z)^{1/2}\rho(x, y)^{1/2}\right), \tag{3.21}$$
$$\rho(z, y) \le C_2 \left(\rho(x, z) + \rho(x, y)\right). \tag{3.22}$$

Part (1) follows from definitions, while part (2) is a consequence of standard ODE theory (dependence of solutions of ODEs on initial conditions and parameters). For the proof of part (3) the reader should see [150], p. 476. □

Let us set

$$y = (z, t) = \Theta(\xi, \eta), \quad (\xi, \eta) \in \Omega.$$

Definition 3.7. A C^∞ function $f(\xi, y)$, $(\xi, y) \in V \times \Theta(\Omega)$, is said to be of *type* O^k if for each compact set $K \subset V$ there is a constant $C_K > 0$ such that

$$|f(\xi, y)| \le C_K |y|^k,$$

for any $\xi \in K$. Here $|y|$ is the Heisenberg norm of y. Moreover, $O^k\mathcal{E}$ denotes an operator involving linear combinations of the indicated derivations, with coefficients of type O^k. □

Summing up the information in Theorem 3.5, the identity (3.20), and Theorem 3.6 above, we may state the following theorem:

Theorem 3.7. *Let* $\{W_\alpha\}$ *be a pseudo-Hermitian frame, defined on the open set* $V \subseteq M$*. Then there is an open set* $\Omega \subseteq V \times V$*, such that* $(\xi, \xi) \in \Omega$ *for any* $\xi \in V$*, and there is a* C^∞ *map* $\Theta : \Omega \to \mathbf{H}_n$ *such that*

(1) $\Theta(\xi, \eta) = -\Theta(\eta, \xi) = \Theta(\eta, \xi)^{-1}$*, for any* $(\xi, \eta) \in \Omega$.
(2) Set $\Theta_\xi(\eta) := \Theta(\xi, \eta)$*. Then* Θ_ξ *is a diffeomorphism of a neighborhood* Ω_ξ *of* ξ *onto a neighborhood of the origin in* $\mathbf{H}_n$.

(3) The following identities hold:

$$(\Theta_\xi^{-1})^*\theta = \theta_0 + O^1 dt + \sum_{j=1}^{n}\left(O^2 dz^j + O^2 d\overline{z}^j\right),$$

$$(\Theta_\xi^{-1})^*\left(\theta \wedge (d\theta)^n\right) = (1 + O^1)\theta_0 \wedge (d\theta_0)^n,$$

$$(\Theta_\xi)_* W_j = Z_j + O^1\mathcal{E}\Big(\frac{\partial}{\partial z}\Big) + O^2\mathcal{E}\Big(\frac{\partial}{\partial t}\Big),$$

$$(\Theta_\xi)_* T = \frac{\partial}{\partial t} + O^1\mathcal{E}\Big(\frac{\partial}{\partial z}, \frac{\partial}{\partial t}\Big),$$

$$(\Theta_\xi)_*\Delta_b = \mathcal{L}_0 + \mathcal{E}\Big(\frac{\partial}{\partial z}\Big) + O^1\mathcal{E}\Big(\frac{\partial}{\partial t}, \frac{\partial^2}{\partial z^2}\Big) + O^2\mathcal{E}\Big(\frac{\partial}{\partial z}, \frac{\partial}{\partial t}\Big) + O^3\mathcal{E}\Big(\frac{\partial^2}{\partial t^2}\Big).$$

Here $\partial/\partial z$ denotes any of the derivations $\partial/\partial z^j$, $\partial/\partial \overline{z}^j$ and

$$\mathcal{L}_0 = -\tfrac{1}{2}\sum_{j=1}^{n}(Z_j\overline{Z}_j + \overline{Z}_j Z_j).$$

Actually the last identity in Theorem 3.7 follows from an additional piece of information, the identity (A.6) below. The uniformity with respect to ξ of bounds on functions of type O^k is not stated explicitly in our preparation (preceding Theorem 3.7), yet follows easily from the fact that the coefficients are C^∞.

Note that when $M = \mathbf{H}_n$ and $\theta = \theta_0$ one may take $\Theta(\xi, \eta) = \xi^{-1}\eta$, and the terms with O^k coefficients (in Theorem 3.7) vanish identically. Hence these terms may be viewed as error terms (and Θ as an approximate group multiplication). Indeed, in the case of an arbitrary strictly pseudoconvex CR manifold, these terms have a higher homogeneity with respect to dilations. Precisely, let us set

$$T^\delta(z, t) = (\delta^{-1}z, \delta^{-2}t), \ (z, t) \in \mathbf{H}_n, \ \delta > 0.$$

Consider a compact set $K \subset V$ and $r > 0$. For sufficiently small $\delta > 0$ and any $\xi \in K$ one has

$$T^\delta\Theta_\xi(\Omega_\xi) \supset B_r := \{y \in \mathbf{H}_n : |y| \le r\}.$$

Then, for any $\xi \in K$ and $y \in B_r$,

$$\begin{aligned} \left[\left(T^{\delta}\Theta_{\xi}\right)^{-1}\right]^{*}\theta &= \delta^{2}\left(1+\delta O^{2}\right)\theta_{0}\,,\\ \left[\left(T^{\delta}\Theta_{\xi}\right)^{-1}\right]^{*}\theta\wedge(d\theta)^{n} &= \delta^{2n+2}\left(1+\delta O^{1}\right)\theta_{0}\wedge(d\theta_{0})^{n}\,,\\ \left(T^{\delta}\Theta_{\xi}\right)_{*}W_{j} &= \delta^{-1}\left(Z_{j}+\delta O^{1}\mathcal{E}\Big(\frac{\partial}{\partial z}\Big)+\delta^{2}O^{2}\mathcal{E}\Big(\frac{\partial}{\partial t}\Big)\right),\\ \left(T^{\delta}\Theta_{\xi}\right)_{*}\Delta_{b} &= \delta^{-2}\left(\mathcal{L}_{0}+\mathcal{E}\Big(\frac{\partial}{\partial z}\Big)+\delta O^{1}\mathcal{E}\Big(\frac{\partial}{\partial t},\frac{\partial^{2}}{\partial z^{2}}\Big)\right.\\ &\qquad\left.+\,\delta^{2}O^{2}\mathcal{E}\Big(\frac{\partial}{\partial z}\cdot\frac{\partial}{\partial t}\Big)+\delta^{3}O^{3}\mathcal{E}\Big(\frac{\partial^{2}}{\partial t^{2}}\Big)\right). \end{aligned} \tag{3.23}$$

Here an O^k may depend on δ, yet its derivatives are bounded by multiples of the frame constants, uniformly as $\delta \to 0$.

Definition 3.8. The term *frame constants* is used to mean bounds on finitely many derivatives of the coefficients in the $O^k\mathcal{E}$ terms in Theorem 3.7. $\square$

We wish to prove (ii) in Theorem 3.4, i.e., $\lambda(M) \le \lambda(S^{2n+1})$.

Lemma 3.4. *The class of test functions defining $\lambda(\mathbf{H}_n)$ can be restricted to C^∞ functions with compact support.*

To this end, let $\psi \in C_0^\infty(\mathbf{H}_n)$ such that $\psi \ge 0$ and $\int_{\mathbf{H}_n}\psi(y)dy = 1$. Let us set

$$\psi_\delta(x) = \delta^{-(2n+2)}\psi(\delta^{-1}x),$$

where $\delta^{-1}x$ is short for $T^\delta(x)$. Moreover, let u be a test function satisfying

$$\int_{\mathbf{H}_n}|u|^p\theta_0\wedge(d\theta_0)^n = 1,\quad Z_ju\in L^2(\mathbf{H}_n),\quad 1\le j\le n.$$

Since Z_j is left invariant, it follows that

$$Z_j(\psi_\delta^* u) = \psi_\delta^* Z_j u.$$

Next, one checks easily that $\psi_\delta^* u \in C^\infty(\mathbf{H}_n)$ and $\psi_\delta^* u \to u$ in $L^p(\mathbf{H}_n)$, $\psi_\delta^* Z_j u \to Z_j u$ in $L^2(\mathbf{H}_n)$, as $\delta \to 0$. Therefore one can restrict the class of test functions to $C^\infty(\mathbf{H}_n)$.

Consider $\phi \in C^\infty(\mathbf{H}_n)$ such that $\phi(x) = 1$ for $|x| < 1$, $\phi(x) = 0$ for $|x| > 2$, and $0 \le \phi(x) \le 1$, for any $x \in \mathbf{H}_n$. Next, let us set $\phi^\delta(x) = \phi(\delta x)$. Then $Z_j\phi^\delta$ is supported in $\delta^{-1} \le |x| \le 2\delta^{-1}$ and there is a constant $C > 0$ such that

$$\left|Z_j\phi^\delta\right| \le C\delta,\quad 1\le j\le n.$$

Consequently

$$\int_{\mathbf{H}_n} \left|Z_j\left(\phi^\delta u\right)\right|^2 \theta_0 \wedge (d\theta_0)^n = \int_{\mathbf{H}_n} \left|\left(Z_j\phi^\delta\right)u + \phi^\delta Z_j u\right|^2 \theta_0 \wedge (d\theta_0)^n$$
$$\leq \int_{\mathbf{H}_n} \left[\left(1+\frac{1}{s}\right)\left|Z_j\phi^\delta\right|^2 |u|^2 + (1+s)|\phi^\delta|^2|Z_j u|^2\right]\theta_0 \wedge (d\theta_0)^n$$
$$\leq C^2\left(1+\frac{1}{s}\right)\int_{\mathbf{H}_n} \delta^2\chi_\delta |u|^2\theta_0 \wedge (d\theta_0)^n + (1+s)\int_{\mathbf{H}_n} |Z_j u|^2\theta_0 \wedge (d\theta_0)^n$$

for any $s > 0$. Here

$$\chi_\delta(x) = \begin{cases} 1, & \text{on } \delta^{-1} \leq |x| \leq 2\delta^{-1}, \\ 0, & \text{otherwise.} \end{cases}$$

Since

$$\int_{\mathbf{H}_n} \chi_\delta(x)dx = C_n\delta^{-(2n+2)},$$

by the Hölder inequality and $(2n+2)(1-\frac{2}{p}) = 2$ it follows that

$$\int_{\mathbf{H}_n} \delta^2\chi_\delta|u|^2 dx \leq \left(\int_{\mathbf{H}_n} |u|^p\chi_\delta dx\right)^{2/p} \delta^2 \left(\int_{\mathbf{H}_n} \chi_\delta(x)dx\right)^{1-2/p}$$
$$= C_n^{1-2/p}\left(\int_{\mathbf{H}_n} |u|^p\chi_\delta dx\right)^{2/p}.$$

This integral tends to zero as $\delta \to 0$ (since $u \in L^p(\mathbf{H}_n)$). Finally, for s and δ sufficiently small we get

$$\limsup_{\delta\to 0} \int_{\mathbf{H}_n} \sum_{j=1}^n \left|Z_j(\phi^\delta u)\right|^2 \theta_0 \wedge (d\theta_0)^n \leq \int_{\mathbf{H}_n} \sum_{j=1}^n |Z_j u|^2\theta_0 \wedge (d\theta_0)^n .$$

Also, let us note that

$$\lim_{\delta\to 0} \int_{\mathbf{H}_n} |\phi^\delta u|^p\theta_0 \wedge (d\theta_0)^n = \int_{\mathbf{H}_n} |u|^p\theta_0 \wedge (d\theta_0)^n;$$

hence one may restrict the class of test functions to $C_0^\infty(\mathbf{H}_n)$. □

To see that $\lambda(M) \leq \lambda(\mathbf{H}_n)$, let $u \in C_0^\infty(\mathbf{H}_n)$ be such that $B_{\theta_0}(u) = 1$ and $A_{\theta_0}(u) < \lambda(\mathbf{H}_n) + \epsilon$. Let us set

$$u_{(\delta)}(x) := \delta^{-n}u(\delta^{-1}x).$$

Let $\xi \in M$ and let Θ_ξ be a Folland–Stein normal coordinate chart at ξ (cf. Theorem 3.7). Set

$$v_{(\delta)}(\eta) := u_{(\delta)}(\Theta_\xi(\eta)).$$

For δ sufficiently small, $\operatorname{supp}(u_{(\delta)}) \subset \Theta_\xi(\Omega_\xi)$; hence $v_{(\delta)}$ has compact support in Ω_ξ. Let us extend $v_{(\delta)}$ to a function in $C^\infty(M)$ by setting $v_{(\delta)} = 0$ outside Ω_ξ. It is immediate that

$$B_{\theta_0}(u_{(\delta)}) = B_{\theta_0}(u) = 1,$$
$$A_{\theta_0}(u_{(\delta)}) = A_{\theta_0}(u) < \lambda(\mathbf{H}_n) + \epsilon.$$

Moreover

$$\int_{\mathbf{H}_n} |u_{(\delta)}|^2 \theta_0 \wedge (d\theta_0)^n = \delta^2 \int_{\mathbf{H}_n} |u|^2 \theta_0 \wedge (d\theta_0)^n \to 0,$$

as $\delta \to 0$. Then (3.23) implies that

$$\lim_{\delta \to 0} B_\theta(v_{(\delta)}) = 1,$$
$$\lim_{\delta \to 0} A_\theta(v_{(\delta)}) = A_\theta(u) < \lambda(\mathbf{H}_n) + \epsilon,$$

with $\epsilon > 0$ arbitrary; hence $\lambda(M) \leq \lambda(S^{2n+1})$. □

3.3 A Sobolev-type lemma

To establish the regularity results for $\Delta_b u = f$, and then for the CR Yamabe equation, we need to prepare the following analogue of the classical Sobolev lemma:

Theorem 3.8. *Set $X_j = \operatorname{Re}(Z_j)$ and $X_{j+n} = \operatorname{Im}(Z_j)$, $1 \leq j \leq n$. There is a constant $C_n > 0$ such that*

$$\left(\int_{\mathbf{H}_n} |f|^p \theta_0 \wedge (d\theta_0)^n \right)^{2/p} \leq C_n \int_{\mathbf{H}_n} \sum_{j=1}^{2n} |X_j f|^2 \theta_0 \wedge (d\theta_0)^n \tag{3.24}$$

for any $f \in C_0^\infty(\mathbf{H}_n)$, where $p = 2 + 2/n$.

The proof of Theorem 3.8 requires the fundamental solution to the operator $\mathcal{L}_0$. We shall present a more general result (cf. Theorem 3.9 below) exhibiting the fundamental solutions for the family of Folland–Stein operators

$$\mathcal{L}_\alpha = -\tfrac{1}{2} \sum_{\mu=1}^{n} \left(T_\mu T_{\overline{\mu}} + T_{\overline{\mu}} T_\mu \right) + i\,\alpha \frac{\partial}{\partial t},$$

(for $\alpha \in \mathbf{C}$ *admissible*; see the definitions below). The results within the harmonic analysis on the Heisenberg group that we need are classical and at least partially presented in book form (cf. E.M. Stein [385]).

Our first task is to write (1.146) for the Heisenberg group $M = \mathbf{H}_n$ (carrying the standard strictly pseudoconvex CR structure). Let $\psi \in \Omega^{0,q+1}(\mathbf{H}_n)$. The Tanaka–Webster connection of $\mathbf{H}_n$ is flat; hence (1.146) becomes

$$\Box_b \psi = -\left(h^{\lambda\bar{\mu}} \nabla_\lambda \nabla_{\bar{\mu}} \psi_{\bar{\alpha}_1 \cdots \bar{\alpha}_{q+1}} + 2i(q+1)\nabla_0 \psi_{\bar{\alpha}_1 \cdots \bar{\alpha}_{q+1}}\right) \theta^{\bar{\alpha}_1} \wedge \cdots \wedge \theta^{\bar{\alpha}_{q+1}} .$$

Also

$$\Gamma^{\bar{\alpha}}_{\mu\bar{\beta}} = \Gamma^{\bar{\alpha}}_{\bar{\mu}\bar{\beta}} = \Gamma^{\bar{\alpha}}_{0\bar{\beta}} = 0$$

and hence we may replace the covariant derivatives by ordinary derivatives

$$\nabla_0 \psi_{\bar{\alpha}_1 \cdots \bar{\alpha}_{q+1}} = \frac{1}{(q+1)!} T(\psi_{\bar{\alpha}_1 \cdots \bar{\alpha}_{q+1}}),$$
$$\nabla_\lambda \nabla_{\bar{\mu}} \psi_{\bar{\alpha}_1 \cdots \bar{\alpha}_{q+1}} = \frac{1}{(q+1)!} T_\lambda T_{\bar{\mu}} \psi_{\bar{\alpha}_1 \cdots \bar{\alpha}_{q+1}} ,$$

where T_λ are given by (1.25) and $T = \partial/\partial t$. Next, using the commutation formula $[T_\alpha, T_{\bar{\beta}}] = -2i\delta_{\alpha\beta} T$ we may calculate $\Box_b \psi$ as follows:

$$\Box_b \psi = -\frac{1}{(q+1)!} \left\{\delta^{\lambda\mu} T_\lambda T_{\bar{\mu}} \psi_{\bar{\alpha}_1 \cdots \bar{\alpha}_{q+1}} + 2i(q+1)T\left(\psi_{\bar{\alpha}_1 \cdots \bar{\alpha}_{q+1}}\right)\right\} \theta^{\bar{\alpha}_1} \wedge \cdots \wedge \theta^{\bar{\alpha}_{q+1}}$$
$$= -\frac{1}{(q+1)!} \sum_{\lambda=1}^{n} \Big\{ \tfrac{1}{2} T_\lambda T_{\bar{\lambda}} \psi_{\bar{\alpha}_1 \cdots \bar{\alpha}_{q+1}} + \tfrac{1}{2}(T_{\bar{\lambda}} T_\lambda - 2iT)\psi_{\bar{\alpha}_1 \cdots \bar{\alpha}_{q+1}}$$
$$+ 2i(q+1)T\left(\psi_{\bar{\alpha}_1 \cdots \bar{\alpha}_{q+1}}\right) \Big\} \theta^{\bar{\alpha}_1} \wedge \cdots \wedge \theta^{\bar{\alpha}_{q+1}} .$$

Therefore, we may adopt the following definition:

Definition 3.9. Let $T_{\bar{\lambda}}$ be the Lewy operators on the Heisenberg group $\mathbf{H}_n$ and $T = \partial/\partial t$. The differential operators $\mathcal{L}_\alpha$ ($\alpha \in \mathbf{C}$), given by

$$\mathcal{L}_\alpha = -\tfrac{1}{2} \sum_{\lambda=1}^{n} \left(T_\lambda T_{\bar{\lambda}} + T_{\bar{\lambda}} T_\lambda\right) + i\alpha T,$$

are called the *Folland–Stein operators.* $\Box$

Proposition 3.1. *The Kohn–Rossi operator of* $(\mathbf{H}_n, \theta_0)$ *is given by*

$$\Box_b \psi = \frac{1}{(q+1)!} \left(\mathcal{L}_{n-2(q+1)} \psi_{\bar{\alpha}_1 \cdots \bar{\alpha}_{q+1}}\right) \theta^{\bar{\alpha}_1} \wedge \cdots \wedge \theta^{\bar{\alpha}_{q+1}} , \tag{3.25}$$

for any $\psi \in \Omega^{0,q+1}(\mathbf{H}_n)$.

A fundamental solution for $\mathcal{L}_0$ has been determined by G.B. Folland [146]. Precisely, there is a constant $c_0 \neq 0$ such that $\varphi_0 : \mathbf{H}_n \setminus \{0\} \to \mathbf{R}$ given by $\varphi_0(x) = |x|^{-2n}$ satisfies $\mathcal{L}_0 \varphi_0 = c_0 \delta$, where δ is the Dirac distribution (concentrated in zero). Following G.B. Folland and E.M. Stein [150], we obtain fundamental solutions for the operators $\mathcal{L}_\alpha$. Let us consider

$$\varphi_\alpha(z, t) = |(z, t)|^{-2n} f\left(\frac{t}{|(z, t)|^2}\right)$$

and look for the unknown function $f(\omega)$ such that $\mathcal{L}_\alpha \varphi_\alpha = 0$ on $\mathbf{H}_n \setminus \{0\}$. Let us set $\omega = t/|(z,t)|^2$. Then

$$\frac{\partial \omega}{\partial t} = |x|^{-2}(1-\omega^2), \quad \frac{\partial \omega}{\partial \bar{z}^j} = -t|x|^{-6}\|z\|^2 z^j \,,$$

where $x = (z,t)$. Using (1.25) one may write $\mathcal{L}_\alpha$ in the following form:

$$\mathcal{L}_\alpha = -\sum_{j=1}^{n} \left\{ \frac{\partial^2}{\partial z^j \partial \bar{z}^j} + |z^j|^2 \frac{\partial^2}{\partial t^2} - i \frac{\partial}{\partial t} \left(z^j \frac{\partial}{\partial z^j} - \bar{z}^j \frac{\partial}{\partial \bar{z}^j} \right) \right\} + i\alpha \frac{\partial}{\partial t} \,.$$

Taking into account

$$\frac{\partial \varphi_\alpha}{\partial z^j} = -|x|^{-2(n+2)} \|z\|^2 \bar{z}^j [n f(\omega) + \omega f'(\omega)],$$
$$\frac{\partial \varphi_\alpha}{\partial t} = -|x|^{-2(n+1)} [n\omega f(\omega) + (\omega^2 - 1) f'(\omega)],$$

one obtains

$$\begin{aligned}\frac{\partial^2 \varphi_\alpha}{\partial t^2} &= |x|^{-2(n+2)} \Big\{ n[(n+2)\omega^2 - 1] f(\omega) \\ &\quad + (2n+3)\omega(\omega^2-1) f'(\omega) + (\omega^2-1)^2 f''(\omega) \Big\} \,, \\ \frac{\partial^2 \varphi_\alpha}{\partial \bar{z}^j \partial z^j} &= |x|^{-2(n+4)} \|z\|^2 |z^j|^2 [(n+1)\omega f'(\omega) + \omega^2 f''(\omega)] \\ &\quad - |x|^{-2(n+2)} [\|z\|^2 + |z^j|^2 - (n+2)\|z\|^4 |z^j|^2 |x|^{-4}][n f(\omega) + \omega f'(\omega)].\end{aligned}$$

Then one may write $\mathcal{L}_\alpha \varphi_\alpha = 0$ as

$$\begin{aligned}&|x|^{-6}\|z\|^6[(n+1)\omega f'(\omega) + \omega^2 f''(\omega)] \\ &\quad - |x|^{-2}\|z\|^2[n+1-(n+2)\|z\|^4|x|^{-4}][n f(\omega) + \omega f'(\omega)] \\ &\quad + \|z\|^2|x|^{-2}\{n[(n+2)\omega^2-1]f(\omega) + (2n+3)\omega(\omega^2-1)f'(\omega) \\ &\qquad + (\omega^2-1)^2 f''(\omega)\} + i\alpha[n\omega f(\omega) + (\omega^2-1)f'(\omega)] = 0.\end{aligned}$$

Finally, using $1-\omega^2 = |x|^{-4}\|z\|^4$, the above ODE may be written as

$$\begin{aligned}(1-\omega^2)^{3/2} f''(\omega) - (1-\omega^2)^{1/2}[(n+1)\omega + i\alpha(1-\omega^2)^{1/2}] f'(\omega) \\ + in\alpha\omega f(\omega) = 0. \quad (3.26)\end{aligned}$$

Since $1-\omega^2 = |x|^{-4}\|z\|^4 > 0$ we may set $\omega = \cos\theta$, $0 \le \theta \le \pi$, and $g(\theta) = f(\cos\theta)$. Consequently (3.26) becomes

$$(\sin\theta) g''(\theta) + (n\cos\theta + i\alpha\sin\theta) g'(\theta) + in\alpha(\cos\theta) g(\theta) = 0,$$

or

$$\left[(\sin\theta)\frac{d}{d\theta}+n\cos\theta\right]\left[\frac{d}{d\theta}+i\alpha\right]g(\theta)=0. \tag{3.27}$$

Note that $g(\theta)=ce^{-i\alpha\theta}$ are bounded solutions of (3.27). Since $e^{-i\theta}=\omega-i\sqrt{1-\omega^2}$ we obtain

$$f(\omega)=c\Big(\frac{t-i\|z\|^2}{|x|^2}\Big)^{\alpha}.$$

Let $c=i^{\alpha}$. Then

$$\varphi_\alpha(z,t)=\big(\|z\|^2-it\big)^{-(n+\alpha)/2}\big(\|z\|^2+it\big)^{-(n-\alpha)/2}. \tag{3.28}$$

We are in a position to prove the following theorem:

Theorem 3.9. (G.B. Folland and E.M. Stein [150])

$$\mathcal{L}_\alpha\varphi_\alpha=c_\alpha\delta,$$

where c_α is given by

$$c_\alpha=\frac{2^{2-2n}\pi^{n+1}}{\Gamma\left(\frac{n+\alpha}{2}\right)\Gamma\left(\frac{n-\alpha}{2}\right)}.$$

Proof. Let $\epsilon>0$ and let us set

$$\rho_\epsilon(z,t)=\|z\|^2+\epsilon^2-it,\quad \varphi_{\alpha,\epsilon}=\rho_\epsilon^{-(n+\alpha)/2}\bar{\rho}_\epsilon^{-(n-\alpha)/2}.$$

Then $\varphi_{\alpha,\epsilon}$ is a C^∞ function on $\mathbf{H}_n$, and we may prove the following lemma:

Lemma 3.5. $\varphi_{\alpha,\epsilon}\to\varphi_\alpha$ *as* $\epsilon\to 0$, *in distribution sense.*

Indeed, let $\psi\in C_0^\infty(\mathbf{H}_n)$. We must show that

$$\lim_{\epsilon\to 0}\int_{\mathbf{H}_n}\varphi_{\alpha,\epsilon}(x)\psi(x)dV(x)=\int_{\mathbf{H}_n}\varphi_\alpha(x)\psi(x)dV(x), \tag{3.29}$$

where $dV(x)=2^{-n}dx$ and dx is the Lebesgue measure on $\mathbf{R}^{2n+1}$. Clearly $\varphi_{\alpha,\epsilon}(x)\to\varphi_\alpha(x)$ as $\epsilon\to 0$, at any $x\in\mathbf{H}_n$. Let us set $K=\operatorname{supp}(\psi)$ and $\Gamma_\psi=\sup_K|\psi|>0$. Let us recall that for any $a,s\in\mathbf{C}$ one has

$$|a^s|=|a|^{\operatorname{Re}(s)}e^{-\operatorname{Im}(s)\arg(a)},$$

where $\arg:\mathbf{C}\to[-\pi,\pi)$. Let $x\in K$. Then

$$\begin{aligned}|\varphi_{\alpha,\epsilon}(x)\psi(x)|&\le\Gamma_\psi|\varphi_{\alpha,\epsilon}(x)|=\Gamma_\psi|\rho_\epsilon(x)|^{-n}e^{\operatorname{Im}(\alpha)\arg(\rho_\epsilon(x))}\\&\le\Gamma_\psi|(\|z\|^2+\epsilon^2)^2+t^2|^{-n/2}e^{\operatorname{Im}(\alpha)\pi}\le\Gamma_\psi e^{\pi\operatorname{Im}(\alpha)}|x|^{-2n}.\end{aligned}$$

On the other hand, for any $u\in\mathbf{H}_n$ we have

$$|u| \leq 1 \Longrightarrow \|u\| \leq |u| \leq \|u\|^{1/2}.$$

Hence

$$\begin{aligned}\int_K |x|^{-2n} dV &= \int_{K\cap\{|x|\leq 1\}} |x|^{-2n} dV + \int_{K\cap\{|x|>1\}} |x|^{-2n} dV \\ &\leq A + \int_{K\cap\{|x|\leq 1\}} |x|^{-2n} dV \leq A + \int_{K\cap\{|x|\leq 1\}} \|x\|^{-2n} 2^{-n} dx \\ &= A + 2^{-n} \int_0^1 \left(\int_{\|x\|=\rho} \rho^{-2n} d\sigma(x) \right) d\rho = A + 2^{-n} \omega_{2n} ,\end{aligned}$$

for some constant $A > 0$. Here ω_N is the "area" of the sphere S^N. Hence one may apply Lebesgue's convergence theorem to obtain (3.29). □

Next, let us set

$$\psi_{\alpha,\epsilon} = \mathcal{L}_\alpha \varphi_{\alpha,\epsilon} .$$

Lemma 3.6. *$\psi_{\alpha,\epsilon} \to c_\alpha \delta$, as $\epsilon \to 0$.*

Let $a \in \mathbf{C}$. Note that

$$T_j(\rho_\epsilon^a) = 2a\bar{z}^j \rho_\epsilon^{a-1}, \quad T_j(\bar{\rho}_\epsilon^a) = T_{\bar{j}}(\rho_\epsilon^a) = 0, \quad T(\rho_\epsilon^a) = -ia\rho_\epsilon^{a-1} ,$$

where $T_j = \partial/\partial z^j + i\bar{z}^j \partial/\partial t$ and $T = \partial/\partial t$. Let us write $\rho = \rho_\epsilon$, for simplicity. Then, for any $a, b \in \mathbf{C}$, we have

$$\begin{aligned}T_j T_{\bar{j}}(\rho^a \bar{\rho}^b) &= 2b\rho^a \bar{\rho}^{b-1} + 4ab\rho^{a-1}\bar{\rho}^{b-1}|z^j|^2, \\ T(\rho^a \bar{\rho}^b) &= i\rho^{a-1}\bar{\rho}^{b-1}(-a\bar{\rho} + b\rho).\end{aligned}$$

Due to the commutation formula $[T_j, T_{\bar{j}}] = -2iT$ the operator

$$\mathcal{L}_\alpha = -\tfrac{1}{2} \sum_{j=1}^n (T_j T_{\bar{j}} + T_{\bar{j}} T_j) + i\alpha T$$

may be also written as

$$\mathcal{L}_\alpha = -\sum_{j=1}^n T_j T_{\bar{j}} + i(\alpha - n)T.$$

In particular for

$$a = -\frac{n+\alpha}{2} , \qquad b = -\frac{n-\alpha}{2}$$

we have

$$\psi_{\alpha,\epsilon} = \mathcal{L}_\alpha \varphi_{\alpha,\epsilon} = \mathcal{L}_\alpha \left(\rho_\epsilon^{-(n+\alpha)/2} \bar{\rho}_\epsilon^{-(n-\alpha)/2} \right)$$
$$= \rho^{-(n+\alpha+2)/2} \bar{\rho}^{-(n-\alpha+2)/2} \Big\{ n(n-\alpha)\rho - (n^2-\alpha^2)\|z\|^2$$
$$- \tfrac{1}{2}(\alpha-n)^2\rho + \tfrac{1}{2}(n^2-\alpha^2)\bar{\rho} \Big\}.$$

Next, since $\rho = \|z\|^2 + \epsilon^2 - it$, we may write $\psi_{\alpha,\epsilon}$ as

$$\psi_{\alpha,\epsilon}(z,t) = \epsilon^2(n^2-\alpha^2)\big(\|z\|^2+\epsilon^2-it\big)^{-(n+\alpha+2)/2}\big(\|z\|^2+\epsilon^2+it\big)^{-(n-\alpha+2)/2},$$

whence

$$\psi_{\alpha,\epsilon} = \epsilon^{-2n-2}\psi_{\alpha,1} \circ \delta_{1/\epsilon};$$

hence

$$\int_{\mathbf{H}_n} \psi_{\alpha,\epsilon} dV = \int_{\mathbf{H}_n} \psi_{\alpha,1} dV. \tag{3.30}$$

This may be checked by a change of variable $(z', t') = \delta_{1/r}(z,t)$ under the first integral sign in (3.30) (since $dV' = \epsilon^{-2n-2}dV$). Let us set

$$c_\alpha = \int_{\mathbf{H}_n} \psi_{\alpha,1} dV. \tag{3.31}$$

Let $\varphi \in C_0^\infty(\mathbf{H}_n)$ be an arbitrary test function. We must show that

$$\lim_{\epsilon\to 0} \int_{\mathbf{H}_n} \psi_{\alpha,\epsilon}\varphi dV = c_\alpha \varphi(0).$$

Using (3.30) we may perform the following estimates:

$$\left| \int_{\mathbf{H}_n} \psi_{\alpha,\epsilon}\varphi dV - c_\alpha\varphi(0) \right| = \left| \int_{\mathbf{H}_n} \psi_{\alpha,\epsilon}\varphi dV - \varphi(0)\int_{\mathbf{H}_n} \psi_{\alpha,1} dV \right|$$
$$= \left| \int_{\mathbf{H}_n} \psi_{\alpha,\epsilon}(\varphi - \varphi(0)) dV \right| \le \int_{\mathbf{H}_n} |\psi_{\alpha,\epsilon}|\,|\varphi-\varphi(0)| dV \le \Gamma_\varphi \int_K |\psi_{\alpha,\epsilon}| dV \to 0,$$

as $\epsilon \to 0$. Here $\Gamma_\varphi = \sup_K |\varphi - \varphi(0)|$ and $K = \text{supp}(\varphi)$. □

Summing up, we have proved that

$$\varphi_{\alpha,\epsilon} \to \varphi_\alpha\,, \quad \mathcal{L}_\alpha\varphi_{\alpha,\epsilon} \to c_\alpha\delta,$$

as $\epsilon \to 0$, in the sense of distributions, where c_α is given by (3.31).

At this point we may show that $\mathcal{L}_\alpha\varphi_\alpha = c_\alpha\delta$. Indeed, let $u \in C_0^\infty(\mathbf{H}_n)$. Then

$$\int_{\mathbf{H}_n} (\mathcal{L}_\alpha\varphi_\alpha) u\, dV = \int_{\mathbf{H}_n} \varphi_\alpha(\mathcal{L}_\alpha^* u) dV$$
$$= \lim_{\epsilon\to 0}\int_{\mathbf{H}_n} \varphi_{\alpha,\epsilon}(\mathcal{L}_\alpha^* u) dV = \lim_{\epsilon\to 0}\int_{\mathbf{H}_n} (\mathcal{L}_\alpha\varphi_{\alpha,\epsilon}) u\, dV = c_\alpha\delta(u),$$

by the Lagrange-Green identity. □

To finish the proof of Theorem 3.9 we compute the constants (3.31). Let us set

$$\beta = \tfrac{1}{2}(n+\alpha+2), \quad \gamma = \tfrac{1}{2}(n-\alpha+2).$$

Let $A(z)$ be a positive function of z. Then

$$\begin{aligned}\int_{\mathbf{R}} (A-it)^{-\beta}(A+it)^{-\gamma}dt &= A^{-\beta-\gamma}\int_{\mathbf{R}}\left(1-i\frac{t}{A}\right)^{-\beta}\left(1+i\frac{t}{A}\right)^{-\gamma}dt \\ &= A^{-n-1}\int_{\mathbf{R}}(1-i\tau)^{-\beta}(1+i\tau)^{-\gamma}d\tau.\end{aligned}$$

In particular, for $A = \|z\|^2+1$, since $dV = 2^{-n}dx\,dy\,dt$ (with $z^j = x^j + iy^j$, $1 \leq j \leq n$) we have

$$\begin{aligned}c_\alpha = \int_{\mathbf{H}_n}\psi_{\alpha,1}dV &= \int_{\mathbf{H}_n}(n^2-\alpha^2)(A-it)^{-\beta}(A+it)^{-\gamma}dV \\ &= 2^{-n}(n^2-\alpha^2)\int_{\mathbf{R}^{2n}}\left(\int_{\mathbf{R}}(A-it)^{-\beta}(A+it)^{-\gamma}dt\right)dx\,dy \\ &= 2^{-n}(n^2-\alpha^2)\int_{\mathbf{R}^{2n}}A^{-n-1}\left(\int_{\mathbf{R}}(1-i\tau)^{-\beta}(1+i\tau)^{-\gamma}d\tau\right)dx\,dy,\end{aligned}$$

that is,

$$\begin{aligned}c_\alpha = 2^{-n}(n^2-\alpha^2)&\left(\int_{\mathbf{C}^n}(\|z\|^2+1)^{-n-1}dx\,dy\right) \\ &\times\left(\int_{\mathbf{R}}(1-it)^{-\beta}(1+it)^{-\gamma}dt\right). \quad (3.32)\end{aligned}$$

The integrals in (3.32) may be computed as follows. First

$$\begin{aligned}\int_{\mathbf{R}^{2n}}(\|z\|^2+1)^{-n-1}dx\,dy &= \int_0^\infty\left(\int_{\|z\|=\rho}(\rho^2+1)^{-n-1}d\sigma(x,y)\right)d\rho \\ = \omega_{2n-1}\int_0^\infty(\rho^2+1)^{-n-1}\rho^{2n-1}d\rho &= \tfrac{1}{2}\omega_{2n-1}\int_1^\infty s^{-n-1}(s-1)^{n-1}ds,\end{aligned}$$

where we set $s = \rho^2+1$. Moreover, $\omega_{2n-1} = 2\pi^n/\Gamma(n)$ and (by setting $\sigma = 1/s$) we get

$$\int_{\mathbf{R}^{2n}}(\|z\|^2+1)^{-n-1}dx\,dy = \frac{\pi^n}{\Gamma(n)}\int_0^1(1-\sigma)^{n-1}d\sigma = \frac{\pi^n}{\Gamma(n+1)}$$

(because of $\Gamma(n) = (n-1)!$). The second integral is more difficult to compute. Assume for the time being that $\alpha \in [-n, n]$. Then $\beta \geq 1$, $\gamma \geq 1$. We shall make use of the identity

$$\int_0^\infty e^{-sx}x^{\gamma-1}dx = \Gamma(\gamma)s^{-\gamma}, \quad (3.33)$$

which holds for any complex number s in the domain

$$\{s \in \mathbf{C} : \mathrm{Re}(s) > 0\}.$$

In particular for $s = 1 + it$ we get

$$(1+it)^{-\gamma}\Gamma(\gamma) = \int_0^\infty e^{-ixt} e^{-x} x^{\gamma-1} dx,$$

which may be written as

$$(1+it)^{-\gamma}\Gamma(\gamma) = \hat{f}(t), \tag{3.34}$$

where f is defined by

$$f(x) = \begin{cases} e^{-x}x^{\gamma-1} & x > 0, \\ 0 & x \le 0, \end{cases}$$

and $\hat{f}$ is its Fourier transform

$$\hat{f}(t) = \int_{\mathbf{R}} e^{-ixt} f(x) dx.$$

Similarly (i.e., again by (3.33)) we obtain

$$(1-it)^{-\beta}\Gamma(\beta) = \hat{g}(t), \tag{3.35}$$

where

$$g(y) = \begin{cases} 0 & y \ge 0, \\ e^{|y|}|y|^{\beta-1} & y < 0. \end{cases}$$

Note that

$$\int_{\mathbf{R}} \hat{f}(t)\hat{g}(t) dt = 2\pi \int_{\mathbf{R}} f(x) g(-x) dx. \tag{3.36}$$

The 2π factor appears due to our choice of definition for $\hat{f}$ (in comparison with the usual one; cf. [147], p. 20) together with Proposition 0.26 in [147], p. 23. Let us multiply (3.34)–(3.35) term by term and integrate over $\mathbf{R}$ in the resulting identity. Then (by (3.36)) we obtain

$$\Gamma(\beta)\Gamma(\gamma)\int_{\mathbf{R}} (1-it)^{-\beta}(1+it)^{-\gamma} dt = 2\pi \int_0^\infty e^{-2x} x^n dx,$$

or (by (3.33) for $s = 2$ and $\gamma = n + 1$)

$$\int_{\mathbf{R}} (1-it)^{-\beta}(1+it)^{-\gamma} dt = \frac{\pi\Gamma(n+1)2^{-n}}{\Gamma(\beta)\Gamma(\gamma)}, \tag{3.37}$$

for any $\alpha \in [-n, n]$. Returning to general $\alpha \in \mathbf{C}$, we may write

$$\int_{\mathbf{R}} (1 - it)^{-\beta}(1 + it)^{-\gamma} dt = \int_{\mathbf{R}} (1 - it)^{-(n+\alpha+2)/2}(1 + it)^{-(n-\alpha+2)/2} dt$$
$$= \int_{\mathbf{R}} (1 + t^2)^{-(n+2)/2}(1 - it)^{-\alpha/2}(1 + it)^{\alpha/2} dt$$

and

$$(1 - it)^{-\alpha/2}(1 + it)^{\alpha/2} = e^{-\alpha/2 \log(1-it)} e^{\alpha/2 \log(1+it)} = e^{i\alpha \arctan t},$$

that is,

$$\int_{\mathbf{R}} (1 - it)^{-\beta}(1 + it)^{-\gamma} dt = \int_{\mathbf{R}} (1 + t^2)^{-(n+\alpha)/2} e^{i\alpha \arctan t} dt;$$

hence $\alpha \mapsto \int_{\mathbf{R}} (1 - it)^{-\beta}(1 + it)^{-\gamma} dt$ is a holomorphic function on $\mathbf{C}$. Thus

$$\int_{\mathbf{R}} (1 - it)^{-\beta}(1 + it)^{-\gamma} dt - \frac{2^{-n}\pi\Gamma(n + 1)}{\Gamma(\beta)\Gamma(\gamma)}$$

is a holomorphic function of α and therefore vanishes identically (otherwise, by Proposition 4.1 in [89], p. 41, it would have a discrete set of zeros; yet (by (3.37)) it vanishes on $[-n, n]$, a contradiction). Summing up, we have computed both integrals in (3.32), so that c_α may be written

$$c_\alpha = \frac{2^{2-2n}\pi^{n+1}}{\Gamma\left(\frac{n+\alpha}{2}\right)\Gamma\left(\frac{n-\alpha}{2}\right)}$$

and Theorem 3.9 is completely proved. □

Note that $c_\alpha = 0$ if and only if $\alpha \in \{\pm n, \pm(n + 2), \pm(n + 4), \dots\}$.

Definition 3.10. We call $\alpha \in \mathbf{C}$ *admissible* if $c_\alpha \neq 0$. □

Let $\alpha \in \mathbf{C}$ be admissible. Then

$$\Phi_\alpha = \frac{1}{c_\alpha}\varphi_\alpha$$

is a fundamental solution for $\mathcal{L}_\alpha$ with source at 0, i.e., $\mathcal{L}_\alpha \Phi_\alpha = \delta$. □

Before attacking the proof of the announced Sobolev-type lemma, we also need to analyze the solutions to the equation $\mathcal{L}_\alpha f = g$. Let $\mathcal{D}'(\mathbf{H}_n)$ be the space of distributions, and $\mathcal{E}'(\mathbf{H}_n)$ the space of distributions of compact support on $\mathbf{H}_n$.

Definition 3.11. Given $f, g : \mathbf{H}_n \to \mathbf{C}$ the *convolution product* of f, g is given by

$$(f * g)(x) = \int_{\mathbf{H}_n} f(y)g(y^{-1}x)dV(y).$$ □

Then we have also

$$(f * g)(x) = \int_{\mathbf{H}_n} f(xy^{-1})g(y)dV(y).$$

The proof is an elementary exercise involving a change of variables via the diffeomorphism $h : \mathbf{H}_n \to \mathbf{H}_n$, $h(y) = y^{-1}x$. If $x = (w, s)$, $y = (z, t)$ and $w^j = u^j + iv^j$, $z^j = x^j + iy^j$, then h is given by

$$h^j(x, y, t) = u^j - x^j, \quad h^{n+j}(x, y, t) = v^j - y^j,$$
$$h^{2n+1}(x, y, t) = s - t + 2\sum_{j=1}^{n}(x^j v^j - y^j u^j),$$

with the Jacobian

$$\det \begin{bmatrix} -\delta^i_j & 0 & 0 \\ 0 & -\delta^i_j & 0 \\ 2v^j & -2u^j & -1 \end{bmatrix} = -1.$$ □

Let us set $\tilde{g}(x) = g(x^{-1})$ for any $x \in \mathbf{H}_n$. Then

$$\begin{aligned}\int_{\mathbf{H}_n} (f * g)(x)h(x)dV(x) &= \int_{\mathbf{H}_n} \left(\int_{\mathbf{H}_n} f(y)g(y^{-1}x)dV(y)\right) h(x)dV(x) \\ &= \int_{\mathbf{H}_n} f(y) \left(\int_{\mathbf{H}_n} h(x)g(y^{-1}x)dV(x)\right) dV(y) \\ &= \int_{\mathbf{H}_n} f(y) \left(\int_{\mathbf{H}_n} h(x)\tilde{g}(x^{-1}y)dV(x)\right) dV(y);\end{aligned}$$

hence

$$\int_{\mathbf{H}_n} (f * g)(x)h(x)dV(x) = \int_{\mathbf{H}_n} f(y)(h * \tilde{g})(y)dV(y), \tag{3.38}$$

whenever both sides make sense. □

Let $f \in C_0^\infty(\mathbf{H}_n)$ and $\alpha \in \mathbf{C}$ admissible. Let us set as before $\Phi_\alpha = \varphi_\alpha / c_\alpha$, where φ_α is given by (3.28).

Definition 3.12. Let us define $K_\alpha : C_0^\infty(\mathbf{H}_n) \to C^\infty(\mathbf{H}_n)$ by setting

$$K_\alpha f = f * \Phi_\alpha .$$ □

Proposition 3.2. (G.B. Folland and E.M. Stein [150])
For any $f \in C_0^\infty(\mathbf{H}_n)$ and any admissible value of $\alpha \in \mathbf{C}$ we have

$$\mathcal{L}_\alpha K_\alpha f = K_\alpha \mathcal{L}_\alpha f = f.$$

Proof. $\mathcal{L}_\alpha$ is an invariant differential operator, i.e.,

$$\mathcal{L}_\alpha(g \circ L_x) = (\mathcal{L}_\alpha g) \circ L_x\,,$$

for any $x \in \mathbf{H}_n$ (where $L_x : \mathbf{H}_n \to \mathbf{H}_n$ is given by $L_x(y) = xy$, for any $y \in \mathbf{H}_n$) and any $g \in C^\infty(\mathbf{H}_n)$. Consequently

$$\begin{aligned}
(\mathcal{L}_\alpha K_\alpha f)(x) &= (\mathcal{L}_\alpha)_x \int_{\mathbf{H}_n} f(y)\Phi_\alpha(y^{-1}x)dV(y) \\
&= \int_{\mathbf{H}_n} \left(\mathcal{L}_\alpha(\Phi_\alpha \circ L_{y^{-1}})\right)(x)dV(y) \\
&= \int_{\mathbf{H}_n} f(y)\,(\mathcal{L}_\alpha \Phi_\alpha)\,(y^{-1}x)dV(y) \\
&= \int_{\mathbf{H}_n} f(xy^{-1})\,(\mathcal{L}_\alpha \Phi_\alpha)\,(y)dV(y) \\
&= (\mathcal{L}_\alpha \Phi_\alpha)\,(g_x) = \delta(g_x) = g_x(0) = f(x),
\end{aligned}$$

where g_x is defined by $g_x(y) = f(xy^{-1})$ for any $y \in \mathbf{H}_n$. We therefore have

$$\mathcal{L}_\alpha K_\alpha f = f. \tag{3.39}$$

To prove the remaining identity in Proposition 3.2, let $g \in C_0^\infty(\mathbf{H}_n)$. Note that $c_{-\alpha} = c_\alpha$ hence if α is admissible, then $-\alpha$ is admissible as well. Since (3.39) holds for any admissible value of α and any test function f, we have

$$\int_{\mathbf{H}_n} g(x)f(x)dV(x) = \int_{\mathbf{H}_n} (\mathcal{L}_{-\alpha}K_{-\alpha}g)\,(x)f(x)dV(x).$$

We shall make use of the following lemma:

Lemma 3.7. $\mathcal{L}_\alpha^* = \mathcal{L}_{\bar\alpha}$.

The adjoint $\mathcal{L}_\alpha^*$ of $\mathcal{L}_\alpha$ is defined by

$$\int_{\mathbf{H}_n} (\mathcal{L}_\alpha \varphi)\overline{\psi} dV = \int_{\mathbf{H}_n} \varphi \overline{\mathcal{L}_\alpha^* \psi} dV.$$

We shall prove Lemma 3.7 later on. For the time being, by Lemma 3.7, we have

$$\begin{aligned}
\int_{\mathbf{H}_n} (\mathcal{L}_{-\alpha}K_{-\alpha}g)\,f dV &= \int_{\mathbf{H}_n} (\mathcal{L}_{-\alpha}K_{-\alpha}g)\,\overline{\overline{f}} dV \\
&= \int_{\mathbf{H}_n} (K_{-\alpha}g)\,\overline{\mathcal{L}_{-\alpha}^* \overline{f}} dV = \int_{\mathbf{H}_n} (K_{-\alpha}g)\,\overline{\mathcal{L}_{-\bar\alpha} \overline{f}} dV.
\end{aligned}$$

But

$$\begin{aligned}
\overline{\mathcal{L}_{-\bar\alpha}} &= \Big(-\tfrac{1}{2}\sum_{j=1}^n \left(T_j T_{\bar j} + T_{\bar j} T_j\right) + i(-\bar\alpha)T\Big)^{-} \\
&= -\tfrac{1}{2}\sum_{j=1}^n \left(T_{\bar j} T_j + T_j T_{\bar j}\right) + i\alpha T = \mathcal{L}_\alpha;
\end{aligned}$$

hence

$$\overline{\mathcal{L}_{-\bar{\alpha}}} = \mathcal{L}_\alpha .$$

Summing up, we have (by (3.38))

$$\begin{aligned}\int_{\mathbf{H}_n} gf\,dV &= \int_{\mathbf{H}_n} (K_{-\alpha}g)\,(\mathcal{L}_\alpha f)\,dV \\ &= \int_{\mathbf{H}_n} (g * \Phi_{-\alpha})\,(\mathcal{L}_\alpha f)\,dV = \int_{\mathbf{H}_n} g\left((\mathcal{L}_\alpha f) * \tilde{\Phi}_{-\alpha}\right)dV.\end{aligned}$$

Finally, because of

$$\tilde{\Phi}_{-\alpha}(x) = \Phi_{-\alpha}(x^{-1}) = \frac{1}{c_\alpha}\varphi_{-\alpha}(x^{-1}) = \frac{1}{c_\alpha}\varphi_\alpha(x) = \Phi_\alpha(x)$$

we have

$$\int_{\mathbf{H}_n} gf\,dV = \int_{\mathbf{H}_n} g((\mathcal{L}_\alpha f) * \Phi_\alpha)dV = \int_{\mathbf{H}_n} g(K_\alpha \mathcal{L}_\alpha f)dV;$$

hence $f = K_\alpha \mathcal{L}_\alpha f$, and Proposition 3.2 is completely proved. □

At this point we may prove Lemma 3.7. Indeed, Lemma 3.7 follows from $T_j^* = -T_{\bar{j}}$, $T_{\bar{j}}^* = -T_j$ (hence $(T_j T_{\bar{j}})^* = T_j T_{\bar{j}}$) and from

$$(i(\alpha - n)T)^* = i(\bar{\alpha} - n)T.$$

In turn, the proof of these identities is mere computation. For instance

$$\begin{aligned}\int_{\mathbf{H}_n} (T_j\varphi)\overline{\psi}\,dV &= \int_{\mathbf{H}_n} \left(\left(\frac{\partial}{\partial z^j} + i\bar{z}^j\frac{\partial}{\partial t}\right)\varphi\right)\overline{\psi}\,dV \\ &= -\int_{\mathbf{H}_n} \varphi\frac{\partial\overline{\psi}}{\partial z^j}dV - i\int_{\mathbf{H}_n} \varphi\frac{\partial}{\partial t}\left(\bar{z}^j\overline{\psi}\right)dV \\ &- \int_{\mathbf{H}_n} \varphi\overline{\left(-\frac{\partial\psi}{\partial\bar{z}^j} + iz^j\frac{\partial\psi}{\partial t}\right)}\,dV.\end{aligned}$$

The operator K_α is clearly continuous from $C_0^\infty(\mathbf{H}_n)$ to $C^\infty(\mathbf{H}_n)$ with the usual topologies; hence we may consider the induced operator $(K_\alpha)' : \mathcal{E}'(\mathbf{H}_n) \to \mathcal{D}'(\mathbf{H}_n)$ given by

$$\left((K_\alpha)'\,T\right)(\varphi) = T(K_\alpha\varphi),$$

for any $T \in \mathcal{E}'(\mathbf{H}_n)$ and $\varphi \in C_0^\infty(\mathbf{H}_n)$. As is well known, there are natural inclusions $C_0^\infty(\mathbf{H}_n) \to \mathcal{E}'(\mathbf{H}_n)$ and $C^\infty(\mathbf{H}_n) \to \mathcal{D}'(\mathbf{H}_n)$, e.g., if $\varphi \in C_0^\infty(\mathbf{H}_n)$ then one may define $\underline{\varphi} : C^\infty(\mathbf{H}_n) \to \mathbf{C}$ by $\underline{\varphi}(\psi) = \int_{\mathbf{H}_n} \varphi\psi\,dV$. The following diagram is commutative:

$$\begin{array}{cccc} & C_0^\infty(\mathbf{H}_n) & \longrightarrow & \mathcal{E}'(\mathbf{H}_n) & \\ K_\alpha & \downarrow & & \downarrow & (K_{-\alpha})' \\ & C^\infty(\mathbf{H}_n) & \longrightarrow & \mathcal{D}'(\mathbf{H}_n) & \end{array}$$

(where horizontal arrows are natural inclusions), that is, $(K_{-\alpha})'$ extends K_α to $\mathcal{E}'(\mathbf{H}_n)$. Indeed (by (3.38)),

$$\begin{aligned}\left((K_{-\alpha})'\,\underline{\varphi}\right)(u) &= \underline{\varphi}(K_{-\alpha}u) \\ &= \int_{\mathbf{H}_n}\varphi\,(K_{-\alpha}u)\,dV = \int_{\mathbf{H}_n}\varphi(u*\Phi_{-\alpha})dV = \int_{\mathbf{H}_n}u\left(\varphi*\tilde{\Phi}_{-\alpha}\right)dV.\end{aligned}$$

But $\tilde{\Phi}_{-\alpha} = \Phi_\alpha$; hence

$$\left((K_{-\alpha})'\,\underline{\varphi}\right)(u) = \int_{\mathbf{H}_n}u\,(\varphi*\Phi_\alpha)\,dV = \int_{\mathbf{H}_n}u\,(K_\alpha\varphi)\,dV = \underline{K_\alpha\varphi}(u)$$

for any $\varphi, u \in C_0^\infty(\mathbf{H}_n)$. Similarly, it is easy to see that the following diagrams are commutative:

$$\begin{array}{cccc} & C_0^\infty(\mathbf{H}_n) & \longrightarrow & \mathcal{E}'(\mathbf{H}^n) & \\ \mathcal{L}_\alpha & \downarrow & & \downarrow & (\mathcal{L}_{-\alpha})' \\ & C_0^\infty(\mathbf{H}_n) & \longrightarrow & \mathcal{E}'(\mathbf{H}_n) & \end{array}$$

and

$$\begin{array}{cccc} & C^\infty(\mathbf{H}^n) & \longrightarrow & \mathcal{D}'(\mathbf{H}_n) & \\ \mathcal{L}_\alpha & \downarrow & & \downarrow & (\mathcal{L}_{-\alpha})' \\ & C^\infty(\mathbf{H}_n) & \longrightarrow & \mathcal{D}'(\mathbf{H}_n) & \end{array}$$

We agree to denote both the operator K_α (respectively $\mathcal{L}_\alpha$) and its extension $(K_{-\alpha})'$ (respectively $(\mathcal{L}_{-\alpha})'$) by the same symbol K_α (respectively by $\mathcal{L}_\alpha$).

Corollary 3.1. (G.B. Folland and E.M. Stein [150])
Let $\alpha \in \mathbf{C}$ be admissible. Then

(i) For any $F \in \mathcal{E}'(\mathbf{H}_n)$ we have

$$\mathcal{L}_\alpha K_\alpha F = K_\alpha \mathcal{L}_\alpha F = F.$$

(ii) $\mathcal{L}_\alpha$ is locally solvable, i.e., for any $g \in \mathcal{E}'(\mathbf{H}_n)$ there is $f \in \mathcal{D}'(\mathbf{H}_n)$ such that $\mathcal{L}_\alpha f = g$.
(iii) The equation $\mathcal{L}_\alpha f = 0$ has no nontrivial solutions in $\mathcal{E}'(\mathbf{H}_n)$.

Proof. To prove (i) we may perform (by Proposition 3.2) the following calculation:

$$\begin{aligned}(\mathcal{L}_\alpha K_\alpha F)\,(\varphi) &= \left((\mathcal{L}_{-\alpha})'\,(K_{-\alpha})'\,F\right)(\varphi) \\ &= \left((K_{-\alpha}\mathcal{L}_{-\alpha})'\,F\right)(\varphi) = F\,(K_{-\alpha}\mathcal{L}_{-\alpha}\varphi) = F(\varphi),\end{aligned}$$

for any $\varphi \in C_0^\infty(\mathbf{H}_n)$, etc. Next, (ii) follows from (i) by setting $f = K_\alpha g$. Finally, to prove (iii) we assume that $\mathcal{L}_\alpha f = 0$, for some $f \in \mathcal{E}'(\mathbf{H}_n)$, $f \neq 0$. Then we apply K_α and use (i) to get $0 = K_\alpha\mathcal{L}_\alpha f = f$, a contradiction. □

By a result of G.B. Folland and E.M. Stein (cf. Proposition 7.5 in [150], p. 444), $\mathcal{L}_\alpha$ is hypoelliptic if and only if α is admissible. Hence the $\mathcal{L}_\alpha$ furnish an example[3] of a family of operators of the form $A + \alpha B$, where A is second-order hypoelliptic and B is first-order, that are hypoelliptic for all but an infinite discrete set of values of the parameter α. According to a result by L. Hörmander [213], this phenomenon cannot occur for operators with real coefficients. □

Let us return to the Sobolev-type lemma.

Proof of Theorem 3.8. By Theorem 3.9, let

$$\Phi_0(x) = a_n |x|^{-2n}, \quad a_n := \frac{1}{c_0} = \frac{2^{2n-2}\Gamma(\frac{n}{2})^2}{\pi^{n+1}},$$

be the fundamental solution of $\mathcal{L}_0 = -\sum_{j=1}^{2n} X_j^2$ (i.e., $\mathcal{L}_0 \Phi_0 = \delta$). Next, let us consider

$$K_0 : C_0^\infty(\mathbf{H}_n) \to C_0^\infty(\mathbf{H}_n), \quad K_0 f = f * \Phi_0,$$

where $*$ is the convolution product. Then (by Proposition 3.2)

$$K_0 \mathcal{L}_0 f = f,$$

for any $f \in C_0^\infty(\mathbf{H}_n)$. Also, since X_j is left invariant, if $g \in C_0^\infty(\mathbf{H}_n)$ then

$$K_0(X_j g) = g * (X_j \Phi_0);$$

hence

$$f = K_0 \mathcal{L}_0 f = (\mathcal{L}_0 f) * \Phi_0 = -\sum_{j=1}^{2n} (X_j f) * (X_j \Phi_0).$$

Note that $X_j \Phi_0$ is homogeneous of degree $-2n - 1$. In particular

$$|(X_j \Phi_0)(x)| \le C |x|^{-2n-1}.$$

Now Theorem 3.8 follows from the following theorem:

[3] Generalized by F. De Mari, M.M. Peloso, and F. Ricci [114] (a generalization further investigated by D. Müller, M.M. Peloso, and F. Ricci [318]) as follows. Let $J := \begin{pmatrix} 0 & I_n \\ -I_n & 0 \end{pmatrix}$ and recall that $S \in \mathrm{Sp}(n, \mathbf{C})$ if $S^t J S = J$ (where S^t is the transpose). Moreover, set $P_n^+ := \{S \in \mathrm{Sp}(n, \mathbf{C}) : S = S^t, \ \mathrm{Re}(S) \ge 0\}$. Given $S \in P_n^+$ and $\alpha \in \mathbf{C}$, the second-order differential operator $\square_{S,\alpha}$, on the Heisenberg group $\mathbf{H}_n$, defined by

$$\square_{S,\alpha} := -\tfrac{1}{4}(X_1, \dots, X_n, Y_1, \dots, Y_n)\, S\, (X_1, \dots, X_n, Y_1, \dots, Y_n)^t + i\alpha T$$

is left-invariant and homogeneous with respect to the dilations δ_s. Clearly, $\square_{I_{2n},\alpha}$ are the Folland–Stein operators $\mathcal{L}_\alpha$ (treated above). D. Müller, M.M. Peloso, and F. Ricci discuss (cf. op. cit.) the solvability of $\square_{S,\alpha}$ (certain operators $\square_{S,\alpha}$ turn out to be solvable, while their transposes are not). Another reason that we mention the work by D. Müller et al. is that no (more refined) regularity results for $\square_{S,\alpha}$ are known (e.g., there is no $\square_{S,\alpha}$-analogue to Theorem 9.5 in [150], p. 457).

Theorem 3.10. (E.M. Stein [383])
If F is a regular homogeneous distribution of degree λ, with $-2n-2 < \lambda < 0$, then the map $\varphi \mapsto \varphi * F$ extends to a bounded map from L^p to L^q, where $1/q = 1/p - \lambda/(2n+2) - 1$, provided that $1 < p < q < \infty$, and from L^1 to $L^{-(2n+2)/\lambda - \epsilon}_{loc}$ for any $\epsilon > 0$.

Indeed, if $0 < \alpha < 2n+2$ and $|H(x)| \le C|x|^{-2n-2+\alpha}$ then (by Theorem 3.10) the map $g \mapsto g * H$ extends to a bounded map $L^r(\mathbf{H}_n) \to L^2(\mathbf{H}_n)$, where $\frac{1}{s} = \frac{1}{r} - \frac{\alpha}{2n+2}$ and $1 < r < s < \infty$. Now let us set $\alpha = 1$, $r = 2$, and $s = p$. □

For the sake of completeness, we end this section with a discussion of homogeneous distributions, as appearing in Theorem 3.10.

Definition 3.13. Let $x \in \mathbf{H}_n$. The *translation operators*

$$\tau_x, \ \tau^x : C^\infty(\mathbf{H}_n) \to C^\infty(\mathbf{H}_n)$$

are defined by setting

$$(\tau_x f)(y) = f(x^{-1}y), \quad (\tau^x f)(y) = f(yx^{-1}),$$

for any $f \in C^\infty(\mathbf{H}_n)$ and any $y \in \mathbf{H}_n$. Also we define the *reflection operator*

$$J : C^\infty(\mathbf{H}_n) \to C^\infty(\mathbf{H}_n)$$

by setting $(Jf)(y) = f(y^{-1})$, for any $y \in \mathbf{H}_n$. □

Definition 3.14. For $\varphi \in C_0^\infty(\mathbf{H}_n)$ and $G \in \mathcal{D}'(\mathbf{H}_n)$ we consider the C^∞ functions $G * \varphi$ and $\varphi * G$ given by

$$(G * \varphi)(x) = G(J\tau_x\varphi), \quad (\varphi * G)(x) = G(J\tau^x\varphi).$$

For $v \in T_0(\mathbf{H}_n)$, i.e., for any vector v tangent to $\mathbf{H}_n$ at the origin, we consider the distribution $D_v \in \mathcal{E}'(\mathbf{H}_n)$ given by

$$D_v f = -(d_0 f)v. \qquad \square$$

Let L_v and R_v be the left-invariant and right-invariant extensions of v, respectively. Then $L_v f = f * D_v$ and $R_v f = D_v * f$. For $1 \le \alpha \le n$ let us set

$$v_\alpha = \frac{1}{2}\frac{\partial}{\partial x^\alpha}\Big|_0, \qquad v_{\alpha+n} = \frac{1}{2}\frac{\partial}{\partial y^\alpha}\Big|_0,$$
$$D_j = D_{v_j}, \quad L_j = L_{v_j}, \quad R_j = R_{v_j},$$

where $\{v_j\} = \{v_\alpha, v_{\alpha+n}\}$. With this notation,

$$X_\alpha = 2L_\alpha, \quad Y_\alpha = 2L_{\alpha+n}, \quad T = \frac{\partial}{\partial t} = [L_{\alpha+n}, L_\alpha].$$

Definition 3.15. For a function f on $\mathbf{H}_n$ and $r > 0$ we define

$$f_r(x) = f(D_r(x)), \quad x \in \mathbf{H}_n ,$$

where D_r is the dilation $D_r(z,t) = (rz, r^2 t)$. We say that f is *homogeneous of degree* λ if $f_r = r^\lambda f$, $r > 0$. □

The notion of homogeneity extends to distributions as follows.

Definition 3.16. Let us set

$$f^r(x) = r^{-2n-2} f(\delta_{1/r}(x)), \quad r > 0, \quad x \in \mathbf{H}_n .$$

Next, let $F \in \mathcal{D}'(\mathbf{H}_n)$ and $r > 0$. We say that F is *homogeneous of degree* λ if

$$F(\varphi^r) = r^\lambda F(\varphi),$$

for any $\varphi \in C_0^\infty(\mathbf{H}_n)$ and any $r > 0$. □

Proposition 3.3. *If $F \in \mathcal{D}'(\mathbf{H}_n)$ is homogeneous of degree λ then $L_j F$ and $R_j F$ are homogeneous of degree $\lambda - 1$, $1 \le j \le 2n$.*

The proof is elementary. One is mainly interested in functions and distributions that are homogeneous of degree $-2n-2$, since this is the exponent on the edge of integrability at 0 and ∞. The first step is to define a notion of *mean value* for such functions. We shall need the following result:

Proposition 3.4. (G.B. Folland and E.M. Stein [150])
Let f be a homogeneous function of degree $-2n-2$ that is locally integrable away from the origin. There is a constant μ_f such that

$$\int_{\mathbf{H}_n} f(x) g(|x|) dV(x) = \mu_f \int_0^\infty \frac{g(r)}{r} dr \tag{3.40}$$

for all measurable functions g on $(0,\infty)$ such that both integrals are defined.

To prove Proposition 3.4 let us set

$$A_f(r) = \begin{cases} \displaystyle\int_{1 \le |x| \le r} f(x) dV(x) & \text{for } r \ge 1, \\ \displaystyle -\int_{r \le |x| \le 1} f(x) dV(x) & \text{for } 0 < r < 1. \end{cases}$$

Then A_f is continuous on $(0,\infty)$ and $A_f(rs) = A_f(r) + A_f(s)$ (because of $dV(\delta_\epsilon(x)) = \epsilon^{2n+2} dV(x)$). Hence $A_f(r) = \mu_f \log r$ for some (unique) constant μ_f, i.e., Proposition 3.4 is proved when g is the characteristic function of an interval. By forming linear combinations and passing to limits one may complete the proof for an arbitrary measurable function g. □

Definition 3.17. The constant μ_f in (3.40) is referred to as the *mean value* of f. □

In particular, let $f(x) = |x|^{-2n-2}$ and let $C_0 = \mu_f$ be its mean value. Also, let $g(r)$ be r^λ times the characteristic function of the interval (a, b) with $0 < a < b < \infty$. Then (3.40) leads to

$$\int_{a \le |x| \le b} |x|^{\lambda-2n-2} dV(x) = \begin{cases} C_0 \lambda^{-1}(b^\lambda - a^\lambda), & \lambda \ne 0, \\ C_0 \log \frac{b}{a}, & \lambda = 0. \end{cases} \tag{3.41}$$

Definition 3.18. Let f be a homogeneous function of degree $-2n-2$, continuous away from the origin, and with $\mu_f = 0$. Then f gives rise to a distribution PVf defined by

$$(PVf)(\varphi) = \lim_{\epsilon \to 0} \int_{|x| \ge \epsilon} f(x)\varphi(x) dV(x)$$

for any $\varphi \in C_0^\infty(\mathbf{H}_n)$. □

Proposition 3.5.
PVf is a homogeneous distribution of degree $-2n-2$.

The proof is left as an exercise to the reader.

Definition 3.19. A distribution F is *regular* if there is a function f that is C^∞ on $\mathbf{H}_n \setminus \{0\}$ such that

$$F(\varphi) = \int f \varphi dV,$$

for any $\varphi \in C_0^\infty(\mathbf{H}_n \setminus \{0\})$. □

The following is a result of G.B. Folland and E.M. Stein [150]:

Theorem 3.11. *If F is a regular homogeneous distribution of degree λ that agrees with $f \in C^\infty(\mathbf{H}_n \setminus \{0\})$ on $\mathbf{H}_n \setminus \{0\}$, then (i) f is homogeneous of degree λ, (ii) if $\lambda > -2n-2$ then $\underline{f} \in \mathcal{D}'(\mathbf{H}_n)$ and $F = \underline{f}$, and (iii) if $\lambda = -2n-2$ then $\mu_f = 0$ and $F = PVf + C\delta$, for some constant C.*

The proof of Theorem 3.11 is beyond the scope of this book. See Proposition 8.5. in [150], p. 447. Therefore, one is entitled to adopt the following definition:

Definition 3.20. A regular homogeneous distribution of degree $-2n-2$ is called a *PV distribution.* □

Note that by a slight abuse of terminology, one refers to the Dirac δ as a PV distribution. PV distributions play the role of the classical singular integral kernels on the group $\mathbf{H}_n$. Indeed, the following analogue of the Calderón–Zygmund theorem holds:

Theorem 3.12. (A.W. Knapp, E.M. Stein, R. Coifman, and G. Weiss [240], [105])
If F is a PV distribution then the map $\varphi \mapsto \varphi * F$, $\varphi \in C_0^\infty(\mathbf{H}_n)$, extends to a bounded transformation on $L^p(\mathbf{H}_n)$, $1 < p < \infty$.

The proof of Theorem 3.12 is omitted. See A.W. Knapp and E.M. Stein [240], for $p = 2$ and A. Korányi and S. Vági [258], or R. Coifman and G. Weiss [105], for the remaining values of p. Kernels of higher homogeneity are known to satisfy a similar property; cf. Theorem 3.10 above.

3.4 Embedding results

Let us consider now the inequality (3.24) for real-valued functions $f \in C_0^\infty(\mathbf{H}_n)$. As shown above, the class of test functions defining $\lambda(\mathbf{H}_n)$ may be restricted to $C_0^\infty(\mathbf{H}_n)$; hence finding $\lambda(\mathbf{H}_n)$ is equivalent to finding the best constant C_n in (3.24). In particular, Theorem 3.8 is equivalent to $\lambda(\mathbf{H}_n) > 0$.

Let U be a relatively compact open subset of a normal coordinate neighborhood $\Omega_\xi \subset M$, as in Theorem 3.7. With the notation there, if $X_j = \mathrm{Re}(W_j)$, $X_{j+n} = \mathrm{Im}(W_j)$, we set

$$X^\alpha = X_{\alpha_1} \cdots X_{\alpha_k}, \quad \alpha = (\alpha_1, \ldots, \alpha_k), \quad 1 \le \alpha_j \le 2n.$$

If $\alpha = (\alpha_1, \ldots, \alpha_\ell)$ set $\ell(\alpha) = \ell$. Consider the norms

$$\|f\|_{S_k^p(U)} = \sup_{\ell(\alpha) \le k} \|X^\alpha f\|_{L^p(U)}, \tag{3.42}$$

where

$$\|g\|_{L^p(U)} = \left(\int_U |g|^p \theta \wedge (d\theta)^n \right)^{1/p}.$$

Definition 3.21.
The *Folland–Stein space* $S_k^p(U)$ is the completion of $C_0^\infty(U)$ under the norm (3.42). □

Hölder spaces suited to (regularity properties of) Δ_b may be introduced as well. Consider the distance function

$$\rho(\xi, \eta) = |\Theta(\xi, \eta)|$$

(Heisenberg norm) on U.

Definition 3.22. Let $0 < \beta < 1$ and let

$$\Gamma_\beta(U) = \{ f \in C^0(\overline{U}) : |f(x) - f(y)| \le C \rho(x, y)^\beta \}$$

with the norm

$$\|f\|_{\Gamma_\beta(U)} = \sup_{x \in U} |f(x)| + \sup_{x, y \in U} \frac{|f(x) - f(y)|}{\rho(x, y)^\beta}.$$

Next, if $k \in \mathbf{Z}$, $k \ge 1$, and $k < \beta < k + 1$, consider

$$\Gamma_\beta(U) = \{ f \in C^0(\overline{U}) : X^\alpha f \in \Gamma_{\beta - k}(U), \, \ell(\alpha) \le k \}$$

with the norm

$$\|f\|_{\Gamma_\beta(U)} = \sup_{x \in U} |f(x)| + \sup_{\substack{x, y \in U \\ \ell(\alpha) \le k}} \frac{|(X^\alpha f)(x) - (X^\alpha f)(y)|}{\rho(x, y)^{\beta - k}}.$$

□

All norms above depend on the choice of pseudo-Hermitian frame.

Definition 3.23. Let M be a compact strictly pseudoconvex CR manifold and let $\{U_j : 1 \leq j \leq s\}$ be a finite open cover of M. Let $\{\varphi_j\}$ be a C^∞ partition of unity subordinate to $\{U_j\}$. We define

$$S_k^p(M) = \{f \in L^1(M) : \varphi_j f \in S_k^p(U),\ 1 \leq j \leq s\}$$

and

$$\Gamma_\beta(M) = \{f \in C^0(M) : \varphi_j f \in \Gamma_\beta(U_j),\ 1 \leq j \leq s\}. \qquad \square$$

Then we have the following embedding:

Theorem 3.13. (G.B. Folland and E.M. Stein [150])
$S_k^r(M) \subset L^s(M)$, where $\dfrac{1}{s} = \dfrac{1}{r} - \dfrac{k}{2n+2}$ and $1 < r < s < \infty$.

The proof of Theorem 3.13 relies on the theory of singular integral operators, i.e., specifically on Lemma 3.9 below. We need some preparation. Let M be a strictly pseudoconvex CR manifold, as before, and $\{T_\alpha\}$ a pseudo-Hermitian frame on some open set $V \subseteq M$. By eventually shrinking V we may assume that V is a relatively compact subset of M. Moreover, as a matter of notation, keeping in mind that our analysis is purely local, we shall write at times M instead of V.

Definition 3.24. A *singular integral operator* on M is an operator of the form

$$Af = \lim_{\epsilon \to 0} A_\epsilon f,$$

where

$$(A_\epsilon f)(x) = \int_{\{\rho(x,y)>\epsilon\}} K(x,y) f(y) dy$$

and K is a *singular integral kernel*, i.e., it satisfies to the following requirements: (1) $K \in C^\infty(M \times M \setminus \Delta)$, where Δ is the diagonal in $M \times M$, (2) the support of K is contained in $\Omega \cap \{(x,y) : \rho(x,y) \leq 1\}$, and (3) when $\rho(x,y)$ is sufficiently small $K(x,y) = k(\Theta(y,x))$, where k is regular homogeneous of degree $-2n-2$ with mean value zero. $\square$

Theorem 3.14. (G.B. Folland and E.M. Stein)
Singular integral operators are bounded on L^p, $1 < p < \infty$.

The proof (cf. [150], pp. 479–486) is quite involved. However, for our needs in this book only a step in the proof of Theorem 3.14 is actually required. First one needs to establish the following analogue of the mean value zero property.

Lemma 3.8. *Let K be a singular integral kernel. Then*

$$\int_{\{\epsilon<\rho(x,y)<\delta\}} K(x,y) dy \leq C(\delta - \epsilon),$$

the constant C being independent of x.

By a result in Chapter 1,

$$\Psi = 2^n i^{n^2} n! \theta \wedge \theta^1 \wedge \cdots \wedge \theta^n \wedge \theta^{\bar{1}} \wedge \cdots \wedge \theta^{\bar{n}} .$$

Also, if $\Psi_0 = \theta_0 \wedge (d\theta_0)^n$ then

$$\Psi_0 = 2^n i^{n^2} n! \theta_0 \wedge dw^1 \wedge \cdots \wedge dw^n \wedge d\bar{w}^1 \wedge \cdots \wedge d\bar{w}^n ,$$

where $(w, \tau) = (\xi + i\eta, \tau)$ are the natural coordinates on $\mathbf{H}_n$. We have

$$\Theta_x^* \Psi = f \Psi_0 ,$$

for some $f \in C^\infty(V_x)$. Then the identity

$$\theta \wedge \theta^1 \wedge \cdots \wedge \theta^n \wedge \theta^{\bar{1}} \wedge \cdots \wedge \theta^{\bar{n}} (T, T_1, \ldots, T_n, T_{\bar{1}}, \ldots, T_{\bar{n}}) = \frac{1}{(2n+1)!}$$

together with

$$(\Theta_x)_* T_\alpha = \frac{\partial}{\partial w^\alpha} + \sum_{\beta=1}^n \left(O^1 \frac{\partial}{\partial w^\beta} + O^1 \frac{\partial}{\partial \bar{w}^\beta} \right) + O^2 \frac{\partial}{\partial \tau}$$

(by Theorem 3.5) lead to

$$f = 1 + O^1 .$$

Let us prove Lemma 3.8. We may assume that δ is small enough so that $K(x, y) = k(\Theta_y(x)) = k(-\Theta_x(y))$ over the region of integration. Then (since $\mu_k = 0$)

$$\begin{aligned} \int_{\epsilon < \rho(x,y) < \delta} K(x, y) dy &= \int_{\epsilon < |\xi| < \delta} k(-\xi)(1 + O^1) dV(\xi) \\ &= \int_{\epsilon < |\xi| < \delta} k(-\xi) O^1 dV(\xi) \leq C \int_{\epsilon < |\xi| < \delta} |\xi|^{-2n-1} dV(\xi) \leq C(\delta - \epsilon) \end{aligned}$$

by (3.41). The uniformity in x is due to the assumption that V is relatively compact, so the error term in the measure is uniformly O^1.

Lemma 3.9. *Let $f \in C_0^\infty$ and A a singular integral operator. Assume that $A_\epsilon f$ is Cauchy in L^p $(1 \leq p \leq \infty)$ as $\epsilon \to 0$. Then Af exists as a strong L^p limit $(1 \leq p \leq \infty)$.*

Proof. Let $\epsilon < \delta < 1$. Then

$$\begin{aligned} (A_\epsilon f - A_\delta f)(x) &= \int_{\epsilon < \rho(x,y) < \delta} K(x, y) f(y)\, dy \\ &= \int_{\epsilon < \rho(x,y) < \delta} K(x, y)(f(y) - f(x))\, dy + f(x) \int_{\epsilon < \rho(x,y) < \delta} K(x, y)\, dy. \end{aligned}$$

By Lemma 3.8, the second term is $\le C(\delta-\epsilon)f(x)$. Also, note that $f(y)-f(x) = O^1$ as a function of y. Therefore, the first term is

$$\le \int_{\epsilon<|\xi|<\delta} |\xi|^{-2n-1} dV(\xi) \le C(\delta-\epsilon).$$

Hence $\{A_\epsilon f\}$ is Cauchy in the uniform norm; hence (since V has finite volume) it is Cauchy in L^p. □

Proof of Theorem 3.13. By Lemma 3.9 there exist operators A_j, $0 \le j \le 2n$, of the form

$$A_j f(x) = \int_M K_j(x,y) f(y) dV(y),$$

$$\left|K_j(x,y)\right| \le \begin{cases} C\rho(x,y)^{-2n-1}, & (x,y)\in\Omega, \\ C, & \text{otherwise}, \end{cases}$$

such that

$$f = \sum_{j=1}^{2n} A_j X_j f + A_0 f,$$

for any $f \in S_1^r(M)$. Since $f,\ X_j f \in L^r(M)$ it follows (by the analogue of Theorem 3.10 for M, rather than $\mathbf{H}_n$) that $f \in L^s(M)$. The proof may be completed by induction over k. □

We end this section by mentioning the following result (the proof of which requires the theory of pseudodifferential operators associated with the subelliptic operator Δ_b (cf. [324]) and a calculus allowing one to define Folland–Stein spaces $S_k^p(M)$ for fractional values of k), as has been announced in [227].

Theorem 3.15. *If* $1 < r < s < \infty$ *and* $\frac{1}{s} > \frac{1}{r} - \frac{1}{2n+2}$ *then the unit ball in* $S_1^r(M)$ *is compact in* $L^s(M)$.

3.5 Regularity results

Let U be a relatively compact open set in a normal coordinate neighborhood Ω_ξ, and $(z,t) = \Theta_\xi$, $\xi \in U$.

Definition 3.25. Given $0 < \beta < 1$ the (standard) *Hölder space* $\Lambda_\beta(U)$ is defined by

$$\Lambda_\beta(U) = \{f \in C^0(\overline{U}) : |f(x)-f(y)| \le C\|x-y\|^\beta\},$$

$$\|f\|_{\Lambda_\beta(U)} = \sup_{x\in U}|f(x)| + \sup_{x,y\in U} \frac{|f(x)-f(y)|}{\|x-y\|^\beta}.$$

Moreover, if $k < \beta < k+1$, with $k \in \mathbf{Z}$, $k \ge 1$, then we set

$$\Lambda_\beta(U) = \{f \in C^0(\overline{U}) : \left(\frac{\partial}{\partial x}\right)^\alpha f \in \Lambda_{\beta-k}(U),\ |\alpha| \le k\}.$$

□

Theorem 3.16. *For any $\beta \in (0, \infty) \setminus \mathbf{Z}$ and any $1 < r < \infty$, and any $k \in \mathbf{Z}$, $k \geq 1$, there is a constant $C > 0$ such that for any $f \in C_0^\infty(U)$:*

$$\begin{aligned}
&(1) \quad \|f\|_{\Gamma_\beta(U)} \leq C\|f\|_{S_k^r(U)}, \quad \frac{1}{r} = \frac{k-\beta}{2n+2};\\
&(2) \quad \|f\|_{\Lambda_{\beta/2}(U)} \leq C\|f\|_{\Gamma_\beta(U)};\\
&(3) \quad \|f\|_{S_2^r(U)} \leq C\left(\|\Delta_b f\|_{L^r(U)} + \|f\|_{L^r(U)}\right);\\
&(4) \quad \|f\|_{\Gamma_{\beta+2}(U)} \leq C\left(\|\Delta_b f\|_{\Gamma_\beta(U)} + \|f\|_{\Gamma_\beta(U)}\right).
\end{aligned}$$

The constant C depends only on the frame constants.

Theorem 3.16 with Δ_b replaced by $\Box_b$ is due to G.B. Folland and E.M. Stein; cf. Theorems 20.1 and 21.1 in [150], p. 515 and p. 517 respectively. The arguments there apply to Δ_b with only minor modifications. By a partition of unity argument it follows that the estimates in Theorem 3.16 hold when U is replaced by a compact strictly pseudoconvex CR manifold M. The estimates in Theorem 3.16 yield the following regularity result:

Theorem 3.17. (G.B. Folland and E.M. Stein [150])
Let $u, v \in L^1_{loc}(U)$ so that $\Delta_b u = v$ in the sense of distributions on U. Then, for any $\eta \in C_0^\infty(U)$:

(1) *if $v \in L^r(U)$, $n+1 < r \leq \infty$, then $\eta u \in \Gamma_\beta(U)$, where $\beta = 2 - \frac{2n+2}{r}$;*
(2) *if $v \in S_k^r(U)$, $1 < r < \infty$, $k = 0, 1, 2, \ldots$, then $\eta u \in S_{k+2}^r(U)$;*
(3) *if $v \in \Gamma_\beta(U)$, $\beta \in (0, \infty) \setminus \mathbf{Z}$, then $\eta u \in \Gamma_{\beta+2}(U)$.*

The proof of Theorem 3.17 is imitative of that of Theorem A.15. The following result is a variant of results in [444], and [409]

Theorem 3.18.
Let U be as in Theorem 3.16. Assume that $f \in L^{n+1}(U)$ and $f \in L^p(U)$, with $p = 2 + \frac{2}{n}$ and $u \geq 0$, and $\Delta_b u + fu = 0$ in the sense of distributions on U. Then $\eta u \in L^s(U)$ for any $\eta \in C_0^\infty(U)$ and any $s < \infty$.

The reader may see the appendix in [227] for a proof.

Theorem 3.19. *Under the hypothesis of Theorem 3.18, if additionally $f \in L^s(U)$, for some $s > n+1$, then u is Hölder continuous in U and there is $\beta > 0$ such that*

$$\|u\|_{\Gamma_\beta(K)} \leq C$$

for any compact set $K \subset U$, where the constant C depends only on the set K, on the numbers $\|f\|_{L^s(U)}$ and $\|u\|_{L^p(U)}$, and on the frame constants.

Proof. Let $\eta_j \in C_0^\infty(U)$ such that $\eta_j = 1$ on K and

$$\operatorname{supp}(\eta_{j+1}) \subseteq \{x \in U : \eta_j(x) = 1\}, \quad j \geq 1.$$

Then

$$fu \in L^q(U), \quad \frac{1}{q} = \frac{1}{p} + \frac{1}{s},$$

by the Hölder inequality. Next $\eta_1 u \in S_2^q(U)$, by (2) in Theorem 3.17. Thus

$$\eta_1 u \in L^{p_1}(U), \quad \frac{1}{p_1} = \frac{1}{q} - \frac{2}{2n+2} = \frac{1}{p} - \left[\frac{1}{n+1} - \frac{1}{s}\right]$$

by Theorem 3.13. An iteration of this argument leads to

$$\eta_k u \in L^{p_k}(U), \quad \frac{1}{p_k} = \frac{1}{p} - k\left[\frac{1}{n+1} - \frac{1}{s}\right],$$

for any k for which $1/p_k > 0$. If k is the largest possible then $p_k > n+1$ and we get Hölder regularity

$$\eta_{k+1} u \in \Gamma_\beta(U), \quad \beta = 2 - \frac{2n+2}{p_k},$$

by (1) of Theorem 8.10 (the bound on $\|u\|_{\Gamma_\beta(K)}$ follows from Theorem 3.16). □

The following Poincaré-type inequality was established in [225] (the proof there is for $q = 2$, yet as indicated in [225], p. 521, the same proof does work for $1 < q < \infty$):

Theorem 3.20. *Let U be as above and $B_r \subset U$ a ball of radius r with respect to the distance function ρ. Then for any f with $\|df\|_\theta \in L^q(B_r)$, $1 < q < \infty$, the following inequality holds:*

$$\int_{B_r} |f - f_{B_r}|^q \theta \wedge (d\theta)^n \le C r^q \int_{B_r} \|df\|_\theta^q \theta \wedge (d\theta)^n,$$

where $C > 0$ is a constant depending only on the frame constants.

Here we have adopted the following definition:

Definition 3.26.

$$f_A := \frac{\int_A f\theta \wedge (d\theta)^n}{\int_A \theta \wedge (d\theta)^n}$$

is the *average value* of f on A. □

Using Theorem 3.20 as the main ingredient, one may adapt J. Moser's proof (cf. [317]) of the Harnack inequality for uniformly elliptic operators to yield the following result:

Theorem 3.21. *Under the hypothesis of Theorem* 3.19, *if additionally $f \in L^\infty(U)$ then*

$$\max_{x \in K} u(x) \le C \min_{x \in K} u(x),$$

where C is a constant depending on the same bounds as in Theorem 3.19 *and in addition on $\|f\|_{L^\infty(U)}$.*

Note that the Poincaré-type inequality above implies the following interpolation inequality for the spaces S_1^q:

Proposition 3.6. *If* $u \in L^1(U)$ *and* $\|du\|_\theta \in L^q(U)$, $1 < q < \infty$, *then* $u \in S_1^q(U)$ *and*

$$\|u\|_{S_1^q(U)} \le C\left(\| \|du\|_\theta \|_{L^q(U)} + \|u\|_{L^1(U)}\right),$$

where the constant C *depends only on the frame constants.*

Proof. It suffices to estimate $\|u\|_{L^q(U)}$:

$$\begin{aligned} \|u\|_{L^q(U)} &\le C\left(\|u - u_U\|_{L^q(U)} + \|u_U\|_{L^q(U)}\right), \\ \|u_U\|_{L^q(U)} &= C\|u_U\|_{L^1(U)} \le C\left(\|u - u_U\|_{L^1(U)} + \|u\|_{L^1(U)}\right) \\ &\le C\left(\|u - u_U\|_{L^q(U)} + \|u\|_{L^1(U)}\right) \end{aligned}$$

(where u_U is the average value of u on U) and one may apply Theorem 3.20 to end the proof. □

At this point, we may establish regularity results for the CR Yamabe equation:

Theorem 3.22. (D. Jerison and J.M. Lee [227])
Let U *be a relatively compact open subset of a normal coordinate neighborhood. Let* $f, g \in C^\infty(U)$, $u \in L^r(U)$ *for some* $r > p$, *and* $u \ge 0$ *on* U. *Assume that*

$$\Delta_b u + gu = f u^{q-1}$$

in the sense of distributions on U, *for some* q, $2 \le q \le p$. *Then* $u \in C^\infty(U)$, $u > 0$, *and if* $K \subset U$ *is a compact set, its norm* $\|u\|_{C^k(K)}$ *depends only on* K, $\|u\|_{L^r(U)}$, $\|f\|_{C^k(K)}$, $\|g\|_{C^k(K)}$, *and the frame constants (but not on* q*).*

Proof. Let $h = f u^{q-2} - g \in L^{r/(q-2)}(U)$. Then

$$h \in L^s(U), \quad s = \frac{r}{p-2} > n+1,$$

by the Hölder inequality. Also $\|h\|_{L^s(U)}$ depends only on the bounds stated in Theorem 3.22. Let us choose a compact set K_1 such that $K \subset K_1 \subset U$. Then $u \in \Gamma_\beta(K_1)$ for some $\beta > 0$ (by Theorem 3.19). Then, by Theorem 3.21, u is bounded away from zero by a constant depending on the same bounds. Since $\Gamma_\beta(K_1)$ is an algebra, $u^\alpha \in \Gamma_\beta(K_1)$ for any $\alpha \in \mathbf{R}$. Therefore, by eventually replacing K_1 with a smaller set denoted by the same symbol, $h \in \Gamma_\beta(K_1)$ and (4) in Theorem 3.16 yields $u \in \Gamma_{\beta+2}(K_1)$. Finally, (2) in Theorem 3.16 and induction lead to $u \in C^k(K)$ for any k. □

Corollary 3.2. *Let* U, f, g, *and* u *be as in Theorem 3.22, except that* $r = p$ *instead of* $r > p$. *Then* $u > 0$ *and* $u \in C^\infty(U)$.

Proof. Consider $h = g - f u^{q-2}$ and K_1 as in the proof of Theorem 3.22. The proof of Corollary 3.2 is carried out in three steps: (1) $h \in L^{n+1}(K_1)$, (2) $u \in L^s(K_1)$, $s < \infty$ (by Theorem 3.18), and (3) $u > 0$ and $u \in C^\infty$ (by Theorem 3.22). □

3.6 Existence of extremals

We shall prove now (iii) in Theorem 3.4. To do so, we consider the following perturbed variational problem. Let M be a compact strictly pseudoconvex CR manifold and θ a contact form on M. For each q, $2 \le q \le p$, let us set

$$\lambda_q = \inf\{A_\theta(\phi) : \phi \in S_1^2(M),\ B_{\theta,q} = 1\}, \tag{3.43}$$

$$B_{\theta,q}(\phi) = \int_M |\phi|^q \theta \wedge (d\theta)^n .$$

Theorem 3.23. (D. Jerison and J.M. Lee [227])
If $2 \le q < p$ there is a positive C^∞ solution u_q to

$$b_n \Delta_b v + \rho v = \lambda_q v^{q-1} \tag{3.44}$$

satisfying $A_\theta(u_q) = \lambda_q$ and $B_{\theta,q}(u_q) = 1$.

Proof. Let ϕ_j be a minimizing sequence for (3.43), i.e.,

$$A_\theta(\phi_j) \to \lambda_q, \quad j \to \infty, \quad B_{\theta,q}(\phi_j) = 1.$$

By replacing ϕ_j with $|\phi_j|$ one may assume that $\phi_j \ge 0$. Since $\{A_\theta(\phi_j)\}$ and $\{B_{\theta,q}(\phi_j)\}$ are bounded, and $\{\phi_j\}$ is bounded in $S_1^2(M)$, there is a subsequence converging weakly in $S_1^2(M)$ to some $\phi \in S_1^2(M)$. By Theorem 3.15, this subsequence converges in the L^q-norm; hence $B_{\theta,q}(\phi) = 1$. Next (by the Hölder inequality)

$$\int_M \rho\phi_j^2 \theta \wedge (d\theta)^n \to \int_M \rho\phi^2 \theta \wedge (d\theta)^n, \quad j \to \infty,$$

and then $A_\theta(\phi) \le \lambda_q$. Yet λ_q is an infimum; hence we must have equality $A_\theta(\phi) = \lambda_q$. Furthermore, since $\phi \ge 0$, a standard variational argument shows that ϕ satisfies (3.44), sense of distributions. Applying Theorem 3.13 we see that $\phi \in L^p(U)$. Finally, by Corollary 3.2, $\phi > 0$ and $\phi \in C^\infty$. □

Now we analyze the behavior of u_q as $q \to p$. We shall need the following lemma:

Lemma 3.10. *Let θ be a contact form on M so that $\int_M \theta \wedge (d\theta)^n = 1$.*

(1) If $\lambda_q < 0$ for some q, then $\lambda_q < 0$ for any $q \ge 2$ and $q \mapsto \lambda_q$ is a nondecreasing function.
(2) If $\lambda_q \ge 0$ for some $q \ge 2$ (and hence for all, in view of (1)), then $q \mapsto \lambda_q$ is a nonincreasing function, and it is continuous from the left.

Proof. Assume that $\lambda_q < 0$ for some q. Let $s \ge 2$ be arbitrary. Given $\epsilon > 0$ such that $\lambda_q + \epsilon < 0$ let ϕ be a C^∞ function such that

$$B_{\theta,s}(\phi) = 1, \quad A_\theta(\phi) < \lambda_q + \epsilon.$$

Let $\alpha \in \mathbf{R}$ and set $\psi = \alpha\phi$. Then

$$B_{\theta,q}(\psi) = \alpha^s B_{\theta,s}(\phi), \quad A_\theta(\psi) = \alpha^2 A_\theta(\phi).$$

Choose now $\alpha := B_{\theta,s}(\phi)^{-1/s}$ such that

$$B_{\theta,s}(\psi) = 1, \quad A_\theta(\psi) < 0.$$

Therefore $\lambda_s < 0$. If $s \le q$ then $\alpha \ge 1$ (by the Hölder inequality and $\int_M \theta \wedge (d\theta)^n = 1$). Hence

$$A_\theta(\psi) \le \lambda_q + \epsilon,$$

i.e., λ_q is nondecreasing (as a function of q). □

When $\lambda_q \ge 0$ the same argument yields $s \ge q \Longrightarrow \lambda_s \le \lambda_q$. As s approaches q, α approaches 1; hence λ_q is continuous from the left. □

Theorem 3.24. (D. Jerison and J.M. Lee [227])
Let θ be such that $\int_M \theta \wedge (d\theta)^n = 1$ and assume that $\lambda(M) < \lambda(S^{2n+1})$. There is a sequence $q_j \le p$ such that $q_j \to p$, $j \to \infty$, and such that u_{q_j} converges in $C^k(M)$, for any k, to some $u \in C^\infty(M)$ with $u > 0$ and

$$b_n \Delta_b u + \rho u = \lambda(M) u^{p-1},$$
$$A_\theta(u) = \lambda(M), \quad B_{\theta,p}(u) = 1.$$

Proof. We distinguish two cases: (I) $\lambda(M) < 0$, and (II) $\lambda(M) \ge 0$. In the first case, let $2 \le q < p$ and $\phi \in S_1^2(M)$. Then

$$\int_M \left(\langle du_q, d\phi\rangle + \rho u_q \phi\right) \theta \wedge (d\theta)^n = \int_M \lambda_q u_q^{q-1} \phi \, \theta \wedge (d\theta)^n,$$

where u_q is furnished by Theorem 3.23. Set $\phi = u_q^{q-1}$ in this equation. Since $\lambda_q < 0$, by Lemma 3.10

$$\frac{q-1}{2} \int_M u_q^{q-2} \|du_q\|_\theta^2 \theta \wedge (d\theta)^n \le \int_M |\rho u_q^q| \theta \wedge (d\theta)^n$$

Let us set $w_q := u_q^{q/2}$. Then

$$\int_M \|dw_q\|_\theta^2 \theta \wedge (d\theta)^n \le C \int_M w_q^2 \theta \wedge (d\theta)^n = C \int_M u_q^q \theta \wedge (d\theta)^n = C.$$

By Theorem 3.13

$$\int_M w_q^p \theta \wedge (d\theta)^n \le C \int_M \left(\|dw_q\|_\theta^2 + w_q^2\right) \theta \wedge (d\theta)^n;$$

hence

$$\int_M w_q^p \theta \wedge (d\theta)^n \le C.$$

Let $q_0 > 2$ and $r := \frac{q_0 p}{2} > p$. Then $\|u_q\|_{L^r(M)}$ is uniformly bounded as $q \to p,\ q \ge q_0$. Then (by Theorem 3.22) $\{u_q\}$ is uniformly bounded in $C^k(M)$; hence there is a subsequence u_{q_j} convergent in $C^k(M)$, for each k, the limit u of which will satisfy

$$b_n \Delta_b u + \rho u = \lambda u^{p-1},$$
$$A_\theta(u) = \lambda, \quad B_{\theta,p}(u) = 1, \quad u > 0, \quad u \in C^\infty(M),$$

where $\lambda = \lim_{j\to\infty} \lambda_{q_j}$. By Lemma 3.10 we get $\lambda \le \lambda(M)$; hence $\lambda = \lambda(M)$ (by the definition of $\lambda(M)$). □

Let us look at case II. Then

$$\lim_{q \to p} \lambda_q = \lambda_p = \lambda(M)$$

by Lemma 3.10. We distinguish two subcases:

(a) $\sup_M \|du_{q_j}\|_\theta$ is uniformly bounded, for some sequence $q_j \to p,\ j \to \infty$, or
(b) $\sup_M \|du_q\|_\theta \to \infty$ for $q \to p$.

We shall show that case II(b) actually doesn't occur. Let $\xi_q \in M$ be a point such that

$$\sup_M \|du_q\|_\theta = \|du_q\|_\theta(\xi_q).$$

Let Θ_{ξ_q} be normal coordinates at ξ_q, as furnished by Theorem 3.7. Moreover, let U be a neighborhood of the origin in $\mathbf{H}_n$, contained in the image of Θ_{ξ_q}, for all q. Next, identify U, via Θ_{ξ_q}, with a neighborhood of ξ_q with coordinates $(z,t) = \Theta_{\xi_q}$. Let us set

$$(\tilde z, \tilde t) = T^\delta(z,t) = (\delta^{-1} z, \delta^{-2} t),$$
$$\tilde\theta_0 = d\tilde t + i \sum_{j=1}^n \left(\tilde z^j d\bar{\tilde z}^j - \bar{\tilde z}^j d\tilde z^j \right) = \delta^{-2} \left[(T^\delta)^{-1} \right]^* \theta_0 .$$

On the set $\delta^{-1} U$, with coordinates $(\tilde z, \tilde t)$, let

$$h_q(\tilde z, \tilde t) = \delta^{2/(q-2)} u_q(\delta \tilde z, \delta^2 \tilde t),$$

where $\delta = \delta_q > 0$ is chosen such that

$$\|dh_q\|_{\tilde\theta_0}(0) = 1.$$

Since $\theta_{\xi_q} = (\theta_0)_0$ and

$$\|\omega\|^2_{\delta^{-2}\theta} = \delta^2 \|\omega\|^2_\theta,$$

for any 1-form ω, it follows that

$$\|dh_q\|_{\tilde\theta_0}(0) = \|(T^\delta)^* dh_q\|_{\delta^{-2}\theta_0}(0) = \delta^{1+2/(q-2)} \|du_q\|_\theta(\xi_q).$$

In particular, if $q \to p$ then $\delta \to 0$; hence $\delta^{-1} U$ tends to the whole of $\mathbf{H}_n$. □

Let us consider now the contact form

$$\theta_q = \delta^{-2}\left[\left(T^{\delta}\right)^{-1}\right]^{*}\theta$$

defined on $\delta^{-1}U$, and let us set

$$\mathcal{L}_q := \Delta_b^{\theta_q} = \delta^2 \Delta_b^{[(T^{\delta})^{-1}]^{*}\theta}$$

(here Δ_b^{θ} is the sub-Laplacian of (M, θ)). Then

$$b_n \mathcal{L}_q h_q + \rho_q \delta_q^2 h_q = \lambda_q h_q^{q-1},$$

where ρ_q is the scalar curvature of θ_q. Note that $|\rho_q| \leq \|\rho\|_{L^\infty(M)}$. Since M is compact we may assume (by passing to a subsequence, if necessary) that ξ_q converges to some $\xi \in M$. At the same time, if $\{W_1^q, \dots, W_n^q\}$ is the pseudo-Hermitian frame used to build Θ_{ξ_q}, we may assume that $\{W_1^q, \dots, W_n^q\}$ converges in C^k, for each k, to some frame $\{W_1, \dots, W_n\}$. Then

$$Z_j^q := \delta(T^{\delta})_* W_j, \quad 1 \leq j \leq n,$$

is a pseudo-Hermitian frame for $(\delta^{-1}U, \theta_q)$. By looking at the error terms in the expression (3.23) of W_j^q one may show that Z_j^q converges in $C^k(B_R)$ to Z_j, for each k and any $R > 0$. Similarly (again by (3.23)) θ_q (respectively $\mathcal{L}_q$) converges uniformly in $C^k(B_R)$ to θ_0 (respectively to $\mathcal{L}_0$). □

Let $R > 0$ be fixed. For q sufficiently close to p one has $B_{3R} \subset \delta_q^{-1}U$. Let $\eta \in C_0^\infty(B_{2R})$ such that $\eta = 1$ on B_R. Then

$$\begin{aligned}\mathcal{L}_q(\eta h_q) &= \eta \mathcal{L}_q h_q - 2L_{\theta_q}^{*}(d\eta, dh_q) + (\mathcal{L}_q \eta) h_q \\ &= \eta(-\rho_q \delta_q^2 h_q + \lambda_q h_q^{q-1}) - 2L_{\theta_q}^{*}(d\eta, dh_q) + (\mathcal{L}_q \eta) h_q .\end{aligned}$$

Note that $\|dh_q\|_{\theta_q}$ is bounded by 1 in B_{2R} (it attains its maximum ($= 1$) at the origin). Also

$$\int_{|(\tilde{z},\tilde{t})|<R} |h_q(\tilde{z}, \tilde{t})|^q \, d\tilde{z} d\tilde{t} = \delta_q^{2q/(q-2)-(2n+2)} \int_{|(z,t)|<\delta_q R} |u_q(z, t)|^q \, dz\, dt. \tag{3.45}$$

If $q < p$ then $2q/(q-2) > 2n + 2$ hence $\delta_q^{2q/(q-2)-(2n+2)} < 1$ as $q \to p$. Moreover (by (3.23))

$$dz\, dt = C_n(1 + \delta O^1)\theta \wedge (d\theta)^n$$

on $B_{2\delta R}$. Thus

$$h_q \in L^q(B_{2R}, d\tilde{z}\, d\tilde{t})$$

with uniform bounds on the norm. In particular $h_q \in L^1(B_{2R}, d\tilde{z}\, d\tilde{t})$, uniformly as $q \to p$, a fact that together with the uniform bound on $\|dh_q\|_{\theta_q}$ yields

$$h_q \in S_1^r(B_{2R}, \theta_q), \quad r < \infty,$$

with uniform bounds on the norm (by Proposition 3.6). Thus (by Theorem 3.13) ηh_q is uniformly bounded in $L^r(B_{2R})$ for any r, and then (by Theorem 3.22) uniformly bounded in $C^k(B_R)$ for each k.

Consider now a subsequence $q_j \to p$ such that h_{q_j} converges in $C^1(B_R)$. Define a function u on $\mathbf{H}_n$ by first choosing a subsequence convergent in $C^1(B_1)$, and then a subsequence convergent in $C^1(B_2)$, and so on. Then $u \geq 0$, $u \in C^1(\mathbf{H}_n)$, and $u \not\equiv 0$ (since $\|du\|_{\theta_0}(0) = 1$). As $\theta_{q_j} \to \theta_0$,

$$\int_{\mathbf{H}_n} \left[b_n L^*_{\theta_0}(du, d\phi) - \lambda(M) u^{p-1}\phi \right] \theta_0 \wedge (d\theta_0)^n = 0, \tag{3.46}$$

for any $\phi \in C_0^\infty(\mathbf{H}_n)$. Let us set

$$\|u\|_p = \left(\int_{\mathbf{H}_n} u^p \theta_0 \wedge (d\theta_0)^n \right)^{1/p}.$$

We claim that

$$\|u\|_p \leq 1. \tag{3.47}$$

Indeed, as $\theta_q \wedge (d\theta_q)^n \to \theta_0 \wedge (d\theta_0)^n$ uniformly on compact sets, the constraint $\int_M u_q^q \theta \wedge (d\theta)^n = 1$ and (3.45) yield $\int_{B_R} u^p \theta_0 \wedge (d\theta_0)^n = 1$ with $R > 0$ arbitrary, and (3.47) is proved. □

Next, we claim that

$$\int_{\mathbf{H}_n} \|du\|^2_{\theta_0} \theta_0 \wedge (d\theta_0)^n \leq C < \infty. \tag{3.48}$$

Indeed

$$\begin{aligned}
\int_{B_R} \|du\|^2_{\theta_0} \theta_0 \wedge (d\theta_0)^n &= \lim_{j\to\infty} \int_{B_R} \|dh_{q_j}\|^2_{\theta_{q_j}} \theta_{q_j} \wedge (d\theta_{q_j})^n \\
&= \lim_{j\to\infty} \int_{B_{\delta R}} \delta^{2q/(q-2)} \|du_{q_j}\|^2_\theta \delta^{-2}\theta \wedge (\delta^{-2} d\theta)^n \\
&\leq \limsup_{j\to\infty} \int_M \|du_{q_j}\|^2_\theta \theta \wedge (d\theta)^n,
\end{aligned}$$

which is bounded. Here $\delta = \delta_{q_j}$. □

At this point, we may conclude the proof of Theorem 3.24. By the estimates (3.47)–(3.48) there is a sequence $\phi_j \in C_0^\infty(\mathbf{H}_n)$ converging to u in the norms associated with (3.47) and (3.48). Thus (by (3.46))

$$b_n \int_{\mathbf{H}_n} \|du\|^2_{\theta_0} \theta_0 \wedge (d\theta_0)^n = \lambda(M) \|u\|_p^p.$$

Now, the function

$$\tilde{u} = \frac{u}{\|u\|_p}$$

satisfies the constraint $\|\tilde{u}\|_p = 1$; yet (by (3.47) and $p \geq 2$)

$$b_n \int_{\mathbf{H}_n} \|d\tilde{u}\|^2_{\theta_0} \theta_0 \wedge (d\theta_0)^n = \lambda(M) \frac{\|u\|_p^p}{\|u\|_p^2} \leq \lambda(M) < \lambda(\mathbf{H}_n),$$

contradicting the definition of $\lambda(\mathbf{H}_n)$. Hence case II.b doesn't occur, and the proof of Theorem 3.24 is complete. □

3.7 Uniqueness and open problems

Another important problem, related to the CR Yamabe problem, is to decide whether a compact contact manifold (M, θ) of constant pseudo-Hermitian scalar curvature is unique. As it turns out, the answer depends on the sign of the CR invariant $\lambda(M)$. Precisely, one may state the following theorem:

Theorem 3.25. (D. Jerison and J.M. Lee [227])
If $\lambda(M) \leq 0$ then any two contact forms of constant pseudo-Hermitian scalar curvature are homothetic.

Lemma 3.11. *Given (M, θ) with $\rho =$ const the sign of ρ is a CR invariant.*

Indeed, let $\hat{\theta} = u^{p-2}\theta$ be another contact form such that $\hat{\rho} =$ const as well. Then

$$b_n \Delta_b u + \rho u = \hat{\rho} u^{p-1}.$$

Integration over M together with $\int_M (\Delta_b u)\theta \wedge (d\theta)^n = \int_M \langle d1, du\rangle \theta \wedge (d\theta)^n = 0$ yields either $\rho = \hat{\rho} = 0$ or

$$\frac{\rho}{\hat{\rho}} = \frac{\int_M u^{p-1}\theta \wedge (d\theta)^n}{\int_M u\theta \wedge (d\theta)^n} > 0.$$ □

Let us return to the proof of Theorem 3.25. Let us assume that $\lambda(M) < 0$. By the Jerison–Lee theorem, there is θ of scalar curvature $\rho = \lambda(M)$. Let $\hat{\theta} = u^{p-2}\theta$ be of constant scalar curvature $\hat{\rho}$. Then $\hat{\rho} < 0$ (by Lemma 3.11); hence, by multiplying u by a suitable constant, one may assume that $\rho = \hat{\rho}$. Then

$$b_n \Delta_b u + \rho u = \rho u^{p-1}.$$

To prove the statement in Theorem 3.25 it suffices to show that $u \equiv 1$. Since Δ_b is degenerate elliptic, it satisfies a weak maximum principle (see the next section). If $x \in M$ is a point where $u(x) = \sup_M u$ then $\Delta_b u(x) \geq 0$; hence $u(x)^{p-1} - u(x) \leq 0$, which yields $u \leq 1$. Similarly, if $y \in M$ is such that $u(y) = \inf_M u$, then $u(y) \geq 1$; hence $u \geq 1$. Therefore $u \equiv 1$. □

Let us look now at the case $\lambda(M) = 0$. By the Jerison–Lee theorem there is θ with $\rho = 0$ and by the first part of the proof of Theorem 3.25 any other $\hat{\theta} = u^{p-2}\theta$ with $\hat{\rho} =$ const has $\hat{\rho} = 0$. Thus $b_n \Delta_b u = 0$ which yields $\int_M \|du\|^2_\theta \theta \wedge (d\theta)^n = 0$. It follows that $du = f\theta$ for some $f \in C^\infty(M)$. Differentiation then gives $0 = (df) \wedge \theta + f d\theta$, an identity that restricted to $H(M)$ yields $f d\theta|_{H(M)\otimes H(M)} = 0$, or $f \equiv 0$, i.e., $u =$ const. □

When $\lambda(M) > 0$, as opposed to the situation in Theorem 3.25, the solution to the CR Yamabe problem may be not unique. For instance, if $M = S^{2n+1}$ and θ_1 is the standard contact form then $\Phi^*\theta_1$ has constant pseudo-Hermitian scalar curvature, for any $\Phi \in \mathrm{Aut}_{CR}(S^{2n+1})$, yet in general $\Phi^*\theta_1$ and θ_1 are not homothetic. Are these solutions extremals for the problem (3.6) for the sphere? As shown in [227], one may observe first that extremals do exist:

Theorem 3.26. (D. Jerison and J.M. Lee [227])
There is $u \in C^\infty(S^{2n+1})$, $u > 0$, such that the infimum $\lambda(S^{2n+1})$, in (3.6) *with $\theta := u^{p-2}\theta_1$, is attained.*

In view of Theorem 3.26 and the results of M. Obata [330], one expects that *the contact forms $\{\Phi^*\theta_1 : \Phi \in \mathrm{Aut}_{CR}(S^{2n+1})\}$ are the only contact forms on S^{2n+1} that have constant pseudo-Hermitian scalar curvature and therefore*

$$\lambda(S^{2n+1}) = \frac{n(n+1)}{2}\left[\int_{S^{2n+1}} \theta_1 \wedge (d\theta_1)^n\right]^{2/p}$$

as conjectured in [227].

Let θ be a contact form on S^{2n+1} with $\rho = $ const. Then $\hat{\theta} = (F^{-1})^*\theta$ is a contact form on $\mathbf{H}_n$ with $\hat{\rho} = $ const. Let $u \in C^\infty(\mathbf{H}_n)$, $u > 0$, so that $\hat{\theta} = u^{p-2}\theta_0$. We have

$$\int_{\mathbf{H}_n} u^p \theta_0 \wedge (d\theta_0)^n = \int_{\mathbf{H}_n} \hat{\theta} \wedge (d\hat{\theta})^n = \int_{S^{2n+1}} \theta \wedge (d\theta)^n < \infty;$$

hence $u \in L^p(\mathbf{H}_n)$. As in the proof of Theorem 3.25, one may multiply u by a constant to get $\rho = \frac{n(n+1)}{2}$; hence

$$4\Delta_b u = n^2 u^{p-1} \tag{3.49}$$

on $\mathbf{H}_n$. On the other hand, if $\Phi \in \mathrm{Aut}_{CR}(S^{2n+1})$ and $\theta = \Phi^*\theta_1$ then a calculation shows that

$$u(z,t) = C\left|t + i|z|^2 + z \cdot \overline{\mu} + \lambda\right|^{-n}, \tag{3.50}$$

where $C > 0$, $\lambda \in \mathbf{C}$, $\mathrm{Im}(\lambda) > 0$, and $\mu \in \mathbf{C}^n$. Therefore, the conjecture above is true whenever the following statement is true: *If $u \in L^p(\mathbf{H}_n)$ is a positive C^∞ solution to* (3.49) *then u is of the form* (3.50). □

The full solution to the question above was given in [229] by following the ideas in Obata's approach (cf. [330]) to the Riemannian counterpart of the problem, which we briefly recall. If g_0 is the standard metric on the sphere S^m and $g = \varphi^2 g_0$ is any conformally equivalent metric then $\varphi^{-2}g$ is an Einstein metric and consequently the traceless Ricci tensor $B(g)$ of g may be computed in terms of the Hessian $\nabla^2\varphi$. Moreover the first Bianchi identity together with the assumption that g has constant scalar curvature imply that

$$\mathrm{div}\left(B(g)^{ij}\varphi_i\partial_j\right) = \varphi|B(g)|^2 \tag{3.51}$$

and integration over S^m yields $B(g) = 0$, i.e., g is an Einstein metric. One is left with the problem of describing the Einstein metrics conformal to g_0, which is easy to solve. We emphasize that Obata's argument leads from *one* nonlinear equation satisfied by φ (the condition that $\varphi^2 g_0$ has constant scalar curvature) to a whole *system* of equations (namely $B(\varphi^2 g_0) = 0$). In the CR case one looks (cf. [229]) for an analog to (3.51). The finding in [229] is a rather complicated identity (involving not only the norm $|B^2|$ of the traceless Ricci tensor $B_{\alpha\bar\beta} = R_{\alpha\bar\beta} - (\rho/n)h_{\alpha\bar\beta}$, cf. Definition 5.11 in Chapter 5, but also the norm $|A|^2$ of the pseudo-Hermitian torsion and higher-order terms such as $|\operatorname{div} A|^2$) playing a role similar to (3.51) in Riemannian geometry (cf. [229], pp. 8–10).

The results of D. Jerison and J.M. Lee presented in this chapter can be formally compared to the partial completion of the proof of the Riemannian Yamabe conjecture by T. Aubin [24]. The remaining cases should, by analogy, be solved using a CR version of the positive mass theorem. Unfortunately, especially because the theory of CR minimal surfaces does not exist at the present stage of research, a CR version of positive mass theorem is not available yet. However, as observed by N. Gamara and R. Yacoub [164], besides T. Aubin's and R. Schoen's proof of the Riemannian Yamabe conjecture, yet another proof due to A. Bahri [27], of the Yamabe conjecture is available (it exploits techniques related to the theory of critical points at infinity) and may be generalized to the CR category (since no use of minimal surfaces or the positive mass theorem is required). N. Gamara and R. Yacoub obtain the following result (related to N. Gamara [163], and to Z. Li, *Webster scalar curvature problem on CR manifolds*, preprint):

Theorem 3.27. (N. Gamara and R. Yacoub [164])
Let M be an orientable compact real $(2n+1)$-dimensional CR manifold, locally CR equivalent to the sphere S^{2n+1}. Let θ be a contact form on M. Then the problem $\Delta_b^c u = u^{1+2/n}$, $u > 0$, on M, admits a solution.

Here Δ_b^c is the CR invariant Laplacian of (M, θ).

3.8 The weak maximum principle for Δ_b

For the convenience of the reader we present briefly the *weak maximum principle* for degenerate elliptic operators (following mainly J.M. Bony [73]). For an extensive treatment (of the theory of degenerate elliptic operators), see, e.g., N. Shimakura [372], pp. 183–224, and again [73]. Let $\Omega \subset \mathbf{R}^n$ be a domain and let L be a second-order differential operator, with real-valued C^∞ coefficients defined in Ω:

$$Lu(x) = \sum_{i,j=1}^n a_{ij}(x)\frac{\partial^2 u}{\partial x_i \partial x_j} + \sum_{i=1}^n a_i(x)\frac{\partial u}{\partial x_i} + a(x)u(x).$$

We assume from now on that L possesses the following three additional properties:

(1) the quadratic form $a_{ij}(x)$ is positive at any x, but not necessarily positive definite, i.e., for any $x \in \Omega$ and any $\xi \in \mathbf{R}^n$,

$$\sum_{i,j=1}^{n} a_{ij}(x)\xi_i\xi_j \geq 0,$$

(2) one has $a(x) \leq 0$ for any $x \in \Omega$,
(3) there exist C^∞ vector fields $X_1, \ldots, X_r, Y$ on Ω such that

$$Lu = \sum_{\alpha=1}^{r} X_\alpha^2 u + Yu + au.$$

Definition 3.27. L is referred to as a *degenerate elliptic* operator. □

See, e.g., J.M. Bony [73]. For instance, consider the operator

$$L = \frac{\partial^2}{\partial x_1^2} + \left(\sum_{j=2}^{n} x_1^{j-1} \frac{\partial}{\partial x_j} \right)^2 .$$

Then $a_i = 0$, $a = 0$, and (a_{ij}) is the matrix

$$\begin{pmatrix} 1 & 0 & \cdots & 0 \\ 0 & x_1^2 & \cdots & x_1^n \\ \vdots & \vdots & & \vdots \\ 0 & x_1^n & \cdots & x_1^{2n-2} \end{pmatrix}.$$

Let H be the hyperplane $x_1 = 0$. If $x \in H$ then

$$\sum_{i,j=1}^{n} a_{ij}(x)\xi^i\xi^j = (\xi^1)^2 \geq 0,$$

while for $x \in \mathbf{R}^n \setminus H$ one has

$$\sum_{ij} a_{ij}(x)\xi^i\xi^j = (\xi^1)^2 + x_1^{-2}\left(\sum_{\alpha=2}^{n} x_1^\alpha \xi^\alpha \right)^2 \geq 0.$$

Also $L = \sum_{j=1}^{n} X_j^2$, where $X_1 = \partial/\partial x_1$ and $X_\alpha = x_1^{\alpha-1}\partial/\partial x_\alpha$ for $2 \leq \alpha \leq n$; hence L is a degenerate elliptic operator. □

Degenerate elliptic operators of second order were investigated in several papers essentially devoted to the Dirichlet problem: existence, uniqueness, and regularity of solutions (cf., e.g., O.A. Oleinik [333], J.J. Kohn and L. Nirenberg [251], J.M. Bony [71]), and to hypoellipticity (cf. L. Hörmander [213]). In [73] one is mainly interested in different properties appearing in the classical potential theory. The degenerate elliptic operators satisfy the following weak form of the maximum principle:

Theorem 3.28. *If a C^2 function u achieves in x_0 a nonnegative local maximum then $(Lu)(x_0) \leq 0$. If moreover this maximum is (strictly) positive and $a(x_0) < 0$ then $(Lu)(x_0) < 0$.*

Proof. Set

$$X_\alpha = a^i_\alpha \partial/\partial x^i, \quad 1 \leq \alpha \leq r.$$

Then the hypothesis (3) amounts to $a_{ij} = \sum_{\alpha=1}^r a^i_\alpha a^j_\alpha$. Since x_0 is a local maximum point, the Hessian $\left[\frac{\partial^2 u}{\partial x^i \partial x^j}(x_0)\right]$ is negative semidefinite. Thus

$$\sum_{i,j=1}^n a_{ij}(x_0)\frac{\partial^2 u}{\partial x^i \partial x^j}(x_0) = \sum_{i,j,\alpha} \frac{\partial^2 u}{\partial x^i \partial x^j}(x_0) a^i_\alpha a^j_\alpha \leq 0.$$

This, together with the property (2), i.e., $a(x_0) \leq 0$, and $\frac{\partial u}{\partial x^i}(x_0) = 0$, yields $(Lu)(x_0) \leq 0$. □

4

Pseudoharmonic Maps

By a well-known result of A. Lichnerowicz [287], any holomorphic map of compact Kähler manifolds is a stable harmonic map. In odd real dimension, the closest analogue of a Kählerian manifold seems to be a strictly pseudoconvex CR manifold (eventually with vanishing pseudo-Hermitian torsion, for some fixed contact form). CR manifolds appear mainly as boundaries of smooth domains in $\mathbf{C}^{n+1}$ and a holomorphic map of a neighborhood of a domain $\Omega \subset \mathbf{C}^{n+1}$ into a neighborhood of a domain $\Omega' \subset \mathbf{C}^{N+1}$, preserving boundaries, gives rise to a CR map $\partial\Omega \to \partial\Omega'$. On the other hand, when Ω and Ω' are strictly pseudoconvex domains, the Levi forms of their boundaries extend to Riemannian metrics (the *Webster metrics*) and it is a natural question whether a CR map $\partial\Omega \to \partial\Omega'$ is harmonic with respect to these metrics. It comes as some surprise that the answer is, in general, negative. Precisely, by a result of H. Urakawa [413], given two strictly pseudoconvex CR manifolds M and M', endowed with contact forms θ and θ' (such that the corresponding Levi forms are positive definite), a CR map $\phi : M \to M'$ is harmonic (with respect to the Webster metrics g_θ and $g_{\theta'}$) if and only if $T(f) = 0$, where f is the unique C^∞ function with $\phi^*\theta' = f\theta$ and T is the characteristic direction of $d\theta$. Even worse, a harmonic map of (M, g_θ) into $(M', g_{\theta'})$ is, in general, unstable. Indeed, by a result of Y.L. Xin [437], any stable harmonic map of the sphere S^{2n+1} to a Riemannian manifold is a constant map. It is natural to ask how CR maps fit into our theory of *pseudoharmonicity* (see the definitions below).

A fundamental ingredient in the study of harmonic maps from a Kählerian manifold to a Riemannian manifold of nonpositive curvature is the *Siu–Sampson formula* (cf. [374]–[361]). In an attempt to recover this formula for C^∞ maps $\phi : M \to N$ from a strictly pseudoconvex CR manifold M to a Riemannian manifold (N, h), R. Petit considered (cf. [346]) the following (analogue of the) second fundamental form (of a C^∞ map between Riemannian manifolds)

$$\beta(\phi)(X, Y) = \left(\phi^{-1}\nabla^h\right)_X (d\phi)Y - (d\phi)\nabla_X Y.$$

Here $\phi^{-1}\nabla^h$ is the connection in the pullback bundle $\phi^{-1}TN \to M$, induced by the Levi-Civita connection ∇^h of (N, h), and ∇ is the Tanaka–Webster connection of (M, θ), for a fixed choice of contact 1-form θ on M. Since one interesting application (cf. Theorem 5.6 in [346]), the (pseudo-Hermitian) analogue of the Siu–Sampson formula (cf. Propositions 3.1 and 3.2 in [346]) led to a description of the curvature properties of a strictly pseudoconvex CR manifold (of *small* pseudo-Hermitian torsion, $\|\tau\| < n/\sqrt{2}$) similar to that in Sasakian geometry (cf. [62], p. 65) where $\tau = 0$.

As a natural continuation of R. Petit's ideas, we restrict $\beta(\phi)$ to the maximal complex distribution of M and take the trace with respect to the Levi form. The resulting ϕ-vector field $\tau(\phi; \theta, \nabla^h) \in \Gamma^\infty(\phi^{-1}TN)$ is the *tension* field of ϕ (with respect to the data (θ, ∇^h)). Smooth maps with $\tau(\phi; \theta, \nabla^h) = 0$ are then pseudo-Hermitian analogues of harmonic maps (of Riemannian manifolds), and are referred to as *pseudoharmonic* in the sequel. It is with this sort of maps that Chapter 4 is mainly concerned. We show that a C^∞ map $\phi : M \to N$ is pseudoharmonic if and only if it satisfies the PDE system

$$\Delta_b \phi^i + 2\, h^{\alpha\bar\beta} T_\alpha(\phi^j) T_{\bar\beta}(\phi^k) \big({\Gamma'}^i_{jk} \circ \phi \big) = 0,$$

whose principal part is the sub-Laplacian Δ_b of (M, θ). As to the geometric interpretation of pseudoharmonicity, we discover the following phenomenon. Let $K(M) = \Lambda^{n+1,0}(M)$ be the canonical bundle over the strictly pseudoconvex CR manifold M. Let $C(M) = (K(M) \setminus \{0\})/\mathbf{R}_+$ be the canonical circle bundle over M and $\pi : C(M) \to M$ the projection. Given a contact 1-form θ on M, let F_θ be the Fefferman metric associated with θ; cf. Chapter 2 of this book. We show that $\phi : M \to N$ is pseudoharmonic if and only if its vertical lift $\phi \circ \pi$ is harmonic as a map of $(C(M), F_\theta)$ into (N, h).

From the point of view of variational calculus, pseudoharmonicity may be looked at as follows. For any compact strictly pseudoconvex pseudo-Hermitian manifold (M, θ), consider

$$E(\phi) = \tfrac{1}{2} \int_M \operatorname{trace}_{G_\theta} \big(\pi_H \phi^* h \big)\, \theta \wedge (d\theta)^n .$$

We show that the critical points of $E(\phi)$ are precisely the pseudoharmonic maps. We derive the second variation formula for $E(\phi)$ and consider the corresponding notion of *stability*. When the target space is a Riemannian manifold of nonpositive curvature, any pseudoharmonic map is shown to be stable. Let N be a totally umbilical real hypersurface of a real space form $M^{m+1}(c)$, of mean curvature $\|H\|$ (a constant a posteriori). If $(m-2)\|H\|^2 + (m-1)c > 0$ then any nonconstant pseudoharmonic map $\phi : M \to N \cap V$ (where $V \subset M^{m+1}(c)$ is a simple and convex open subset) is shown to be unstable. The results in Chapter 4 are based on the works by E. Barletta et al. [43], and E. Barletta [32, 33].

4.1 CR and pseudoharmonic maps

Let $(M, T_{1,0}(M))$ be a nondegenerate CR manifold (of CR dimension n), θ a contact form on M, and T the characteristic direction of $d\theta$. Let (N, D') be a manifold with linear connection. Let $\phi : M \to N$ be a C^∞ map and $\phi^{-1}TN \to M$ the pullback of $T(N)$ by ϕ. Let $\phi^{-1}D'$ be the connection in $\phi^{-1}TN$ induced by D'. This is most easily described in local coordinates, as follows.

Definition 4.1. The *natural lift* $\tilde{Y} : \phi^{-1}(V) \to \phi^{-1}TN$ of a tangent vector field $Y : V \to T(N)$ (with $V \subseteq N$ open) is given by $\tilde{Y}(x) = Y(\phi(x)),\ x \in \phi^{-1}(V)$. □

Let $\{T_\alpha\}$ be a local frame of $T_{1,0}(M)$, defined on the open set $U \subseteq M$, and $\{\theta^\alpha\}$ the corresponding admissible coframe, i.e., $\theta^\alpha(T_\beta) = \delta^\alpha_\beta$, $\theta^\alpha(T_{\bar\beta}) = 0$, $\theta^\alpha(T) = 0$. Let (V, y^i) be a local coordinate system on N with $\phi(U) \subseteq V$ and let Y_i be the natural lift of $\partial/\partial y^i$, so that $d\phi = \phi^i_A \theta^A \otimes Y_i$, for some C^∞ functions ϕ^i_A on U, where $A \in \{0, 1, \dots, n, \bar{1}, \dots, \bar{n}\}$ and $\theta^0 = \theta$.

Definition 4.2. The *induced* connection $\phi^{-1}D'$ in $\phi^{-1}TN \to M$ is defined by

$$\left(\phi^{-1}D'\right)_{T_A} Y_j = \phi^i_A \left(\Gamma'^k_{ij} \circ \phi\right) Y_k,$$

where $T_0 = T$ and Γ'^i_{jk} are the local coefficients of D' with respect to (V, y^i). □

Moreover, let ∇^ϕ be the connection in $T^*(M) \otimes \phi^{-1}TN$ determined by

$$\nabla^\phi_X(\omega \otimes s) = (\nabla_X \omega) \otimes s + \omega \otimes (\phi^{-1}D')_X s, \quad X \in \mathcal{X}(M).$$

As usual, for any bilinear form B on $T(M)$, we denote by $\pi_H B$ the restriction of B to $H(M) \otimes H(M)$.

Definition 4.3. Let us consider the ϕ-tensor field on M given by

$$\tau(\phi; \theta, D') = \text{trace}_{G_\theta}\{\pi_H \nabla^\phi d\phi\} \in \Gamma^\infty(\phi^{-1}TN).$$

We say that ϕ is *pseudoharmonic*, with respect to the data (θ, D'), if $\tau(\phi; \theta, D') = 0$. □

Theorem 4.1. *Let M, N be nondegenerate CR manifolds and θ, θ' contact forms on M and N, respectively. Let ∇' be the Tanaka–Webster connection of (N, θ'). Then for any CR map $\phi : M \to N$,*

$$\tau(\phi; \theta, \nabla') = 2n\tilde{J}'(d\phi)T. \tag{4.1}$$

In particular, a CR map ϕ is pseudoharmonic with respect to the data (θ, ∇') if and only if ϕ is a pseudo-Hermitian map, i.e., $\phi^*\theta' = c\theta$, for some $c \in \mathbf{R}$. Here $\tilde{J}' : \phi^{-1}TN \to \phi^{-1}TN$ is the natural lift of $J' : T(N) \to T(N)$, i.e., $J'_x Y_x = J'_{\phi(x)} Y_{\phi(x)}$, $Y \in \mathcal{X}(N)$, $x \in M$.

Proof. To compute $\tau(\phi; \theta, \nabla')$ we need

$$(\nabla^\phi_X d\phi)Y = \left(\phi^{-1}\nabla'\right)_X (d\phi)Y - (d\phi)\nabla_X Y.$$

Let $\{X_\alpha, JX_\alpha\}$ be a local G_θ-orthonormal frame of $H(M)$, i.e.,

$$G_\theta(X_\alpha, X_\beta) = \epsilon_\alpha \delta_{\alpha\beta},$$
$$\epsilon_1 = \cdots = \epsilon_r = -\epsilon_{r+1} = \cdots = -\epsilon_{r+s} = 1, \quad r + s = n.$$

Then

$$\tau(\phi;\theta,\nabla') = \sum_{\alpha=1}^{n} \epsilon_\alpha \left\{ \left(\nabla^\phi_{X_\alpha} d\phi\right) X_\alpha + \left(\nabla^\phi_{JX_\alpha} d\phi\right) JX_\alpha \right\}.$$

We have

$$\nabla_{JX} JX = -\nabla_X X + J[JX, X] + JT_\nabla(JX, X),$$

for any $X \in H(M)$, because of $\nabla J = 0$. On the other hand (cf., e.g., (5) in [120], p. 174), $\pi_H T_\nabla = 2(d\theta) \otimes T$; hence $J\pi_H T_\nabla = 0$, and the previous identity becomes

$$\nabla_{JX} JX = -\nabla_X X + J[JX, X]. \tag{4.2}$$

Similarly,

$$\left(\phi^{-1}\nabla'\right)_{JX} (d\phi) JX = \tilde{J}' \left(\phi^{-1}\nabla'\right)_{JX} (d\phi) X$$

(by $(d\phi)JX = \tilde{J}'(d\phi)X$, since ϕ is a CR map). Also

$$\tilde{T}_{\nabla'}((d\phi)X, (d\phi)Y) = 2d(\phi^*\theta')(X, Y) T' \circ \phi$$

(where $\tilde{T}_{\nabla'}$ is the ϕ-vector field induced by $T_{\nabla'}$) leads to

$$\left(\phi^{-1}\nabla'\right)_X (d\phi) Y = \left(\phi^{-1}\nabla'\right)_Y (d\phi) X + (d\phi)[X, Y] + 2d(\phi^*\theta')(X, Y)\tilde{T}',$$

for any $X, Y \in H(M)$. Then (by $\tilde{J}'\tilde{T}' = 0$)

$$\left(\phi^{-1}\nabla'\right)_{JX} (d\phi) JX + \left(\phi^{-1}\nabla'\right)_X (d\phi) X = \tilde{J}'(d\phi)[JX, X]. \tag{4.3}$$

Finally (by (4.2)–(4.3))

$$\begin{aligned}\tau(\phi;\theta,\nabla') &= \sum_{\alpha=1}^{n} \epsilon_\alpha \left\{ \tilde{J}'(d\phi)[JX_\alpha, X_\alpha] - (d\phi)J[JX_\alpha, X_\alpha] \right\} \\ &= \sum_{\alpha=1}^{n} \epsilon_\alpha \tilde{J}' d\phi \left\{ [JX_\alpha, X_\alpha] - \pi_{H(M)}[JX_\alpha, X_\alpha] \right\},\end{aligned}$$

where $\pi_{H(M)} : T(M) \to H(M)$ is the projection associated with the direct sum decomposition $T(M) = H(M) \oplus \mathbf{R}T$. Hence

$$\tau(\phi;\theta,\nabla') = \sum_{\alpha=1}^{n} \epsilon_\alpha \tilde{J}'\theta\left([JX_\alpha, X_\alpha]\right)(d\phi)T$$

and

$$\sum_{\alpha=1}^{n} \epsilon_\alpha \theta([JX_\alpha, X_\alpha]) = -2\sum_{\alpha=1}^{n} \epsilon_\alpha (d\theta)(JX_\alpha, X_\alpha) = 2\sum_{\alpha=1}^{n} \epsilon_\alpha G_\theta(X_\alpha, X_\alpha) = 2n,$$

which leads to (4.1). Now sufficiency follows from Corollary 3.2 in [413], p. 236. Indeed, for a CR map $\phi : M \to N$ the property $\phi^*\theta' = c\theta$, for some $c \in \mathbf{R}$, is equivalent (cf. op. cit.) to $(d\phi)T = c\tilde{T}'$ and then $\tau(\phi;\theta,\nabla') = 0$ by (4.1) and $J'T' = 0$.

Conversely, since ϕ is a CR map, $\phi^*\theta' = \lambda\theta$, for some $\lambda \in C^\infty(M)$. To have Theorem 4.1, we only have to see that pseudoharmonicity yields that λ is constant. Indeed, we have

$$d(\phi^*\theta') = (d\lambda) \wedge \theta + \lambda d\theta,$$

so that

$$T(\lambda)\theta - d\lambda = T \rfloor d(\phi^*\theta'). \tag{4.4}$$

Assume ϕ to be pseudoharmonic. Then (by (4.1))

$$(d_x\phi)T_x = \lambda(x)T'_{\phi(x)}, \quad x \in M.$$

Consequently

$$\left(T \rfloor d(\phi^*\theta')\right)_x = \lambda(x)(d\theta')_{\phi(x)}(T'_{\phi(x)}, (d_x\phi)\cdot) = 0$$

and (by (4.4)) $d\lambda = 0$ on $H(M)$. In particular, λ is a real-valued CR function, hence a constant, by the following lemma (see also Theorem 2.1 in [215]):

Lemma 4.1. *Let M be a connected nondegenerate CR manifold and $\lambda : M \to \mathbf{R}$ a C^∞ solution of $\overline{\partial}_b\lambda = 0$. Then λ must be constant.*

Proof. Let $\{T_\mu\}$ be a local frame of $T_{1,0}(M)$. The tangential Cauchy–Riemann equations (satisfied by λ) read $T_{\overline{\mu}}(\lambda) = 0$. Then $T_\mu(\lambda) = 0$, by complex conjugation, hence $[T_\mu, T_{\overline{\mu}}](\lambda) = 0$. Let θ be a contact form on M, and (by the nondegeneracy assumption) T the characteristic direction of $d\theta$. Let Γ^A_{BC} be the local coefficients of the Tanaka–Webster connection (of (M,θ)) with respect to $\{T_\mu, T_{\overline{\mu}}, T\}$. By the purity axiom

$$\Gamma^{\overline{\mu}}_{\alpha\overline{\beta}}T_{\overline{\mu}} - \Gamma^{\mu}_{\overline{\beta}\alpha}T_\mu - [T_\alpha, T_{\overline{\beta}}] = 2i h_{\alpha\overline{\beta}}T,$$

where $h_{\alpha\overline{\beta}} = L_\theta(T_\alpha, T_{\overline{\beta}})$. Then $T(\lambda) = 0$. $\square$

4.2 A geometric interpretation

The main purpose of this section is to show that a map is pseudoharmonic if and only if its vertical lift is harmonic with respect to the Fefferman metric (cf. Theorem 4.2). We start by establishing the following result:

Proposition 4.1. *Let $\phi : M \to N$ be a C^∞ map of a nondegenerate CR manifold M into a C^∞ manifold N. Let θ be a contact form on M, and D' a torsion-free linear connection on N. Then ϕ is pseudoharmonic, with respect to the data (θ, D'), if and only if*

$$\Delta_b \phi^i + 2 h^{\alpha\overline{\beta}} T_\alpha(\phi^j) T_{\overline{\beta}}(\phi^k) \left({\Gamma'}^i_{jk} \circ \phi \right) = 0, \tag{4.5}$$

for some local frame $\{T_\alpha\}$ of $T_{1,0}(M)$ and some local coordinate system (V, y^i) on N, where $\phi^i = y^i \circ \phi$.

Proof. Let $\{X_\alpha, JX_\alpha\}$ be a G_θ-orthonormal local frame of $H(M)$, i.e., $G_\theta(X_\alpha, X_\beta) = \epsilon_\alpha \delta_{\alpha\beta}$. As usual, we set $Z_\alpha = \frac{1}{\sqrt{2}}\{X_\alpha - \sqrt{-1} J X_\alpha\}$, so that for any bilinear form φ on $H(M)$,

$$\text{trace}_{G_\theta}\{\varphi\} = \sum_{\alpha=1}^{n} \epsilon_\alpha \{\varphi(Z_\alpha, Z_{\overline{\alpha}}) + \varphi(Z_{\overline{\alpha}}, Z_\alpha)\}.$$

Let $\{\theta_\alpha\}$ be the admissible coframe associated with $\{Z_\alpha\}$. The identities $\nabla\theta = 0$ and $\nabla\theta^\alpha = -\omega^\alpha_\beta \otimes \theta^\beta$ lead to

$$\nabla^\phi_Z d\phi = \{Z(\phi^i_A)\theta^A - \phi^i_\alpha \omega^\alpha_\beta(Z)\theta^\beta - \phi^i_{\overline{\alpha}} \omega^{\overline{\alpha}}_{\overline{\beta}}(Z)\theta^{\overline{\beta}}\} \otimes Y_i + (\phi^i_A \theta^A) \otimes (\phi^{-1} D')_Z Y_i,$$

for any $Z \in T(M) \otimes \mathbf{C}$. Hence

$$\text{trace}_{G_\theta}\{\pi_H \nabla^\phi d\phi\} =$$

$$\sum_{\mu=1}^{n} \epsilon_\mu \{Z_\mu(\phi^i_{\overline{\mu}}) - \phi^i_{\overline{\alpha}} \Gamma^{\overline{\alpha}}_{\mu\overline{\mu}} + Z_{\overline{\mu}}(\phi^i_\mu) - \phi^i_\alpha \Gamma^\alpha_{\overline{\mu}\mu} + (\phi^j_\mu \phi^k_{\overline{\mu}} + \phi^k_\mu \phi^j_{\overline{\mu}}) {\Gamma'}^i_{jk} \circ \phi\} Y_i,$$

for any linear connection D' on N. When D' is torsion-free one may express the tension field of ϕ, with respect to the data (θ, D'), since

$$\tau(\phi; \theta, D') = \left\{ \Delta_b \phi^i + \sum_{\mu=1}^{n} \epsilon_\mu Z_\mu(\phi^j) Z_{\overline{\mu}}(\phi^k) ({\Gamma'}^i_{jk} \circ \phi) \right\} Y_i.$$

Finally, if $\{T_\alpha\}$ is an arbitrary local frame of $T_{1,0}(M)$ then $Z_\alpha = U^\beta_\alpha T_\beta$, for some C^∞ functions U^β_α with $\sum_\beta \epsilon_\beta U^\lambda_\beta U^{\overline{\mu}}_{\overline{\beta}} = h^{\lambda\overline{\mu}}$, so that (4.5) is proved. □

To state the main result of Chapter 4, we need the Fefferman metric of (M, θ). Assume M to be strictly pseudoconvex and θ chosen such that the Levi form L_θ is positive definite. The Fefferman metric of (M, θ) is expressed by

$$F_\theta = \pi^* \tilde{G}_\theta + 2(\pi^*\theta) \odot \sigma, \tag{4.6}$$
$$\sigma = \frac{1}{n+2} \left\{ d\gamma + \pi^* \left(i\omega^\alpha_\alpha - \frac{i}{2} h^{\alpha\overline{\beta}} dh_{\alpha\overline{\beta}} - \frac{\rho}{4(n+1)} \theta \right) \right\};$$

cf. Chapter 2 of this book. Let (N, h) be a Riemannian manifold, where h denotes the Riemannian metric, and $\Phi : (C(M), F_\theta) \to (N, h)$ a C^∞ map.

Definition 4.4. The *energy* of Φ over a compact domain $D \subseteq C(M)$ is

$$\mathbf{E}(\Phi; D) = \frac{1}{2} \int_D \operatorname{trace}_{F_\theta} \left(\Phi^* h\right) d\mathrm{vol}(F_\theta),$$

where $d\mathrm{vol}(F_\theta)$ is the volume element of $(C(M), F_\theta)$. Then Φ is *harmonic* if for any compact domain $D \subseteq C(M)$, it is an extremal of the energy $\mathbf{E}(\,\cdot\,; D)$ with respect to all variations of Φ supported in D. □

Therefore, Φ is harmonic if and only if it satisfies the Euler–Lagrange equations

$$\Box \Phi^i + \sum_{p,q=1}^{2n+2} g^{pq} \left(\Gamma'^i_{jk} \circ \Phi\right) \frac{\partial \Phi^j}{\partial u^p} \frac{\partial \Phi^k}{\partial u^q} = 0, \tag{4.7}$$

for some local coordinate system (U, x^a) on M, respectively (V, y^i) on N (such that $\Phi(U) \subseteq V$), with $\Phi^i = y^i \circ \Phi$. See, e.g., [124] for an elementary introduction to the theory of harmonic maps and the first variation formula. Here $\Box$ is the wave operator (the Laplace–Beltrami operator associated with the Lorentz metric F_θ). One endows $C(M)$ with the induced local coordinate system $(\pi^{-1}(U), u^p)$, with $u^a = x^a \circ \pi$ and $u^{2n+2} = \gamma$. Also $[g^{pq}] = [g_{pq}]^{-1}$ and $g_{pq} = F_\theta(\partial_p, \partial_q)$, where ∂_p is short for $\partial / \partial u^p$.

Theorem 4.2. *Let M be a strictly pseudoconvex CR manifold and θ a contact form such that L_θ is positive definite. Let F_θ be the Fefferman metric of (M, θ) and (N, h) a Riemannian manifold. Then a C^∞ map $\phi : M \to N$ is pseudoharmonic, with respect to the data (θ, ∇^h), if and only if its vertical lift $\Phi = \phi \circ \pi : (C(M), F_\theta) \to (N, h)$ is a harmonic map.*

Proof. Let $\Phi : C(M) \to N$ be an S^1-invariant C^∞ map. Then Φ descends to a C^∞ map $\phi : M \to N$ (so that $\Phi = \phi \circ \pi$). Again by a result in Chapter 2, the wave operator $\Box$ pushes forward to a differential operator $\pi_* \Box : C^\infty(M) \to C^\infty(M)$ (given by $(\pi_* \Box) u = \Box(u \circ \pi)$, for any $u \in C^\infty(M)$) and $\pi_* \Box = \Delta_b$. Hence, the S^1-invariant map Φ is harmonic if and only if (by (4.7))

$$(\Delta_b \phi^i) \circ \pi + \sum_{a,c=1}^{2n+1} g^{ac} \left(\Gamma'^i_{jk} \circ \phi \circ \pi\right) \left(\frac{\partial \phi^j}{\partial x^a} \circ \pi\right) \left(\frac{\partial \phi^k}{\partial x^c} \circ \pi\right) = 0. \tag{4.8}$$

Let $\{T_\alpha\}$ be a local frame of $T_{1,0}(M)$ and $\{\theta^\alpha\}$ the corresponding admissible coframe. Then $\{\pi^* \theta^A, \sigma\}$ is a local frame of $T^*(C(M)) \otimes \mathbf{C}$. Let $\{V_A, \Sigma\}$ be dual to $\{\pi^* \theta^A, \sigma\}$. Relabel the variables x^a, $1 \leq a \leq 2n+1$, since x^A, $A \in \{0, 1, \dots, n, \overline{1}, \dots, \overline{n}\}$, where $x^{\overline{\alpha}} = x^{\alpha + n}$, $x^0 = x^{2n+1}$. Then $T_A = \lambda^B_A \partial / \partial x^B$, for some C^∞ functions $\lambda^B_A : U \to \mathbf{C}$. It follows that

$$V_A = (\lambda^B_A \circ \pi) \frac{\partial}{\partial u^B}, \quad \Sigma = (n+2) \frac{\partial}{\partial \gamma}.$$

Set $\mu = \lambda^{-1}$. Since $L_\theta = 2 h_{\alpha\overline{\beta}} \theta^\alpha \odot \theta^{\overline{\beta}}$ we get (by (4.6))

$$g = \begin{bmatrix} h_{\alpha\overline{\beta}}(\mu^\alpha_A \mu^{\overline{\beta}}_B + \mu^{\overline{\beta}}_A \mu^\alpha_B) & \frac{1}{n+2}\mu^0_A \\ \frac{1}{n+2}\mu^0_B & 0 \end{bmatrix}, \tag{4.9}$$

with respect to the frame $\{\partial/\partial u^A, \partial/\partial \gamma\}$. The inverse of (4.9) is denoted by

$$\begin{bmatrix} g^{AB} & g^{A,2n+2} \\ g^{2n+2,B} & g^{2n+2,2n+2} \end{bmatrix},$$

and a calculation shows that

$$g^{AB}\mu^\alpha_A \mu^{\overline{\beta}}_B = h^{\alpha\overline{\beta}}, \quad g^{AB}\mu^0_A = 0, \quad g^{AB}\mu^\alpha_A \mu^\beta_B = 0.$$

Consequently

$$g^{AB}\frac{\partial \phi^j}{\partial x^A}\frac{\partial \phi^k}{\partial x^B} = h^{\alpha\overline{\beta}}\{T_\alpha(\phi^j)T_{\overline{\beta}}(\phi^k) + T_{\overline{\beta}}(\phi^j)T_\alpha(\phi^k)\}. \tag{4.10}$$

Hence (4.8) yields (4.5). □

4.3 The variational approach

In this section, we first introduce an energy functional, similar to the Dirichlet energy functional in the theory of harmonic maps, by means of the following definition:

Definition 4.5. Let $\phi : M \to N$ be a C^∞ map of a compact strictly pseudoconvex CR manifold M into a Riemannian manifold (N, h). The *energy* of ϕ is given by

$$E(\phi) = \tfrac{1}{2}\int_M \text{trace}_{G_\theta}\left(\pi_H \phi^* h\right) \theta \wedge (d\theta)^n.$$

Here θ is a contact form on M with L_θ positive definite. □

Theorem 4.3. *The critical points of E are precisely the C^∞ maps that are pseudoharmonic with respect to the data (θ, ∇^h). Moreover, let $\{\phi_t\}_{|t|<\epsilon}$ be a smooth 1-parameter variation of ϕ $(\phi_0 = \phi)$ and set*

$$\Phi : (-\epsilon, \epsilon) \times M \to N, \quad \Phi(t, x) = \phi_t(x), \quad x \in M, \; |t| < \epsilon,$$

$$V \in \Gamma^\infty(\phi^{-1}TN), \quad V_x = (d_{(0,x)}\Phi)\frac{\partial}{\partial t}\bigg|_{(0,x)}.$$

Then

$$\frac{d}{dt}\{E(\phi_t)\}_{t=0} = -\int_M \tilde{h}(V, \tau(\phi; , \theta, \nabla^h))\,\theta \wedge (d\theta)^n. \tag{4.11}$$

Here $\tilde{h}$ is the Riemannian bundle metric induced by h in $\phi^{-1}TN$. Also $\tilde{\nabla} = \phi^{-1}\nabla^h$ and $\tilde{R}^h$ is the ϕ-tensor field induced by R^h (the curvature of ∇^h).

Proof. Let F_θ be the Fefferman metric of (M, θ). Set $\mathbf{E} = \mathbf{E}(\,\cdot\,; C(M))$ for simplicity. Note that

$$\int_{C(M)} (f \circ \pi) d\mathrm{vol}(F_\theta) = 2\pi \int_M f\, \theta \wedge (d\theta)^n$$

(integration along the fiber) for any $f \in C^\infty(M)$. Then (by (4.10))

$$\begin{aligned}
\mathbf{E}(\phi \circ \pi) &= \tfrac{1}{2} \int_{C(M)} \sum_{a,c=1}^{2n+1} g^{ac} \Big(\frac{\partial \phi^j}{\partial x^a} \circ \pi \Big) \Big(\frac{\partial \phi^k}{\partial x^c} \circ \pi \Big) (g'_{jk} \circ \phi \circ \pi) d\mathrm{vol}(F_\theta) \\
&= \int_{C(M)} (h^{\alpha\overline{\beta}} \circ \pi)\, (T_\alpha(\phi^j) \circ \pi)\, (T_{\overline{\beta}}(\phi^k) \circ \pi)\, (g'_{jk} \circ \phi)\, d\mathrm{vol}(F_\theta) \\
&= 2\pi \int_M h^{\alpha\overline{\beta}} T_\alpha(\phi^j) T_{\overline{\beta}}(\phi^k) (g'_{jk} \circ \phi)\, \theta \wedge (d\theta)^n,
\end{aligned}$$

where $g'_{jk} = h(\partial/\partial y^j, \partial/\partial y^k)$. On the other hand,

$$\mathrm{trace}_{G_\theta} \left(\pi_H \phi^* h \right) = 2 h^{\alpha\overline{\beta}} T_\alpha(\phi^j) T_{\overline{\beta}}(\phi^k) g'_{jk} \circ \phi,$$

hence

$$\mathbf{E}(\phi \circ \pi) = 2\pi E(\phi). \tag{4.12}$$

Let $\phi : M \to N$ be pseudoharmonic, with respect to the data (θ, ∇^h). Then (by Theorem 4.2) $\phi \circ \pi$ is a critical point of $\mathbf{E}$. If ϕ_t is a 1-parameter variation of ϕ, then $\phi_t \circ \pi$ is a 1-parameter variation of $\phi \circ \pi$, and (by (4.12)) it follows that ϕ is a critical point of E. To prove the converse, one may no longer use (4.12) (since vertical lifts of 1-parameter variations do not lead to arbitrary 1-parameter variations of $\phi \circ \pi$). However, if ϕ is a critical point of E then (by the first variation formula (4.11)) $\tau(\phi; \theta, \nabla^h) = 0$. □

Let us prove (4.11). Let $\Phi^{-1} TN \to (-\epsilon, \epsilon) \times M$ be the pullback of $T(N)$ by Φ. Let $\{X_j : 1 \leq j \leq 2n\}$ be a local G_θ-orthonormal frame of $H(M)$. The product manifold $(-\epsilon, \epsilon) \times M$ is endowed with the Riemannian metric $dt \otimes dt + g_\theta$. Of course, one may think of X_j as (orthonormal) vector fields tangent to $(-\epsilon, \epsilon) \times M$. Then

$$\frac{d}{dt} \mathrm{trace}_{G_\theta} \left\{ \pi_H \phi_t^* h \right\} = \frac{\partial}{\partial t} \sum_{j=1}^{2n} \left(\Phi^* h \right) (X_j, X_j) = 2 \sum_{j=1}^{2n} \tilde{h}(\tilde{\nabla}_{\partial/\partial t} (d\Phi) X_j, (d\Phi) X_j),$$

where $\tilde{\nabla} = \Phi^{-1} \nabla^h$ (and $\tilde{h}$ is induced by h in $\Phi^{-1} TN$). Moreover (since ∇^h is torsion-free),

$$\begin{aligned}
\frac{d}{dt} \mathrm{trace}_{G_\theta} \left\{ \pi_H \phi_t^* h \right\} &= 2 \sum_{j=1}^{2n} \tilde{h}(\tilde{\nabla}_{X_j} (d\Phi) \frac{\partial}{\partial t}, (d\Phi) X_j) \\
&= 2 \sum_{j=1}^{2n} \Big[X_j (\tilde{h}(d\Phi) \frac{\partial}{\partial t}, (d\Phi) X_j)) - \tilde{h}((d\Phi) \frac{\partial}{\partial t}, \tilde{\nabla}_{X_j} (d\Phi) X_j) \Big].
\end{aligned}$$

Let $X_t \in \Gamma^\infty(H(M))$ be defined by

$$G_\theta(X_t, Y)_x = \tilde{h}((d\Phi)\frac{\partial}{\partial t}, (d\Phi)Y)_{(t,x)}, \quad x \in M, \quad |t| < \epsilon,$$

for any $Y \in \Gamma^\infty(H(M))$. Then

$$\frac{d}{dt}\text{trace}_{G_\theta}\left\{\pi_H \phi_t^* h\right\} = 2\sum_{j=1}^{2n}\left\{X_j(G_\theta(X_t, X_j)) - \tilde{h}(d\Phi)(\frac{\partial}{\partial t}, \tilde{\nabla}_{X_j}(d\Phi)X_j)\right\}$$

$$= 2\sum_{j=1}^{2n}\left\{G_\theta(\nabla_{X_j}X_t, X_j) + G_\theta(X_t, \nabla_{X_j}X_j) - \tilde{h}((d\Phi)\frac{\partial}{\partial t}, \tilde{\nabla}_{X_j}(d\Phi)X_j)\right\}$$

by $\nabla g_\theta = 0$. Let us compute the divergence of X_t with respect to the volume form Ψ. We have

$$\text{div}(X_t) = \sum_{j=1}^{2n} G_\theta(\nabla_{X_j}X_t, X_j).$$

Finally, let us integrate over M in

$$\frac{d}{dt}\text{trace}_{G_\theta}\left\{\pi_H \phi_t^* h\right\} = 2\,\text{div}(X_t) - 2\sum_{j=1}^{2n}\tilde{h}((d\Phi)\frac{\partial}{\partial t}, \tilde{\nabla}_{X_j}(d\Phi)X_j - (d\Phi)\nabla_{X_j}X_j),$$

and set $t = 0$ in the resulting identity. This leads to (4.11). □

Theorem 4.4. *Under the hypothesis of Theorem 4.3, let $\{\phi_{s,t}\}_{-\epsilon<s,t<\epsilon}$ be a smooth 2-parameter variation of $\phi : M \to N$ $(\phi_{0,0} = \phi)$ and let us set*

$$\Phi : (-\epsilon, \epsilon) \times (-\epsilon, \epsilon) \times M \to N,$$

$$\Phi(s, t, x) = \phi_{s,t}(x), \quad x \in M, \quad -\epsilon < s, t < \epsilon,$$

$$V_x = (d_{(0,0,x)}\Phi)\left.\frac{\partial}{\partial t}\right|_{(0,0,x)}, \qquad W_x = (d_{(0,0,x)}\Phi)\left.\frac{\partial}{\partial s}\right|_{(0,0,x)}, \qquad x \in M.$$

Then, for any pseudoharmonic (with respect to the data (θ, ∇^h)) map $\phi : M \to N$,

$$\frac{\partial^2}{\partial s \partial t}\left\{E(\phi_{s,t})\right\}_{s=t=0} =$$
$$\int_M \left\{\langle\tilde{\nabla}V, \tilde{\nabla}W\rangle - \sum_{j=1}^{2n}\tilde{h}(\tilde{R}^h(V, (d\phi)X_j)(d\phi)X_j, W)\right\}\theta \wedge (d\theta)^n, \quad (4.13)$$

for any G_θ-orthonormal (local) frame $\{X_j\}$ in $H(M)$.

Proof. We have

$$\frac{\partial^2}{\partial s \partial t}\left\{E(\phi_{s,t})\right\}_{s=t=0} = \tfrac{1}{2}\frac{\partial}{\partial s}\left\{\int_M \frac{\partial}{\partial t}\sum_{j=1}^{2n}\left(\Phi^* h\right)(X_j, X_j)\,\theta\wedge(d\theta)^n\right\}_{s=t=0}$$

$$= \frac{\partial}{\partial s}\left\{\int_M \left[X_j((d\Phi)\frac{\partial}{\partial t}, (d\Phi)X_j)) - \tilde{h}((d\Phi)\frac{\partial}{\partial t}, \tilde{\nabla}_{X_j}(d\Phi)X_j)\right]\theta\wedge(d\theta)^n\right\}_{s=t=0}.$$

Let $X_{s,t} \in \Gamma^\infty(H(M))$ be defined by

$$G_\theta(X_{s,t}, Y) = \tilde{h}((d\Phi)\frac{\partial}{\partial t}, (d\Phi)Y)\circ\alpha_{s,t},$$
$$\alpha_{s,t} : M \to (-\epsilon, \epsilon)^2 \times M,\ \ \alpha_{s,t}(x) = (s,t,x),\ \ x \in M,$$

for any $Y \in \Gamma^\infty(H(M))$. Let us set

$$\beta(X, Y) = \tilde{\nabla}_X(d\Phi)Y - (d\Phi)\nabla_X Y.$$

Then

$$\frac{\partial^2}{\partial s \partial t}\left\{E(\phi_{s,t})\right\}_{s=t=0}$$
$$= \frac{\partial}{\partial s}\left\{\int_M \sum_{j=1}^{2n}\left[X_j(G_\theta(X_{s,t}, X_j)) - \tilde{h}\Big((d\Phi)\frac{\partial}{\partial t}, \tilde{\nabla}_{X_j}(d\Phi)X_j\Big)\right]\Psi\right\}_{s=t=0}$$
$$= \frac{\partial}{\partial s}\left\{\int_M \left[\operatorname{div}(X_{s,t}) - \tilde{h}\Big((d\Phi)\frac{\partial}{\partial t}, \sum_{j=1}^{2n}\beta(X_j, X_j)\Big)\right]\Psi\right\}_{s=t=0}$$
$$= -\left\{\int_M \tilde{h}\Big(\tilde{\nabla}_{\partial/\partial s}(d\Phi)\frac{\partial}{\partial t}, \sum_{j=1}^{2n}\beta(X_j, X_j)\Big)\Psi\right\}_{s=t=0}$$
$$\quad - \left\{\int_M \tilde{h}\Big((d\Phi)\frac{\partial}{\partial t}, \tilde{\nabla}_{\partial/\partial s}\sum_{j=1}^{2n}\beta(X_j, X_j)\Big)\Psi\right\}_{s=t=0}$$
$$= -\int_M \tilde{h}\Big(\big(\tilde{\nabla}_{\partial/\partial s}(d\Phi)\frac{\partial}{\partial t}\big)_{s=t=0}, \tau(\phi; \theta, \nabla^h)\Big)\Psi$$
$$\quad - \left\{\int_M \tilde{h}\Big((d\Phi)\frac{\partial}{\partial t}, \sum_{j=1}^{2n}\left[\tilde{\nabla}_{\partial/\partial s}\tilde{\nabla}_{X_j}(d\Phi)X_j - \tilde{\nabla}_{\partial/\partial s}(d\Phi)\nabla_{X_j}X_j\right]\Big)\Psi\right\}_{s=t=0}.$$

Note that

$$\tilde{\nabla}_{\partial/\partial s}\tilde{\nabla}_{X_j}(d\Phi)X_j = \tilde{\nabla}_{X_j}\tilde{\nabla}_{\partial/\partial s}(d\Phi)X_j + \tilde{R}(\frac{\partial}{\partial s}, X_j)(d\Phi)X_j,$$

since $[\partial/\partial s, X_j] = 0$. Therefore, if ϕ is pseudoharmonic,

$$\frac{\partial^2}{\partial s \partial t}\left\{E(\phi_{s,t})\right\}_{s=t=0} = -\int_M \tilde{h}\Big((d\Phi)\frac{\partial}{\partial t}, \sum_{j=1}^{2n}\Big[\tilde{\nabla}_{X_j}\tilde{\nabla}_{\partial/\partial s}(d\Phi)X_j$$

$$+ \tilde{R}\Big(\frac{\partial}{\partial s}, X_j\Big)(d\Phi)X_j - \tilde{\nabla}_{\partial/\partial s}(d\Phi)\nabla_{X_j}X_j\Big]\Big)_{s=t=0} \Psi$$

$$= -\int_M \sum_{j=1}^{2n}\Big[X_j\Big(\tilde{h}\Big((d\Phi)\frac{\partial}{\partial t}, \tilde{\nabla}_{X_j}(d\Phi)\frac{\partial}{\partial s}\Big)\Big) - \tilde{h}\Big(\tilde{\nabla}_{X_j}(d\Phi)\frac{\partial}{\partial t}, \tilde{\nabla}_{X_j}(d\Phi)\frac{\partial}{\partial s}\Big)$$

$$+ \tilde{h}\Big(\tilde{R}\Big(\frac{\partial}{\partial s}, X_j\Big)(d\Phi)X_j - \tilde{\nabla}_{\partial/\partial s}(d\Phi)\nabla_{X_j}X_j, (d\Phi)\frac{\partial}{\partial t}\Big)\Big]_{s=t=0} \Psi.$$

Let $X_\phi \in \Gamma^\infty(H(M))$ be defined by

$$G_\theta(X_\phi, Y) = \tilde{h}(\tilde{\nabla}_Y W, V),$$

for any $Y \in \Gamma^\infty(H(M))$. Then

$$\sum_{j=1}^{2n}\Big[X_j\Big(\tilde{h}\Big((d\Phi)\frac{\partial}{\partial t}, \tilde{\nabla}_{X_j}(d\Phi)\frac{\partial}{\partial s}\Big)\Big) - \tilde{h}\Big(\tilde{\nabla}_{\partial/\partial s}(d\Phi)\nabla_{X_j}X_j, (d\Phi)\frac{\partial}{\partial t}\Big)\Big]_{s=t=0}$$

$$= \sum_{j=1}^{2n}\big[X_j(G_\theta(X_\phi, X_j)) - G_\theta(X_\phi, \nabla_{X_j}X_j)\big] = \sum_{j=1}^{2n} G_\theta(\nabla_{X_j}X_\phi, X_j) = \operatorname{div}(X_\phi)$$

and (by Green's lemma) (4.13) is proved. □

As a consequence of (4.13) we get an immediate pseudo-Hermitian analogue of the notion of stability (of a harmonic map between Riemannian manifolds).

Definition 4.6. Given a pseudoharmonic map $\phi : M \to N$, we set

$$I(V, V) = \int_M \Big\{\|\tilde{\nabla}V\|^2 - \sum_{j=1}^{2n}\tilde{h}\big(\tilde{R}^h(V, (d\phi)X_j)(d\phi)X_j, V\big)\Big\}\, \theta \wedge (d\theta)^n,$$

and call ϕ *stable* if $I(V, V) \geq 0$ for any $V \in \Gamma^\infty(\phi^{-1}TN)$. □

Then, as well as in Riemannian geometry, we have the following result:

Proposition 4.2. *Any pseudoharmonic map, of a strictly pseudoconvex CR manifold into a Riemannian manifold of nonpositive sectional curvature, is stable.*

Next, our purpose in this section is to prove Theorem 4.5 below. To this end, we need some preparation. Let $j : N \subset (M^{m+1}(c), \overline{h})$ be a real hypersurface, and let $h = j^*\overline{h}$ be its first fundamental form. Let $\tan_x$ and nor_x be the canonical projections associated with the direct sum decomposition

$$T_x(M^{m+1}(c)) = T_x(N) \oplus T_x(N)^\perp, \quad x \in N.$$

Let ξ be a unit normal field on N. Then $\mathrm{nor}(V) = \overline{h}(V, \xi)\xi$, for any $V \in \mathcal{X}(M^{m+1}(c))$. We shall need the Gauss and Weingarten equations (of the immersion j)

$$\overline{\nabla}_X Y = \nabla^N_X Y + B(X, Y), \tag{4.14}$$

$$\overline{\nabla}_X \eta = -A_\eta X + \nabla^\perp_X \eta, \tag{4.15}$$

where ∇^N, B, A_η, and $\nabla^\perp$ are respectively the induced connection, the second fundamental form (of j), the Weingarten operator (associated with the normal section η), and the normal connection ($\nabla^\perp \xi = 0$). Next, let $\{X_j : 1 \leq j \leq 2n\}$ be a local G_θ-orthonormal frame in $H(M)$, defined on an open subset $U \subseteq M$.

It is well known that any nonconstant harmonic map of a Riemannian manifold into a sphere is unstable. This result carries over easily to the case of pseudoharmonic maps.

Theorem 4.5. *Let M be a strictly pseudoconvex, compact CR manifold and N an orientable totally umbilical real hypersurface, of mean curvature vector H, regularly embedded in a real space form $M^{m+1}(c)$. Let $V \subseteq M^{m+1}(c)$ be a simple and convex*[1] *open subset, $N \cap V \neq \emptyset$. If $(m-2)\|H\|^2 + (m-1)c > 0$ then any nonconstant pseudoharmonic (with respect to the data (θ, ∇^N)) map $\phi : M \to N \cap V$ is unstable.*

Proof. Let $\phi : M \to N \cap V$ be a pseudoharmonic map, with respect to the data (θ, ∇^N). Let $\{V_a : 1 \leq a \leq m+1\}$ be a parallel (i.e., $\overline{\nabla} V_a = 0$) local $\overline{h}$-orthonormal frame on V. To build such a frame one merely starts with an orthonormal frame $\{v_a\} \subset T_p(M^{m+1}(c))$ at some point $p \in V$ and considers the vector field V_a obtained by parallel translation of v_a along geodesics issuing at p. We wish to compute

$$I(\tan(V_a), \tan(V_a)) =$$
$$\sum_{j=1}^{2n} \int_M \left\{ \|\tilde{\nabla}_{X_j} \tan(V_a)\|^2 - \tilde{h}(\tilde{R}^N(\tan(V_a), (d\phi)X_j)(d\phi)X_j, \tan(V_a)) \right\} \theta \wedge (d\theta)^n,$$

where $\tilde{\nabla} = \phi^{-1}\nabla^N$. Then (by (4.14)–(4.15))

$$\begin{aligned} \tilde{\nabla}_{X_j} \tan(V_a) &= \nabla^N_{(d\phi)X_j} \tan(V_a) = \tan\left(\overline{\nabla}_{(d\phi)X_j} \tan(V_a)\right) \\ &= \tan\left(\overline{\nabla}_{(d\phi)X_j}(V_a - \mathrm{nor}(V_a))\right) = -\tan\left(\overline{\nabla}_{(d\phi)X_j} \mathrm{nor}(V_a)\right) \\ &= A_{\mathrm{nor}(V_a)}(d\phi)X_j. \end{aligned}$$

Using $\|X\|^2 = \sum_{a=1}^{m+1} \overline{h}(X, V_a)^2$ we may compute

$$\|\tilde{\nabla}_{X_j} \tan(V_a)\|^2 = \|A_{\mathrm{nor}(V_a)}(d\phi)X_j\|^2 = \sum_{b=1}^{m+1} \overline{h}(A_{\mathrm{nor}(V_a)}(d\phi)X_j, V_b)^2$$

$$= \sum_{b=1}^{m+1} (A_{nor(V_a)}(d\phi)X_j, \tan(V_b)) = \sum_{b=1}^{m+1} \overline{h}(B((d\phi)X_j, \tan(V_b)), \mathrm{nor}(V_a))^2.$$

[1] In the sense of [241], Vol. I, p. 149.

Assume from now on that N is totally umbilical in $M^{m+1}(c)$, i.e., $B = h \otimes H$ (in particular $\|H\| = \text{const}$, by the Codazzi equation). Then

$$\|\tilde{\nabla}_{X_j} \tan(V_a)\|^2 = \sum_{b=1}^{m+1} h((d\phi)X_j, \tan(V_b))^2 \, \overline{h}(H, \text{nor}(V_a))^2$$
$$= \|(d\phi)X_j\|^2 \, \overline{h}(H, \text{nor}(V_a))^2.$$

For any normal section η one has $\|\eta\|^2 = \overline{h}(\eta, \xi)^2$. Thus

$$\overline{h}(H, \text{nor}(V_a)) = \|\text{nor}(V_a)\|^2 \, \|H\|^2,$$

and the last identity becomes

$$\|\tilde{\nabla}_{X_j} \tan(V_a)\|^2 = \|(d\phi)X_j\|^2 \, \|\text{nor}(V_a)\|^2 \, \|H\|^2;$$

hence

$$\sum_{j=1}^{2n} \|\tilde{\nabla}_{X_j} \tan(V_a)\|^2 = \text{trace}_{G_\theta}\{\pi_H \phi^* h\} \, \|\text{nor}(V_a)\|^2 \, \|H\|^2.$$

Since $\|\xi\| = 1$ and $\overline{h}(V_a, V_b) = \delta_{ab}$, it follows that $\sum_{a=1}^{m+1} \|\text{nor}(V_a)\|^2 = 1$. Therefore

$$\sum_{a=1}^{m+1} \sum_{j=1}^{2n} \|\tilde{\nabla}_{X_j} \tan(V_a)\|^2 = \|H\|^2 \, \text{trace}_{G_\theta}\{\pi_H \phi^* h\}.$$

The Gauss equation (cf., e.g., [91])

$$R^N(X, Y)Z = (c + \|H\|^2)\{h(Y, Z)X - h(X, Z)Y\}$$

leads to

$$\tilde{h}(\tilde{R}^N(\tan(V_a), (d\phi)X_j)(d\phi)X_j, \tan(V_a))$$
$$= (c + \|H\|^2)\{\|(d\phi)X_j\|^2 \, \|\tan(V_a)\|^2 - \tilde{h}(\tan(V_a), (d\phi)X_j)^2\}.$$

Next $\|\tan(V_a)\|^2 + \|\text{nor}(V_a)\|^2 = 1$ yields $\sum_{a=1}^{m+1} \|\tan(V_a)\|^2 = m$; hence

$$\sum_{j=1}^{2n} \tilde{h}(\tilde{R}^N(\tan(V_a), (d\phi)X_j)(d\phi)X_j, \tan(V_a))$$
$$= (c + \|H\|^2)\{\|\tan(V_a)\|^2 \, \text{trace}_{G_\theta}\Big\{\pi_H \phi^* h\} - \sum_{j=1}^{2n} \overline{h}((d\phi)X_j, V_a)^2\Big\},$$

or

$$\sum_{a=1}^{m+1}\sum_{j=1}^{2n} \tilde{h}(\tilde{R}^N(\tan(V_a), (d\phi)X_j)(d\phi)X_j, \tan(V_a))$$
$$= (c + \|H\|^2)\Big\{m \operatorname{trace}_{G_\theta}\{\pi_H\phi^*h\} - \sum_{j=1}^{2n}\|(d\phi)X_j\|^2\Big\}$$
$$= (m-1)(c + \|H\|^2)\operatorname{trace}_{G_\theta}\{\pi_H\phi^*h\}.$$

Summing up

$$\sum_{a=1}^{m+1} I(\tan(V_a), \tan(V_a)) = -2\Big[(m-2)\|H\|^2 + (m-1)c\Big]\ E(\phi) < 0.$$

Theorem 4.5 is proved. □

Pseudoharmonic maps are solutions to the nonlinear subelliptic PDE system (4.5). Other subelliptic PDEs, such as the the CR Yamabe equation, were successfully studied by exploiting results in harmonic analysis on the Heisenberg group; cf. Chapter 3 of this book. Let us also remark that as well as (4.5), the CR Yamabe equation is the projection on M of a nonelliptic PDE (whose principal part is the wave operator on $C(M)$).

Let $\phi : M \to N$ be a pseudoharmonic map, with respect to the data (θ, ∇^h). Let $\{X_p\} = \{X_\alpha, JX_\alpha\}$ be a local orthonormal frame of $H(M)$, defined on a coordinate neighborhood $(\tilde{U}, \varphi)$ in M, and (V, y^i) a local coordinate system on N such that $\phi(\tilde{U}) \subseteq V$. Since $\Delta_b = H$ the equations (4.5)) may be written

$$H\phi^i + \sum_{p=1}^{2n} X_p(\phi^j)X_p(\phi^k)({\Gamma'}^i_{jk} \circ \phi) = 0,$$

where $H = -\sum_{p=1}^{2n} X_p^* X_p$ is the Hörmander operator (cf. Chapter 2 of this book). We wish to emphasize the formal analogy between $\phi \circ \varphi^{-1} : \overline{\Omega} \to V$ and a subelliptic harmonic map, in the sense of J. Jost and C.J. Xu [234] ($\Omega = \varphi(U)$, $U \subset\subset \tilde{U}$). Nevertheless, two differences occur. First, J. Jost and C.-J. Xu discuss (cf. op. cit.) Hörmander systems of vector fields $\{X_p\}$ defined on an open set in $\mathbf{R}^N$ (the dimension N is arbitrary, while $N = 2n + 1$ in CR geometry). Second, the formal adjoints of the X_p's, and therefore the Hörmander operator, are built with respect to the Euclidean metric of $\mathbf{R}^N$ (while we use the volume form Ψ arising from the given contact structure). See also Z.R. Zhou [449]. On the other hand, there is an increasing literature regarding subelliptic equations on domains in $\mathbf{R}^N$ (cf., e.g., [230], [340], [362]–[363], [441], and [443]). Encouraged by the progress there we may state the following **conjecture** (whose Riemannian counterpart is a result by J. Eells and M.J. Ferreira [131]): *Let M be a nondegenerate CR manifold, θ a contact form on M, and (N, h) a Riemannian manifold. Let $\mathcal{H}$ be a homotopy class of C^∞ maps $\phi : M \to N$. Then there are $u \in C^\infty(M)$ and $\phi \in \mathcal{H}$ such that ϕ is pseudoharmonic with respect to the data* $(\exp(u)\,\theta, \nabla^h)$. □

We end this section by looking at some examples of (globally defined) pseudoharmonic maps and by stating a few related open problems.

Example 3. Let M be a nondegenerate CR manifold. For a C^∞ map $\Phi : M \to (\mathbf{R}^m, h_0)$,

$$\tau(\Phi; \theta, D^0) = \left(\Delta_b \Phi^i\right) Y_i$$

(where D^0 is the Levi-Civita connection of the Euclidean metric h_0).

Proposition 4.3. *Φ is pseudoharmonic (with respect to (θ, D^0)) if and only if Φ^i are harmonics of the sub-Laplacian.*

Example 4. Let M be a strictly pseudoconvex CR manifold, θ a contact form on M with L_θ positive definite, and g_θ the Webster metric of (M, θ). Let $\phi : M \to (N, h)$ be an isometric ($\phi^* h = g_\theta$) immersion of (M, g_θ) into (N, h). Then

$$\tau(\phi; \theta, \nabla^h) = (2n + 1) H(\phi) - \alpha(\phi)(T, T), \tag{4.16}$$

where $\dim(M) = 2n + 1$ and $H(\phi) = \frac{1}{2n+1} \text{trace}_{g_\theta} \alpha(\phi)$ is the mean curvature vector of ϕ.

Proposition 4.4. *Any pseudo-Hermitian immersion of (M, θ) into a strictly pseudoconvex CR manifold (N, θ') is pseudoharmonic with respect to (θ, ∇^h), with $h := g_{\theta'}$ the Webster metric of (N, θ').*

This follows from the description of pseudo-Hermitian immersions performed in detail in Chapter 6. We anticipate a few notions and results that enable us to give a proof of Proposition 4.4. Let $M, \ N$ be two CR manifolds. A *CR immersion* $\phi : M \to N$ is an immersion and a CR map (cf. [424]). Let $\theta, \ \theta'$ be contact forms on $M, \ N$, respectively. A CR map $\phi : M \to N$ is *isopseudo-Hermitian* if $\phi^* \theta' = \theta$ (cf. [219]). Let T' be the characteristic direction of $d\theta'$. A *pseudo-Hermitian immersion* is an isopseudo-Hermitian CR immersion $\phi : M \to N$ such that T' is normal to $\phi(M)$ (cf. [120] [36]). See [120], p. 190, and Theorem 12, p. 196, for examples of pseudo-Hermitian immersions. Let us prove Proposition 4.4. Given $X, Y \in \mathcal{X}(M)$ one has

$$\beta(\phi)(X, Y) = \nabla^h_{\phi_* X} \phi_* Y - \phi_* \nabla_X Y.$$

On the other hand, the Levi-Civita connection ∇^{g_θ} of the Webster metric and the Tanaka–Webster connection ∇ are related (cf. Chapter 1 of this book) by

$$\nabla^{g_\theta} = \nabla - (A + d\theta) \otimes T + \tau \otimes \theta + 2\theta \odot J,$$

where $A(X, Y) = g_\theta(X, \tau Y)$. Using the Gauss formula

$$\nabla^h_{\phi_* X} \phi_* Y = \phi_* \nabla^{g_\theta}_X Y + \alpha(\phi)(X, Y)$$

we obtain

$$\beta(\phi)(X,Y) = \alpha(\phi)(X,Y) - \{A(X,Y) + (d\theta)(X,Y)\}\phi_* T + \theta(Y)\phi_*\tau X + \theta(X)\phi_* JY + \theta(Y)\phi_* JX. \quad (4.17)$$

Let us restrict (4.17) to $H(M) \otimes H(M)$ and take traces (with respect to G_θ); then $\text{trace}_{g_\theta} \tau = 0$ and $\tau(T) = 0$ yield (4.16). □

Assume $h = g_{\theta'}$ and let ϕ be a pseudo-Hermitian immersion. Then

$$\alpha(\phi)(T,T) = \nabla^{g_{\theta'}}_{T'} T' - \phi_* \nabla^{g_\theta}_T T = 0$$

and (by Theorem 7 in [120], p. 189) $H(\phi) = 0$ hence $\tau(\phi; \theta, \nabla^{g_{\theta'}}) = 0$. □

Example 5. Let $\phi : (M, g_\theta) \to (S^m(r), h)$ be an isometric immersion, where $h := \iota^* h_0$ and $\iota : S^m(r) \subset \mathbf{R}^{m+1}$. Set $\Phi := \iota \circ \phi : M \to \mathbf{R}^{m+1}$. Then we have the following result:

Proposition 4.5. *ϕ is pseudoharmonic with respect to (θ, ∇^h) if and only if*

$$\Delta_b \Phi = \lambda \Phi + \alpha(\phi)(T,T) + T^2(\Phi),$$

for some $\lambda \in C^\infty(M)$. If additionally $m = 2\ell + 1$ and $\phi_* T = T_0$ [the characteristic direction of $(S^{2\ell+1}(r), \frac{\sqrt{-1}}{2}\iota^*(\bar\partial - \partial)|z|^2)$] then ϕ is pseudoharmonic if and only if

$$\Delta_b \Phi = -\frac{2(n+1)}{r^2}\Phi,$$

i.e., Φ^i are eigenfunctions of the sub-Laplacian corresponding to the eigenvalue $2(n+1)r^{-2}$.

To prove this statement, note that (by Theorem 10.1 in [124], p. 84)

$$H(\phi) = \frac{1}{2n+1} \tan_{S^m(r)} \Delta\Phi;$$

hence (by (4.16)) ϕ is pseudoharmonic (with respect to (θ, ∇^h)) if and only if

$$\tan_{S^m(r)} \Delta\Phi = \alpha(\phi)(T,T),$$

i.e.,

$$\Delta\Phi = \alpha(\phi)(T,T) + \lambda\Phi,$$

for some $\lambda \in C^\infty(M)$, or, by a formula in [186] [cf., e.g., (68) in [120], p. 194, i.e., $\Delta_b = \Delta + T^2$ on $C^\infty(M)$ (*Greenleaf's formula*)]

$$\Delta_b \Phi = \alpha(\phi)(T,T) + \lambda\Phi + T^2(\Phi).$$

Assume from now on that $m = 2\ell + 1$ and $\phi_* T = T_0$. Then

$$\alpha(\phi)(T,T) = \nabla^h_{T_0} T_0 - \phi_* \nabla^{g_\theta}_T T = 0.$$

Also

$$T_0 = \frac{1}{r} J_0 \Big(x^j \frac{\partial}{\partial x^j} + y^j \frac{\partial}{\partial y^j} \Big) = \frac{1}{r} \Big(x^j \frac{\partial}{\partial y^j} - y^j \frac{\partial}{\partial x^j} \Big),$$

so that $\sum_{A=1}^{2\ell+2} T(\Phi^A)\partial/\partial x^A = T_0$ yields

$$T(\Phi^j) = -\frac{1}{r}\Phi^{j+\ell+1}, \quad T(\Phi^{j+\ell+1}) = \frac{1}{r}\Phi^j;$$

hence

$$T^2(\Phi) = -r^{-2}\Phi.$$

Then ϕ is pseudoharmonic if and only if $\Delta_b \Phi = (\lambda - r^{-2})\Phi$, for some $\lambda \in C^\infty(M)$. Yet (by $\Delta\Phi = \lambda\Phi$ and the proof of Theorem 10.2 in [124], p. 86) $\lambda = -(2n+1)r^{-2}$. □

Example 6. By Example 5, pseudoharmonicity is related to the spectrum of Δ_b [very much as minimality in submanifold theory is related to the spectrum of the Laplacian on a Riemannian manifold; cf. T. Takahashi [395], whose work led to the (still growing) theory of submanifolds of finite type, cf., e.g., [92]]. Since Δ_b is subelliptic, by a result of A. Menikoff and J. Söstrand [300], Δ_b has a discrete spectrum tending to $+\infty$. Since explicit minimal immersions among spheres may be built (cf., e.g., Theorem 10.4 in [124], p. 92) by using orthonormal systems of eigenfunctions of the Laplacian, one may ask whether eigenfunctions of Δ_b may be produced in any effective way. For instance, if $M = S^{2n+1}$ then $\Delta v = -k(k+2n)v$, where $v = H|_{S^{2n+1}}$ with $H \in \mathcal{H}_k$ (the space of all harmonic polynomials $H : \mathbf{R}^{2n+2} \to \mathbf{R}$, homogeneous of degree k), and all eigenfunctions of Δ are obtained in this way (cf. [59], p. 160–162). Let $H \in \mathcal{H}_2$, i.e., $H = \sum_{i,j=1}^{n+1}(a_{ij}x_i x_j + b_{ij}x_i y_j + c_{ij}y_i y_j)$ with $\sum_{i=1}^{n+1}(a_{ii}+c_{ii}) = 0$. Set $T_0 = \sum_{j=1}^{n+1}\big(x^j\partial/\partial x^j - y^j\partial/\partial x^j\big)$ (hence $\iota_* T = T_0$, where T is the standard contact vector of S^{2n+1}). Then

$$H = \sum_{i=1}^{n+1} a_{ii}(x_i^2 + y_i^2) + \sum_{i<j}(a_{ij}+a_{ji})(x_i x_j + y_i y_j) + \sum_{i,j=1}^{n+1} b_{ij}x_i y_j, \tag{4.18}$$

$$\sum_{i=1}^{n+1} a_{ii} = 0, \quad b_{ij} = -b_{ji},$$

are all harmonic polynomials $H \in \mathcal{H}_2$ satisfying $T_0^2(H) = 0$. Set

$$\mathcal{P}_0 := \mathrm{Eigen}(\Delta; 4(n+1)) \cap \mathrm{Ker}(T^2).$$

By the calculation above, $\mathcal{P}_0 = \{H|_{S^{2n+1}} : H \text{ given by (4.18)}\}$. Also $\mathcal{P}_0 \subseteq \mathrm{Eigen}(\Delta_b; 4(n+1))$, by Greenleaf's formula. See [59], p. 160. The restriction to the sphere $C^\infty(\mathbf{R}^{2n+2}) \to C^\infty(S^{2n+1})$ descends to an isomorphism $\mathcal{H}_2 \to \mathrm{Eigen}(\Delta; 4(n+1))$; hence $\dim_{\mathbf{R}} \mathcal{P}_0 = n(n+2)$. More generally, set

$$\mathcal{P}_\lambda := \text{Eigen}(\Delta; 4(n+1)) \cap \text{Ker}(T^2 - \lambda I),$$
$$N_\lambda := \dim_{\mathbf{R}} \mathcal{P}_\lambda, \quad \lambda \in \mathbf{R} \setminus \{-2\}.$$

(For $\lambda = -2$ one has $\mathcal{H}_2 \cap \text{Ker}(T_0^2 - \lambda I) = (0)$.) Then

$$\mathcal{P}_\lambda \subseteq \text{Eigen}(\Delta_b; 4(n+1) - \lambda)$$

and

$$N_\lambda = \begin{cases} \dfrac{n(n+1)}{2}, & \lambda \in \mathbf{R} \setminus \{-4, -2, 0\}, \\ (n+1)(n+2), & \lambda = -4, \\ n(n+2), & \lambda = 0. \end{cases}$$

Let $\{v_1, \ldots, v_{n(n+2)}\}$ be a basis of $\mathcal{P}_0$ and set $\Phi = (v_1, \ldots, v_{n(n+2)}) : S^{2n+1} \to \mathbf{R}^{n(n+2)}$. Then $\Delta_b \Phi = -4(n+1)\Phi$. *If* Φ *is an isometric immersion* [of (S^{2n+1}, g_θ) into $(\mathbf{R}^{n(n+2)}, h_0)$] *then* (by Theorem 6.2 in [124], p. 45, Greenleaf's formula, and $T^2(\Phi) = 0$) $\Delta_b \Phi = (2n+1) H(\Phi)$; *hence* Φ *is normal to* S^{2n+1}. Therefore $F(x) := \|\Phi(x)\|^2_{\mathbf{R}^{n(n+2)}}$ is a constant function, i.e., $F(x) = c = \text{const}$, $x \in S^{2n+1}$, i.e., $\Phi(S^{2n+1}) \subset S^{n(n+2)-1}(\sqrt{c})$. In view of Theorem 10.4 in [124], p. 92 (cf. also [395]), it is an open question whether $S^{2n+1} \to S^{n(n+2)-1}(\sqrt{c})$ may be adjusted into a pseudoharmonic map.

We close with a few open problems. A smooth curve $\gamma : I \to M$ in a nondegenerate CR manifold, defined on an open interval I containing the origin, is a *parabolic geodesic* of M if (i) $\dot\gamma(0) \in H(M)_{\gamma(0)}$ and (ii) $(\nabla_{\dot\gamma}\dot\gamma)_{\gamma(t)} = 2cT_{\gamma(t)}$, $t \in I$, for some $c \in \mathbf{R}$ (cf. [228]; see also [120], pp. 191–192). *Compute the form* $\beta(\phi)$ *of a map* $\phi : M \to N$ *carrying parabolic geodesics to parabolic geodesics.*

Let X be a unit ($g_\theta(X, X) = 1$) vector field on a strictly pseudoconvex CR manifold M. Then X is a map of M into $U(M)$, the (total space of the) tangent sphere bundle over (M, g_θ). $U(M)$ has a well-known almost CR structure (integrable if, for instance, $M = S^{2n+1}$; see, e.g., [37]). Also $U(M)$ carries a natural Riemannian metric (induced by the *Sasaki metric* on $T(M)$) arising from g_θ. In view of [69] one may ask, *when is X a pseudoharmonic (a CR, a pseudo-Hermitian) map*?

Let $\phi : M \to N$ be a continuous map of a strictly pseudoconvex CR manifold M into a Riemannian manifold N. We call ϕ a *pseudoharmonic morphism* if there is a contact form θ on M, with L_θ positive definite, such that for any local harmonic function $v : V \to \mathbf{R}$ ($V \subseteq N$ open, $\Delta v = 0$) one has $\phi^{-1}(V) \neq \emptyset$ and the pullback $v \circ \phi$ satisfies $\Delta_b(v \circ \phi) = 0$ in $\phi^{-1}(V)$. If this is the case, note that $\square(v \circ \phi \circ \pi) = \Delta_b(v \circ \phi) \circ \pi = 0$; hence $\phi \circ \pi$ is a harmonic morphism of $(C(M), F_\theta)$ into N. The problem to *prove a CR analogue of the Fuglede–Ishihara theorem* (cf. [159], [217]) that harmonic morphisms are precisely the horizontally weakly conformal harmonic maps has recently been dealt with by E. Barletta [33]. The results are described in detail in the next section.

Already dealing with "horizontal conformality" leads to interesting problems. Let $\phi : M \to N$ be a smooth map and set $\Phi = \phi \circ \pi$. Then the restriction of the Fefferman metric F_θ to $\operatorname{Ker}(d\Phi)$ is indefinite. Indeed, if S is the vector field on $C(M)$ tangent to the S^1-action then $S \in \operatorname{Ker}(d\Phi)$ and $F_\theta(S,S)=0$. Is $\operatorname{Ker}(d\Phi)$ at least nondegenerate in $(T(C(M)), F_\theta)$?

Proposition 4.6. *(a)* $\operatorname{Ker}(d\Phi) = \left[\pi^*\operatorname{Ker}(d\phi)\right] \oplus \mathbf{R}S$.
(b) The restriction of F_θ to $\pi^*\operatorname{Ker}(d\phi)$ is positive definite if and only if T is transverse to $\operatorname{Ker}(d\phi)$.

Here $\left[\pi^*\operatorname{Ker}(d\phi)\right]_u$ is the preimage of $\operatorname{Ker}(d_{\pi(u)}\phi)$ by $d_u\pi : \operatorname{Ker}(\sigma_u) \to T_{\pi(u)}(M)$, $u \in C(M)$ (a linear isomorphism, since σ is a connection 1-form in the principal S^1-bundle $\pi : C(M) \to M$). The statement (a) follows easily from definitions. To prove "$\Longrightarrow$" in (b), assume that $T_x \in \operatorname{Ker}(d_x\phi)$ for some $x \in M$. Let $u \in \pi^{-1}(x)$ and let $A \in \operatorname{Ker}(\sigma_u)$ be the unique vector with $(d_u\pi)A = T_x$. Thus $A \in \left[\pi^*\operatorname{Ker}(d\phi)\right]_u$. Next,

$$0 = \tilde{G}_\theta(T,T)_x = F_{\theta,u}(A,A); \qquad \text{(by } A \in \operatorname{Ker}(\sigma_u) \text{ and (4.6));}$$

hence $A=0$, i.e., $T_x = 0$, a contradiction. To prove "$\Longleftarrow$" in (b), let $A \in \pi^*\operatorname{Ker}(d\phi)$ and $X := \pi_* A,\ X = X_H + \theta(X)T$, where $X_H \in H(M)$. Then

$$F_\theta(A,A) = G_\theta(X_H, X_H) \geq 0$$

and $=0$ if and only if $X_H = 0$, i.e., $\operatorname{Ker}(d\phi) \ni X = \theta(X)T$, which gives $\theta(X)=0$ (since T is transverse to $\operatorname{Ker}(d\phi)$). Thus $X=0$, i.e., $A=0$. □

Definition 4.7. Let h be a Riemannian metric on N. We say that ϕ is *horizontally pseudoconformal* if (i) T is transverse to $\operatorname{Ker}(d\phi)$ and (ii) for any $u \in C(M)$ there is $\lambda(u) \in \mathbf{R}$ such that

$$\lambda(u)^2 F_{\theta,u}(A,B) = (\Phi^* h)_u(A,B),$$

for any $A, B \in \mathcal{H}_u$, where $\mathcal{H}_u$ is the orthogonal complement (with respect to $F_{\theta,u}$) of $\mathcal{V}_u := \left[\pi^*\operatorname{Ker}(d\phi)\right]_u$. □

By condition (i) $\mathcal{V}_u$ is, in particular, nondegenerate in $(T_u(C(M)), F_{\theta,u})$. In particular, the *dilation* $\lambda : C(M) \to \mathbf{R}$ of ϕ is a continuous function (not necessarily smooth). If $\hat\theta = e^f\theta,\ f \in C^\infty(M)$, let $\hat T$ be the characteristic direction of $d\hat\theta$. Since $F_{\hat\theta} = e^{f\circ\pi} F_\theta$ it follows (by (b)) that T is transverse to $\operatorname{Ker}(d\phi)$ if and only if $\hat T$ is transverse to $\operatorname{Ker}(d\phi)$. Also $\hat\lambda^2 F_{\hat\theta} = \Phi^* h$ on $\mathcal{H}\otimes\mathcal{H}$, where $\hat\lambda := e^{f\circ\pi)}\lambda$. We have thus proved the following result:

Proposition 4.7.
Horizontal pseudoconformality is a CR-invariant property.

An interesting question is *whether pseudoharmonic maps of S^{2n+1} into a Riemannian manifold are unstable*. In view of Theorem 3.1 of [437], p. 611, the answer is expected to rely on a CR analogue of the Weitzenböck formula (for $\phi^{-1}TN$-valued forms on S^{2n+1}). Only a CR version of the Bochner formula (due to [186] and presented in detail in Chapter 9 of this book) is known so far.

4.4 Hörmander systems and harmonicity

In this section, given a Hörmander system $X = \{X_1, \dots, X_m\}$ on a domain $\Omega \subseteq \mathbf{R}^n$ we show that any *subelliptic harmonic morphism* ϕ from Ω into a ν-dimensional Riemannian manifold (N, h) is a (smooth) subelliptic harmonic map (in the sense of J. Jost and C.-J. Xu [234]). Also ϕ is a submersion provided that $\nu \le m$ and X has rank m. If $\Omega = \mathbf{H}_n$ (the Heisenberg group) and $X = \{\frac{1}{2}(L_\alpha + L_{\overline{\alpha}}), \frac{1}{2i}(L_\alpha - L_{\overline{\alpha}})\}$, where $L_{\overline{\alpha}} = \partial/\partial\overline{z}^\alpha - iz^\alpha\partial/\partial t$ is the Lewy operator, then a smooth map $\phi : \Omega \to (N, h)$ is a subelliptic harmonic morphism if and only if $\phi \circ \pi : (C(\mathbf{H}_n), F_{\theta_0}) \to (N, h)$ is a harmonic morphism, where $S^1 \to C(\mathbf{H}_n) \xrightarrow{\pi} \mathbf{H}_n$ is the canonical circle bundle and F_{θ_0} is the Fefferman metric of $(\mathbf{H}_n, \theta_0)$. For any S^1-invariant weak solution Φ to the harmonic map equation on $(C(\mathbf{H}_n), F_{\theta_0})$,

$$\Box\Phi^i + F_{\theta_0}^{ab}\left(\left|\begin{matrix} i \\ jk \end{matrix}\right| \circ \pi\right)\frac{\partial\Phi^j}{\partial u^a}\frac{\partial\Phi^k}{\partial u^b} = 0, \quad 1 \le i \le \nu,$$

the base map ϕ (i.e., $\phi \circ \pi = \Phi$) is shown to be a weak subelliptic harmonic map. We obtain a regularity result for *weak* harmonic morphisms from $(C(\{x_1 > 0\}), F_{\theta(k)})$ into a Riemannian manifold, where $F_{\theta(k)}$ is the Fefferman metric associated with the system of vector fields $X_1 = \partial/\partial x_1$, $X_2 = \partial/\partial x_2 + x_1^k\, \partial/\partial x_3$ $(k \ge 1)$ on $\Omega = \mathbf{R}^3 \setminus \{x_1 = 0\}$.

J. Jost and C.-J. Xu studied (cf. [234]) the existence and regularity of weak solutions $\phi : \Omega \to (N, h)$ to the nonlinear subelliptic system

$$H\phi^i + \sum_{a=1}^{m}\left(\left|\begin{matrix} i \\ jk \end{matrix}\right| \circ \phi\right)X_a(\phi^j)X_a(\phi^k) = 0, \quad 1 \le i \le \nu\,, \tag{4.19}$$

where $H = \sum_{a=1}^m X_a^* X_a$ is the Hörmander operator associated with a system $X = \{X_1, \dots, X_m\}$ of smooth vector fields on an open set $\Omega \subseteq \mathbf{R}^n$ satisfying the Hörmander condition on Ω and (N, h) is a Riemannian manifold. See Definitions 4.10 and 4.11 below. Also $\left|\begin{smallmatrix} i \\ jk \end{smallmatrix}\right|$ are the Christoffel symbols associated with the metric h. If $\omega \subset \Omega$ is a smooth domain such that $\partial\omega$ is noncharacteristic for X, the result of J. Jost and C.-J. Xu (cf. op. cit., Theorems 1 and 2, pp. 4641–4644) is that the Dirichlet problem for (4.19), with boundary data having values in regular balls of (N, h), may be solved and the solution is continuous on ω, up to the boundary. Any such map is then smooth by a result of C.-J. Xu and C. Zuily (cf. [443]), who studied higher regularity of continuous solutions to a quasilinear subelliptic systems including (4.20).

Definition 4.8. Solutions (smooth a posteriori) to (4.19) are *subelliptic harmonic maps* and (4.19) is the *subelliptic harmonic map system.* □

See also Z-R. Zhou [449]. Clearly, if $X_a = \partial/\partial x^a$, $1 \le a \le n$, then a subelliptic harmonic map is an ordinary harmonic map (Ω is thought of as a Riemannian manifold, with the Euclidean metric). An important class of harmonic maps are *harmonic morphisms*, i.e., smooth maps of Riemannian manifolds pulling back local harmonic functions to harmonic functions. That these are indeed harmonic maps is a classical result of T. Ishihara (cf. [217]), actually holding in general for harmonic morphisms between semi-Riemannian manifolds (cf. B. Fuglede [160]). In the present section we extend the notion of a harmonic morphism to the context of systems of vector fields and generalize the Fuglede–Ishihara theorem. The results are due to E. Barletta [32].

Definition 4.9. A localizable[2] map $\phi : \Omega \to (N, h)$ is a (weak) *subelliptic harmonic morphism* if for any $v : V \to \mathbf{R}$, with $V \subseteq N$ open and $\Delta_N v = 0$ in V, one has (i) $v \circ \phi \in L^1_{\text{loc}}(U)$, for any open set $U \subset \Omega$ such that $\phi(U) \subset V$, and (ii) $H(v \circ \phi) = 0$, in the distributional sense. □

The main result in this section is the following:

Theorem 4.6. (E. Barletta [32])
Let $X = \{X_1, \ldots, X_m\}$ be a Hörmander system on a domain $\Omega \subseteq \mathbf{R}^n$ and (N, h) a ν-dimensional Riemannian manifold. If $\nu > m$ there are no subelliptic harmonic morphisms of Ω into (N, h), except for the constant maps. If $\nu \leq m$ then any subelliptic harmonic morphism $\phi : \Omega \to (N, h)$ is an actually smooth subelliptic harmonic map and there is a smooth function $\lambda : \Omega \to [0, +\infty)$ such that

$$\sum_{a=1}^{m} (X_a \phi^i)(x)(X_a \phi^j)(x) = \lambda(x)\delta^{ij}, \qquad 1 \leq i, j \leq \nu \,, \tag{4.20}$$

for any $x \in \Omega$ and any normal coordinate system (V, y^i) at $\phi(x) \in N$, where $\phi^i = y^i \circ \phi$. In particular if $x \in U = \phi^{-1}(V)$ is such that $\lambda(x) \neq 0$ then the matrix $[(X_a \phi^i)(x)]$ has maximal rank; hence ϕ is a C^∞ submersion provided that $\{X_1, \ldots, X_m\}$ are independent at any $x \in \Omega$.

When $\Omega = \mathbf{H}_n$, the Heisenberg group, and $X = \{X_\alpha, Y_\alpha : 1 \leq \alpha \leq n\}$, where $X_\alpha = (1/2)\partial/\partial x^\alpha + y^\alpha \partial/\partial t$ and $Y_\alpha = J X_\alpha$, we relate subelliptic harmonic morphisms to harmonic morphisms (from a certain Lorentzian manifold).

Theorem 4.7. (E. Barletta [32])
Let $\mathbf{H}_n = \mathbf{C}^n \times \mathbf{R}$ be the Heisenberg group endowed with the standard strictly pseudoconvex CR structure and the contact form $\theta_0 = dt + i \sum_{\alpha=1}^n (z^\alpha d\bar{z}^\alpha - \bar{z}^\alpha dz^\alpha)$. Consider the Fefferman metric

$$F_{\theta_0} = \pi^* G_{\theta_0} + \frac{2}{n+2} (\pi^* \theta_0) \odot (d\gamma)$$

on $C(\mathbf{H}_n) = \left(\Lambda^{n+1,0}(\mathbf{H}_n) \setminus \{0\}\right) / \mathbf{R}_+$, where $\pi : C(\mathbf{H}_n) \to \mathbf{H}_n$ is the projection and γ a fiber coordinate on $C(\mathbf{H}_n)$. Then a smooth map $\phi : \mathbf{H}_n \to (N, h)$ is a subelliptic harmonic morphism, with respect to the system of vector fields $X = \{X_\alpha, Y_\alpha\}$, if and only if $\phi \circ \pi : (C(\mathbf{H}_n), F_{\theta_0}) \to (N, h)$ is a harmonic morphism.

[2] In the sense of [233], p. 434, i.e., for any $x_0 \in \Omega$ there is an open neighborhood $U \subset \Omega$ of x_0 and a coordinate neighborhood (V, y^i) on N such that $\phi(U) \subset V$.

The main ingredient in the proof is the relationship between the Laplace–Beltrami operator $\Box$ of the Fefferman metric F_{θ_0} and the Hörmander operator H on $\mathbf{H}_n$. This was presented in Chapter 2 of this book (cf. also J.M. Lee [271]), where $\Box$ is related to the sub-Laplacian Δ_b of the given strictly pseudoconvex CR manifold, yet seems to be unknown in the literature on PDEs. We emphasize the relationship between subelliptic and hyperbolic PDEs by providing a short direct proof that for the Heisenberg group, $\pi_* \Box = -2H$, where

$$\Box f = \frac{1}{2}\sum_{\alpha=1}^{n}\left(\frac{\partial^2 f}{\partial (u^\alpha)^2} + \frac{\partial^2 f}{\partial (u^{\alpha+n})^2}\right) + 2(|z|^2 \circ \pi)\frac{\partial^2 f}{\partial (u^{2n+1})^2}$$
$$+ 2u^{\alpha+n}\frac{\partial^2 f}{\partial u^\alpha \partial u^{2n+1}} - 2u^\alpha \frac{\partial^2 f}{\partial u^{\alpha+n}\partial u^{2n+1}} + 2(n+2)\frac{\partial^2 f}{\partial u^{2n+1}\partial u^{2n+2}},$$

for any $f \in C^2(\mathbf{H}_n)$. Here $u^A = x^A \circ \pi$, $1 \le A \le 2n+1$, and $u^{2n+2} = \gamma$, where $(x^A) = (z^\alpha = x^\alpha + i y^\alpha, t)$ are coordinates on $\mathbf{H}_n$.

4.4.1 Hörmander systems

Let $\Omega \subseteq \mathbf{R}^n$ be an open set and $X = \{X_1, \dots, X_m\}$ a system of C^∞ vector fields on Ω.

Definition 4.10. We say that X satisfies the *Hörmander condition* (or that X is a *Hörmander system*) on Ω if the vector fields $X_1, \dots, X_m$ together with their commutators up to some fixed length r span the tangent space $T_x(\Omega)$, at each $x \in \Omega$. $\Box$

A commutator of the form $[X_a, X_b]$ has length 2 (and by convention each X_a has length 1). If $X_a = b_a^A(x)\partial/\partial x^A$ then we set $X_a^* f = -\partial(b_a^A f)/\partial x^A$, for any $f \in C_0^1(\Omega)$. Our convention as to the range of indices is $a, b, \dots \in \{1, \dots, m\}$ and $A, B, \dots \in \{1, \dots, n\}$.

Definition 4.11. The *Hörmander operator* is

$$Hu = -\sum_{a=1}^{m} X_a^* X_a u = \sum_{A,B=1}^{n} \frac{\partial}{\partial x^A}\left(a^{AB}(x)\frac{\partial u}{\partial x^B}\right),$$

where $a^{AB}(x) = \sum_{a=1}^{m} b_a^A(x) b_a^B(x)$. $\Box$

The reader should observe that although we employ the same terminology, H is distinct from the operator in Definition 2.2. Indeed the formal adjoint X_a^* is defined with respect to the Euclidean metric (while X_a^* in (2.8) is built with respect to the volume form arising from a fixed contact form). The matrix a^{AB} is symmetric and positive semidefinite; yet it may fail to be definite; hence in general H is not elliptic (H is a degenerate elliptic operator).

Example 1. (Cf. [234], p. 4634) The system of vector fields

$$X_1 = \partial/\partial x^1, \;\; X_2 = \partial/\partial x^2 + (x^1)^k \partial/\partial x^3 \quad (k \ge 0) \tag{4.21}$$

satisfies the Hörmander system on $\mathbf{R}^3$ with $r = k + 1$. We have $X_a^* = -X_a$, $a \in \{1, 2\}$; hence the Hörmander operator is

$$Hu = \frac{\partial^2 u}{\partial (x^1)^2} + \frac{\partial^2 u}{\partial (x^2)^2} + (x^1)^{2k} \frac{\partial^2 u}{\partial (x^3)^2} + 2(x^1)^k \frac{\partial^2 u}{\partial x^2 \partial x^3} . \tag{4.22}$$

As we shall see later, there is a CR structure $\mathcal{H}(k)$ on $\Omega = \mathbf{R}^3 \setminus \{x^1 = 0\}$ such that the (rank-2) distribution $\mathcal{D}$ spanned by the X_a's is precisely the Levi (or maximally complex) distribution of $(\Omega, \mathcal{H}(k))$.

Example 2. Let $\mathbf{H}_n = \mathbf{C}^n \times \mathbf{R}$ be the Heisenberg group with coordinates $(z, t) = (z^1, \ldots, z^n, t)$ and set $z^\alpha = x^\alpha + i y^\alpha$, $1 \leq \alpha \leq n$. Consider the Lewy operators $L_{\overline{\alpha}} = \partial/\partial \overline{z}^\alpha - i z^\alpha \, \partial/\partial t$, $1 \leq \alpha \leq n$, and the system of vector fields $X := \{X_\alpha, X_{\alpha+n} : 1 \leq \alpha \leq n\}$, $X_{\alpha+n} = J X_\alpha$, where

$$X_\alpha = \tfrac{1}{2} \left(L_\alpha + L_{\overline{\alpha}} \right), \tag{4.23}$$

and $L_\alpha = \overline{L_{\overline{\alpha}}}$. As in Chapter 1 of this book, the Heisenberg group is thought of as a CR manifold (of hypersurface type) with the standard CR structure

$$T_{1,0}(\mathbf{H}_n)_x = \sum_{\alpha=1}^n \mathbf{C} L_{\alpha,x} , \quad x \in \mathbf{H}_n .$$

Also, J is the complex structure in the (real rank $2n$) distribution $H(\mathbf{H}_n) = \operatorname{Re} e \left(T_{1,0}(\mathbf{H}_n) \oplus T_{0,1}(\mathbf{H}_n) \right)$, i.e., $J(Z + \overline{Z}) = i(Z - \overline{Z})$, for any $Z \in T_{1,0}(\mathbf{H}_n)$. Since $[L_\alpha, L_{\overline{\alpha}}] = -2i \delta_{\alpha\beta} T$ (with $T = \partial/\partial t$), (4.23) is a Hörmander system on $\mathbf{H}_n$, with $r = 2$. Next $X_a^* = -X_a$ and the corresponding Hörmander operator is

$$Hu = \tfrac{1}{4} \sum_{\alpha=1} \left\{ \frac{\partial^2 u}{\partial (x^\alpha)^2} + \frac{\partial^2 u}{\partial (y^\alpha)^2} \right\} + y^\alpha \frac{\partial^2 u}{\partial x^\alpha \partial t} + x^\alpha \frac{\partial^2 u}{\partial y^\alpha \partial t} + |z|^2 \frac{\partial^2 u}{\partial t^2} . \tag{4.24}$$

We end this section with a discussion of the function spaces $W_X^{k,p}(\Omega)$ (which the reader will meet again in applications in Section 4.4.4).

Let $U \subseteq \mathbf{R}^n$ be an open set and $X = \{X_1, \ldots, X_m\}$ a Hörmander system on U. Let Ω be a bounded domain in $\mathbf{R}^n$ such that $\Omega \subset\subset U$. Let $k \in \mathbf{Z}$, $k \geq 1$, and $p \in \mathbf{R}$, $p \geq 1$. We define the function spaces

$$W_X^{k,p}(\Omega) = \{ f \in L^p(\Omega) : X^J f \in L^p(\Omega), \quad \forall J = (j_1, \ldots, j_s), \quad |J| := s \leq k \} .$$

If $J = (j_1, \ldots, j_s)$ and $1 \leq j_i \leq m$ then $X^J f := X_{j_1} \cdots X_{j_s} f$ (and derivatives are intended in the sense of distribution theory). As a matter of notation, we admit that $|J| = 0$ (i.e., J is the empty multi-index) and then $X^J f := f$. Also we set

$$\| f \|_{W_X^{k,p}(\Omega)} = \left(\sum_{|J| \leq k} \| X^J f \|_{L^p(\Omega)}^p \right)^{1/p}, \quad W_X^k(\Omega) := W_X^{k,2}(\Omega).$$

Theorem 4.8. (C.-J. Xu [440])
(1) If $1 \le p < \infty$ then $W_X^{k,p}(\Omega)$ is a separable[3] *Banach space.*
(2) If $1 < p < \infty$ then $W_X^{k,p}(\Omega)$ is reflexive.[4]
(3) $W_X^k(\Omega)$ is a separable Hilbert space.

If $J = (j_1, \dots, j_s)$ with $1 \le j_i \le m$ then we denote by $(X^J)^*$ the adjoint of X^J on $C_0^\infty(\Omega)$, i.e.,

$$\int_\Omega (X^J u)\, v\, dx = \int_\Omega u\, (X^J)^* v\, dx, \qquad \forall\, u,\, v \in C_0^\infty(\Omega).$$

Then the function spaces $W_X^{k,p}(\Omega)$ may be also described as

$$W_X^{k,p}(\Omega) = \Big\{ f \in L^p(\Omega) : \exists\, g_J \in L^p(\Omega) \quad \text{such that} \\ \int_\Omega f\, \big((X^J)^*\varphi\big)\, dx = \int_\Omega g_J\, \varphi\, dx, \quad \forall\, \varphi \in C_0^\infty(\Omega),\ \forall\, |J| \le k \Big\}.$$

Of course g_J is unique (up to a zero measure set) and commonly denoted by $X^J f$.

Proof of Theorem 4.8. Let $\{u_\nu\}$ be a Cauchy sequence in $W_X^{k,p}(\Omega)$. Then for any $\epsilon > 0$, there is $\nu_\epsilon \ge 1$ such that

$$\|X^J u_\nu - X^J u_\mu\|^p_{L^p(\Omega)} \le \|u_\nu - u_\mu\|^p_{W_X^{k,p}(\Omega)} < \epsilon^p$$

for any $\nu, \mu \ge \nu_\epsilon$, i.e., $\{X^J u_\nu\}$ is a Cauchy sequence in $L^p(\Omega)$, for any $|J| \le k$. Therefore, there is $u^J \in L^p(\Omega)$ such that $X^J u_\nu \to u^J$ in $L^p(\Omega)$, since $\nu \to \infty$. Since

$$\left| \int_\Omega (X^J u_\nu)\, \psi\, dx - \int_\Omega u^J\, \psi\, dx \right| \le \|X^J u_\nu - u^J\|_{L^p(\Omega)} \|\psi\|_{L^q(\Omega)}, \qquad \frac{1}{p} + \frac{1}{q} = 1,$$

it follows that $\int_\Omega X^J u_\mu\, \psi\, dx \to \int_\Omega u^J\, \psi\, dx$ as $\nu \to \infty$, for any $\psi \in C_0^\infty(\Omega)$. Hence, by letting $\nu \to \infty$ in

$$\int_\Omega u_\nu\, (X^J)^*\varphi\, dx = \int_\Omega (X^J u_\nu)\, \varphi\, dx$$

we obtain

$$\int_\Omega u^0\, (X^J)^*\varphi\, dx = \int_\Omega u^J\, \varphi\, dx,$$

---

[3] A topological space X is *separable* if there is a countable dense subset $A \subseteq X$.

[4] If X is a Banach space, let X^* be its normed dual (itself a Banach space). Consider the linear isometry $\phi : X \to X^{**}$ given by $\langle \phi x, x^* \rangle = \langle x^*, x \rangle$, for any $x \in X,\ x^* \in X^*$. Then ϕ is an isometric isomorphism of X onto a closed subspace of X^{**} and X is *reflexive* if $\phi(X) = X^{**}$.

where 0 is the empty multi-index. Thus $u^0 \in W_X^{k,p}(\Omega)$ and $X^J u^0 = u^J$ and

$$\|u_\nu - u^0\|^p_{W_X^{k,p}(\Omega)} = \sum_{|J|\le k} \|X^J(u_\nu - u^0)\|^p_{L^p(\Omega)} = \sum_{|J|\le k} \|X^J u_\nu - u^J\|^p_{L^p(\Omega)} \to 0$$

for $\nu \to \infty$, i.e., $u_\nu \to u^0$ in $W_X^{k,p}(\Omega)$. This proves that $W_X^{k,p}(\Omega)$ is Banach. Any L^p space with $1 < p < \infty$ is reflexive. Thus the product space

$$E := \prod_{|J|\le k} L^p(\Omega) = \big\{\{f_J\}_{|J|\le k} : f_J \in L^p(\Omega),\ |J| \le k\big\},$$

$$\|\{f_J\}_{|J|\le k}\|_E := \Big(\sum_{|J|\le k} \|f_J\|^p_{L^p(\Omega)}\Big)^{1/p},$$

is a reflexive Banach space. Moreover, the map

$$T : W_X^{k,p}(\Omega) \to E, \quad T(u) := X^J u, \quad u \in W_X^{k,p}(\Omega),$$

is a linear isometry. Since $T(W_X^{k,p}(\Omega))$ is a closed subspace of E and $T(W_X^{k,p}(\Omega))$ is reflexive, it follows that $W_X^{k,p}(\Omega)$ is reflexive as well. The proof of the separability may be obtained in a similar way.

4.4.2 Subelliptic harmonic morphisms

Proposition 4.8. *A weak subelliptic harmonic morphism $\phi : \Omega \to (N, h)$ is actually smooth.*

Indeed, let $x \in \Omega$ and $p = \phi(x) \in N$. Let (V, y^i) be a local system of harmonic coordinates at p (cf., e.g., [60], p. 143), i.e., $p \in V$ and $\Delta_N y^i = 0$ in V, where Δ_N is the Laplace–Beltrami operator of (N, h). Since ϕ is localizable, we may consider an open neighborhood U of x such that $\phi(U) \subset V$. Then $y^i \circ \phi \in L^1_{\text{loc}}(U)$ and $H(y^i \circ \phi) = 0$. Moreover, it is a well-known fact that H is hypoelliptic, i.e., if $Hu = f$ in the distributional sense, and f is smooth, then u is smooth, too. Hence $y^i \circ \phi \in C^\infty(U)$. □

To show that ϕ is a subelliptic harmonic map we need the following lemma:

Lemma 4.2. (T. Ishihara [217])
Let (N, h) be a ν-dimensional Riemannian manifold and $C_i,\ C_{ij} \in \mathbf{R}$ a system of constants such that $C_{ij} = C_{ji}$ and $\sum_{i=1}^{\nu} C_{ii} = 0$. Let $p \in N$. Then there is a normal coordinate system (V, y^i) in p and a harmonic function $v : V \to \mathbf{R}$ such that

$$\frac{\partial v}{\partial y^i}(p) = C_i, \quad v_{i,j}(p) = C_{ij}\ .$$

Lemma 4.2 is referred to as *Ishihara's lemma*. Here $v_{i,j}$ are the second-order covariant derivatives

$$v_{i,j} = \frac{\partial^2}{\partial y^i \partial y^j} - \left| \begin{matrix} k \\ ij \end{matrix} \right| \frac{\partial v}{\partial y^k} .$$

Let $i_0 \in \{1, \dots, \nu\}$ be a fixed index and consider the constants $C_i = \delta_{i i_0}$ and $C_{ij} = 0$. By Ishihara's lemma there is a local harmonic function $v : V \to \mathbf{R}$ such that

$$\frac{\partial v}{\partial y^i}(p) = \delta_{i i_0}, \qquad v_{i,j}(p) = 0.$$

A calculation shows that

$$X_a(v \circ \phi) = \frac{\partial v}{\partial y^j} X_a(\phi^j),$$

$$H(v \circ \phi) = (H\phi^j) \frac{\partial v}{\partial y^j} - \sum_{a=1}^{m} (X_a\phi^j)(X_a\phi^k) \left\{ v_{j,k} + \left| \begin{matrix} i \\ jk \end{matrix} \right| \frac{\partial v}{\partial y^i} \right\} . \tag{4.25}$$

Then (by (4.25)),

$$0 = H(v \circ \phi)(x) = (H\phi^{i_0})(x) - \sum_{a=1}^{m} (X_a\phi^j)(x)(X_a\phi^k)(x) \left| \begin{matrix} i_0 \\ jk \end{matrix} \right| (p).$$

To prove (4.20) in Theorem 4.6 consider the constants $C_{ij} \in \mathbf{R}$ such that $C_{ij} = C_{ji}$ and $\sum_{i=1}^{\nu} C_{ii} = 0$. Let $x \in \Omega$ and $p = \phi(x) \in N$. By Ishihara's lemma there is a normal coordinate system (V, y^i) in p and a local harmonic function v on V such that

$$\frac{\partial v}{\partial y^i}(p) = 0, \quad v_{i,j}(p) = C_{ij} .$$

Since ϕ is a subelliptic harmonic morphism (again by (4.25)),

$$0 = H(v \circ \phi)(x) = - \sum_{a=1}^{m} (X_a\phi^j)(x)(X_a\phi^k)(x) C_{jk},$$

that is,

$$C_{jk} X^{jk}(x) = 0, \tag{4.26}$$

where

$$X^{jk} := \sum_{a=1}^{m} (X_a\phi^j)(X_a\phi^k).$$

The identity (4.26) may be also written as

$$\sum_{i \neq j} C_{ij} X^{ij}(x) + \sum_i C_{ii} \left\{ X^{ii}(x) - X^{11}(x) \right\} = 0. \tag{4.27}$$

Now let us choose the constants C_{ij} such that $C_{ij} = 0$ for any $i \neq j$ and

$$C_{ii} = \begin{cases} 1, & i = i_0, \\ -1, & i = 1, \\ 0, & \text{otherwise}, \end{cases}$$

where $i_0 \in \{2, \dots, \nu\}$ is a fixed index. Then (4.27) gives

$$X^{i_0 i_0}(x) - X^{11}(x) = 0,$$

that is,

$$X^{11}(x) = X^{22}(x) = \cdots = X^{\nu\nu}(x),$$

and (4.27) becomes

$$\sum_{i \neq j} C_{ij} X^{ij}(x) = 0. \tag{4.28}$$

Let us fix $i_0, j_0 \in \{1, \dots, \nu\}$ such that $i_0 \neq j_0$, otherwise arbitrary, and set

$$C_{ij} = \begin{cases} 1, & i = i_0\,,\ j = j_0 or i = j_0\,,\ j = i_0, \\ 0, & \text{otherwise}. \end{cases}$$

Then (4.28) implies that $X^{i_0 j_0}(x) = 0$. Let us set

$$\lambda := X^{11} = \sum_{a=1}^{m} (X_a \phi^1)^2 \in C^\infty(U),$$

where $U = \phi^{-1}(V) \subset \Omega$. Summing up the results obtained so far, we have

$$\sum_{a=1}^{m} (X_a \phi^i)(x)(X_a \phi^j)(x) = \lambda(x) \delta^{ij}\ ,$$

which is (4.20), and in particular

$$\nu\, \lambda(x) = \sum_{a,i} (X_a \phi^i)(x)^2\ .$$

Therefore, we have built a global C^∞ function $\lambda : \Omega \to [0, +\infty)$. Indeed, if $(V, \varphi = (y^1, \dots, y^\nu))$ and $(V', \varphi' = (y'^1, \dots, y'^\nu))$ are two normal coordinate systems at $p = \phi(x)$ and $F = \varphi' \circ \varphi^{-1}$, then the identities

$$X_a \phi'^i = \frac{\partial F^i}{\partial \xi^j} X_a \phi^j, \quad \sum_k \frac{\partial F^k}{\partial \xi^i}(p) \frac{\partial F^k}{\partial \xi^j}(p) = \delta_{ij}$$

yield

$$\sum_i (X_a\phi'^i)(x)^2 = \sum_j (X_a\phi^j)(x)^2 .$$

Assume that there is $x_0 \in \Omega$ such that $\lambda(x_0) \neq 0$ and consider

$$v^i := \Big((X_1\phi^i)(x_0), \ldots, (X_m\phi^i)(x_0) \Big) \in \mathbf{R}^m, \quad 1 \leq i \leq \nu .$$

Clearly $v^i \neq 0$, for any i, and $v^i \cdot v^j = 0$, for any $i \neq j$. Consequently $\text{rank}[(X_a\phi^i)(x_0)] = \nu$; hence $\nu \leq m$. Thus, whenever $\nu > m$ it follows that $\lambda = 0$, i.e., $X_a\phi^i = 0$, and then the commutators of the X_a's, up to length r, annihilate ϕ^i. Since $X = \{X_1, \ldots, X_m\}$ is a Hörmander system and Ω is connected, it follows that $\phi^i = \text{const}$.

4.4.3 The relationship to hyperbolic PDEs

We recall (cf., e.g., [433]) the following definition:

Definition 4.12. A smooth map $\Phi : M \to N$ of semi-Riemannian manifolds is a *harmonic morphism* if for any local harmonic function $v : V \to \mathbf{R}$ on N, the pullback $v \circ \Phi$ is harmonic on M, i.e., $\Delta(v \circ \Phi) = 0$ in $U = \Phi^{-1}(V)$. Here Δ is the Laplace–Beltrami operator of M. □

In the context of Example 2, we shall relate the subelliptic harmonic morphisms $\phi : \mathbf{H}_n \to N$ to harmonic morphisms from the Lorentzian manifold $(C(\mathbf{H}_n), F_{\theta_0})$. Recall that $\theta_0 = dt + i\sum_{\alpha=1}^n (z^\alpha d\overline{z}^\alpha - \overline{z}^\alpha dz^\alpha)$ is a contact form on $\mathbf{H}_n$, i.e., $\theta_0 \wedge (d\theta_0)^n$ is a volume form. Let us consider the 1-form $\sigma = \frac{1}{n+2} d\gamma$ on $C(\mathbf{H}_n)$ and let us set

$$F_{\theta_0} = \pi^* G_{\theta_0} + 2(\pi^*\theta_0) \odot \sigma.$$

By the results in Chapter 2, F_{θ_0} is a Lorentz metric on $C(\mathbf{H}_n)$, the Fefferman metric of $(\mathbf{H}_n, \theta_0)$. With respect to the local coordinates $(u^a) = (u^A, \gamma)$, where $u^A = x^A \circ \pi$, the Fefferman metric may be written as

$$F_{\theta_0} = 2\sum_{\alpha=1}^{n} \left[(du^\alpha)^2 + (du^{\alpha+n})^2 \right] + \frac{2}{n+2}\left[du^{2n+1} + 2\sum_{\alpha=1}^{n} (u^\alpha du^{\alpha+n} - u^{\alpha+n} du^\alpha) \right] \odot du^{2n+2} . \tag{4.29}$$

We wish to compute the Laplace–Beltrami operator

$$\Box f = \frac{1}{\sqrt{|F|}} \frac{\partial}{\partial u^a}\left(\sqrt{|F|}\, F_{\theta_0}^{ab} \frac{\partial f}{\partial u^b} \right), \qquad f \in C^2(C(\mathbf{H}_n)).$$

A calculation shows that

$$F := \det[(F_{\theta_0})_{ab}] = -\left(\frac{2^n}{n+2} \right)^2, \tag{4.30}$$

$$
F_{\theta_0}^{ab} : \begin{pmatrix}
1/2 & \cdots & 0 & 0 & \cdots & 0 & u^{n+1} & 0 \\
\vdots & & \vdots & \vdots & & \vdots & \vdots & \vdots \\
0 & \cdots & 1/2 & 0 & \cdots & 0 & u^{2n} & 0 \\
0 & \cdots & 0 & 1/2 & \cdots & 0 & -u^1 & 0 \\
\vdots & & \vdots & \vdots & & \vdots & \vdots & \vdots \\
0 & \cdots & 0 & 0 & \cdots & 1/2 & -u^n & 0 \\
u^{n+1} & \cdots & u^{2n} & -u^1 & \cdots & -u^n & 2|z|^2 \circ \pi & n+2 \\
0 & \cdots & 0 & 0 & \cdots & 0 & n+2 & 0
\end{pmatrix}; \tag{4.31}
$$

hence

$$
\begin{aligned}
\Box f = {} & \tfrac{1}{2} \sum_{\alpha=1}^{n} \left[\frac{\partial^2 f}{\partial (u^\alpha)^2} + \frac{\partial^2 f}{\partial (u^{\alpha+n})^2} \right] + 2(n+2) \frac{\partial^2 f}{\partial u^{2n+1} \partial u^{2n+2}} \\
& + 2u^{\alpha+n} \frac{\partial^2 f}{\partial u^\alpha \partial u^{2n+1}} - 2u^\alpha \frac{\partial^2 f}{\partial u^{\alpha+n} \partial u^{2n+1}} + 2(|z| \circ \pi)^2 \frac{\partial^2 f}{\partial (u^{2n+1})^2} \, .
\end{aligned} \tag{4.32}
$$

Proposition 4.9. *Let $c \in \pi^{-1}(0) \subset C(\mathbf{H}_n)$. Then $[F_{\theta_0}^{ab}(c)]$ has spectrum $\{1/2,\ n+2,\ -n-2\}$ (with multiplicities $\{2n,\ 1,\ 1\}$, respectively). The corresponding eigenspaces are*

$$
\text{Eigen}(1/2) = \sum_{j=1}^{2n} \mathbf{R} e_j \, , \quad \text{Eigen}(\pm(n+2)) = \mathbf{R}(0, \ldots, 0, 1, \pm 1).
$$

Here $\{e_1, \ldots, e_{2n+2}\} \subset \mathbf{R}^{2n+2}$ is the canonical linear basis. Consequently, under the coordinate transformation

$$
\begin{cases}
w^j = \sqrt{2}\, u^j, \quad 1 \le j \le 2n, \\
w^{2n+1} = \dfrac{1}{\sqrt{2(n+2)}} \left(u^{2n+1} + \gamma \right), \\
w^{2n+2} = \dfrac{1}{\sqrt{2(n+2)}} \left(u^{2n+1} - \gamma \right),
\end{cases}
$$

(4.32) goes over to the canonical hyperbolic form

$$
(\Box f)(c) = \sum_{A=1}^{2n+1} \frac{\partial^2 f}{\partial (w^A)^2}(c) - \frac{\partial^2 f}{\partial (w^{2n+2})^2}(c).
$$

The unit circle S^1 acts freely on $C(\mathbf{H}_n)$ by $R_w([\omega]) = [\omega] \cdot w := [w\omega],\ w \in S^1$. Then $u^A \circ R_w = u^A$ and $\gamma \circ R_w = \gamma + \arg(w) + 2k\pi$, for some $k \in \mathbf{Z}$; hence $R_w^* F_{\theta_0} = F_{\theta_0}$, i.e., $S^1 \subset \text{Isom}(C(\mathbf{H}_n), F_{\theta_0})$. As is well known, this yields $\Box^{R_w} = \Box$, where we set $\Box^\psi f := (\Box f^{\psi^{-1}})^\psi$ and $f^{\psi^{-1}} := f \circ \psi$, for any diffeomorphism ψ of $C(\mathbf{H}_n)$ into itself. Therefore,

$$
\pi_* \Box : C^\infty(\mathbf{H}_n) \to C^\infty(\mathbf{H}_n), \quad (\pi_* \Box) u := (\Box (u \circ \pi))\tilde{}\,,
$$

is well defined, where for a given S^1 invariant function f on $C(\mathbf{H}_n)$, $\tilde{f}$ denotes the corresponding base map. Finally, a calculation based on (4.24) and (4.32) leads to

$$(\pi_* \Box) u = 2\, H u, \quad u \in C^2(\mathbf{H}_n). \tag{4.33}$$

At this point, Theorem 4.7 is proved. For given a local harmonic function $v : V \to \mathbf{R}$ on N and $\phi : \mathbf{H}_n \to N$ a subelliptic harmonic morphism then (by (4.33)) $0 = 2\, H(v \circ \phi) = (\pi_* \Box)(v \circ \phi)$; hence $\Box(v \circ \Phi) = 0$, i.e., $\Phi = \phi \circ \pi$ is a harmonic morphism.

Example 1. (*continued*) Theorem 4.7 applies, with only minor modifications, to subelliptic harmonic morphisms from $\Omega = \mathbf{R}^3 \setminus \{x^1 = 0\}$, with respect to the Hörmander system (4.23). $\mathbf{R}^3$ is a CR manifold with the CR structure $\mathcal{H}(k)$ spanned by

$$Z := 2\,\frac{\partial}{\partial z} - i \left(\frac{z + \bar{z}}{2} \right)^k \frac{\partial}{\partial t},$$

where $z = x^1 + i x^2$ and $t = x^3$. Next

$$\theta(k) = dt + \frac{i}{2} \left(\frac{z + \bar{z}}{2} \right)^k (dz - d\bar{z})$$

is a pseudo-Hermitian structure on $\mathbf{R}^3$ with the Levi form

$$L_{\theta(k)}(Z, \overline{Z}) = -k \left(\frac{z + \bar{z}}{2} \right)^{k-1};$$

hence $(\mathbf{R}^3, \mathcal{H}(0))$ is Levi flat, while $(\Omega, \mathcal{H}(k))$ is nondegenerate, for any $k \geq 1$. Moreover, if $k \geq 1$, each connected component $\Omega^+ = \{x^1 > 0\}$ and $\Omega^- = \{x^1 < 0\}$ is strictly pseudoconvex. The Tanaka–Webster connection of (Ω, θ_k) is given by

$$\Gamma^{\bar{1}}_{1\bar{1}} = 0, \quad \Gamma^1_{11} = \frac{2(k-1)}{z + \bar{z}}, \quad \Gamma^1_{01} = 0.$$

In particular, the Tanaka–Webster connection of $(\Omega, \theta(k))$ has pseudo-Hermitian scalar curvature $R = -\frac{k-1}{k}\,(2/(z + \bar{z}))^{k+1}$, and one may explicitly compute the Fefferman metric $F_{\theta(k)}$ of $(\Omega^{\pm}, \theta(k))$. Also, the pseudo-Hermitian torsion vanishes (i.e., $A^{\bar{1}}_1 = 0$). Note that $(\Omega, \theta(1))$ is Webster flat. Set $\nabla^H u = \pi_{\mathcal{D}} \nabla u$, where ∇u is the gradient of $u \in C^\infty(\Omega)$ with respect to the Webster metric and $\pi_H : T(\Omega) \to \mathcal{D}$ the projection with respect to the direct sum decomposition $T(\Omega) = \mathcal{D} \oplus \mathbf{R}\partial/\partial t$. That is, $\nabla^H u = u^1 Z + u^{\bar{1}} \overline{Z}$, where $u^1 = h^{1\bar{1}} u_{\bar{1}}$ and $u_1 = Z(u)$. The sub-Laplacian is $\Delta_b u = \operatorname{div}(\nabla^H u)$, where the divergence is taken with respect to the volume form $\theta(k) \wedge (d\theta(k))^n$. Since

$$\operatorname{div}(Z) = \frac{k-1}{x_1}, \quad h^{1\bar{1}} = -\frac{1}{k}\, x_1^{1-k} \qquad (x_1 = x^1),$$

one has $\Delta_b = -\frac{1}{k}\, x_1^{1-k}(Z\overline{Z} + \overline{Z}Z)$; hence (by (4.22)) $\Delta_b = -\frac{1}{k}\, x_1^{1-k}\, H$ on C^∞ functions. By a result in Chapter 2, the Laplace–Beltrami operator $\Box$ of the Fefferman metric of $(\Omega^{\pm}, \theta(k))$ is related to the sub-Laplacian by $\pi_* \Box = \Delta_b$; hence

$$(\pi_* \Box) u = -\frac{1}{k}\, x_1^{1-k} H u, \qquad u \in C^\infty(\Omega^{\pm}). \tag{4.34}$$

Proposition 4.10.
For any subelliptic harmonic morphism $\phi : \Omega^{\pm} \to N$, with respect to the Hörmander system (4.22)*, the map $\Phi = \phi \circ \pi : C(\Omega^{\pm}) \to N$ is a harmonic morphism, with respect to the Fefferman metric of* $(\Omega^{\pm}, \theta(k))$.

For the converse, one may see our next section.

4.4.4 Weak harmonic maps from $C(\mathbf{H}_n)$

From now on, we assume that N is covered by one coordinate chart $\varphi = (y^1, \ldots, y^\nu) : N \to \mathbf{R}^\nu$ and $\Phi^i := y^i \circ \Phi$.

Definition 4.13. A map $\Phi : C(\mathbf{H}_n) \to N$ is said to satisfy *weakly* the harmonic map system

$$\Box \Phi^i + F^{ab}_{\theta_0} \left(\left| \begin{matrix} i \\ jk \end{matrix} \right| \circ \pi \right) \frac{\partial \Phi^j}{\partial u^a} \frac{\partial \Phi^k}{\partial u^b} = 0, \quad 1 \leq i \leq \nu\,, \tag{4.35}$$

if $\Phi^i \in L^2(C(\mathbf{H}_n)),\ X(\Phi^i) \in L^2(\mathcal{U})$, for any smooth vector field X on $\mathcal{U} \subseteq C(\mathbf{H}_n)$ open, and

$$\sum_{i=1}^{\nu} \left\{ \int_{C(\mathbf{H}_n)} \Phi^i \, \Box \varphi^i \, d\,\mathrm{vol}(F_{\theta_0}) + \int_{C(\mathbf{H}_n)} F^{ab}_{\theta_0} \left(\left| \begin{matrix} i \\ jk \end{matrix} \right| \circ \pi \right) \frac{\partial \Phi^j}{\partial u^a} \frac{\partial \Phi^k}{\partial u^b} \varphi^i \, d\,\mathrm{vol}(F_{\theta_0}) \right\} = 0,$$

for any $\varphi \in C^\infty_0(C(\mathbf{H}_n), \mathbf{R}^\nu)$. □

Of course, the prescription $X(\Phi^i) \in L^2(\mathcal{U})$ means that there is $g^i_X \in L^2(\mathcal{U})$ such that

$$\int_{\mathcal{U}} g^i_X \varphi \, d\,\mathrm{vol}(F_{\theta_0}) = \int_{\mathcal{U}} \Phi^i \, X^* \varphi \, d\,\mathrm{vol}(F_{\theta_0}),$$

where X^* is the formal adjoint of X with respect to the L^2-inner product $(u, v) = \int uv\, d\mathrm{vol}(F_{\theta_0})$, for $u, v \in C^\infty(C(\mathbf{H}_n))$, at least one of compact support. A posteriori, the element g^i_X is uniquely determined a.e. and denoted by $X(\Phi^i)$.

Earlier in this chapter, given a strictly pseudoconvex CR manifold M, one related smooth harmonic maps from $C(M)$ (with the Fefferman metric corresponding to a fixed choice of contact form on M) to smooth pseudoharmonic maps from M (as argued there, these are locally formally similar to J. Jost and C.-J. Xu's subelliptic harmonic maps). Here we wish to attack the same problem for weak solutions (of the harmonic, respectively subelliptic harmonic, map equations). We recall the Sobolev-type spaces

$$W^{1,2}_X(\Omega) = \{u \in L^2(\Omega) : X_a u \in L^2(\Omega),\ 1 \leq a \leq m\},$$

adapted to a system of vector fields $X = \{X_1, \ldots, X_m\}$ on $\Omega \subseteq \mathbf{R}^n$ (the $X_a u$'s are understood in the distributional sense). See also Section 4.4.1 of this book.

Definition 4.14. $\phi : \Omega \to N$ is a *weak solution* to (4.19) if $\phi^i \in W_X^{1,2}(\Omega)$ and

$$\sum_{i=1}^{\nu} \left\{ \int_\Omega \sum_{a=1}^{m} (X_a \phi^i)\,(X_a \varphi^i)\,dx - \int_\Omega \sum_{a=1}^{m} \left(\left| \begin{matrix} i \\ jk \end{matrix} \right| \circ \phi \right) (X_a \phi^j)(X_a \phi^k) \varphi^i \,dx \right\} = 0,$$

for any $\varphi \in C_0^\infty(\Omega, \mathbf{R}^\nu)$. □

See, e.g., [234], p. 4641.

Lemma 4.3. *Let $\Phi : C(\mathbf{H}_n) \to N$ be an S^1-invariant map and $\phi = \tilde{\Phi}$ the corresponding base map. If $\Phi^i \in L^2(C(\mathbf{H}_n))$ and $Y(\Phi^i) \in L^2(\mathcal{U})$, for any smooth vector field Y on $\mathcal{U} \subseteq C(\mathbf{H}_n)$, then $\phi^i \in W_X^{1,2}(\mathbf{H}_n)$.*

One has $\|\Phi^i\|_{L^2(C(\mathbf{H}_n))} = 2\pi \|\phi^i\|_{L^2(\mathbf{H}_n)}$; hence $\phi^i \in L^2(\mathbf{H}^n)$. In particular $\phi^i \in L^1_{\rm loc}(\mathbf{H}_n)$ and $\Phi^i \in L^1_{\rm loc}(C(\mathbf{H}_n))$. Let $\varphi \in C_0^\infty(\mathbf{H}_n)$. Then $\varphi \circ \pi \in C_0^\infty(C(\mathbf{H}_n))$ (because S^1 is compact) and

$$\begin{aligned} \int_{C(\mathbf{H}_n)} \Phi^i \Box(\varphi \circ \pi)\, d\mathrm{vol}(F_{\theta_0}) &= -2 \int_{C(\mathbf{H}_n)} (\phi^i\ H\varphi) \circ \pi \ d\mathrm{vol}(F_{\theta_0}) \qquad \text{(by (4.33))} \\ &= -4\pi \int_{\mathbf{H}_n} \phi^i\ H\varphi\ \theta_0 \wedge (d\theta_0)^n \\ &= -4\pi \int_{\mathbf{H}_n} \sum_{a=1}^{2n} (X_a \phi^i)(X_a \varphi)\, \theta_0 \wedge (d\theta_0)^n. \end{aligned}$$

On the other hand,

$$\begin{aligned} \int_{\mathbf{H}_n} \phi^i (X_\alpha^* \varphi) dx &= - \int_{\mathbf{H}_n} \phi^i (X_\alpha \varphi) dx \\ &= -\frac{1}{2\pi} \int_{C(\mathbf{H}_n)} \Phi^i (X_\alpha \varphi) \circ \pi\ dx\, d\gamma = -\frac{1}{2\pi} \int \Phi^i \hat{X}_\alpha (\varphi \circ \pi)\, dx\, d\gamma \end{aligned}$$

(since $(\partial/\partial u^a)^* = -\partial/\partial u^a$)

$$= \frac{1}{2\pi} \int \Phi^i \left(\hat{X}_a\right)^* (\varphi \circ \pi)\, dx\, d\gamma = \frac{1}{2\pi} \int \left(\hat{X}_a \Phi^i\right) (\varphi \circ \pi)\, dx\, d\gamma.$$

The notation dx (respectively, $dx\, d\gamma$) is short for $\theta_0 \wedge (d\theta_0)^n$ (respectively, for $d\,\mathrm{vol}(F_{\theta_0})$). Also, we set

$$\hat{X}_\alpha := \frac{1}{2} \frac{\partial}{\partial u^\alpha} + u^{\alpha+n} \frac{\partial}{\partial u^{2n+1}}, \qquad \hat{X}_{\alpha+n} = \frac{1}{2} \frac{\partial}{\partial u^{\alpha+n}} - u^\alpha \frac{\partial}{\partial u^{2n+1}}.$$

The Jacobian of the right translation R_w with $w \in S^1$ is the unit matrix, hence for any $\psi \in C_0^\infty(C(\mathbf{H}_n))$,

$$\int \left(\hat{X}_a \Phi^i\right) \psi\, dx\, d\gamma = - \int \Phi^i\ \hat{X}_a \psi\, dx\, d\gamma = - \int (\Phi^i \circ R_w)\ (\hat{X}_a \psi) \circ R_w\, dx\, d\gamma$$

(since $\hat{X}_a$ is right-invariant)

$$= -\int \left(\Phi^i \circ R_w\right)(x,\gamma)\, \left((d_{(x,\gamma)} R_w)\hat{X}_a\right)(\psi)(x,\gamma)$$

(since Φ^i is S^1-invariant)

$$= -\int \Phi^i\, \hat{X}_a(\psi \circ R_w)\, dx\, d\gamma = \int (\hat{X}_a \Phi^i)\,(\psi \circ R_w)\, dx\, d\gamma$$
$$= \int \left((\hat{X}_a \Phi^i) \circ R_{w^{-1}}\right) \psi\, dx\, d\gamma;$$

hence $\hat{X}_a \Phi^i = (\hat{X}_a \Phi^i) \circ R_{w^{-1}}$, i.e., there is an element of $L^2(\mathbf{H}_n)$, which we denote by $X_a \phi^i$, such that

$$\hat{X}_a \Phi^i = (X_a \phi^i) \circ \pi\ .$$

We may conclude (by Fubini's theorem) that

$$\int \phi^i\, X_a^* \varphi\, dx = \int (X_a \phi^i)\, \varphi\, dx\ ,$$

i.e., $X_a \phi^i$ is indeed the weak derivative of ϕ^i. Lemma 4.3 is proved. □

At this point we may establish the following theorem:

Theorem 4.9. *Let $\phi : \mathbf{H}_n \to N$ be a map such that $\Phi := \phi \circ \pi$ satisfies weakly the harmonic map system* (4.35). *Then ϕ is a weak solution to the subelliptic harmonic map system* (4.19).

Combining the regularity results in [234] and [443] with Theorem 4.9 we obtain the following result:

Corollary 4.1. *Let N be a complete Riemannian manifold of sectional curvature $\leq \kappa^2$. Let $\Phi : C(\mathbf{H}_n) \to N$ be a bounded S^1-invariant weak solution to the harmonic map equation* (4.35) *such that $\Phi(C(\mathbf{H}_n))$ is contained in a regular[5] ball of N. Then Φ is smooth.*

The statement in Theorem 4.9 follows from the preceding calculations and the identity

$$\left(\left|\begin{matrix} i \\ jk \end{matrix}\right| \circ \Phi\right) \frac{\partial \Phi^j}{\partial u^a} \frac{\partial \Phi^k}{\partial u^b} F^{ab}_{\theta_0}$$
$$= \left(\left|\begin{matrix} i \\ jk \end{matrix}\right| \circ \Phi\right) \left\{ 2 \sum_{a=1}^{2n} (\hat{X}_a \Phi^j)(\hat{X}_a \Phi^k) + (n+2)\left(\frac{\partial \Phi^j}{\partial \gamma} \frac{\partial \Phi^k}{\partial u^{2n+1}} + \frac{\partial \Phi^j}{\partial u^{2n+1}} \frac{\partial \Phi^k}{\partial \gamma} \right) \right\}$$

[5] That is, a ball $B(p,\nu) = \{q \in N : d_N(q,p) \leq \nu\}$ such that $\nu < \min\{\pi/(2\kappa),\, i(p)\}$, where $i(p)$ is the injectivity radius of p (cf. [234], p. 4644).

(itself a consequence of (4.31)) provided we show that $\partial \Phi^j / \partial \gamma = 0$, as a distribution. Indeed, given $\varphi \in C_0^\infty(C(\mathbf{H}_n))$ set $C := \sup_{C(\mathbf{H}_n)} |\partial \varphi / \partial \gamma|$, $\Gamma := \operatorname{supp}(\varphi)$, and $\Gamma_H = \pi(\Gamma)$. Then, given $\phi_\nu^j \in C_0^\infty(\mathbf{H}_n)$ such that $\phi^j = L^2 - \lim_{\nu \to \infty} \phi_\nu^j$,

$$\left| \int_{C(\mathbf{H}_n)} (\phi_\nu^j \circ \pi) \frac{\partial \varphi}{\partial \gamma} d\mathrm{vol}(F_{\theta_0}) - \int_{C(\mathbf{H}_n)} (\phi^j \circ \pi) \frac{\partial \varphi}{\partial \gamma} d\mathrm{vol}(F_{\theta_0}) \right| \\ \leq 2\pi C \operatorname{vol}(\Gamma_H)^{1/2} \|\phi_\nu^j - \phi^j\|_{L^2(\mathbf{H}_n)};$$

hence (since (4.30) implies $(\partial/\partial\gamma)^* = -\partial/\partial\gamma$)

$$\begin{aligned} \frac{\partial \Phi^j}{\partial \gamma}(\varphi) &= - \int_{C(\mathbf{H}_n)} (\phi^j \circ \pi) \frac{\partial \varphi}{\partial \gamma} d\mathrm{vol}(F_{\theta_0}) \\ &= - \lim_{\nu \to \infty} \int_{C(\mathbf{H}_n)} (\phi_\nu^j \circ \pi) \frac{\partial \varphi}{\partial \gamma} d\mathrm{vol}(F_{\theta_0}) = 0, \end{aligned}$$

by Green's lemma and $\operatorname{div}(\partial/\partial\gamma) = 0$, again as a consequence of (4.30).

Example 1. (*continued*) As shown by Proposition 4.8, as a consequence of the hypoellipticity of the Hörmander operator together with the existence of local harmonic coordinates on the target manifold, there is no notion of *weak* subelliptic harmonic morphism (of course, this is true for harmonic morphisms between Riemannian manifolds as well). Nevertheless, in the context of the Hörmander system (4.22) it makes sense (since $\Box$ is nonelliptic) to adopt the following definition:

Definition 4.15. We say that a localizable map $\Phi : (C(\Omega^\pm), F_{\theta(k)}) \to (N, h)$ is a *weak harmonic morphism* if, for any local harmonic function $v : V \to \mathbf{R}$ on N one has $v \circ \Phi \in L^1_{\mathrm{loc}}(\mathcal{U})$, for any $\mathcal{U} \subseteq C(\Omega^\pm)$ open such that $\Phi(\mathcal{U}) \subset V$, and $\Box(v \circ \Phi) = 0$ in the distributional sense. □

Then we may prove the following regularity result (and converse of Proposition 4.10):

Proposition 4.11. *If $\Phi : C(\Omega^\pm) \to N$ is an S^1-invariant weak harmonic morphism then the base map $\phi = \tilde{\Phi} : \Omega^\pm \to N$ is a smooth subelliptic harmonic morphism (in particular, Φ is smooth).*

Let $\varphi \in C_0^\infty(\Omega^\pm)$. Then

$$0 = \Box(v \circ \Phi)(\varphi \circ \pi) = \int_{C(\Omega_\pm)} (v \circ \Phi) \Box(\varphi \circ \pi) \, d\mathrm{vol}(F_{\theta(k)})$$

(by (4.34) and Fubini's theorem)

$$= -\frac{2\pi}{k} \int_{\Omega^\pm} (v \circ \phi)(x) x_1^{1-k} (H\varphi)(x) \, dx = -\frac{2\pi}{k} H(x_1^{1-k} \, v \circ \phi)(\varphi),$$

i.e., $H(x_1^{1-k} \, v \circ \phi) = 0$ in the distributional sense [here dx is short for $\theta(k) \wedge (d\theta(k))^n$]. Hence there is $f \in C^\infty(U)$ such that $v \circ \phi = x_1^{k-1} f$, i.e., $v \circ \phi$ is smooth (here $U \subseteq \Omega^\pm$ is any open set such that $\phi(U) \subset V$). Then, again by (4.34), $H(v \circ \phi) = 0$. □

4.5 Generalizations of pseudoharmonicity

Owing to the ideas of M. Ara (cf. [22]) and K. Uhlenbeck (cf. [410]), in this section we consider *F-pseudoharmonic maps*, i.e., critical points of the energy

$$E_F(\phi) = \int_M F\left(\tfrac{1}{2}\text{trace}_{G_\theta}(\pi_H \phi^* h)\right) \theta \wedge (d\theta)^n ,$$

on the class of smooth maps $\phi : M \to N$ from a (compact) strictly pseudoconvex CR manifold (M, θ) to a Riemannian manifold (N, h), where θ is a contact form and $F : [0, \infty) \to [0, \infty)$ is a C^2 function such that $F'(t) > 0$. F-pseudoharmonic maps generalize both J. Jost and C.-J. Xu's subelliptic harmonic maps (the case $F(t) = t$, cf. [234]) and P. Hajlasz and P. Strzelecki's subelliptic p-harmonic maps (the case $F(t) = (2t)^{p/2}$, cf. [192]).

The present section is organized as follows. We obtain the first variation formula for $E_F(\phi)$. We investigate the relationship between F-pseudoharmonicity and pseudoharmonicity, by exploiting the analogy between CR and conformal geometry (cf. M. Ara [22] for the Riemannian counterpart). We consider *pseudoharmonic morphisms* from a strictly pseudoconvex CR manifold and show that any pseudoharmonic morphism is a pseudoharmonic map (the CR analogue of T. Ishihara's theorem, cf. [217]). We give a geometric interpretation of F-pseudoharmonicity in terms of the Fefferman metric of (M, θ). The results we report on belong to E. Barletta [33].

Let $(M, T_{1,0}(M))$ be a strictly pseudoconvex CR manifold, of CR dimension n, and θ a contact form on M such that the Levi form G_θ is positive definite.

Definition 4.16. Let $F : [0, \infty) \to [0, \infty)$ be a C^2 function such that $F'(t) > 0$. For a smooth map $\phi : (M, \theta) \to (N, h)$ and a compact domain $D \subseteq M$ we consider the *energy* function

$$E_F(\phi; D) = \int_D F\left(\tfrac{1}{2}\,\text{trace}_{G_\theta}(\pi_H \phi^* h)\right) \theta \wedge (d\theta)^n . \tag{4.36}$$

Here (N, h) is a Riemannian manifold. Then ϕ is *F-pseudoharmonic* if for any compact domain $D \subseteq M$, it is an extremal of the energy $E_F(\cdot\,; D)$ with respect to all variations of ϕ supported in D. □

For $F(t) = t$, (4.36) is the energy function in Definition 4.5 (and extremals were referred to as pseudoharmonic maps). The cases $F(t) = (2t)^{p/2}$ $(p \geq 4)$ and $F(t) = \exp(t)$ (familiar in the theory of harmonic maps of Riemannian manifolds and their generalizations, such as p-harmonic maps, cf., e.g., P. Baird and S. Gudmundson [28], L.F. Cheung and P.F. Leung [101], or exponentially harmonic maps, cf. M.C. Hong [210], S.E. Koh [245]) have not been studied from the point of view of CR and pseudo-Hermitian geometry. However, if $F(t) = (2t)^{p/2}$ and $\phi : (M, \theta) \to (S^m, h_0)$ is F-pseudoharmonic [where S^m is the unit sphere in $\mathbf{R}^{m+1}$ and h_0 the standard Riemannian metric on S^m] then given a local coordinate system (U, φ) on M, the map $\phi \circ \varphi^{-1} : \Omega \to S^m$ $(\Omega = \varphi(U))$ is formally similar to a subelliptic p-harmonic map in the sense of P. Hajlasz and P. Strzelecki [192]. We emphasize, however, that our notion is distinct from theirs. Indeed, P. Hajlasz and P. Strzelecki (cf. op. cit.) employ the Euclidean metric on Ω (rather than the volume form $\theta \wedge (d\theta)^n$) to define the relevant operators (formal adjoints, the Hörmander operator, etc.).

Let us look at the following *first variation formula* [stated for simplicity in the case M is compact (and then one adopts the shorter notation $E_F(\phi) := E_F(\phi; M)$)].

Theorem 4.10. (E. Barletta [33])
Let M be a compact strictly pseudoconvex CR manifold, of CR dimension n, and θ a contact form on M such that the Levi form G_θ is positive definite. Let (N, h) be a Riemannian manifold. Let $F : [0, \infty) \to [0, \infty)$ be a C^2 map such that $F'(s) > 0$ and set $\rho(s) := F'(s/2)$. Let $\{\phi_t\}_{|t|<\epsilon}$ be a 1-parameter variation of a smooth map $\phi = \phi_0 : M \to N$. Then

$$\frac{d}{dt}\{E_F(\phi_t)\}_{t=0} = -\int_M \tilde{h}\,(V, \tau_F(\phi; \theta, h))\,\theta \wedge (d\theta)^n,$$

where $Q := \mathrm{trace}_{G_\theta}(\pi_H \phi^* h)$ and

$$\tau_F(\phi; \theta, h) := \sum_{a=1}^{2n} \left[(\phi^{-1}\nabla^N)_{X_a}(\rho(Q)\phi_* X_a) - \rho(Q)\phi_* \nabla_{X_a} X_a \right].$$

Here $\{X_a\}$ is a local G_θ-orthonormal frame of $H(M)$. Also we set $\tilde{M} := (-\epsilon, \epsilon) \times M$ and

$$\Phi : \tilde{M} \to N, \quad \Phi(t, x) := \phi_t(x), \quad x \in M, \ |t| < \epsilon,$$

$$V_x := (d_{(0,x)}\Phi)\left.\frac{\partial}{\partial t}\right|_{(0,x)} \in T_{\phi(x)}(N), \quad x \in M.$$

Then ϕ is F-pseudoharmonic if

$$\tau_F(\phi; \theta, h) = 0. \tag{4.37}$$

Moreover, for each smooth map $\phi : M \to N$ the tension field

$$\tau_F(\phi; \theta, h) \subset \Gamma^\infty(\phi^{-1}TN)$$

is also given by

$$\tau_F(\phi; \theta, h) = \left\{ \mathrm{div}(\rho(Q)\nabla^H \phi^i) + \sum_{a=1}^{2n} \rho(Q) \left(\left| \begin{matrix} i \\ jk \end{matrix} \right| \circ \phi \right) X_a(\phi^j) X_a(\phi^k) \right\} Y_i \tag{4.38}$$

on $U := \phi^{-1}(V)$, where (V, y^i) is a local coordinate system on N, $\phi^j := y^j \circ \phi$, and $Y_j(x) := (\partial/\partial y^j)(x)$, $x \in U$, $1 \le j \le m$.

The proof is relegated to the next section. Here

$$\left| \begin{matrix} i \\ jk \end{matrix} \right| = h^{i\ell}|jk, \ell|, \quad |ij, k| = \frac{1}{2}\left(\frac{\partial h_{ik}}{\partial y^j} + \frac{\partial h_{jk}}{\partial y^i} - \frac{\partial h_{ij}}{\partial y^k} \right),$$

are the Christoffel symbols of (N, h). As a consequence of (4.38) the Euler–Lagrange equations (4.37) (the *F-pseudoharmonic map equation*) for $\phi : (M, \theta) \to (\mathbf{R}^m, h_0)$ may be written

$$\operatorname{div}(\rho(Q)\nabla^H \phi^j) = 0, \quad 1 \le j \le m. \tag{4.39}$$

Here h_0 is the natural flat metric on $\mathbf{R}^m$. The reader should compare to (4.39) to the equation (0.1) considered by K. Uhlenbeck [410].

The relationship between F-pseudoharmonicity and pseudoharmonicity is clarified by the following theorem:

Theorem 4.11. (E. Barletta [33])
Let a CR manifold (M, θ), a Riemannian manifold (N, h), and a C^2 function $F : [0, \infty) \to [0, \infty)$ be as in Theorem 4.10. Then

$$\tau_F(\phi; \theta, h) = F'\left(\frac{Q}{2}\right)^{1+1/n} \tau\left(\phi;\ F'\left(\frac{Q}{2}\right)^{1/n} \theta\,,\ h\right).$$

Thus $\phi : (M, \theta) \to (N, h)$ is an F-pseudoharmonic map if and only if $\phi : (M, F'(Q/2)^{1/n}\,\theta) \to (N, h)$ is a pseudoharmonic (with respect to the data $(F'(Q/2)^{1/n}\,\theta\,,\ h)$) map.

The tension field $\tau(\phi; \theta, h)$ in Theorem 4.11 is obtained from $\tau_F(\phi; \theta, h)$ for $F(t) = t$.

4.5.1 The first variation formula

Let $\{Z_1, \dots, Z_n\}$ be a local frame of $T_{1,0}(M)$, defined on an open set $U \subseteq M$, such that $L_\theta(Z_\alpha, Z_{\overline{\beta}}) = \delta_{\alpha\beta}$ (with $Z_{\overline{\beta}} = \overline{Z}_\beta$). We recall that for any bilinear form B on $T(M)$,

$$\operatorname{trace}_{G_\theta}(\pi_H B) = \sum_{\alpha=1}^{n} \{B(X_\alpha, X_\alpha) + B(JX_\alpha, JX_\alpha)\}\,,$$

where $X_\alpha := \frac{1}{\sqrt{2}}(Z_\alpha + Z_{\overline{\alpha}})$ (hence $G_\theta(X_\alpha, X_\beta) = \delta_{\alpha\beta}$). If X is a tangent vector field on M, $\phi_* X$ denotes the section in $\phi^{-1}TN \to M$ given by $(\phi_* X)(x) := (d_x\phi)X_x \in T_{\phi(x)}(N) = (\phi^{-1}TN)_x$, $x \in M$. Note that

$$\operatorname{trace}_{G_\theta}(\pi_H \phi^* h) = \sum_{a=1}^{2n} \tilde{h}(\phi_* X_a, \phi_* X_a) \ge 0;$$

hence the definition of $E_F(\phi; D)$ makes sense (remember that $F(t)$ is defined only for $t \ge 0$). Here $\{X_a : 1 \le a \le 2n\} := \{X_\alpha, JX_\alpha : 1 \le \alpha \le n\}$.

Let $\phi : M \to N$ be a smooth map. Then ϕ is F-pseudoharmonic if

$$\frac{d}{dt}\{E_F(\phi_t)\}_{t=0} = 0,$$

for any compactly supported 1-parameter variation $\{\phi_t\}_{|t|<\epsilon}$ of $\phi_0 = \phi$. To write the first variation formula we set $\tilde{M} := (-\epsilon, \epsilon) \times M$ and

$$\Phi : \tilde{M} \to N, \quad \Phi(t, x) := \phi_t(x), \quad x \in M, \ |t| < \epsilon.$$

Also

$$V_x := (d_{(0,x)}\Phi) \left.\frac{\partial}{\partial t}\right|_{(0,x)} \in T_{\phi(x)}(N), \quad x \in M.$$

Then $V \in \Gamma^\infty(\phi^{-1}TN)$. Moreover, let

$$\mathbf{V} := \Phi_* \frac{\partial}{\partial t} \in \Gamma^\infty(\Phi^{-1}TN)$$

(hence $\mathbf{V}_{(0,x)} = V_x$). Let $\tilde{\nabla} := \Phi^{-1}\nabla^N$ be the connection in $\Phi^{-1}TN \to \tilde{M}$ induced by ∇^N (the Levi-Civita connection of (N, h)). We have

$$\begin{aligned}
\frac{d}{dt}\{E_F(\phi_t)\} &= \frac{d}{dt}\int_M F\left(\tfrac{1}{2}\,\mathrm{trace}_{G_\theta}\left(\pi_H \phi_t^* h\right)\right)\Psi \\
&= \int_M \frac{d}{dt} F\Big(\tfrac{1}{2}\sum_{a=1}^{2n}(\phi_t^* h)(X_a, X_a)\Big)\Psi,
\end{aligned}$$

where as usual $\Psi = \theta \wedge (d\theta)^n$. Let $\alpha_t : M \to \tilde{M}$ be given by $\alpha_t(x) := (t, x), \ x \in M$. If X is a tangent vector field on M we set $\tilde{X}_{(t,x)} := (d_x\alpha_t)X_x$. The symbol $\tilde{h}$ denotes the bundle metric $\Phi^{-1}h$ (induced by h) in $\Phi^{-1}TN \to \tilde{M}$ as well. Then

$$(\phi_t^* h)(X, Y) = \tilde{h}(\Phi_*\tilde{X}, \Phi_*\tilde{Y}) \circ \alpha_t.$$

Set

$$Q_t := \mathrm{trace}_{G_\theta}\left(\pi_H \phi_t^* h\right) \in C^\infty(M), \ |t| < \epsilon.$$

Then

$$\begin{aligned}
\frac{d}{dt}\{E_F(\phi_t)\} &= \tfrac{1}{2}\int_M F'\Big(\frac{Q_t}{2}\Big)\sum_{a=1}^{2n}\frac{d}{dt}\Big\{\tilde{h}(\Phi_*\tilde{X}_a, \Phi_*\tilde{X}_a)\circ\alpha_t\Big\}\Psi \\
&= \int_M F'\Big(\frac{Q_t}{2}\Big)\sum_{a=1}^{2n}\tilde{h}(\tilde{\nabla}_{\partial/\partial t}\Phi_*\tilde{X}_a, \Phi_*\tilde{X}_a)\Psi \\
&= \int_M F'\Big(\frac{Q_t}{2}\Big)\sum_{a=1}^{2n}\tilde{h}(\tilde{\nabla}_{\tilde{X}_a}\Phi_*\frac{\partial}{\partial t}, \Phi_*\tilde{X}_a)\Psi,
\end{aligned}$$

since $\tilde{\nabla}\tilde{h} = 0$, ∇^N is torsion-free, and $[\tilde{X}, \partial/\partial t] = 0$. Next

$$\frac{d}{dt}\{E_F(\phi_t)\} = \int_M F'\Big(\frac{Q_t}{2}\Big)\sum_a\Big\{\tilde{X}_a(\tilde{h}(\mathbf{V}, \Phi_*\tilde{X}_a)) - \tilde{h}(\mathbf{V}, \tilde{\nabla}_{\tilde{X}_a}\Phi_*\tilde{X}_a)\Big\}\Psi.$$

For $|t| < \epsilon$ fixed, let $X_t \in \Gamma^\infty(H(M))$ be defined by

$$G_\theta(X_t, Y) = \tilde{h}(\mathbf{V}, \Phi_* \tilde{Y}) \circ \alpha_t, \qquad Y \in \Gamma^\infty(H(M)).$$

Also, if $f \in C^\infty(\tilde{M})$ and $|t| < \epsilon$ we set $f_t := f \circ \alpha_t \in C^\infty(M)$, so that

$$X(f_t) = \tilde{X}(f) \circ \alpha_t ,$$

for any $X \in T(M)$. Therefore

$$\frac{d}{dt}\{E_F(\phi_t)\} = \int_M \rho(Q_t) \sum_a \left\{ X_a(G_\theta(X_t, X_a)) - \tilde{h}(\mathbf{V}, \tilde{\nabla}_{\tilde{X}_a} \Phi_* \tilde{X}_a) \right\} \Psi$$

$$= \int_M \rho(Q_t) \sum_a \left\{ G_\theta(\nabla_{X_a} X_t, X_a) + G_\theta(X_t, \nabla_{X_a} X_a) - \tilde{h}(\mathbf{V}, \tilde{\nabla}_{\tilde{X}_a} \Phi_* \tilde{X}_a) \right\} \Psi,$$

where $\rho(s) := F'(s/2)$ and ∇ is the Tanaka–Webster connection of (M, θ) (so that $\nabla G_\theta = 0$). As usual we compute the divergence by the formula

$$\operatorname{div}(X) = \operatorname{trace}\{Y \mapsto \nabla_Y X\}.$$

Then

$$\frac{d}{dt}\{E_F(\phi_t)\} = \int_M \rho(Q_t) \left\{ \operatorname{div}(X_t) - - \sum_a \tilde{h}(\mathbf{V}, \tilde{\nabla}_{\tilde{X}_a} \Phi_* \tilde{X}_a - \Phi_* \widetilde{\nabla_{X_a} X_a}) \right\} \Psi$$

$$= \int_M \left\{ \operatorname{div}(\rho(Q_t) X_t) - X_t(\rho(Q_t)) - \rho(Q_t) \sum_a \tilde{h}(\mathbf{V}, \tilde{\nabla}_{\tilde{X}_a} \Phi_* \tilde{X}_a - \Phi_* \widetilde{\nabla_{X_a} X_a}) \right\} \Psi,$$

because of $\operatorname{div}(fX) = f \operatorname{div}(X) + X(f)$. Since ϕ_t is compactly supported, so is X_t. Therefore (by Green's lemma)

$$\frac{d}{dt}\{E_F(\phi_t)\} = -\int_M \Big\{ G_\theta(X_t, \nabla^H \rho(Q_t))$$

$$+ \tilde{h}(\mathbf{V}, \rho(Q_t) \sum_a (\tilde{\nabla}_{\tilde{X}_a} \Phi_* \tilde{X}_a - \Phi_* \widetilde{\nabla_{X_a} X_a})) \Big\} \Psi$$

$$= -\int_M \tilde{h}\left(\mathbf{V}, \sum_a \left(\tilde{\nabla}_{\tilde{X}_a} (\rho(Q_t) \Phi_* \tilde{X}_a) - \rho(Q_t) \Phi_* \widetilde{\nabla_{X_a} X_a} \right) \right) \Psi .$$

The last equality holds because of

$$\sum_{a=1}^{2n} \tilde{\nabla}_{\tilde{X}_a} \left(\rho(Q_t) \Phi_* \tilde{X}_a \right) = \sum_a \left(\tilde{X}_a(\rho(Q_t)) \Phi_* \tilde{X}_a + \rho(Q_t) \tilde{\nabla}_{\tilde{X}_a} \Phi_* \tilde{X}_a \right)$$

$$= \Phi_* \nabla^H \rho(Q_t) + \rho(Q_t) \sum_a \tilde{\nabla}_{\tilde{X}_a} \Phi_* \tilde{X}_a .$$

Note that

$$\left((\Phi^{-1}\nabla^N)_X \Phi_* Y\right)_{(0,x)} = \left((\phi^{-1}\nabla^N)_X \phi_* Y\right)_x .$$

We are left with the proof of (4.38). We have

$$\begin{aligned}
\operatorname{div}(\rho(Q)\nabla^H \phi^j) &= \operatorname{trace}\{Y \mapsto \nabla_Y(\rho(Q)\nabla^H\phi^j)\} \\
&= \sum_{a=1}^{2n} G_\theta(\nabla_{X_a}(\rho(Q)\nabla^H\phi^j), X_a) + \theta(\nabla_T(\rho(Q)\nabla^H\phi^j)) \\
&= \sum_{a=1}^{2n} \left[X_a(G_\theta(\rho(Q)\nabla^H\phi^j, X_a)) - G_\theta(\rho(Q)\nabla^H\phi^j, \nabla_{X_a}X_a)\right] \\
&\qquad + T(\theta(\rho(Q)\nabla^H\phi^j)) ,
\end{aligned}$$

because of $\nabla g_\theta = 0$ and $\nabla T = 0$. Also (since $H(M)$ is ∇-parallel)

$$G_\theta(\nabla^H\phi^j, \nabla_{X_a}X_a)Y_j = g_\theta(\nabla\phi^j, \nabla_{X_a}X_a)Y_j = (\nabla_{X_a}X_a)(\phi^j)\, Y_j = \phi_*\nabla_{X_a}X_a;$$

hence

$$\operatorname{div}\big(\rho(Q)\nabla^H\phi^j\big)Y_j = \sum_{a=1}^{2n}\Big\{X_a\big(\rho(Q)X_a(\phi^j)\big)Y_j - \rho(Q)\phi_*\nabla_{X_a}X_a\Big\}.$$

Consequently (by the definition of $\tau_F(\phi;\theta,h)$)

$$\tau_F(\phi;\theta,h) = \operatorname{div}(\rho(Q)\nabla^H\phi^j)Y_j + \rho(Q)\sum_a\{\tilde\nabla_{X_a}\phi_*X_a - (X_a^2\phi^j)Y_j\}$$

and the following calculation leads to (4.38)

$$\begin{aligned}
\tilde\nabla_{X_a}\phi_*X_a - (X_a^2\phi^j)Y_j &= \tilde\nabla_{X_a}(X_a(\phi^j)Y_j) - (X_a^2\phi^j)Y_j \\
&= X_a(\phi^j)X_a(\phi^k)\left(\nabla^N_{\partial/\partial y^k}\frac{\partial}{\partial y^j}\right)\circ\phi \\
&\quad - \left(\left|\begin{matrix} i \\ jk \end{matrix}\right|\circ\phi\right) X_a(\phi^j)X_a(\phi^k)Y_i
\end{aligned}$$

Theorem 4.10 is proved. □

To prove Theorem 4.11 we need to derive the transformation law for $\tau(\phi;\theta,h)$ under a change of contact form $\hat\theta = e^{2u}\theta$, $u \in C^\infty(M)$. Set

$$\beta_\phi(X,Y) := (\phi^{-1}\nabla^N)_X\phi_*Y - \phi_*\nabla_X Y ,$$

for any $X, Y \in T(M)$, so that

$$\tau(\phi;\theta,h) = \operatorname{trace}_{G_\theta}\left(\pi_H\beta_\phi\right).$$

Consequently

$$\tau_F(\phi;\theta,h) = F'\left(\frac{Q}{2}\right)\tau(\phi;\theta,h) + \phi_*\nabla^H F'\left(\frac{Q}{2}\right). \tag{4.40}$$

If $\{Z_\alpha\}$ is a local orthonormal frame of $T_{1,0}(M)$ we set $\hat{Z}_\alpha := e^{-u} Z_\alpha$. Note that

$$\tau(\phi;\theta,h) = \sum_{\alpha=1}^{n} \left\{\beta_\phi(Z_\alpha, Z_{\overline{\alpha}}) + \beta_\phi(Z_{\overline{\alpha}}, Z_\alpha)\right\}.$$

Therefore

$$\tau(\phi;\hat{\theta},h) = \sum_{\alpha=1}^{n} \{\tilde{\nabla}_{\hat{Z}_\alpha}\phi_*\hat{Z}_{\overline{\alpha}} - \phi_*\hat{\nabla}_{\hat{Z}_\alpha}\hat{Z}_{\overline{\alpha}} + \tilde{\nabla}_{\hat{Z}_{\overline{\alpha}}}\phi_*\hat{Z}_\alpha - \phi_*\hat{\nabla}_{\hat{Z}_{\overline{\alpha}}}\hat{Z}_\alpha\},$$

where $\tilde{\nabla} = \phi^{-1}\nabla^N$ and $\hat{\nabla}$ is the Tanaka–Webster connection of $(M,\hat{\theta})$. Set $\nabla_{Z_A} Z_B = \Gamma^C_{AB} Z_C$, where $A, B, \ldots \in \{1,\ldots,n,\overline{1},\ldots,\overline{n},0\}$ and $Z_0 := T$. Then

$$\hat{\nabla}_Z \overline{W} = \nabla_Z \overline{W} - 2L_\theta(Z,\overline{W})\nabla^{0,1}u\,, \quad Z, W \in T_{1,0}(M)$$

(where $\nabla^{0,1}u := \pi_{0,1}\nabla u$ and $\pi_{0,1} : T(M)\otimes \mathbf{C} \to T_{0,1}(M)$ is the projection (associated with $T(M)\otimes\mathbf{C} = T_{1,0}(M)\oplus T_{0,1}(M)\oplus \mathbf{C}T$)) leads to

$$e^{2u}\,\tau(\phi;\hat{\theta},h) = \tau(\phi;\theta,h) + 2n\,\phi_*\nabla^H u\,.$$

Set $\lambda := e^u$. Then

$$\tau(\phi;\lambda^2\theta,h) = \lambda^{-2(n+1)}\left\{\lambda^{2n}\tau(\phi;\theta,h) + \phi_*\nabla^H(\lambda^{2n})\right\}. \tag{4.41}$$

This is the transformation law we were looking for. Setting $\lambda := \rho(Q)^{1/(2n)}$, the formulas (4.40)–(4.41) lead to the identity in Theorem 4.11. □

4.5.2 Pseudoharmonic morphisms

Pseudo-Hermitian maps are known to be examples of pseudoharmonic maps $\phi: (M,\theta)\to(N,g_{\theta_N})$, where g_{θ_N} is the Webster metric of (N,θ_N), and these are also the only CR maps that are pseudoharmonic; cf. Theorem 4.1. New examples, as obtained in this section, are the *pseudoharmonic morphisms*.

Definition 4.17. Let M be a nondegenerate CR manifold and θ a contact form on M. Let $\phi: M\to N$ be a smooth map into a Riemannian manifold (N,h). We say that ϕ is a *pseudoharmonic morphism* if for each local harmonic function $v: V\to\mathbf{R}$ ($V\subseteq N$ open, $\Delta_N v = 0$, where Δ_N is the Laplace–Beltrami operator of (N,h)) one has $\Delta_b(v\circ\phi) = 0$ in $U := \phi^{-1}(V)$. □

Theorem 4.12. (E. Barletta [33])
Let M be a nondegenerate CR manifold, of CR dimension n, and θ a contact form on M. Let (N,h) be an m-dimensional Riemannian manifold. If $m > n$ there is no pseudoharmonic morphism of (M,θ) into (N,h), except for the constant maps. If $m \leq$

n then any pseudoharmonic morphism $\phi : (M, \theta) \to (N, h)$ is a pseudoharmonic map and a C^∞ submersion and there is a unique C^∞ function $\lambda : M \to [0, +\infty)$ such that

$$g_\theta^*(d_H\phi^i, d_H\phi^j)_x = 2\lambda(x)\delta^{ij}, \quad 1 \le i, j \le m, \tag{4.42}$$

for any $x \in M$ and any normal coordinate system (V, y^i) at $\phi(x) \in N$.

The Riemannian counterpart of Theorem 4.12 is a result of T. Ishihara [217] (thought of as foundational for the theory of harmonic morphisms; cf., e.g., J.C. Wood [433]). To prove Theorem 4.12 we make use of T. Ishihara's lemma (cf. Lemma 4.2 in this book). The proof of Theorem 4.12 is similar to that of Theorem 4.6.

Let $\phi : (M, \theta) \to (N, h)$ be a pseudoharmonic morphism. Let $x \in M$ be an arbitrary point and set $p := \phi(x) \in N$. Let $i_0 \in \{1, \dots, m\}$ be a fixed index and set $C_i := \delta_{ii_0}$ and $C_{ij} := 0$, $1 \le i, j \le m$. By Ishihara's lemma, given normal coordinates (V, y^i) at p, there is a harmonic function $v : V \to \mathbf{R}$. Then $\Delta_b(v \circ \phi) = 0$ in $U := \phi^{-1}(V)$, by the definition of pseudoharmonic morphisms. Let $\{Z_\alpha\}$ be an orthonormal frame of $T_{1,0}(M)$ on U. Note that

$$(v \circ \phi)_{\alpha,\overline{\beta}} = (v_j \circ \phi)\phi^j_{\alpha,\overline{\beta}} + \phi^i_\alpha \phi^j_{\overline{\beta}} \left(v_{i,j} + \left| \begin{matrix} k \\ ij \end{matrix} \right| v_k \right) \circ \phi, \qquad v_i := \frac{\partial v}{\partial y^i},$$

$$\Delta_b(v \circ \phi) = \sum_\alpha \{(v \circ \phi)_{\overline{\alpha},\alpha} + (v \circ \phi)_{\alpha,\overline{\alpha}}\},$$

yield

$$\Delta_b(v \circ \phi) = (\Delta_b \phi^j)(v_j \circ \phi) + 2 \sum_\alpha \phi^j_\alpha \phi^k_{\overline{\alpha}} \left(v_{j,k} + \left| \begin{matrix} i \\ jk \end{matrix} \right| v_i \right) \circ \phi . \tag{4.43}$$

Let us apply (4.43) at the preferred point x:

$$0 = (\Delta_b \phi^{i_0})(x) + 2 \sum_\alpha \phi^j_\alpha(x) \phi^k_{\overline{\alpha}}(x) \left| \begin{matrix} i_0 \\ jk \end{matrix} \right| (p) = \tau(\phi; \theta, h)^{i_0}_x ,$$

i.e., ϕ is a pseudoharmonic map.

Let us consider now $C_{ij} \in \mathbf{R}$ such that $C_{ij} = C_{ji}$ and $\sum_{i=1}^m C_{ii} = 0$. By Ishihara's lemma there is $v : V \to \mathbf{R}$ such that $\Delta_N v = 0$, $v_i(p) = 0$ and $v_{i,j}(p) = C_{ij}$. Since ϕ is a pseudoharmonic morphism (by (4.43)),

$$0 = \Delta_b(v \circ \phi)(x) = \sum_\alpha \phi^j_\alpha(x) \phi^k_{\overline{\alpha}}(x) C_{jk} .$$

Set

$$X^{jk} := \sum_\alpha \phi^j_\alpha \phi^k_{\overline{\alpha}} ,$$

so that

$$C_{jk} X^{jk}(x) = 0.$$

Thus

$$\sum_{i \neq j} C_{ij} X^{ij}(x) + \sum_{i} \left(X^{ii}(x) - X^{11}(x) \right) C_{ii} = 0. \tag{4.44}$$

Let us choose for a moment $C_{ij} = 0$ for any $i \neq j$ and

$$C_{ii} = \begin{cases} 1, & i = i_0, \\ -1, & i = 1, \\ 0, & \text{otherwise}, \end{cases}$$

for $i_0 \in \{2, \ldots, m\}$ fixed. Then (4.44) yields

$$X^{i_0 i_0}(x) - X^{11}(x) = 0,$$

i.e.,

$$X^{11}(x) = \cdots = X^{mm}(x),$$

and (4.44) becomes

$$\sum_{i \neq j} C_{ij} X^{ij}(x) = 0.$$

Again we may fix $i_0, j_0 \in \{1, \ldots, m\}$ with $i_0 \neq j_0$ and choose

$$C_{ij} = \begin{cases} 1, & i = i_0, \quad j = j_0, \\ 0, & \text{otherwise}, \end{cases}$$

to get

$$X^{i_0 j_0}(x) = 0.$$

We have proved that $X^{ij}(x) = 0$ for any $i \neq j$. If we set

$$\lambda := X^{11} = \sum_{\alpha=1}^{n} \phi^1_\alpha \phi^1_{\bar{\alpha}} \in C^\infty(U),$$

then

$$\sum_{\alpha=1}^{n} \phi^i_\alpha(x) \phi^j_{\bar{\alpha}}(x) = \lambda(x) \delta^{ij} . \tag{4.45}$$

The contraction of i, j now leads to

$$m\lambda = \sum_{\alpha, i} |\phi^i_\alpha|^2 \geq 0;$$

hence $\lambda : M \to [0, +\infty)$ is a C^∞ function. To complete the proof of Theorem 4.12 assume that there is $x \in M$ such that $\lambda(x) \neq 0$ and set

$$w^i := (\phi^i_1(x), \ldots, \phi^i_n(x)) \in \mathbf{C}^n .$$

Clearly $w^i \neq 0$ and (by (4.45)) $i \neq j \Longrightarrow w^i \cdot \overline{w^j} = 0$, that is, the rows of $[\phi^i_\alpha(x)] = (w^1, \ldots, w^m)^t$ are mutually orthogonal nonzero vectors in $\mathbf{C}^n$. Hence $\text{rank}[\phi^i_\alpha(x)] = m$ and then $m \leq n$. Therefore $m > n$ implies $\lambda = 0$; hence $\phi^i_\alpha = 0$. Thus $\phi^i_{\overline{\alpha}} = 0$ (by complex conjugation) or $\overline{\partial}_b \phi^i = 0$, i.e., ϕ^i is an $\mathbf{R}$-valued CR function on a nondegenerate CR manifold; hence $\phi^i = \text{const}$, $1 \leq i \leq m$, i.e., ϕ is a constant map. □

4.5.3 The geometric interpretation of F-pseudoharmonicity

The notion of a F-pseudoharmonic map admits the following geometric interpretation (in terms of the Fefferman metric):

Theorem 4.13. (E. Barletta [33])
Let M be a compact strictly pseudoconvex CR manifold, θ a contact form on M, (N, h) a Riemannian manifold, and $F : [0, +\infty) \to [0, +\infty)$ a C^2 function, as in Theorem 4.10. Let $S^1 \to C(M) \xrightarrow{\pi} M$ be the canonical circle bundle and F_θ the Fefferman metric of (M, θ). Let $\phi : M \to N$ be a smooth map. Then $\phi : (M, \theta) \to (N, h)$ is an F-pseudoharmonic map if and only if its vertical lift $\phi \circ \pi : (C(M), F_\theta) \to (N, h)$ is an F-harmonic map in the sense of M. Ara, [22], i.e., a critical point of the energy

$$\mathbf{E}(\Phi) = \int_{C(M)} F\left(\tfrac{1}{2} \operatorname{trace}_{F_\theta} \Phi^* h\right) \, d\operatorname{vol}(F_\theta),$$

on the class of all smooth functions $\Phi : C(M) \to N$. Here $d\operatorname{vol}(F_\theta)$ is the natural volume form on the Lorentzian manifold $(C(M), F_\theta)$.

Let $(M, T_{1,0}(M))$ be a strictly pseudoconvex CR manifold, of CR dimension n, and θ a contact form on M with G_θ positive definite. Let $F(t) \geq 0$ be a C^2 map defined for $t \geq 0$, such that $F'(t) > 0$. For simplicity, assume for the rest of this section that M is compact; hence $C(M)$ is compact as well.

Definition 4.18. A smooth map $\Phi : (C(M), F_\theta) \to (N, h)$ is said to be *F-harmonic* if it is a critical point of

$$\mathbf{E}_F(\Phi) = \int_{C(M)} F\left(\tfrac{1}{2} \operatorname{trace}_{F_\theta}(\Phi^* h)\right) \, d\operatorname{vol}(F_\theta),$$

i.e., if $\tau_F(\Phi; F_\theta, h) = 0$, where (cf. M. Ara [22])

$$\tau_F(\Phi; F_\theta, h) := F'\left(\tfrac{1}{2} \operatorname{trace}_{F_\theta}(\Phi^* h)\right) \tau(\Phi; F_\theta, h) + \Phi_* \left\{\nabla F'\left(\tfrac{1}{2} \operatorname{trace}_{F_\theta}(\Phi^* h)\right)\right\}. \tag{4.46}$$

Here $\tau(\Phi; F_\theta, h)$ is the ordinary tension field of $\Phi : (C(M), F_\theta) \to (N, h)$ and $\nabla : C^\infty(C(M)) \to \mathcal{X}(C(M))$ the gradient operator with respect to the Fefferman metric F_θ. □

At this point we may prove Theorem 4.13. Let (U, x^A) be a local coordinate system on M (the convention as to the range of indices is $A, B, \ldots \in \{0, 1, \ldots, n, \overline{1}, \ldots, \overline{n}\}$) and let us set $u^A := x^A \circ \pi : \pi^{-1}(U) \to \mathbf{R}$. Moreover, let $g_{ab} = F_\theta(\partial/\partial u^a \, , \, \partial/\partial u^b)$ be the local components of the Fefferman metric with respect to this coordinate system, and $[g^{ab}] = [g_{ab}]^{-1}$, where $a, b, \ldots \in \{0, 1, \ldots, n, \overline{1}, \ldots, \overline{n}, 2n+2\}$ and $u^{2n+2} = \gamma$. Let $\phi : M \to N$ be a smooth map and let us set $\Phi := \phi \circ \pi$. Then

$$\operatorname{trace}_{F_\theta}\left(\Phi^* h\right) = g^{ab} \frac{\partial \Phi^j}{\partial u^a} \frac{\partial \Phi^k}{\partial u^b} (h_{jk} \circ \Phi) = g^{AB} \left(\frac{\partial \phi^j}{\partial x^A} \circ \pi \right) \left(\frac{\partial \phi^k}{\partial x^B} \circ \pi \right) (h_{jk} \circ \Phi)$$

(because $\partial \pi^A / \partial \gamma = 0$) and

$$g^{AB} \left(\frac{\partial \phi^j}{\partial x^A} \circ \pi \right) \left(\frac{\partial \phi^k}{\partial x^B} \circ \pi \right) = h^{\alpha\overline{\beta}} \left\{ T_\alpha(\phi^j) T_{\overline{\beta}}(\phi^k) + T_{\overline{\beta}}(\phi^j) T_\alpha(\phi^k) \right\} \circ \pi;$$

hence

$$\operatorname{trace}_{F_\theta}\left(\Phi^* h\right) = 2h^{\alpha\overline{\beta}} T_\alpha(\phi^j) T_{\overline{\beta}}(\phi^k) \circ \pi = \left\{ \operatorname{trace}_{G_\theta}\left(\pi_H \phi^* h\right) \right\} \circ \pi = Q \circ \pi.$$

Therefore (by (4.46))

$$\tau_F(\Phi; F_\theta, h) = F'\left(\tfrac{1}{2}\, Q \circ \pi\right) \tau(\Phi; F_\theta, h) + \Phi_* \nabla \left\{ F'\left(\tfrac{1}{2}\, Q \circ \pi\right) \right\},$$

i.e., $\Phi = \phi \circ \pi$ is F-harmonic if and only if

$$(\rho(Q) \circ \pi) \tau(\Phi; F_\theta, h) + \Phi_* \nabla(\rho(Q) \circ \pi) = 0, \tag{4.47}$$

where $\rho(s) := F'(s/2)$. It is well known that under a conformal change of metric, the tension tensor field transforms as

$$\tau(\Phi; \lambda^2 F_\theta, h) = \lambda^{-(2n+2)} \left\{ \lambda^{2n} \tau(\Phi; F_\theta, h) + \Phi_* \nabla(\lambda^{2n}) \right\}, \tag{4.48}$$

for any $\lambda \in C^\infty(C(M))$. By Theorem 4.11, ϕ is F-pseudoharmonic, if and only if $\tau(\phi; \rho(Q)^{1/n}\theta, h) = 0$, that is (by Theorem 4.1) if and only if $\phi \circ \pi$ is harmonic with respect to the Fefferman metric $\mathcal{F}_{\rho(Q)^{1/n}\theta}$, i.e.,

$$\tau(\phi \circ \pi; \mathcal{F}_{\rho(Q)^{1/n}\theta}, h) = 0.$$

Finally [by (4.48) with $\lambda := (\rho(Q) \circ \pi)^{1/(2n)}$] ϕ is F-pseudoharmonic if and only if (4.47) holds. □

4.5.4 Weak subelliptic F-harmonic maps

Earlier in this chapter we introduced F-pseudoharmonic maps (see Definition 4.16 in Section 4.5) as critical points $\phi : M \to N$ of the functional

$$E_F(\phi) = \int_M F\left(\tfrac{1}{2} \operatorname{trace}_{G_\theta}(\pi_H \phi^* h)\right) \theta \wedge (d\theta)^n, \tag{4.49}$$

where M is a compact strictly pseudoconvex CR manifold and (N, h) a Riemannian manifold. Also $F : [0, +\infty) \to [0, +\infty)$ is a C^2 function such that $F'(t) > 0$. Moreover, it has been shown (cf. Theorem 4.10 in Chapter 4) that the Euler–Lagrange equations of the variational principle $\delta E_F(\phi) = 0$ are

$$\operatorname{div}(\rho(Q)\nabla^H \phi^i) + \sum_{a=1}^{2n} \rho(Q)(\left| \begin{matrix} i \\ j\ell \end{matrix} \right| \circ \phi)(X_a\phi^j)(X_a\phi^\ell) = 0, \tag{4.50}$$

$$\rho(t) := F'(t/2), \qquad Q := \operatorname{trace}_{G_\theta}(\pi_H \phi^* h).$$

When $N = S^m$ (the standard sphere in $\mathbf{R}^{m+1}$) the equations (4.50) become

$$-\sum_{a=1}^{2n} X_a^*(\rho(Q)X_a\phi^i) = \rho(Q)\phi^i|X\phi|^2, \quad 1 \leq i \leq m, \tag{4.51}$$

$$|X\phi|^2 := \sum_{a=1}^{2n}\sum_{A=1}^{m+1} X_a(\phi^A)^2, \quad \phi^A := y^A \circ \phi,$$

where (y^A) are the Cartesian coordinates on $\mathbf{R}^{m+1}$. To see that (4.50) implies (4.51) one takes into account the local expression of the divergence operator div with respect to the local G_θ-orthonormal frame $\{X_a\}$ of $H(M)$ on U. The adjoint X_a^* is with respect to $\Psi = \theta \wedge (d\theta)^n$, i.e., $\int uX_a^*v \;\Psi = -\int (X_a u)v \;\Psi$ for any $u \in C_0^\infty(U)$ and $v \in C^\infty(U)$. Taking into account the constraint $\sum_{A=1}^{m+1} \phi_A^2 = 1$ (where $\phi_A = \phi^A$) it follows that ϕ_{m+1} satisfies (4.50) as well.

Our scope in Section 4.5.4 is to look at local properties of weak solutions to (4.50). At the present time the theory is rather incomplete, and we deal only with the problem in which $\phi : \Omega \to S^m$, for some bounded domain $\Omega \subset \mathbf{R}^n$. If this is the case the local frame $\{X_a : 1 \leq a \leq 2n\}$ of $H(M)$ is replaced by an arbitrary Hörmander system $\{X_1, \dots, X_k\}$ on $\mathbf{R}^n$, and then (4.49) and (4.51) become

$$E_F(\phi) = \int_\Omega F(\tfrac{1}{2}|X\phi|^2)dx,$$

$$-X^* \cdot (\rho(|X\phi|^2)X\phi) = \rho(|X\phi|^2)\phi|X\phi|^2. \tag{4.52}$$

Here if $V = (V_1, \dots, V_{2n})$ is a vector field we adopt the notation $X^* \cdot V = \sum_{a=1}^{2n} X_a^* V_a$. When $F(t) := (2t)^{p/2}$, $t \geq 0$, and $m = 1$ the left-hand side of (4.52) becomes $\mathcal{L}_p\phi$, where $\mathcal{L}_p u \equiv -X^* \cdot (|Xu|^{p-2})Xu)$ is the *subelliptic p-Laplacian* in [85]. L. Capogna, D. Danielli, and N. Garofalo were (cf. op. cit.) the first to study regularity properties to (a single equation) $\mathcal{L}_p u = 0$.

In Section 4.5.4 we also recall the "subelliptic technicalities" that are necessary to understand the proof of P. Hájlasz and P. Strzelecki result in [192] (which carries over to the more general case of (4.52), cf. E. Barletta et al. [42.1]). Rudiments of subelliptic theory were presented in Chapter 3, as needed for solving the CR Yamabe problem. Here we give additional details, trying to emphasize (in the spirit of P. Hájlasz [191]) the results that hold in the general setting of metric spaces endowed with a Borel measure.

Let $X := \{X_1, \ldots, X_k\}$ be a Hörmander system with smooth real coefficients defined on $\mathbf{R}^n$ (see Definition 4.10 in Chapter 4 of this book). We shall need the following definition:

Definition 4.19. An absolutely continuous curve $C : [0, T] \to \mathbf{R}^n$ is said to be *admissible* if $\dot{C}(t) = \sum_{a=1}^{k} f_a(t) X_a(C(t))$, for some functions $f_a(t)$ satisfying $\sum_{a=1}^{k} f_a(t)^2 \leq 1$. □

If the vector fields X_a are not linearly independent at some point then the coefficients f_a are not unique.

Definition 4.20. The *Carnot–Carathéodory* distance $d_X(x, y)$, $x, y \in \mathbf{R}^n$, is the infimum of all $T > 0$ for which there is an admissible curve $C : [0, T] \to \mathbf{R}^n$ such that $C(0) = x$ and $C(T) = y$. Balls with respect to d_X are referred to as *Carnot–Carathéodory balls*. □

The main ingredient in the proof that d_X is indeed a distance function on $\mathbf{R}^n$ is a classical result of W.L. Chow [103], according to which any two points in $\mathbf{R}^n$ may be joined by an admissible curve.

Example 7. Let $\mathbf{H}_n$ be the Heisenberg group (cf. Section 1.1.5 in Chapter 1) and

$$X_j = \frac{\partial}{\partial x^j} + 2y^j \frac{\partial}{\partial t}, \qquad Y_j = \frac{\partial}{\partial y^j} - 2x^j \frac{\partial}{\partial t}. \tag{4.53}$$

Then $X = \{X_j, Y_j : 1 \leq j \leq n\}$ is a Hörmander system on $\mathbf{R}^{2n+1}$ (cf. also Example 2 in Section 4.4.1 of this book). Let d_X be the corresponding Carnot–Carathéodory distance and set $|x|_X = d_X(0, x)$. Then (i) $|\cdot|_X : \mathbf{H}_n \to [0, +\infty)$ is continuous, (ii) $|x^{-1}|_X = |x|_X$, (iii) $|D_\delta x|_X = \delta |x|_X$, (iv) $|x|_X = 0$ if and only if $x = 0$. Here D_δ is the dilation by the factor $\delta > 0$ (see Definition 1.11). The metric d_X is left invariant. Consequently $d_X(x, y) = |x^{-1} y|_X$. Also, d_X commutes with dilations, i.e., $d_X(D_\delta x, D_\delta y) = \delta(x, y)$. It may be shown (cf. J. Mitchell [305]) that the Hausdorff dimension of $\mathbf{H}_1$ with respect to d_X equals 4. Another homogeneous (with respect to dilations) norm $|\cdot|$ on $\mathbf{H}_n$ was considered in Definition 1.12 (the Heisenberg norm). This satisfies (i)–(iv) above, and $d(x, y) = |x^{-1} y|$ is a distance function on $\mathbf{H}_n$. The metrics d_X and d are equivalent.[6] □

Several relevant properties of the Carnot–Carathéodory metric were discovered[7] by A. Nagel, E.M. Stein, and S. Wainger [323].

---

[6] This is an easy exercise following from the fact that both d_X and d are left invariant and commute with dilations.

[7] While our presentation doesn't aim to be accurate from the historical point of view, we should nevertheless mention the work by E. Lanconelli [266], in which a geometric approach (based on the properties of the integral curves of vector fields associated with the given partial differential operator) is used to study the Hölder continuity of weak solutions to certain *strongly degenerate* equations (for which a prototype is $L_\alpha u \equiv \partial^2 u / \partial x^2 + |x|^{2\alpha} \partial^2 u / \partial y^2 = 0$ $(\alpha > 0)$). The paper [266] was published in 1983 (while [323] appeared in 1985).

Theorem 4.14. (A. Nagel, E.M. Stein, and S. Wainger [323])
For every open set $\Omega \subset\subset \mathbf{R}^n$ *there are constants* $C_i > 0$ *and* $\lambda \in (0, 1]$ *such that*

$$C_1 \|x - y\| \le d_X(x, y) \le C_2 \|x - y\|^{\lambda} \tag{4.54}$$

for any $x, y \in \Omega$.

Here $\|\cdot\|$ is the Euclidean norm. In particular the identity map $1 : \mathbf{R}^n \to \mathbf{R}^n$, $1(x) = x$, gives a homeomorphism $(\mathbf{R}^n, d_X) \approx (\mathbf{R}^n, d_0)$, where d_0 is the Euclidean metric on $\mathbf{R}^n$. As another consequence, if Ω is bounded with respect to d_0 then it is bounded with respect to d_X as well. The converse is false, in general (if for instance $X_1 = x_1^2 \, \partial/\partial x_1$ then it may be shown that the Carnot–Carathéodory distance to infinity is finite). To avoid any difficulties of the sort we assume throughout that in addition to being C^∞, the coefficients of the vector fields $\{X_1, \ldots, X_k\}$ are globally Lipschitz on $\mathbf{R}^n$. Indeed, if this is the case then (by a result of N. Garofalo and D.M. Nhieu [166]) a subset of $\mathbf{R}^n$ is bounded with respect to d_X if and only if it is bounded with respect to d_0.

Example 7. (*continued*) For the Hörmander system (4.53) on $\mathbf{H}_1$ it may be proved directly that for any bounded set $\Omega \subset \mathbf{H}_1$ there is a constant $C > 0$ such that

$$\frac{1}{C} \|x - y\| \le d_X(x, y) \le C \, \|x - y\|^{1/2}$$

for any $x, y \in \Omega$. □

Definition 4.21. Let X be a metric space with the distance function ρ. Let μ be a Borel measure on X such that μ is finite on bounded sets. We say that μ is *doubling* on $\Omega \subset X$ if there is a constant $C_d \ge 1$ such that

$$\mu(B(x, 2r)) \le C_d \, \mu(B(x, r))$$

for any $x \in \Omega$ and any $0 < r < 5 \operatorname{diam}(\Omega)$. C_d is referred to as the *doubling constant.* □

Here $B(x, R) = \{y \in X : \rho(x, y) < R\}$. Under our assumptions the Lebesgue measure is doubling on any bounded open subset of $\mathbf{R}^n$ with respect to the Carnot–Carathéodory metric. This is made precise in the following theorem:

Theorem 4.15. (A. Nagel, E.M. Stein, and S. Wainger [323])
Let $X = \{X_1, \ldots, X_k\}$ *be a Hörmander system on* $\mathbf{R}^n$, *with globally Lipschitz coefficients. Then for every bounded open set* $\Omega \subset \mathbf{R}^n$ *there is a constant* $C > 1$ *such that*

$$|B_X(x, 2r)| \le |B_X(x, r)| \tag{4.55}$$

for any $x \in \Omega$ *and any* $0 < r \le 5 \operatorname{diam}(\Omega)$.

Here the ball $B_X(x, R)$ and the diameter are meant with respect to d_X. Also if $A \subset \mathbf{R}^n$ is a Lebesgue measurable set then $|A|$ denotes its Lebesgue measure.

Example 7. (*continued*) The Lebesgue measure of a Carnot–Carathéodory ball $B_X(x, r)$ in the lowest-dimensional Heisenberg group $(\mathbf{H}_1, d_X)$ is

$$|B_X(x, r)| = Cr^4$$

for some constant $C > 0$ and any $x \in \mathbf{H}_1$, $r > 0$. □

Definition 4.22. Let $H = \sum_{a=1}^{k} X_a^* X_a$ be the Hörmander operator. A function $G(x, y)$ defined for $(x, y) \in \Omega \times \Omega$, smooth off $\Delta = \{(x, y) \in \Omega \times \Omega : x = y\}$, is said to be a *fundamental solution* for H if

$$\left(H \int_\Omega G(\cdot, y)\varphi(y)dy\right)(x) = \varphi(x) \tag{4.56}$$

for any $\varphi \in C_0^\infty(\Omega)$. □

By a result of A. Sánchez-Calle [363], for any bounded domain $\Omega \subset \mathbf{R}^n$ there is[8] a fundamental solution of the Hörmander operator H, possessing growth properties that extend naturally those of the fundamental solution for the classical Laplacian. Precisely, if G is a fundamental solution for H in Ω and $n \geq 3$ then

$$\frac{1}{C} \frac{d_X(x, y)^2}{|B_X(x, d_X(x, y))|} \leq G(x, y) \leq C \frac{d_X(x, y)^2}{|B_X(x, d_X(x, y))|}$$

and (for $n \geq 2$)

$$|X_a G(x, y)| \leq C \frac{d_X(x, y)}{|B_X(x, d_X(x, y))|}, \tag{4.57}$$

$$|X_a X_b G(x, y)| \leq \frac{C}{|B_X(x, d_X(x, y))|} \tag{4.58}$$

for any $x, y \in \Omega$. Here it is irrelevant whether differentiation is performed with respect to x or y. Only the estimates on the derivatives of G are needed in the sequel.

Example 8. Let us consider the differential operator $\mathcal{L}_0$ (cf. Definition 3.9 for $\alpha = 0$). The Hörmander operator (associated with the Hörmander system (4.53)) is given by $H = 4\mathcal{L}_0 = \sum_{j=1}^{n}(X_j^2 + Y_j^2)$. Let us recall (cf. Theorem 3.9 in Chapter 5, for $\alpha = 0$) that $\mathcal{L}_0\left(|x|^{-2n}\right) = c_0\delta$, where $c_0 = (2^{2-2n}\pi^{n+1}/\Gamma(\frac{n}{2}))^2$. A fundamental solution for H is given by $G(x, y) = w(xy^{-1})$, where

$$w(x) = \frac{C}{|x|^{2n}} = \frac{C}{(\|z\|^4 + t^2)^{n/2}}, \qquad x = (z, t) \in \mathbf{H}_n,$$

where $C = 1/(4c_0)$. □

Let us multiply both sides of (4.56) by $u \in C_0^\infty(\Omega)$ and integrate with respect to x. Next, we integrate by parts (in the left-hand side) to obtain the following result:

[8] The existence of $G(x, y)$ is well known to follow from the hypoellipticity of H and Bony's maximum principle (cf. J.M. Bony [73]).

Proposition 4.12. *Let $\Omega \subset \mathbf{R}^n$ be a bounded domain. Then*

$$u(x) = \int_\Omega X_y G(y, x) \cdot X_y u(y)\, dy \tag{4.59}$$

for any $u \in C_0^\infty(\Omega)$ and any $x \in \Omega$.

Definition 4.23. A function $\varphi(x)$ is said to be a *cut-off* function if $\varphi(x) = 1$ for any point x in a Carnot–Carathéodory ball $B_X(x_0, r)$, $\varphi(x) = 0$ for any $x \in \mathbf{R}^n \setminus B(x_0, 2r)$, and there is a constant $C > 0$ such that $|X\varphi| \le C/r$. □

Smooth cut-off functions were built by G. Citti, N. Garofalo, and E. Lanconelli [104] (by exploiting the fact that a fundamental solution G has decay properties similar to those of d_X):

Theorem 4.16. (G. Citti, N. Garofalo, and E. Lanconelli [104])
For any open bounded set $\Omega \subset \mathbf{R}^n$ there is a constant such that for any $x \in \Omega$ and any $0 < r \le \operatorname{diam}(\Omega)$ there is a function $\varphi \in C_0^\infty(B_X(x, r))$ such that $0 \le \varphi \le 1$, $\varphi = 1$ on $B_X(x, r/2)$ and $|X\varphi| \le C/r$.

Let $\Omega \subseteq \mathbf{R}^n$ be an open set and $W_X^{1,p}(\Omega)$ the Sobolev-type spaces associated with the Hörmander system X on $\mathbf{R}^n$, i.e.,

$$W_X^{1,p}(\Omega) = \{u \in L^p(\Omega) : X_a u \in L^p(\Omega), \quad a \in \{1, 2, \ldots, k\}\}$$

equipped with the norm $\|u\|_{W_X^{1,p}(\Omega)} = \|u\|_{L^p(\Omega)} + \|Xu\|_{L^p(\Omega)}$ (derivatives are meant in the distributional sense). See also Section 4.4.1 of Chapter 4. The case $p = 2$ was considered in Section 4.4.4 of this book. By a result of B. Franchi, R. Serapioni, and F. Serra Cassano [155], $C^\infty(\Omega)$ is dense in $W_X^{1,p}(\Omega)$.

Definition 4.24. Let $\Omega \subset \mathbf{R}^n$ be a bounded open set. A number D is called a *homogeneous dimension* relative to Ω (with respect to X) if there is a constant $C > 0$ such that

$$\frac{|B_X(x, r)|}{|B_X(x_0, r_0)|} \ge C \left(\frac{r}{r_0}\right)^D \tag{4.60}$$

for any Carnot–Carathéodory ball $B_0 = B(x_0, r_0)$ of center $x_0 \in \Omega$ and radius $0 < r_0 \le \operatorname{diam}(\Omega)$, and any Carnot–Carathéodory ball $B = B(x, r)$ of center $x \in B_0$ and radius $0 < r \le r_0$. □

The doubling property (cf. Theorem 4.15 above) is known to imply the existence of homogeneous dimensions. Precisely, this holds in the general context of metric spaces endowed with Borel measures.

Theorem 4.17. (R. Coifman and G. Weiss [105])
Let (X, ρ) be a metric space and μ a Borel measure on X that is finite on bounded sets. Let $\Omega \subset X$ be an open bounded set and assume that μ is doubling on Ω, with doubling constant $C_d \ge 1$. Then

$$\frac{\mu(B(x,r))}{\mu(\Omega)} \geq \left(\frac{r}{2\,\mathrm{diam}(\Omega)}\right)^{\log_2 C_d}$$

for any $x \in \Omega$ *and any* $0 < r \leq \mathrm{diam}(\Omega)$.

Clearly, if D is a homogeneous dimension (relative to Ω) then any $D' \geq D$ is a homogeneous dimension as well.

Example 9. For the Heisenberg group $\mathbf{H}_1$ the smallest homogeneous dimension is known to be $D = 4$, so that $w(x)$ in Example 8 (with $n = 1$) may be written[9] as $w(x) = C|x|^{2-D}$. □

When $F(t) = (2t)^{2p}$ $(t \geq 0)$ the subelliptic F-harmonic map equations become

$$-X^* \cdot (|X\phi|^{p-2} X\phi) = |X\phi|^p \phi, \tag{4.61}$$

a form that is well suited for introducing the notion of weak solution $\phi \in W_X^{1,p}(\Omega, S^m)$.

Definition 4.25. (P. Hájlasz and P. Strzelecki [192])
Let $1 < p < \infty$. An element $\phi = (\phi_1, \ldots, \phi_{m+1}) \in W_X^{1,p}(\Omega, S^m)$ (that is, each $\phi_A \in W_X^{1,p}(\Omega)$ and $\sum_{A=1}^{m+1} \phi_A^2 = 1$) is a *weak solution* to the system (4.61) if

$$\int_\Omega |X\phi|^{p-2}(X\phi_A) \cdot (X\psi)\, dx = \int_\Omega |X\phi|^p \phi_A \psi\, dx \tag{4.62}$$

for any $1 \leq A \leq m+1$ and any $\psi \in C_0^\infty(\Omega)$. Any such weak solution[10] is referred to as a *weak subelliptic p-harmonic* map. □

Definition 4.26. An element $\phi \in W_X^{1,p}(\Omega, S^m)$ is a *weak solution* to (4.52) if $\rho(|X\phi|^2)(X\phi_A) \cdot (X\psi)$, and $\rho(|X\phi|^2)|X\phi|^2\phi_A\psi$ are in $L^1(\Omega)$ and

$$\int_\Omega \rho(|X\phi|^2)(X\phi_A) \cdot (X\psi)\, dx = \int_\Omega \rho(|X\phi|^2)|X\phi|^2 \phi_A \psi\, dx$$

for any $1 \leq A \leq m+1$ and any $\psi \in C_0^\infty(\Omega)$. Any such weak solution[11] is called a *weak subelliptic F-harmonic* map. □

---

[9] The form of $w(x)$ is reminiscent of the fundamental solution to the classical Laplacian (cf., e.g., (2.12) in [177], p. 17) and suggests that the (smallest) homogeneous dimension plays (within the analysis of the Hörmander operator) the role of the Euclidean dimension (in the analysis of the Laplacian).

[10] Note that (by $|\phi| \leq 1$ and Hölder's inequality)

$$\left|\int_\Omega |X\phi|^p \phi_A \psi\, dx\right| \leq C \int_\Omega |X\phi|^p |\phi|\, dx \leq C\|X\phi\|^p_{L^p(\Omega)},$$

$$\left|\int_\Omega |X\phi|^{p-2}(X\phi_A) \cdot (X\psi)\, dx\right| \leq \int_\Omega |X\phi|^{p-1}|X\psi|\, dx$$

$$\leq C \int_\Omega |X\phi|^{p-1} dx \leq C\mu(\Omega)^{1/p}\|X\phi\|^{p/q}_{L^p(\Omega)};$$

hence both integrals in (4.62) make sense.

[11] Under the hypothesis of Theorem 4.19 (that is, $\rho(t) \leq Kt^p$ for some $0 < p < (D-2)/2$) one has (by Hölder's inequality (since $D/(2p+1) > 1$))

Theorem 4.18. (P. Hájlasz and P. Strzelecki [192])
Let $X = \{X_a = b_a^A(x)\partial/\partial x^A : 1 \le a \le k\}$ *be a Hörmander system on* $\mathbf{R}^n$ *such that* $b_a^A \in C^\infty(\mathbf{R}^n) \cap \mathrm{Lip}(\mathbf{R}^n)$. *Let* $\Omega \subset \mathbf{R}^n$ *be an open bounded set and* D *be a homogeneous dimension relative to* Ω, *with respect to* X. *Then every weak subelliptic* D*-harmonic map* $\phi \in W_X^{1,D}(\Omega, S^m)$ *is locally Hölder continuous.*

For the case of an arbitrary $F(t)$ ($F \in C^2$, $F(t) \ge 0$, $F'(t) > 0$) only the following less-precise result is known:

Theorem 4.19. (E. Barletta et al. [42.1])
Let X, Ω, *and* D *be as in Theorem* 4.18. *Assume that* $t^p/K \le \rho(t) \le K t^p$ *for some constant* $K \ge 1$ *and some* $0 < p < (D-2)/2$. *Let* $\phi \in W_X^{1,D}(\Omega, S^m)$ *be a weak solution to the system* (4.52). *Let* $R_0 > 0$ *and* $\Omega_1 \subset\subset \Omega$ *such that* $B(x, 2R_0/\tau) \subset \Omega$ *for any* $x \in \Omega_1$, $\tau := 1/200$. *Then there are* $\lambda \in [1/2, 1)$ *and* $0 < r_0 \le R_0$ *such that*

$$I_p(r) := \int_{B_X(x,r)} |X\phi|^{2(p+1)}(y)\, dy \le C\, r^\gamma, \qquad \gamma := (\log\lambda)/(\log\tau),$$

for any $x \in \Omega_1$ *and any* $0 < r \le r_0$. *Consequently, if* $\tau^D < \lambda < \tau^{D-2(p+1)}$ *then* $\phi \in S^{0,\alpha}_{\mathrm{loc}}(\Omega)$ *with* $\alpha := 1 + (\gamma - D)/(2p+2)$.

The Hölder-like spaces (associated with the given Hörmander system) in Theorem 4.19 are given by

$$S^{0,\alpha}(\Omega) = \left\{ f \in L^\infty(\Omega) : \sup_{x,y\in\Omega} \frac{|f(x)-f(y)|}{d_X(x,y)^\alpha} < \infty \right\}, \qquad 0 < \alpha \le 1.$$

In the sequel we prove Theorem 4.19 (for a complete proof of Theorem 4.18 the reader may see [192], pp. 353–359).

We shall need the Poincaré inequality. This is known to follow easily from the Sobolev inequality. Generalizations (to the context of systems of vector fields) of the Sobolev embedding theorem have been obtained by several authors:[12]

Theorem 4.20. *Let* X *be a Hörmander system on* $\mathbf{R}^n$ *with globally Lipschitz coefficients. Let* $\Omega \subset \mathbf{R}^n$ *be an open bounded set and* D *a homogeneous dimension relative to* Ω. *For each* $1 \le p < D$ *there is a constant* $C > 0$ *such that for any Carnot–Carathéodory ball* $B = B_X(x, r)$ *with* $x \in \Omega$ *and* $0 < r \le \mathrm{diam}(\Omega)$,

$$\left(\frac{1}{|B|}\int_B |u - u_B|^{p^*} dx\right)^{1/p^*} \le Cr\left(\frac{1}{|B|}\int_B |Xu|^p dx\right)^{1/p}, \tag{4.63}$$

where $p^* = Dp/(D-p)$ *and* $u_B = (1/|B|)\int_B u\, dx$.

$$\left|\int_\Omega \rho(|X\phi|^2)(X\phi_A)\cdot(X\psi)\, dx\right| \le C\int_\Omega |X\phi|^{2p+1}\, dx \le C\mu(\Omega)^{D/(D-2p-1)}\|X\phi\|^{2p+1}_{L^D(\Omega)},$$

so that $\rho(|X\phi|^2)(X\phi_A)\cdot(X\psi)$ is integrable. Similarly $\rho(|X\phi|^2)|X\phi|^2\phi_A\psi \in L^1(\Omega)$.

12 The statement of Theorem 4.20 is taken from B. Franchi, G. Lu, and R.L. Wheeden [154], and L. Capogna, D. Danielli, and N. Garofalo [86]. However, versions of Theorem 4.20 are also due to M. Biroli and U. Mosco [61], B. Franchi [153], B. Franchi and E. Lanconelli [152], G. Lu [289], N. Garofalo, and D.M. Nhieu [167], and D. Jerison [225].

The inequality (4.63) implies the Poincaré inequality

$$\Big(\int_B |u - u_B|^p dx\Big)^{1/p} \le Cr\Big(\int_B |Xu|^p dx\Big)^{1/p} \tag{4.64}$$

for any $1 \le p < \infty$. To prove (4.64) one should distinguish two cases: (I) $1 \le p < D$ and (II) $1 \le p < \infty$ is arbitrary. In the first case one applies the Hölder inequality.[13] In the second case we may choose $D' > \max\{D, p\}$ and proceed as in case I (as emphasized above, any $D' > D$ is another homogeneous dimension relative to Ω).

Proof of Theorem 4.19. If $C > 0$ is a constant then $C\,B(x_0, r)$ is the ball $B(x_0, Cr)$. Also, by $r \approx s$ we mean $r/C \le s \le Cr$, for some $C \ge 1$. To start the study of weak solutions ϕ to (4.52) with the constraint $\sum_{i=1}^{m+1} \phi_i^2 = 1$, we set $V_{i,a} := \rho(Q) X_a \phi_i$, $1 \le a \le k$, and $V_i = (V_{i,1}, \ldots, V_{i,k})$. Then

$$V_i = \sum_{j=1}^{m+1} \phi_j (\phi_j V_i - \phi_i V_j),$$

merely as a consequence of the constraint. Next, we set $E_{i,j} := \phi_j V_i - \phi_i V_j$ and then (4.52) implies

$$X^* \cdot E_{i,j} = 0, \quad 1 \le i, j \le m+1. \tag{4.65}$$

Indeed, for any $\psi \in C_0^\infty$,

$$\begin{aligned}
\int_\Omega (X^* \cdot (\phi_i V_j))\psi\, dx &= \sum_{a=1}^k \int_\Omega X_a^*(\phi_i V_{j,a})\psi\, dx = -\sum_a \int_\Omega \phi_i V_{j,a} X_a \psi\, dx \\
&= -\sum_a \int \rho(Q)(X_a \phi_j)\,[X_a(\phi_i \psi) - \psi X_a \phi_i]\, dx \\
&= \sum_a \int X_a^*(\rho(Q) X_a \phi_j)\phi_i \psi\, dx \\
&\quad + \sum_a \int \psi \rho(Q)(X_a \phi_i)(X_a \phi_j) dx \\
&= \int \rho(Q)\Big[-Q\phi_i\phi_j + \sum_a (X_a\phi_i)(X_a\phi_j)\Big]\psi\, dx. \\
&\hspace{20em} \text{(by (4.52))}
\end{aligned}$$

Hence $X^* \cdot (\phi_i V_j)$ is symmetric in i, j, which yields (4.65). The identity (4.65) implies the following result:

---

[13] Note that $1/(p^*/p) + 1/D = 1$; hence (by Hölder's inequality)

$$\int_B |u - u_B|^p dx \le |B|^{1/D} \Big(\int_B |u - u_B|^{p^*} dx\Big)^{p/p^*}$$

and then one may use (4.63).

Lemma 4.4. (The duality inequality)
Let $\Omega \subset \mathbf{R}^n$ be a bounded domain and $X_a = b_a^A \partial/\partial x^A$ a Hörmander system on $\mathbf{R}^n$ with $b_a^A(x)$ globally Lipschitz. Let $R_0 > 0$ and $\Omega_1 \subset\subset \Omega$ such that $B(x, 400R_0) \subset \Omega$, for any $x \in \Omega_1$. Let $B = B(x_0, r)$, $x_0 \in \Omega_1$, be a ball such that $0 < r \le R_0$, and $\varphi \in W_X^{1,D}(B)$ a function of compact support. Then

$$\left| \int_B X^* \cdot (\phi_j E_{i,j}) \varphi \, dx \right| \le CK \|X\varphi\|_{L^D(B)} \left(\|X\phi\|_{L^{2p+2}(100B)} \right)^{2p+2} , \tag{4.66}$$

for some constant $C = C(\Omega_1, D, C_d, R_0) > 0$, provided that $\rho(t) \le Kt^p$, $\quad t \ge 0$, for some $K > 0$ and $0 < p < (D-2)/2$.

Aside from some additional technical difficulties (e.g., one applies twice the fractional integration theorem), the proof of Lemma 4.4 is similar to that of Lemma 3.2 in P. Hájlasz and P. Strzelecki [192], p. 354.

To prove Theorem 4.19 we fix $\Omega_1 \subset\subset \Omega$ and $R_0 > 0$ as in Lemma 4.4. Taking the dot product of $V_i = \sum_{j=1}^{m+1} \phi_j E_{i,j}$ with X^* we get

$$X^* \cdot (\rho(Q) X\phi_i) = \sum_{j=1}^{m+1} X^* \cdot (\phi_j E_{i,j})$$

(a consequence of the constraint alone) and integrating over $2B$ [where $B = B(x, r)$, $x \in \Omega_1$, $0 < r < R_0$] against $\psi_i := \eta(\phi_i - (\phi_i)_{2B})$, where $0 \le \eta \le 1$ is a smooth cut-off function such that $\eta = 1$ on B, $\eta = 0$ on $\Omega \setminus 2B$, and $|X\eta| \le C/r$, we obtain

$$\int X^* \cdot (\rho(Q) X\phi_i) \psi_i \, dx = \sum_{j=1}^{m+1} \int X^* \cdot (\phi_j E_{i,j}) \psi_i \, dx \, . \tag{4.67}$$

The left-hand side may be also written

$$\begin{aligned} \int X^* \cdot (\rho(Q) X\phi_i) \psi_i \, dx &= - \int \rho(Q)(X\phi_i) \cdot (X\psi_i) \, dx \\ &= - \int \rho(Q)(X\phi_i) \cdot \left[(X\eta)(\phi_i - (\phi_i)_{2B}) + \eta(X\phi_i) \right] dx; \end{aligned}$$

hence (4.67) becomes (by summing over $1 \le i \le m+1$)

$$\begin{aligned} \int_{2B} \eta Q \rho(Q) \, dx + \sum_{i=1}^{m+1} \int_{2B} \rho(Q)(\phi_i - (\phi_i)_{2B})(X\phi_i) \cdot (X\eta) \, dx \\ = - \sum_{i,j} \int_{2B} X^* \cdot (\phi_j E_{i,j}) \psi_i \, dx \, . \end{aligned}$$

Consequently

$$\int_B Q\rho(Q)\,dx \le \int_{2B} \eta Q\rho(Q)\,dx$$
$$\le \sum_i \int_{2B} \rho(Q)\left|\phi_i - (\phi_i)_{2B}\right| |X\phi_i|\,|X\eta|\,dx + \sum_{i,j}\left|I_{i,j}\right| ,$$

where

$$I_{i,j} := \int_{2B} X^* \cdot (\phi_j E_{i,j})\psi_i \,dx .$$

Moreover, by $|X\phi_i| \le |X\phi| = Q^{1/2}$ and by the Hölder inequality (with $1/[2(p+1)] + 1/q = 1$),

$$\int_B Q\rho(Q)\,dx \le \sum_i \int_{2B} Q^{1/2}\rho(Q)\,|X\eta|\,\left|\phi_i - (\phi_i)_{2B}\right| dx + \sum_{i,j}\left|I_{i,j}\right|$$
$$\le \sum_{i,j}\left|I_{i,j}\right| + \sum_i \left(\int_{2B} \left|\phi_i - (\phi_i)_{2B}\right|^{2(p+1)}\right)^{1/[2(p+1)]}$$
$$\times \left(\int_{2B\setminus B} \left(Q^{1/2}\rho(Q)|X\eta|\right)^{2(p+1)/(2p+1)}\right)^{(2p+1)/[2(p+1)]} .$$

At this point, we may apply the Poincaré inequality

$$\left(\int_{2B} |u - u_{2B}|^s dx\right)^{1/s} \le Cr\left(\int_{2B} |Xu|^s dx\right)^{1/s}, \quad 1 \le s < \infty,$$

and Lemma 4.4 (with φ replaced by ψ_i) to obtain

$$\int_B Q\rho(Q)\,dx \le C\sum_i \left(\int_{2B} |X\phi_i|^{2(p+1)}\,dx\right)^{1/[2(p+1)]}$$
$$\times \left(\int_{2B\setminus B} \left(Q^{1/2}\rho(Q)\right)^{2(p+1)/(2p+1)} dx\right)^{(2p+1)/[2(p+1)]} + \sum_{i,j}\left|I_{i,j}\right|$$
$$\le C\left(\int_{2B} Q^{p+1}\,dx\right)^{1/[2(p+1)]} \left(\int_{2B\setminus B} \left(Q\rho(Q)^2\right)^{(p+1)/(2p+1)} dx\right)^{(2p+1)/[2(p+1)]}$$
$$+ C\sum_i \|X\psi_i\|_{L^D(2B)} \left(\|X\phi\|_{L^{2p+2}(200B)}\right)^{2p+2} .$$

By $\rho(t) \le Kt^p$ we have

$$\left(\int_{2B\setminus B} \left(Q\rho(Q)^2\right)^{(p+1)/(2p+1)} dx\right)^{(2p+1)/[2(p+1)]}$$
$$\le K\left(\int_{2B\setminus B} |X\phi|^{2(p+1)}\,dx\right)^{(2p+1)/[2(p+1)]} .$$

Now we may use $t^p/K \leq \rho(t)$ to estimate $\int Q\rho(Q)dx$ from below, and the inequality

$$\sum_i \|X\psi_i\|_{L^D(2B)} \leq C\|X\phi\|_{L^D(2B)} \tag{4.68}$$

to obtain

$$I_p(r) \leq \\ C\big[I_p(2r)^{1/(2p+2)}(I_p(2r) - I_p(r))^{(2p+1)/(2p+2)} + \|X\phi\|_{L^D(2B)} I_p(200r)\big], \tag{4.69}$$

where

$$I_p(r) := \int_{B(x,r)} |X\phi|^{2p+2}\,dx\ .$$

As to (4.68), it follows from

$$\begin{aligned}
\sum_i \|X\psi_i\|_{L^D(2B)} &\leq \sum_i (\|(X\eta)(\phi_i - (\phi_i)_{2B})\|_{L^D(2B)} + \|\eta X\phi_i\|_{L^D(2B)}) \\
&= \sum_i \left(\int_{2B} |X\eta|^D \left|\phi_i - (\phi_i)_{2B}\right|^D dx\right)^{1/D} + \sum_i \left(\int_{2B} |\eta|^D |X\phi_i|^D\,dx\right)^{1/D} dx \\
&\leq \frac{C}{r} \sum_i \left(\int_{2B} \left|\phi_i - (\phi_i)_{2B}\right|^D dx\right)^{1/D} + \sum_i \left(\int_{2B} |X\phi_i|^D\,dx\right)^{1/D} dx \\
&\leq C\left(\int_{2B} |X\phi|^D\,dx\right)^{1/D} dx \quad \text{(by the Poincaré inequality).}
\end{aligned}$$

Using (4.69) we may establish the following lemma:

Lemma 4.5. *There are* $r_0 > 0$ *and* $\lambda \in [1/2, 1)$ *such that*

$$I_p(r) \leq \lambda I_p(200r), \tag{4.70}$$

for any $0 < r \leq r_0$.

The proof is by contradiction. Assume that for any $r_0 > 0$ and any $\lambda \in [1/2, 1)$ there is $0 < r \leq r_0$ such that $\lambda I_p(200r) < I_p(r)$. Then (by (4.69))

$$\begin{aligned}
&\lambda I_p(200r) < I_p(r) \\
&\quad \leq C\left[I_p(2r)^{1/(2p+2)}(I_p(2r) - I_p(r))^{(2p+1)/(2p+2)} + \|X\phi\|_{L^D(2B)} I_p(200r)\right] \\
&\quad\quad \leq C\left[I_p(200r)(1-\lambda)^{(2p+1)/(2p+2)} + \|X\phi\|_{L^D(2B)} I_p(200r)\right].
\end{aligned}$$

That is,

$$\tfrac{1}{2} \leq \lambda < C\left[(1-\lambda)^{(2p+1)/(2p+2)} + \|X\phi\|_{L^D(2B)}\right] .$$

Consequently, for any $r_0 > 0$ there is $0 < r \le r_0$ such that

$$\left(\frac{1}{2C}\right)^D \le \int_{2B} |X\phi|^D \, dx \, .$$

Indeed, let $\lambda_\nu \in [1/2, 1)$, $\lambda_\nu \to 1$ as $\nu \to \infty$, and $0 < r_\nu \le r_0$ correspondingly. By eventually passing to a subsequence, we may assume that $r_\nu \to r_\infty$ as $\nu \to \infty$, for some $0 \le r_\infty \le r_0$. Let us take $\nu \to \infty$ in $1/2 < C[(1-\lambda_\nu)^{(2p+1)/(2p+2)} + (\int_{B(x,2r_\nu)} |X\phi|^D \, dy)^{1/D}]$. Then we may use the Vitali absolute continuity of the integral to conclude that either $r_\infty > 0$ and then we get the desired inequality, or $r_\infty = 0$ and then $1/2 \le 0$, a contradiction. In particular, for $r_0 = 1/k$ there is $0 < r \le 1/k$ such that

$$\left(\frac{1}{2C}\right)^D \le \int_{B(x,2r)} |X\phi|^D \, dy \le \int_{B(x,2/k)} |X\phi|^D \, dy$$

and (again using absolute continuity) the last integral goes to 0 as $k \to \infty$, a contradiction. Lemma 4.5 is proved. □

The inequality (4.70) may be written $I_p(\tau r) \le \lambda I_p(r)$, where $\tau = 1/200$. Therefore

$$I_p(\tau^m r) \le \lambda^m I_p(r),$$

for any integer $m \ge 1$. The following argument (leading to the estimate (4.71)) is standard. Details are for the sake of completeness. The family of intervals $\{(\tau^m, \tau^{m-1}] : m \ge 1\}$ is a cover of $(0, 1]$; hence $\tau^m < r/r_0 \le \tau^{m-1}$, for some $m \ge 1$. Now $r \le \tau^{m-1} r_0$ yields

$$I_p(r) \le I_p(\tau^{m-1} r_0) \le \lambda^{m-1} I_p(r_0).$$

Let us set $\gamma := (\log \lambda)/(\log \tau)$ (then $0 < \gamma < 1$, because of $\lambda \ge 1/2 > \tau$). On the other hand, $r/r_0 \ge \tau^m$ yields

$$\left(\frac{r}{r_0}\right)^\gamma > \tau^{m\gamma} = \tau^{(\log \lambda^m)/(\log \tau)} = \lambda^m \, .$$

Then $\lambda^{m-1} < (r/r_0)^\gamma / \lambda$, whence

$$I_p(r) \le \lambda^{m-1} I_p(r_0) < \frac{1}{\lambda}\left(\frac{r}{r_0}\right)^\gamma I_p(r_0) = C r^\gamma$$

(where $C = I_p(r_0)/(\lambda r_0^\gamma)$). We have obtained

$$\int_{B(x,r)} |X\phi|^{2(p+1)}(y) \, dy \le C r^\gamma \, , \tag{4.71}$$

which is the Caccioppoli-type estimate sought. To complete the proof of Theorem 4.19 we need to recall (cf. Proposition 2.1 in C.-J. Xu and C. Zuily [443], p. 326) the following result. Let $u \in L^2(\Omega)$. Then the following two conditions are equivalent: (i) $u \in S^{0,\alpha}_{\mathrm{loc}}(\Omega)$, and (ii) there are constants $r_0 > 0$ and $C > 0$ such that for any $0 < r \leq r_0$ and any $x \in \Omega$ such that $B(x, 2r) \subset \Omega$ one has

$$\int_{B(x,r)} |u(y) - u_{B(x,r)}|^2 \, dy \leq C|B(x,r)|r^{2\alpha} .$$

By the Poincaré inequality

$$\begin{aligned}\int_{B(x,r)} |\phi_i(y) - (\phi_i)_{B(x,r)}|^2 dy &\leq Cr^2 \int_{B(x,r)} |X\phi|^2 \, dy \\ &\leq Cr^2 \left(\int_{B(x,r)} |X\phi|^{2p+2} dy \right)^{1/(p+1)} |B(x,r)|^{p/(p+1)}\end{aligned}$$

by the Hölder inequality (with $1/(p+1) + 1/q = 1$), which in turn

$$\leq Cr^2 |B(x,r)|^{p/(p+1)} r^{\gamma/(p+1)} \leq C|B(x,r)|r^{2\alpha}$$

by (4.71) and by the definition of homogeneous dimension, where $\alpha := 1 + (\gamma - D)/(2p+2)$. Now $\alpha > 0$ provided that $p > (D-2)/2 - \gamma/2$, and $\alpha \leq 1$ when $D > \gamma$, which holds since D is tacitly assumed to be large (usually D is larger than the Euclidean dimension). Theorem 4.19 is proved.

We end this section by proving the duality inequality. It suffices to prove Lemma 4.4 for $\varphi \in C_0^\infty(B)$. Since the proof is fairly long, we organize it in several steps, as follows. For any bounded domain $\Omega \subset \mathbf{R}^n$ and any $u \in C_0^\infty(\Omega)$ one has the representation formula (4.59) (see Proposition 4.12 above). By Theorem 4.16 above, we may construct a smooth cut-off function η_0 such that $\eta_0 = 1$ on $2B$, $\eta_0 = 0$ on $\Omega \setminus 4B$, and $|X\eta_0| \leq C/\mathrm{diam}(B)$. The diameter of a set is meant with respect to d_X. Then, using (4.59) (with $u = \varphi$)

$$\begin{aligned}\int_B X^* \cdot (\phi_j E_{i,j})\varphi \, dx &= \int_B X^* \cdot (\phi_j E_{i,j})\varphi\eta_0 \, dx \\ &= \iint_{(x,y)\in B\times B} X^* \cdot (\phi_j E_{i,j})(x)\eta_0(x) \left(X_y G(y,x)\right) \cdot \left(X_y\varphi(y)\right) \, dx \, dy;\end{aligned}$$

hence

$$\int_B X^* \cdot (\phi_j E_{i,j})\varphi \, dx = \int_B A \cdot (X\varphi) \, dy ,$$

where

$$A(y) := \int_B X^* \cdot (\phi_j E_{i,j})(x)\eta_0(x) X_y G(y,x) \, dx.$$

Our **Step 1** is to establish *a bound on* $|A_a(y)|$. To this end we need the Whitney decomposition of $\Omega_y := \Omega \setminus \{y\}$. The Whitney decomposition of an open set in $\mathbf{R}^n$ has been generalized to the context of metric spaces endowed with doubling measures by R.A. Macías and C. Segovia [292]. We recall a few details, as follows. Let (X, ρ) be a metric space and $\Omega \subset X$ an open set, $X \setminus \Omega \neq \emptyset$. Let μ be a measure on X that is doubling on Ω, with the doubling constant $C_d \geq 1$.

Theorem 4.21. (R.A. Macías and C. Segovia [292])
Let $x \in \Omega$ and $r_x = \text{dist}(x, X \setminus \Omega)/1000$, so that $\{B(x, r_x) : x \in \Omega\}$ is a covering of Ω. Let $\{B(x_i, r_i) : i \in I\}$ be a maximal subfamily of mutually disjoint balls. Then

$$\Omega = \bigcup_{i \in I} B(x_i, 3r_i) \tag{4.72}$$

and there is $N \geq 1$ such that any point of Ω belongs to at most N balls $B(x_i, 6r_i)$.

Proof. (4.72) follows from the maximality of $\{B(x_i, r_i) : i \in I\}$. The existence of $N \geq 1$ in the second statement follows from the doubling property (N depends only on C_d). □

Definition 4.27. The decomposition (4.72) is called a *Whitney decomposition* of Ω. □

To a Whitney decomposition of Ω into balls one may associate a partition of unity as follows. Let $0 \leq \psi \leq 1$ be a smooth function such that $\psi(t) = 1$ for $0 \leq t \leq 1$ and $\psi(t) = 0$ for $t \geq 4/3$. Let us set

$$\varphi_i(x) = \psi(\rho(x, x_i)/(3r_i)).$$

Then $\varphi_i(y) = 1$ for any $y \in B(x_i, 3r_i)$, $\varphi_i(y) = 0$ for any $y \in X \setminus B(x_i, 4r_i)$, and each φ_i is Lipschitz, with a Lipschitz constant of the form C/r_i. Finally, we set

$$\theta_i(x) = \frac{\varphi_i(x)}{\sum_{k \in I} \varphi_k(x)}.$$

By Theorem 4.21 at most N terms in the sum above are nonzero, so that $\theta_i(x)$ is well defined.

Proposition 4.13. (R.A. Macías and C. Segovia [292])
$\sum_{i \in I} \theta_i(x) = 1$ *for any $x \in \Omega$ and* $\text{supp}(\theta_i) \subset B(x_i, 6r_i)$. *Also θ_i is a Lipschitz function with a Lipschitz constant of the form C/r_i, where the constant C depends only on C_d.*

Let us go back to the proof of Step 1. Let $y \in B$ and let $\{\theta_\alpha\}_{\alpha \in I}$ be a smooth partition of unity associated with a Whitney decomposition of $\Omega_y := \Omega \setminus \{y\}$. Precisely as in Theorem 4.21, for $x \in \Omega_y$ we set $r_x := d_X(x, \mathbf{R}^n \setminus \Omega_y)/1000$ and choose, among $\{B(x, r_x)\}_{x \in \Omega_y}$, a maximal family of mutually disjoint balls $\{B(x_\alpha, r_\alpha)\}_{\alpha \in I}$ (hence $\Omega_y = \bigcup_{\alpha \in I} B(x_\alpha, 3r_\alpha)$ and there is $N \geq 1$ such that each $x \in \Omega$ belongs to at most N balls $B(x_\alpha, 6r_\alpha)$). Then (by Proposition 4.13) we may consider a family of smooth functions $\{\theta_\alpha\}_{\alpha \in I}$ such that $0 \leq \theta_\alpha \leq 1$, $\sum_{\alpha \in I} \theta_\alpha = 1$ on Ω_y, $\text{supp}(\theta_\alpha) \subset B_\alpha := B(x_\alpha, 6r_\alpha)$, and $|X\theta_\alpha| \leq C/r_\alpha$. The bounds on the gradients follow from Theorem 4.16). Then

$$\begin{aligned} A_a(y) &= \sum_{\alpha \in I} \int_{B_\alpha} X^* \cdot (\phi_j E_{i,j})(x) \eta_0(x) \theta_\alpha(x) X_{a,y} G(y, x)\, dx \\ &= \sum_{\alpha \in I} \int_{B_\alpha} X^* \cdot \left[\phi_j - (\phi_j)_{B_\alpha}\right] E_{i,j}(x) \eta_0(x) \theta_\alpha(x) X_{a,y} G(y, x)\, dx, \quad \text{(by (4.65))} \end{aligned}$$

where $(\phi_j)_{B_\alpha} := (1/|B_\alpha|)\int_{B_\alpha} \phi_j(x)dx$. Next

$$A_a(y) = -\sum_{\alpha\in I}\int_{B_\alpha} \left[\phi_j - (\phi_j)_{B_\alpha}\right] E_{i,j}(x)\cdot X_x\left(\eta_0(x)\theta_\alpha(x)X_{a,y}G(y,x)\right) dx\,.$$

Using the estimates (4.57)–(4.58) and $|X\eta_0(x)| \le Cd_X(x,y)^{-1}$ and $|X\theta_\alpha(x)| \le Cd_C(x,y)^{-1}$, $\alpha \in I$, we obtain

$$\left|X_{b,x}(\eta_0(x)\theta_\alpha(x)X_{a,y}G(y,x))\right| \le \frac{C}{|B(y,d_X(x,y))|};$$

hence

$$|A_a(y)| \le C\sum_{\alpha\in I}\int_{B_\alpha}\frac{|\phi_j(x)-(\phi_j)_{B_\alpha}||E_{i,j}|}{|B(y,d_X(x,y))|}\,dx\,. \tag{4.73}$$

Note that

$$|B(y,d_X(x,y))| \ge C\,|B_\alpha|\,, \quad x\in B_\alpha\,. \tag{4.74}$$

Indeed, $d_X(x,x_\alpha) < 6r_\alpha$. On the other hand, $d_X(y,x_\alpha) \ge 1000r_\alpha$, from $y \in \mathbf{R}^n\setminus\Omega_y$ and the definition of r_α. Hence

$$1000r_\alpha \le d_X(y,x_\alpha) \le d_X(x,y) + d_X(x,x_\alpha) \le d_X(x,y) + 6r_\alpha,$$

that is, $r_\alpha \le d_X(x,y)/994$ or $6r_\alpha \le d_X(x,y)$. This yields $|B(y,6r_\alpha)| \le |B(y,d_X(x,y))|$. Moreover

$$\frac{|B(y,6r_\alpha)|}{|B(x_\alpha,6r_\alpha)|} \ge C,$$

as a consequence of (4.60) (which holds by Theorem 4.17). Combining the last two inequalities leads to (4.74). Let us consider the set of indices

$$J := \{\alpha\in I\,;\,\mathrm{supp}(\theta_\alpha)\cap 4B \ne \emptyset\}.$$

By (4.73)–(4.74) and the Hölder inequality (with $1/\nu^* + 1/\beta = 1$)

$$\begin{aligned}|A_a(y)| &\le C\sum_{\alpha\in J}\frac{1}{|B_\alpha|}\int_{B_\alpha}\left|\phi_j(x)-(\phi_j)_{B_\alpha}\right|\,\left|E_{i,j}\right|\,dx\\ &\le C\sum_{\alpha\in J}\left(\frac{1}{|B_\alpha|}\int_{B_\alpha}\left|\phi_j(x)-(\phi_j)_{B_\alpha}\right|^{\nu^*}dx\right)^{1/\nu^*}\left(\frac{1}{|B_\alpha|}\int_{B_\alpha}\left|E_{i,j}\right|^{\beta}dx\right)^{1/\beta}.\end{aligned}$$

Next, we need to apply the Sobolev inequality (4.63). Precisely, given $1 \le p < D$ there is a constant $C > 0$ such that for any ball $B = B(x,r)$ with $x\in\Omega$ and $0 < r \le \mathrm{diam}(\Omega)$ one has

$$\left(\frac{1}{B}\int_B |u-u_B|^{p^*}\,dx\right)^{1/p^*} \le Cr\left(\frac{1}{B}\int_B |Xu|^p\,dx\right)^{1/p}, \qquad p^* = \frac{Dp}{D-p}$$

(where D is a homogeneous dimension of Ω relative to X). Let us choose $\nu^* := D\nu/(D-\nu)$ (hence $\beta = \nu^*/(\nu^*-1) = D\nu/[D(\nu-1)+\nu]$) with $1 \le \nu < D$. Then (since $X_a\phi_j \in L^\nu$)

$$|A_a(y)| \le C\sum_{\alpha\in J} r_\alpha \left(\frac{1}{|B_\alpha|}\int_{B_\alpha} |X\phi_j|^\nu dx\right)^{1/\nu}\left(\frac{1}{|B_\alpha|}\int_{B_\alpha} \left|E_{i,j}\right|^\beta\,dx\right)^{1/\beta}.$$

Note that (by the definition of $E_{i,j}$) one has $|E_{i,j}| \le 2\rho(Q)|X\phi|$. Therefore, using also $\rho(Q) \le KQ^p$,

$$|A_a(y)| \le CK\sum_{\alpha\in J} r_\alpha \left(\frac{1}{|B_\alpha|}\int_{B_\alpha} |X\phi|^\nu dx\right)^{1/\nu}\left(\frac{1}{|B_\alpha|}\int_{B_\alpha} |X\phi|^{\beta(2p+1)} dx\right)^{1/\beta}.$$

The second integral converges if $\beta \le D/(2p+1)$. Later on, we shall choose ν (and this will produce a limitation on p). Given $\alpha \in J$, there is $k \in \mathbf{Z}$ such that $x_\alpha \in B(y,2^{k-1}) \setminus B(y,2^{k-2})$. Let us observe (together with P. Hájlasz and P. Strzelecki [192], p. 356) that $B_\alpha = B(x_\alpha, 6r_\alpha) \subset B(y,2^k)$. Moreover, $r_\alpha \approx 2^k$; hence, by applying (4.60) with $x_0 = y$, $r_0 = 2^k$, and $x = x_\alpha$, $r = 6r_\alpha$,

$$\frac{|B(x_\alpha, 6r_\alpha)|}{\left|B(y,2^k)\right|} \ge C\left(\frac{6r_\alpha}{2^k}\right)^D,$$

we get $|B_\alpha| \approx |B(y,2^k)|$. In the end, when $2^{k-2} \ge \operatorname{diam}(8B)$ the set $\{\alpha \in J : x_\alpha \in B(y,2^{k-1}) \setminus B(y,2^{k-2})\}$ is empty. Therefore

$$|A_a(y)| \le C \sum_{2^k \le 4\operatorname{diam}(8B)} 2^k \left(\frac{1}{|B(y,2^k)|}\int_{B(y,2^k)} |X\phi|^\nu dx\right)^{1/\nu} \times \left(\frac{1}{|B(y,2^k)|}\int_{B(y,2^k)} |X\phi|^{\beta(2p+1)} dx\right)^{1/\beta}. \tag{4.75}$$

Our **Step 2** is to *rewrite the estimate* (4.75) *in terms of Riesz potentials* and apply the fractional integration theorem. We recall its essentials, in the framework of metric spaces with Borel measures.

Definition 4.28. Let (X,ρ) be a metric space and let μ be a Borel measure on X such that $\mu(B) > 0$ for any ball $B \subset X$. Given a bounded open set $A \subset X$ and the numbers $q > 0$, $\sigma \ge 1$, and $h > 0$ we set

$$(J_{h,q}^{\sigma,A} g)(x) = \sum_{2^k \le 2\sigma\operatorname{diam}(A)} 2^{kh}\left(\frac{1}{|B(x,2^k)|}\int_{B(x,2^k)} |g(z)|^q\,dz\right)^{1/q}. \tag{4.76}$$

We call $J_{h,q}^{\sigma,A}$ an *abstract Riesz potential operator*. □

Theorem 4.22. (P. Hajlasz and P. Koskela [190]) *Let (X, ρ) be a metric space and let μ be a Borel measure on X such that $\mu(B) > 0$ for any ball $B \subset X$. Let $A \subset X$ be a bounded open set. Assume that μ is doubling on*

$$V = \{x \in X : \operatorname{dist}(x, A) < 2\sigma \operatorname{diam}(A)\}.$$

Moreover, let us assume that there are constants $b > 0$ and $D_1 > 0$ such that

$$\mu(B(x, r)) \geq b \left(\frac{r}{\operatorname{diam}(A)}\right)^{D_1} \mu(A)$$

for any $x \in A$ and any $0 < r \leq 2\sigma \operatorname{diam}(A)$. Let $h > 0$ and $0 < q < s < D_1/h$. Then

$$\|J_{h,q}^{\sigma,A} g\|_{L^{s^*}(A,\mu)} \leq C \left(\frac{\operatorname{diam}(A)}{\mu(A)^{D_1}}\right)^h \|g\|_{L^s(V,\mu)},$$

where $s^* = sD_1(D_1 - hs)$ and the constant $C > 0$ depends only on h, σ, q, s, b, D_1, and C_d.

Let us go back to Step 2. With the notation adopted in Definition 4.28 we may rewrite (4.75) as

$$|A_a(y)| \leq C \left(J_{1/2,\nu}^{2,8B}|X\phi|(y)\right) \left(J_{1/2,\beta}^{2,8B}|X\phi|^{2p+1}(y)\right). \tag{4.77}$$

Since **Step 3** we may *end the proof of Lemma* 4.4. By the Hölder inequality (with $1/D + 1/D' = 1$)

$$\begin{aligned}\left|\int_B X^* \cdot (\phi_j E_{i,j})\varphi \, dx\right| &\leq \sum_a \|X\varphi\|_{L^D(B)} \left(\int_B |A_a(y)|^{D'} dy\right)^{1/D'} \\ &\leq C\|X\varphi\|_{L^D(B)} \left(\int_B (J_\alpha|X\phi|)^{D'} \left(J_\beta|X\phi|^{2p+1}\right)^{D'} dy\right)^{1/D'}\end{aligned}$$

(by (4.77) in Step 2)

$$\leq C\|X\varphi\|_{L^D(B)} \left(\|(J_\nu|X\phi|)^{D'}\|_{L^{s^*/D'}(8B)} \cdot \|(J_\beta|X\phi|^{2p+1})^{D'}\|_{L^{r'}(8B)}\right)^{1/D'}$$

(again by the Hölder inequality, with $D'/s^* + 1/r' = 1$). At this point we may apply (twice) Theorem 4.22 (with $A = 8B$, $D_1 = D$, $\sigma = 2$, $h = 1/2$ and $q = \nu$ (respectively $q = \beta$)). Let us set

$$J_q := J_{1/2,q}^{2,8B}$$

for simplicity. Now, on the one hand,

$$\|J_\nu|X\phi|\|_{L^{s^*}(8B)} \leq C \left(\frac{\operatorname{diam}(8B)}{|8B|^{1/D}}\right)^{1/2} \|X\phi\|_{L^s(V)}, \qquad s^* = \frac{2Ds}{2D - s},$$

and on the other,

$$\|J_\beta |X\phi|^{2p+1}\|_{L^{r^*}(8B)} \leq C\left(\frac{\text{diam}(8B)}{|8B|^{1/D}}\right)^{1/2} \||X\phi|^{2p+1}\|_{L^r(V)}, \qquad r^* = \frac{2Dr}{2D-r}.$$

We wish to have $r^* = s^* D'/(s^* - D') = 2Ds/[2s(D-1) - 2D + s]$; hence we must take $r := s/(s-1)$ and require that $0 < \nu < s$ and $0 < \beta < s/(s-1)$. Summing up (by $\|g^n\|_{L^{m/n}} = (\|g\|_{L^m})^n$),

$$\left|\int_B X^* \cdot (\phi_j E_{i,j})\varphi \, dx\right| \leq C\frac{\text{diam}(8B)}{|8B|^{1/D}} \|X\varphi\|_{L^D(B)} \|X\phi\|_{L^s(V)} \||X\phi|^{2p+1}\|_{L^{s/(s-1)}(V)},$$

and the integrals in the right-hand member are convergent if

$$D/(D-2p-1) \leq s \leq D. \tag{4.78}$$

At this point, we choose $s := 2(p+1)$ (hence $s/(s-1) = 2(p+1)/(2p+1)$). The inequalities (4.78) are satisfied (because $0 < p < (D-2)/2$). With this choice of s we must have $\beta = D\nu/[D(\nu-1)+\nu] < 2(p+1)/(2p+1)$; hence

$$2D(p+1)/[D+2(p+1)] < \nu < 2(p+1)$$

(again, such a choice of ν is possible because $p < (D-2)/2$). Finally, note that $\||X\phi|^{2p+1}\|_{L^{s/(s-1)}(V)} = \left(\|X\phi\|_{L^{2(p+1)}(V)}\right)^{2p+1}$ and $V \subset 100B$; hence

$$\left|\int_B X^* \cdot (\phi_j E_{i,j})\varphi \, dx\right| \leq C\frac{\text{diam}(8B)}{|8B|^{1/D}} \|X\varphi\|_{L^D(B)} \left(\|X\phi\|_{L^{2(p+1)}(100B)}\right)^{2(p+1)}.$$

To end the proof of Lemma 4.4, let $R_0 > 0$ and consider a relatively compact subset $\Omega_1 \subset\subset \Omega$ such that $B(x, 400R_0) \subset \Omega$, for any $x \in \Omega_1$. For any $0 < r \leq R_0$, from the definition of the homogeneous dimension,

$$|8B| \geq C\left(\frac{8r}{400R_0}\right)^D |B(x, 400R_0)| \geq C(8r)^D(400R_0)^{s_d - D}\frac{|\Omega|}{(2\,\text{diam}(\Omega))^{s_d}}$$

where $s_d := \log_2 C_d$ and $C_d \geq 1$ is the doubling constant (on Ω, relative to the Lebesgue measure). In the end, $\text{diam}(8B)/|8B|^{1/D} \leq C$, for some constant $C = C(\Omega_1, D, C_d, R_0) > 0$. The inequality (4.66) is proved.

5

Pseudo-Einsteinian Manifolds

Let $(M, T_{1,0}(M))$ be a nondegenerate CR manifold (of hypersurface type) of CR dimension n. A contact form θ on M is said to be *pseudo-Einstein* if the pseudo-Hermitian Ricci tensor $R_{\alpha\overline{\beta}}$ of (the Tanaka–Webster connection of) (M, θ) is proportional to the Levi form, i.e.,

$$R_{\alpha\overline{\beta}} = \frac{\rho}{n} h_{\alpha\overline{\beta}}$$

on U, for any (local) frame $\{T_\alpha : 1 \leq \alpha \leq n\}$ in $T_{1,0}(M)$ on U. Here $\rho = h^{\alpha\overline{\beta}} R_{\alpha\overline{\beta}}$ is the pseudo-Hermitian scalar curvature (cf. Chapter 1 of this book). A nondegenerate CR manifold admitting some pseudo-Einsteinian contact form is a *pseudo-Einsteinian manifold*.

The pseudo-Einsteinian condition was considered for the first time by J.M. Lee; cf. [270]. It is of course an analogue of the Einstein condition in Riemannian geometry, yet less rigid than the Einstein condition: indeed, the Bianchi identities no longer imply $\rho = \text{const}$ (due to the presence of torsion terms).

The natural problem, given a compact nondegenerate CR manifold, is to find a pseudo-Einsteinian contact form. As we shall shortly see, the problem has a local aspect, and a global aspect as well. Note that any contact form on a 3-dimensional nondegenerate CR manifold is actually pseudo-Einsteinian (since the pseudo-Hermitian Ricci tensor has but one component $R_{1\overline{1}}$); hence we may assume that M has CR dimension $n \geq 2$.

In the sequel, we describe J.M. Lee's results (cf. [270]) in some detail, together with some results in [121] and [36, 37].

5.1 The local problem

In this section we address the problem of the existence of pseudo-Einsteinian contact forms in some neighborhood of an arbitrary point of M. The known results in this direction follow from a theorem of J.M. Lee (cf. op. cit.) and the positive (local) embeddability results available in contemporary science.

Theorem 5.1. (J.M. Lee [270]) *Let M be a nondegenerate CR manifold, of CR dimension n. If M admits a nowhere-vanishing closed $(n+1,0)$-form then M admits a (global) pseudo-Einsteinian contact form. Conversely, if M is a pseudo-Einsteinian manifold then for any point of M, there is an open neighborhood U and a nowhere-vanishing closed section in $K(M) = \Lambda^{n+1,0}(M)$ defined on U.*

Clearly, any real hypersurface $j : M \subset \mathbf{C}^{n+1}$ admits a globally defined, nowhere-vanishing closed $(n+1,0)$-form ($\eta = j^*\left(dz^1 \wedge \cdots \wedge dz^{n+1}\right)$); hence we have the following corollary:

Corollary 5.1. *Let M be a nondegenerate CR manifold (of hypersurface type). If M admits a CR embedding in $\mathbf{C}^{n+1}$ then M is a pseudo-Einstein manifold.*

Also, by the results of L. Boutet de Monvel [77] (any compact, strictly pseudoconvex CR manifold is locally embeddable in $\mathbf{C}^{n+1}$) and M. Kuranishi [263], T. Akahori [2] (the same is true in the noncompact case if the CR dimension is ≥ 3) one has the following:

Corollary 5.2. *Let M be a strictly pseudoconvex CR manifold. If either M is compact or* $\dim(M) \geq 7$ *then in a neighborhood of each point of M there is a pseudo-Einsteinian contact form.*

When M is 5-dimensional or the Levi form is not definite there may be no global closed sections of $K(M)$ and manifolds with closed sections are not necessarily embeddable (cf. H. Jacobowitz [222]).

To prove Theorem 5.1, let us examine the relationship between sections in the canonical bundle $K(M)$ and pseudo-Hermitian structures. Let θ be a contact form on M and $h_{\alpha\overline{\beta}}$ the components of the Levi form L_θ, with respect to a (local) frame $\{T_\alpha\}$ in $T_{1,0}(M)$, defined on the open set U. There is $P : U \to \mathrm{GL}(n, \mathbf{C})$ such that

$$[h_{\alpha\overline{\beta}}] = P \cdot \mathrm{diag}(\lambda_1, \ldots, \lambda_n) \cdot \overline{P}^{\,t}$$

on U, where $\mathrm{Spec}(L_{\theta,x}) = \{\lambda_1(x), \ldots, \lambda_n(x)\}$, $x \in M$, are the eigenvalues of the Levi form. Then

$$\det(h_{\alpha\overline{\beta}}) = \lambda_1 \cdots \lambda_n \cdot |\det P|^2 ;$$

hence $\mathrm{sign}\{\det(h_{\alpha\overline{\beta}})\} = (-1)^s$, provided that L_θ has signature (r, s).

Definition 5.1. A contact form θ is said to be *volume-normalized* with respect to a section $\omega \in \Gamma^\infty(K(M))$ if

$$2^n i^{n^2} n! (-1)^s \theta \wedge (T \rfloor \omega) \wedge (T \rfloor \overline{\omega}) = \theta \wedge (d\theta)^n ,$$

where T is the characteristic direction of $d\theta$. □

We set $K^0(M) := K(M) \setminus \{\text{zero section}\}$ and assume that there is $\zeta \in \Gamma^\infty(K^0(M))$. By (the proof of) Lemma 2.2 (in Chapter 2) there is a C^∞ function $\lambda : M \to \mathbf{R}^*$ such that $\mathrm{sign}(\lambda(x)) = (-1)^s$ and

$$2^n i^{n^2} n! \theta \wedge (T \rfloor \zeta) \wedge (T \rfloor \overline{\zeta}) = \lambda\, \theta \wedge (d\theta)^n .$$

Let us set $\mu := (-1)^s \lambda$. Then θ is volume-normalized with respect to $\omega := \mu^{-1/2} \zeta$.

Proposition 5.1. (J.M. Lee [270]) *Let θ be a contact form on M. Then θ is pseudo-Einsteinian if and only if for any $x \in M$ there is an open set $U \subseteq M$ with $x \in U$ and a section $\zeta \in \Gamma^\infty(U, K^0(M))$ such that $d\zeta = 0$ and θ is volume-normalized with respect to ζ.*

Proof. To prove the sufficiency let $\{T_\alpha\}$ be an orthonormal frame of $T_{1,0}(M)$, i.e., $h_{\alpha\overline{\beta}} = \epsilon_\alpha \delta_{\alpha\beta}$, $\epsilon_\alpha^2 = 1$. Then $\zeta = f\,\theta \wedge \theta^{1\cdots n}$ for some C^∞ function $f : U \to \mathbf{C}^*$, and

$$(d\theta)^n = 2^n i^{n^2} n!(-1)^s \theta^{1\cdots n\cdot\overline{1}\cdots\overline{n}}.$$

Taking into account

$$\theta \wedge (d\theta)^n = 2^n i^{n^2} n!(-1)^s \theta \wedge (T \rfloor \zeta) \wedge (T \rfloor \overline{\zeta})$$

(since by hypothesis, θ is volume-normalized with respect to ζ) and

$$T \rfloor \zeta = \frac{f}{n+1}\theta^{1\cdots n}, \quad T \rfloor \overline{\zeta} = \frac{\overline{f}}{n+1}\theta^{\overline{1}\cdots\overline{n}},$$

we obtain $|f| = n+1$. Let us set

$$[U^\alpha_\beta] := \operatorname{diag}(f, 1, \ldots, 1), \quad \hat{T}_\alpha := U^\beta_\alpha T_\beta .$$

Let $\{\hat{\theta}^\alpha\}$ be the corresponding admissible coframe, i.e.,

$$\hat{\theta}^1 = f^{-1}\theta^1, \quad \hat{\theta}^\alpha = \theta^\alpha, \quad \alpha \geq 2.$$

Then

$$\zeta = \theta \wedge \hat{\theta}^1 \wedge \cdots \wedge \hat{\theta}^n,$$

and taking the exterior derivative yields

$$\begin{aligned} d\zeta &= (d\theta) \wedge \hat{\theta}^1 \wedge \cdots \wedge \hat{\theta}^n + \sum_{\alpha=1}^{n} (-1)^\alpha \theta \wedge \hat{\theta}^1 \wedge \cdots \wedge (d\hat{\theta}^\alpha) \wedge \cdots \wedge \hat{\theta}^n \\ &= \sum_{\alpha=1}^{n} (-1)^\alpha \theta \wedge \hat{\theta}^1 \wedge \cdots \wedge \left(\hat{\theta}^\beta \wedge \hat{\omega}^\alpha_\beta + \theta \wedge \hat{\tau}^\alpha\right) \wedge \cdots \wedge \hat{\theta}^n \\ &= \sum_{\alpha} (-1)^\alpha \theta \wedge \hat{\theta}^1 \wedge \cdots \wedge \left(\hat{\theta}^\alpha \wedge \hat{\omega}^\alpha_\alpha\right) \wedge \cdots \wedge \hat{\theta}^n = -\hat{\omega}^\alpha_\alpha \wedge \zeta, \end{aligned}$$

i.e.,

$$d\zeta = -\hat{\omega}^\alpha_\alpha \wedge \zeta.$$

Yet $d\zeta = 0$; hence $\hat{\omega}^\alpha_\alpha$ is a $(1,0)$-form. The components $\hat{h}_{\alpha\overline{\beta}}$ of the Levi form L_θ, with respect to the new frame $\{\hat{T}_\alpha\}$, are given by

$$\hat{h}_{\alpha\overline{\beta}} = \hat{\epsilon}_\alpha \delta_{\alpha\beta}, \quad \hat{\epsilon}_\alpha := \begin{cases} (n+1)^2 \epsilon_1, & \alpha = 1, \\ \epsilon_\alpha, & \alpha \geq 2. \end{cases}$$

On the other hand, as a consequence of $\nabla g_\theta = 0$,

$$d\hat{h}_{\alpha\overline{\beta}} = \hat{\omega}_\alpha^\gamma \hat{h}_{\gamma\overline{\beta}} + \hat{h}_{\alpha\overline{\gamma}} \hat{\omega}_{\overline{\beta}}^{\overline{\gamma}},$$

or

$$\hat{\omega}_\alpha^\beta \hat{\epsilon}_\beta \delta_{\beta\gamma} + \hat{\omega}_{\overline{\gamma}}^{\overline{\beta}} \hat{\epsilon}_\alpha \delta_{\alpha\beta} = 0,$$

i.e.,

$$\hat{\epsilon}_\gamma \hat{\omega}_\alpha^\gamma + \hat{\epsilon}_\alpha \hat{\omega}_{\overline{\gamma}}^{\overline{\alpha}} = 0$$

(no sums), which may be also written as

$$\hat{\omega}_\alpha^\beta + \hat{\epsilon}_\alpha \hat{\epsilon}^\beta \hat{\omega}_{\overline{\beta}}^{\overline{\alpha}} = 0,$$

where $\hat{\epsilon}^\alpha := 1/\hat{\epsilon}_\alpha$. Contraction of α and β now gives

$$\hat{\omega}_\alpha^\alpha + \hat{\omega}_{\overline{\alpha}}^{\overline{\alpha}} = 0;$$

hence $\hat{\omega}_\alpha^\alpha$ is pure imaginary. Therefore (since $\hat{\omega}_\alpha^\alpha$ is a $(1,0)$-form, as previously shown)

$$\hat{\omega}_\alpha^\alpha = iu\theta,$$

for some real-valued $u \in C^\infty(U)$. On the other hand, by (1.89) in Chapter 1,

$$d\hat{\omega}_\alpha^\alpha = \hat{R}_{\lambda\overline{\mu}} \hat{\theta}^\lambda \wedge \hat{\theta}^{\overline{\mu}} + \hat{\varphi} \wedge \theta, \quad \hat{\varphi} := \hat{W}_{\alpha\lambda}^\alpha \hat{\theta}^\lambda - \hat{W}_{\alpha\overline{\lambda}}^\alpha \hat{\theta}^{\overline{\lambda}},$$
$$d\hat{\omega}_\alpha^\alpha = i\, du \wedge \theta + iu\, d\theta\,;$$

hence

$$\hat{R}_{\lambda\overline{\mu}} = 2u\hat{h}_{\lambda\overline{\mu}}\,,$$

i.e., θ is pseudo-Einsteinian. □

Conversely, assume θ to be pseudo-Einsteinian. Let $\{T_\alpha\}$ such that $h_{\alpha\overline{\beta}} = \epsilon_\alpha \delta_{\alpha\beta}$. Consider the (locally defined) $(n+1,0)$-form

$$\zeta_0 := \theta \wedge \theta^1 \wedge \cdots \wedge \theta^n\,.$$

Then (by a computation similar to that of $d\zeta$ above)

$$d\zeta_0 = -\omega_\alpha^\alpha \wedge \zeta_0\,. \tag{5.1}$$

We need the following lemma:

Lemma 5.1. *The contact form θ is pseudo-Einsteinian if and only if the 1-form $\omega_\alpha^\alpha + \frac{i}{2n}\rho\theta$ is closed, for any frame $\{T_\alpha\}$ of $T_{1,0}(M)$ on $U \subseteq M$.*

We shall prove Lemma 5.1 later on. By Lemma 5.1, by eventually restricting further the open set U, there is a real-valued function $\varphi \in C^\infty(U)$ such that

$$\omega_\alpha^\alpha + \frac{i\rho}{2n}\theta = i\, d\varphi.$$

Here, we have also exploited the fact that ω_α^α is pure imaginary (a fact following from $\omega_{\alpha\overline{\beta}} + \omega_{\overline{\beta}\alpha} = 0$). Note that

$$\begin{aligned} d(e^{i\varphi}\zeta_0) &= e^{i\varphi}\{i\, d\varphi \wedge \zeta_0 + d\zeta_0\} \\ &= e^{i\varphi}\left\{i\, d\varphi - \omega_\alpha^\alpha\right\} \wedge \zeta_0 = e^{i\varphi}\frac{i\rho}{2n}\theta \wedge \zeta_0 = 0 \end{aligned}$$

(by (5.1)); hence $\zeta := e^{i\varphi}\zeta_0$ is closed and

$$\begin{aligned} 2^n i^{n^2} n!(-1)^s \theta \wedge (T \rfloor \zeta) \wedge (T \rfloor \overline{\zeta}) &= \frac{(-1)^s}{(n+1)^2}\frac{1}{\det(h_{\alpha\overline{\beta}})}\theta \wedge (d\theta)^n \\ &= \frac{1}{(n+1)^2}\theta \wedge (d\theta)^n, \end{aligned}$$

i.e., $\omega := (n+1)\zeta$ is closed and θ is volume-normalized with respect to ω. Proposition 5.1 is completely proved. □

5.2 The divergence formula

We first define two differential operators d_b and d_b^c on functions.

Definition 5.2. For any C^∞ function $f : M \to \mathbf{C}$ we set

$$d_b f := (\overline{\partial}_b + \partial_b) f, \quad d_b^c := i(\overline{\partial}_b - \partial_b) f.$$ □

Consequently, on any nondegenerate CR manifold on which a contact form θ has been fixed,

$$df = f_0 \theta + d_b f,$$

where $f_0 = T(f)$, and T is the characteristic direction of (M, θ). Given a $(1,0)$-form $\sigma = \sigma_\alpha \theta^\alpha$ we set as usual

$$\sigma_{\alpha,\overline{\beta}} := (\nabla_{T_{\overline{\beta}}}\sigma)T_\alpha, \quad \sigma_{\alpha,}{}^{\beta} := h^{\beta\overline{\gamma}}\sigma_{\alpha,\overline{\gamma}},$$

and define the differential operator δ_b on $(1,0)$-forms as follows.

Definition 5.3. Let $\delta_b : \Omega^{1,0}(M) \to C^\infty(M) \otimes \mathbf{C}$ be the differential operator defined by $\delta_b(\sigma_\alpha\theta^\alpha) := \sigma_{\alpha,}{}^{\alpha}$. □

It is an easy exercise that the definition of $(\delta_b\sigma)_x$ doesn't depend on the choice of local frame $\{T_\alpha\}$ on $U \ni x$, i.e., $\delta_b\sigma$ is a globally defined smooth function. Assume $\sigma \in \Gamma_0^\infty(\Lambda^{1,0}(M))$, i.e., that σ has compact support.

Proposition 5.2. *Let M be a nondegenerate CR manifold, θ a contact form on M, and σ a $(1,0)$-form on M. Then*

$$\int_M (\delta_b \sigma)\, \theta \wedge (d\theta)^n = 0.$$

Proposition 5.2 is referred to as the *divergence formula*. Indeed, let $\sigma^\sharp$ be the vector field given by

$$g_\theta(\sigma^\sharp, X) = \sigma(X),$$

for any $X \in \mathcal{X}(M)$. Then $\sigma^\sharp = \sigma^{\overline{\alpha}} T_{\overline{\alpha}}$ and

$$\operatorname{div}(\sigma^\sharp) = T_{\overline{\alpha}}(\sigma^{\overline{\alpha}}) + \sigma^{\overline{\beta}} \Gamma^{\overline{\alpha}}_{\overline{\alpha}\overline{\beta}} = \sigma^{\overline{\alpha}}{}_{,\overline{\alpha}} = \sigma_{\alpha,}{}^{\alpha} = \delta_b \sigma.$$

Clearly $\delta_b \sigma = \operatorname{div}(\sigma^\sharp)$ yields the divergence formula. □

Let ∂_b^* be the formal adjoint of ∂_b, i.e.,

$$(\partial_b^* \sigma, u) = (\sigma, \partial_b u)$$

for any $(1,0)$-form σ and any $u \in C_0^\infty(M)$. Then (by Green's lemma)

$$(\partial_b^* \sigma, u) = \int_M \sigma_\alpha u^\alpha \, \Psi = \int_M \left\{ \operatorname{div}(u \sigma^{\overline{\beta}} T_{\overline{\beta}}) - u \sigma^{\overline{\beta}} \Gamma^{\overline{\alpha}}_{\overline{\alpha}\overline{\beta}} - u T_{\overline{\beta}}(\sigma^{\overline{\beta}}) \right\} \Psi = (-\delta_b \sigma, u).$$

We have proved the following result:

Proposition 5.3. *On any nondegenerate CR manifold M one has $\partial_b^* = -\delta_b$. In particular $d_b^* = -\delta_b - \overline{\delta}_b$.*

5.3 CR-pluriharmonic functions

Definition 5.4. A real-valued function $u \in C^\infty(M)$ is said to be *CR-pluriharmonic* if for any point $x \in M$, there is an open neighborhood U of x in M and a real-valued function $v \in C^\infty(U)$ such that $\overline{\partial}_b(u + iv) = 0$. □

In other words, CR-pluriharmonic functions are, locally, real parts of CR-holomorphic functions. On a nondegenerate CR manifold, the following description of CR-pluriharmonic functions is available:

Lemma 5.2. *Let M be a nondegenerate CR manifold, of hypersurface type, and $u \in C^\infty(M)$ a real-valued function. Let θ be a contact form on M.*

(1) u is CR-pluriharmonic if and only if for any $x \in M$ there is an open neighborhood $U \subseteq M$ of X and a real-valued function $\lambda \in C^\infty(U)$ such that the 1-form $\eta := d_b^c u + \lambda\theta$ is closed.

(2) *There is a (globally defined) real-valued function $v \in C^\infty(M)$ such that $\overline{\partial}_b(u + iv) = 0$ if and only if there is a real-valued function $\lambda \in C^\infty(M)$ such that the 1-form $\eta := d_b^c u + \lambda\theta$ is exact.*

Proof. Let us prove the implication $\Longrightarrow$ in statement (1). To this end, let $x \in M$ and $v \in C^\infty(U)$, $U \ni x$, such that $u + iv$ is CR-holomorphic in U. Then

$$0 = i\overline{\partial}_b(u + iv) - i\partial_b(u - iv) = d_b^c u - \overline{\partial}_b v - \partial_b v = d_b^c u - dv + v_0\theta,$$

i.e.,

$$d_b^c u + v_0\theta = dv.$$

Let us set $\lambda := v_0 \in C^\infty(U)$. Then $d_b^c u + \lambda\theta$ is closed. □

Next, we prove the implication $\Longleftarrow$ in statement (1). If $d(d_b^c u + \lambda\theta) = 0$ then (by the Poincaré lemma) there is $V \subseteq U$ such that $x \in V$ and there is $v \in C^\infty(V)$ such that

$$d_b^c u + \lambda\theta = dv,$$

which may be written as

$$iu_{\overline{\beta}}\theta^{\overline{\beta}} - iu_\beta\theta^\beta + \lambda\theta = v_\beta\theta^\beta + v_{\overline{\beta}}\theta^{\overline{\beta}} + v_0\theta.$$

Comparing the (0, 1) components we find that

$$iu_{\overline{\beta}} = v_{\overline{\beta}},$$

i.e., $\overline{\partial}_b(u + iv) = 0$. □

Let us prove the implication $\Longrightarrow$ in statement (2). If there is $v \in C^\infty(M)$ such that $\overline{\partial}_b(u + iv) = 0$ then

$$d_b^c u + v_0\theta = dv = \text{exact}. \qquad \square$$

Conversely, to prove $\Longleftarrow$, let us assume that $d_b^c u + \lambda\theta = \text{exact} = dv$, for some λ. Then $iu_{\overline{\beta}} = v_{\overline{\beta}}$ in a neighborhood of any point of M, i.e., $\overline{\partial}_b(u + iv) = 0$. Lemma 5.2 is completely proved. □

Lemma 5.3. *Let M be a nondegenerate CR manifold of dimension ≥ 5 and $\xi \in \Omega^2(M)$ a closed ($d\xi = 0$) complex two-form such that $\xi|_{H(M)\otimes H(M)} = 0$. Then $\xi = 0$.*

Proof. Since $\xi|_{H(M)\otimes H(M)} = 0$ there is a complex 1-form $\sigma \in \Omega^1(M)$ such that $\xi = \sigma \wedge \theta$. Yet ξ is closed; hence

$$0 = d\xi = (d\sigma) \wedge \theta - \sigma \wedge d\theta.$$

Consequently

$$\sigma \wedge d\theta|_{H(M)\otimes H(M)\otimes H(M)} = 0.$$

Let $\alpha, \beta, \gamma \in \{1, \dots, n\}$ be chosen such that $\{T_\alpha, T_\beta, T_{\overline{\gamma}}\}$ are linearly independent at each point of U and let us set

$$(A_1, A_2, A_3) := (\alpha, \beta, \overline{\gamma}).$$

Then

$$\begin{aligned}0 &= \tfrac{1}{6} \sum_{\pi \in \pi_3} \epsilon(\pi)(\sigma \otimes d\theta)(T_{A_{\pi(1)}}, T_{A_{\pi(2)}}, T_{A_{\pi(3)}}) \\ &= \tfrac{1}{3}\left(\sigma(T_\alpha)(d\theta)(T_\beta, T_{\overline{\gamma}}) + \sigma(T_\beta)(d\theta)(T_{\overline{\gamma}}, T_\alpha) + \sigma(T_{\overline{\gamma}})(d\theta)(T_\alpha, T_\beta)\right);\end{aligned}$$

or

$$0 = i\sigma_\alpha h_{\beta\overline{\gamma}} - i\sigma_\beta h_{\alpha\overline{\gamma}}\,.$$

Let us contract by $h^{\overline{\gamma}\mu}$ followed by a contraction of the indices μ and β. We get $(n-1)\sigma_\alpha = 0$; hence by the assumption on the dimension ($2n+1 > 3$), $\sigma_\alpha = 0$. Thus $\sigma|_{H(M)} = 0$ and consequently $\xi = 0$. □

Proposition 5.4. (J.M. Lee [271]) *Let M be a nondegenerate CR manifold of dimension* $\dim(M) > 3$. *Let $u \in C^\infty(M)$. Then u is CR-pluriharmonic if and only if for any (local) frame $\{T_\alpha\}$ of $T_{1,0}(M)$ on U, there is a function $\mu \in C^\infty(U) \otimes \mathbf{C}$ such that*

$$u_{\alpha\overline{\beta}} = \mu h_{\alpha\overline{\beta}}$$

in U, where $u_{AB} := (\nabla_{T_B} du)T_A$.

Proof. To prove the implication $\Longrightarrow$, let u be a CR-pluriharmonic function. Then, for any $x \in M$, there is an open neighborhood U of x in M and a function $\lambda \in C^\infty(U)$ such that $d(d_b^c u + \lambda\theta) = 0$, i.e.,

$$\begin{aligned}0 &= d(iu_{\overline{\alpha}}\theta^{\overline{\alpha}} - iu_\alpha\theta^\alpha + \lambda\theta) \\ &= i(du_{\overline{\alpha}}) \wedge \theta^{\overline{\alpha}} + iu_{\overline{\alpha}}d\theta^{\overline{\alpha}} - i(du_\alpha) \wedge \theta^\alpha - iu_\alpha d\theta^\alpha + (d\lambda) \wedge \theta + \lambda\, d\theta \\ &= iT_\beta(u_{\overline{\alpha}})\,\theta^\beta \wedge \theta^{\overline{\alpha}} + iT_{\overline{\beta}}(u_{\overline{\alpha}})\theta^{\overline{\beta}} \wedge \theta^{\overline{\alpha}} \\ &\quad + iT(u_{\overline{\alpha}})\,\theta \wedge \theta^{\overline{\alpha}} + iu_{\overline{\alpha}}\left\{\theta^{\overline{\beta}} \wedge \omega^{\overline{\alpha}}_{\overline{\beta}} + \theta \wedge \tau^{\overline{\alpha}}\right\} \\ &\qquad - iT_\beta(u_\alpha)\,\theta^\beta \wedge \theta^\alpha - iT_{\overline{\beta}}(u_\alpha)\,\theta^{\overline{\beta}} \wedge \theta^\alpha - iT(u_\alpha)\,\theta \wedge \theta^\alpha \\ &\qquad\quad - iu_\alpha\left\{\theta^\beta \wedge \omega^\alpha_\beta + \theta \wedge \tau^\alpha\right\} + \lambda_\alpha\,\theta^\alpha \wedge \theta + \lambda_{\overline{\alpha}}\theta^{\overline{\alpha}} \wedge \theta + 2i\lambda h_{\alpha\overline{\beta}}\theta^\alpha \wedge \theta^{\overline{\beta}}\,.\end{aligned}$$

In particular, the $(1, 1)$ component must vanish, i.e.,

$$\left\{T_\alpha(u_{\overline{\beta}}) - \Gamma^{\overline{\gamma}}_{\alpha\overline{\beta}}u_{\overline{\gamma}} + T_{\overline{\beta}}(u_\alpha) - \Gamma^{\gamma}_{\overline{\beta}\alpha}u_\gamma + 2\lambda h_{\alpha\overline{\beta}}\right\}\theta^\alpha \wedge \theta^{\overline{\beta}} = 0;$$

hence

$$u_{\overline{\beta}\alpha} + u_{\alpha\overline{\beta}} + 2\lambda h_{\alpha\overline{\beta}} = 0. \tag{5.2}$$

On the other hand, the identity

$$(\nabla_X du)Y = (\nabla_Y du)X - T_\nabla(X, Y)(u)$$

leads to

$$u_{\alpha\overline{\beta}} = u_{\overline{\beta}\alpha} + 2ih_{\alpha\overline{\beta}}u_0 \,.$$

Using the commutation formula to replace $u_{\overline{\beta}\alpha}$ in (5.2) we obtain

$$u_{\alpha\overline{\beta}} = (iu_0 - \lambda)h_{\alpha\overline{\beta}} \,.$$

□

Conversely, in order to prove $\Longleftarrow$, let us assume that $u_{\alpha\overline{\beta}} = \mu h_{\alpha\overline{\beta}}$ on U and set $\lambda := iu_0 - \mu$. By taking complex conjugates we get $u_{\overline{\alpha}\beta} = \overline{\mu}h_{\overline{\alpha}\beta}$. Hence, again by the commutation formula for $u_{\alpha\overline{\beta}}$,

$$\mu - 2iu_0 = \overline{\mu} \,.$$

Consequently $\overline{\lambda} = \lambda$, i.e., λ is real-valued, as it should be. Taking into account the identity

$$d(u_\alpha\theta^\alpha) = u_{\beta\alpha}\theta^\alpha \wedge \theta^\beta - u_{\alpha\overline{\beta}}\theta^\alpha \wedge \theta^{\overline{\beta}} + \theta \wedge \left(u_{\alpha 0}\theta^\alpha + u_\alpha A^\alpha_{\overline{\beta}}\theta^{\overline{\beta}}\right)$$

and the commutation formula

$$u_{\alpha\beta} = u_{\beta\alpha}$$

we derive

$$d(d_b^c u + \lambda\theta) = i\left\{u_{\overline{\beta}\alpha} + u_{\alpha\overline{\beta}} + 2\lambda\, h_{\alpha\overline{\beta}}\right\}\theta^\alpha \wedge \theta^{\overline{\beta}} + \theta \wedge \phi, \tag{5.3}$$
$$\phi := (iu_{\overline{\beta}}A^{\overline{\beta}}_\alpha - iu_{\alpha 0} - \lambda_\alpha)\theta^\alpha + (iu_{\overline{\alpha}0} - iu_\beta A^\beta_{\overline{\alpha}} - \lambda_{\overline{\alpha}})\theta^{\overline{\alpha}} \,.$$

Before going any further, let us collect the commutation formulas:

$$u_{\alpha\beta} = u_{\beta\alpha} \,,$$
$$u_{\overline{\beta}\alpha} = u_{\alpha\overline{\beta}} - 2ih_{\alpha\overline{\beta}}u_0 \,,$$
$$u_{\beta 0} = u_{0\beta} - u_{\overline{\alpha}}A^{\overline{\alpha}}_\beta \,.$$

By the second commutation formula, (5.3) may be also written as

$$d\eta = 2i\left\{u_{\alpha\overline{\beta}} + (\lambda - iu_0)h_{\alpha\overline{\beta}}\right\}\theta^\alpha \wedge \theta^{\overline{\beta}} + \theta \wedge \phi. \tag{5.4}$$

A remark regarding the first part of the proof is in order. If u is CR-pluriharmonic, then in addition to the expression of $u_{\alpha\overline{\beta}}$ we also get $\theta \wedge \phi = 0$; hence

$$u_{\alpha 0} = i\lambda_\alpha + u_{\overline{\beta}}A^{\overline{\beta}}_\alpha \,.$$

Let us return to the proof of the opposite implication. Under the assumptions made (5.4) becomes

$$d\eta = \theta \wedge \phi.$$

At this point, we may apply Lemma 5.3 for $\xi := d\eta$ to conclude that $\xi = 0$; hence by Lemma 5.2, u is CR-pluriharmonic. Proposition 5.4 is completely proved. □

In particular

$$\begin{aligned} u_{\alpha 0} &= i\lambda_\alpha + u_{\overline{\beta}} A^{\overline{\beta}}_{\alpha} = i T_\alpha(iu_0 - \mu) + u_{\overline{\beta}} A^{\overline{\beta}}_{\alpha} \\ &= -T_\alpha(u_0) - i\mu_\alpha + u_{\overline{\beta}} A^{\overline{\beta}}_{\alpha} = -u_{0\alpha} - i\mu_\alpha + u_{\overline{\beta}} A^{\overline{\beta}}_{\alpha} = -u_{\alpha 0} - i\mu_\alpha \, , \end{aligned}$$

i.e.,

$$u_{\alpha 0} = -\frac{i}{2}\mu_\alpha \, .$$

For the sake of completeness, let us examine now the case $n = 1$. If this is the case, then *any* function u satisfies $u_{1\overline{1}} = \mu h_{1\overline{1}}$ for some μ; hence to characterize CR-pluriharmonic functions one should use directly Lemma 5.2.

Proposition 5.5. *Let M be a 3-dimensional nondegenerate CR manifold and $P : C^\infty(M) \to \Gamma^\infty(T^*(M) \otimes \mathbf{C})$ the differential operator given by*

$$Pu := \left(u_{\overline{1}}{}^{\overline{1}}{}_{1} + 2i A_{11} u^{1} \right) \theta^{1} \, .$$

Then u is CR-pluriharmonic if and only if $Pu = 0$.

Proof. Assume that u is CR-pluriharmonic. Then $d(d_b^c u + \lambda\theta) = 0$ for some $\lambda \in C^\infty(U)$, by Lemma 5.2. That is,

$$\begin{gathered} 2i \left\{ u_{1\overline{1}} + (\lambda - iu_0) h_{1\overline{1}} \right\} \theta^1 \wedge \theta^{\overline{1}} + \theta \wedge \phi = 0, \\ \phi = (-iu_{10} + iu_{\overline{1}} A^{\overline{1}}_{1} - \lambda_1)\theta^1 + (iu_{\overline{1}0} - iu_1 A^{1}_{\overline{1}} - \lambda_{\overline{1}})\theta^{\overline{1}} \, , \end{gathered}$$

or

$$\begin{aligned} u_{1\overline{1}} &= (iu_0 - \lambda) h_{1\overline{1}} \, , \\ -iu_{10} + iu_{\overline{1}} A^{\overline{1}}_{1} - \lambda_1 &= 0. \end{aligned} \tag{5.5}$$

The commutation formula $u_{\overline{1}1} = u_{1\overline{1}} - 2ih_{1\overline{1}} u_0$ yields

$$iu_0 = \tfrac{1}{2} \left(u_1{}^1 - u_{\overline{1}}{}^{\overline{1}} \right);$$

hence $\lambda = iu_0 - u_1{}^1$ may be written

$$\lambda = -\tfrac{1}{2} \left(u_1{}^1 + u_{\overline{1}}{}^{\overline{1}} \right).$$

Let us set

$$\begin{aligned} u_{ABC} &:= (\nabla^3 u)(T_C, T_B, T_A) \\ &= (\nabla_{T_C} \nabla^2 u)(T_B, T_A) - (\nabla^2 u)(\nabla_{T_C} T_B, T_A) - (\nabla^2 u)(T_B, \nabla_{T_C} T_A), \\ (\nabla^2 u)(X, Y) &:= (\nabla_X du)Y. \end{aligned}$$

Then

$$u_{\beta\overline{\mu}\alpha} = T_\alpha(u_{\beta\overline{\mu}}) - \Gamma^{\overline{\nu}}_{\alpha\overline{\mu}} u_{\beta\overline{\nu}} - \Gamma^{\nu}_{\alpha\beta} u_{\nu\overline{\mu}},$$

or, using

$$T_\alpha(h^{\gamma\overline{\nu}}) = -h^{\gamma\overline{\rho}} h^{\epsilon\overline{\nu}} T_\alpha(h_{\epsilon\overline{\rho}})$$

we get

$$T_\alpha(u_\beta{}^\gamma) = u_\beta{}^\gamma{}_\alpha + u_\nu{}^\gamma \Gamma^\nu_{\alpha\beta} - u_\beta{}^\epsilon \Gamma^\gamma_{\alpha\epsilon};$$

hence when $n = 1$,

$$T_1(u_1{}^1) = u_1{}^1{}_1\,.$$

Consequently

$$\lambda_1 = -\tfrac{1}{2}\left(u_1{}^1{}_1 + u_{\overline{1}}{}^{\overline{1}}{}_1\right).$$

At this point we need the commutation formula (cf. Chapter 9)

$$f_{\beta\overline{\gamma}A} - 2i f_{0A} h_{\beta\overline{\gamma}} = f_{\overline{\gamma}\beta A},$$

whence

$$f_\alpha{}^\alpha{}_A - 2ni f_{0A} = f^\alpha{}_{\alpha A}\,.$$

Then, for $A = \beta$,

$$f_\alpha{}^\alpha{}_\beta = f^\alpha{}_{\alpha\beta} + 2ni f_{0\beta},$$

or

$$f_\alpha{}^\alpha{}_\beta = f_{\overline{\alpha}}{}^{\overline{\alpha}}{}_\beta + 2ni\left\{f_{\beta 0} + f_{\overline{\alpha}} A^{\overline{\alpha}}_\beta\right\}.$$

For $n = 1$,

$$u_1{}^1{}_1 = u_{\overline{1}}{}^{\overline{1}}{}_1 + 2i\left\{u_{10} + u_{\overline{1}} A^{\overline{1}}_1\right\};$$

hence λ_1 may be written

$$\lambda_1 = -u_{\overline{1}}{}^{\overline{1}}{}_1 - i\left(u_{10} + u_{\overline{1}} A^{\overline{1}}_1\right),$$

and the second of the identities (5.5) becomes

$$u_{\bar{1}}{}^{\bar{1}}{}_{1} + 2iu_{\bar{1}}A_{1}^{\bar{1}} = 0. \qquad \square$$

Conversely, if $Pu = 0$ then

$$u_{\bar{1}}{}^{\bar{1}}{}_{1} + 2iu_{\bar{1}}A_{1}^{\bar{1}} = 0,$$

and by letting $\lambda := -\frac{1}{2}\left(u_1{}^1 + u_{\bar{1}}{}^{\bar{1}}\right)$ in the identity

$$d\eta = 2i\left\{u_{1\bar{1}} + (\lambda - iu_0)h_{1\bar{1}}\right\}\theta^1 \wedge \theta^{\bar{1}} + \theta \wedge \phi,$$
$$\phi := (-iu_{10} + iu_{\bar{1}}A_1^{\bar{1}} - \lambda_1)\theta^1 + (iu_{\bar{1}0} - iu_1 A_{\bar{1}}^{1} - \lambda_{\bar{1}})\theta^{\bar{1}},$$

we get $d\eta = 0$; hence by Lemma 5.2, u is CR-pluriharmonic. □

We end this section by discussing a few well-known matters, perhaps classical at least within the Italian mathematical culture, related to the subject of CR-pluriharmonic functions and suitable for shedding light on the latter. Let Ω be a domain in $\mathbf{C}^n$ and U an open set such that $U \supseteq \overline{\Omega}$. Assume that $\partial\Omega$ is a smooth hypersurface in $\mathbf{C}^n$.

Definition 5.5. A C^2 function $u : U \to \mathbf{R}$ is *pluriharmonic* if $\partial\overline{\partial}u = 0$ in U. □

Let $u : U \to \mathbf{R}$ be a pluriharmonic function. If U is simply connected, there is a function w, holomorphic in U ($\overline{\partial}w = 0$ in U), such that $\mathrm{Re}(w) = u$. Thus $f := w|_{\partial\Omega}$ is a CR function on $\partial\Omega$. Moreover, $u|_{\partial\Omega} = \mathrm{Re}(f)$; hence $U := u|_{\partial\Omega}$ is a CR-pluriharmonic function on $\partial\Omega$, in the sense adopted in this chapter. Thus CR-pluriharmonic functions may be thought of as boundary values of pluriharmonic functions. On the other hand, one of the antique problems in analysis is to *characterize traces on* $\Sigma := \partial\Omega$ *of pluriharmonic functions*. We present one of the first contributions in this direction, in two complex variables, belonging to L. Amoroso (cf. [11]) (a new proof of which is due to G. Fichera (cf. [141])), together with a generalization of the result to an arbitrary number n of complex variables ($n \geq 2$) (cf. [142]). We close with a characterization of boundary values of pluriharmonic functions on the unit ball in $\mathbf{C}^n$, due to E. Bedford (cf. [52]) and related to the result by L. Amoroso.

1 L. Amoroso's theorem. The real part of a function w that is holomorphic ($\overline{\partial}w = 0$) in Ω is pluriharmonic. If Ω is simply connected the converse is true (i.e., any pluriharmonic function is the real part of some function holomorphic in Ω). The first to consider the previously stated problem was H. Poincaré (cf. [349]): *given* U *on* Σ, *find* u *pluriharmonic in* Ω *whose trace on* Σ *coincides with* U (the Dirichlet problem for pluriharmonic functions). As H. Poincaré observed, any pluriharmonic function is in particular harmonic in Ω and, since a harmonic function is determined by its boundary values, one may not assign U arbitrarily if one wants the given harmonic function to be additionally pluriharmonic. The difficulty of the problem (i.e., *find necessary and sufficient conditions on* U *such that* U *is the boundary values of a pluriharmonic function*) was emphasized by T. Levi-Civita (cf. [281]). L. Amoroso (cf. op. cit.) was the first to solve the problem. Although later F. Severi showed (cf. [369]) that the conditions found by L. Amoroso were overdetermined, the work by L. Amoroso remains of great importance, and according to G. Fichera (cf. op. cit.), insufficient credit is given to L. Amoroso in the existing literature on functions of several complex variables. See also G. Tomassini [404].

Let $\rho \in C^m(\mathbf{R}^4)$, with $m \geq 1$, such that $\Omega = \{\rho > 0\}$, $\Sigma = \{\rho = 0\}$, and $\mathbf{R}^4 \setminus \overline{\Omega} = \{\rho < 0\}$, and let us assume that $D\rho(x) \neq 0$, for any $x \in \Sigma$.

Definition 5.6. A linear differential operator L, of order m, with continuous coefficients is *tangential* to Σ if $Lu = 0$ on Σ for any $u \in C^m(\mathbf{R}^4)$ that satisfies $u|_\Sigma = 0$. □

L. Amoroso (cf. op. cit.) characterized traces of pluriharmonic functions in terms of the following invariant. Let ν be the inward unit normal vector on Σ.

Definition 5.7. The *Levi invariant* of Σ is

$$\mathcal{L}(\rho) := \sum_{i,j=1}^{2} \frac{\partial^2 \rho}{\partial z_i \partial \overline{z}_j} \lambda_i \overline{\lambda}_j,$$

where

$$\lambda_1 := \frac{\partial \rho}{\partial z_2}, \quad \lambda_2 := -\frac{\partial \rho}{\partial z_1}. \qquad \square$$

In the language of pseudo-Hermitian structures, introduced by S. Webster ([422]) and adopted by us throughout this book, if $j : \Sigma \subset \mathbf{C}^n$ is the inclusion and $\theta := j^*\{\frac{i}{2}(\overline{\partial} - \partial)\rho\}$, then

$$\mathcal{L}(\rho) = 2L_\theta(T_1, T_{\overline{1}}),$$

where L_θ is the Levi form and T_1, the generator of the CR structure $T_{1,0}(\Sigma)$, is given by

$$j_* T_1 = \frac{\partial \rho}{\partial z_1}\frac{\partial}{\partial z_2} - \frac{\partial \rho}{\partial z_2}\frac{\partial}{\partial z_1}.$$

As such, the Levi invariant $\mathcal{L}_\theta := 2L_\theta(T_1, T_{\overline{1}})$ makes sense on an arbitrary 3-dimensional CR manifold M on which a pseudo-Hermitian structure has been fixed. If T_1 is defined on the open set $U \subseteq M$ then $\mathcal{L}_\theta \in C^\infty(U)$ and the Levi invariant transforms as $\mathcal{L}_{\hat{\theta}} = \lambda \mathcal{L}_\theta$, under a transformation $\hat{\theta} = \lambda\theta$ of pseudo-Hermitian structure.

Theorem 5.2. (L. Amoroso [11])
Assume that Ω admits a defining function $\rho \in C^2(\Omega)$ such that $\mathcal{L}(\rho) \neq 0$ everywhere on Σ. Then there is a tangential (relative to Σ) second-order linear differential operator D such that

$$\frac{\partial u}{\partial \nu} = \frac{1}{\mathcal{L}(\rho)} Du, \tag{5.6}$$

for any $u \in C^2(\overline{\Omega})$ which is pluriharmonic in Ω.

Proof. (G. Fichera [141]) Let $\omega \subset \Omega$ be a simply connected subdomain. Let $u \in C^2(\overline{\Omega})$ be such that $\partial\overline{\partial}u = 0$ in Ω. Then $\partial\overline{\partial}u = 0$ in ω, hence there is a function $w(z)$ holomorphic in ω such that $u = \text{Re}(w)$. Let us set $v := \text{Im}(w)$ (in particular, v is pluriharmonic in ω, as the real part of the holomorphic function iw). Then $w = u + iv$ and $\overline{\partial}w = 0$ yield

$$\frac{\partial v}{\partial \overline{z}_j} = i\frac{\partial u}{\partial \overline{z}_j}. \tag{5.7}$$

Let us set

$$L := j_* T_{\overline{1}} = \frac{\partial \rho}{\partial z_1}\frac{\partial}{\partial z_2} - \frac{\partial \rho}{\partial z_2}\frac{\partial}{\partial z_1}. \tag{5.8}$$

Then L, $\overline{L}$ are tangential first-order linear differential operators. By the Cauchy–Riemann equations (5.7)

$$Lv = \rho_{\overline{z}_1}\frac{\partial v}{\partial \overline{z}_2} - \rho_{\overline{z}_2}\frac{\partial v}{\partial \overline{z}_1} = i\left(\rho_{\overline{z}_1}\frac{\partial u}{\partial \overline{z}_2} - \rho_{\overline{z}_2}\frac{\partial u}{\partial \overline{z}_1}\right) = iLu,$$

i.e.,

$$Lv = iLu, \quad \overline{L}v = -i\overline{L}u. \tag{5.9}$$

Let us set

$$\Lambda := \left[\overline{L}, L\right] = \overline{L}L - L\overline{L} = a_1\frac{\partial}{\partial \overline{z}_1} + a_2\frac{\partial}{\partial \overline{z}_2} - \overline{a}_1\frac{\partial}{\partial z_1} - \overline{a}_2\frac{\partial}{\partial z_2}, \tag{5.10}$$

$$\begin{aligned} M &:= \overline{L}L + L\overline{L} \\ &= 2\Big(\rho_{z_1}\rho_{\overline{z}_1}\frac{\partial^2}{\partial z_2 \partial \overline{z}_2} - \rho_{z_1}\rho_{\overline{z}_2}\frac{\partial^2}{\partial \overline{z}_1 \partial z_2} - \rho_{\overline{z}_1}\rho_{z_2}\frac{\partial^2}{\partial z_1 \partial \overline{z}_2} + \rho_{z_2}\rho_{\overline{z}_2}\frac{\partial^2}{\partial z_2 \partial \overline{z}_2}\Big) \\ &\quad + a_1\frac{\partial}{\partial \overline{z}_1} + a_2\frac{\partial}{\partial \overline{z}_2} + \overline{a}_1\frac{\partial}{\partial z_1} + \overline{a}_2\frac{\partial}{\partial z_2}, \end{aligned} \tag{5.11}$$

where

$$a_1 := \lambda_1\frac{\partial^2 \rho}{\partial \overline{z}_2 \partial z_1} + \lambda_2\frac{\partial^2 \rho}{\partial \overline{z}_2 \partial z_2}, \quad a_2 := -\lambda_1\frac{\partial^2 \rho}{\partial \overline{z}_1 \partial z_1} - \lambda_2\frac{\partial^2 \rho}{\partial \overline{z}_1 \partial z_2}.$$

Clearly Λ, M are tangential linear differential operators, of the first and second order, respectively. Note that (by (5.9))

$$\Lambda v = \overline{L}Lv - L\overline{L}v = \overline{L}(iLu) - L(-i\overline{L}u) = i(\overline{L}L + L\overline{L})u = iMu,$$

i.e.,

$$\Lambda v = iMu$$

in $\overline{\omega}$, which, in view of (5.10)–(5.11) becomes

$$\rho_{z_1}\rho_{\bar z_1}\frac{\partial^2 u}{\partial z_2\partial \bar z_2} - \rho_{z_1}\rho_{\bar z_2}\frac{\partial^2 u}{\partial \bar z_1\partial z_2} - \rho_{\bar z_1}\rho_{z_2}\frac{\partial^2 u}{\partial z_1\partial \bar z_2} + \rho_{z_2}\rho_{\bar z_2} = 0,$$

hence (5.11) leads to

$$a_1\frac{\partial u}{\partial \bar z_1} + a_2\frac{\partial u}{\partial \bar z_2} + \bar a_1\frac{\partial u}{\partial z_1} + \bar a_2\frac{\partial u}{\partial z_2} = Mu \tag{5.12}$$

in $\overline{\omega}$. Since ω may be chosen around an arbitrary point of Ω, (5.12) holds in $\overline{\Omega}$. Note that

$$a_1 = \lambda_1\rho_{z_1\bar z_2} + \lambda_2\rho_{z_2\bar z_2}, \quad a_2 = -\lambda_1\rho_{z_1\bar z_1} - \lambda_2\rho_{\bar z_1 z_2}$$

may be written as

$$a_1 = -\frac{\mathcal{L}(\rho)}{|\lambda_1|^2+|\lambda_2|^2}\frac{\partial\rho}{\partial z_1} + p_1, \quad a_2 = -\frac{\mathcal{L}(\rho)}{|\lambda_1|^2+|\lambda_2|^2}\frac{\partial\rho}{\partial z_2} + p_2, \tag{5.13}$$

where

$$p_1 := \lambda_1\rho_{z_1\bar z_2} + \lambda_2\rho_{z_2\bar z_2} - \frac{\lambda_2\mathcal{L}(\rho)}{|\lambda_1|^2+|\lambda_2|^2},$$
$$p_2 := -\lambda_2\rho_{\bar z_1 z_2} - \lambda_1\rho_{z_1\bar z_1} + \frac{\lambda_1\mathcal{L}(\rho)}{|\lambda_1|^2+|\lambda_2|^2}.$$

Moreover, note that

$$Q := p_1\frac{\partial}{\partial \bar z_1} + p_2\frac{\partial}{\partial \bar z_2}, \quad \overline{Q} = \bar p_1\frac{\partial}{\partial z_1} + \bar p_2\frac{\partial}{\partial z_2}, \tag{5.14}$$

are tangential differential operators. Since Q is a vector field (a priori tangent to $\mathbf{C}^2$) it suffices to check that $Q\rho = 0$. Indeed,

$$\begin{aligned}
Q\rho &= p_1\rho_{\bar z_1} + p_2\rho_{\bar z_2}\\
&= -\Big(\lambda_1\rho_{z_1\bar z_2} + \lambda_2\rho_{z_2\bar z_2} - \frac{\lambda_2\mathcal{L}(\rho)}{|\lambda_1|^2+|\lambda_2|^2}\Big)\bar\lambda_2\\
&\quad + \Big(-\lambda_2\rho_{\bar z_1 z_2} - \lambda_1\rho_{z_1\bar z_1} + \frac{\lambda_1\mathcal{L}(\rho)}{|\lambda_1|^2+|\lambda_2|^2}\Big)\bar\lambda_1\\
&= -\lambda_1\bar\lambda_2\rho_{z_1\bar z_2} - \lambda_2\bar\lambda_2\rho_{z_2 z_2} - \lambda_2\bar\lambda_1\rho_{\bar z_1 z_2} - \lambda_1\bar\lambda_1\rho_{z_1\bar z_2}\\
&\quad + \frac{|\lambda_2|^2\mathcal{L}(\rho)}{|\lambda_1|^2+|\lambda_2|^2} + \frac{|\lambda_1|^2\mathcal{L}(\rho)}{|\lambda_1|^2+|\lambda_2|^2} = 0.
\end{aligned}$$

□

Let us set

$$\nabla u := u_{z_1}\frac{\partial}{\partial z_1} + u_{z_2}\frac{\partial}{\partial z_2} + u_{\bar z_1}\frac{\partial}{\partial \bar z_1} + u_{\bar z_2}\frac{\partial}{\partial \bar z_2}.$$

Since $\nu = \|\nabla\rho\|^{-1}\nabla\rho$,

$$\frac{\partial u}{\partial\nu} = \langle\nabla u, \nu\rangle = \frac{1}{\|\nabla\rho\|}\big(u_{z_1}\rho_{\bar z_1} + u_{z_2}\rho_{\bar z_2} + u_{\bar z_1}\rho_{z_1} + u_{\bar z_2}\rho_{z_2}\big);$$

hence (by (5.13))

$$a_1 u_{\overline{z}_1} + a_2 u_{\overline{z}_2} + \overline{a}_1 u_{z_1} + \overline{a}_2 u_{z_2} =$$
$$-\frac{\mathcal{L}(\rho)}{|\lambda_1|^2 + |\lambda_2|^2}\left(\rho_{z_1} u_{\overline{z}_1} \rho_{z_2} u_{\overline{z}_2} + \rho_{\overline{z}_1} u_{z_1} + \rho_{\overline{z}_2} u_{z_2}\right) + p_1 u_{\overline{z}_1} + p_2 u_{\overline{z}_2} + \overline{p}_1 u_{z_1} + \overline{p}_2 u_{z_2},$$

or (by (5.12))

$$Mu = -\frac{\mathcal{L}(\rho)}{\|\lambda\|^2}\|\nabla\rho\|\frac{\partial u}{\partial \nu} + Qu + \overline{Q}u.$$

Finally, since $\|\nabla\rho\|^2 = 2\|\lambda\|^2$, the last equation may be written

$$\frac{\partial u}{\partial \nu} = \frac{1}{\mathcal{L}(\rho)} Du$$

on Σ, where

$$D := -\tfrac{1}{2}\|\nabla\rho\|\left(L\overline{L} + \overline{L}L - Q - \overline{Q}\right),$$

and Theorem 5.2 is completely proved. □

2 G. Fichera's theorem. Let $\Delta \subset \mathbf{R}^{2n}$ be a $(2n-1)$-cell $(n > 2)$ of class C^ℓ $(\ell \geq 1)$, i.e., $\Delta := \phi(T^{2n-1})$ where $T^k := \{t \in \mathbf{R}^k : t_\alpha \geq 0,\ \sum_{\alpha=1}^k t_\alpha \leq 1\}$ is the standard simplex in $\mathbf{R}^k$ and $\phi : T^{2n-1} \to \mathbf{R}^{2n}$, $\phi(t_1, \ldots, t_{2n-1}) =: (x_1, y_1, \ldots, x_n, y_n)$ satisfies (1) $\phi \in C^\ell(T^{2n-1})$, (2) $\frac{\partial(x_1,y_1,\ldots,x_n,y_n)}{\partial(t_1,\ldots,t_{2n-1})}$ has rank $2n-1$ at any point of T^{2n-1}, and (3) ϕ is injective. That is, a $(2n-1)$-cell in $\mathbf{R}^{2n}$ is a regularly embedded (into $\mathbf{R}^{2n}$) standard $(2n-1)$-simplex.

Definition 5.8. Let $\Omega \subset \mathbf{R}^{2n}$ be a domain such that $\overline{\Omega} \supset \Delta$. A linear differential operator of order ℓ,

$$D = \sum_{|\mu|=0}^{\ell} a_\mu(z) D^\mu, \quad a_\mu \in C^0(\overline{\Omega}),$$

is *tangential* to Δ if $Du|_\Delta = 0$ for any $u \in C^\ell(\overline{\Omega})$ with $u|_\Delta = 0$. We say that D is a *Cauchy–Riemann operator* in Ω if for any open set $A \subseteq \Omega$ and any function w, holomorphic in A, we have $Dw = 0$ in A. Moreover, we say that D is a *pluriharmonic operator* in Ω if for any open set $A \subseteq \Omega$ and any function u, pluriharmonic in A, we have $Du = 0$ in A. Also D is a *real operator* if the a_μ are real-valued. □

Proposition 5.6. *Any pluriharmonic operator is a Cauchy–Riemann operator. Conversely, any real Cauchy–Riemann operator is a (real) pluriharmonic operator.*

Let us consider again the Dirichlet problem for pluriharmonic functions (i.e., given $U \in C^0(\Sigma)$ find $u \in C^0(\overline{\Omega})$, pluriharmonic in Ω, such that $u|_\Sigma = U$). Each pluriharmonic function (in Ω) is harmonic (in Ω) hence uniqueness in the Dirichlet problem is obvious. Therefore, the Dirichlet problem admits the following reformulation: *find necessary and sufficient conditions on U such that its harmonic extension in Ω is pluriharmonic*. G. Fichera (cf. [143]–[145]) has formulated two approaches to the Dirichlet problem. The so-called *local approach* consists in finding, for any $x \in \Sigma$, an open neighborhood $x \in A \subset \Sigma$ and (differential) conditions satisfied by U such that U is the trace on A of a pluriharmonic function (in Ω). The *global approach* is to determine a set Ψ of functions defined on Σ such that the following statements are equivalent: (1) U is the trace on Σ of a pluriharmonic (in Ω) function, and (2) $\int_\Sigma U\psi\, d\sigma = 0$ for any $\psi \in \Psi$.

As shown by G. Fichera (cf. op. cit.), the local approach leads to a solution of the following problem: *given U on a* $(2n-1)$*-cell* $\Delta \subset \Sigma$ *find u pluriharmonic in some domain* $\Omega^+ \subseteq \Omega$ *with* $\partial\Omega^+ \supset \Delta$ *such that* $u|_\Delta = U$. The local approach uses tangential Cauchy–Riemann operators as its main tool.

Let $M^n_\ell(\Delta)$ be the set of all real Cauchy–Riemann operators, of order ℓ, tangential to Δ.

Definition 5.9. Two operators in $M^n_\ell(\Delta)$ are *equivalent* if their restrictions to Δ coincide. □

The same symbol $M^n_\ell(\Delta)$ will denote the quotient space, modulo this equivalence relation, with the natural structure of a $C^0(\overline{\Delta})$-module (actually an algebra). Clearly $M^n_\ell(\Delta)$ has finite rank.

Definition 5.10. Assume that $\Sigma \in C^2$ and let $z \in \Sigma$. We say that Σ satisfies the *E.E. Levi condition* at z if there is $Z \in T_{1,0}(\Sigma)_z$ such that $L_{\theta,z}(Z, \overline{Z}) \neq 0$ (where L_θ is the Levi form corresponding to the pseudo-Hermitian structure $\theta = j^*[\frac{i}{2}(\overline{\partial} - \partial)\rho]$). □

In other words, if Σ fails to satisfy the E.E. Levi condition at each $z \in \Sigma$ then Σ is Levi flat. Since L_θ changes conformally the E.E. Levi condition is a CR-invariant condition.

The following result, the proof of which will not be given here, is an important instrument in G. Fichera's local approach (to the Dirichlet problem for pluriharmonic functions):

Theorem 5.3. (G. Fichera [142])

(i) Let $\Delta \in C^1$. *Then* rank $M^n_1(\Delta) = 0$.
(ii) Let $\Delta \in C^2$. *Let us assume that the E.E. Levi condition holds at each point of* $\overline{\Delta}$. *Then we have* rank $M^n_2(\Delta) = n^2 - 2n$. *In particular* rank $M^2_2(\Delta) = 0$.
(iii) Let $\Delta \in C^3$. *Let us assume that the E.E. Levi condition holds at each point of* $\overline{\Delta}$. *Let us set*

$$P^n_3(\Delta) := \Big\{ D \in M^n_3(\Delta) : D = \sum_{j=1}^{s} f_j L_j D_j,\ s \in \mathbf{Z},\ s \geq 1,$$
$$f_j \in C^0(\overline{\Delta}) \text{ (real-valued)},$$
$$L_j \text{ a real first-order tangential (to } \Delta\text{) operator},\quad D_j \in M^n_2(\Delta),\ \ 1 \leq j \leq s \Big\}.$$

Then there is a $C^0(\overline{\Delta})$*-module* $Q_3^n(\Delta)$ *such that*

$$M_3^n(\Delta) = P_3^n(\Delta) \oplus Q_3^n(\Delta)$$

and rank $Q_3^n(\Delta) > n$ *(in particular* rank $Q_3^2(\Delta) = 3$).

Let $\rho \in C^2(\mathbf{R}^{2n})$ be such that (1) $\rho|_\Delta = 0$ and $D\rho \neq 0$ everywhere on Δ (where D is the ordinary gradient), and (2) for any $z \in \Delta$, there are $\lambda_1, \ldots, \lambda_n \in \mathbf{C}$ such that

$$\sum_{j=1}^{n} \frac{\partial \rho}{\partial z_j}(z)\lambda_j = 0, \tag{5.15}$$

$$\mathcal{L}(\rho, \lambda) := \sum_{j,k=1}^{n} \frac{\partial^2 \rho}{\partial z_j \partial \overline{z}_k}(z)\lambda_j \overline{\lambda}_k \neq 0. \tag{5.16}$$

Let Ω be a domain such that $\overline{\Omega} \supset \Delta$ and ν the inward unit normal on Σ. Let $z^0 \in \Delta$ be fixed and assume, to fix ideas, that $\frac{\partial \rho}{\partial z_1}(z^0) \neq 0$. We may state the following generalization of L. Amoroso's result (i.e., of Theorem 5.2):

Theorem 5.4. (G. Fichera [142])
Let $u \in C^2(\Omega \cup \Delta)$ *be pluriharmonic in* Ω*. There is a neighborhood* A *of* z^0 *in* Δ *such that*

$$\frac{\partial u}{\partial \nu} = \frac{1}{\mathcal{L}(\rho, \lambda(z))} Du \tag{5.17}$$

in A*, for some tangential (to* Δ*) second-order differential operator, where*

$$\lambda_1(z) := -\frac{1}{\rho_{z_1}(z^0)} \sum_{j=2}^{n} \lambda_j \rho_{z_j}(z),$$

$$\lambda_j(z) := \frac{1}{\rho_{z_1}(z^0)} \lambda_j \rho_{z_1}(z), \quad 2 \leq j \leq n,$$

and $\lambda := (\lambda_1, \ldots, \lambda_n)$ *is some vector in* $\mathbf{C}^n$ *satisfying* (5.15)–(5.16) *with* $z = z^0$.

Proof. Let $\lambda_1, \ldots, \lambda_n \in \mathbf{C}$ such that (5.15)–(5.16) hold at $z = z^0$. Without loss of generality, we may assume $\lambda_2 \neq 0$ (since a consequence of (5.16)). Let us set $\alpha := 1/\rho_{z_1}(z^0)$ and consider the matrix $a \in \mathcal{M}_n(\mathbf{C})$ given by

$$a := \begin{bmatrix} \alpha & \lambda_1 - \alpha & 0 & 0 & \cdots & 0 \\ 0 & \lambda_2 & 0 & 0 & \cdots & 0 \\ 0 & \lambda_3 & 1 & 0 & \cdots & 0 \\ \vdots & \vdots & \vdots & \vdots & & \vdots \\ 0 & \lambda_n & 0 & 0 & \cdots & 1 \end{bmatrix}.$$

Then $\det(a) = \lambda_2 \alpha \neq 0$; hence a is invertible. Let us set $A := a^{-1}$. Then A gives rise to a biholomorphism $\zeta = Az$. Let $\tilde{\Delta} := A(\Delta)$ be the transform of Δ by this biholomorphism and set $\zeta^0 := Az^0 \in \tilde{\Delta}$. Let $r := \rho \circ a$. Then $r(\zeta) = 0$ for any $\zeta \in \tilde{\Delta}$. Let us set $\mu := A\lambda$. Since

$$r(\zeta_1, \dots, \zeta_n) = \rho(a_1^j \zeta_j, \dots, a_n^j \zeta_j)$$

we have

$$\frac{\partial r}{\partial \zeta_j}(\zeta^0) = \sum_{k=1}^n a_k^j \rho_{z_k}(z^0).$$

Thus

$$\sum_{j=1}^n r_{\zeta_j}(\zeta^0)\mu_j = 0, \quad \sum_{j,k=1}^n r_{\zeta_j \overline{\zeta}_k}(\zeta^0)\mu_j \overline{\mu}_k \neq 0. \tag{5.18}$$

From the definitions of a and μ we get

$$\mu_1 = \mu_2 = 1, \quad \mu_3 = \cdots = \mu_n = 0.$$

Note that

$$r_{\zeta_1}(\zeta^0) = \sum_{k=1}^n a_k^1 \rho_{z_k}(z^0) = \alpha \rho_{z_1}(z^0) = 1$$

(in a_j^i the index i is the column index). Also, the first of the formulas (5.18) may be explicitly written

$$r_{\zeta_1}(\zeta^0) + r_{\zeta_2}(\zeta^0) = 0.$$

Let us set

$$\mu_1(\zeta) := -r_{\zeta_2}(\zeta), \quad \mu_2(\zeta) := r_{\zeta_1}(\zeta), \quad \mu_3(\zeta) := \cdots = \mu_n(\zeta) := 0.$$

Note that

$$\sum_{j=1}^n \frac{\partial r}{\partial \zeta_j}(\zeta)\mu_j(\zeta) = 0.$$

Also

$$\mu_j(\zeta^0) = \mu_j, \quad 1 \le j \le n;$$

hence by (the second of the formulas (5.18) and by) continuity, there is an open neighborhood $\tilde{I}(\zeta^0)$ of ζ^0 in $\tilde{\Delta}$ such that

$$\mathcal{L}(r, \mu(\zeta)) \neq 0,$$

for any $\zeta \in \tilde{I}(\zeta^0)$. Consider now the first-order differential operators (tangential to $\tilde{\Delta}$)

$$L := \frac{\partial r}{\partial \overline{\zeta}_1}\frac{\partial}{\partial \overline{\zeta}_2} - \frac{\partial r}{\partial \overline{\zeta}_2}\frac{\partial}{\partial \overline{\zeta}_1}, \quad \overline{L} = \frac{\partial r}{\partial \zeta_1}\frac{\partial}{\partial \zeta_2} - \frac{\partial r}{\partial \zeta_2}\frac{\partial}{\partial \zeta_1}.$$

Since our considerations are local, we may assume without loss of generality that Ω is simply connected (otherwise choose a simply connected subdomain $\omega \subset \Omega$ such that $\overline{\omega} \ni z^0$). Let w be a holomorphic function in Ω such that $\text{Re}(w) = u$ and let us set $v := \text{Im}(w)$. Next, let us set $\tilde{\Omega} := A(\Omega)$ and $\tilde{w} := w \circ a$. Clearly $\tilde{w}$ is holomorphic in $\tilde{\Omega}$ (since a is a biholomorphism); hence by the Cauchy–Riemann equations in $\tilde{\Omega}$,

$$L\tilde{v} = iL\tilde{u}, \quad \overline{L}\tilde{v} = -i\overline{L}\tilde{u}\,, \tag{5.19}$$

where $\tilde{w} = \tilde{u} + i\tilde{v}$ are the real and imaginary parts of $\tilde{w}$. Let us set $\Lambda := [\overline{L}, L]$ and $M := \overline{L}L + L\overline{L}$. Then (by (5.19)) $\Lambda\tilde{v} = iM\tilde{u}$ in $\tilde{\Omega}$; hence (by a calculation similar to that in the previous section)

$$a_1\frac{\partial\tilde{u}}{\partial\overline{\zeta}_1} + a_2\frac{\partial\tilde{u}}{\partial\overline{\zeta}_2} + \overline{a}_1\frac{\partial\tilde{u}}{\partial\zeta_1} + \overline{a}_2\frac{\partial\tilde{u}}{\partial\zeta_2} = M\tilde{u}$$

in $\tilde{\Omega}$ and, since $\tilde{u} \in C^2(\tilde{\Omega} \cup \tilde{\Delta})$, in $\tilde{\Delta}$ as well. This is formally similar to (5.12), and the proof of Theorem 5.4 may be completed along the lines of that of Theorem 5.2. □

3 E. Bedford's theorem. Consider the tangential first-order differential operators (complex vector fields) on S^{2n-1}

$$\mathcal{L}_{ij} = \overline{\zeta}_i\frac{\partial}{\partial\zeta_j} - \overline{\partial}_j\frac{\partial}{\partial\zeta_i}, \quad 1 \le i, j \le n.$$

These operators extend naturally in the interior of the unit ball B^n.

Theorem 5.5. (E. Bedford [52]) *Let L be one of the operators $\mathcal{L}_{ij}$, or their complex conjugates. If $u \in C^3(S^{2n-1})$ then $\overline{L}\,\overline{L}\,L(u) = 0$ if and only if u extends to a pluriharmonic function in B^n.*

The proof is a rather simple consequence of the fact that given a function $u \in C^1(S^{2n-1})$ one has

$$(L_\zeta u) * P(z,\zeta) = L_z[u * P(z,\zeta)]$$

for $\zeta \in S^{2n-1}$ and $z \in B^n$, where $P(z,\zeta)$ is the Poisson kernel of B^n. The reader may see [52], p. 21, for details. See also P. De Bartolomeis et al. [115].

5.4 More local theory

At this point we may prove Lemma 5.1. We recall (see Chapter 1 of this book)

$$\Omega_\alpha^\beta = R_\alpha{}^\beta{}_{\lambda\overline{\mu}}\theta^\lambda \wedge \theta^{\overline{\mu}} + W^\beta_{\alpha\lambda}\theta^\lambda \wedge \theta - W^\beta_{\alpha\overline{\lambda}}\theta^{\overline{\lambda}} \wedge \theta,$$

where

$$\Omega_\alpha^\beta := \Pi_\alpha^\beta - 2i\theta_\alpha \wedge \tau^\beta + 2i\tau_\alpha \wedge \theta^\beta ,$$
$$\Pi_\alpha^\beta := d\omega_\alpha^\beta - \omega_\alpha^\gamma \wedge \omega_\gamma^\beta .$$

Since $R_{\lambda\overline{\mu}} = R_\alpha{}^\alpha{}_{\lambda\overline{\mu}}$ we may derive

$$\Omega_\alpha^\alpha = R_{\lambda\overline{\mu}}\theta^\lambda \wedge \theta^{\overline{\mu}} + W_{\alpha\lambda}^\alpha \theta^\lambda \wedge \theta - W_{\alpha\overline{\lambda}}^\alpha \theta^{\overline{\lambda}} \wedge \theta.$$

Also, note that $\Omega_\alpha^\alpha = d\omega_\alpha^\alpha$; hence

$$d\omega_\alpha^\alpha = R_{\alpha\overline{\beta}}\theta^\alpha \wedge \theta^{\overline{\beta}} + W_{\alpha\beta}^\alpha \theta^\beta \wedge \theta - W_{\alpha\overline{\beta}}^\alpha \theta^{\overline{\beta}} \wedge \theta. \tag{5.20}$$

Assume now that θ is pseudo-Einsteinian, i.e., $R_{\alpha\overline{\beta}} = (\rho/n)h_{\alpha\overline{\beta}}$. Substitution in (5.20) gives

$$d\omega_\alpha^\alpha = -\frac{i\rho}{2n}\, d\theta + \left(W_{\alpha\beta}^\alpha \theta^\beta - W_{\alpha\overline{\beta}}^\alpha \theta^{\overline{\beta}} \right) \wedge \theta,$$

or

$$d(\omega_\alpha^\alpha + \frac{i\rho}{2n}\theta) = \left\{ W_{\alpha\beta}^\alpha \theta^\beta - W_{\alpha\overline{\beta}}^\alpha \theta^{\overline{\beta}} + \frac{i}{2n} d\rho \right\} \wedge \theta. \tag{5.21}$$

Let us set

$$\xi := d\left(\omega_\alpha^\alpha + \frac{i\rho}{2n}\theta \right)$$

and note that $\xi|_{H(M)\otimes H(M)} = 0$. Then, by Lemma 5.3, $\xi = 0$, i.e., $\omega_\alpha^\alpha + \frac{i\rho}{2n}\theta$ is closed. □

To prove the converse, assume $\omega_\alpha^\alpha + \frac{i\rho}{2n}\theta$ to be closed. By (5.20)

$$d\omega_\alpha^\alpha = R_{\lambda\overline{\mu}}\theta^\lambda \wedge \theta^{\overline{\mu}} + \varphi\theta, \quad \varphi := W_{\alpha\lambda}^\alpha \theta^\lambda - W_{\alpha\overline{\lambda}}^\alpha \theta^{\overline{\lambda}} .$$

Then our assumption yields

$$0 = d\left(\omega_\alpha^\alpha + \frac{i\rho}{2n}\theta \right) = \frac{i}{2n} d(R\theta) + R_{\alpha\overline{\beta}}\theta^\alpha \wedge \theta^{\overline{\beta}} + \varphi \wedge \theta,$$

or

$$R_{\alpha\overline{\beta}}\theta^\alpha \wedge \theta^{\overline{\beta}} + \varphi \wedge \theta = -\frac{i\rho}{2n} d\theta - \frac{i}{2n}(d\rho) \wedge \theta. \tag{5.22}$$

Let us apply (5.22) to $(T_\alpha, T_{\overline{\beta}})$. It follows that $R_{\alpha\overline{\beta}} = (\rho/n)h_{\alpha\overline{\beta}}$, i.e., θ is pseudo-Einsteinian. Lemma 5.1 is completely proved. □

In particular

$$\varphi \wedge \theta = -\frac{i}{2n}(d\rho) \wedge \theta.$$

Applying this identity to (T_λ, T), respectively to $(T_{\overline{\lambda}}, T)$, we get

$$W_{\alpha\lambda}^\alpha = -\frac{i}{2n}\rho_\lambda, \quad W_{\alpha\overline{\lambda}}^\alpha = \overline{W_{\alpha\lambda}^\alpha} .$$

5.5 Topological obstructions

5.5.1 The first Chern class of $T_{1,0}(M)$

One of the main purposes of the present section is to establish the following proposition:

Proposition 5.7. (J.M. Lee [270])
If (M, θ) is pseudo-Einsteinian then $c_1(T_{1,0}(M)) = 0$.

Here $c_1(T_{1,0}(M)) \in H^2(M; \mathbf{R})$ is the first Chern class of $T_{1,0}(M)$. Before proving Proposition 5.7, let us look at Chern classes of an arbitrary complex vector bundle over a CR manifold. Let $\pi : E \to M$ be a rank-r complex vector bundle over M. The kth Chern class $c_k(E) \in H^{2k}(M; \mathbf{R})$ is the cohomology class $c_k(E) = [\gamma_k]$, where the $2k$-form γ_k on M is determined by

$$\det\left(I_r - \frac{1}{2\pi i}\Omega\right) = p^*\left(1 + \gamma_1 + \cdots + \gamma_r\right).$$

Here $p : L(E) \to M$ is the principal $\mathrm{GL}(r, \mathbf{C})$-bundle of all frames in the fibers of E (so that $(L(E) \times \mathbf{C}^r)/\mathrm{GL}(r, \mathbf{C}) \simeq E$, a bundle isomorphism). Also

$$\Omega \in \Gamma^\infty(\Lambda^2 T^*(L(E)) \otimes \mathbf{gl}(r, \mathbf{C}))$$

is the curvature 2-form of an arbitrary, fixed connection

$$D : \Gamma^\infty(E) \to \Gamma^\infty(T^*(M) \otimes E).$$

If $R^D(X, Y) = [D_X, D_Y] - D_{[X,Y]}$ is the curvature tensor of D then (cf., e.g., [241], vol. I)

$$\Omega(X^H, Y^H)_u u^{-1} s(p(u)) = u^{-1}(R^D(X, Y)s)_{p(u)}, \tag{5.23}$$

for any $X, Y \in \mathcal{X}(M)$, $u \in L(E)$, and $s \in \Gamma^\infty(E)$. Also X^H denotes the horizontal lift (with respect to D) of X to $L(E)$.

Let $\{s_1, \ldots, s_r\}$ be a local frame in E, defined on the open set U. By eventually restricting the open set U, we may assume the existence of a local frame $\{T_\alpha\}$ in $T_{1,0}(M)$, defined on U as well. Let $\{E^i_j\}$ be a linear basis of $\mathbf{gl}(r, \mathbf{C}) \simeq \mathbf{C}^{r^2}$ and let us set

$$\Omega = \Omega^i_j \otimes E^j_i, \quad \Omega^i_j \in \Gamma^\infty(\Lambda^2 T^* L(E)).$$

If $u \in L(E)_x$, $x \in M$, i.e., $u : \mathbf{C}^r \to E_x$ (a $\mathbf{C}$-linear isomorphism) we set $\sigma_j := u(e_j)$, where $\{e_j\}$ is the canonical linear basis in $\mathbf{C}^r$. We have (by (5.23))

$$u^{-1}\left(R^D(T_A, T_B)s_j\right)_x = \Omega(T^H_A, T^H_B)_u u^{-1} s_j(x). \tag{5.24}$$

On the other hand, we set

$$R^D(T_A, T_B)s_j =: R^k_{jAB} s_k$$

and $\sigma := [\sigma^k_j] \in \mathrm{GL}(r, \mathbf{C})$, where $\sigma_j = \sigma^k_j s_k(x)$. Then (5.24) may be written

$$R^k_{jAB}(x)(\sigma^{-1})^i_k e_i = \Omega^\ell_k(T^H_A, T^H_B)_u E^k_\ell \cdot \left((\sigma^{-1})^i_j e_i\right),$$

or, by taking into account that $E^j_k \cdot e_i = \delta_{ki} e^j$,

$$\Omega^\ell_i(T^H_A, T^H_B)_u = \sigma^j_\ell R^k_{jAB}(x)(\sigma^{-1})^i_k . \tag{5.25}$$

Let $\zeta^i_j : p^{-1}(U) \to \mathbf{C}$ be fiber coordinates on $L(E)$, i.e., $\zeta^i_j(u) := \sigma^i_j$. Since $\mathrm{Ker}(p_*) \rfloor \Omega = 0$, the identity (5.25) may be written

$$\Omega^\ell_i = \zeta^j_\ell (R^k_{jAB} \circ p)(\zeta^{-1})^i_k (p^*\theta^A) \wedge (p^*\theta^B). \tag{5.26}$$

We recall (cf. again [241], vol. II)

$$p^*(\gamma_k) = C_k \sum \delta^{j_1 \cdots j_k}_{i_1 \cdots i_k} \Omega^{i_1}_{j_1} \wedge \cdots \wedge \Omega^{i_k}_{j_k},$$

where

$$C_k := \frac{(-1)^k}{(2\pi i)^k k!}$$

and the sum is taken over all ordered subsets $(i_1, \ldots, i_k)$ in $\{1, \ldots, r\}$ and all permutations $(j_1, \ldots, j_k)$ of $(i_1, \ldots, i_k)$, while $\delta^{i_1 \cdots i_k}_{j_1 \cdots j_k}$ is the sign of the permutation $(i_1, \ldots, i_k) \mapsto (j_1, \ldots, j_k)$. In particular

$$p^*(\gamma_1) = C_1 \sum \delta^{j_1}_{i_1} \Omega^{i_1}_{j_1} = C_1 \sum_{i=1}^{r} \Omega^i_i .$$

Now (5.26) yields

$$\Omega^i_i = p^* \left(R^i_{iAB} \theta^A \wedge \theta^B\right).$$

Note that p^* is injective (and $C_1 = -1/(2\pi i)$). We obtain

$$\gamma_1 = -\frac{1}{2\pi i} R^j_{jAB} \theta^A \wedge \theta^B . \tag{5.27}$$

The computation of the (representatives of the) Chern classes $c_k(E)$, for $k \geq 2$, is usually more involved. For instance

$$p^*\gamma_2 = C_2 \sum \delta^{j_1 j_2}_{i_1 i_2} \Omega^{i_1}_{j_1} \wedge \Omega^{i_2}_{j_2} = C_2 \sum_{(i_1, i_2) \in \{1, \ldots, r\}^2} \left(\Omega^{i_1}_{i_1} \wedge \Omega^{i_2}_{i_2} - \Omega^{i_1}_{i_2} \wedge \Omega^{i_2}_{i_1}\right),$$

i.e.,

$$p^*\gamma_2 = C_2 \sum_{i=1}^{r} \sum_{j=1}^{r} \left(\Omega^i_i \wedge \Omega^j_j - \Omega^i_j \wedge \Omega^j_i\right).$$

A careful examination of Chern classes of CR-holomorphic vector bundles over CR manifolds is still missing from the contemporary literature, even for well-known examples (cf. the many examples in Chapter 1). If the given CR-holomorphic bundle $(E, \overline{\partial}_E) \to M$ carries a Hermitian structure and M is nondegenerate then one may use, at least in principle, the Tanaka connection to compute the representatives γ_k. While we leave this as an open problem, let us look at the main example, which is of course $E = T_{1,0}(M)$. The Tanaka–Webster connection ∇ descends to a connection in $T_{1,0}(M)$; hence (by (5.27))

$$c_1(T_{1,0}(M)) = \Big[-\frac{1}{2\pi i} R_\alpha{}^\alpha{}_{AB} \theta^A \wedge \theta^B \Big].$$

By a result in Chapter 1

$$R_\alpha{}^\alpha{}_{\lambda\mu} = 0, \quad R_\alpha{}^\alpha{}_{\overline{\lambda}\overline{\mu}} = 0, \quad R_\alpha{}^\alpha{}_{0\overline{\mu}} = W^\alpha_{\alpha\overline{\mu}}, \quad R_\alpha{}^\alpha{}_{\lambda 0} = W^\alpha_{\alpha\lambda}.$$

We may conclude that the first Chern class of $T_{1,0}(M)$ is represented by

$$\gamma_1 = -\frac{1}{2\pi i} \Big\{ R_{\lambda\overline{\mu}} \theta^\lambda \wedge \theta^{\overline{\mu}} + W^\alpha_{\alpha\lambda} \theta^\lambda \wedge \theta - W^\alpha_{\alpha\overline{\mu}} \theta^{\overline{\mu}} \theta^{\overline{\mu}} \wedge \theta \Big\}.$$

At this point we may prove Proposition 5.7. Assume that θ is pseudo-Einsteinian, i.e.,

$$R_{\alpha\overline{\mu}} = \frac{\rho}{n} h_{\alpha\overline{\mu}}, \quad W^\alpha_{\alpha\lambda} = -\frac{i}{2n} \rho_\lambda, \quad W^\alpha_{\alpha\overline{\mu}} = \frac{i}{2n} \rho_{\overline{\mu}};$$

hence

$$\begin{aligned}
\gamma_1 &= -\frac{1}{2\pi i} \Big\{ \frac{\rho}{n} h_{\lambda\overline{\mu}} \theta^\lambda \wedge \theta^{\overline{\mu}} - \frac{i}{2n} \rho_\lambda \theta^\lambda \wedge \theta - \frac{i}{2n} \rho_{\overline{\mu}} \theta^{\overline{\mu}} \wedge \theta \Big\} \\
&= -\frac{1}{2n\pi i} \Big\{ \rho h_{\lambda\overline{\mu}} \theta^\lambda \wedge \theta^{\overline{\mu}} - \frac{i}{2} (d\rho) \wedge \theta \Big\} \\
&= -\frac{1}{2n\pi i} \Big\{ \frac{1}{2i} \rho \, d\theta + \frac{1}{2i} (d\rho) \wedge \theta \Big\} = \frac{1}{4n\pi} d(R\theta).
\end{aligned}$$

We have proved the following result:

Proposition 5.8. *For any pseudo-Einsteinian manifold, the first Chern class of $T_{1,0}(M)$ is represented by*

$$\gamma_1 = \frac{1}{4n\pi} d(\rho\theta).$$

This is exact; hence $c_1(T_{1,0}(M)) = 0$. □

Note that for any pseudo-Einsteinian contact form θ, the curvature form of the Tanaka–Webster connection of (M, θ) satisfies

$$\Omega^j_j = \frac{1}{2ni} p^* d(\rho\theta).$$

5.5.2 The traceless Ricci tensor

Let M be a nondegenerate CR manifold and θ a contact form on M.

Definition 5.11. The *traceless Ricci tensor* $B_{\alpha\overline{\beta}}$ is given by

$$B_{\alpha\overline{\beta}} := R_{\alpha\overline{\beta}} - \frac{\rho}{n} h_{\alpha\overline{\beta}} . \qquad \square$$

Let us recall that (by a result in Chapter 2) under a transformation $\hat{\theta} = e^{2u}\theta$, the contracted connection 1-forms (of the Tanaka–Webster connection) transform as

$$\hat{\omega}_\alpha^\alpha = \omega_\alpha{}^\alpha + (n+2)(u_\alpha\theta^\alpha - u^\alpha\theta_\alpha) + i\left\{2(n+1)u_\alpha u^\alpha + \tfrac{1}{2}\Delta_b u\right\}\theta + n\,du.$$

Consequently

$$\begin{aligned} d\hat{\omega}_\alpha^\alpha &= d\omega_\alpha{}^\alpha + (n+2)d(u_\alpha\theta^\alpha - u^\alpha\theta_\alpha) \\ &\quad + i\left\{2(n+1)u_\alpha u^\alpha + \tfrac{1}{2}\Delta_b u\right\} d\theta + i d\left(2(n+1)u_\alpha u^\alpha + \tfrac{1}{2}\Delta_b u\right) \wedge \theta, \end{aligned}$$

or, by a curvature formula in Chapter 1,

$$\begin{aligned} &\hat{R}_{\lambda\overline{\mu}}\theta^\lambda \wedge \theta^{\overline{\mu}} + \hat{W}^\alpha_{\alpha\lambda}\theta^\lambda \wedge \theta - \hat{W}^\alpha_{\alpha\overline{\mu}}\theta^{\overline{\mu}} \wedge \theta \\ &\quad = R_{\lambda\overline{\mu}}\theta^\lambda \wedge \theta^{\overline{\mu}} + W^\alpha_{\alpha\lambda}\theta^\lambda \wedge \theta - W^\alpha_{\alpha\overline{\mu}}\theta^{\overline{\mu}} \wedge \theta + (n+2)d(u_\alpha\theta^\alpha - u^\alpha\theta_\alpha) \\ &\qquad + i\big(2(n+1)u_\alpha u^\alpha + \tfrac{1}{2}\Delta_b u\big)d\theta + i d\big(2(n+1)u_\alpha u^\alpha + \tfrac{1}{2}\Delta_b u\big) \wedge \theta; \end{aligned}$$

hence

$$\begin{aligned} (\hat{R}_{\lambda\overline{\mu}} - R_{\lambda\overline{\mu}})\theta^\lambda \wedge \theta^{\overline{\mu}} &\equiv -\left\{\Delta_b u + 4(n+1)u_\alpha u^\alpha\right\} h_{\lambda\overline{\mu}}\theta^\lambda \wedge \theta^{\overline{\mu}} \\ &\quad + (n+2)\left\{d(u_\alpha\theta^\alpha) - d(u_{\overline{\alpha}}\theta^{\overline{\alpha}})\right\} \mod \theta, \end{aligned}$$

or

$$\begin{aligned} (\hat{R}_{\lambda\overline{\mu}} - R_{\lambda\overline{\mu}})\theta^\lambda \wedge \theta^{\overline{\mu}} &= -\left\{\Delta_b u + 4(n+1)u_\alpha u^\alpha\right\} h_{\lambda\overline{\mu}}\theta^\lambda \wedge \theta^{\overline{\mu}} \\ &\quad - (n+2)\left\{u_{\lambda\overline{\mu}}\theta^\lambda \wedge \theta^{\overline{\mu}} + u_{\overline{\mu}\lambda}\theta^\lambda \wedge \theta^{\overline{\mu}}\right\} \mod \theta^\alpha \wedge \theta^\beta,\ \theta^{\overline{\alpha}} \wedge \theta^{\overline{\beta}},\ \theta, \end{aligned}$$

whence

$$\hat{R}_{\lambda\overline{\mu}} = R_{\lambda\overline{\mu}} - (n+2)(u_{\lambda\overline{\mu}} + u_{\overline{\mu}\lambda}) - \left\{\Delta_b u + 4(n+1)u_\alpha u^\alpha\right\} h_{\lambda\overline{\mu}} .$$

Using the commutation formula for $u_{\lambda\overline{\mu}}$, we have

$$\hat{R}_{\lambda\overline{\mu}} = R_{\lambda\overline{\mu}} - 2(n+1)(u_{\lambda\overline{\mu}} - i h_{\lambda\overline{\mu}} u_0) - \left\{\Delta_b u + 4(n+1)u_\alpha u^\alpha\right\} .$$

Contraction with $h^{\lambda\overline{\mu}}$ leads to

$$e^{2u}\hat{\rho} = \rho - (n+2)\left(u_\lambda{}^\lambda + u_{\overline{\lambda}}{}^{\overline{\lambda}}\right) - n\left(\Delta_b u + 4(n+1)u_\alpha u^\alpha\right)$$

and since

$$\Delta_b u = u_\lambda{}^\lambda + u_{\overline{\lambda}}{}^{\overline{\lambda}}$$

we get

$$e^{2u}\hat{\rho} = \rho - 2(n+1)\Delta_b u - 4n(n+1)u_\alpha u^\alpha .$$

Therefore

$$\hat{B}_{\lambda\overline{\mu}} = \hat{R}_{\lambda\overline{\mu}} - \frac{\hat{\rho}}{n}\hat{h}_{\lambda\overline{\mu}} = R_{\lambda\overline{\mu}} - (n+2)(u_{\lambda\overline{\mu}} + u_{\overline{\mu}\lambda})$$
$$- \left\{\Delta_b u + 4(n+1)u_\alpha u^\alpha\right\} h_{\lambda\overline{\mu}} - \frac{1}{n}h_{\lambda\overline{\mu}}\left\{\rho - 2(n+1)\Delta_b u - 4n(n+1)u_\alpha u^\alpha\right\},$$

i.e.,

$$\hat{B}_{\lambda\overline{\mu}} = B_{\lambda\overline{\mu}} - (n+2)(u_{\lambda\overline{\mu}} + u_{\overline{\mu}\lambda}) + \frac{n+2}{n}(u_\alpha{}^\alpha + u_{\overline{\alpha}}{}^{\overline{\alpha}})h_{\lambda\overline{\mu}} .$$

Let us set

$$(Qu)_{\alpha\overline{\beta}} := u_{\alpha\overline{\beta}} - \frac{1}{n}u_\gamma{}^\gamma h_{\alpha\overline{\beta}} .$$

We have already proved that u is CR-pluriharmonic if and only if $Qu = 0$; cf. Proposition 5.4. Then

$$\hat{B}_{\lambda\overline{\mu}} = B_{\lambda\overline{\mu}} - (n+2)\left\{(Qu)_{\lambda\overline{\mu}} + (Qu)_{\overline{\mu}\lambda}\right\}.$$

Yet, again by the commutation formula for second-order covariant derivatives

$$(Qu)_{\overline{\mu}\lambda} = u_{\overline{\mu}\lambda} - \frac{1}{n}u_{\overline{\gamma}}{}^{\overline{\gamma}}h_{\overline{\mu}\lambda} = u_{\lambda\overline{\mu}} - 2ih_{\lambda\overline{\mu}}u_0 - \frac{1}{n}h_{\lambda\overline{\mu}}\left\{u^{\overline{\gamma}}{}_{\overline{\gamma}} - 2inu_0\right\} = (Qu)_{\lambda\overline{\mu}} ,$$

i.e.,

$$(Qu)_{\lambda\overline{\mu}} = (Qu)_{\overline{\mu}\lambda}$$

and we have proved the following:

Proposition 5.9. *The traceless Ricci tensor transforms, under a transformation of the contact form $\hat{\theta} = e^{2u}\theta$, as*

$$\hat{B}_{\lambda\overline{\mu}} = B_{\lambda\overline{\mu}} - 2(n+2)(Qu)_{\lambda\overline{\mu}}. \tag{5.28}$$

In particular, $\hat{\theta}$ is pseudo-Einsteinian if and only if

$$Qu = \frac{1}{2(n+2)}B.$$

Let $\mathcal{P}$ be the sheaf of CR-pluriharmonic functions (on open sets of M). If a pseudo-Einsteinian contact form is available, the remaining pseudo-Einsteinian contact forms may be parameterized by elements in $\mathcal{P}(M)$, as follows.

Theorem 5.6. (J.M. Lee [270]) *If (M, θ) is pseudo-Einsteinian then*

$$\{e^{2u}\theta : u \in \mathcal{P}(M)\}$$

is the set of all pseudo-Einsteinian contact forms on M.

Proof. Let θ be pseudo-Einsteinian, i.e., $B = 0$. Let $u \in C^\infty(M)$ and set $\hat{\theta} := e^{2u}\theta$. Then $\hat{\theta}$ is pseudo-Einsteinian if and only if $Qu = \frac{1}{2(n+2)} B$; yet $B = 0$; hence $\hat{\theta}$ is pseudo-Einsteinian if and only if u is CR-pluriharmonic. □

5.5.3 The Lee class

When the given nondegenerate CR manifold is locally embeddable (as a real hypersurface in $\mathbf{C}^{n+1}$), a precise description of the obstruction to the existence of global pseudo-Einstein contact forms is available, in terms of a cohomology class $\gamma(M)$, with coefficients in $\mathcal{P}$, referred to hereinafter as the *Lee class* of M.

Theorem 5.7. (J.M. Lee [270]) *Let M be a locally realizable nondegenerate CR manifold. There exists a CR-invariant cohomology class $\gamma(M) \in H^1(M; \mathcal{P})$ such that $\gamma(M) = 0$ if and only if M admits a global pseudo-Einsteinian contact form.*

Proof. Let θ be a contact form on M. By hypothesis, for any $x \in M$ there is an open neighborhood $U \ni x$ and a CR immersion $\psi : U \to \mathbf{C}^{n+1}$. Let us set

$$\zeta := \psi^*(dz^1 \wedge \cdots \wedge dz^{n+1}) \in \Gamma^\infty(U, K^0(M)).$$

Then $d\zeta = 0$ and (by Lemma 2.2 in Chapter 2) there is a unique C^∞ function $\lambda : U \to (0, \infty)$ such that

$$\lambda\theta \wedge (d\theta)^n = 2^n i^{n^2} n!(-1)^s \theta \wedge (T \rfloor \zeta) \wedge (T \rfloor \overline{\zeta}).$$

If $\hat{\theta} = e^{2u}\theta$ then

$$d\hat{\theta} = e^{2u}\,(2(du) \wedge \theta + d\theta);$$

hence

$$\hat{\theta} \wedge (d\hat{\theta})^n = e^{2(n+1)u}\theta \wedge (d\theta)^n$$

and consequently

$$\hat{\theta} \wedge (\hat{T} \rfloor \zeta) \wedge (\hat{T} \rfloor \overline{\zeta}) = e^{-2u}\theta \wedge (T \rfloor \zeta) \wedge (T \rfloor \overline{\zeta}).$$

In particular for

$$u := \frac{\log \lambda}{2(n+2)}$$

we have $\lambda = e^{2(n+2)u}$ and then

$$\begin{aligned}\hat{\theta} \wedge (d\hat{\theta})^n &= e^{2(n+1)u}\theta \wedge (d\theta)^n = e^{-2u}\lambda\theta \wedge (d\theta)^n \\ &= e^{-2u} n! 2^n i^{n^2} (-1)^s \theta \wedge (T \rfloor \zeta) \wedge (T \rfloor \overline{\zeta}) \\ &= n! 2^n i^{n^2} (-1)^s \hat{\theta} \wedge (\hat{T} \rfloor \zeta) \wedge (\hat{T} \rfloor \overline{\zeta}),\end{aligned}$$

i.e., $\hat{\theta}$ is volume-normalized with respect to ζ. Since ζ is also closed, we may apply Proposition 5.1 to conclude that $\hat{\theta}$ is a pseudo-Einstein contact form on U. Therefore, there is a locally finite open covering $\{U_i : i \in I\}$ of M and a family $\{\theta_i : i \in I\}$ of local contact forms such that each (U_i, θ_i) is pseudo-Einsteinian. Then

$$\theta_i = e^{2u_{ij}}\theta_j$$

on $U_i \cap U_j$, for some $u_{ij} \in C^\infty(U_i \cap U_j)$. By Theorem 5.6, $u_{ij} \in \mathcal{P}(U_i \cap U_j)$. Also, clearly

$$\begin{aligned} u_{ij} + u_{ji} &= 0 \quad \text{on } U_i \cap U_j, \\ u_{ij} + u_{jk} + u_{ki} &= 0 \quad \text{on } U_i \cap U_j \cap U_k. \end{aligned}$$

Let us set $\mathcal{U} := \{U_i\}_{i \in I}$ and let $N(\mathcal{U})$ be the nerve of $\mathcal{U}$. Let $f \in \mathcal{C}^1(N(\mathcal{U}), \mathcal{P})$ be the 1-cochain defined by

$$f(\sigma) := u_{ij} \in \mathcal{P}(\cap\sigma), \quad \sigma = (U_i U_j) \in \mathcal{S}^1(N(\mathcal{U})).$$

Consider the coboundary operator

$$\delta^1 : \mathcal{C}^1(N(\mathcal{U}), \mathcal{P}) \to \mathcal{C}^2(N(\mathcal{U}), \mathcal{P}).$$

Then, for $\sigma = (U_i U_j U_k)$,

$$\begin{aligned}(\delta^1 f)(\sigma) &= \sum_{\alpha=0}^{2} \rho_{\sigma^\alpha,\sigma} f(\sigma^\alpha) \\ &= \rho_{\sigma^0,\sigma} f(\sigma^0) - \rho_{\sigma^1,\sigma} f(\sigma^1) + \rho_{\sigma^2,\sigma} f(\sigma^2).\end{aligned}$$

Here we have adopted the following notation. If $\sigma = \Delta(i_0, \dots, i_p)$ is a p-simplex, we set $\cap\sigma := \cap_{j=0}^{p} U_{i_j}$ and

$$\sigma^j := \Delta(i_0, \dots, i_{j-1}, i_{j+1}, i_p) \in \mathcal{S}^{p-1}(N(\mathcal{U})).$$

Also, if $\mathcal{F}$ is a given presheaf on M then

$$\rho_{\sigma^j,\sigma} : \mathcal{F}(\cap\sigma^j) \to \mathcal{F}(\cap\sigma)$$

denotes the restriction map. We use the notation and conventions (as to Čech cohomology) in [178]. Thus

$$(\delta^1 f)(\sigma) = u_{jk}\big|_{\cap\sigma} - u_{ik}\big|_{\cap\sigma} + u_{ij}\big|_{\cap\sigma} = 0,$$

i.e., $f \in Z^1(N(\mathcal{U}), \mathcal{P})$. We may then set by definition

$$\Gamma := [f] \in H^1(N(\mathcal{U}), \cap P),$$

and

$$\gamma(M) := \langle \Gamma \rangle \in H^1(M, \cap P) = \lim_{\mathcal{U}} H^1(N(\mathcal{U}), \mathcal{P}).$$

Definition 5.12. $\gamma(M)$ is called the *Lee class* of M. □

First, let us show that the Lee class depends only on the CR structure. Indeed, let $\tilde{U} := \{\tilde{U}_j\}_{j \in J}$ be another open covering of M, together with pseudo-Einstein contact forms $\tilde{\theta}_j$ on $\tilde{U}_j$. By passing to a common refinement if necessary, one may assume without loss of generality that $I = J$ and $U_i = \tilde{U}_i$, for any $i \in I$. Then

$$\tilde{\theta}_i = e^{2h_i} \theta_i$$

on U_i, for some $h_i \in \mathcal{P}(U_i)$. Thus

$$\tilde{u}_{ij} - u_{ij} = h_i - h_j \, ,$$

i.e.,

$$\tilde{f} = f - \delta h;$$

hence $\tilde{\Gamma} = \Gamma$. Here $h \in C^0(N(\mathcal{U}), \mathcal{P})$ is defined by $h(\Delta(i)) := h_i \in \mathcal{P}(U_i)$. Next, let us show that $\gamma(M) = 0$ if and only if M admits a global pseudo-Einsteinian contact form. The implication "$\Longleftarrow$" is left as an exercise to the reader. To prove "$\Longrightarrow$" assume that $0 = \gamma(M) = \langle \Gamma \rangle$, i.e., there is a refinement $\mathcal{V}$ of $\mathcal{U}$ $(\mathcal{V} < \mathcal{U})$ such that[1] $\phi_{\mathcal{U}\mathcal{V}} \Gamma = 0$. Thus $[\tilde{\phi} f] = 0$, i.e.,

$$\tilde{\phi} f = \delta h. \tag{5.29}$$

Let $V, V' \in \mathcal{V}$ such that $V \cap V' \neq \emptyset$ and let us set

$$V \subset \phi(V) =: U_i, \quad V' \subset \phi(V') =: U_j \, ,$$

for some $i, j \in I$. Then $U_i \cap U_j \neq \emptyset$ and $\theta_i = e^{2u_{ij}} \theta_j$ on V. Let us set

$$\sigma := (V, V') \in \mathcal{S}^1(N(\mathcal{V})).$$

Then (by (5.29))

---

[1] Again, we are consistent with the conventions in [178], i.e., if $\phi : \mathcal{V} \to \mathcal{U}$ is a map such that $V \subseteq \phi(V)$, for any $V \in \mathcal{V}$, then the naturally induced map of simplicial complexes $\phi_* N(\mathcal{V}) \to N(\mathcal{U})$ induces a map $\tilde{\phi}$ on cochains, and then a map ϕ^* on cohomology

$$(\phi^*)^p : H^p(N(\mathcal{U}), \mathcal{P}) \to H^p(N(\mathcal{V}), \mathcal{P}).$$

This depends only on $\mathcal{V}$ and $\mathcal{U}$ and is commonly denoted by $\phi_{\mathcal{U}\mathcal{V}}$.

$$u_{ij}\big|_{\cap\sigma} = \rho^1 f(\phi^1\sigma) = (\delta^0 h)(\sigma) = h(V')\big|_{\cap\sigma} - h(V)|_{\cap\sigma}\,;$$

hence

$$e^{2h(V)}\theta_i = e^{2h(V')}\theta_j$$

on $V \cap V'$. Let us set $\theta_V := \theta_i|_V$. Since $U \mapsto \Gamma^\infty(U, H(M)^\perp)$ is a sheaf, there is $\theta \in \Gamma^\infty(M, H(M)^\perp)$ such that $\theta|_V = e^{2h(V)}\theta_V$ for any $V \in \mathcal{V}$. Now Theorem 5.6 implies that θ is pseudo-Einsteinian in each $V \in \mathcal{V}$, and hence in M. □

5.6 The global problem

In this section we examine certain sufficient conditions under which the global existence problem for pseudo-Einsteinian contact forms admits a positive solution. As we have previously shown, for a given strictly pseudoconvex CR manifold M the first Chern class of its CR structure $T_{1,0}(M)$ is an obstruction to the existence of globally defined pseudo-Einsteinian contact forms. The following conjecture, referred to as the *Lee conjecture* in this book, has been proposed in [270]:

Conjecture. *Any compact strictly pseudoconvex CR manifold M whose CR structure has a vanishing first Chern class* ($c_1(T_{1,0}(M)) = 0$) *admits a global pseudo-Einsteinian contact form.*

Already with the work of J.M. Lee (cf. op. cit.) the conjecture was known to be true when either M admits some contact form of positive semidefinite pseudo-Hermitian Ricci tensor or M has *transverse symmetry* (a notion to be defined shortly). Successively we shall examine a result in [121], i.e., show that M admits a global pseudo-Einsteinian contact form, provided it admits some pseudo-Hermitian structure whose characteristic direction T is regular (in the sense of R. Palais [336]). Both transverse symmetry and regularity will be seen to yield zero pseudo-Hermitian torsion ($\tau = 0$). Yet, examples of compact strictly pseudoconvex CR manifolds admitting global pseudo-Einsteinian contact forms of nonzero pseudo-Hermitian torsion will be built, thus hinting at a wider validity of the Lee conjecture.

Let θ be a contact form on a compact, strictly pseudoconvex CR manifold M. Assume that $c_1(T_{1,0}(M)) = 0$. Then

$$\frac{i}{2\pi} d\omega_\alpha^\alpha = \gamma_1 = -d\lambda,$$

for some global 1-form λ on M. Since ω_α^α is pure imaginary, we may take λ to be real. Let us set $\sigma := i\lambda$. Then

$$R_{\alpha\overline{\beta}}\theta^\alpha \wedge \theta^{\overline{\beta}} + W^\alpha_{\alpha\mu}\theta^\mu \wedge \theta - W^\alpha_{\alpha\overline{\mu}}\theta^{\overline{\mu}} \wedge \theta = 2\pi\, d\sigma. \tag{5.30}$$

If we write σ locally as

$$\sigma = \sigma_{\overline{\alpha}}\theta^{\overline{\alpha}} - \sigma_\alpha\theta^\alpha + i\sigma_0\theta,$$

then

$$d\sigma = \sigma_{\overline{\beta},\overline{\alpha}}\theta^{\overline{\alpha}} \wedge \theta^{\overline{\beta}} - \sigma_{\overline{\alpha},\beta}\theta^{\overline{\alpha}} \wedge \theta^{\beta} + \theta \wedge (\sigma_{\overline{\alpha},0}\theta^{\overline{\alpha}} + \sigma_{\overline{\alpha}} A^{\overline{\alpha}}_{\beta}\theta^{\beta}) - \sigma_{\beta,\alpha}\theta^{\alpha} \wedge \theta^{\beta}$$
$$+ \sigma_{\alpha,\overline{\beta}}\theta^{\alpha} \wedge \theta^{\overline{\beta}} - \theta \wedge (\sigma_{\alpha,0}\theta^{\alpha} + \sigma_{\alpha} A^{\alpha}_{\overline{\beta}}\theta^{\overline{\beta}}) + i(d\sigma_0) \wedge \theta + i\sigma_0 d\theta.$$

Hence (from (5.30))

$$\begin{aligned} R_{\alpha\overline{\beta}} &= 2\pi \left(\sigma_{\overline{\beta},\alpha} + \sigma_{\alpha,\overline{\beta}} - 2\sigma_0 h_{\alpha\overline{\beta}}\right), \\ \frac{1}{2\pi} W^{\alpha}_{\alpha\mu} &= -\sigma_{\overline{\alpha}} A^{\overline{\alpha}}_{\mu} + \sigma_{\mu,0} + i T_{\mu}(\sigma_0), \\ -\frac{1}{2\pi} W^{\alpha}_{\alpha\overline{\mu}} &= -\sigma_{\overline{\mu},0} + \sigma_{\alpha} A^{\alpha}_{\overline{\mu}} + i T_{\overline{\mu}}(\sigma_0). \end{aligned} \tag{5.31}$$

Also

$$\sigma_{\overline{\beta},\overline{\alpha}}\theta^{\overline{\alpha}} \wedge \theta^{\overline{\beta}} - \sigma_{\beta,\alpha}\theta^{\alpha} \wedge \theta^{\beta} = 0;$$

hence

$$\sigma_{\beta,\alpha} = \sigma_{\alpha,\beta}\,. \tag{5.32}$$

Let us set $\hat{\theta} = e^{2u}\theta$, with $u \in C^{\infty}(M)$. Then $\hat{\theta}$ is pseudo-Einsteinian, i.e., $\hat{B}_{\lambda\overline{\mu}} = 0$ if and only if (by (5.28))

$$B_{\lambda\overline{\mu}} - (n+2)(u_{\lambda\overline{\mu}} + u_{\overline{\mu}\lambda}) + \frac{n+2}{n}\left(u_{\alpha}{}^{\alpha} + u_{\overline{\alpha}}{}^{\overline{\alpha}}\right) h_{\lambda\overline{\mu}} = 0$$

if and only if (by (5.31))

$$\begin{aligned} &(n+2)(u_{\lambda\overline{\mu}} + u_{\overline{\mu}\lambda}) \\ &\quad = 2\pi(\sigma_{\lambda,\overline{\mu}} + \sigma_{\overline{\mu},\lambda}) + \left\{\frac{n+2}{n}\left(u_{\alpha}{}^{\alpha} + u_{\overline{\alpha}}{}^{\overline{\alpha}}\right) - \frac{\rho}{n} - 4\pi\sigma_0\right\} h_{\lambda\overline{\mu}}\,. \end{aligned} \tag{5.33}$$

We shall use now J.J. Kohn's "Hodge theory" for the $\overline{\partial}_b$ complex; cf. [246].

Theorem 5.8. (J.J. Kohn [246])
If M is a compact, strictly pseudoconvex CR manifold and $\eta \in \Omega^{0,1}(M)$ a smooth $(0,1)$-form on M such that $\overline{\partial}_b\eta = 0$ then there is $f \in C^{\infty}(M) \otimes \mathbf{C}$ such that

$$\overline{\partial}_b f - \eta \in \operatorname{Ker}(\square_b).$$

Let us set $\gamma := \overline{\partial}_b f - \eta$ (hence $\square_b\gamma = 0$). Then

$$0 = (\square_b\gamma, \gamma) = 2\left(\overline{\partial}_b^*\overline{\partial}_b\gamma + \overline{\partial}_b\overline{\partial}_b^*\gamma, \gamma\right) = 2\|\overline{\partial}_b\gamma\|^2 + 2\|\overline{\partial}_b^*\gamma\|^2;$$

hence

$$\overline{\partial}_b \gamma = 0, \quad \overline{\partial}_b^* \gamma = 0. \tag{5.34}$$

Since

$$\overline{\partial}_b(\sigma_{\overline{\alpha}} \theta^{\overline{\alpha}}) = \sigma_{\overline{\beta},\overline{\alpha}} \theta^{\overline{\alpha}} \wedge \theta^{\overline{\beta}}$$

the identity (5.32) yields

$$\overline{\partial}_b(\sigma_{\overline{\alpha}} \theta^{\overline{\alpha}}) = 0.$$

In particular for

$$\eta := \frac{2\pi}{n+2} \sigma_{\overline{\alpha}} \theta^{\overline{\alpha}} \in \Omega^{0,1}(M)$$

we have $\overline{\partial}_b \eta = 0$; hence by the result in [246] quoted above, there is $f \in C^\infty(M) \otimes \mathbf{C}$ such that

$$\gamma := \frac{n+2}{2\pi} \left(\overline{\partial}_b f - \eta \right) \in \operatorname{Ker}(\Box_b).$$

With respect to a local frame, taking into account (5.34) as well, we get

$$f_{\overline{\alpha}} = \frac{2\pi}{n+2} \left(\sigma_{\overline{\alpha}} + \gamma_{\overline{\alpha}} \right), \tag{5.35}$$

$$\gamma_{\overline{\alpha},\overline{\beta}} = \gamma_{\overline{\beta},\overline{\alpha}}, \quad \gamma_{\overline{\alpha},}{}^{\overline{\alpha}} = 0. \tag{5.36}$$

Summing up, we have proved the following:

Lemma 5.4. *Let M be a compact, strictly pseudoconvex CR manifold, with $c_1(T_{1,0}(M)) = 0$. Let θ be a contact form on M. Then there are a 1-form $\sigma \in \Omega^1(M)$ and a $(0,1)$-form $\gamma \in \Omega^{0,1}(M)$ and a function $f = u + iv \in C^\infty(M) \otimes \mathbf{C}$ such that the identities* (5.30) *and* (5.35)–(5.36) *are satisfied.*

Let us set, from now on $\gamma_\alpha := \overline{\gamma_{\overline{\alpha}}}$.

Lemma 5.5. *Let u and γ be as in Lemma* 5.4. *Then $\hat{\theta} := e^{2u}\theta$ is pseudo-Einsteinian if and only if*

$$\gamma_{\alpha,\overline{\beta}} + \gamma_{\overline{\beta},\alpha} = 0.$$

Proof. Taking covariant derivatives in

$$u_{\overline{\alpha}} + i v_{\overline{\alpha}} = \frac{2\pi}{n+2} \left(\sigma_{\overline{\alpha}} + \gamma_{\overline{\alpha}} \right)$$

we obtain

$$
\begin{aligned}
(n+2)(u_{\overline{\alpha}\beta} + i v_{\overline{\alpha}\beta}) &= 2\pi(\sigma_{\overline{\alpha},\beta} + \gamma_{\overline{\alpha},\beta}),\\
(n+2)(u_{\beta\overline{\alpha}} - i v_{\beta\overline{\alpha}}) &= 2\pi(\sigma_{\beta,\overline{\alpha}} + \gamma_{\beta,\overline{\alpha}}),
\end{aligned}
$$

whence, by adding up these two identities,

$$
(n+2)(u_{\overline{\alpha}\beta} + u_{\beta\overline{\alpha}}) = i(n+2)(v_{\beta\overline{\alpha}} - v_{\overline{\alpha}\beta}) + 2\pi(\sigma_{\overline{\alpha},\beta} + \sigma_{\beta,\overline{\alpha}} + \gamma_{\overline{\alpha},\beta} + \gamma_{\beta,\overline{\alpha}}),
$$

or, by applying the commutation formula for $v_{\alpha\overline{\beta}}$,

$$
(n+2)(u_{\overline{\alpha}\beta} + u_{\beta\overline{\alpha}}) = 2\pi(\sigma_{\overline{\alpha},\beta} + \sigma_{\beta,\overline{\alpha}} + \gamma_{\overline{\alpha},\beta} + \gamma_{\beta,\overline{\alpha}}) - 2(n+2)v_0 h_{\beta\overline{\alpha}}\,. \tag{5.37}
$$

Hence (by (5.33)) $\hat{\theta}$ is pseudo-Einsteinian if and only if

$$
2\pi\left(\gamma_{\beta,\overline{\alpha}} + \gamma_{\overline{\alpha},\beta}\right) - 2(n+2)v_0 h_{\beta\overline{\alpha}} = \left(\frac{(n+2)}{n}\Delta_b u - \frac{\rho}{n} - 4\pi\sigma_0\right)h_{\beta\overline{\alpha}}\,,
$$

i.e., if and only if

$$
\begin{aligned}
\gamma_{\beta,\overline{\alpha}} + \gamma_{\overline{\alpha},\beta} &= \lambda h_{\beta\overline{\alpha}},\\
\lambda &:= \frac{n+2}{\pi}v_0 + \frac{n+2}{2n\pi}\Delta_b u - \frac{\rho}{2n\pi} - 2\sigma_0,
\end{aligned}
$$

or, by contraction with $h^{\beta\overline{\alpha}}$,

$$
\gamma_{\beta,}{}^{\beta} + \gamma_{\overline{\alpha},}{}^{\overline{\alpha}} = n\lambda;
$$

hence (by (5.36)) $\lambda = 0$ and Lemma 5.5 is completely proved. □

Lemma 5.6. *Let M be a compact, strictly pseudoconvex CR manifold, and θ a contact form on M. Assume that* $\dim(M) \geq 5$. *If the pseudo-Hermitian Ricci tensor $R_{\alpha\overline{\beta}}$ of (M,θ) is positive semidefinite, then for any* $(0,1)$*-form $\gamma_{\overline{\alpha}}\theta^{\overline{\alpha}} \in \Omega^{0,1}(M)$,*

$$
\Box_b(\gamma_{\overline{\alpha}}\theta^{\overline{\alpha}}) = 0 \Longrightarrow \begin{cases} \gamma_{\overline{\alpha},\beta} = 0,\\ \gamma_{\overline{\alpha},\beta} = 0,\\ \gamma_{\alpha,0} - \quad \frac{i}{n} R_{\alpha\beta}\gamma^{,\beta}\,. \end{cases}
$$

Proof. Let us set $\gamma := \gamma_{\overline{\alpha}}\theta^{\overline{\alpha}}$. As seen before, $\Box_b\gamma = 0$ yields $\gamma_{\overline{\alpha},}{}^{\overline{\alpha}} = 0$. This may be written $h^{\overline{\alpha}\beta}\gamma_{\overline{\alpha},\beta} = 0$, and by differentiating covariantly and taking into account the commutation formula

$$
\gamma_{\overline{\alpha},\overline{\beta}\gamma} - \gamma_{\overline{\alpha},\gamma\overline{\beta}} = i h_{\overline{\beta}\gamma}\gamma_{\overline{\alpha},0} + R_{\overline{\alpha}}{}^{\overline{\rho}}{}_{\overline{\beta}\gamma}\gamma_{\overline{\rho}}
$$

we have

$$
0 = h^{\overline{\alpha}\beta}\gamma_{\overline{\alpha},\beta\overline{\gamma}} = h^{\overline{\alpha}\beta}\left(\gamma_{\overline{\alpha},\overline{\gamma}\beta} + i h_{\overline{\gamma}\beta}\gamma_{\overline{\alpha},0} - R_{\overline{\alpha}}{}^{\overline{\rho}}{}_{\overline{\gamma}\beta}\gamma_{\overline{\rho}}\right) = \gamma_{\overline{\alpha},\overline{\gamma}}{}^{\overline{\alpha}} + i\gamma_{\overline{\gamma},0} - R_{\overline{\alpha}}{}^{\overline{\rho}}{}_{\overline{\gamma}}{}^{\overline{\alpha}}\gamma_{\overline{\rho}}.
$$

Yet $\gamma_{\overline{\alpha},\overline{\gamma}} = \gamma_{\overline{\gamma},\overline{\alpha}}$; hence

$$\gamma_{\overline{\gamma},\overline{\alpha}}{}^{\overline{\alpha}} = -i\gamma_{\overline{\gamma},0} + R_{\lambda\overline{\gamma}}\gamma^{\lambda}\,. \tag{5.38}$$

To get (5.38) we also rely on the following calculation [taking into account the symmetry property (1.98) of the curvature tensor (of the Tanaka–Webster connection); cf. Chapter 1]:

$$R_{\overline{\alpha}}{}^{\overline{\rho}}{}_{\overline{\gamma}}{}^{\overline{\alpha}}\gamma_{\overline{\rho}} = h^{\overline{\rho}\lambda}h^{\overline{\alpha}\mu}R_{\overline{\alpha}\lambda\overline{\gamma}\mu} = h^{\overline{\rho}\lambda}h^{\overline{\alpha}\mu}R_{\overline{\alpha}\mu\overline{\gamma}\lambda} = R_{\overline{\alpha}}{}^{\overline{\alpha}}{}_{\overline{\gamma}\lambda}h^{\overline{\rho}\lambda} = R_{\overline{\gamma}\lambda}h^{\overline{\rho}\lambda}\,.$$

On the other hand, the identity

$$\gamma_{\alpha,\beta\overline{\gamma}} - \gamma_{\alpha,\overline{\gamma}\beta} = ih_{\beta\overline{\gamma}}\gamma_{\alpha,0} + R_{\alpha}{}^{\mu}{}_{\beta\overline{\gamma}}\gamma_{\mu} \tag{5.39}$$

may be written

$$\gamma_{\alpha,\beta\overline{\gamma}}h^{\beta\overline{\gamma}} - \gamma_{\alpha,\overline{\gamma}}{}^{\overline{\gamma}} = in\gamma_{\alpha,0} + R_{\alpha}{}^{\mu\overline{\gamma}}{}_{\overline{\gamma}}\gamma_{\mu}\,.$$

Yet $\gamma_{\alpha,\beta} = \gamma_{\beta,\alpha}$; hence

$$\gamma_{\beta,\alpha\overline{\gamma}}h^{\beta\overline{\gamma}} - \gamma_{\alpha,\overline{\gamma}}{}^{\overline{\gamma}} = in\gamma_{\alpha,0} + R_{\alpha}{}^{\mu}{}_{\beta}{}^{\beta}\gamma_{\mu},$$

and using once again the commutation formula we started with, we obtain

$$h^{\beta\overline{\gamma}}\left(\gamma_{\beta,\overline{\gamma}\alpha} + ih_{\alpha\overline{\gamma}}\gamma_{\beta,0} + R_{\beta}{}^{\mu}{}_{\alpha\overline{\gamma}}\gamma_{\mu}\right) - \gamma_{\alpha,\overline{\gamma}}{}^{\overline{\gamma}} = in\gamma_{\alpha,0} + R_{\alpha}{}^{\mu\overline{\gamma}}{}_{\overline{\gamma}}\gamma_{\mu},$$

or

$$\gamma_{\beta,\overline{\gamma}\alpha}h^{\beta\overline{\gamma}} + i\gamma_{\alpha,0} + R_{\beta}{}^{\mu}{}_{\alpha}{}^{\beta}\gamma_{\mu} - \gamma_{\alpha,\overline{\gamma}}{}^{\overline{\gamma}} = in\gamma_{\alpha,0} + R_{\alpha}{}^{\mu\overline{\gamma}}{}_{\overline{\gamma}}\gamma_{\mu},$$

or

$$\underbrace{\gamma_{\beta,}{}^{\beta}{}_{\alpha}}_{=0} - \gamma_{\alpha,\overline{\gamma}}{}^{\overline{\gamma}} = i(n-1)\gamma_{\alpha,0} + \left(R_{\alpha}{}^{\mu}{}_{\beta}{}^{\beta} - R_{\beta}{}^{\mu}{}_{\alpha}{}^{\beta}\right)\gamma_{\mu}\,. \tag{5.40}$$

On the other hand [by (1.98) in Chapter 1],

$$R_{\alpha}{}^{\mu}{}_{\beta}{}^{\beta} - R_{\beta}{}^{\mu}{}_{\alpha}{}^{\beta} = h^{\mu\overline{\lambda}}h^{\beta\overline{\sigma}}\left[R_{\alpha\overline{\lambda}\beta\overline{\sigma}} - R_{\beta\overline{\lambda}\alpha\overline{\sigma}}\right] = 0;$$

hence (5.40) becomes

$$\gamma_{\overline{\alpha},\gamma}{}^{\gamma} = i(n-1)\gamma_{\overline{\alpha},0}\,. \tag{5.41}$$

At this point, we may perform the following calculation:

$$\begin{aligned}
&\gamma_{\overline{\beta},\alpha}\gamma^{\overline{\beta},\alpha} + (n-1)\gamma_{\overline{\beta},\overline{\gamma}}\gamma^{\overline{\beta},\overline{\gamma}} + (n-1)R_{\alpha\overline{\beta}}\gamma^{\alpha}\gamma^{\overline{\beta}} \\
&\text{(by (5.38))} \quad = \gamma_{\overline{\beta},\alpha}\gamma^{\overline{\beta},\alpha} + (n-1)\gamma_{\overline{\beta},\overline{\gamma}}\gamma^{\overline{\beta},\overline{\gamma}} + (n-1)\gamma^{\overline{\beta}}\left[\gamma_{\overline{\beta},\overline{\alpha}}{}^{\overline{\alpha}} + i\gamma_{\overline{\beta},0}\right] \\
&\text{(by (5.41))} \quad = \gamma_{\overline{\beta},\alpha}\gamma^{\overline{\beta},\alpha} + (n-1)\gamma_{\overline{\beta},\overline{\gamma}}\gamma^{\overline{\beta},\overline{\gamma}} + (n-1)\gamma^{\overline{\beta}}\gamma_{\overline{\beta},\overline{\gamma}}{}^{\overline{\gamma}} + \gamma^{\overline{\beta}}\gamma_{\overline{\beta},\gamma}{}^{\gamma} \\
&\qquad = \gamma_{\overline{\beta},\alpha}\gamma^{\overline{\beta},\alpha} + \gamma_{\overline{\beta},\alpha}{}^{\alpha}\gamma^{\overline{\beta}} + (n-1)\left[\rho_{\overline{\beta},\overline{\gamma}}\gamma^{\overline{\beta},\overline{\gamma}} + \gamma_{\overline{\beta},\overline{\gamma}}{}^{\overline{\gamma}}\gamma^{\overline{\beta}}\right] \\
&\qquad = \lambda_{\alpha,}{}^{\alpha} + (n-1)\mu_{\overline{\gamma},}{}^{\overline{\gamma}}
\end{aligned}$$

where

$$\lambda_\alpha := \gamma_{\overline{\beta},\alpha}\gamma^{\overline{\beta}}, \quad \mu_{\overline{\gamma}} := \gamma_{\overline{\beta},\overline{\gamma}}\gamma^{\overline{\beta}}.$$

By the divergence formula,

$$\int_M \left[\gamma_{\overline{\beta},\alpha}\gamma^{\overline{\beta},\alpha} + (n-1)\gamma_{\overline{\beta},\overline{\gamma}}\gamma^{\overline{\beta},\overline{\gamma}} + +(n-1)R_{\alpha\overline{\beta}}\gamma^\alpha\gamma^{\overline{\beta}}\right]\theta \wedge (d\theta)^n = 0. \quad (5.42)$$

Yet by hypothesis $R_{\alpha\overline{\beta}}\gamma^\alpha\gamma^{\overline{\beta}} \geq 0$; hence

$$\gamma_{\overline{\beta},\alpha} = 0, \quad \gamma_{\overline{\beta},\overline{\gamma}} = 0.$$

Then (5.39) becomes

$$ih_{\beta\overline{\gamma}}\gamma_{\alpha,0} + R_\alpha{}^\mu{}_{\beta\overline{\gamma}}\gamma_\mu = 0$$

and contraction with $h^{\beta\overline{\gamma}}$ leads to

$$in\gamma_{\alpha,0} + R_\alpha{}^\mu{}_\beta{}^\beta\gamma_\mu = 0.$$

Yet (by using twice (1.98))

$$R_\alpha{}^\mu{}_\beta{}^\beta = R_{\alpha\overline{\lambda}}h^{\mu\overline{\lambda}};$$

hence

$$in\gamma_{\alpha,0} + R_{\alpha\overline{\lambda}}\gamma^{\overline{\lambda}} = 0,$$

and Lemma 5.6 is completely proved. □

As a byproduct, we obtain the following:

Proposition 5.10. *Let M be a compact, strictly pseudoconvex CR manifold, of dimension* $\dim(M) \geq 5$. *Let θ be a contact form on M. If the pseudo-Hermitian Ricci tensor $R_{\alpha\overline{\beta}}$ of (M, θ) is positive definite then*

$$H^{0,1}_{\overline{\partial}_b}(M) = 0.$$

Proof. By J.J. Kohn's "Hodge theory" for $\overline{\partial}_b$.

Theorem 5.9. (J.J. Kohn [246]) *Any cohomology class in the Kohn–Rossi cohomology group $H^{0,1}_{\overline{\partial}_b}(M)$ has a unique smooth representative γ with $\Box_b\gamma = 0$.*

Then (5.42) yields $R_{\alpha\overline{\beta}}\gamma^\alpha\gamma^{\overline{\beta}} = 0$; hence $\gamma = 0$. □

To state the main result of this section we need the following definition:

Definition 5.13. We say that M has *transverse symmetry* if M admits a 1-parameter group of CR automorphisms transverse to the Levi distribution. □

Precisely, if M has transverse symmetry, there is a tangent vector field $V \in \mathcal{X}(M)$, transverse to $H(M)$, whose 1-parameter transformation group $\{\varphi_t\}_{t\in\mathbf{R}}$ consists of CR automorphisms $\varphi_t : M \to M$ (since M is compact, the local integration of V furnishes, of course, a *global* 1-parameter group of *global* automorphisms).

Theorem 5.10. (J.M. Lee [270])
Let M be a compact, strictly pseudoconvex CR manifold whose CR structure has a vanishing first Chern class ($c_1(T_{1,0}(M)) = 0$). Let us suppose that (at least) one of the following assumptions is satisfied

(i) M admits a contact form θ with $R_{\alpha\overline{\beta}}$ positive semidefinite.
(Ii) M has transverse symmetry.

Then M admits a globally defined pseudo-Einsteinian contact form.

Proof. To prove (i), let θ be a contact form with $R_{\alpha\overline{\beta}}$ positive semidefinite. By Lemma 5.4 there are a 1-form $\sigma \in \Omega^1(M)$, a $(0,1)$-form $\gamma \in \Omega^{0,1}(M)$, and a function $f = u + iv \in C^\infty(M) \otimes \mathbf{C}$ such that

$$R_{\alpha\overline{\beta}}\theta^\alpha \wedge \theta^{\overline{\beta}} + W^\alpha_{\alpha\mu}\theta^\mu \wedge \theta - W^\alpha_{\alpha\overline{\mu}}\theta^{\overline{\mu}} \wedge \theta = 2\pi\, d\sigma\,,$$

$$f_{\overline{\alpha}} = \frac{2\pi}{n+2}(\sigma_{\overline{\alpha}} + \gamma_{\overline{\alpha}})\,, \quad \gamma_{\overline{\alpha},\overline{\beta}} = 0, \quad \gamma_{\overline{\alpha}}{}^{\overline{\alpha}} = 0.$$

By Lemma 5.6 we get $\gamma_{\overline{\beta},\alpha} = 0$ and (by complex conjugation) $\gamma_{\beta,\overline{\alpha}} = 0$. Finally, by Lemma 5.5, $e^{2u}\theta$ is pseudo-Einsteinian. □

Let us prove (ii). Let θ_1 be a contact form on M and let us set $\lambda := \theta_1(V)$. Since $T(M) = H(M)\oplus\mathbf{R}V$ it follows that $\lambda(x) \neq 0$, for any $x \in M$. Let us set $\theta := (1/\lambda)\theta_1$ (hence $\theta(V) = 1$). The Lie derivative of θ, in the direction V, is given by

$$(\mathcal{L}_V\theta)_x = \lim_{t\to 0}\frac{1}{t}\left\{\theta_x - (\varphi_t^*\theta)_x\right\} = \lim_{t\to 0}\frac{1}{t}\{\theta_x - \lambda_t(x)\theta_x\} = u(x)\theta_x\,,$$

for some $\lambda_t \in C^\infty(M)$ (since φ_t is a CR map), where

$$u(x) := \lim_{t\to 0}\frac{1}{t}(1 - \lambda_t(x)).$$

Then

$$u\theta = \mathcal{L}_V\theta = (\iota_V\, d + d\,\iota_V)\,\theta = V \rfloor d\theta + \underbrace{d(V \rfloor \theta)}_{=0},$$

and applying both members to T,

$$u = u\theta(T) = (d\theta)(V, T) = 0$$

since $T \rfloor d\theta = 0$. We have proved that

$$\theta(V) = 1, \quad V \rfloor d\theta = 0,$$

relations that uniquely describe the characteristic direction of $d\theta$, i.e., we may conclude that $V = T$, hence that T is an infinitesimal CR automorphism of M. By S. Webster's Theorem 1.5 in Chapter 1, (the Tanaka–Webster connection of) θ has zero pseudo-Hermitian torsion, i.e., $\tau = 0$. Hence the identity

$$\rho_\gamma - R_{\gamma\overline{\sigma},}{}^{\overline{\sigma}} = i(n-1)A_{\alpha\gamma,}{}^{\alpha}$$

becomes

$$\rho_\gamma = R_{\gamma\overline{\sigma},}{}^{\overline{\sigma}}\,. \tag{5.43}$$

Also, the commutation formula

$$\sigma_{\alpha,\overline{\beta}\overline{\gamma}} - \sigma_{\alpha,\overline{\gamma}\overline{\beta}} = ih_{\alpha\overline{\beta}}A_{\overline{\gamma}\overline{\rho}}\sigma^{\overline{\rho}} - ih_{\alpha\overline{\gamma}}A_{\overline{\beta}\overline{\rho}}\sigma^{\overline{\rho}}$$

(true for any $(1,0)$-form $\sigma_\alpha\theta^\alpha$ on M) yields (since $A_{\alpha\beta} = 0$) $u_{\alpha\overline{\beta}\overline{\gamma}} = u_{\alpha\overline{\gamma}\overline{\beta}}$, or, by contraction with $h^{\overline{\gamma}\alpha}$,

$$u_{\alpha\overline{\beta}}{}^{\alpha} = u_{\alpha}{}^{\alpha}{}_{\overline{\beta}}\,. \tag{5.44}$$

Let γ be as in Lemma 5.4. By (5.37)

$$2\pi(\gamma_{\beta,\overline{\alpha}} + \gamma_{\overline{\alpha},\beta}) = (n+2)(u_{\beta\overline{\alpha}} + u_{\overline{\alpha}\beta}) - 2\pi(\sigma_{\beta,\overline{\alpha}} + \sigma_{\overline{\alpha},\beta}) + 2(n+2)v_0 h_{\beta\overline{\alpha}}\,.$$

At this point, we may perform the following calculation:

$$\begin{aligned}
2\pi(\gamma_{\overline{\beta},\alpha} + \gamma_{\alpha,\overline{\beta}})(\gamma^{\overline{\beta},\alpha} + \gamma^{\alpha,\overline{\beta}}) &= 2\pi\gamma^{\overline{\beta},\alpha}(\gamma_{\overline{\beta},\alpha} + \gamma_{\alpha,\overline{\beta}}) + 2\pi\overline{\gamma^{\overline{\beta},\alpha}(\gamma_{\overline{\beta},\alpha} + \gamma_{\alpha,\overline{\beta}})}\\
&= 4\pi\,\mathrm{Re}\{\gamma^{\overline{\beta},\alpha}(\gamma_{\overline{\beta},\alpha} + \gamma_{\alpha,\overline{\beta}})\}\\
= 2\,\mathrm{Re}\{\gamma^{\overline{\beta},\alpha}(n+2)(u_{\alpha\overline{\beta}} + u_{\overline{\beta}\alpha}) &- 2\pi\gamma^{\overline{\beta},\alpha}(\sigma_{\alpha,\overline{\beta}} + \sigma_{\overline{\beta},\alpha}) + 2(n+2)v_0 h_{\alpha\overline{\beta}}\gamma^{\overline{\beta},\alpha}\}.
\end{aligned}$$

Yet

$$2\pi(\sigma_{\alpha,\overline{\beta}} + \sigma_{\overline{\beta},\alpha}) = R_{\alpha\overline{\beta}} + 4\pi\sigma_0 h_{\alpha\overline{\beta}}\,.$$

Thus

$$\begin{aligned}
&2\pi(\gamma_{\overline{\beta},\alpha} + \gamma_{\alpha,\overline{\beta}})(\gamma^{\overline{\beta},\alpha} + \gamma^{\alpha,\overline{\beta}})\\
&= 2\,\mathrm{Re}\,\gamma^{\overline{\beta},\alpha}\left[(n+2)\big(2u_{\alpha\overline{\beta}} - ih_{\alpha\overline{\beta}}u_0\big) - R_{\alpha\overline{\beta}} - 4\pi\sigma_0 h_{\alpha\overline{\beta}} + 2(n+2)v_0 h_{\alpha\overline{\beta}}\right]\\
&= 2\,\mathrm{Re}\,\gamma^{\overline{\beta},\alpha}\big[2(n+2)u_{\alpha\overline{\beta}} - R_{\alpha\overline{\beta}}\big] + 2\,\mathrm{Re}\,\underbrace{\gamma^{\overline{\beta},\alpha}h_{\alpha\overline{\beta}}}_{=\gamma_{\alpha,}{}^{\alpha}=0}[(n+2)(2v_0 - iu_0) - 4\pi\sigma_0]\,,
\end{aligned}$$

i.e.,

$$2\pi(\gamma_{\overline{\beta},\alpha} + \gamma_{\alpha,\overline{\beta}})(\gamma^{\overline{\beta},\alpha} + \gamma^{\alpha,\overline{\beta}}) = 2\,\mathrm{Re}\,\gamma^{\overline{\beta},\alpha}\big[2(n+2)u_{\alpha\overline{\beta}} - R_{\alpha\overline{\beta}}\big]. \tag{5.45}$$

On the other hand,

$$
\begin{aligned}
\gamma^{\overline{\beta},\alpha} u_{\alpha\overline{\beta}} &= \left(\gamma^{\overline{\beta}} u_{\alpha\overline{\beta}}\right)^{,\alpha} - \gamma^{\overline{\beta}} u_{\alpha\overline{\beta}}{}^{\alpha} \\
\text{(by (5.44))} \qquad &= \left(\gamma^{\overline{\beta}} u_{\alpha\overline{\beta}}\right)^{,\alpha} - \gamma^{\overline{\beta}} u_{\alpha}{}^{\alpha}{}_{\overline{\beta}} \\
\text{(because of } \gamma^{\overline{\beta}}_{,\overline{\beta}} = 0) \qquad &= \left(\gamma^{\overline{\beta}} u_{\alpha\overline{\beta}}\right)^{,\alpha} - \left(\gamma^{\overline{\beta}} u_{\alpha}{}^{\alpha}\right)_{,\overline{\beta}}
\end{aligned}
$$

i.e.,

$$
\gamma^{\overline{\beta},\alpha} u_{\alpha\overline{\beta}} = \left(\gamma^{\overline{\beta}} u_{\alpha\overline{\beta}}\right)^{,\alpha} - \left(\gamma^{\overline{\beta}} u_{\alpha}{}^{\alpha}\right)_{,\overline{\beta}} . \tag{5.46}
$$

Moreover,

$$
\begin{aligned}
\gamma^{\overline{\beta},\alpha} R_{\alpha\overline{\beta}} &= \left(\gamma^{\overline{\beta}} R_{\alpha\overline{\beta}}\right)^{,\alpha} - \gamma^{\overline{\beta}} R_{\alpha\overline{\beta},}{}^{\alpha} \\
\text{(by (5.43))} \qquad &= \left(\gamma^{\overline{\beta}} R_{\alpha\overline{\beta}}\right)^{,\alpha} - \gamma^{\overline{\beta}} \rho_{\overline{\beta}}
\end{aligned}
$$

i.e.,

$$
\gamma^{\overline{\beta},\alpha} R_{\alpha\overline{\beta}} = \left(\gamma^{\overline{\beta}} R_{\alpha\overline{\beta}}\right)^{,\alpha} - \left(\gamma^{\overline{\beta}} \rho\right)_{,\overline{\beta}} . \tag{5.47}
$$

Substitution from (5.46)–(5.47) into (5.45) leads to

$$
\begin{aligned}
&2\pi(\gamma_{\overline{\beta},\alpha} + \gamma_{\alpha,\overline{\beta}})(\gamma^{\overline{\beta},\alpha} + \gamma^{\alpha,\overline{\beta}}) = \\
&\qquad 2\,\mathrm{Re}\left\{2(n+2)\left[\left(\gamma^{\overline{\beta}} u_{\alpha\overline{\beta}}\right)^{,\alpha} - \left(\gamma^{\overline{\beta}} u_{\alpha}{}^{\alpha}\right)_{,\overline{\beta}}\right] - \left(\gamma^{\overline{\beta}} R_{\alpha\overline{\beta}}\right)^{,\alpha} + \left(\gamma^{\overline{\beta}} \rho\right)_{,\overline{\beta}}\right\}.
\end{aligned}
$$

Finally, let us integrate over M and use the divergence formula. We get $\gamma_{\overline{\beta},\alpha} + \gamma_{\alpha,\overline{\beta}} = 0$; hence (by Lemma 5.5) $e^{2u}\theta$ is pseudo-Einsteinian. □

5.7 The Lee conjecture

In this section, we deal with the Lee conjecture,[2] as stated in the preceding section. We solve in the affirmative the Lee conjecture for compact strictly pseudoconvex CR manifolds with a *regular* (in the sense of R. Palais [336]) contact vector. The regularity assumption leads (via the Boothby–Wang theorem [75] and B. O'Neill's fundamental equations of a submersion [334]) to zero pseudo-Hermitian torsion (and we may apply a result of J.M. Lee [270]).

Let $(M, T_{1,0}(M))$ be a CR manifold of CR dimension n. In the previous sections we formulated the following natural problem: assuming that M is nondegenerate, find a pseudo-Hermitian structure θ such that (M, θ) is pseudo-Einsteinian. It turned out that the solution to the local problem is intimately related to the question of embeddability, and then to classical results by M. Kuranishi [263] and T. Akahori [2], while

[2] A CR analogue of the Calabi problem.

there are a number of obstructions to (the solution of) the global problem, such as the first Chern class of the given CR structure. In the defense of the Lee conjecture we construct an example of a compact strictly pseudoconvex CR manifold that carries a globally pseudo-Einsteinian contact form with nonvanishing pseudo-Hermitian torsion. This is obtained as a quotient of the Heisenberg group $\mathbf{H}_n$ by a discrete group of CR automorphisms (and is the CR analogue of the construction of H. Hopf [211], endowing $S^{2n-1} \times S^1$ with a complex structure). Precisely, we construct a family $\mathbf{H}_n(s)$, $0 < s < 1$, of compact strictly pseudoconvex CR manifolds such that each $\mathbf{H}_n(s)$ satisfies the Lee conjecture. We endow $\mathbf{H}_n(s)$ with the contact form (5.51). Our construction is reminiscent of W.C. Boothby's Hermitian metric (cf. [74]) on a complex Hopf manifold (cf. also [125]).

5.7.1 Quotients of the Heisenberg group by properly discontinuous groups of CR automorphisms

Let $D_s : \mathbf{H}_n \setminus \{0\} \to \mathbf{H}_n \setminus \{0\}$, $s > 0$, be the parabolic dilations, i.e., $D_s(z,t) = (sz, s^2 t)$. If $m \in \mathbf{Z}$, $m > 0$, we set $D_s^m = D_s \circ \cdots \circ D_s$ (m factors). Also, we set $D_s^{-m} = \delta_{1/s}^m$ and $D_s^0 = I$. Consider the discrete group $G_s = \{D_s^m : m \in \mathbf{Z}\}$.

Theorem 5.11. ([121]) *Let $0 < s < 1$ and $n > 1$. Then G_s acts freely on $\mathbf{H}_n \setminus \{0\}$ as a properly discontinuous group of CR automorphisms of $\mathbf{H}_n \setminus \{0\}$. The quotient space $\mathbf{H}_n(s) = (\mathbf{H}_n \setminus \{0\})/G_s$ is a compact strictly pseudoconvex CR manifold of CR dimension n.*

Proof. Clearly $\delta_m^s x = x$ for some $x \in \mathbf{H}_n \setminus \{0\}$ yields $m = 0$. Thus the action of G_s on $\mathbf{H}_n \setminus \{0\}$ is free.

Let $|x|$ be the Heisenberg norm on $\mathbf{H}_n$. Let $x_0 \in \mathbf{H}_n \setminus \{0\}$ and set $U_\epsilon(x_0) = \{x \in \mathbf{H}_n \setminus \{0\} : |x - x_0| < \epsilon\}$, $\epsilon > 0$. Let $\|x\|$ be the Euclidean norm on $\mathbf{H}_n \simeq \mathbf{R}^{2n+1}$. By a result in Chapter 1, for any $x \in \mathbf{H}_n$ with $|x| \leq 1$ one has

$$\|x\| \leq |x| \leq \|x\|^{1/2}.$$

Thus the sets $U_\epsilon(x)$, $x \in \mathbf{H}_n \setminus \{0\}$, $0 < \epsilon < 1$, form a fundamental system of neighborhoods in $\mathbf{H}_n \setminus \{0\}$. To show that G_s is properly discontinuous, given $x_0 \in \mathbf{H}_n \setminus \{0\}$ one needs to choose $\epsilon > 0$ such that

$$\delta_s^m(U_\epsilon(x_0)) \cap U_\epsilon(x_0) = \emptyset, \tag{5.48}$$

for any $m \in \mathbf{Z}$, $m \neq 0$. By Lemma 8.9 in [150], p. 449, there exists $\gamma \geq 1$ such that $|x + y| \leq \gamma(|x| + |y|)$ for any $x, y \in \mathbf{H}_n$. Consequently

$$|x| - \gamma|y| \leq \gamma|x - y|, \tag{5.49}$$

for any $x, y \in \mathbf{H}_n$. Let

$$\xi_m = |\delta_s^m(x_0) - x_0|$$

for $m \in \mathbf{Z}$. Since G_s acts freely on $\mathbf{H}_n \setminus \{0\}$, it follows that $\xi_m \geq 0$ and $\xi_m = 0 \Longleftrightarrow m = 0$. Next, since $0 < s < 1$, one obtains

$$0 \leq m_1 < m_2 \Longrightarrow \xi_{m_1} < \xi_{m_2}, \quad \xi_{-m_1} < \xi_{-m_2}.$$

Therefore

$$\xi_m \geq \min\{\xi_1, \xi_{-1}\} = \xi_1$$

for any $m \in \mathbf{Z}$, $m \neq 0$. Let us set $N = 2\gamma + 1$. Choose $0 < \epsilon < \frac{1}{N}\xi_1$. Let $x \in U_\epsilon(x_0)$. Then

$$|\delta_s^m(x) - \delta_s^m(x_0)| = s^m |x - x_0| < s^m \epsilon < \epsilon$$

shows that

$$\delta_s^m(U_\epsilon(x_0)) \subseteq U_\epsilon(\delta_s^m(x_0)). \tag{5.50}$$

Using (5.49)–(5.50) we have the estimates

$$\begin{aligned} \gamma|x_0 - \delta_s^m(x)| &= \gamma|x_0 - \delta_s^m(x_0) - (\delta_s^m(x) - \delta_s^m(x_0))| \\ &\geq |x_0 - \delta_s^m(x_0)| - \gamma|\delta_s^m(x) - \delta_s^m(x_0)| \\ &> \xi_m - \gamma\epsilon \geq \xi_1 - \gamma\epsilon > N\epsilon - \gamma\epsilon = (\gamma + 1)\epsilon, \end{aligned}$$

so that

$$|x_0 - D_s^m(x)| > \frac{\gamma + 1}{\gamma}\epsilon > \epsilon.$$

This shows that $\delta_s^m(x) \notin U_\epsilon(x_0)$, for any $x \in U_\epsilon(x_0)$, $m \in \mathbf{Z}$, $m \neq 0$, so that (5.48) holds. □

Let $\pi : \mathbf{H}_n \setminus \{0\} \to \mathbf{H}_n(s)$ be the natural map. Let

$$\Sigma^{2n} = \{x \in \mathbf{H}_n : |x| = 1\}.$$

Then Σ^{2n} is a compact real hypersurface in $\mathbf{H}_n$. The map $\mathbf{H}_n(s) \to \Sigma^{2n} \times S^1$ defined by

$$\pi(x) \mapsto \left(\frac{z}{|x|}, \ \frac{t}{|x|^2}, \ \exp\left(\frac{2\pi i \log|x|}{\log s}\right)\right)$$

is a diffeomorphism, $x = (z, t) \in \mathbf{H}_n \setminus \{0\}$. Thus $\mathbf{H}_n(s)$ is compact. Since π is a local diffeomorphism, $\mathbf{H}_n(s)$ inherits the structure of a CR manifold (of hypersurface type) of CR dimension n. Let $(U, z^1, \ldots, z^n, t)$ be the natural local coordinate system on $\mathbf{H}_n(s)$, $z^\alpha = x^\alpha + iy^\alpha$. Let us set

$$\theta = |x|^{-2}\left\{dt + 2\sum_{\alpha=1}^{n}\left(x^\alpha dy^\alpha - y^\alpha dx^\alpha\right)\right\} \tag{5.51}$$

on U. The right-hand member of (5.51) is G_s-invariant and thus defines a global 1-form on $\mathbf{H}_n(s)$. Let $\{\theta^\alpha\}$ be dual to T_α, where $T_\alpha = \partial/\partial z^\alpha + i\overline{z}^\alpha \partial/\partial t$ on U. The Levi form associated with (5.51) is given by

$$L_{\dot\theta} = |x|^{-2}\delta_{\alpha\beta}\theta^\alpha \wedge \theta^{\overline{\beta}}$$

on U. Thus $\mathcal{H}_s := T_{1,0}(\mathbf{H}_n(s))$ is strictly pseudoconvex. Our Theorem 5.11 is completely proved. □

Let

$$\gamma_s := \gamma(\mathbf{H}_n(s)) \in H^1(\mathbf{H}_n(s), \mathcal{P})$$

be the Lee class of $\mathbf{H}_n(s)$, a CR invariant of $\mathbf{H}_n(s)$, as previously shown. Let $\{(U_i, z_i^\alpha, t_i^\alpha)\}_{i\in I}$ be an atlas on $\mathbf{H}_n(s)$ such that for any $i, j \in I$ with $U_i \cap U_j \neq \emptyset$ the coordinate transformation reads

$$z_j^\alpha = s^{m_{ji}} z_i^\alpha, \quad t_j = s^{2m_{ji}} t_i, \tag{5.52}$$

for some $m_{ji} \in \mathbf{Z}$. Let us define

$$\theta_i = dt_i + 2\sum_{\alpha=1}^{n} \left(x_i^\alpha dy_i^\alpha - y_i^\alpha dx_i^\alpha\right)$$

on U_i, $i \in I$. Each (U_i, θ_i) is a strictly pseudoconvex CR manifold with a vanishing Ricci tensor (in particular θ_i is pseudo-Einsteinian). As a consequence of (5.52) one has

$$\theta_j = \exp(2m_{ji} \log s)\theta_i$$

on $U_i \cap U_j$. Let $c = (2m_{ij} \log s) \in Z^1(N(\mathcal{U}), \mathbf{R})$ be the corresponding cocycle, where $N(\mathcal{U})$ is the nerve of $\mathcal{U} = \{U_i\}_{i\in I}$. If

$$i : C^1(N(\mathcal{U}), \mathbf{R}) \to C^1(N(\mathcal{U}), \mathcal{P})$$

is the natural cochain map then γ_s is the image of $[c]$ via

$$i_* : H^1(\mathbf{H}_n(s), \mathbf{R}) \to H^1(\mathbf{H}_n(s), \mathcal{P}).$$

We are going to show that (5.51) is globally pseudo-Einsteinian, so that $\gamma_s = 0$. Yet $c \neq 0$ (since $\ker(i_*) \neq 0$). Indeed $[c]$ corresponds (under the isomorphism $H^1_{dR}(\mathbf{H}_n(s)) \simeq H^1(\mathbf{H}_n(s), \mathbf{R})$) to the de Rham cohomology class $[\omega]$ of the 1-form $\omega = d \log |x|^{-1}$ (which is not exact).[3] Also $\gamma_s = 0$ yields $c_1(\mathcal{H}_s) = 0$. We may show that actually all Chern classes of $\mathcal{H}_s$ vanish (by constructing a flat connection D in $\mathcal{H}_s$). We do this in the following more general setting.

[3] Note that $d \log |x|^{-1}$ is G_s-invariant, so that ω is globally defined.

Let $(M, T_{1,0}(M))$ be a nondegenerate CR manifold. Let $u \in C^\infty(M)$ be a real-valued smooth function on M. Let $\{T_\alpha\}$ be a frame in $T_{1,0}(M)$ defined on some open set $U \subseteq M$. Let

$$\hat{\theta} = e^{2u}\theta\,, \quad \hat{\theta}^\alpha = \theta^\alpha + 2iu^\alpha\theta,$$
$$\hat{T} = e^{-2u}\{T - 2iu^\beta T_\beta + 2iu^{\overline{\beta}} T_{\overline{\beta}}\},$$

where

$$u^\alpha = h^{\alpha\overline{\beta}} u_{\overline{\beta}}, \; u_{\overline{\beta}} = T_{\overline{\beta}}(u), \; u^{\overline{\sigma}} = \overline{u^\sigma}.$$

Note that with these choices, one has

$$\hat{T} \rfloor \hat{\theta} = 1, \quad \hat{T} \rfloor d\hat{\theta} = 0, \quad \hat{T} \rfloor \hat{\theta}^\alpha = 0.$$

Then $G_{\hat{\theta}} = e^{2u} G_\theta$; hence $\hat{h}_{\alpha\overline{\beta}} = e^{2u} h_{\alpha\overline{\beta}}$, where $\hat{h}_{\alpha\overline{\beta}} = l_{\hat{\theta}}(T_\alpha, T_{\overline{\beta}})$.

Let $(M, T_{1,0}(M))$ be a nondegenerate CR manifold admitting a real, closed (globally defined) 1-form ω. Let $B = \omega^\sharp$, where $\sharp$ denotes raising of indices with respect to g_θ. Next, set $B^{1,0} = \pi_+ B$. Locally, if

$$\omega = \omega_\alpha \theta^\alpha + \omega_{\overline{\alpha}} \theta^{\overline{\alpha}} + \omega_0 \theta,$$

where $\omega_{\overline{\alpha}} = \overline{\omega_\alpha}$, then

$$B^{1,0} = h^{\alpha\overline{\beta}} \omega_{\overline{\beta}} T_\alpha.$$

By the Poincaré lemma, there exists an open covering $\{U_i\}_{i \in I}$ of M and a family $\{u_i\}_{i \in I}$ of $\mathbf{R}$-valued functions $u_i \in C^\infty(U_i)$ such that

$$\omega\big|_{U_i} = du_i, \; i \in I.$$

Let us set $\theta_i = \exp(2u_i)\theta|_{U_i}$. By applying the identities

$$\begin{aligned} \hat{\Gamma}^\sigma_{\beta\alpha} &= \Gamma^\sigma_{\beta\alpha} + 2u_\beta \delta^\sigma_\alpha + 2u_\alpha \delta^\sigma_\beta\,, \\ \hat{\Gamma}^\sigma_{\overline{\beta}\alpha} &= \Gamma^\sigma_{\overline{\beta}\alpha} - 2u^\sigma h_{\overline{\beta}\alpha}\,, \\ e^{2u}\hat{\Gamma}^\sigma_{\hat{0}\alpha} &= \Gamma^\sigma_{0\alpha} + 2u_0 \delta^\sigma_\alpha - 4iu_\alpha u^\sigma + 2iu_{\alpha,}{}^\sigma + 2i\Gamma^\sigma_{\overline{\mu}\alpha} u^{\overline{\mu}} - 2i\Gamma^\sigma_{\mu\alpha} u^\mu\,, \end{aligned} \tag{5.53}$$

to $u = u_i$ it follows that the Tanaka–Webster connections of the nondegenerate CR manifolds (U_i, θ_i), $i \in I$, glue up to a (globally defined) linear connection D expressed by

$$\begin{aligned} D_Z W &= \nabla_Z W + 2\{\omega(Z)W + \omega(W)Z\}, \\ D_{\overline{Z}} W &= \nabla_{\overline{Z}} W - 2L_\theta(\overline{Z}, W)B^{1,0}, \\ D_T W &= \nabla_T W + 2i\nabla_W B^{1,0} + 4i\omega(W)B^{1,0} + 4i\|B^{1,0}\|^2 W, \\ D_Z T_\omega &= 2\omega(Z)T_\omega\,, \quad D_{T_\omega} T_\omega = 2\omega(T)T_\omega\,, \end{aligned} \tag{5.54}$$

for any $Z, W \in T_{1,0}(M)$. Here ∇ is the Tanaka–Webster connection of (M, θ) and

$$T_\omega = T - 2iB^{1,0} + 2iB^{0,1}.$$

Note that T_ω is transversal to $H(M)$ (so that the formulas (5.54) define D everywhere on $T(M)$). In analogy with I. Vaisman [415], we make the following definition:

Definition 5.14. We call D the *Weyl connection* of (M, θ, ω). □

Theorem 5.12. *Let* $0 < s < 1$ *and* $n > 1$. *Then (i) all Chern classes of* $\mathcal{H}_s$ *vanish, and (ii) the contact form* (5.51) *is pseudo-Einsteinian and has a nonvanishing pseudo-Hermitian torsion.*

Proof. Let $M = \mathbf{H}_n(s)$ with the C^∞ atlas $\{U_i, z_i^\alpha, t_i\}_{i\in I}$ as above. Let $u_i \in C^\infty(U_i)$ be defined by $u_i = \log|x_i|$, $x_i = (z_i, t_i)$. Then (by (5.52)) we have

$$u_j - u_i = m_{ji} \log s = \text{const}$$

on $U_i \cap U_j$. Consequently, the local 1-forms du_i glue up to a real (closed) global 1-form ω on $\mathbf{H}_n(s)$. The Tanaka–Webster connections of the local pseudo-Hermitian structures $\{\theta_i\}_{i\in I}$ are flat, so that the Weyl connection D of $(\mathbf{H}_n(s), \theta, \omega)$ (with θ given by (5.51)) is flat. Since $DJ = 0$ the Weyl connection is reducible to a (flat) connection in $\mathcal{H}_s$. By the Chern–Weil theorem the characteristic ring of $\mathcal{H}_s$ must vanish. □

Let $(M, T_{1,0}(M), \theta)$ be a nondegenerate CR manifold. Let us set $\hat{\theta} = e^{2u}\theta$. As a consequence of (5.53) one has

$$\hat{A}_{\alpha\beta} = A_{\alpha\beta} + 2iu_{\alpha,\beta} - 4iu_\alpha u_\beta. \tag{5.55}$$

At this point we may prove (ii) in Theorem 5.12. Indeed, we may apply (5.55) with $u = \log|x|^{-1}$, $A_{\alpha\beta} = 0$, and $\omega_\beta^\alpha = 0$. If $T_\alpha = \partial/\partial z^\alpha + i\overline{z}^\alpha \partial/\partial t$ then

$$u_\alpha = -\tfrac{1}{2}|x|^{-4}\overline{z}_\alpha\phi,$$
$$T_\alpha(u_\beta) = |x|^{-8}\phi^2\overline{z}_\alpha\overline{z}_\beta,$$

where $\phi(z, t) = \|z\|^2 + it$. Finally, since ϕ is CR-holomorphic, (5.55) yields

$$\hat{A}_{\alpha\beta} = 2iT_\alpha(u_\beta) - 4iu_\alpha u_\beta = i|x|^{-8}\overline{z}_\alpha\overline{z}_\beta\phi^2,$$

so that (5.51) has nonvanishing pseudo-Hermitian torsion. □

Let $(M, T_{1,0}(M))$ be a nondegenerate CR manifold of CR dimension n and $\hat{\theta} = e^{2u}\theta$. Then the pseudo-Hermitian Ricci tensors $R_{\alpha\overline{\beta}}$, $\hat{R}_{\alpha\overline{\beta}}$ of θ, $\hat{\theta}$ are related by

$$\hat{R}_{\alpha\overline{\beta}} = R_{\alpha\overline{\beta}} - (n+2)(u_{\alpha,\overline{\beta}} + u_{\overline{\beta},\alpha}) - \left[u_{\rho,}{}^{\rho} + u_{\overline{\rho},}{}^{\overline{\rho}} + 4(n+1)u_\rho u^\rho\right] h_{\alpha\overline{\beta}}. \tag{5.56}$$

If $M = \mathbf{H}_n(s)$ and θ is given by (5.51) then we may apply (5.56) with $R_{\alpha\overline{\beta}} = 0$, $u = \log|x|^{-1}$, $h_{\alpha\overline{\beta}} = \delta_{\alpha\beta}$, and $\omega_\beta^\alpha = 0$. Then

$$u_{\rho,}{}^{\rho} = -\frac{n}{2}|x|^{-4}\phi, \quad u_\rho u^\rho = \tfrac{1}{4}|x|^{-4}\|z\|^2, \quad u_{\alpha,\overline{\beta}} = -\tfrac{1}{2}|x|^{-4}\phi\delta_{\alpha\beta},$$

so that (5.56) yields

$$\hat{R}_{\alpha\overline{\beta}} = (n+1)|x|^{-2}\|z\|^2\hat{h}_{\alpha\overline{\beta}},$$

which means that (5.51) is pseudo-Einsteinian. Our Theorem 5.12 is completely proved. □

A remark is in order. Let $\mathbf{R}^* \simeq \{(0,t) : t \in \mathbf{R}\setminus\{0\}\} \subset \mathbf{H}_n \setminus \{0\}$. The pseudo-Hermitian Ricci curvature of the contact form (5.51) vanishes on $\pi(\mathbf{R}^*)$, so that Proposition 6.4 in [270], p. 175, does not apply. □

5.7.2 Regular strictly pseudoconvex CR manifolds

Let M be an m-dimensional differentiable manifold.

Definition 5.15. A local chart (U,φ) on M is *cubical* (of breadth $2a$ centered at $x \in M$) if $\varphi(x) = (0,\dots,0)$ and $\varphi(U) = \{(t^1,\dots,t^m) \in \mathbf{R}^m : |t^j| < a, 1 \le j \le m\}$. □

Definition 5.16. Let (U,φ), $\varphi = (x^1,\dots,x^m)$, be a cubical local chart on M. Let $1 \le p \le m$ and $t = (t^{p+1},\dots,t^m) \in \mathbf{R}^{m-p}$ such that $|t^{p+j}| < a$, $1 \le j \le m-p$. The p-dimensional *slice* Σ_t of (U,φ) is given by

$$\Sigma_t = \{y \in U : x^{p+j}(y) = t^{p+j},\ 1 \le j \le m-p\}.$$ □

Let $(M, T_{1,0}(M), \theta, T)$ be a nondegenerate CR manifold of CR dimension n.

Definition 5.17. T is *regular* if M admits a C^∞ atlas $\{(U,x^i)\}$ such that the intersection of U with any maximal integral curve of T is a 1-dimensional slice of (U,x^i). □

Let $\langle T\rangle$ be the distribution spanned by T, i.e., $\langle T\rangle_x = \mathbf{R}T_x$, $x \in M$.

Theorem 5.13. (Theorem VIII and Theorem X in [336], pp. 19–20)
If T is regular then the quotient space[4] $M/\langle T\rangle$ admits a natural manifold structure with respect to which the canonical projection $\pi : M \to M/\langle T\rangle$ is differentiable.

Theorem 5.14. *Let $(M, T_{1,0}(M))$ be a compact strictly pseudoconvex CR manifold. If M admits a contact form whose contact vector is regular then M admits a global pseudo-Einstein structure.*

To prove Theorem 5.14 we need to recall the essentials of the Boothby–Wang theorem (cf. [75]). Since T is regular, its maximal integral curves are closed subsets of M (cf. Theorem VII [336], p. 18). But M is compact, so that each maximal integral curve is homeomorphic to S^1. Let λ be the period of T, i.e.,

[4] That is, the space of all maximal integral curves of T.

$$\lambda(x) = \inf\{t > 0 : \varphi_t(x) = x\}, \quad x \in M,$$

where $\{\varphi_t\}_{t\in\mathbf{R}}$ is the (global) 1-parameter group generated by T. We may assume that $\lambda = 1$ (otherwise, since $\lambda = \text{const} > 0$ (by an argument in [400]) we may replace T by $(1/\lambda)T$). Then, by the Boothby–Wang theorem, T generates a free and effective action of S^1 on M. Next M becomes the total space of a principal S^1-bundle $\pi : M \to B$, where $B = M/\langle T\rangle$. The projection map of any principal bundle is in particular a submersion (and we may apply the results in [334]). Let g_θ be the Webster metric. Let d/dt be the generator of Lie algebra $L(S^1) \simeq \mathbf{R}$. Then $\theta \otimes (d/dt)$ is a connection 1-form in $S^1 \to M \longrightarrow B$. Let us set

$$h_\theta(X, Y)_u = g_\theta(X^H, Y^H)_x,$$

where $x \in \pi^{-1}(u)$, $u \in B$, and $X, Y \in T_u(B)$. Here X^H denotes the horizontal lift of X with respect to $\theta \otimes (d/dt)$. The definition of $h_\theta(X, Y)_u$ does not depend on the choice of x in $\pi^{-1}(u)$. It follows that $\pi : M \to B$ is a Riemannian submersion from (M, g_θ) onto (B, h_θ). Let P, Q be the fundamental tensors of π (cf. [334], p. 460), that is,

$$P_X Y = h\tilde{\nabla}_{vX} vY + v\tilde{\nabla}_{vX} hY, \tag{5.57}$$

$$Q_X Y = h\tilde{\nabla}_{hX} hY + v\tilde{\nabla}_{hX} vY, \tag{5.58}$$

for any $X, Y \in T(M)$. Here $\tilde{\nabla}$ denotes the Levi-Civita connection of (M, g_θ). Moreover, $h = \pi_H$ and $vX = \theta(X)T$ are the canonical projections associated with $T(M) = H(M) \oplus \mathbf{R}T$. Let us substitute from

$$\tilde{\nabla} = \nabla - (d\theta + A) \otimes T + \tau \otimes \theta + \theta \odot J \tag{5.59}$$

into (5.58). Since $JT = 0$, $\tau T = 0$, $\nabla T = 0$ and $H(M)$ is parallel with respect to ∇, our (5.58) becomes

$$\begin{aligned} Q_X Y &= -\{(d\theta)(X, Y) + A(X, Y)\}T, \\ Q_X T &= \tau(X) + JX, \\ Q_T X &= 0, \quad Q_T T = 0, \end{aligned} \tag{5.60}$$

for any $X, Y \in H(M)$. Since τ is self-adjoint, by a result in [334], p. 460, Q is skew-symmetric on horizontal vectors. Clearly the Levi distribution $H(M)$ coincides with the horizontal distribution of the Riemannian submersion $\pi : M \to B$. Then the first of the formulas (5.60) yields $A = 0$ and thus there is $u \in C^\infty(M)$ such that $\exp(2u)\theta$ is globally pseudo-Einsteinian. The proof of Theorem 5.14 is complete. □

A couple of remarks are in order.

(i) Let us substitute from (5.59) into (5.57). This procedure leads to $P = 0$. Consequently the fibers of the submersion $\pi : M \to B$ are totally geodesic in (M, g_θ).

(ii) By a result of G. Gigante [175], p. 151, and by the proof of Theorem 5.14, any compact strictly pseudoconvex symmetric (in the sense of [175], p. 150) CR manifold is a Sasakian manifold.

5.7.3 The Bockstein sequence

Let $(M, T_{1,0}(M))$ be a nondegenerate CR manifold, and θ a contact form. Let $\mathcal{E}_{CR}$ be the sheaf of (local) CR-holomorphic functions on M.

Proposition 5.11. *There is a short exact sequence*

$$0 \to \mathbf{R} \xrightarrow{j} \mathcal{E}_{CR} \xrightarrow{\eta} \mathcal{P} \to 0, \tag{5.61}$$

where

$$j_U : \mathbf{R} \to \mathcal{E}_{CR}(U), \quad j_U(c) = ic,$$

and

$$\eta_U : \mathcal{E}_{CR}(U) \to \mathcal{P}(U), \quad \eta_U(f) = \mathrm{Re}(f),$$

for any $c \in \mathbf{R}$ and $f \in \mathcal{E}_{CR}(U)$.

Indeed, let $\sigma_x \in \mathrm{Ker}(\eta_x)$, $x \in M$. That is, there are an open set $U \subset M$, $x \in U$, and a real-valued function $v \in C^\infty(U)$ such that $[iv]_x = \sigma_x$ and $\overline{\partial}_b v = 0$. Then $\partial_b v = 0$ (by complex conjugation) and $dv = T(v)\theta$. Exterior differentiation gives

$$0 = (dT(v)) \wedge \theta + T(v) d\theta = (dT(v)) \wedge \theta + 2i T(v) h_{\alpha\overline{\beta}} \theta^\alpha \wedge \theta^{\overline{\beta}}.$$

Let us apply this to the pair $(T_\alpha, T_{\overline{\beta}})$ to yield $0 = iT(v)h_{\alpha\overline{\beta}}$, which (by contraction with $h^{\alpha\overline{\beta}}$) gives $T(v) = 0$, i.e., there is an open set $V \subset U$, $x \in V$, and a constant $c \in \mathbf{R}$ such that $v = c$ on V. Thus $\sigma_x = [ic]_x = j_x(c)$. □

Consider the Bockstein exact sequence

$$\cdots \to H^1(M, \mathbf{R}) \to H^1(M, \mathcal{E}_{CR}) \xrightarrow{\eta_*} H^1(M, \mathcal{P}) \xrightarrow{b} H^2(M, \mathbf{R}) \to \cdots$$

associated with (5.61). If M is compact and strictly pseudoconvex one may try to show that (i) $b(\gamma(M)) = c_1(T_{1,0}(M))$ and (ii) $\mathrm{Im}(\eta_*) = 0$ (by the results in the previous sections, this would imply the Lee conjecture). The example $\mathbf{H}_n(s)$ kills a hope to solve the Lee conjecture along the lines indicated above. Indeed, the map $r : \mathbf{H}_n \setminus \{0\} \to \Sigma^{2n}$ defined by

$$r(x) = \delta_{|x|^{-1}}(x), \quad x \in \mathbf{H}_n \setminus \{0\},$$

is a deformation retract. Thus, by $\mathbf{H}_n(s) \simeq \Sigma^{2n} \times S^1$ and the Küneth formula, it follows that

$$H^2(\mathbf{H}_n(s), \mathbf{R}) = H^2(\Sigma^{2n}, \mathbf{R}) = H^2(\mathbf{H}_n \setminus \{0\}, \mathbf{R}) = H^2(S^{2n}, \mathbf{R}) = 0,$$

and the Bockstein sequence yields

$$\mathrm{Im}(\eta_*) = H^1(\mathbf{H}_n(s), \mathcal{P}).$$

5.7.4 The tangent sphere bundle

Let M be a Riemannian manifold and $U(M)$ its tangent sphere bundle. The natural almost complex structure $\tilde{J}$ of $T(M)$ induces an almost CR structure $\mathcal{H}$ on $U(M)$ (as a real hypersurface of $T(M)$). Although $\tilde{J}$ is rarely integrable (in fact, only when M is locally Euclidean; cf. P. Dombrowski [116]) $\mathcal{H}$ may turn out to be a CR structure. For instance, if M is a space form then $\mathcal{H}$ is integrable.

Theorem 5.15. (E. Barletta et al. [37])
Let M be an n-dimensional Riemannian manifold and $\mathcal{H}$ the natural almost CR structure of $U(M)$. The following statements are equivalent:

(i) $(U(M), \mathcal{H})$ is a strictly pseudoconvex CR manifold (of CR dimension $n-1$) whose Tanaka–Webster connection has a vanishing pseudo-Hermitian torsion.
(ii) M is an elliptic space form $M^n(c)$ of sectional curvature $c = 1$.

The natural pseudo-Hermitian structure of $U(M^n(1))$ is globally pseudo-Einsteinian. In particular $U(M^n(1))$ has positive pseudo-Hermitian scalar curvature and the first Chern class of its CR structure $\mathcal{H}$ vanishes.

As a corollary, the first statement in Theorem 5.15 yields a short proof of a result by Y. Tashiro [402] (the contact vector of $U(M)$ is Killing if and only if $M = M^n(1)$). The proof of the second statement in Theorem 5.15 relies on a result by E.T. Davies and K. Yano [113].

We need a brief preparation on the geometry of the tangent bundle over a Riemannian manifold. Let (M, G) be an n-dimensional Riemannian manifold. Let D be the Levi-Civita connection of (M, G).

Definition 5.18. (W. Barthel [48]) A C^∞ distribution on $T(M)$,

$$N : v \in T(M) \mapsto N_v \subset T_v(T(M)),$$

is called a *nonlinear connection* on M if

$$T_v(T(M)) = N_v \oplus [\mathrm{Ker}(d_v \Pi)] , \tag{5.62}$$

for any $v \in T(M)$. Here $\Pi : T(M) \to M$ is the natural projection. □

D gives rise to a nonlinear connection N on $T(M)$. Precisely, let (U, x^i) be a local coordinate system on M and $(\Pi^{-1}(U), x^i, y^i)$ the naturally induced local coordinates on $T(M)$. Let $\Gamma^i_{jk}(x)$ be the coefficients of D (with respect to (U, x^i)) and let us set

$$\frac{\delta}{\delta x^i} = \frac{\partial}{\partial x^i} - N_i^j \frac{\partial}{\partial y^j} ,$$

where

$$N^i_j(x, y) = \Gamma^i_{jk}(x) y^k .$$

Then $\{\delta/\delta x^i : 1 \leq i \leq n\}$ is a local frame of N (on $\Pi^{-1}(U)$).

Definition 5.19. For each $v \in T(M)$ let $\beta_v : T_x(M) \to N_v$ be the *horizontal lift*, i.e., the inverse of $d_v\Pi : N_v \to T_x(M)$, where $x = \Pi(v)$. □

Locally, one has

$$\beta \frac{\partial}{\partial x^i} = \frac{\delta}{\delta x^i}.$$

Definition 5.20. The *vertical lift* $\gamma_v : T_x(M) \to \operatorname{Ker}(d_v\Pi)$ is given by

$$\gamma_v(w) = \frac{dC}{dt}(0),$$

for any $w \in T_x(M)$. Here $C : (-\epsilon, \epsilon) \to T_x(M)$ is the curve given by $C(t) = v + tw, \ |t| < \epsilon$. □

Locally

$$\gamma \frac{\partial}{\partial x^i} = \frac{\partial}{\partial y^i},$$

so that γ is a bundle isomorphism. Let $Q_v : T_v(T(M)) \to \operatorname{Ker}(d_v\Pi)$ be the natural projection (associated with (5.62)).

Definition 5.21. (P. Dombrowski [116])
The *Dombrowski map* $K_v : T_v(T(M)) \to T_x(M)$ is given by $K = \gamma^{-1} \circ Q$. □

Locally

$$K \frac{\delta}{\delta x^i} = 0, \quad K \frac{\partial}{\partial y^i} = \frac{\partial}{\partial x^i}.$$

Next, we shall need the following definition:

Definition 5.22. The *Sasaki metric* $\tilde{g}$ on $T(M)$ is given by

$$\tilde{g}(V, W) = G(KV, KW) + G(\Pi_* V, \Pi_* W)$$

for any $V, W \in T(T(M))$. □

This is a Riemannian metric on $T(M)$ and the distributions N and $\operatorname{Ker}(\Pi_*)$ are orthogonal (with respect to $\tilde{g}$).

Let us set $U(M)_x = \{v \in T_x(M) : G_x(v, v) = 1\}$. The disjoint union $U(M)$ of $U(M)_x$ for all $x \in M$ is a real hypersurface of $T(M)$ and the total space of an S^{n-1}-bundle $\pi : U(M) \to M$. The portion of $U(M)$ over U is given by the equation

$$g_{ij}(x) y^i y^j = 1, \tag{5.63}$$

where the g_{ij} are the components of G with respect to (U, x^i). Note that

$$N_v \subset T_v(U(M)),$$
$$\operatorname{Ker}(d_v\pi) = T_v(U(M)) \cap \operatorname{Ker}(d_v\Pi),$$

for any $v \in U(M)$.

Definition 5.23. Let $\tilde{J}$ be the natural almost complex structure of $T(M)$ given by $\tilde{J} \circ \beta = \gamma$ and $\tilde{J} \circ \gamma = -\beta$. □

Locally

$$\tilde{J}\frac{\delta}{\delta x^i} = \frac{\partial}{\partial y^i}, \quad \tilde{J}\frac{\partial}{\partial y^i} = -\frac{\delta}{\delta x^i}.$$

Let us set

$$T_{1,0}(U(M))_v = T^{1,0}(T(M))_v \cap [T_v(U(M)) \otimes_{\mathbf{R}} \mathbf{C}],$$

where $T^{1,0}(T(M)) \subset T(T(M)) \otimes \mathbf{C}$ is the eigenbundle of $\tilde{J}$ corresponding to the eigenvalue i.

Proposition 5.12. *$\mathcal{H} = T_{1,0}(U(M))$ is an almost CR structure on $U(M)$, i.e., $\mathcal{H} \cap \overline{\mathcal{H}} = (0)$.*

By a result of P. Dombrowski [116], if D is flat then $(T(M), \tilde{J})$ is a complex manifold.

Corollary 5.3. *If (M, G) is locally Euclidean then $\mathcal{H}$ is integrable.*

For instance $U(\mathbf{R}^n)$ is a CR manifold (of CR dimension $n-1$). In general, $\mathcal{H}$ is only an almost CR structure, and the problem of studying the corresponding Cauchy–Riemann pseudocomplex is open.

Let $P = \beta \circ K$. Then $P_v : \operatorname{Ker}(d_v \Pi) \to N_v$ is a linear isomorphism. We shall need the following lemma:

Lemma 5.7. *The maximal complex distribution $H(U(M))$ of the almost CR manifold $(U(M), \mathcal{H})$ is given by*

$$H(U(M))_v = \operatorname{Ker}(d_v \pi) \oplus [P_v \operatorname{Ker}(d_v \pi)]$$

for any $v \in U(M)$.

Proof. Let $E(\iota) \to U(M)$ be the normal bundle of the immersion $\iota : U(M) \hookrightarrow T(M)$. Let us set $\nu = y^i \partial/\partial y^i$. Then ν is a (globally defined) unit normal ($\nu_v \in E(\iota)_v$) on $U(M)$. Let us set $\xi' = -\tilde{J}\nu$. Then ξ' is tangent to $U(M)$. Locally, we have $\xi' = y^i \delta/\delta x^i$. Let η' be the real 1-form on $U(M)$ given by

$$\eta'(V) = g'(V, \xi'),$$

for any $V \in T(T(M))$, where $g' = \iota^* \tilde{g}$ is the metric induced on $U(M)$ by the Sasaki metric $\tilde{g}$ of $T(M)$. Note that $H(U(M)) = \operatorname{Ker}(\eta')$. Also $\operatorname{Ker}(\pi_*) \subset \operatorname{Ker}(\eta')$. Let us set $y_i = g_{ij} y^j$. Then $\eta'(\delta/\delta x^i) = y_i$. At this point Lemma 5.7 follows from the fact that a vertical tangent vector $X = B^i \partial/\partial y^i$ is tangent to $U(M)$ if and only if $g_{ij} B^i y^j = 0$ (by taking into account (5.63)). □

Let us set $\tilde{U} = \{y_m \neq 0\} \subset \pi^{-1}(U)$. The portion of $\mathrm{Ker}(\pi_*)$ over $\tilde{U}$ is the span of $\{Y_\alpha : 1 \leq \alpha \leq n-1\}$, where

$$Y_\alpha = \frac{\partial}{\partial y^\alpha} - A_\alpha \frac{\partial}{\partial y^n}, \quad A_\alpha = \frac{y_\alpha}{y_n}.$$

Let us set $\varphi V = \tan\{\tilde{J} V\}$ for any $V \in T(U(M))$. Here $\tan_v : T_v(T(M)) \to T_v(U(M))$ is the natural projection associated with the decomposition

$$T_v(T(M)) = t_v(U(M)) \oplus E(\iota)_v\,,$$

for any $v \in U(M)$. The restriction J of φ to $H(U(M))$ and the complex structure $J_{U(M)}$ of $H(U(M))$ actually coincide. Note that $JX_\alpha = Y_\alpha$, where

$$X_\alpha = \frac{\delta}{\delta x^\alpha} - A_\alpha \frac{\delta}{\delta x^n}.$$

Thus (by Lemma 5.7) $\mathcal{H} = T_{1,0}(U(M))$ is (locally) the span of $\{T_\alpha\}$, where $T_\alpha = X_\alpha - iY_\alpha$. Let $R^i_{jk\ell}$ be the components of the curvature tensor field R of D (with respect to (U, x^i)). Let us set

$$R^i_{k\ell} = R^i_{jk\ell} y^j.$$

Note that

$$\left[\frac{\delta}{\delta x^i}, \frac{\delta}{\delta x^j}\right] = -R^k_{ij} \frac{\partial}{\partial y^k} \tag{5.64}$$

(so that the Pfaffian system $dy^i + N^i_j(x, y)dx^j = 0$ is integrable if and only if $R^k_{ij} = 0$). We shall prove the following lemma:

Lemma 5.8. *The almost CR structure $\mathcal{H}$ on $U(M)$ is integrable if and only if*

$$R^i_{\alpha\beta} + A_\alpha R^i_{\beta n} + A_\beta R^i_{n\alpha} = 0 \tag{5.65}$$

on $\tilde{U}$, for any local coordinate neighborhood (U, x^i) on M.

Proof. The generators X_α, Y_α satisfy the commutation formulas

$$[X_\alpha, X_\beta] = \{X_\beta(A_\alpha) - X_\alpha(A_\beta)\}\frac{\delta}{\delta x^n} - \{R^i_{\alpha\beta} + A_\alpha R^i_{\beta n} + A_\beta R^i_{n\alpha}\}\frac{\partial}{\partial y^i},$$

$$[Y_\alpha, Y_\beta] = \{Y_\beta(A_\alpha) - Y_\alpha(A_\beta)\}\frac{\partial}{\partial y^n},$$

$$[X_\alpha, Y_\beta] = Y_\beta(A_\alpha)\frac{\delta}{\delta x^n} - X_\alpha(A_\beta)\frac{\partial}{\partial y^n} + \{Y_\beta(N^i_\alpha) - A_\alpha Y_\beta(N^i_n)\}\frac{\partial}{\partial y^i}.$$

These follow from (5.64) together with the identities

$$\left[\frac{\delta}{\delta x^i}, \frac{\partial}{\partial y^j}\right] = \Gamma^k_{ij}\frac{\partial}{\partial y^k}, \quad \left[\frac{\partial}{\partial y^i}, \frac{\partial}{\partial y^j}\right] = 0.$$

A straightforward calculation shows that

$$X_\alpha(A_\beta) = X_\beta(A_\alpha), \quad Y_\alpha(A_\beta) = Y_\beta(A_\alpha),$$
$$Y_\alpha(N^i_\beta) - A_\beta Y_\alpha(N^i_n) = Y_\beta(N^i_\alpha) - A_\alpha Y_\beta(N^i_n),$$

so that

$$[T_\alpha, T_\beta] = -\{R^i_{\alpha\beta} + A_\alpha R^i_{\beta n} + A_\beta R^i_{n\alpha}\}\frac{\partial}{\partial y^i}. \tag{5.66}$$

Finally, from (5.66) and from

$$T_\alpha = \frac{\delta}{\delta x^\alpha} - i\frac{\partial}{\partial y^\alpha} - A_\alpha\left(\frac{\delta}{\delta x^n} - i\frac{\partial}{\partial y^n}\right)$$

we see that $[T_\alpha, T_\beta] \in \mathcal{H}$ if and only if (5.65) holds on $\tilde{U}$. □

Let X be a real $(2m+1)$-dimensional manifold. Let (φ, ξ, η) be an almost contact structure on X. The restriction of φ to $\mathrm{Ker}(\eta)$ is a complex structure. Let us extend φ to $\mathrm{Ker}(\eta) \otimes \mathbf{C}$ and let us set $T_{1,0}(X) = \mathrm{Eigen}(i)$. Then $T_{1,0}(X)$ is an almost CR structure (of CR dimension m) on X. If (φ, ξ, η) is normal (in the sense of [62], p. 48) then (by a result of [214]) $T_{1,0}(X)$ is integrable. Going back to $X = U(M)$, set $\xi = 2\xi'$ and $\eta = (1/2)\eta'$. Then (φ, ξ, η) is an almost contact structure on $U(M)$. Next, set $g = (1/4)g'$. Then (φ, ξ, η, g) is a contact metric structure on $U(M)$ (in the sense of [62], p. 25). By a result of Y. Tashiro [402], if $M = M^n(1)$ (i.e., (M, G) has constant sectional curvature 1) then (φ, ξ, η, g) is a Sasakian structure on $U(M)$. In particular it is normal, so that by applying the theorem of S. Ianus quoted above, we have proved the following:

Proposition 5.13. (Y. Tashiro [402])
$U(M^n(1))$ is a CR manifold.

Our Lemma 5.8 may be used to indicate examples of Riemannian manifolds M other than those covered by Y. Tashiro's theorem, for which $U(M)$ is CR.

Definition 5.24. ([420]) (M, G) is a Riemannian manifold of *quasi-constant curvature* if its curvature tensor field is given by

$$R^\ell_{ijk} = c\{\delta^\ell_j g_{ki} - \delta^\ell_k g_{ji}\} + b\{(\delta^\ell_j v_k - \delta^\ell_k v_j)v_i + (v_j g_{ki} - v_k g_{ji})v^\ell\}, \tag{5.67}$$

for some real-valued functions $b, c \in C^\infty(M)$ and some unit tangent vector field $V = v^i \partial/\partial x^i$ on M. □

If M is a Riemannian manifold of quasi-constant curvature we write $M = M^n_{c,b}(V)$.

Proposition 5.14. (E. Barletta et al. [37]) *Let M be an n-dimensional Riemannian manifold. Then*

$$R^\ell_{jk} y_i + R^\ell_{ki} y_j + R^\ell_{ij} y_k = 0 \tag{5.68}$$

on $U(M)$ is a sufficient condition for the integrability of the natural almost CR structure $\mathcal{H}$ of $U(M)$. If $M = M^n_{c,b}(V)$ then (5.68) holds if and only if either $n = 2$ or $n \geq 3$ and $b = 0$. In particular $U(M^n(c))$ is CR for any space form $M^n(c)$.

Proof. Clearly (5.68) yields (5.65), so that the first statement follows from Lemma 5.8. Next (by taking into account (5.67))

$$R^\ell_{jk} y_i + R^\ell_{ki} y_j + R^\ell_{ij} y_k = bf T^\ell_{jik},$$

where

$$T^\ell_{jik} = \delta^\ell_j y_i v_k - \delta^\ell_i y_j v_k + \delta^\ell_i y_k v_j - \delta^\ell_j y_k v_i + \delta^\ell_k y_j v_i - \delta^\ell_k y_i v_j$$

and $f : U(M) \to \mathbf{R}$ is $f = v_i y^i$. Note that

$$T^\ell_{jik} + T^\ell_{ijk} = T^\ell_{jik} + T^\ell_{jki} = T^\ell_{jik} + T^\ell_{kij} = 0.$$

Clearly $n = 2$ or $b = 0$ yields (5.68). □

Conversely, assume that (5.68) holds. Then

$$bf T^j_{jik} v^i = 0,$$

or

$$(n-2) bf (f v_k - y_k) = 0.$$

If $n \geq 3$ then $b = 0$. Indeed, if $b(x) \neq 0$ for some $x \in M$ then one may choose $u \in U(M)_x$ such that $\{u, V_x\}$ span a 2-plane and $G_x(u, V_x) \neq 0$. Then $f(u) \neq 0$ and $f(u) v^i(x) - u^i \neq 0$, a contradiction. □

Let us consider a C^∞ manifold X carrying the contact metric structure (φ, ξ, η, g). Let us set $T = -\xi$ and $\theta = -\eta$. By the contact condition ($d\eta = \Phi$ where $\Phi(V, W) = g(V, \varphi W)$) one has $T \rfloor d\theta = 0$. Assume $T_{1,0}(X) = \text{Eigen}(i)$ to be a CR structure on X. Again by the contact condition $g_\theta = g$, so that $(X, T_{1,0}(X))$ is a strictly pseudoconvex CR manifold. Let ∇ and ∇^θ be respectively the Tanaka–Webster connection and the Levi-Civita connection of (X, g). Then $\nabla T = 0$; hence

$$\nabla^\theta \xi = -\varphi - \tau. \tag{5.69}$$

In particular, the almost contact metric structure (φ, ξ, η, g) of $X = U(M)$ satisfies the contact condition (cf. [62], p. 133), so that the considerations above may be applied to compute the pseudo-Hermitian torsion τ of $(U(M), \eta)$. Precisely, we establish the following:

Lemma 5.9. *Assume $\mathcal{H} = T_{1,0}(U(M))$ to be a CR structure. Then*

$$\tau \frac{\delta}{\delta x^i} = H^k_i (\delta^\ell_k - y_k y^\ell) \frac{\partial}{\partial y^\ell}, \tag{5.70}$$

$$\tau \frac{\partial}{\partial y^i} = H^k_i \frac{\delta}{\delta x^k}, \tag{5.71}$$

where

$$H^k_i = R^k_{ij} y^j + y_i y^k - \delta^k_i.$$

Proof. Let us use

$$\tau = -\tfrac{1}{2} J_X \circ (\mathcal{L}_T J_X)$$

for $X = U(M)$. Then (5.70)–(5.71) may be gotten from the following identities:

$$\begin{aligned}
\left[\xi, \frac{\delta}{\delta x^j}\right] &= 2\left\{N^i_j \frac{\delta}{\delta x^i} - y^i R^k_{ij} \frac{\partial}{\partial y^k}\right\},\\
\left[\xi, \frac{\partial}{\partial y^j}\right] &= 2\left\{-\frac{\delta}{\delta x^j} + N^k_j \frac{\partial}{\partial y^k}\right\},\\
[\xi, \nu] &= -\xi,\\
\xi(y_j) &= 2y^i y^k\, |jk, i|,\\
\tan\left(\frac{\partial}{\partial y^i}\right) &= (\delta^k_i - y_i y^k)\frac{\partial}{\partial y^k},
\end{aligned}$$

where $\xi = 2y^i \delta/\delta x^i$ and $\nu = y^i \partial/\partial y^i$.

At this point we may prove Theorem 5.15. Assume (ii) holds. Then (5.68) holds on $U(M)$, so that $U(M)$ is a CR manifold. On the other hand, $M = M^n(1)$ yields

$$R^k_{ij} y^j = \delta^k_i - y_i y^k, \tag{5.72}$$

so that (by Lemma 5.9) $\tau = 0$. Conversely, assume (i) holds. Then Lemma 5.9 yields (5.72). Let $x \in M$ and $X, v \in T_x(M)$ be two unit tangent vectors such that $G_x(X, v) = 0$. Let us set $X = X^i \partial/\partial x^i$. Let us apply (5.72) at v and contract with X^i in the resulting identity. This procedure leads to

$$R^k_{\ell ij}(x) X^i v^j v^\ell = X^k$$

(since $X^i v_i = 0$), or

$$R_x(X, v)v = X,$$

which (by taking the inner product with X) yields constant sectional curvature 1. □

Let us show that Y. Tashiro's theorem ([402]) follows from the first statement of our Theorem 5.15. Indeed, if $M = M^n(1)$ then (with the arguments above) it makes sense to consider the Tanaka–Webster connection, and $\tau = 0$ yields normality by a result in [120]. Thus $U(M)$ is Sasakian (and any Sasakian structure is in particular K-contact). Conversely, if the contact structure (φ, ξ, η, g) of $U(M)$ is K-contact then $\nabla^\theta \xi = -\varphi$ (by (8) in [62], p. 64), which together with (5.69) yields $\tau = 0$ and Theorem 5.15 applies. □

When $M^n(1)$ is compact $U(M^n(1))$ is a compact strictly pseudoconvex CR manifold; hence (by a result of L. Boutet de Monvel [77]) $U(M^n(1))$ is locally embeddable (as a real hypersurface in $\mathbf{C}^n$). It is natural to ask whether $U(M^n(1))$ is globally pseudo-Einsteinian.

Let $\tilde{K}$ be the Ricci tensor field of the Sasaki metric $\tilde{g}$ on $T(M)$. By a result of E.T. Davies and K. Yano [113], one has

$$\tilde{K}\left(\frac{\delta}{\delta x^j}, \frac{\delta}{\delta x^k}\right) = R_{jk} + \tfrac{1}{4} y^m \left\{ R^r_{\ ji} R^i_{\ krm} + R^r_{\ ki} R^i_{\ jrm} \right\}, \tag{5.73}$$

$$\tilde{K}\left(\frac{\partial}{\partial y^j}, \frac{\delta}{\delta x^k}\right) = \tfrac{1}{2} y^\ell \left\{ \nabla_i R^i_{k\ell j} + R^i_{k\ell j} \frac{\partial}{\partial x^i} (\log \sqrt{\Delta}) \right\}, \tag{5.74}$$

$$\tilde{K}\left(\frac{\partial}{\partial y^j}, \frac{\partial}{\partial y^k}\right) = \tfrac{1}{4} R^\ell_{irk} R^i_{\ell js} y^r y^s, \tag{5.75}$$

where R_{jk} denotes the Ricci curvature of (M, G) and $\Delta = \det[g_{ij}]$. We need the Gauss equation (cf. [241], vol. II, p. 23) of $U(M)$ in $(T(M), \tilde{g})$:

$$\begin{aligned} &\tilde{R}(X, Y, Z, W) \\ &\quad = R(X, Y, Z, W) + g'(X, a_\nu Z) g'(Y, a_\nu W) - g'(X, a_\nu W) g'(Y, a_\nu Z), \end{aligned} \tag{5.76}$$

for any $X, Y, Z, W \in T(U(M))$. Here a_ν is the shape operator of ι. Taking traces in (5.76) leads to

$$K^\theta(X, Y) = \tilde{K}(X, Y) + g'(a_\nu X, a_\nu Y) - g'(X, a_\nu Y)\|\mu\| - \tilde{R}(X, \nu, Y, \nu), \tag{5.77}$$

where K^θ is the Ricci curvature of $((M), g')$ and μ is the mean curvature vector of ι. At this point, a calculation based on the identities (2) in [62], p. 130, shows that

$$\tilde{R}(X, \nu, Y, \nu) = 0,$$

for any $X, Y \in T(U(M))$. Next, by taking into account

$$a_\nu \frac{\delta}{\delta x^i} = 0, \quad a_\nu X = -X,$$

for any $X \in \mathrm{Ker}(\pi_*)$; cf. [62], p. 132, we obtain

$$\begin{aligned} K^\theta\left(\frac{\delta}{\delta x^i}, \frac{\delta}{\delta x^j}\right) &= \tilde{K}\left(\frac{\delta}{\delta x^i}, \frac{\delta}{\delta x^j}\right), \\ K^\theta(X, Y) &= \tilde{K}(X, Y) + (1 + \|\mu\|) g'(X, Y), \\ K^\theta\left(X, \frac{\delta}{\delta x^j}\right) &= \tilde{K}\left(X, \frac{\delta}{\delta x^j}\right), \end{aligned} \tag{5.78}$$

for any $X, Y \in \mathrm{Ker}(\pi_*)$. Note that $g(X_\alpha, X_{\beta}) = 2h_{\alpha\overline{\beta}}$, where $h_{\alpha\overline{\beta}}$ is the Levi form of $(U(M), \theta)$. If $M = \mathbf{R}^n$ then (5.73)–(5.75) and (5.78) lead to

$$K^\theta_{\alpha\overline{\beta}} = 2(1 + \|\mu\|) h_{\alpha\overline{\beta}},$$

where $K^\theta_{\alpha\overline{\beta}} = K^\theta(T_\alpha, T_{\overline{\beta}})$. By a result in Chapter 1 of this book

$$K^\theta_{\alpha\overline{\beta}} = K_{\alpha\overline{\beta}} - \tfrac{1}{2} h_{\alpha\overline{\beta}},$$

where $K_{\alpha\overline{\beta}}$ is the (pseudo-Hermitian) Ricci tensor (of $(U(M), \theta)$).

Corollary 5.4. *$U(\mathbf{R}^n)$ is globally pseudo-Einsteinian.*

Similarly, if $M = M^n(1)$ then (5.73)–(5.75) may be written as

$$\tilde{K}\left(\frac{\delta}{\delta x^j}, \frac{\delta}{\delta x^k}\right) = \frac{2n-3}{2} g_{jk} - \frac{n-2}{2} y_j y_k,$$
$$\tilde{K}\left(\frac{\partial}{\partial y^j}, \frac{\partial}{\partial y^k}\right) = \tfrac{1}{2}(g_{jk} - y_j y_k),$$
$$\tilde{K}\left(\frac{\partial}{\partial y^j}, \frac{\delta}{\delta x^k}\right) = \tfrac{1}{2}(y^i g_{jk} - \delta^i_j)\frac{\partial}{\partial x^i}(\log\sqrt{\Delta})$$

on $U(M^n(1))$, which together with (5.78) furnish

$$K^\theta_{\alpha\overline{\beta}} = 2(n + \|\mu\|)h_{\alpha\overline{\beta}}.$$

Thus

$$\rho = 2n(n + \|\mu\|) + \frac{n}{2} > 0. \qquad \square$$

Let $c_1(\mathcal{H}) \in H^2(U(M^n(1)), \mathbf{R})$ be the first Chern class of $\mathcal{H}$. Since $U(M^n(1))$ is pseudo-Einsteinian, we may conclude that $c_1(\mathcal{H}) = 0$. $\square$

5.8 Pseudo-Hermitian holonomy

Let $(M, T_{1,0}(M))$ be a strictly pseudoconvex CR manifold, of CR dimension n. Let θ be a contact form on M such that the Levi form L_θ is positive definite. Let T be the characteristic direction of (M, θ). Let $\mathrm{GL}(2n+1, \mathbf{R}) \to \mathcal{L}(M) \longrightarrow M$ be the principal bundle of all linear frames tangent to M. For each $x \in M$, let $B(\theta)_x$ consist of all $\mathbf{R}$-linear isomorphisms $u : \mathbf{R}^{2n+1} \to T_x(M)$ such that

$$u(e_0) = T_x, \quad u(e_\alpha) \in H(M)_x, \quad u(e_{\alpha+n}) = J_x\, u(e_\alpha),$$
$$g_{\theta,x}(u(e_\alpha), u(e_\beta)) = \delta_{\alpha\beta}, \quad g_{\theta,x}(u(e_\alpha), u(e_{\beta+n})) = 0,$$

where J is the complex structure in $H(M)$ and g_θ is the Webster metric of (M, θ). Also $\{e_0, e_\alpha, e_{\alpha+n}\} \subset \mathbf{R}^{2n+1}$ is the canonical linear basis.

Proposition 5.15. *$B(\theta) \to M$ is a* $\mathrm{U}(n) \times 1$*-structure on M, i.e., a principal* $\mathrm{U}(n) \times 1$*-subbundle of $\mathcal{L}(M)$.*

On a strictly pseudoconvex CR manifold, there are two natural families of holonomy groups one may consider: the holonomy of the Levi-Civita connection of (M, g) and the holonomy of the Tanaka–Webster connection. The Tanaka–Webster connection ∇ of (M, θ) gives rise to a connection Γ in $B(\theta)$. Let $\Phi^0(u)$ be the restricted holonomy group of Γ, with reference point $u \in B(\theta)$.

Definition 5.25. We call $\Phi^0(u)$ the *pseudo-Hermitian holonomy group* of (M, θ) at u. $\square$

A systematic study of the (pseudo-Hermitian) holonomy of a CR manifold is still missing in the present-day mathematical literature. In this section we establish a pseudo-Hermitian analogue of a result by H. Iwamoto [218].

Theorem 5.16. (D.E. Blair et al. [68])
Let M be a real $(2n+1)$-dimensional Sasakian manifold, with the structure tensors (φ, ξ, η, g). The pseudo-Hermitian holonomy groups of (M, η) are contained in $\mathrm{SU}(n) \times 1$ if and only if the Tanaka–Webster connection of (M, η) is Ricci flat $(Ric = 0)$.

Proof. M is thought of as a strictly pseudoconvex CR manifold carrying a contact form with vanishing Webster torsion $(\tau = 0)$ and g is its Webster metric. For any $u \in B(\theta)$, $\Phi^0(u) \subset \mathrm{U}(n) \times 1$. Let Ω be the curvature 2-form of Γ. By Lemma 1 in [241], Vol. II, p. 151, given an ideal $\mathbf{h}$ of $L(\mathrm{U}(n) \times 1)$, the Lie algebra of $\mathrm{U}(n) \times 1$, one has the following:

Proposition 5.16. $L(\Phi^0(u)) \subset \mathbf{h}$ *if and only if* Ω *is* $\mathbf{h}$*-valued.*

Throughout, $L(G)$ is the Lie algebra of the Lie group G. Let $E^i_j \in \mathbf{gl}(2n+1, \mathbf{R})$ be the matrix with 1 in the jth row and ith column and 0 at all other entries. Then $\Omega = \Omega^i_j \otimes E^j_i$. A basis of $L(\mathrm{U}(n) \times 1)$ is

$$\{E^{\beta+1}_{\alpha+1} - E^{\alpha+1}_{\beta+1} + E^{\beta+n+1}_{\alpha+n+1} - E^{\alpha+n+1}_{\beta+n+1},\ E^{\beta+1}_{\alpha+n+1} - E^{\alpha+n+1}_{\beta+1} + E^{\alpha+1}_{\beta+n+1} - E^{\beta+n+1}_{\alpha+1}\};$$

hence

$$\Omega^i_1 = 0, \quad \Omega^1_j = 0,$$
$$\Omega^{\alpha+1}_{\beta+1} = \Omega^{\alpha+n+1}_{\beta+n+1} = \Phi^\alpha_\beta - \Phi^\beta_\alpha,$$
$$\Omega^{\alpha+n+1}_{\beta+1} = -\Omega^{\alpha+1}_{\beta+n+1} = \Psi^\alpha_\beta + \Psi^\beta_\alpha,$$

for some scalar 2-forms Φ^α_β, Ψ^α_β on $B(\theta)$. Since $\mathrm{SU}(n) = \mathrm{O}(n) \cap \mathrm{SL}(n, \mathbf{C})$ it follows that Ω is $L(\mathrm{SU}(n) \times 1)$-valued if and only if $\Psi^\alpha_\alpha = 0$. On the other hand, using the identity

$$2u\left(\Omega(X^\Gamma, Y^\Gamma)_u\, u^{-1}(Z_x)\right) = (R(X, Y)Z)_x, \quad u \in B(\theta)_x, \tag{5.79}$$

we may compute the forms Ψ^α_α in terms of $R_A{}^B{}_{CD}$. Here X, Y, Z are vector fields on M and X^Γ is the Γ-horizontal lift of X. Let $x \in M$ and let $\{X_\alpha, JX_\alpha, T\}$ be a cross-section in $B(\theta)$, defined on some open neighborhood U of x. Let us set $\xi_\alpha = \frac{1}{\sqrt{2}}(X_\alpha - iJX_\alpha)$ (hence $g_{\alpha\overline{\beta}} = \delta_{\alpha\beta}$). Let $u = (x, \{X_{\alpha,x}, \varphi_x X_{\alpha,x}, T_x\})$ and note that $u^{-1}(\xi_{\gamma,x}) = \frac{1}{\sqrt{2}}(e_\gamma - i\, e_{\gamma+n})$. Then (5.79) leads to

$$\left(R(X, Y)\xi_\gamma\right)_x = 2\{\Phi^\alpha_\gamma - \Phi^\gamma_\alpha + i(\Psi^\alpha_\gamma + \Psi^\gamma_\alpha)\}(X^\Gamma, Y^\Gamma)_u \xi_{\alpha,x}\ ,$$

because of $E^i_j e_k = \delta^i_k e_j$. Take the inner product with $\xi_{\overline{\alpha}}$ and contract α and γ in the resulting identity. We obtain

$$4i\,\Psi_\alpha^\alpha(X^\Gamma, X^\Gamma)_u = \sum_{\alpha=1}^{n} g(R(X,Y)\xi_\alpha, \xi_{\overline{\alpha}})_x. \tag{5.80}$$

The curvature form Ω is horizontal; hence $L(\mathrm{U}(n) \times 1)^* \rfloor \psi_\alpha^\alpha = 0$, where A^* is the fundamental vertical vector field associated with the left-invariant vector field A. Also

$$4i\,\Psi_\alpha^\alpha(\xi_\lambda^\Gamma, \xi_\mu^\Gamma)_u = \sum_\alpha R_\alpha{}^\sigma{}_{\lambda\mu} g_{\sigma\overline{\alpha}} = R_\alpha{}^\alpha{}_{\lambda\mu} = 0.$$

Similarly

$$4i\,\Psi_\alpha^\alpha(T^\Gamma, \xi_\lambda^\Gamma) = R_\alpha{}^\alpha{}_{0\lambda} = \sum_\alpha S_{\alpha\overline{\alpha}}^{\overline{\lambda}} = 0.$$

Finally (again by (5.80))

$$R_{\lambda\overline{\mu}}(x) = 4i\,\Psi_\alpha^\alpha(\xi_\lambda^\Gamma, \xi_{\overline{\mu}}^\Gamma)_u. \tag{5.81}$$

Since Ψ_α^α is a real form, (5.81) shows that $\Psi_\alpha^\alpha = 0$ if and only if $R_{\lambda\overline{\mu}} = 0$. Yet when $\tau = 0$ the only nonzero components of Ric are $R_{\lambda\overline{\mu}}$ (cf. also Lemma 5.10). □

Note that the hypothesis $\tau = 0$ was not fully used in the proof of Theorem 5.16 (only $S = 0$ was actually needed). Therefore, we have obtained the following result:

Theorem 5.17. *Let M be a strictly pseudoconvex CR manifold, of CR dimension n, and θ a contact form with parallel Webster torsion ($\nabla\tau = 0$). Then the Tanaka–Webster connection ∇ of (M,θ) has pseudo-Hermitian holonomy contained in* $\mathrm{SU}(n) \times 1$ *if and only if the pseudo-Hermitian Ricci tensor of (M,θ) vanishes ($R_{\alpha\overline{\beta}} = 0$).*

5.9 Quaternionic Sasakian manifolds

The closest odd-dimensional analogue of Kählerian manifolds seems to be Sasakian manifolds. On the other hand, real $4m$-dimensional Riemannian manifolds whose holonomy group is contained in $\mathrm{Sp}(m)$ (the so-called *hyperkählerian* manifolds) or in $\mathrm{Sp}(m)\mathrm{Sp}(1)$ (the *quaternionic-Kähler* manifolds) are quaternion analogues, and by a well-known result of M. Berger [58], any hyperkählerian manifold is Ricci flat, while any quaternion-Kähler manifold is Einstein, (provided that $m \geq 2$). Cf. also S. Ishihara [216].

It is then a natural question whether a Sasakian counterpart of quaternionic-Kähler manifolds may be devised, with the expectation of producing new examples of pseudo-Einsteinian contact forms.

Evidence on the existence of such a notion may be obtained as follows. Recall (cf., e.g., [60], p. 403) the following proposition:

Proposition 5.17. *A Riemannian manifold (M^{4m}, g) is a quaternionic-Kähler manifold if and only if there is a covering of M^{4m} by open sets U_i and for each i, two* *almost complex structures F and G on U_i such that (a) g is Hermitian with respect to F and G on U_i, (b) $FG = -GF$, (c) the covariant derivatives (with respect to the Levi-Civita connection of (M^{4m}, g)) of F and G are linear combinations of F, G and $H := FG$, and (d) for any $x \in U_i \cap U_j$ the linear space of endomorphisms of $T_x(M^{4m})$ spanned by F, G, and H is the same for both i and j.*

In an attempt to unify the treatment of *quaternionic* submanifolds, and of *totally real* submanifolds of a quaternionic-Kähler manifold (cf. S. Funbashi [161], S. Marchiafava [293], A. Martinez [294], A. Martinez, J.D. Pérez, and F.G. Santos [295], G. Pitis [348], Y. Shibuya [371]) M. Barros, B.-Y. Chen, and F. Urbano introduced (cf. [47]) the notion of *quaternionic CR submanifold* of a quaternionic-Kähler manifold, as follows.

Definition 5.26. Let N be a real submanifold of a quaternionic-Kähler manifold M^{4m}. A C^∞ distribution $H(N)$ on N is a *quaternionic distribution* if for any $x \in N$ and any i such that $x \in U_i \subseteq M^{4m}$ one has $F(H(N)_x) \subseteq H(N)_x$, $G(H(N)_x) \subseteq H(N)_x$ [and then, of course, $H_x(H(N)_x) \subseteq H(N)_x$]. A submanifold N of a quaternionic-Kähler manifold is a *quaternionic CR submanifold* if it is endowed with a quaternionic distribution $H(N)$ such that its orthogonal complement $H(N)^\perp$ in $T(N)$ satisfies $F(H(N)_x^\perp) \subseteq T(N)_x^\perp$, $G(H(N)_x^\perp) \subseteq T(N)_x^\perp$ and $H(H(N)_x^\perp) \subseteq T(N)_x^\perp$, for any $x \in U_i$ and any i. □

Here $T(N)^\perp \to N$ is the normal bundle (of the given immersion of N in M^{4m}). Let us also recall (cf., e.g., [60], p. 398) the following result:

Proposition 5.18. *A Riemannian manifold (M^{4m}, g) is hyperkählerian if and only if there exist on M^{4m} two complex structures F and G compatible with g and such that (a) F and G are parallel, i.e., g is a Kählerian metric for both F and G, and (b) $FG = -GF$.*

Given a quaternionic CR submanifold $(N, H(N))$ of a hyperkählerian manifold (M^{4m}, g, F, G), by a theorem of D.E. Blair and B-Y. Chen [64], the complex structures F and G induce two CR structures on N (provided N is *proper*, i.e., $H(N) \neq 0$ and $H(N)^\perp \neq 0$) such that $H(N)$ is the Levi distribution for both. Taking this situation as a model one may produce the following notion of abstract (i.e., not embedded) *hyper CR manifold.*

Definition 5.27. Let $(M, T_{1,0}(M))$ be a CR manifold of type (n, k) where $n = 2m$ (hence $\dim_{\mathbf{R}} M = 4m + k$) and $k \geq 1$. Let $H(M)$ be its Levi distribution and

$$F : H(M) \to H(M), \quad F(Z + \overline{Z}) = i(Z - \overline{Z}), \quad Z \in T_{1,0}(M),$$

its complex structure. We say that $(M, T_{1,0}(M))$ is a *hyper CR manifold* if it possesses two additional CR structures, say $T_{1,0}(M)'$ and $T_{1,0}(M)''$, with the same Levi distribution $H(M)$, such that the corresponding complex structures $G, H : H(M) \to H(M)$ satisfy

$$\begin{aligned} F^2 = G^2 = H^2 &= -I, \\ FG = -GF &= H, \\ GH = -HG &= F, \\ HF = -FH &= G. \end{aligned} \tag{5.82}$$

□

Definition 5.28. A *quaternionic CR manifold* is by definition a real $(4m + k)$-dimensional manifold M, $k \geq 1$, endowed with a real rank-$4m$ subbundle $H(M) \subset T(M)$ and a real rank-3 subbundle $E \to M$ of $H(M)^* \otimes H(M) \to M$ such that for any $x \in M$ there is an open neighborhood U of x and a local frame $\{F, G, H\}$ of E on U satisfying the identities (5.82). □

A priori, the notions of a hyper CR manifold, or a quaternionic CR manifold, seem not to be direct analogues of the notions of hyperkählerian and quaternionic-Kähler manifolds, since there is no counterpart of the metric structure there. However, in complex analysis one is interested in the metric structure arising from the complex structure, e.g., the Levi form of a given CR manifold, extending (for nondegenerate CR structures) to a semi-Riemannian metric (the Webster metric).

Let M^{4m+1} be a hyper CR manifold such that $(M^{4m+1}, T_{1,0}(M))$ is nondegenerate, and let θ be a fixed contact form.

Definition 5.29. We say that θ is a *hyper contact form* if $\nabla G = 0$ and $\nabla H = 0$, where ∇ is the Tanaka–Webster connection of (M^{4m+1}, θ). □

More generally, let $(M^{4m+1}, H(M), E)$ be a quaternionic CR manifold of the following sort: M^{4m+1} carries a nondegenerate CR structure $T_{1,0}(M)$ whose Levi distribution is $H(M)$ and for any $x \in M^{4m+1}$ there is an open neighborhood U and a local frame of E on U of the form $\{F, G, H\}$, where F is the (restriction to U of the) complex structure in $H(M)$ associated with $T_{1,0}(M)$ and satisfying the identities (5.82). Such a local frame of E at x will be referred to hereinafter as an *F-frame*.

Definition 5.30. A contact form θ on $(M^{4m+1}, T_{1,0}(M))$ is said to be a *quaternionic contact form* if for any $x \in M^{4m+1}$ there is an open neighborhood U and an F-frame $\{F, G, H\}$ of E on U such that

$$\begin{aligned} (d\theta)(FX, FY) + (d\theta)(X, Y) &= 0, \\ (d\theta)(GX, GY) + (d\theta)(X, Y) &= 0, \\ (d\theta)(HX, HY) + (d\theta)(X, Y) &= 0, \end{aligned} \tag{5.83}$$

for any $X, Y \in H(M)$, and moreover,

$$\begin{aligned} \nabla_X F &= 0, \\ \nabla_X G &= p(X)H, \\ \nabla_X H &= -p(X)G, \end{aligned} \tag{5.84}$$

for some 1-form p on U and any $X \in T(M)$, where ∇ is the Tanaka–Webster connection of (M^{4m+1}, F, θ). □

Note that the first row identities in (5.83)–(5.84) are written for uniformity's sake (and are automatically satisfied, one as a consequence of the formal integrability property of $T_{1,0}(M)$, and the other by the very construction of ∇).

Definition 5.31. A quaternionic CR manifold carrying a quaternionic contact form θ is said to be a *quaternionic Sasakian manifold.* □

This is motivated by Theorem 5.18 below, according to which the Webster torsion of θ vanishes ($\tau = 0$), i.e., the underlying Riemannian metric is indeed Sasakian.

Theorem 5.18. (D.E. Blair et al. [68])
Let (M^{4m+1}, θ) be a quaternionic Sasakian manifold. Then $\tau = 0$, i.e., the Webster metric g of (M^{4m+1}, θ) is a Sasakian metric. Moreover, either the Tanaka–Webster connection of (M^{4m+1}, θ) is Ricci flat, or $m = 1$. If this is the case (i.e., $m = 1$) then (M^5, θ) is pseudo-Einsteinian if and only if $4p + \rho\,\theta$ is a closed 1-form on U, for any F-frame of E on U satisfying (5.83)–(5.84).

A few remarks are in order. By Theorem 5.16 any quaternionic Sasakian manifold M^{4m+1} of dimension ≥ 9 has pseudo-Hermitian holonomy contained in $\mathrm{SU}(2m) \times 1$. By a result in this very chapter, the first Chern class of the CR structure of M^{4m+1} must vanish ($c_1(T_{1,0}(M^{4m+1})) = 0$). Finally, let $M^n(1)$ be a real space form of sectional curvature 1. By a result in this chapter the pseudo-Hermitian Ricci tensor of $U(M^n(1))$ is given by $R_{\alpha\overline{\beta}} = [\frac{1}{2} + 2(n + \|\mu\|)]g_{\alpha\overline{\beta}}$, where μ is the mean curvature vector of $U(M^n(1))$ in $T(M^n(1))$. Therefore (by Theorem 5.18) $U(M^{2m+1}(1))$ admits no quaternionic Sasakian structure for $m \geq 2$. □

To prove Theorem 5.18, let (M^{4m+1}, θ) be a quaternionic Sasakian manifold and $\{F, G, H\}$ a (local) F-frame on U satisfying (5.82)–(5.84). Let g be the Webster metric of (M^{4m+1}, θ). Then

$$\begin{cases} g(FX, FY) = g(X, Y), \\ g(GX, GY) = g(X, Y), \\ g(HX, HY) = g(X, Y), \end{cases} \tag{5.85}$$

for any $X, Y \in H(M)$. The first identity is obvious. The second, for instance, follows from

$$\begin{aligned} g(GX, GY) &= (d\theta)(GX, FGY) = (d\theta)(GX, HY) \\ &= -(d\theta)(G^2X, GHY) = (d\theta)(X, FY) = g(X, Y), \end{aligned}$$

by the definition of the Webster metric g. We shall need the following curvature identities

$$[R(X, Y), F] = 0, \tag{5.86}$$

$$[R(X, Y), G] = \alpha(X, Y)H, \tag{5.87}$$

$$[R(X, Y), H] = -\alpha(X, Y)G, \tag{5.88}$$

for any $X, Y \in T(M)$, where $\alpha := 2\,dp$. The first identity is a consequence of $\nabla F = 0$. The second, for instance, follows from

$$\begin{aligned}[R(X,Y),G]Z &= R(X,Y)GZ - G\,R(X,Y)Z \\ &= \nabla_X(\nabla_Y G)Z - \nabla_Y(\nabla_X G)Z - (\nabla_{[X,Y]}G)Z \\ &\quad + (\nabla_X G)\nabla_Y Z - (\nabla_Y G)\nabla_X Z \\ &= 2(dp)(X,Y)HZ + p(Y)(\nabla_X H)Z - p(X)(\nabla_Y H)Z \\ &= \alpha(X,Y)HZ,\end{aligned}$$

for any $X, Y \in T(M)$ and $Z \in H(M)$. □

Let us take the inner product of (5.88),

$$[R(X,Y),H]Z = -\alpha(X,Y)\,GZ, \quad Z \in H(M),$$

with GZ to obtain

$$\alpha(X,Y)\|Z\|^2 = g(HZ, R(X,Y)GZ) + g(R(X,Y)Z, FZ). \tag{5.89}$$

Consider a local orthonormal frame of $H(M)$ on U of the form

$$\{X_i : 1 \le i \le 4m\} = \{X_a, FX_a, GX_a, HX_a : 1 \le a \le m\}.$$

Let us set $Z = X_i$ in (5.89) and sum over i to obtain

$$4m\,\alpha(X,Y) = \sum_{i=1}^{4m}\{g(HX_i, R(X,Y)GX_i) + g(R(X,Y)X_i, FX_i)\}. \tag{5.90}$$

Since

$$\{(GX_i, HX_i) : 1 \le i \le 4m\} = \{(\epsilon_i X_i, \epsilon_i FX_i) : 1 \le i \le 4m\},$$

where $\epsilon_i \in \{\pm 1\}$, the equation (5.90) becomes

$$2m\,\alpha(X,Y) = \sum_{i=1}^{4m} g(R(X,Y)X_i, FX_i). \tag{5.91}$$

We shall need the first Bianchi identity,

$$\sum_{XYZ}\{R(X,Y)Z + T_\nabla(T_\nabla(X,Y),Z) + (\nabla_X T_\nabla)(Y,Z)\} = 0,$$

for any $X, Y, Z \in T(M)$. Throughout $\sum_{XYZ}$ denotes the cyclic sum over X, Y, Z. Also, we recall that

$$T_\nabla(X,Y) = 2(d\theta)(X,Y)T,$$

for any $X, Y \in H(M)$. Therefore (by $\nabla T = 0$ and $\nabla\Omega = 0$)

$$\sum_{XYZ} \{R(X,Y)Z - 2\Omega(X,Y)\,\tau Z\} = 0, \tag{5.92}$$

for any $X, Y, Z \in H(M)$. Let us set $Z = FX_i$ in (5.92), and take the inner product with X_i in the resulting identity. Next, sum over i to obtain

$$\begin{aligned} -\,2m\,\alpha(X,Y) + \sum_{i=1}^{4m} \{g(X_i, R(Y, FX_i)X) + g(X_i, R(FX_i, X)Y)\} \\ = 2\Omega(X,Y)\,\mathrm{trace}(\tau F) + 2\Omega(Y, F\tau X) + 2\Omega(F\tau Y, X). \end{aligned} \tag{5.93}$$

Note that $\mathrm{trace}(\tau F) = 0$, because $\tau\, T_{1,0}(M) \subseteq T_{0,1}(M)$, and

$$\Omega(Y, F\tau X) + \Omega(F\tau Y, X) = 0,$$

by the symmetry property of $A(X,Y) = g(\tau X, Y)$, with the corresponding simpler from of (5.93). From the curvature theory developed in Chapter 1 of this book we recall that

$$\begin{aligned} \tilde{R}(X,Y)Z = R(X,Y)Z + (LX \wedge LY)Z + 2\Omega(X,Y)\varphi Z \\ - g(S(X,Y), Z)\xi + \eta(Z)S(X,Y) - 2g((\eta \wedge \mathcal{O})(X,Y), Z)\xi \\ + 2\eta(Z)(\eta \wedge \mathcal{O})(X,Y), \end{aligned} \tag{5.94}$$

for any $X, Y, Z \in T(M)$, where

$$L = \varphi - \tau, \quad \mathcal{O} = \tau^2 - 2\varphi\,\tau - I,$$

and $(X \wedge Y)Z = g(X,Z)Y - g(Y,Z)X$. Here $\tilde{R}$ is the curvature of (M, g). In particular, if $X, Y, Z \in H(M)$,

$$\tilde{R}(X,Y)Z = R(X,Y)Z + (LX \wedge LY)Z + 2\Omega(X,Y)\varphi Z - g(S(X,Y), Z)\xi\,.$$

Let us take the inner product with $W \in H(M)$ to obtain

$$\begin{aligned} \tilde{R}(W,Z,X,Y) \\ = g(R(X,Y)Z, W) + g((LX \wedge LY)Z, W) - 2\Omega(X,Y)\Omega(Z,W), \end{aligned} \tag{5.95}$$

for any $X, Y, Z, W \in H(M)$, where $\tilde{R}(W,Z,X,Y) = g(\tilde{R}(X,Y)Z, W)$ is the Riemann–Christoffel 4-tensor of (M, g). Exploiting the well-known symmetry

$$\tilde{R}(W,Z,X,Y) = \tilde{R}(X,Y,W,Z),$$

the identity (5.95) furnishes

$$\begin{aligned} g(R(X,Y)Z, W) \\ = g(R(W,Z)Y, X) + g((LW \wedge LZ)Y, X) - g((LX \wedge LY)Z, W). \end{aligned} \tag{5.96}$$

Let us replace the vectors (X, Y, Z, W) by the vectors (FX_i, X, Y, X_i) in the formula (5.96). We obtain

$$\sum_{i=1}^{4m} g(R(FX_i, X)Y, X_i) = \sum_{i=1}^{4m} g(R(X_i, Y)X, FX_i) + g(X, LY)\,\mathrm{trace}(FL) \\ - g(X, LFLY) + g(LX, Y)\,\mathrm{trace}(LF) - g(LFLX, Y).$$

Substitution into (5.93) gives

$$2m\,\alpha(X, Y) = \sum_{i=1}^{4m} \{g(R(Y, FX_i)X, X_i) - g(R(Y, X_i)X, FX_i)\} \\ + g(X, LY)\,\mathrm{trace}(FL) - g(X, LFLY) \\ + g(LX, Y)\,\mathrm{trace}(LF) - g(LFLX, Y),$$

and by observing that

$$\{(FX_i, X_i) : 1 \le i \le 4m\} = \{(\lambda_i X_i, \mu_i FX_i) : 1 \le i \le 4m\}, \\ \lambda_i,\ \mu_i \in \{\pm 1\}, \quad \lambda_i \mu_i = -1,$$

we obtain

$$2m\,\alpha(X, Y) = 2\sum_i g(R(X_i, Y)X, FX_i) + \text{terms},$$

or (replacing X by FX)

$$2\,\mathrm{Ric}(X, Y) = 2m\,\alpha(FX, Y) \\ + g(FX, LFY) - g(FX, LY)\,\mathrm{trace}(FL) \\ + g(LFLFX, Y) - g(LFX, Y)\,\mathrm{trace}(LF), \tag{5.97}$$

for any $X, Y \in H(M)$. Next, let us take the inner product of (5.86),

$$[R(X, Y), F]Z = 0, \quad X, Y, Z \in H(M),$$

with GZ, so that

$$g(R(X, Y)FZ, GZ) + g(R(X, Y)Z, HZ) = 0.$$

Let us set $Z = X_i$ and sum over i to obtain

$$\sum_{i=1}^{4m} \left\{g(R(X, Y)FX_i, GX_i) + g(R(X, Y)X_i, HX_i)\right\} = 0$$

and observe that

$$\{(FX_i, GX_i) : 1 \le i \le 4m\} = \{(\epsilon_i X_i, \epsilon_i HX_i) : 1 \le i \le 4m\}, \quad \epsilon_i \in \{\pm 1\}.$$

Therefore

$$\sum_{i=1}^{4m} g(R(X, Y)X_i, HX_i) = 0. \tag{5.98}$$

Let us set $Z = HX_i$ in (5.92) and take the inner product with X_i in the resulting identity. Then (by (5.98))

$$\sum_{i=1}^{4m} \{g(R(Y, HX_i)X, X_i) + g(R(HX_i, X)Y, X_i)\}$$
$$= 2\Omega(X, Y)\,\text{trace}(\tau H) + 2\Omega(Y, H\tau X) + 2\Omega(H\tau Y, X). \tag{5.99}$$

Now replace (X, Y, Z, W) by (HX_i, X, Y, X_i) in (5.96) to obtain

$$g(R(HX_i, X)Y, X_i) = g(R(X_i, Y)X, HX_i) + g((LX_i \wedge LY)X, HX_i) - g((LHX_i \wedge LX)Y, X_i)$$

and substitute into (5.99). Also observe that

$$\{(HX_i, X_i) : 1 \le i \le 4m\} = \{(\lambda_i X_i, \mu_i HX_i) : 1 \le i \le 4m\},$$
$$\lambda_i, \mu_i \in \{\pm 1\}, \quad \lambda_i \mu_i = -1;$$

hence

$$2\sum_{i=1}^{4m} g(R(X_i, Y)X, HX_i) - g(LHLY, X) + g(LY, X)\,\text{trace}(HL)$$
$$- g(LHLX, Y) + g(LX, Y)\,\text{trace}(LH)$$
$$= 2\Omega(X, Y)\,\text{trace}(\tau H) + 2\Omega(Y, H\tau X) + 2\Omega(H\tau Y, X). \tag{5.100}$$

The inner product of (5.88),

$$R(X, Y)HZ = HR(X, Y)Z - \alpha(X, Y)GZ,$$

with X_i gives

$$g(R(X, Y)HZ, X_i) = g(HR(X, Y)Z, X_i) - \alpha(X, Y)g(GZ, X_i),$$

or, replacing (X, Z) by (X_i, X),

$$g(R(X_i, Y)HX, X_i) = g(H\,R(X_i, Y)X, X_i) - \alpha(X_i, Y)g(GX, X_i),$$

and taking the sum over i we have

$$\sum_{i=1}^{4m} g(R(X_i, Y)X, HX_i) = -\alpha(GX, Y) - \text{Ric}(HX, Y),$$

i.e., (by (5.100) and replacing X by HX)

$$\begin{aligned} 2\,\mathrm{Ric}(X, Y) &= 2\,\alpha(FX, Y) + g(LHLY, HX) \\ &\quad - g(LY, HX)\,\mathrm{trace}(HL) + g(LHLHX, Y) - g(LHX, Y)\,\mathrm{trace}(LH) \\ &\quad + 2\Omega(HX, Y)\,\mathrm{trace}(\tau H) + 2\Omega(Y, H\tau HX) + 2\Omega(H\tau Y, HX), \quad (5.101) \end{aligned}$$

for any $X, Y \in H(M)$. To compute the pseudo-Hermitian Ricci curvature, set first $X = \xi_\lambda$ and $Y = \xi_{\overline{\mu}}$ in (5.97). We obtain

$$R_{\lambda\overline{\mu}} = i\, m\, \alpha_{\lambda\overline{\mu}} + \tfrac{1}{2}\big(A_\lambda^{\overline{\beta}}\, A_{\overline{\beta}\overline{\mu}} - A_{\alpha\lambda}\, A_{\overline{\mu}}^{\alpha}\big).$$

The torsion terms vanish (by $A_{\alpha\beta} = A_{\beta\alpha}$); hence

$$R_{\lambda\overline{\mu}} = i\, m\, \alpha_{\lambda\overline{\mu}}. \qquad (5.102)$$

Let us set also $X = \xi_\lambda$ and $Y = \xi_\mu$ in (5.97) and note that

$$\begin{aligned} g(F\xi_\lambda, LFL\xi_\mu) &= -2i\, A_{\lambda\mu} = g(LFLF\xi_\lambda, \xi_\mu), \\ g(F\xi_\lambda, L\xi_\mu) &= i\, A_{\lambda\mu} = g(LF\xi_\lambda, \xi_\mu), \\ \mathrm{trace}(FL) &= -4m = \mathrm{trace}(LF), \end{aligned}$$

i.e.,

$$R_{\lambda\mu} = im\,\alpha_{\lambda\mu} + 2i(2m-1)\, A_{\lambda\mu}. \qquad (5.103)$$

Taking traces in (5.94), we obtain the following:

Lemma 5.10. *Let M^{2n+1} be a nondegenerate CR manifold, on which a contact form θ has been fixed. Let g be the Webster metric of (M^{2n+1}, θ). Then*

$$\begin{aligned} &R_{\alpha\overline{\beta}} = 2(g_{\alpha\overline{\beta}} - R^g_{\alpha\overline{\beta}}), \quad R_{\alpha\beta} = i(n-1)\, A_{\alpha\beta}, \\ &R_{0\beta} = S^{\overline{\alpha}}_{\overline{\alpha}\beta}, \quad R_{\alpha 0} = R_{00} = 0, \end{aligned}$$

for any local frame $\{\xi_\alpha\}$ in $T_{1,0}(M)$. Here $R^g_{\alpha\overline{\beta}} = \widetilde{\mathrm{Ric}}(\xi_\alpha, \xi_{\overline{\beta}})$ and $\widetilde{\mathrm{Ric}}$ is the Ricci tensor of (M^{2n+1}, g). Also $S^A_{BC}\xi_A := S(\xi_B, \xi_C)$ with $A, B, \ldots \in \{0, 1, \ldots, n, \overline{1}, \ldots, \overline{n}\}$ and $T_0 = T$.

Combining (5.103) and Lemma 5.10 (with $n = 2m$) gives $m\,\alpha_{\lambda\mu} + (2m-1)\, A_{\lambda\mu} = 0$. Yet α is skew, while the Webster torsion is symmetric, hence $\alpha_{\lambda\mu} = 0$ and $A_{\lambda\mu} = 0$. Thus g is a Sasakian metric. As another consequence of $\tau = 0$, (5.101) becomes

$$2\,\mathrm{Ric}(X, Y) = 2\,\alpha(FX, Y) + g(FHFY, HX) + g(FHFHX, Y)$$

(since $\mathrm{trace}(HL) = \mathrm{trace}(HF) = \mathrm{trace}(G) = 0$) and then (by (5.85))

$$\mathrm{Ric}(X, Y) = \alpha(FX, Y),$$

for any $X, Y \in H(M)$. Consequently $R_{\lambda\overline{\mu}} = i\,\alpha_{\lambda\overline{\mu}}$ and by (5.102) we get $(m-1)\, R_{\lambda\overline{\mu}} = 0$; hence either $m = 1$ or the pseudo-Hermitian Ricci curvature vanishes. Therefore, if $m \geq 2$ then (by Lemma 5.10) $\mathrm{Ric} = 0$. □

Let us look now at the case $m = 1$. As a consequence of (5.89) we may write

$$\alpha(FX, X)\|Z\|^2 = g(R(FX, X)Z, FZ) + g(R(FX, X)GZ, HZ),$$
$$\alpha(HX, GX)\|Z\|^2 = g(R(HX, GX)Z, FZ) + g(R(HX, GX)GZ, HZ).$$

Summing up the last two identities we get

$$\begin{aligned}\alpha(FX, X)\|Z\|^2 + \alpha(HX, GX)\|Z\|^2 &= g(R(FX, X)Z, FZ) + g(R(FX, X)GZ, HZ)\\ &\quad + g(R(HX, GX)Z, FZ) + g(R(HX, GX)GZ, HZ).\end{aligned}$$

Note that the right-hand member of this last identity is symmetric in X, Z. Hence

$$\{\alpha(FX, X) + \alpha(HX, GX)\}\|Z\|^2 = \{\alpha(FZ, Z) + \alpha(HZ, GZ)\}\|X\|^2;$$

or

$$\{\mathrm{Ric}(X, X) + \mathrm{Ric}(GX, GX)\}\|Z\|^2 = \{\mathrm{Ric}(Z, Z) + \mathrm{Ric}(GZ, GZ)\}\|X\|^2 .$$

Let us set $Z = X_i$ and sum over i. We have

$$4\{\mathrm{Ric}(X, X) + \mathrm{Ric}(GX, GX)\} = \sum_{i=1}^{4}\{\mathrm{Ric}(X_i, X_i) + \mathrm{Ric}(GX_i, GX_i)\}\|X\|^2,$$

or, again due to the particular form of our frame,

$$2\{\mathrm{Ric}(X, X) + \mathrm{Ric}(GX, GX)\} = \sum_{i=1}^{4}\mathrm{Ric}(X_i, X_i)\|X\|^2 .$$

Finally, $\mathrm{trace}(\mathrm{Ric}) = 2\rho$; hence (by $T \rfloor \mathrm{Ric} = 0$; cf. Lemma 5.10 with $\tau = 0$)

$$\mathrm{Ric}(X, Y) + \mathrm{Ric}(GX, GY) = \rho\, g(X, Y), \tag{5.104}$$

for any $X, Y \in H(M)$. Note that use was made of the symmetry of Ric on $H(M) \otimes H(M)$, a consequence of Lemma 5.10 as well. It remains to be shown that θ is pseudo-Einsteinian if and only if

$$d(4p + \rho\theta) = 0. \tag{5.105}$$

Due to $FG = -GF$, $G^2 = -I$, and $g(GX, GY) = g(X, Y)$ for any $X, Y \in H(M)$, one has

$$G\xi_\alpha = G_\alpha^{\overline{\beta}}\xi_{\overline{\beta}}, \quad G_\alpha^{\overline{\beta}}G_{\overline{\beta}}^{\lambda} = -\delta_\alpha^\lambda, \quad g_{\alpha\overline{\beta}} = G_\alpha^{\overline{\mu}}G_{\overline{\beta}}^{\lambda}g_{\lambda\overline{\mu}},$$

for some smooth functions $G_\alpha^{\overline{\beta}} : U \to U$, where $G_{\overline{\alpha}}^{\beta} = \overline{G_\alpha^{\overline{\beta}}}$. Let us set $\alpha_{AB} = \alpha(\xi_A, \xi_B)$ and note that $\alpha_{\lambda\mu} = 0$, $\alpha_{\overline{\lambda}\overline{\mu}} = 0$. The identity (5.104) may be written

$$R_{\alpha\overline{\beta}} + i\, G^{\overline{\mu}}_{\alpha} G^{\lambda}_{\overline{\beta}} \alpha_{\lambda\overline{\mu}} = \rho\, g_{\alpha\overline{\beta}}.$$

Consequently θ is pseudo-Einsteinian, i.e., $R_{\alpha\overline{\beta}} = (\rho/2) g_{\alpha\overline{\beta}}$ if and only if

$$i\, \alpha_{\lambda\overline{\beta}} = (\rho/2)\, g_{\alpha\overline{\beta}}. \tag{5.106}$$

Since $d(\rho\theta) = (d\rho) \wedge \theta + \rho\, d\theta$ and θ vanishes on $H(M)$, (5.106) may be written

$$4(dp)(\xi_\alpha, \xi_{\overline{\beta}}) + d(\rho\theta)(\xi_\alpha, \xi_{\overline{\beta}}) = 0,$$

i.e., $d(4p + \rho\theta) = 0$ on $H(M) \otimes H(M)$. Finally, by Lemma 5.3 (established earlier in this chapter) if a closed 2-form Ξ vanishes on $H(M) \otimes H(M)$ then $\Xi = 0$; hence (5.105) holds. □

5.10 Homogeneous pseudo-Einsteinian manifolds

A complete classification of all homogeneous pseudo-Einsteinian manifolds is obtained in [319].

Definition 5.32. A pseudo-Hermitian manifold (M, θ) is said to be G-*homogeneous* if there is a closed subgroup $G \subseteq \mathrm{Psh}(M, \theta)$ such that G is transitive on M. □

Of course, if we fix a point $x_0 \in M$ then we may identify M and G/H, where H is the isotropy group at x_0. The results in [319] are mainly concerned with strictly pseudoconvex CR manifolds M on which a contact form θ has been fixed in such a way that the Levi form L_θ is positive definite and the corresponding characteristic direction T is regular (in the sense of R. Palais [336]). If this is the case then M is principal bundle over the orbit space B of all trajectories of T, with structure group S^1 or **R**. Indeed G-homogeneous pseudo-Hermitian manifolds (M, θ) with L_θ positive definite belong to this class (cf., e.g., Proposition 3.4 in [319], p. 226).

In the terminology of [319] two pseudo-Hermitian manifolds (M, θ) and (M', θ') are *contact homothetic* if there is a C^∞ diffeomorphism such that $f^*\theta' = s\,\theta$ and $(df)T = r\,T'$, for some $r, s \in (0, +\infty)$. Here T and T' are the characteristic directions of $d\theta$ and $d\theta'$, respectively. See [319], p. 230, for the definition of the principal S^1-bundle $\pi^{(1)} : B^{(1)} \to B$ (appearing in Theorem 5.19 below). Also, given a positive integer k, let $\pi^{(k)} : B^{(k)} \to B$ be the kth tensor power of $\pi^{(1)} : B^{(1)} \to B$. Cf. [319], p. 232. $B^{(k)}$ admits a canonical pseudo-Hermitian structure related (via $\pi^{(k)}$) to $(B, \sqrt{k}g)$, where g is the original Kähler–Einstein metric on B. If additionally (M, θ) is pseudo-Einsteinian then B is a G-invariant homogeneous Einstein–Kähler manifold, thus prompting the following result:

Theorem 5.19. (E. Musso [319])
Let M be a strictly pseudoconvex CR manifold and let θ be a contact form on M with L_θ positive definite. Assume that (M, θ) is G-homogeneous and pseudo-Einsteinian. (1) If (M, θ) has positive pseudo-Hermitian scalar curvature ($\rho > 0$) then M is a

principal S^1-bundle over B and the integral Chern class $c_1(M)$ is an integer multiple of $c_1(B^{(1)})$, i.e.,

$$c_1(M) = k\, c_1(B^{(1)})$$

for some $k \in \mathbf{Z}$. Moreover, (M, θ) is contact homothetic to the canonical pseudo-Hermitian manifold $B^{(k)}$. (2) If (M, θ) has negative pseudo-Hermitian scalar curvature ($\rho < 0$) then the orbit space B is an affinely homogeneous Siegel domain of the second kind with its Bergman metric and (M, θ) is contact homothetic to either (i) $B \times S^1$ with the contact form

$$c\, d\gamma + i(\partial - \overline{\partial}) \log K(z, z),$$

where $c > 0$ is a constant, γ is a fiber coordinate, and $K(z, \zeta)$ is the Bergman kernel of B, or to (ii) $B \times \mathbf{R}$ with the contact form

$$c\, dt + i(\partial - \overline{\partial}) \log K(z, z).$$

(3) If $\rho = 0$ then (M, θ) is contact homothetic to either (i) $\mathbf{C}^n \times S^1$ with the contact form $c\, d\gamma + 2y_i dx^i$, or to (ii) $\mathbf{C}^n \times \mathbf{R}$ with the contact form $c\, dt + 2y_i dx^i$.

A deeper analysis of the 3-dimensional case is due to D. Perrone [343]–[344].

Theorem 5.20. (D. Perrone [344])
Let (M, θ) be a simply connected 3-dimensional pseudo-Hermitian manifold such that L_θ is positive definite. Assume that there is a Lie group $G \subseteq \mathrm{Psh}(M, \theta)$ such that (M, θ) is G-homogeneous. Then M is a Lie group and both θ and g_θ are invariant. Moreover, the following classification holds:

(1) If M is unimodular then it is one of the following Lie groups:
  - *(i) the Heisenberg group when $\rho = \|\tau\| = 0$,*
  - *(ii) the 3-sphere group $\mathrm{SU}(2)$ when $4\sqrt{2}\rho > \|\tau\|$,*
  - *(iii) the group $\tilde{E}(2)$, i.e., the universal covering of the group of rigid motions of Euclidean 2-space, when $4\sqrt{2}\rho = \|\tau\| > 0$,*
  - *(iv) the group $\widetilde{\mathrm{SL}}(2, \mathbf{R})$ when $-\|\tau\| \neq 4\sqrt{2}\rho < \|\tau\|$,*
  - *(v) the group $E(1,1)$ of rigid motions of the Minkowski 2-space when $4\sqrt{2}\rho = -\|\tau\| < 0$.*

(2) If M is nonunimodular then its Lie algebra is given by

$$[E_1, E_2] = \alpha E_2 + 2T, \quad [E_1, T] = \gamma E_2, \quad [E_2, T] = 0,$$

where $\alpha \neq 0$, with E_1, $E_2 = J_b E_1 \in H(M)$ and $4\sqrt{2}\rho < \|\tau\|$. Moreover, if $\gamma = 0$ then the pseudo-Hermitian torsion vanishes ($\tau = 0$), i.e., M is a Sasakian manifold, and the pseudo-Hermitian scalar curvature is given by $\rho = -\alpha^2/4$.

Since it may be shown that

$$4\rho - \frac{\|\tau\|}{\sqrt{2}} = \frac{\rho_\theta}{2} + \left(1 - \frac{\|\tau\|}{2\sqrt{2}}\right)^2,$$

where ρ_θ is the (Riemannian) scalar curvature of g_θ, one gets the following corollary: the 3-sphere group SU(2) is the only simply connected 3-dimensional manifold that admits a structure of a homogeneous contact Riemannian manifold (respectively Sasakian manifold) with scalar curvature $\rho_\theta > -2(1 - \|\tau\|/(2\sqrt{2}))^2$ (respectively $\rho_\theta > -2$). As another consequence, the Heisenberg group and $\widetilde{\mathrm{SL}}(2, \mathbf{R})$ are the only simply connected 3-manifolds that admit a unimodular homogeneous contact Riemannian manifold structure of vanishing pseudo-Hermitian scalar curvature ($\rho = 0$).

The theory of homogeneous pseudo-Hermitian manifolds enters, of course, the more general theory of homogeneous CR manifolds, as mentioned in the preface of this book.

Definition 5.33. A CR manifold M is said to be *homogeneous* if there is a Lie group G acting transitively[5] on M as a group of CR automorphisms. □

A more pedantic terminology is to refer to a pair (M, G) consisting of a CR manifold M and a real Lie G group acting on M as above, as a homogeneous CR manifold. Any real analytic nondegenerate CR manifold is homogeneous, because in this case (by a result of N. Tanaka [397]) $\mathrm{Aut}_{CR}(M)$, the group of all CR automorphisms of M, is a real Lie group. The only CR structure (of hypersurface type) on S^{2n+1} $(n \geq 2)$ that admits a transitive action of a Lie group of CR automorphisms is the standard CR structure inherited from $\mathbf{C}^{n+1}$ (cf. R. Lehmann and D. Feldmueller [277]).

Much work has been done on the classification of compact homogeneous CR manifolds. The history of the subject starts with the work of Y. Morimoto and T. Nagano [316], under the assumptions that (1) M is strictly pseudoconvex and the boundary of a domain in a Stein manifold, and (2) the fundamental group $\pi_1(M)$ is finite. An important consequence (cf. H. Samelson [360]) of the assumption (2) is that a maximal compact subgroup $K \subset G$ acts transitively on M. The result by Y. Morimoto and T. Nagano (cf. op. cit.) is that such M is either the sphere $S^{2n+1} \subset \mathbf{C}^{n+1}$ or a finite cover of a sphere bundle in the tangent bundle of a symmetric space of rank one.[6] H. Rossi [354], replaced Y. Morimoto and T. Nagano's assumption (1) by the requirement that (1′) M be strictly pseudoconvex and its real dimension be ≥ 5. This is known to imply (cf., e.g., A. Andreotti and Y.T. Siu [21]) that M is the boundary of a domain in a Stein space Z. Then K acts on Z and has precisely one orbit E. Moreover, either Z is a manifold and then one may apply the Y. Morimoto and T. Nagano result above, or the singular set of Z (a K-stable analytic set) is precisely E, i.e., $E = \{x_0\}$ is a fixed point. H. Rossi (cf. op. cit.) desingularizes, $\pi : \tilde{Z} \to Z$, and shows (by using the K-action) that $Q := \pi^{-1}(x_0)$ is a homogeneous rational manifold and that $\tilde{Z}$ is a K-invariant tube in the normal bundle of Q in $\tilde{Z}$. Therefore M is the boundary of a K-invariant domain in an affine cone over Q. The story of the strictly pseudoconvex case ends up with the work of D. Burns and S. Shnider [79], concerning the 3-dimensional

[5] Certain authors (cf., e.g., H. Azad et al. [26]) assume also that the action of G on M is almost effective.

[6] Such a symmetric space is either a sphere, or a projective space over **R**, **C**, or **H**, or the projective space over the Cayley numbers. Moreover, its tangent bundle possesses a natural K-invariant complex structure as a Stein manifold.

case. Their result is that any 3-dimensional strictly pseudoconvex compact homogeneous CR hypersurface M has a finite $\pi_1(M)$. The case that M is a nondegenerate CR manifold (compact, of hypersurface type) was examined by H. Azad et al. [26]. Any such M is (via the so-called *basic* **g**-*anticanonical fibration*; cf. [26], pp. 130–133) a principal S^1-bundle over a homogeneous rational manifold Q, or a finite covering of a CR hypersurface that is a G-orbit under a linear representation in $\mathbf{C}P^m$. In the S^1-bundle case M is the boundary of a tube in a line bundle over Q with nondegenerate invariant Chern form. In the projective algebraic case the following may be proved. Let G be a linear Lie group in $\mathrm{PSL}_{m+1}(\mathbf{C})$ and let $\hat{G}$ be the smallest complex Lie group containing G. Assume that there is $p \in \mathbf{C}P^m$ such that the orbit $\tilde{M} := G(p)$ is a compact hypersurface in the $\hat{G}$-orbit $\hat{G}(p)$. Then either the semisimple part K^{ss} of K acts transitively on $\tilde{M}$ or $\tilde{M}$ is a CR-product $S^1 \times Q$, where Q is a complex homogeneous rational manifold. In particular, unless $\tilde{M}$ is the Levi flat product $S^1 \times Q$, the fundamental group $\pi_1(\tilde{M})$ is finite. Moreover, if S is the complexification of K^{ss} in $\mathrm{PSL}_{m+1}(\mathbf{C})$ then $\tilde{M}$ is a hypersurface in $\Omega := S(p)$ that is Zariski open in its S-almost homogeneous closure X (and X is a rational algebraic variety). See [26] for the detailed classification, according to whether the **g**-anticanonical fiber is S^1 or finite.

6

Pseudo-Hermitian Immersions

Let M and A be two CR manifolds, of CR dimensions n and $N = n + k$, $k \geq 1$, respectively. A *CR immersion* $f : M \to A$ is an immersion and a CR map. If f is the inclusion then M is a *CR submanifold* of A (a *CR hypersurface* when $k = 1$). For instance, let M^{2n+1} be the intersection between the sphere S^{2n+3} and a transverse complex hypersurface in $\mathbf{C}^{n+2}$. Then M^{2n+1} is a CR hypersurface of S^{2n+3} (in particular M^{2n+1} is strictly pseudoconvex). Let M be a CR submanifold of A. Then M is *rigid* in A if any CR diffeomorphism $F : M \to M'$ onto another CR submanifold M' of A (e.g., F may be the restriction of a biholomorphic mapping) extends to a CR automorphism of A (e.g., if $A = S^{2n+3}$ then F should extend to a fractional linear, or projective, transformation preserving S^{2n+3}).

A theory of CR immersions has been initiated by S. Webster [424]. There it is shown that S^{2n+1} is rigid in S^{2n+3} if $n \geq 2$. Also, if $n \geq 3$ then any CR hypersurface of S^{2n+3} is rigid. The basic idea in [424] is to endow the ambient space S^{2n+3} with the Tanaka–Webster connection (rather than the Levi-Civita connection associated with the canonical Riemannian structure) and obtain CR analogues of the Gauss–Weingarten (respectively Gauss–Ricci–Codazzi) equations (from the theory of isometric immersions between Riemannian manifolds). In the end, these could be used to show that the intrinsic geometry determines the (CR analogue of the) second fundamental form of the given CR immersion.

The main inconvenience of this approach seems to be the nonuniqueness of choice of a canonical connection on the CR submanifold (i.e., the induced and "intrinsic" Tanaka–Webster connection of the submanifold do not coincide, in general). In [120] one compensates for this inadequacy by restricting oneself to a smaller class of CR immersions, as follows. Let $f : M \to A$ be a CR immersion between two strictly pseudoconvex CR manifolds on which contact 1-forms θ and Θ have been fixed. Then

$$f^*\Theta = \lambda\theta$$

for some C^∞ function $\lambda : M \to \mathbf{R}$. If $\lambda \equiv 1$ then f is called *isopseudo-Hermitian* (following the terminology in [219]). An isopseudo-Hermitian immersion $f : M \to A$ is a *pseudo-Hermitian immersion* if $f(M)$ is tangent to the characteristic direction of

(A, Θ). If this is the case then (by a result in [120] and the first section below) f is an isometry (with respect to the Webster metrics of (M, θ) and (A, Θ)). Also, one may use the axiomatic description of the Tanaka–Webster connection to show that the induced and intrinsic connections on M coincide. Moreover, by a result of H. Urakawa (any CR map $f : M \to A$ satisfying $f_* T = \lambda T_A$ for some $\lambda \in C^\infty(M)$ with $T(\lambda) = 0$ is harmonic with respect to the Webster metrics of M and A; cf. Corollary 3.2 in [413], p. 263) any pseudo-Hermitian immersion is actually minimal. Cf. also Theorem 7 in [120].

The scope of Chapter 6 is to present the theory of pseudo-Hermitian immersion together with its application to the problem of the existence of pseudo-Einsteinian pseudo-Hermitian structures (cf. Chapter 5 of this book) on (locally realizable) CR manifolds.

As in [120], our main tool will consist of pseudo-Hermitian analogues of the Gauss and Weingarten equations. In particular, we shall introduce the concept of *normal Tanaka–Webster connection* $\nabla^\perp$ (of a given pseudo-Hermitian immersion between two strictly pseudoconvex CR manifolds). When $\nabla^\perp$ is flat we use the (pseudo-Hermitian analogues of the) Gauss–Ricci–Codazzi equations to relate the pseudo-Hermitian Ricci tensors of the Tanaka–Webster connections of the submanifold and ambient space and prove the following theorem:

Theorem 6.1. (E. Barleta et al. [36])
Let $f : M \to A$ be a pseudo-Hermitian immersion between two strictly pseudoconvex CR manifolds (M, θ) and (A, Θ). If the normal Tanaka–Webster connection is flat (i.e., $R^\perp = 0$) then

$$R_{\alpha\overline{\beta}} = \text{trace}\{Z \mapsto R^A(Z, f_* T_\alpha) f_* T_{\overline{\beta}}\}. \tag{6.1}$$

In particular, if Θ is pseudo-Einsteinian, then θ is pseudo-Einsteinian, too.

As a corollary, we may regard the Lee class $\gamma(M)$ (a cohomology class in the first cohomology group of the given (locally realizable) CR manifold M with coefficients in the sheaf of CR-pluriharmonic functions; cf. Chapter 5) as an obstruction to the existence of pseudo-Hermitian immersions $f : M \to S^{2N+1}$ with a flat normal Tanaka–Webster connection of a strictly pseudoconvex CR manifold M into an odd-dimensional sphere.

The methods we employ are similar to those in B.Y. Chen and H.S. Lue [93] (where holomorphic immersions between Kähler manifolds are dealt with). We exploit the symmetries of the curvature tensor field of the Tanaka–Webster connection (rather than the Riemann–Christoffel tensor field in [93]) and deal with the highly complicated (due to the presence of torsion terms there) of the Bianchi identities (cf., e.g., (6.66)). The key points (leading from (6.78) to (6.1) in Theorem 6.1) are Lemma 6.8 (the $(0, 2)$-tensor field E_a is proportional to the Levi form of the submanifold) and a nontrivial cancellation of torsion terms.

As a byproduct we prove (cf. Theorem 6.10) the nonexistence of pseudo-Hermitian immersions of $H_n(s)$ (cf. Chapter 5) into a Tanaka–Webster flat strictly pseudoconvex CR manifold (e.g., $\mathbf{H}_N$ or $U_{\alpha,\beta}$).

6.1 The theorem of H. Jacobowitz

In this section we refine a result by H. Jacobowitz (any real analytic CR map between real analytic CR manifolds (of hypersurface type) of the same dimension is a local diffeomorphism or a constant map; cf. [219]) to the case in which the dimensions of the source and target CR manifolds are not necessarily equal (cf. Theorem 6.2). The proof parallels closely the proof of the original result in [219]. Lemma 6.2 is however new. Our point of view is that Theorem 6.2 may be regarded as a motivation for the study of CR immersions (submersions) between (abstract) CR manifolds. As a main objective of the present chapter, a study of CR immersions is carried out in the next two sections.

Definition 6.1. Let $\Phi(z, t)$ be an $\mathbf{R}$-valued function (defined on some domain $D \subset \mathbf{H}_n$ with $0 \in D$). Then Φ is said to be of *weight* r if $\Phi(sz, s^2 t) = s^r \Phi(z, t)$, for any $s \in \mathbf{R}$. □

If Φ is real analytic, one may regroup the terms of its Taylor series expansion to obtain a decomposition

$$\Phi = \sum_{\chi=0}^{\infty} \Phi_\chi ,$$

where Φ_χ is of weight χ. Then Φ is of weight greater than or equal to r if and only if $\Phi_\chi = 0$ for $0 \le \chi \le r - 1$.

Let M be an orientable C^∞ nondegenerate CR manifold of CR dimension n. Fix a point $x_0 \in M$.

Lemma 6.1. (H. Jacobowitz [219])
There exist Folland–Stein local coordinates (V, z^α, u) with origin at x_0 and there are functions $k^\alpha(z, u)$, $h^{\alpha\beta}(z, u)$ of weight at least 2 such that

$$\theta = du + \sum_{\alpha=1}^{n} \{(iz^\alpha + k^\alpha) d\overline{z}^\alpha + (-i\overline{z}^\alpha + \overline{k}^\alpha) dz^\alpha\} \tag{6.2}$$

is a pseudo-Hermitian structure on M and

$$\theta^\alpha = dz^\alpha + \sum_{\beta=1}^{n} h^{\alpha\beta} d\overline{z}^\beta \tag{6.3}$$

is a frame of $T_{1,0}(M)^*$ on V.

Cf. also [150], p. 471, and [227], p. 177, and Chapter 4 of this book. The proof of Lemma 6.1 is omitted. Cf. [219], p. 233, for details. Let $f : M \to A$ be a CR map, of class C^∞, between two orientable C^∞ nondegenerate CR manifolds of CR dimensions n and N, respectively. By Lemma 6.1, we may choose Folland–Stein coordinates (Z^j, U) on A with origin at $f(x_0)$, and respectively a pseudo-Hermitian structure Θ on A and a (local) frame $\{\Theta^j\}_{1 \le j \le N}$ of $T_{1,0}(A)^*$ given by

$$\Theta = dU + \sum_{j=1}^{N} \{(iZ^j + K^j) d\overline{Z}^j + (-i\overline{Z}^j + \overline{K}^j) dZ^j\}, \tag{6.4}$$

$$\Theta^j = dZ^j + \sum_{k=1}^{N} H^{jk} d\overline{Z}^k, \tag{6.5}$$

for some functions $K^j(Z, U)$, $H^{jk}(Z, U)$ of weight at least 2. Since f is a CR map we have

$$f^* \Theta^j U \equiv 0 \mod \theta, \theta^\alpha, \tag{6.6}$$

$$f^* \Theta \equiv 0 \mod \theta. \tag{6.7}$$

Using (6.2)–(6.3) the equations (6.6)–(6.7) may be written

$$\begin{aligned} \mathcal{L}^\beta f^j &= k^\beta \frac{\partial f^j}{\partial u} + \sum_{k=1}^{N} H^{jk} \Big(- \mathcal{L}^\beta \overline{f^k} + k^\beta \frac{\partial \overline{f^k}}{\partial u} \Big) \\ &\quad + \sum_{\alpha=1}^{n} h^{\alpha\beta} \left\{ \overline{\mathcal{L}^\alpha} f^j - \overline{k^\alpha} \frac{\partial f^j}{\partial u} + \sum_{k=1}^{N} H^{jk} \Big(\overline{\mathcal{L}^\alpha} f^k - \overline{k^\alpha} \frac{\partial \overline{f^k}}{\partial u} \Big) \right\}, \end{aligned} \tag{6.8}$$

$$\begin{aligned} \mathcal{L}^\alpha f^0 &= k^\alpha \frac{\partial f^0}{\partial u} + \sum_{j=1}^{N} \left\{ (i f^j + K^j) \Big(- \mathcal{L}^\alpha \overline{f^j} + k^\alpha \frac{\partial \overline{f^j}}{\partial u} \Big) \right. \\ &\quad \left. + (i \overline{f^j} + \overline{K^j}) \Big(- \mathcal{L}^\alpha f^j + k^\alpha \frac{\partial \overline{f^j}}{\partial u} \Big) \right\}, \end{aligned} \tag{6.9}$$

where

$$\mathcal{L}^\alpha = \frac{\partial}{\partial \overline{z}^\alpha} - i z^\alpha \frac{\partial}{\partial u}$$

(the Lewy operator) and $f^j = Z^j \circ f$, $f^0 = U \circ f$.

Theorem 6.2. *Let $f : M \to A$ be a real analytic CR map between two orientable real analytic strictly pseudoconvex CR manifolds of CR dimensions n and N, respectively, M connected. Then either f has maximal rank or $f(M)$ is a point.*

Proof. Let $x_0 \in M$. Choose, as above, Folland–Stein coordinates (z^α, u) and (Z^j, U) at x_0 and $f(x_0)$, respectively, pseudo-Hermitian structures θ and Θ, and (local) frames $\{\theta^\alpha\}$ and $\{\Theta^j\}$ given by (6.2)–(6.5). Since the CR structures on M, A are supposed to be real analytic, one may choose k^α, $h^{\alpha\beta}$, K^j, and H^{jk} to be real analytic, too. The proof of Theorem 6.2 follows from a direct inspection of the equations (6.8)–(6.9) and is a refinement of the proof in [219], pp. 234–238. Let

$$f^0 = \sum_{\chi=1}^{\infty} f^0_\chi, \quad f^j = \sum_{\chi=1}^{\infty} f^j_\chi \tag{6.10}$$

where f^0_χ, f^j_χ are of weight χ, for $\chi \geq 1$ (since (Z^j, U) have origin at $f(x_0)$ one has $f^0(0) = 0$, $f^j(0) = 0$ and the Taylor series of f^0, f^j contain no terms of weight zero). Arguments analogous to that in [219] (which we omit) furnish

$$\mathcal{L}^\alpha f^j_1 = 0 \tag{6.11}$$

and

$$\frac{\partial f^j}{\partial \overline{z}^\alpha}(0) = 0. \tag{6.12}$$

Finally, the commutation formula

$$\left[\mathcal{L}^\alpha, \overline{\mathcal{L}^\beta}\right] = 2i\,\delta_{\alpha\beta}\frac{\partial}{\partial u}$$

and (6.11) lead to

$$\delta_{\alpha\beta}\frac{\partial f^0}{\partial u}(0) = \frac{\partial f^j}{\partial z^\beta}(0)\,\overline{\frac{\partial f^j}{\partial \overline{z}^\alpha}}(0). \tag{6.13}$$

Let us set $f^j = X^j + iY^j$ and $z^\alpha = x^\alpha + iy^\alpha$. By (6.12) the (real) Jacobian of f at $z = 0$, $u = 0$ is given by

$$J_{\mathbf{R}}(f)(0) = \begin{pmatrix} X^j_{x^\alpha}(0) & -Y^j_{x^\alpha}(0) & X^j_u(0) \\ Y^j_{x^\alpha}(0) & X^j_{x^\alpha}(0) & Y^j_u(0) \\ 0 & 0 & f^0_u(0) \end{pmatrix}$$

(where subscripts denote partial derivatives). We distinguish two cases. Either $\operatorname{rank} J_{\mathbf{R}}(f)(0) = \min\{2n+1, 2N+1\}$ and then we are done, or

$$\operatorname{rank} J_{\mathbf{R}}(f)(0) < \min\{2n+1, 2N+1\}. \tag{6.14}$$

If (6.14) holds we show that the derivatives of arbitrary order of f^j, f^0 at $z = 0$, $u = 0$ vanish, so that the statement in Theorem 6.2 may be obtained by analytic continuation. □

Lemma 6.2.

$$\frac{\partial f^0}{\partial u}(0) = 0.$$

Proof. Let $\mathcal{M} : \mathbf{C}^n \to \mathbf{C}^N$ be the $\mathbf{C}$-linear map defined by

$$\mathcal{M}e_\alpha = \sum_{j=1}^N \frac{\partial f^j}{\partial z^\alpha}(0)e_j\,,$$

where $\{e_\alpha\}$ and $\{e_j\}$ are the canonical linear bases in $\mathbf{C}^n$ and $\mathbf{C}^N$ respectively. By (6.13), for any $x \in \mathbf{C}^n$ one has

$$\|\mathcal{M}x\|^2 = \langle \overline{\mathcal{M}^t}\mathcal{M}x, x\rangle = \frac{\partial f^0}{\partial u}(0)\|x\|^2. \tag{6.15}$$

The proof of Lemma 6.2 is by contradiction. Assume that $f_u^0(0) \neq 0$. Then $\mathcal{M}$ is injective, by (6.15). We distinguish three cases: $n = N$, $n > N$, and $n < N$. When $n = N$ our Lemma 6.2 follows from [219], p. 237. If $n > N$ the fact that $\mathcal{M}$ is injective already leads to a contradiction. Hence we may assume that $n < N$. Let $(i_1, \dots, i_n) \in \{1, \dots, N\}^n$, $i_1 < \cdots < i_n$. Denote by $\Delta_{i_1 \cdots i_n}$ the square matrix of order $2n$ extracted from $J_{\mathbf{R}}(f)(0)$ as follows:

(1) The first n lines of $\Delta_{i_1\cdots i_n}$ are the lines $i_1, \dots, i_n$ of $J_{\mathbf{R}}(f)(0)$.
(2) The last n lines of $\Delta_{i_1\cdots i_n}$ are the lines $i_1 + N, \dots, i_n + N$ of $J_{\mathbf{R}}(f)(0)$.
(3) From each line chosen at (1)–(2) one has previously eliminated the elements of the last column of $J_{\mathbf{R}}(f)(0)$.

For instance,

$$\Delta_{1\cdots n} = \begin{pmatrix} X^\alpha_{x^\beta}(0) & -Y^\alpha_{x^\beta}(0) \\ Y^\alpha_{x^\beta}(0) & X^\alpha_{x^\beta}(0) \end{pmatrix}.$$

Since $n < N$, our (6.14) yields rank $J_{\mathbf{R}}(f)(0) < 2n + 1$, so that all determinants of order $2n + 1$ of $J_{\mathbf{R}}(f)(0)$ must vanish. Thus

$$f_u^0(0)\, \det\left(\Delta_{i_1\cdots i_n}\right) = 0,$$

and consequently

$$\det\left(\Delta_{i_1\cdots i_n}\right) = 0.$$

Then

$$\left|\det\left[\frac{\partial f^{i_\alpha}}{\partial z^\beta}(0)\right]\right|^2 = \det\left(\Delta_{i_1\cdots i_n}\right) = 0,$$

i.e., rank$(\mathcal{M}) < n$, a contradiction ($\mathcal{M}$ is injective). Our Lemma 6.2 is completely proved. □

Using (6.13) and Lemma 6.2 we obtain

$$\frac{\partial f^j}{\partial z^\alpha}(0) = 0,$$

and the proof of Theorem 6.2 may be completed by using the following fact:

Lemma 6.3. (H. Jacobowitz [219], p. 237)
For any real analytic CR map $f = (f^1, \dots, f^N, f^0)$ with the weight decomposition (6.10), if $f^j_\chi = 0$, $1 \le \chi \le r$, then (i) $f^0_\chi = 0$, $1 \le \chi \le 2r + 1$, and (ii) $f^j_{r+1} = 0$.

A remark is in order. The assumption of strict pseudoconvexity in Theorem 6.2 is motivated by the following example. Let $Q_3 \subset \mathbf{C}^2$ be the hyperquadric $Q_3 = \{(z_1, z_2) : \text{Im}(z_2) - |z_1|^2 = 0\}$. Let $A = \{z \in \mathbf{C}^3 : \text{Im}(z_3) = |z_1|^2 - |z_2|^2\}$ and $f : Q_3 \to A$ the CR map given by $f(z_1, z_2) = (z_1, z_1, 0)$. Then $\text{rank}(f) = 1$ and f is nonconstant. Our Theorem 6.2 does not apply since A is only nondegenerate. □

6.2 The second fundamental form

Let $(M, T_{1,0}(M))$ and $(A, T_{1,0}(A))$ be two CR manifolds of CR dimensions n and $n+k$, $k > 0$, respectively.

Definition 6.2. A *CR immersion* is a CR map $f : M \to A$ such that $\text{rank}(d_x f) = \dim(M)$ at any $x \in M$. □

That is, a CR immersion is an immersion and a CR map. For instance, let M^{2n} be a complex n-dimensional Hermitian manifold and $T^{1,0}(M^{2n})$ its holomorphic tangent bundle. Then $(M, T^{1,0}(M^{2n}))$ is a CR manifold of type $(n, 0)$. Let M be a proper CR submanifold of M^{2n} (in the sense of A. Bejancu [56]; see Chapter 1 for definitions). Then the inclusion $j : M \to M^{2n}$ is a CR immersion.

Definition 6.3. Let $f : M \to A$ be a CR immersion. The *CR normal bundle* of f is defined by

$$\nu_H^{2k}(f) = \mathcal{L}_A / f_* \mathcal{L}_M \,, \tag{6.16}$$

where $\mathcal{L}_M$ and $\mathcal{L}_A$ are the bundles whose total spaces are $H(M)$ and $H(A)$ respectively (i.e., the Levi distributions regarded as vector bundles). □

When A is a strictly pseudoconvex CR manifold the CR normal bundle (of the given CR immersion f) is isomorphic to the orthogonal complement of $f_* \mathcal{L}_M$ in $\mathcal{L}_A$ with respect to the (real) Levi form of A.

Let $(M, T_{1,0}(M), \theta)$ and $(A, T_{1,0}(A), \Theta)$ be two strictly pseudoconvex CR manifolds of CR dimensions n and $n+k$ respectively (thus $\dim(M) = 2n+1$ and $\dim(A) = 2(n+k)+1$).

Definition 6.4. Let $f : M \to A$ be a CR immersion. Then f is *isopseudo-Hermitian* if

$$f^* \Theta = \theta. \tag{6.17}$$

Let $f : M \to A$ be an isopseudo-Hermitian CR immersion. $(T_{1,0}(A), \Theta)$ is strictly pseudoconvex, hence (A, g_Θ) is a Riemannian manifold. Let $\nu^{2k}(f)$ be the normal bundle of f and $E(\nu^{2k}(f))$ its total space. Then

$$T_{f(x)}(A) = [f_* T_x(M)] \oplus E(\nu^{2k}(f))_x \tag{6.18}$$

for any $x \in M$. Note that (6.17) yields

$$f^* d\Theta = d\theta. \tag{6.19}$$

Using the fact that f is a CR map,

$$J_A \circ f_* = f_* \circ J; \tag{6.20}$$

hence

$$f^* G_\Theta = G_\theta. \tag{6.21}$$

Yet in general, the immersion f fails to be isometric (with respect to the Riemannian metrics g_θ and g_Θ) unless $f(M)$ is tangent to the characteristic direction T_A of (A, Θ); cf. our Theorem 6.3 below. Again, since f is a CR map,

$$f_* H(M) \subseteq H(A). \tag{6.22}$$

Let $E(\nu_H^{2k}(f))_x$ be the orthogonal (with respect to $G_{\Theta, f(x)}$) complement of $(d_x f) H(M)_x$ in $H(A)_{f(x)}$, for $x \in M$. Then $E(\nu_H^{2k}(f))$ is the total space of a (real) rank-$2k$ vector bundle isomorphic to $\nu_H^{2k}(f)$ (and denoted by the same symbol in the sequel). Moreover,

$$H(A)_{f(x)} = [(d_x f) H(M)_x] \oplus E(\nu_H^{2k}(f))_x, \tag{6.23}$$

for any $x \in M$. The CR normal bundle $\nu_H^{2k}(f)$ and the normal bundle $\nu^{2k}(f)$ do not coincide (i.e., $E(\nu_H^{2k}(f))_x \neq E(\nu^{2k}(f))_x$, for $x \in M$) in general, unless $f(M)$ is tangent to T_A. Yet we have the following:

Proposition 6.1. *$\nu_H^{2k}(f)$ and $\nu^{2k}(f)$ are canonically isomorphic.*

If N is a manifold, let τ_N be its tangent bundle. To prove Proposition 6.1, we construct a bundle isomorphism

$$\Phi : \mathcal{L}_A / f_* \mathcal{L}_M \to \tau_A / f_* \tau_M$$

by setting

$$\Phi_x(X + f_* H(M)_x) = X + f_* T_x(M),$$

where $X \in H(A)_{f(x)}$ and $x \in M$. Since $\mathcal{L}_A / f_* \mathcal{L}_M$ and $\tau_A / f_* \tau_M$ have the same rank, it is sufficient to check that Φ is a bundle monomorphism. To see this, assume that $X + f_* T_x(M) = 0$ for some $X \in H(A)_{f(x)}$. Then $X \in H(A)_{f(x)} \cap f_* T_x(M)$. Since

$$f_* H(M)_x \subseteq H(A)_{f(x)} \cap f_* T_x(M) \subseteq f_* T_x(M)$$

and $\dim_{\mathbf{R}} H(M)_x = 2n$, $\dim_{\mathbf{R}} T_x(M) = 2n + 1$, it follows that either

$$H(A)_{f(x)} \cap f_* T_x(M) = f_* H(M)_x$$

and then we are done, or

$$H(A)_{f(x)} \cap f_* T_x(M) = f_* T_x(M).$$

Yet since f is isopseudo-Hermitian, the last alternative does not occur. Indeed, if $f_* T_x(M) \subseteq H(A)_{f(x)}$ then $f_* T_x \in H(A)_{f(x)}$ and by (6.17) one has

$$0 = \Theta_{f(x)}(f_* T_x) = (f^*\Theta)_x(T_x) = \theta_x(T_x) = 1,$$

a contradiction. □

Let $\tan_x$, nor_x be the natural projections associated with (6.18). Let us set $T_A^\perp = \mathrm{nor}(T_A)$. Let ∇^A be the Tanaka–Webster connection of (A, Θ). Let $X, Y \in T(M)$. Let us set

$$\nabla_X Y = \tan\{\nabla^A_{f_* X} f_* Y\}, \quad \alpha(f)(X, Y) = \mathrm{nor}\{\nabla^A_{f_* X} f_* Y\}.$$

Then ∇ is a linear connection on M, while $\alpha(f)$ is $C^\infty(M)$-bilinear and has values in $E(\nu^{2k}(f))$. Thus, by (6.18), one obtains the following CR analogue of the Gauss formula (from the theory of isometric immersions between Riemannian manifolds):

$$\nabla^A_{f_* X} f_* Y = f_* \nabla_X Y + \alpha(f)(X, Y). \tag{6.24}$$

Next, if $X \in T(M)$ and $\xi \in E(\nu^{2k}(f))$ then we set

$$a_\xi X = -\tan\{\nabla^A_{f_* X} \xi\}, \quad \nabla^\perp_X \xi = \mathrm{nor}\{\nabla^A_{f_* X} \xi\}.$$

Then a is $C^\infty(M)$-bilinear, while $\nabla^\perp$ is a connection in $\nu^{2k}(f)$. We are led to a CR analogue of the Weingarten formula (from Riemannian geometry)

$$\nabla^A_{f_* X} \xi = -f_* a_\xi X + \nabla_X \xi\,. \tag{6.25}$$

In general, the "induced" connection ∇ in (6.24) does not coincide with the "intrinsic" Tanaka–Webster connection of (M, θ) (unless $T_A^\perp = 0$), nor is $\alpha(f)$ (the CR analogue of the second fundamental form of an isometric immersion between Riemannian manifolds) symmetric.

Theorem 6.3. *Let $f : M \to A$ be an isopseudo-Hermitian CR immersion between two strictly pseudoconvex CR manifolds of CR dimensions n and $n + k$. The following statements are equivalent:*

(i) $f^* g_\Theta = g_\theta$.
(ii) $T_A^\perp = 0$.
(iii) $E(\nu_H^{2k}(f))_x = E(\nu^{2k}(f))_x$, *for any* $x \in M$.

Lemma 6.4. *The following identities hold:*

$$X_A \rfloor \theta = 1 - \|T_A^\perp\|^2, \quad X_A \rfloor d\theta = -f^*\left(T_A^\perp \rfloor d\Theta\right),$$

where $X_A = \tan(T_A)$.

The proof of Lemma 6.4 is straightforward. At this point we may prove Theorem 6.3. To this end, assume (i) holds. For any $X \in T(M)$ one has

$$g_\theta(X_A, X) = g_\Theta(f_* X_A, f_* X) = g_\Theta(T_A, f_* X) = \Theta(f_* X) = \theta(X) = g_\theta(T, X)$$

and thus $X_A = T$. One still has to show that T_A is tangent to $f(M)$. To see this, apply Θ to $T_A = f_* T + T_A^\perp$. This procedure yields $\|T_A^\perp\| = 0$.

Conversely, if (ii) holds, then $X_A = T$ by Lemma 6.4. By (1.16) in Chapter 1 and (6.19) the verification of (i) amounts to checking that

$$g_\Theta(f_* T, f_* X) = 0,$$

for any $X \in H(M)$. This follows from (6.22).

Assume (iii) holds. Since $T_A^\perp \in E(\nu^{2k}(f))$, by (iii) and (6.23) it follows that $T_A^\perp \in H(A)$. Thus $\|T_A^\perp\|^2 = g_\Theta(T_A, T_A^\perp) = 0$. Conversely, if (ii) holds then $T_A = f_* T$ by Lemma 6.4. Let $\xi \in E(\nu_H^{2k}(f))_x$ and $X \in T_x(M)$, for $x \in M$. By (1.20) in Chapter 1 one has

$$X = Y + cT_x, \quad Y \in H(M)_x, \quad c \in \mathbf{R}.$$

Thus

$$g_\Theta(\xi, f_* X) = c g_\Theta(\xi, f_* T) = c g_\Theta(\xi, T_A) = 0$$

and thus $E(\nu_H^{2k}(f))_x \subseteq E(\nu^{2k}(f))_x$, for any $x \in M$. □

Taking into account Theorem 6.3 we define the following central notion.

Definition 6.5. A *pseudo-Hermitian immersion* is an isopseudo-Hermitian CR immersion with the additional property $T_A^\perp = 0$. □

Let $f : M \to A$ be a pseudo-Hermitian immersion. Let $\pi_H \alpha(f)$ be a vector-valued form defined by $(\pi_H \alpha(f))(X, Y) = \alpha(f)(\pi_H X, \pi_H Y)$, for any $X, Y \in T(M)$. Here $\pi_H : T(M) \to H(M)$ is the projection associated with $T(M) = H(M) \oplus \mathbf{R}T$.

Theorem 6.4. *Let M, A be two strictly pseudoconvex CR manifolds and $f : M \to A$ a pseudo-Hermitian immersion. Then*

(i) ∇ is the Tanaka–Webster connection of (M, θ).
(ii) $\pi_H \alpha(f)$ is symmetric.
(iii) a_ξ is $H(M)$-valued and for any $x \in M$, $(a_\xi)_x : H(M)_x \to H(M)_x$ is self adjoint (with respect to $G_{\theta,x}$).

We organize the proof in several steps, as follows.

Step 1. *$H(M)$ is parallel with respect to ∇.*

Let $Y \in H(M)$. Since f is a CR map, $f_* Y \in H(A)$. Then (6.24) and Theorem 6.3 yield

$$\alpha(f)(X, Y) \in E(\nu^{2k}(f)) = E(\nu_H^{2k}(f)) \subset H(A).$$

Thus, again by (6.24), one has $f_* \nabla_X Y \in H(A)$. To complete the proof of Step 1, recall that since f is pseudo-Hermitian, one has

$$(d_x f) H(M)_x = [(d_x f) T_x(M)] \cap H(A)_{f(x)}$$

for any $x \in M$.

Step 2. *J is parallel with respect to ∇.*

This follows from $\nabla^A J_A = 0$, (6.20), (6.24), and Step 1. Moreover, since $\xi \in E(\nu_H^{2k}(f)) \Longrightarrow J_A \xi \in E(\nu_H^{2k}(f))$, besides $\nabla J = 0$ one additionally obtains

$$\alpha(f)(X, JY) = J_A \alpha(f)(X, Y) \tag{6.26}$$

for any $X \in T(M),\ Y \in H(M)$.

Step 3. *g_θ is parallel with respect to ∇.*

This follows from (6.24) since f is an isometric immersion. By Tanaka's theorem (i.e., Theorem 1.3 in Chapter 1), it remains to show that T_∇ is pure.

Step 4. *T_∇ is pure.*

Let $Z \in T_{1,0}(M),\ W \in T(M) \otimes \mathbf{C}$. Since f is a CR map,

$$\pi_+^A T_{\nabla^A}(f_* Z, f_* W) = 0 \tag{6.27}$$

(by Theorem 1.3 and the equivalent formulation of the purity conditions (1.36)–(1.38)), where T_{∇^A} is the torsion of ∇^A. Also $\pi_+^A : T(A) \otimes \mathbf{C} \to T_{1,0}(A)$ is the natural projection. We have

$$\nabla^A_{f_* Z} f_* W = f_* \nabla_Z W + \alpha(f)(Z, W)$$

since both sides are $\mathbf{C}$-linear and coincide (by (6.24)) on real vectors. Then (6.27) may be written

$$\pi_+^A \{ f_* T_\nabla(Z, W) + \alpha(f)(Z, W) - \alpha(f)(W, Z) \} = 0 \tag{6.28}$$

for any $Z \in T_{1,0}(M),\ W \in T(M) \otimes \mathbf{C}$. We have

$$T(A) \otimes \mathbf{C} = [f_* T(M) \otimes \mathbf{C}] \oplus [E(\nu^{2k}(f)) \otimes \mathbf{C}].$$

As noticed above, J_A restricts to a complex structure on $\nu^{2k}(f) (= \nu_H^{2k}(f))$. Extend J_A to $E(\nu^{2k}(f)) \otimes \mathbf{C}$ (by $\mathbf{C}$-linearity) and set

$$E(\nu^{2k}(f))^{1,0} = \mathrm{Eigen}(J_A; i).$$

Since f is a CR map, $f_* \circ \pi_+ = \pi_+^A \circ f_*$ and the restriction of π_+^A to $E(\nu^{2k}(f)) \otimes \mathbf{C}$ is the natural projection $E(\nu^{2k}(f)) \otimes \mathbf{C} \to E(\nu^{2k}(f))^{1,0}$. Finally, note that

$$T_{1,0}(A) = [f_* T_{1,0}(M)] \oplus E(\nu^{2k}(f))^{1,0},$$

so that (6.28) yields $\pi_+ T_\nabla(Z, W) = 0$. Thus part (i) in Theorem 6.4 is proved. □

Note that since $\nabla^A T_A = 0$, $\nabla T = 0$, and $T_A = f_* T$, the identity (6.24) yields

$$\alpha(f)(X, T) = 0, \tag{6.29}$$

for any $X \in T(M)$. Let us consider the (vector-valued) differential 1-form $Q \in \Gamma^\infty(T^*(M) \otimes E(\nu^{2k}(f)))$ defined by

$$QX = \alpha(f)(T, X), \tag{6.30}$$

for any $X \in T(M)$. Then

$$\tau_A f_* X = f_* \tau X + QX. \tag{6.31}$$

Here τ_A denotes the pseudo-Hermitian torsion of ∇^A. At this point (6.24), (1.60), and (6.31) yield

$$\alpha(f)(X, Y) = \alpha(f)(Y, X) + 2(\theta \wedge Q)(X, Y). \tag{6.32}$$

Therefore, in general, $\alpha(f)$ is not symmetric. Yet by (6.32), since θ vanishes on $H(M)$, $\pi_H \alpha(f)$ is symmetric. This proves part (ii) in Theorem 6.4. □

Let us use (6.24)–(6.25) and $\nabla^A g_\Theta = 0$ to obtain

$$g_\Theta(\alpha(f)(X, Y), \xi) = g_\theta(a_\xi X, Y), \tag{6.33}$$

for any $X, Y \in T(M)$, $\xi \in E(\nu^{2k}(f))$. Let us set $Y = T$ in (6.33) and use (6.29). It follows that

$$\theta(a_\xi X) = 0, \tag{6.34}$$

that is, a_ξ is $H(M)$-valued. Then (6.33) and the symmetry of $\pi_H \alpha(f)$ complete the proof of part (iii). □

6.3 CR immersions into $\mathbf{H}_{n+k}$

We may state the following theorem:

Theorem 6.5. *Let $f : M \to A$ be a pseudo-Hermitian immersion between two strictly pseudoconvex CR manifolds M and A of CR dimensions n and $n+k$, respectively. Then f is a minimal isometric immersion. Consequently, there are no pseudo-Hermitian immersions from a compact oriented strictly pseudoconvex CR manifold of CR dimension n into the Heisenberg group $\mathbf{H}_{n+k}$ (carrying the standard strictly pseudoconvex pseudo-Hermitian structure).*

Proof. Let $f : M \to A$ be a pseudo-Hermitian immersion. Then

$$\text{trace}[\pi_H \alpha(f)] = 0. \tag{6.35}$$

Let now $\tilde{\nabla}^A$ be the Levi-Civita connection of (A, g_Θ). We shall need the Gauss formula of (M, g_θ) in (A, g_Θ), i.e.,

$$\tilde{\nabla}^A_{f_* X} f_* Y = f_* \tilde{\nabla}_X Y + \beta(f)(X, Y) \tag{6.36}$$

for any $X, Y \in T(M)$. Here $\beta(f)$ denotes the second fundamental form (of the isometric immersion f), while $\tilde{\nabla}$ is the Levi-Civita connection of (M, g_θ). Since $\tilde{\nabla}^A$, ∇^A are related (cf. (1.61)), the Gauss formula (6.36) and its CR analogue (6.24) furnish

$$\beta(f) = \alpha(f) + Q \otimes \theta. \tag{6.37}$$

By (6.29), $QT = 0$. Thus (6.35), (6.37) lead to

$$\text{trace}[\beta(f)] = 0, \tag{6.38}$$

that is, f is minimal. Next, we need a lemma on the Riemannian structure of $\mathbf{H}_n$ (endowed with the pseudo-Hermitian structure (1.25)). Let us use (1.61) (with $\tau = 0$) to obtain

$$\tilde{\nabla} = \nabla - (d\theta_0) \otimes X_0 + \theta_0 \odot J, \tag{6.39}$$

where $X_0 = \partial/\partial t$. Let us set $\{\partial_i\}_{1 \leq i \leq 2n+1} = \{\partial/\partial x^\alpha, \partial/\partial y^\alpha, \partial/\partial t\}$. By (6.39), we obtain

$$\tilde{\nabla}_X Y = X(Y^i)\partial_i + 3(d\theta_0)(X, Y)X_0 + (\theta_0 \odot J)(X, Y) \tag{6.40}$$

for any $X, Y \in T(\mathbf{H}_n)$, where $Y = Y^i \partial_i$. Let $A = \mathbf{H}_{n+k}$. Let us set $f = (f^i)$, $1 \leq i \leq 2(n+k) + 1$. Then (6.40) and the Gauss formula of M in $\mathbf{H}_{n+k}$ yield

$$(\Delta f^i)\partial_i = \text{trace}[\beta(f)].$$

Here Δ is the Laplace–Beltrami operator (on $C^\infty(M)$) associated with g_θ. Thus, by (6.38), each f^i is harmonic. □

Let us look at a few examples.

(1) Let $n < N$ and consider the natural embedding $f : \mathbf{H}_n \to \mathbf{H}_N$ given by

$$f(z^1, \ldots, z^n, t) = (z^1, \ldots, z^n, 0, \ldots, 0, t).$$

Then f is a CR map and $f(\mathbf{H}_n)$ is tangent to $X_0 = \partial/\partial t$. Also

$$f^*\left\{dt + i\sum_{j=1}^{N}(z^j d\overline{z}^j - \overline{z}^j dz^j)\right\} = dt + i\sum_{\alpha=1}^{n}(z^\alpha d\overline{z}^\alpha - \overline{z}^\alpha dz^\alpha),$$

so that f is isopseudo-Hermitian. Next $\alpha(f) = 0$ (by (6.24)) and f is totally geodesic (with respect to the Webster metrics) by (6.37). □

(2) Let $a_1, \ldots, a_n \in \mathbf{Z}$, $a_j \geq 2$, and

$$X(a_1, \ldots, a_{n+1}) = \{(z^1, \ldots, z^{n+1}) : (z^1)^{a_1} + \cdots + (z^{n+1})^{a_{n+1}} = 0\}.$$

Then $X(a_1, \ldots, a_{n+1})$ is an algebraic hypersurface with one singularity (at the origin) and $X(a_1, \ldots, a_{n+1}) \setminus \{0\}$ is a complex n-dimensional manifold. The *Brieskorn sphere*

$$\Sigma^{2n-1}(a_1, \ldots, a_{n+1}) = [X(a_1, \ldots, a_{n+1}) \setminus \{0\}] \cap S^{2n+1}$$

is a CR manifold and the inclusion $\Sigma^{2n-1}(a_1, \ldots, a_{n+1}) \to S^{2n+1}$ is a CR immersion. If $a_1 = \cdots = a_{n+1} = a$ then $\Sigma^{2n-1}(a, \ldots, a)$ is tangent to the contact vector of S^{2n+1} (and Theorem 6.5 applies). □

Let $f : M \to A$ be a CR immersion. We may give the following geometric interpretation of $\pi_H \alpha(f)$. We start by giving the following definition:

Definition 6.6. (D. Jerison and J.M. Lee [227])
A regular curve $\gamma : I \to M$, defined on some open interval I containing the origin, is a *parabolic geodesic* of the nondegenerate CR manifold (of hypersurface type) M if

(i) $\dot{\gamma}(0) \in H(M)_{\gamma(0)}$.
(ii) There is $c \in \mathbf{R}$ such that

$$(\nabla_{\dot{\gamma}} \dot{\gamma})_{\gamma(t)} = 2c T_{\gamma(t)} \tag{6.41}$$

for any $t \in I$. □

Lemma 6.5. *Let $f : M \to A$ be a pseudo-Hermitian immersion such that for any parabolic geodesic γ of M, $f \circ \gamma$ is a parabolic geodesic of A. Then $\pi_H \alpha(f) = 0$.*

The proof is straightforward. Let $\gamma : I \to M$ be a parabolic geodesic. Then

$$\frac{d}{dt}\theta(\dot{\gamma}) = \frac{d}{dt} g_\theta(\dot{\gamma}, T) = g_\theta(\nabla_{\dot{\gamma}} \dot{\gamma}, T) = 2c g_\theta(T, T) = 2c,$$

so that

$$\dot{\gamma}(t) = \pi_H \dot{\gamma}(f) + 2ct T_{\gamma(t)}\,. \tag{6.42}$$

The converse of Lemma 6.5 does not hold, in general. Indeed, if $\pi_H \alpha(f) = 0$ and γ is a parabolic geodesic of M then (by the pseudo-Hermitian analogue of the Gauss equation) we get only as much as

$$\nabla^A_{f_* \dot{\gamma}} f_* \dot{\gamma} = 2c T_A(f(\gamma(t))) + 2ct\, Q(\pi_H \dot{\gamma}).$$

Nevertheless, if $Q = 0$, i.e., the entire fundamental form $\alpha(f)$ vanishes, then $f \circ \gamma$ is a parabolic geodesic of A. □

As another useful tool in the theory of isometric immersions between Riemannian manifolds, one derives the Gauss–Codazzi–Ricci equations, relating the second fundamental form of the given immersion to the Riemannian curvature of the ambient space and submanifold, respectively. It is of course desirable to have CR, or pseudo-Hermitian, analogues of these equations, relating $\alpha(f)$ to the curvature of the Tanaka–Webster connections ∇^A and ∇, respectively. To this end, let $f : M \to A$ be a pseudo-Hermitian immersion. Using (6.24)–(6.25) and (1.60) one may derive

$$\tan\{R^A(f_*X, f_*Y)f_*Z\} = R(X,Y)Z + a_{\alpha(f)(X,Z)}Y - a_{\alpha(f)(Y,Z)}X, \tag{6.43}$$

$$\begin{aligned}\operatorname{nor}\{R^A(f_*X, f_*Y)f_*Z &= (\nabla_X\alpha(f))(Y,Z) - (\nabla_Y\alpha(f))(X,Z) + \theta(X)\alpha(f)(\tau Y, Z)\\ &\quad - \theta(Y)\alpha(f)(\tau X, Z) + (d\theta)(X,Y)QZ,\end{aligned} \tag{6.44}$$

$$\begin{aligned}g_\Theta(R^A(f_*X, f_*Y)\xi, \eta) &= g_\Theta(R^\perp(X,Y)\xi, \eta)\\ &\quad + g_\theta(a_\xi X, a_\eta Y) - g_\theta(a_\xi Y, a_\eta X),\end{aligned} \tag{6.45}$$

for any $X, Y, Z \in T(M)$ and $\xi, \eta \in E(\nu^{2k}(f))$. Here R^A, R, and $R^\perp$ denote the curvature tensor fields of ∇^A, ∇, and $\nabla^\perp$, respectively. Note that only the CR analogue (6.44) of the Codazzi equation presents a different aspect with respect to its Riemannian counterpart. The additional terms in (6.44) are a manifestation of the pseudo-Hermitian torsion τ_A. By (6.31), τ and Q are its tangential and normal components, respectively.

Definition 6.7. Let M be a CR manifold and $x \in M$. *A parabola tangent to M at x* is a curve $\gamma : \mathbf{R} \to T_x(M)$ given by

$$\gamma(s) = sX + (s^2 + as + b)cT_x, \quad s \in \mathbf{R},$$

for some $X \in H(M)_x$ and $a, b, c \in \mathbf{R}$. □

Let $\{X_\alpha, JX_\alpha\}$ be a holomorphic frame of $H(M)$ defined on some open neighborhood of x. This induces an isomorphism $h_x : T_x(M) \to \mathbf{H}_n$ given by

$$h_x : X + cT_x \longmapsto (z, c), \quad X \in H(M)_x,$$

where $X = z^\alpha T_{\alpha,x} + \overline{z}^\alpha T_{\overline{\alpha},x}$, $z = (z^1, \ldots, z^n) \in \mathbf{C}^n$, $c \in \mathbf{R}$, and $T_\alpha = \frac{1}{2}(X_\alpha - iJX_\alpha)$. Let $\gamma : \mathbf{R} \to T_x(M)$ be a parabola tangent to M at x. If $a = b = 0$ then $h_x(\gamma(s)) = D_s(z,c)$ (where $D_s : \mathbf{H}_n \to \mathbf{H}_n$ is $D_s(z,c) = (sz, s^2c)$, $s > 0$), i.e., parabolas tangent to a CR manifold correspond to parabolic dilations under the natural identification of tangent spaces with the Heisenberg group.

Each D_s, $s \neq 0$, is a CR automorphism of $\mathbf{H}_n$. On the other hand, if $(z,c) \mapsto (sz, (s^2 + as + b)c)$ is a CR map for at least two distinct values of $s \neq 0$, then $a = b = 0$.

Definition 6.8. A curve $\gamma : \mathbf{R} \to \mathbf{H}_n$ given by $\gamma(s) = (z,t) \cdot (sZ, s^2c)$ (Heisenberg product), $s \in \mathbf{R}$, for some $(z,t), (Z,c) \in \mathbf{H}_n$, is called a *parabola* in $\mathbf{H}_n$. □

If $z = 0$, $t = 0$ then γ is a parabola through the origin (neutral element) in $\mathbf{H}_n$.

Lemma 6.6. *Any parabolic geodesic of $\mathbf{H}_n$ is a parabola.*

The proof is straightforward.

Theorem 6.6. *Let M be a strictly pseudoconvex CR manifold of CR dimension n and $f : M \to \mathbf{H}_{n+k}$ a pseudo-Hermitian immersion. Then*

(i) The Tanaka–Webster connection of M has a negative semidefinite Ricci form.
(ii) If $f = (f^i)$ then $\Delta_b f^i = 0$, $1 \leq i \leq 2(n+k)+1$.

(iii) *The pseudo-Hermitian scalar curvature R vanishes if and only if for any parabolic geodesic γ of M, $f \circ \gamma$ is a parabola in $\mathbf{H}_{n+k}$. In particular, if $R = 0$ then M is a Sasakian space-form.*

Here Δ_b is the sub-Laplacian of (M, θ). If $A = \mathbf{H}_{n+k}$ then $\tau_A = 0$ and consequently $\tau = 0$, $Q = 0$, by (6.31). In particular M turns out to be a Sasakian manifold. Next $R^A = 0$, so that (6.43) becomes

$$R(X, Y)Z = a_{\alpha(f)(Y,Z)}X - a_{\alpha(f)(X,Z)}Y.$$

A suitable contraction of indices then leads to

$$\mathrm{Ric}(X, Y) = -\sum_{j=1}^{2k} g_\theta(a_j^2 X, Y), \tag{6.46}$$

where $a_j = a_{\xi_j}$ and $\{\xi_j\}_{1 \leq j \leq 2k}$ is a given (local) orthonormal frame of $\nu^{2k}(f)$. Since $Q = 0$ each a_j is self-adjoint and thus $\mathrm{Ric}(X, X) \leq 0$ for any $X \in T(M)$. Yet $\mathrm{Ric}(X, T) = 0$, i.e., Ric is degenerate.

Let f^i be the components of $f : M \to \mathbf{H}_{n+k}$. Then $\Delta f^i = 0$ by the proof of Theorem 6.5. Also, since $X_0^\perp = 0$, it follows that

$$T(f^{2(n+k)+1}) = 1, \quad T(f^a) = 0, \quad 1 \leq a \leq 2(n+k).$$

At this point (ii) follows from a formula of A. Greenleaf ([186]),

$$\Delta_b u = \Delta u + T(T(u)), \tag{6.47}$$

for any $u \in C^\infty(M)$. A further contraction of indices in (6.46) gives

$$2R = -\sum_{j=1}^{2k} \mathrm{trace}(a_j^2).$$

Then $R = 0$ if and only if $\alpha(f) = 0$. This follows from (6.33). Indeed, since $Q = 0$, there exists a local frame with respect to which $a_j = \mathrm{diag}(\lambda_{j,1}, \ldots, \lambda_{j,2n+1})$. Thus $a_j^2 = \mathrm{diag}(\lambda_{j,1}^2, \ldots, \lambda_{j,2n+1}^2)$. At this point we may apply Lemma 6.5. Finally, if the pseudo-Hermitian scalar curvature vanishes then the Tanaka–Webster connection of M is flat and (1.71) in Chapter 1 becomes

$$\begin{aligned} R^\theta(X, Y)Z = \tfrac{1}{4}\{&\theta(X)g_\theta(Y, Z)X_0 - \theta(Y)g_\theta(X, Z)X_0 \\ &+ \theta(Z)\theta(Y)X - \theta(Z)\theta(X)Y \\ &- (JX \wedge JY)Z + 2(d\theta)(X, Y)JZ\}, \end{aligned} \tag{6.48}$$

for any $X, Y, Z \in T(M)$. An inspection of (6.48) in comparison with (3) in [62], p. 97, shows that M (with the Sasakian structure $(J, X_0, \theta, g_\theta)$) is a Sasakian space-form. Theorem 6.6 is completely proved. □

We close this section with a remark relating Sasakian and pseudo-Hermitian geometry. By a result of S. Ianus [214], any $(2m+1)$-dimensional manifold A carrying a normal almost contact structure (φ, ξ, η) is a CR manifold. Precisely, let $\mathcal{L}$ be the distribution on A given by the Pfaffian equation $\eta = 0$. Then $\varphi^2 = -I$ on $\mathcal{L}$. Let us complexify $\mathcal{L}$ and φ and set $\mathcal{L}^{1,0} = \text{Eigen}(\varphi; i)$. Then $(A, \mathcal{L}^{1,0})$ is a CR manifold (of hypersurface type) of CR dimension m and η is a pseudo-Hermitian structure. Let (φ, ξ, η, g) be a Sasakian structure. Then $G_{-\eta} = g$ on $\mathcal{L}$, so that $(A, \mathcal{L}^{1,0}, -\eta)$ is strictly pseudoconvex. Conversely, we have the following theorem (thus completing a discussion started in Chapter 1):

Theorem 6.7.
Let $(M, T_{1,0}(M), J, \theta, T)$ be a strictly pseudoconvex CR manifold. Then (J, T, θ, g_θ) is a contact metric structure on M. The almost contact structure (J, T, θ) is normal if and only if $\tau = 0$. In particular, the Heisenberg group (carrying the Sasakian structure $(J, -2T, -\frac{1}{2}\theta, \frac{1}{4}g_\theta)$ with θ given by (1.25)) is a Sasakian space-form of φ-sectional curvature $c = -3$.

6.4 Pseudo-Einsteinian structures

As mentioned in the introduction to this chapter, the machinery in the previous sections may be applied to the study of pseudo-Einsteinian structures; cf. our Chapter 5, from which we recall a few basic facts concerning the Lee class. These are used to establish Theorem 6.8.

6.4.1 CR-pluriharmonic functions and the Lee class

Let $f : M \to A$ be a pseudo-Hermitian immersion. Then

$$\overline{\partial}_b f^* \eta = f^* \overline{\partial}_b \eta \tag{6.49}$$

for any $(0, q)$-form η on A. Here the symbol $\overline{\partial}_b$ denotes both the tangential Cauchy–Riemann operators on M and those A. Let $\mathcal{P}_A$ be the sheaf of CR-pluriharmonic functions on A. Assume for the rest of this section that f is a homeomorphism on its image. As a consequence of (6.49), we have the following result:

Proposition 6.2. *If $D \subseteq A$ is open and $v \in \mathcal{P}_A(D)$ then $v \circ f \in \mathcal{P}(V)$, where $V = f^{-1}(D \cap f(M))$.*

We need to recall the construction of the CR-invariant cohomology class $\gamma(A) \in H^1(A, \mathcal{P}_A)$ built in Chapter 5 of this book. Assume from now on that A is locally realizable. Then there is an open covering $\mathcal{D} = \{D_j\}_{j \in \Sigma}$ of A and a pseudo-Einsteinian pseudo-Hermitian structure Θ_j on each D_j, $j \in \Sigma$. If $I_{ij} : D_i \cap D_j \to D_j$ are inclusions, then

$$I_{ij}^* \Theta_j = \exp(2U_{ji}) I_{ji}^* \Theta_i,$$

for some C^∞ functions $U_{ji} : D_i \cap D_j \to \mathbf{R}$. By Theorem 5.6 in Chapter 5, $U_{ji} \in \mathcal{P}_A(D_i \cap D_j)$. Let $N(\mathcal{D})$ be the nerve of $\mathcal{D}$. Let $C \in \mathcal{C}^1(N(\mathcal{D}), \mathcal{P}_A)$ be the 1-cochain mapping each 1-simplex $\sigma = (D_i D_j)$ of $N(\mathcal{D})$ in $U_{ji} \in \mathcal{P}_A(\cap\sigma)$. Then $C \in Z^1(N(\mathcal{D}), \mathcal{P}_A)$, i.e., C so built is a 1-cocycle with coefficients in $\mathcal{P}_A$. Finally $\gamma(A) \in H^1(A, \mathcal{P}_A)$ is the equivalence class of $[C] \in H^1(N(\mathcal{D}), \mathcal{P}_A)$.

Proposition 6.3. *Each pseudo-Hermitian immersion $f : M \to A$ such that $f : M \simeq f(M)$ (a homeomorphism) induces a map on cohomology*

$$f^* : H^p(A, \mathcal{P}_A) \to H^p(M, \mathcal{P}).$$

Let $\mathrm{Cov}(A)$ be the set of all open coverings of A. Let $\Gamma \in H^p(A, \mathcal{P}_A)$. Since

$$H^p(A, \mathcal{P}_A) = \varinjlim H^p(N(\mathcal{D}), \mathcal{P}_A),$$

there is $\mathcal{D} \in \mathrm{Cov}(A)$ and $h \in H^p(N(\mathcal{D}), \mathcal{P}_A)$ such that $\Gamma = [h]$. Let $V_j = f^{-1}(D_j \cap f(M))$ and set $\mathcal{V} = \{V_j\}_{j\in\Sigma}$. Then $\mathcal{V} \in \mathrm{Cov}(M)$. Let us set $f^*\Gamma = [f^*h]$ where

$$f^* : H^p(N(\mathcal{D}), \mathcal{P}_A) \to H^p(N(\mathcal{V}), \mathcal{P})$$

is described as follows. Let $c \in Z^p(N(\mathcal{D}), \mathcal{P}_A)$ be such that $h = [c]$ and set $f^*h = [f^*c]$, where

$$f^* : \mathcal{C}^p(N(\mathcal{D}), \mathcal{P}_A) \to \mathcal{C}^p(N(\mathcal{V}), \mathcal{P})$$

is described as follows. Let $\sigma = (V_{j_0} \cdots V_{j_p})$ be a p-simplex of $N(\mathcal{V})$ and set

$$(f^*c)\sigma = \rho_{f^*\sigma,\sigma} c(f^*\sigma),$$

where $f^*\sigma = (D_{j_0} \cdots D_{j_p})$, while

$$\rho_{f^*\sigma,\sigma} : \mathcal{P}_A(\cap f^*\sigma) \to \mathcal{P}(\cap\sigma), \quad \rho_{f^+\sigma,\sigma}(v) = v \circ f,$$

for any CR-pluriharmonic function $v : D_{j_0} \cap \cdots \cap D_{j_p} \to \mathbf{R}$. It is an elementary matter to check that the definition of f^* doesn't depend (at the various stages) on the choice of representatives. Throughout we use the notation and conventions in [178], pp. 272–275.

Theorem 6.8. (E. Barletta et al. [36])
Let $f : M \to A$ be a pseudo-Hermitian immersion (so that $f : M \to f(M)$ is a homeomorphism) between two strictly pseudoconvex CR manifolds M and A of CR dimensions n and $N = n + k$. Assume that both M, A are locally realizable (e.g., either M, A are compact or $n > 2$). Then

$$f^*\gamma(A) - \gamma(M) \in \mathrm{Ker}(j),$$

where $j : H^1(M, \mathcal{P}) \to H^1(M, \mathcal{E})$ is the map induced on cohomology by the natural sheaf morphism $\mathcal{P} \to \mathcal{E}$ (and $\mathcal{E}$ is the sheaf of C^∞ functions on M). Let us set $\varphi_j = f^*\Theta_j$, $V_j = f^{-1}(D_j \cap f(M))$, $j \in \Sigma$. If each (V_j, φ_j) is pseudo-Einsteinian then $f^*\gamma(A) = \gamma(M)$; in particular, if A admits a global pseudo-Einsteinian structure, then so does M.

Proof. Let $\mathcal{V} = \{U_\alpha\}_{\alpha \in I} \in \mathrm{Cov}(M)$ and $u_{\beta\alpha} \in \mathcal{P}(U_\alpha \cap U_\beta)$ such that $i^*_{\alpha\beta}\theta_\beta = \exp(2u_{\beta\alpha}) i^*_{\beta\alpha}\theta_\alpha$, where $i_{\alpha\beta} : U_\alpha \cap U_\beta \to U_\beta$ are inclusions. Then $\gamma(M) \in H^1(M, \mathcal{P})$ is the equivalence class of $[c] \in H^1(N(\mathcal{U}), \mathcal{P})$, where $c : \Delta(\alpha\beta) \mapsto u_{\beta\alpha}$. Let $\mathcal{W} \in \mathrm{Cov}(M)$ so that $\mathcal{W} < \mathcal{U}$ and $\mathcal{W} < \mathcal{V}$. Let us set $\mathcal{W} = \{W_a\}_{a \in J}$. There are maps $\phi : J \to I$ and $\psi : J \to \Sigma$ such that $W_a \subset U_{\phi(a)} \cap V_{\psi(a)}$ for each $a \in J$. Let us set $\lambda_a = r^*_a \theta_{\phi(a)}$ and $\mu_a = s^*_a \varphi_{\psi(a)}$ where $r_a : W_a \to U_{\phi(a)}$ and $s_a : W_a \to V_{\psi(a)}$ are inclusions. Note that

$$k^*_{ab}\lambda_b = \exp(2h_{ba}) k^*_{ba}\lambda_a, \tag{6.50}$$

where $k_{ab} : W_a \cap W_b \to W_b$ are inclusions and $h_{ba} = u_{\beta\alpha} \circ r_{ab}$ with $\alpha = \phi(a)$ and $\beta = \phi(b)$, and $r_{ab} : W_a \cap W_b \subset U_\alpha \cap U_\beta$. In other words $h_{ba} = \rho_{\phi\sigma,\sigma}(u_{\beta\alpha})$, where $\rho_{\phi\sigma,\sigma} : \mathcal{P}(\cap\phi\sigma) \to \mathcal{P}(\cap\sigma)$ is the restriction map ($\sigma = \Delta(ab) \in N(\mathcal{W})$) and $\phi : N(\mathcal{W}) \to N(\mathcal{U})$ the natural simplicial map. If $\tilde{\phi} : C^1(N(\mathcal{U}, \mathcal{P}) \to C^1(N(\mathcal{W}), \mathcal{P})$ is the induced map on cochains, then $(\tilde{\phi}c)\sigma = h_{ba}$, and if $\phi^* : H^1(N(\mathcal{U}), \mathcal{P}) \to H^1(N(\mathcal{W}), \mathcal{P})$ is the induced map on cohomology then $\phi^* g = [\tilde{\phi}c]$ with $g = [c]$, so that

$$\gamma(M) = [\phi^* g]$$

(one checks that $g \sim \phi^* g$ by looking at $\mathcal{W}$ as a common refinement of itself and $\mathcal{U}$). Both (W_a, λ_a) and (W_a, μ_a) are strictly pseudoconvex CR manifolds, so that

$$\mu_a = \exp(2v_a)\lambda_a \tag{6.51}$$

for some $v_a \in \mathcal{E}(W_a)$. Let $v \in C^1(N(\mathcal{W}), \mathcal{E})$ be given by $v : \Delta(a) \mapsto v_a$. Similarly to (6.50) we have

$$k^*_{ab}\mu_b = \exp(2\tilde{h}_{ba}) k^*_{ba}\mu_a, \tag{6.52}$$

where $\tilde{h}_{ba} = \tilde{u}_{ji} \circ s_{ab}$ with $i = \psi(a)$, $j = \psi(b)$ and $s_{ab} : W_a \cap W_b \subset V_i \cap V_j$. Also $\tilde{u}_{ji} = U_{ji} \circ f_{ij}$ and $f_{ij} : V_i \cap V_j \to D_i \cap D_j$ is induced by f. Finally (6.50)–(6.52) lead to

$$\tilde{h}_{ba} = v_b \circ k_{ab} + h_{ba} - v_a \circ k_{ba}. \tag{6.53}$$

Let $j : C^1(N(\mathcal{W}), \mathcal{P}) \to C^1(N(\mathcal{W}), \mathcal{E})$ be induced by the natural sheaf morphism $\mathcal{P} \to \mathcal{E}$ (i.e., $\mathcal{P}(U) \to \mathcal{E}(U)$ is the inclusion, for each $U \subseteq M$ open). Then (6.53) may be written

$$j\tilde{\psi} f^* C = \delta_{\mathcal{E}} v + j\tilde{\phi}c,$$

where

$$\delta_{\mathcal{E}} : C^1(N(\mathcal{W}), \mathcal{E}) \to C^2(N(\mathcal{W}), \mathcal{E})$$

is the coboundary operator. Consequently

$$j\phi_{\mathcal{VW}} f^* G = j\phi_{\mathcal{UW}} g,$$

where $j : H^1(N(\mathcal{W}), \mathcal{P}) \to H^1(N(\mathcal{W}), \mathcal{E})$. Finally, since j and ϕ^* (respectively j and ψ^*) commute, it follows that $j(f^*\gamma(A) - \gamma(M)) = 0$. Note that in general $\mathrm{Ker}(j) \neq 0$ (because $B^1(N(\mathcal{W}), \mathcal{P}) \subset B^1(N(\mathcal{W}), \mathcal{E})$, strict inclusion). If each μ_a is pseudo-Einsteinian then $v_a \in \mathcal{P}(W_a)$ and (6.53) may be written

$$\tilde{\phi} f^* C = \delta v + \tilde{\phi} c,$$

where $\delta : \mathcal{C}^1(N(\mathcal{W}), \mathcal{P}) \to \mathcal{C}^2(N(\mathcal{W}), \mathcal{P})$ is the coboundary operator. Thus

$$\phi_{\mathcal{VW}} f^* G = \phi_{\mathcal{UW}} g,$$

that is,

$$f^*\gamma(A) = \gamma(M),$$

and Theorem 6.8 is proved. □

6.4.2 Consequences of the embedding equations

In order to prove Theorem 6.1 we shall need the following lemma:

Lemma 6.7. *For any $X, Y \in T(M)$ and any $\xi \in \nu^{2k}(f)$ the following identity holds:*

$$g_\theta(a_\xi JX + Ja_\xi X, Y) = g_\theta(T_{\nabla^A}(f_*X, f_*Y), J_A\xi). \tag{6.54}$$

Using

$$\begin{aligned} g_\theta(a_\xi X, Y) &= g_\Theta(\alpha(f)(X, Y), \xi), \\ \alpha(f)(X, Y) - \alpha(f)(Y, X) &= \mathrm{nor}\{T_{\nabla^A}(f_*X, f_*Y)\}, \\ \alpha(f)(X, JY) &= J_A\alpha(f)(X, Y), \end{aligned} \tag{6.55}$$

we may perform the calculation

$$\begin{aligned} g_\theta(a_\xi JX, Y) &= g_\Theta(\alpha(f)(JX, Y), \xi) \\ &= g_\Theta(\alpha(f)(Y, JX) + T_{\nabla^A}(f_*JX, f_*Y), \xi) \\ &= g_\Theta(J_A\alpha(f)(Y, X) + T_{\nabla^A}(f_*JX, f_*Y), \xi) \\ &= g_\Theta(J_A\alpha(f)(X, Y), \xi) + g_\Theta(T_{\nabla^A}(f_*JX, f_*Y), \xi) \\ &\quad - g_\Theta(J_A\,\mathrm{nor}\{T_{\nabla^A}(f_*X, f_*Y)\}, \xi). \end{aligned}$$

Finally,

$$\begin{aligned} J_A^2 &= -I + \theta \otimes T_A, \\ g_\Theta(J_AX, J_AY) &= g_\Theta(X, Y) - \Theta(X)\Theta(Y), \quad X, Y \in T(A), \end{aligned}$$

lead to (6.54). □

Let $\xi \in \nu^{2k}(f)$ be such that $R^{\perp}(X, Y)\xi = 0$, for any $X, Y \in T(M)$. Then

$$g_\Theta(R^A(f_*X, f_*Y)\xi, \eta) = g_\Theta(R^{\perp}(X, Y)\xi, \eta) + g_\theta(a_\eta Y, a_\xi X) - g_\theta(a_\eta X, a_\xi Y),$$
$$a_{J_A\xi} X = J a_\xi X,$$

furnish

$$R^A(f_*X, f_*Y, \xi, J_A\xi) = g_\theta(J a_\xi Y, a_\xi X) - g_\theta(J a_\xi X, a_\xi Y). \tag{6.56}$$

Throughout, $R(X, Y, Z, W) = g_\theta(R(X, Y)Z, W)$. □
Note that (6.55) may be restated as

$$g_\theta(a_\xi X, Y) = g_\theta(X, a_\xi Y) + g_\Theta(T_{\nabla^A}(f_*X, f_*Y), \xi). \tag{6.57}$$

By (6.57) and $J^2 = -I + \theta \otimes T$ we obtain

$$g_\theta(J a_\xi Y, a_\xi X) = -g_\theta(a_\xi J a_\xi X, Y) - g_\Theta(T_{\nabla^A}(f_*Y, f_* J a_\xi X), \xi). \tag{6.58}$$

Let us replace X by $a_\xi X$ in (6.54) of Lemma 6.7 to obtain

$$g_\theta(a_\xi J a_\xi X, Y) = -g_\theta(J a_\xi^2 X, Y)$$
$$+ g_\Theta(T_{\nabla^A}(f_* a_\xi X, f_*Y), J_A\xi) + g_\Theta(T_{\nabla^A}(f_* J a_\xi X, f_*Y), \xi). \tag{6.59}$$

Substitution from (6.59) into (6.58) now leads to

$$g_\theta(J a_\xi Y, a_\xi X) = g_\theta(J a_\xi^2 X, Y) - g_\Theta(T_{\nabla^A}(f_* a_\xi X, f_*Y), J_A\xi). \tag{6.60}$$

On the other hand, we may replace X by Y and Y by $J a_\xi X$ in (6.57). The resulting identity and (6.59) furnish

$$g_\theta(a_\xi X, a_\xi Y) = -g_\theta(J a_\xi^2 X, Y) + g_\Theta(T_{\nabla^A}(f_* a_\xi X, f_*Y), J_A\xi). \tag{6.61}$$

Finally, by (6.60)–(6.61) the (CR analogue of) Ricci's equation becomes

$$R^A(f_*X, f_*Y, \xi, J_A\xi) = 2g_\theta(J a_\xi^2 X, Y) - 2g_\Theta(T_{\nabla^A}(f_* a_\xi X, f_*Y), J_A\xi), \tag{6.62}$$

for $X, Y \in T(M)$ and $\xi \in \nu^{2k}(f)$ with the property that $R^{\perp}(X, Y)\xi = 0$. Let

$$\{\xi_1, \ldots, \xi_k, J_A\xi_1, \ldots, J_A\xi_k\}$$

be a local orthonormal frame of $\nu^{2k}(f)$. Moreover, let

$$\{E_1, \ldots, E_{2n+1}\}$$

be a local orthonormal frame of $T(M)$, with $E_{2n+1} = T$ and $E_j \in H(M)$, for any $1 \le j \le 2n$. Let

$$K(Z, W) = \text{trace}\{V \to R^A(V, Z)W\}.$$

It is our purpose to compute $K(f_*X, f_*Y)$ for any $X, Y \in T(M)$. To this end, note that

$$\tan\{R^A(f_*X, f_*Y)f_*Z\} = R(X, Y)Z + a_{\alpha(f)(X,Z)}Y - a_{\alpha(f)(Y,Z)}X$$

may be restated as

$$R^A(f_*X, f_*Y, f_*Z, f_*W) = R(X, Y, Z, W) + g_\Theta(\alpha(f)(Y, W), \alpha(f)(X, Z)) - g_\Theta(\alpha(f)(X, W), \alpha(f)(Y, Z)), \quad (6.63)$$

for any $X, Y, Z, W \in T(M)$. To compute traces we use

$$K(f_*X, f_*Y) = \sum_{i=1}^{2n+1} R^A(f_*E_i, f_*X, f_*Y, f_*E_i) + \sum_{a=1}^{k}\{R^A(\xi_a, f_*X, f_*Y, \xi_a) + R^A(J_A\xi_a, f_*X, f_*Y, J_A\xi_a)\}.$$

We may assume that $E_{\alpha+n} = JE_\alpha,\ 1 \le \alpha \le n$. Consequently

$$\sum_{i=1}^{2n+1} \alpha(f)(E_i, E_i) = 0.$$

Of course, here $\alpha(f)$ is not the second fundamental form of f (with respect to the Webster metrics of M and A) but rather its pseudo-Hermitian analogue. Nevertheless, as previously shown, the "true" second fundamental form of f is traceless as well (and f is a minimal isometric immersion). This is natural since pseudo-Hermitian immersions appear to behave very much like holomorphic isometric immersions between Kählerian manifolds. The implications of minimality have been discussed earlier in this chapter. Next (6.63) leads to

$$\mathrm{Ric}(X, Y) = K(f_*X, f_*Y) - \sum_{a=1}^{k}\left\{R^A(\xi_a, f_*X, f_*Y, \xi_a) + R^A(J_A\xi_a, f_*X, f_*Y, J_A\xi_a)\right\} - \sum_{i=1}^{2n+1} g_\Theta(\alpha(f)(X, E_i), \alpha(f)(E_i, Y)), \quad (6.64)$$

for any $X, Y \in T(M)$. □

At this point we may start the proof of Theorem 6.1. We shall need the first Bianchi identity for ∇^A (cf., e.g., S. Kobayashi and K. Nomizu [241], Vol. I, p. 135)

$$\sum_{VZW} R^A(V, Z)W = \sum_{VZW} \{(\nabla^A_V T_{\nabla^A})(Z, W) + T_{\nabla^A}(T_{\nabla^A}(V, Z), W)\}, \quad (6.65)$$

for any $V, Z, W \in T(A)$. Here $\sum_{VZW}$ is the cyclic sum over V, Z, W. Let us set $V = f_*X$, $Z = J_A f_* Y$ and $W = \xi_a$ in (6.65) and take the inner product of the resulting identity with $J_A\xi$. This procedure leads to

$$R^A(f_*X, J_A f_*Y, \xi_a, J_A\xi_a) = E_a(X, Y) \\ + R^A(\xi_a, J_A f_*Y, f_*X, J_A\xi_a) - R^A(\xi_a, f_*X, J_A f_*Y, J_A\xi_a), \quad (6.66)$$

where

$$E_a(X, Y) = g_\Theta((\nabla^A_{f_*X} T_{\nabla^A})(J_A f_*Y, \xi_a), J_A\xi_a) \\ + g_\Theta((\nabla^A_{J_A f_*Y} T_{\nabla^A})(\xi_a, f_*X), J_A\xi_a) + g_\Theta((\nabla^A_{\xi_a} T_{\nabla^A})(f_*X, J_A f_*Y), J_A\xi_a) \\ + g_\Theta(T_{\nabla^A}(T_{\nabla^A}(f_*X, J_A f_*Y), \xi_a) J_A\xi_a) + g_\Theta(T_{\nabla^A}(T_{\nabla^A}(J_A f_*Y, \xi_a), f_*X), J_A\xi_a) \\ + g_\Theta(T_{\nabla^A}(T_{\nabla^A}(\xi_a, f_*X), J_A f_*Y), J_A\xi_a).$$

Note that

$$R^A(V, Z) J_A W = J_A R^A(V, Z) W \quad (6.67)$$

(as a consequence of $\nabla^A J_A = 0$) for any $V, Z, W \in T(A)$. By (6.67) and $\Theta(\xi_a) = 0$ we obtain

$$R^A(\xi_a, f_*X, J_A f_*Y, J_A\xi_a) = R^A(\xi_a, F_*X, f_*Y, \xi_a). \quad (6.68)$$

Next, replace ξ by ξ_a and Y by JY in (6.62) to obtain (provided $R^\perp = 0$)

$$R^A(f_*X, f_*JY, \xi_a, J_A\xi_a) \\ = 2g_\theta(a^2_{\xi_a} X, Y) - 2g_\Theta(T_{\nabla^A}(f_* a_{\xi_a} X, f_*JY), J_A\xi_a), \quad (6.69)$$

for any $X, Y \in T(M)$. At this point we may use (6.68)–(6.69) to write (6.66) as follows:

$$2g_\theta(a^2_{\xi_a} X, Y) - 2g_\Theta(T_\nabla(f_* a_{\xi_a} X, f_*JY), J_A\xi_a) \\ = R^A(\xi_a, J_A f_*Y, f_*X, J_A\xi_a) - R^A(\xi_a, f_*X, f_*Y, \xi_a) + E_a(X, Y), \quad (6.70)$$

for any $X, Y \in T(M)$. To deal with the torsion terms in (6.70) we need the following lemma:

Lemma 6.8. *Let* $T_\alpha = \frac{1}{2}(E_\alpha - iJE_\alpha)$, $1 \le \alpha \le n$. *Then*

$$E_a(T_\alpha, T_{\overline{\beta}}) = i g_\Theta(\tau_A\xi_a, J_A\xi_a) h_{\alpha\overline{\beta}}. \quad (6.71)$$

The proof of Lemma 6.8 is a straightforward consequence of

$$T_{\nabla^A}(Z, W) = T_{\nabla^A}(\overline{Z}, \overline{W}) = 0,$$
$$T_{\nabla^A}(Z, \overline{W}) = 2i G_\Theta(Z, \overline{W}) T_A,$$
$$\tau_A Z \in T_{0,1}(A),$$

for any $Z, W \in T_{1,0}(A)$. □

Lemma 6.9. *For any $X, Y, Z, W \in H(A)$ the following identity holds:*

$$R^A(X,Y,Z,W) = R^A(Z,W,X,Y) - A_\Theta(Y,Z)(d\Theta)(W,X) - A_\Theta(X,W)(d\Theta)(Z,Y) - A_\Theta(W,Y)(d\Theta)(X,Z) - A_\Theta(Z,X)(d\Theta)(Y,W), \tag{6.72}$$

where $A_\Theta(X,Y) = g_\Theta(\tau_A X, Y)$.

For the proof of Lemma 6.9 the reader may see Chapter 1 of this book. Using (6.72) we may compute the first curvature term in (6.70) as

$$R^A(\xi_a, f_*JY, f_*X, J_A\xi_a) = R^A(f_*X, J_A\xi_a, \xi_a, f_*JY) - A_\Theta(f_*JY, f_*X)(d\Theta)(J_A\xi_a, \xi_a) - A_\Theta(\xi_a, J_A\xi_a)(d\Theta)(f_*X, f_*JY), \tag{6.73}$$

for any $X, Y \in H(M)$. Also

$$R^A(f_*X, J_A\xi_a, \xi_a, f_*JY) = -R^A(J_A\xi_a, f_*X, f_*Y, J_A\xi_a). \tag{6.74}$$

Let us substitute from (6.73)–(6.74) into (6.70) and use the identities

$$A_\Theta(f_*X, f_*Y) = A(X,Y),$$
$$(d\Theta)(f_*X, f_*JY) = g_\Theta(X,Y),$$

to obtain

$$2g_\Theta(a^2_{\xi_a}X, Y) - 2g_\Theta(T_{\nabla^A}(f_*a_{\xi_a}X, f_*JY), J_A\xi_a) = R^A(J_A\xi_a, f_*X, f_*Y, J_A\xi_a) - R^A(\xi_a, f_*X, f_*Y, \xi_a) + A(X,JY) - g_\theta(X,Y)A_\Theta(\xi_a, J_A\xi_a) + E_a(X,Y), \tag{6.75}$$

for any $X, Y \in H(M)$. On the other hand (using (6.54)) one may show that

$$\sum_{i=1}^{2n+1} g_\Theta(\alpha(f)(X,E_i), \alpha(f)(E_i,Y)) = \sum_{a=1}^{k}\Big\{2g_\theta(a^2_{\xi_a}X, Y) + g_\Theta(T_{\nabla^A}(f_*Y, f_*a_{\xi_a}X), \xi_a) - g_\Theta(T_{\nabla^A}(f_*Y, f_*Ja_{\xi_a}X), J_A\xi_a)\Big\}. \tag{6.76}$$

Finally, substitution from (6.75)–(6.76) into (6.64) gives

$$\mathrm{Ric}(X,Y) = K(f_*X, f_*Y) - \sum_{a=1}^{k}\Big\{g_\Theta(T_{\nabla^A}(f_*Ja_{xi_a}X, f_*Y), \xi_a) - 2g_\theta(T_{\nabla^A}(f_*a_{\xi_a}, f_*JY), J_A\xi_a) - A(X,JY) + g_\theta(X,Y)A_\Theta(\xi_a, J_A\xi_a) - E_a(X,Y)\Big\}, \tag{6.77}$$

for $X, Y \in H(M)$. Let us extend both sides of (6.77) by **C**-linearity to $H(M) \otimes \mathbf{C}$. It follows that (6.77) holds for any $X, Y \in H(M) \otimes \mathbf{C}$ (since both sides are **C**-linear and coincide on real vectors). Let us set $X = Z,\ Y = \overline{W}$, with $Z, W \in T_{1,0}(M)$. We obtain

$$\mathrm{Ric}(Z, \overline{W}) = K(f_* Z, f_* \overline{W}) + \sum_{a=1}^{k} \{A_\Theta(\xi_a, J_A \xi_a) g_\theta(Z, \overline{W}) - E_a(Z, \overline{W})\}. \quad (6.78)$$

Finally, we set $Z = T_\alpha$ and $W = T_\beta$ in (6.78) and use (6.71) to obtain (6.1). □

6.4.3 The first Chern class of the normal bundle

Let (M, θ) and (A, Θ) be two strictly pseudoconvex CR manifolds and $f : M \to A$ a pseudo-Hermitian immersion. The purpose of the present section is to look at the converse of Theorem 6.1, i.e., it may be asked whether (6.1) yields $R^\perp = 0$. We establish the following weaker result. Let $\nu^{2k}(f) \to M$ be the normal bundle of f. By a result in this chapter, $\nu^{2k}(f)_x \subset H(A)_{f(x)}$, for any $x \in M$, so that J_A descends to a complex structure $J^\perp$ in $\nu^{2k}(f)$. Let us extend $J^\perp$ by complex linearity to $\nu^{2k}(f) \otimes \mathbf{C}$ and let $\nu^{2k}(f)^{1,0}$ be the eigenbundle corresponding to the eigenvalue i.

Theorem 6.9. *Let $f : M \to A$ be a pseudo-Hermitian immersion with the property $R_{\alpha\overline{\beta}} = K_{\alpha\overline{\beta}}$, where $K_{\alpha\overline{\beta}} = K(f_* T_\alpha, f_* T_{\overline{\beta}})$. If the Tanaka–Webster connection of A has parallel pseudo-Hermitian torsion (that is, $\nabla^A \tau_A = 0$) then $c_1(\nu^{2k}(f)^{1,0}) = 0$.*

Throughout, if $E \to M$ is a complex vector bundle then $c_1(E) \in H^2(M, \mathbf{R})$ denotes its first Chern class. To prove Theorem 6.9 we need the following lemma:

Lemma 6.10. *Let $f : M \to A$ be a pseudo-Hermitian immersion. If the ambient space A has parallel pseudo-Hermitian torsion then*

$$(\nabla_X A)(Y, Z) = g_\Theta(\alpha(f)(X, Z), Q(f)Z) + g_\Theta(Q(f)Y, \alpha(f)(X, Z)), \quad (6.79)$$

for any $X, Y, Z \in T(M)$.

The proof of Lemma 6.10 follows from $\nabla^A \tau_A = 0$, the (pseudo-Hermitian analogues of the) Gauss–Weingarten formulas, and

$$\tau_A f_* X = f_* \tau X + Q(f) X$$

in a straightforward manner. □

Let us recall that $c_1(T_{1,0}(M))$ is represented by $(i/(2\pi)) d\omega_\alpha^\alpha$, where

$$d\omega_\alpha^\alpha = R_{\alpha\overline{\beta}} \theta^\alpha \wedge \theta^{\overline{\beta}} + W_{\alpha\beta}^\alpha \theta^\beta \wedge \theta - W_{\overline{\alpha}\overline{\beta}}^{\overline{\alpha}} \theta^{\overline{\beta}} \wedge \theta,$$

$$W_{\beta\gamma}^\alpha = A_{\beta\gamma,\overline{\sigma}} h^{\alpha\overline{\sigma}}, \quad A_{\alpha\beta,\overline{\gamma}} = (\nabla_{T_{\overline{\gamma}}} A)(T_\alpha, T_\beta).$$

Here ω_β^α are the connection 1-forms of ∇. Cf. Chapter 5 of this book. Moreover, let $\{\Theta^1, \dots, \Theta^N\}$ be the admissible coframe corresponding to $\{T_1, \dots, T_n, \zeta_1, \dots, \zeta_k\}$, where $\zeta_a = (1/2)(\xi_a - i J_A \xi_a)$. Then $f^*\Theta^\alpha = \theta^\alpha$ and $f^*\Theta^{\alpha+n} = 0$. Next $c_1(T_{1,0}(A))$ is represented by $(i/(2\pi))d\Omega^j_j$, where Ω^i_j are the connection 1-forms of ∇^A and $(A_\Theta)_{ij,\overline{k}} = 0$ yields

$$d\Omega^j_j = K_{j\overline{k}}\,\Theta^j \wedge \Theta^{\overline{k}}.$$

Finally (6.79) gives $A_{\alpha\beta,\overline{\gamma}} = 0$, so that $f^*c_1(T_{1,0}(A)) = c_1(T_{1,0}(M))$, and the direct sum decomposition

$$T_{1,0}(A)_{f(x)} = [(d_x f)T_{1,0}(M)_x] \oplus \nu^{2k}(f)^{1,0}_x, \quad x \in M,$$

yields $c_1(\nu^{2k}(f)^{1,0}) = 0$. □

Let $f : M \to A$ be a pseudo-Hermitian immersion. Assume that $R^A = 0$ (e.g., $A = \mathbf{H}_N$). Then (6.64) gives

$$\mathrm{Ric}(X, Y) = -\sum_{i=1}^{2n+1} g_\Theta(\alpha(f)(X, E_i), \alpha(f(E_i, Y)),$$

or (by computing traces)

$$2\rho = -\|\alpha(f)\|^2 \le 0. \tag{6.80}$$

Theorem 6.10. (E. Barletta et al. [36])
There is no pseudo-Hermitian immersion of

$$\left(\mathbf{H}_n(s),\ |x|^{-2}\left\{dt + 2\sum_{\alpha=1}^{n}(x^\alpha dy^\alpha - y^\alpha dx^\alpha)\right\}\right)$$

into a Tanaka–Webster flat strictly pseudoconvex CR manifold.

Proof. By a result in Chapter 5 of this book, we have

$$R_{\alpha\overline{\beta}} = (n+1)|x|^{-2}\|z\|^2 h_{\alpha\overline{\beta}}, \tag{6.81}$$

or (by computing traces)

$$\rho = n(n+1)|x|^{-2}\|z\|^2. \tag{6.82}$$

Assume that there is a strictly pseudoconvex CR manifold A with $R^A = 0$ and a pseudo-Hermitian immersion $f : \mathbf{H}_n(s) \to A$. Then (6.82) contradicts (6.80) and Theorem 6.10 is completely proved. □

A remark regarding the analogy with Kählerian geometry (cf. [93], p. 554) is in order. Let $f : M \to A$ be a pseudo-Hermitian immersion. Assume that $c_1(T_{1,0}(M)) = 0$. Then there is a real 1-form η on M such that

$$\Gamma = d\eta, \tag{6.83}$$

where $\Gamma = (i/(2\pi))d\omega^\alpha_\alpha$.

Definition 6.9. A $\mathbf{C}$-valued 2-form on M is a $(1,1)$*-form* if $T \rfloor \eta = 0$ and $\eta(Z, W) = \eta(\overline{Z}, \overline{W}) = 0$ for any $Z, W \in T_{1,0}(M)$. $\square$

Let $\Lambda^{1,1}(M)$ be the bundle of $(1,1)$-forms on M.

Definition 6.10. Let

$$L_\theta : C^\infty(M) \otimes \mathbf{C} \to \Lambda^{1,1}(M)$$

be defined by setting

$$L_\theta f = -f\, d\theta\,,$$

for any C^∞ function $f : M \to \mathbf{C}$. Also, we consider

$$\Lambda_\theta : \Lambda^{1,1}(M) \to C^\infty(M) \otimes \mathbf{C}$$

given by

$$(\Lambda_\theta \psi, f)_\theta = (\psi, L_\theta f)_\theta, \quad \psi \in \Gamma^\infty(\Lambda^{1,1}(M)). \qquad \square$$

Here $(\ ,\ )_\theta$ is the usual L^2 inner product

$$(\phi, \psi)_\theta = \int_M \langle \phi, \psi \rangle\, \theta \wedge (d\theta)^n,$$

for any $(1,1)$-forms ϕ, ψ on M, at least one of compact support, where

$$\langle \phi, \psi \rangle = \phi_{\alpha\overline{\beta}} \psi^{\alpha\overline{\beta}}\,,$$
$$\phi = \phi_{\alpha\overline{\beta}} \theta^\alpha \wedge \theta^{\overline{\beta}}, \quad \psi = \psi_{\alpha\overline{\beta}} \theta^\alpha \wedge \theta^{\overline{\beta}},$$
$$\psi^{\alpha\overline{\beta}} = \overline{\psi^{\overline{\alpha}\beta}}, \quad \psi^{\overline{\alpha}\beta} = \psi_{\lambda\overline{\mu}} h^{\lambda\overline{\alpha}} h^{\overline{\mu}\beta}.$$

We may extend Λ_θ to an operator

$$\Lambda_\theta : \Lambda^2 T^* M \otimes \mathbf{C} \to C^\infty(M) \otimes \mathbf{C}$$

by declaring it to be zero on $\Lambda^{0,2}(M) \oplus \Lambda^{2,0}(M)$. Then

$$\Lambda_\theta \Gamma = -\frac{1}{\pi} \rho,$$

and we may apply Λ_θ to (6.83) to obtain

$$\frac{1}{2\pi} \rho = n\eta_0 + i \operatorname{div}(Z), \tag{6.84}$$

where $\eta = \eta_\alpha \theta^\alpha + \eta_{\overline{\alpha}} \theta^{\overline{\alpha}} + \eta_0 \theta$ and $Z = Z^{\overline{\alpha}} T_{\overline{\alpha}} - Z^\alpha T_\alpha$ with $Z^\alpha = h^{\alpha\overline{\beta}} \eta_{\overline{\beta}}$. The divergence in (6.84) is taken with respect to the volume form $\Psi = \theta \wedge (d\theta)^n$. Therefore, if $\int_M \eta_0 \Psi \geq 0$ then (6.80) gives $\alpha(f) = 0$ and thus $R^\perp = 0$ (as a consequence of the (pseudo-Hermitian analogue of the) Ricci equation) provided that $R^A = 0$. Yet, by a previous result, if (M, θ) is pseudo-Einsteinian, one representative of Γ is $\eta = (1/(2\pi n))\rho\theta$, so that (in view of (6.80)) the hypothesis $\int_M \eta_0 \Psi \geq 0$ is generically not satisfied. Indeed, let η' be another real 1-form such that $\Gamma = d\eta'$. If for instance $H^1(M, \mathbf{R}) = 0$ then $\eta' = \eta + du$ for some C^∞ function $u : M \to \mathbf{R}$ and (6.84) yields

$$\int_M T(u)\Psi = 0,$$

that is,

$$\int_M \eta_0' \Psi = \int_M \eta_0 \Psi \leq 0. \qquad \square$$

Let us look at several examples.

(1) (*Heisenberg groups*) Let $\mathbf{H}_n$ be the Heisenberg group, endowed with a standard strictly pseudoconvex CR structure spanned by $T_\alpha = \partial/\partial z^\alpha + i\overline{z}^\alpha \partial/\partial t$, and the contact form $\theta_0 = dt + i\sum_{\alpha=1}^n \{z_\alpha d\overline{z}_\alpha - \overline{z}_\alpha dz_\alpha\}$, $z^\alpha = z_\alpha$. The map $f : \mathbf{H}_n \to \mathbf{H}_N$, $N = n+k$, $k \geq 1$, induced by the natural inclusion $\mathbf{C}^n \to \mathbf{C}^N$ (i.e., $f(z,t) = (z, 0, t)$, $0 \in \mathbf{C}^k$), is a pseudo-Hermitian immersion with a flat normal Tanaka–Webster connection. $R^\perp = 0$ follows from the Ricci equation.

(2) (*Quotients of the Heisenberg group by discrete groups of dilations*) Let $\mathbf{H}_n(s)$, $0 < s < 1$, carry the CR structure induced by the covering map $\pi : \mathbf{H}_n \setminus \{0\} \to \mathbf{H}_n(s)$, and the contact 1-form θ given by

$$\theta_{\pi(x)} = |x|^{-2}\theta_{0,x} \circ (d_x\pi)^{-1}, \tag{6.85}$$

for any $x \in \mathbf{H}_n \setminus \{0\}$. The map $F : \mathbf{H}_n(s) \to \mathbf{H}_N(s)$ induced by $f : \mathbf{H}_n \setminus \{0\} \to \mathbf{H}_N \setminus \{0\}$ (i.e., $F \circ \pi = \Pi \circ f$, where $\Pi : \mathbf{H}_N \setminus \{0\} \to \mathbf{H}_N(s)$ is the natural covering map) is a pseudo-Hermitian immersion. Indeed, if $H_N(s)$ is endowed with the contact 1-form Θ given by

$$\Theta_{\Pi(X)} = |X|^{-2}\Theta_{0,X} \circ (d_X\Pi)^{-1}, \tag{6.86}$$

$$\Theta_0 = ds + \sum_{j=1}^N \{w_j d\overline{w}_j - \overline{w}_j dw_j\},$$

for any $X \in \mathbf{H}_N \setminus \{0\}$, then $|f(x)| = |x|$ (Heisenberg norms), $x \in \mathbf{H}_n$, yields $F^*\Theta = \theta$ (i.e., F is isopseudo-Hermitian). Moreover, we may write (6.85)–(6.86) as

$$\theta = e^{2u}\theta_0, \quad \Theta = e^{2U}\Theta_0$$

(with $U = \log|X|^{-1}$ and $u = U \circ f$). Therefore, the characteristic directions T and T_A of $(\mathbf{H}_n(s), \theta)$ and $(\mathbf{H}_N(s), \Theta)$ are respectively given by

$$T = e^{-2u}\left\{\frac{\partial}{\partial t} - 2iu^\beta T_\beta + 2iu^{\overline{\beta}} T_{\overline{\beta}}\right\},$$

$$T_A = e^{-2U}\left\{\frac{\partial}{\partial s} - 2iU^j W_j + 2iU^{\overline{j}} W_{\overline{j}}\right\}.$$

Note that

$$U^j(f(x)) = |x|^2 W_{\overline{j}}(U)_{f(x)}$$

and

$$W_{\overline{j}}(U) = -\tfrac{1}{2}|X|^{-4} w_j \overline{\Phi}$$

where $\Phi(w, s) = |w|^2 + is$ (note that $\overline{\Phi}$ is CR-holomorphic). Finally

$$U^\alpha = |x|^2 U_{\overline{\alpha}}, \quad U_{\overline{\alpha}} \circ f = u_{\overline{\alpha}}, \quad U^\alpha \circ f = u^\alpha,$$

and

$$T_{\overline{\alpha}}(u) = -\tfrac{1}{2}|x|^{-4} z_\alpha \overline{\phi}$$

(where $\phi = \Phi \circ f$) yield $f_* T = T_A$. Next, let us compute the curvature of the normal Tanaka–Webster connection $\nabla^\perp$ of F. We perform our task in a more general setting, as follows. Let $f : M \to A$ be a pseudo-Hermitian immersion between (M, θ) and (A, Θ) and set $\hat{\theta} = e^{2u}\theta$, $\hat{\Theta} = e^{2U}\Theta$, with $U \in C^\infty(A)$, $u = U \circ f$. Readily $f^*\hat{\Theta} = \hat{\theta}$. Let us set

$$\hat{T} = e^{-2u}\{T - 2iu^\beta T_\beta + 2iu^{\overline{\beta}} T_{\overline{\beta}}\},$$

where T is the characteristic direction of (M, θ). Since $U^\alpha \circ f = u^\alpha$ we obtain

$$\begin{aligned}\hat{T}_A(f(x)) &= (f_*\hat{T})(f(x)) \\ &\quad + 2ie^{-2u(x)}\{U^{\overline{\alpha+n}}(f(x)) W_{\overline{\alpha+n}}(f(x)) - U^{\alpha+n}(f(x)) W_{\alpha+n}(f(x))\},\end{aligned}$$

for any $x \in M$. Thus $f_*\hat{T} = \hat{T}_A$ (i.e., f is a pseudo-Hermitian immersion from $(M, \hat{\theta})$ into $(A, \hat{\Theta})$) if and only if $\zeta_{\overline{a}}(U) = 0$. Let us look now at the relation between $\nabla^\perp$ and $\hat{\nabla}^\perp$ (the normal Tanaka–Webster connection of $(M, \hat{\theta})$ in $(A, \hat{\Theta})$). Let $\hat{\nu}^{2k}(f)_x$ be the orthogonal complement (with respect to $g_{\hat{\Theta}, f(x)}$) of $(d_x f) T_x(M)$ in $T_{f(x)}(A)$, for any $x \in M$. Then $\hat{\nu}^{2k}(f)_x = \nu^{2k}(f)_x$, although the Webster metrics $g_{\hat{\Theta}}$ and g_Θ are not conformally related.

Assume from now on that f is a pseudo-Hermitian immersion both as a map of (M, θ) into (A, Θ) and as a map of $(M, \dot{\theta})$ into (A, Θ). We need to recall (cf. Chapter 2) the following result:

Lemma 6.11. *Let $(M, T_{1,0}(M), \theta, T)$ be a nondegenerate CR manifold. Then, under a transformation $\hat{\theta} = e^{2u}\theta$, the Christoffel symbols of the Tanaka–Webster connections of $(T_{1,0}(M), \theta)$ and $(T_{1,0}(M), \hat{\theta})$ are related by*

$$\begin{aligned}\hat{\gamma}^\sigma_{\beta\alpha} &= \Gamma^\sigma_{\beta\alpha} + 2u_\beta \delta^\sigma_\alpha + 2u_\alpha \delta^\sigma_\beta, \\ \hat{\Gamma}^\sigma_{\overline{\beta}\alpha} &= \Gamma^\sigma_{\overline{\beta}\alpha} - 2u^\sigma h_{\overline{\beta}\alpha}, \\ e^{2u}\hat{\Gamma}^\sigma_{\hat{0}\alpha} &= \Gamma^\sigma_{0\alpha} + 2u_0\delta^\sigma_\alpha + iu_{\alpha,}{}^\sigma + 2i\Gamma^\sigma_{\overline{\mu}\alpha} u^{\overline{\mu}} - 2i\Gamma^\sigma_{\mu\alpha} u^\mu,\end{aligned}$$

where $u_{\alpha,}{}^\sigma = u_{\alpha,\overline{\beta}} h^{\sigma\overline{\beta}}$.

Using Lemma 6.11, the Weingarten formula, and

$$\hat{\nabla}^A_{f_* X}\xi = -f_*\hat{a}_\xi X + \hat{\nabla}^\perp_X \xi,$$

for $X \in \mathcal{X}(M)$ and $\xi \in \Gamma^\infty(\nu^{2k}(f))$, we obtain

$$\begin{aligned} \hat{\nabla}^\perp_{T_\beta}\zeta_a &= \nabla^\perp_{T_\beta}\zeta_a + 2u_\beta\zeta_a, \\ \hat{\nabla}^\perp_{T_{\overline{\beta}}}\zeta_a &= \nabla^\perp_{T_{\overline{\beta}}}\zeta_a\,, \\ \hat{\nabla}_{\hat{T}}\zeta_a &= \nabla^\perp_{\hat{T}}\zeta_a + 2u_0 e^{-2u}\zeta_a. \end{aligned} \tag{6.87}$$

If $M = \mathbf{H}_n$ and $A = \mathbf{H}_N$ we have $\nabla^A\zeta_a = 0$ and thus $\nabla^\perp\zeta_a = 0$. Thus (by (6.87)) if $M = \mathbf{H}_n(s)$ and $A = \mathbf{H}_N(s)$ the normal Tanaka–Webster connection of F is given by

$$\begin{aligned} \nabla^\perp_{T_\beta}\zeta_a &= 2u_\beta\zeta_a, \\ \nabla^\perp_{T_{\overline{\beta}}}\zeta_a &= 0, \\ \nabla^\perp_{T}\zeta_a &= 2u_0 e^{-2u}\zeta_a, \end{aligned} \tag{6.88}$$

with $u = \log|x|^{-1}$. Next (as a consequence of (6.88)) we may use the identities

$$[T_\alpha, T_\beta] = 0, \quad [T_\alpha, T_{\overline{\beta}}] = -2i\delta_{\alpha\beta}\frac{\partial}{\partial t},$$

and

$$\nabla^\perp_{\partial/\partial t}\zeta_a = 2(u_0 + 2iu_\beta u^\beta)\zeta_a$$

to obtain

$$R^\perp(T_\alpha, T_\beta = \zeta_a = 0, \quad R^\perp(T_{\overline{\alpha}}, T_\beta)\zeta_a = 0, \tag{6.89}$$

and

$$R^\perp(T_\alpha, T_{\overline{\beta}})\zeta_a = \{-2T_{\overline{\beta}}(u_\alpha 9 + 4i\delta_{\alpha\beta}(u_0 + 2iu_\sigma u^\sigma)\}\zeta_a.$$

Finally, taking into account the identities

$$\begin{aligned} u_\alpha = -\tfrac{1}{2}|x|^{-4}\overline{z}_\alpha\phi, \quad & T_{\overline{\beta}}(u_\alpha) = -\tfrac{1}{2}|x|^{-4}\delta_{\alpha\beta}\phi, \\ u_0 = -\tfrac{1}{2}|x|^{-4}t, \quad & \phi\overline{\phi} = |x|^4, \\ u_\sigma u^\sigma = & \tfrac{1}{4}|x|^{-4}\|z\|^2, \end{aligned}$$

it follows that

$$R^\perp(T_\alpha, T_{\overline{\beta}})\zeta_a = -|x|^{-4}\phi\delta_{\alpha\beta}\zeta_a. \tag{6.90}$$

Summing up, the pseudo-Hermitian immersion $F : \mathbf{H}_n(s) \to \mathbf{H}_N(s)$ has (by (6.90)) $R^\perp \neq 0$. However, (6.81) yields $K_{\alpha\beta} = \lambda R_{\alpha\beta}$ with $\lambda = (N+1)/(n+1)$.

(3) (*Pseudo-Siegel domains*) Let $(\alpha, \beta) = (\alpha_1, \ldots, \alpha_n, \beta) \in \mathbf{Z}_+^{n+1}$ be a fixed multi-index and

$$D_{\alpha,\beta} = \Big\{(z_1, \ldots, z_n, w) \in \mathbf{C}^{n+1} : \sum_{j=1}^{n} |z_j|^{2\alpha_j} + Im(w^\beta) - 1 < 0\Big\}$$

(cf. [45]). Then $D_{\mathbf{1},1}$ is the Siegel domain in $\mathbf{C}^{n+1}$ (and $\partial D_{\mathbf{1},1} \simeq \mathbf{H}_n$). Assume that $\beta > 1$ from now on. The boundary $\partial D_{\alpha,\beta}$ of $D_{\alpha,\beta}$ inherits a CR structure (as a real hypersurface of $\mathbf{C}^{n+1}$) spanned by

$$T_j = \frac{\partial}{\partial z_j} - 2i f_j \frac{\partial}{\partial w} \tag{6.91}$$

in some neighborhood of $w \neq 0$ where

$$f_j = \frac{\alpha_j}{\beta} w^{1-\beta} z_j^{\alpha_j - 1} \overline{z}_j^{\alpha_j}.$$

Hence we have the commutation relations

$$\begin{aligned} &[T_j, T_k] = 0, \\ &[T_j, T_{\overline{k}}] = \frac{2i}{\beta}\left\{\frac{\alpha_j^2 |z_j|^{2(\alpha_j-1)}}{w^{\beta-1}} \frac{\partial}{\partial w} + \frac{\alpha_k^2 |z_k|^{2(\alpha_k-1)}}{\overline{w}^{\beta-1}} \frac{\partial}{\partial \overline{w}}\right\}\delta_{jk}. \end{aligned} \tag{6.92}$$

Endow $\partial D_{\alpha,\beta}$ with the pseudo-Hermitian structure $\theta = \theta_{\alpha,\beta}$ given by

$$\theta = \beta w^{\beta-1} dw + \beta \overline{w}^{\beta-1} d\overline{w} + 2i \sum_{j=1}^{n} (g_j dz_j - \overline{g}_j d\overline{z}_j), \tag{6.93}$$

where

$$g_j = \alpha_j z_j^{\alpha_j - 1} \overline{z}_j^{\alpha_j}.$$

Therefore the Levi form of $(\partial D_{\alpha,\beta}, \theta)$ is $\text{diag}(\lambda_1, \ldots, \lambda_n)$, where

$$\lambda_j = 4\alpha_j^2 |z_j|^{2(\alpha_j - 1)}.$$

Therefore, if $\alpha_j > 1,\ 1 \leq j \leq n$, then G_θ is degenerate at each point of $\bigcup_{j=1}^n M_j$, where M_j is the trace of the complex hyperplane $L_j = \{(z, w) : z_j = 0\}$ on the boundary of $D_{\alpha,\beta}$. Next

$$U_{\alpha,\beta} = \partial D_{\alpha,\beta} \setminus \Big[\bigcup_{j=1}^{n} M_j\Big]$$

(an open subset of $\partial D_{\alpha,\beta}$) is a strictly pseudoconvex CR manifold. The characteristic direction T of

$$d\theta = -4i\alpha_j^2 |z_j|^{2(\alpha_j - 1)} dz_j \wedge d\overline{z}^j$$

is given by

$$T = \frac{1}{4\beta|w|^{2(\beta-1)}} \left\{ \overline{w}^{\beta-1} \frac{\partial}{\partial w} + w^{\beta-1} \frac{\partial}{\partial \overline{w}} \right\}.$$

Note that (6.92) may be written

$$j \neq k \Longrightarrow [T_j, T_{\overline{k}}] = 0,$$
$$[T_j, T_{\overline{j}}] = i\lambda_j T.$$

Also

$$[T_j, T] = 0.$$

Using the explicit expressions of the Christoffel symbols in Chapter 1, we derive the (Christoffel symbols of the) Tanaka–Webster connection of $(U_{\alpha,\beta}, \theta)$:

$$\Gamma^s_{jk} = \frac{\partial}{\alpha_j - 1} z_j \delta_{jk} \delta_{js}, \quad \Gamma^s_{\overline{j}k} = 0, \quad \Gamma^s_{0k} = 0. \tag{6.94}$$

Therefore $(U_{\alpha,\beta}, \theta)$ has a vanishing pseudo-Hermitian torsion ($\tau = 0$). As a straightforward consequence of (6.94) the Tanaka–Webster connection of $(U_{\alpha,\beta}, \theta)$ is flat ($R = 0$).

Finally, we look at the structure of the points of weak pseudoconvexity of $\partial D_{\alpha,\beta}$. Let $1 \leq p \leq n$ and set

$$M_{j_1 \cdots j_p} = \partial D_{\alpha,\beta} \cap L_{j_1} \cap \cdots \cap L_{j_p}.$$

Then

$$M_{j_1 \cdots j_p} \simeq \partial D_{\alpha_{j_1 \cdots j_p}, \beta} \subset \mathbf{C}^{n+1-p}$$

(a diffeomorphism), where $\alpha_{j_1 \cdots j_p} = (\alpha_1, \ldots, \hat{\alpha}_{j_1}, \ldots, \hat{\alpha}_{j_p}, \ldots, \alpha_n)$. A natural question is how $M_{j_1 \cdots j_p}$ sits in $\partial D_{\alpha,\beta}$, i.e., equivalently study the geometry of the immersion $f : \partial D_{(\alpha_1, \ldots, \alpha_k), \beta} \to \partial D_{\alpha,\beta}$ induced by the natural map

$$\mathbf{C}^k \times \mathbf{C} \to \mathbf{C}^n \times \mathbf{C}, \ (z, w) \mapsto (z, 0, w), \ 0 \in \mathbf{C}^k, \ 0 < k < n.$$

Using (6.91) one may show that f is a CR immersion. Finally (6.93) yields

$$f^* \theta_{\alpha,\beta} = \theta_{(\alpha_1, \ldots, \alpha_k), \beta},$$

i.e., f is isopseudo-Hermitian.

7

Quasiconformal Mappings

Our scope in Chapter 7 is to report on results by A. Korányi and H.M. Reimann [254]–[256]. The central notion is that of a K*-quasiconformal map*, that is, a contact transformation $f : M \to M'$ of a strictly pseudoconvex CR manifold M, whose differential df is a quasi-isometry as a map $df : H(M) \to H(M')$, with respect to the Levi forms. One of the main results of A. Korányi and H.M. Reimann (cf. op. cit.) is that the existence of smooth K-quasiconformal maps $f : \mathbf{H}_n \to M \subset \mathbf{C}^{n+1}$ from the Heisenberg group is tied to the existence of solutions to the *tangential Beltrami equations*, very much as in the one complex variable counterpart of the theory, cf. L.V. Ahlfors [1]. We also present a generalization of the celebrated Fefferman theorem (cf. [139]) that biholomorphisms of strictly pseudoconvex domains extend smoothly up to the boundary, and therefore induce CR isomorphisms of the boundaries. Here one weakens the hypothesis on the given transformation $F : \Omega \to \Omega$, i.e., one assumes only that F is a symplectomorphism (with respect to the Kähler form associated with the Bergman metric) that extends smoothly up to the boundary. The (weaker) conclusion is that the boundary values of F give rise at least to a contact transformation (cf. Theorem 7.6). The proof relies on the Fefferman asymptotic expansion of the Bergman kernel of Ω. Symplectomorphisms of the Siegel domain are shown to satisfy the (several complex variables) Beltrami system (cf. E. Barletta et al. [41])

$$\frac{\partial f}{\partial \bar{z}^j} = \sum_k d^k_{\bar{j}} \frac{\partial f}{\partial z^k} \,. \tag{7.1}$$

The reader may consult L. Wang [419], for an analysis of the solutions to (7.1). Closing a circle of ideas, the boundary values of solutions to (7.1) satisfy the tangential Beltrami equations.

7.1 The complex dilatation

Let $(M, T_{1,0}(M))$ be an orientable CR manifold (of hypersurface type) of CR dimension n. Let $H(M)$ be its Levi distribution.

Definition 7.1. A C^2 diffeomorphism $f : M \to M'$ of M onto another CR manifold M' is a *contact transformation* if

$$(d_x f)H(M)_x = H(M')_{f(x)},$$

for any $x \in M$. □

Assume from now on that both M and M' are strictly pseudoconvex and let θ and θ' be pseudo-Hermitian structures, on M and M' respectively, whose corresponding Levi forms are positive definite. Let $f : M \to M'$ be a contact transformation. Then

$$f^*\theta' = \lambda_f(\theta, \theta')\,\theta \tag{7.2}$$

for some nowhere-vanishing C^∞ function $\lambda_f(\theta, \theta') : M \to \mathbf{R}$.

Proposition 7.1. *For any contact transformation $f : M \to M'$ of strictly pseudoconvex CR manifolds, the property*

$$\lambda_f(\theta, \theta') > 0 \tag{7.3}$$

(when achieved) is a CR invariant property.

Indeed, if $\hat{\theta}$ and $\hat{\theta}'$ are pseudo-Hermitian structures on M and M' such that the corresponding Levi forms are positive definite, then there are C^∞ functions $u : M \to \mathbf{R}$ and $u' : M' \to \mathbf{R}$ such that $\hat{\theta} = e^u\theta$ and $\hat{\theta}' = e^{u'}\theta'$. Let $v := u' \circ f$. Then (by (7.2))

$$\lambda_f(\hat{\theta}, \hat{\theta}') = e^{v-u}\lambda_f(\theta, \theta'),$$

and Proposition 7.1 is proved. □

We shall need the following lemma:

Lemma 7.1. *Let M and M' be two strictly pseudoconvex CR manifolds and $f : M \to M'$ a contact transformation. Assume that f possesses the CR invariant property* (7.3). *Then $(d_x f)\overline{Z}$ does not belong to $T_{1,0}(M')_{f(x)}$, for any $Z \in T_{1,0}(M)_x$, $Z \neq 0$, and any $x \in M$.*

Proof. The proof is by contradiction. Assume that

$$(d_x f)\overline{Z} \in T_{1,0}(M')_{f(x)},$$

for some $Z \in T_{1,0}(M)_x$, $Z \neq 0$, and some $x \in M$. Let θ and θ' be pseudo-Hermitian structures on M and M' respectively such that the corresponding Levi forms L_θ and $L_{\theta'}$ are positive definite. Then

$$L_{\theta', f(x)}((d_x f)\overline{Z}, (d_x f)Z) > 0.$$

On the other hand,

$$\begin{aligned}
L_{\theta', f(x)}((d_x f)\overline{Z}, (d_x f)Z) &= -i(d\theta')_{f(x)}((d_x f)\overline{Z}, (d_x f)Z) \\
&= -i(f^*d\theta')_x(\overline{Z}, Z) = -id(\lambda_f\theta)_x(\overline{Z}, Z) \\
&= -i\left\{(d\lambda_f) \wedge \theta + \lambda_f\, d\theta\right\}_x (\overline{Z}, Z) \\
&= i\lambda_f\,(d\theta)_x(Z, \overline{Z}) = -\lambda_f L_{\theta,x}(Z, \overline{Z}) < 0,
\end{aligned}$$

a contradiction. Here λ_f is short for $\lambda_f(\theta, \theta')$. □

Definition 7.2. Let $T_{1,0}(M)_f \subset H(M) \otimes \mathbf{C}$ be defined by setting

$$T_{1,0}(M)_f = \{Z \in H(M) \otimes \mathbf{C} : (df)Z \in T_{1,0}(M')\}.$$ □

Lemma 7.2. *Let $f : M \to M'$ be a contact transformation as in Lemma 7.1. Then there is a* **C***-antilinear bundle map $\mu : T_{1,0}(M) \to T_{1,0}(M)$ such that*

$$T_{1,0}(M)_f = \{Z - \overline{\mu Z} : Z \in T_{1,0}(M)\}.$$

Proof. Let $\pi'_{0,1} : H(M') \otimes \mathbf{C} \to T_{0,1}(M')$ be the natural projection. Then

$$T_{1,0}(M)_f = \operatorname{Ker}(\pi'_{0,1} \circ (df)).$$

Let $\{T_\alpha\}$ be a frame of $T_{1,0}(M)$ on U, and $\{T'_\alpha\}$ a frame of $T_{1,0}(M')$ on U', such that $f(U) \subseteq U'$. Then

$$(df)T_\alpha = f_\alpha^\beta T'_\beta + f_\alpha^{\bar\beta} T'_{\bar\beta}, \quad (df)T_{\bar\alpha} = f_{\bar\alpha}^\beta T'_\beta + f_{\bar\alpha}^{\bar\beta} T'_{\bar\beta},$$

for some $f_\alpha^\beta, f_\alpha^{\bar\beta} : U \to \mathbf{C}$. Here $f_{\bar\alpha}^\beta = \overline{f_\alpha^{\bar\beta}}$ and $f_{\bar\alpha}^{\bar\beta} = \overline{f_\alpha^\beta}$. The matrix $\left[f_{\bar\alpha}^\beta\right]$ is nonsingular at each point of U. Indeed, if $\det\left(f_{\bar\alpha}^\beta\right) = 0$ at some $x_0 \in U$ then

$$f_{\bar\alpha}^{\bar\beta}(x_0)\overline{\zeta^\alpha} = 0, \quad 1 \le \beta \le n,$$

for some $(\zeta^1, \dots, \zeta^n) \in \mathbf{C}^n \setminus \{0\}$. Let us set $Z = \zeta^\alpha T_{\alpha,x_0} \in T_{1,0}(M)_{x_0}$. Then

$$(d_{x_0} f)\overline{Z} = \overline{\zeta^\alpha}(d_{x_0} f)T_{\bar\alpha,x_0} = \overline{\zeta^\alpha} f_{\bar\alpha}^\beta(x_0) T'_{\beta, f(x_0)} \in T_{1,0}(M')_{f(x_0)},$$

in contradiction to Lemma 7.1.

Let $\mu : T_{1,0}(M) \to T_{1,0}(M)$ be defined by setting

$$\mu\, T_\alpha = \mu_{\bar\alpha}^\beta T_\beta$$

followed by **C**-antilinear extension, where $\mu_{\bar\alpha}^\beta$ is given by

$$\mu_{\bar\alpha}^\beta f_\beta^\gamma = f_{\bar\alpha}^\gamma .$$

Finally, note that $T_\alpha - \overline{\mu T_\alpha} \in T_{1,0}(M)_f$. Indeed

$$\begin{aligned} \pi'_{0,1} \circ (df)(T_\alpha - \overline{\mu T_\alpha}) &= \pi'_{0,1} \circ (df)(T_\alpha - \mu_\alpha^{\bar\beta} T_{\bar\beta}) \\ &= \pi'_{0,1}(f_\alpha^\beta T'_\beta + f_\alpha^{\bar\beta} T'_{\bar\beta} - \mu_\alpha^{\bar\beta}(f_{\bar\beta}^\gamma T'_\gamma + f_{\bar\beta}^{\bar\gamma} T'_{\bar\gamma})) \\ &= (f_\alpha^{\bar\gamma} - \mu_\alpha^{\bar\beta} f_{\bar\beta}^{\bar\gamma}) T'_{\bar\gamma} = 0, \end{aligned}$$

where $\mu_\alpha^{\bar\beta} = \overline{\mu_{\bar\alpha}^\beta}$. □

Definition 7.3. The bundle map $\mu : T_{1,0}(M) \to T_{1,0}(M)$ (determined by f via Lemma 7.2) is called the *complex dilatation* of f. □

When danger of confusion may arise, we denote μ by μ_f. Note that

$$(d\theta)(Z, \overline{\mu W}) + (d\theta)(\overline{\mu Z}, W) = 0,$$

for any $Z, W \in T_{1,0}(M)$. Indeed, let $Z \in T_{1,0}(M)$. Then (by Lemma 7.2) $Z - \overline{\mu Z} \in T_{1,0}(M)_f$; hence (by the definition of $T_{1,0}(M)_f$) $(df)(Z - \overline{\mu Z}) \in T_{1,0}(M')$. By the integrability of the CR structure $T_{1,0}(M)$, the 2-form $d\theta$ vanishes on complex vectors of the same type. Hence

$$\begin{aligned} 0 &= (d\theta')((df)(Z - \overline{\mu Z}), (df)(W - \overline{\mu W})) \\ &= \lambda_f \, (d\theta)(Z - \overline{\mu Z}, W - \overline{\mu W}) \\ &= -\lambda_f \left((d\theta)(Z, \overline{\mu W}) + (d\theta)(\overline{\mu Z}, W) \right). \end{aligned}$$

A couple of remarks are in order.

(1) Let $f : M \to M'$ be a contact transformation of strictly pseudoconvex CR manifolds possessing the property (7.3). Then f is a CR map if and only if $\mu_f = 0$.

(2) If $f : M \to M'$ is a CR map then $f^* G_{\theta'} = \lambda_f G_\theta$ on $H(M) \otimes H(M)$. This is no longer true when f is only contact. Nevertheless, one may change the complex structure on $H(M)$ in a suitable way. That is, let $J_f : H(M) \to H(M)$ be given by

$$J_{f,x} = (d_x f)^{-1} \circ J'_{f(x)} \circ (d_x f)$$

for any $x \in M$. Also, let us set

$$G_f(X, Y) = (d\theta)(X, J_f Y)$$

for any $X, Y \in H(M)$. Then $f^* G_{\theta'} = \lambda_f G_f$ on $H(M) \otimes H(M)$.

Proposition 7.2. *Let $f_i : M \to M_i$, $i = 1, 2$, be two contact transformations of strictly pseudoconvex CR manifolds, both possessing the CR invariant property* (7.3). *If f_i, $i \in \{1, 2\}$, have the same complex dilatation, i.e., $\mu_{f_1} = \mu_{f_2}$, then $\phi = f_2 \circ f_1^{-1} : M_1 \to M_2$ is a CR diffeomorphism.*

Proof. Let us choose contact forms θ, θ_i on M, M_i, respectively, and adopt the notation

$$\lambda_i = \lambda_{f_i}(\theta, \theta_i) \ , \ i \in \{1, 2\},$$

so that

$$f_i^* \theta_i = \lambda_i \theta \ , \ \lambda_i > 0.$$

It is easy to see that

$$\phi^* \theta_2 = \frac{\lambda_2}{\lambda_1} \theta_1;$$

hence

$$\lambda_\phi = \lambda_\phi(\theta_1, \theta_2) = \frac{\lambda_1}{\lambda_2} > 0.$$

In particular ϕ is a contact transformation of M_1 into M_2 possessing the CR invariant property (7.3). Let then $\mu_\phi : T_{1,0}(M_1) \to T_{1,0}(M_1)$ be its complex dilatation. Since the maps f_i have the same dilatation,

$$\mathrm{Ker}(\pi_{0,1}^{(1)} \circ (df_1)) = \mathrm{Ker}(\pi_{0,1}^{(2)} \circ (df_2)),$$

where $\pi_{0,1}^{(i)} : H(M_i) \otimes \mathbf{C} \to T_{0,1}(M_i)$ are the natural projections. Then

$$\begin{aligned}\{Z - \overline{\mu_\phi Z} : Z \in T_{1,0}(M_1)\} &= \mathrm{Ker}(\pi_{0,1}^{(2)} \circ (d\phi)) \\ &\subseteq (df_1)\left(\mathrm{Ker}(\pi_{0,1}^{(2)} \circ (df_2))\right) = (df_1)\left(\mathrm{Ker}(\pi_{0,1}^{(1)} \circ (df_1))\right) \\ &= (df_1)\{Z \in H(M) \otimes \mathbf{C} : (df_1)Z \in T_{1,0}(M_1)\} \subseteq T_{1,0}(M_1).\end{aligned}$$

Consequently, for any $Z \in T_{1,0}(M_1)$ we get $Z - \overline{\mu_\phi Z} \in T_{1,0}(M_1)$. It follows that

$$\overline{\mu_\phi Z} \in T_{1,0}(M_1) \cap T_{0,1}(M_1) = (0);$$

hence $\mu_\phi = 0$. □

We close this section by relating the notion of *Beltrami differential* (such as introduced by L. Lempert [278]) to the notion of complex dilatation.

Let M be a 3-dimensional CR manifold. Let $\{g_\zeta : \zeta \in S^1\}$ be a positive contact action of S^1 on M. Let $p \in M$ and $X \in H(M)_p \otimes \mathbf{C}$ be such that $\{X, \overline{X}\}$ are linearly independent. Since M is 3-dimensional, $H(M)_p \otimes \mathbf{C}$ has complex dimension 2; hence $\{X, \overline{X}\}$ span $H(M)_p \otimes \mathbf{C}$. Since we have $T_{0,1}(M)_p \subset H(M)_p \otimes \mathbf{C}$ it follows that there is $(\alpha, \beta) \in \mathbf{C}^2_*$ with

$$\alpha \overline{X} + \beta X \in T_{0,1}(M)_p\,.$$

Then $\overline{\alpha} X + \overline{\beta X} \in T_{1,0}(M)_p$; hence $\{\alpha \overline{X} + \beta X,\ \overline{\alpha} X + \overline{\beta}\ \overline{X}\}$ are linearly independent. As an immediate consequence $|\alpha| \neq |\beta|$. Indeed, if $|\alpha| = |\beta| = r$ then $\alpha = re^{i\varphi}$ and $\beta = re^{i\psi}$ $(r \neq 0)$ and

$$\alpha \overline{X} + \beta X = r(e^{i\varphi}\overline{X} + e^{i\psi}X) = re^{i(\varphi+\psi)}(e^{-i\varphi}X + e^{-i\psi}\overline{X}) = e^{i(\varphi+\psi)}(\overline{\alpha}X + \overline{\beta}\ \overline{X}),$$

i.e., $\{\alpha\overline{X} + \beta X,\ \overline{\alpha}X + \overline{\beta}\ \overline{X}\}$ are linearly dependent, a contradiction. □

Consequently, either $|\alpha| < |\beta|$ or $|\alpha| > |\beta|$. We adopt the following definition:

Definition 7.4. X is said to be a $(1, 0)$*-like vector* if

$$\{(\alpha, \beta) \in \mathbf{C}^2_* : \alpha\overline{X} + \beta X \in T_{0,1}(M)_p\} \subseteq \{(\alpha, \beta) \in \mathbf{C}^2 : |\alpha| > |\beta|\}. \qquad \square$$

Here $\mathbf{C}^2_* = \mathbf{C}^2 \setminus \{(0,0)\}$. Since $g_\zeta : M \to M$ is a contact map for each $\zeta \in S^1$, one has

$$(d_p g_\zeta)X, \ (d_p g_\zeta)\overline{X} \in H(M)_{g_\zeta(p)} \otimes \mathbf{C},$$

and $\{(d_p g_\zeta)X, \ (d_p g_\zeta)\overline{X}\}$ are linearly independent (since $d_p g_\zeta$ is a linear isomorphism). Therefore, there is $(a, b) \in \mathbf{C}^2_*$ such that

$$a(d_p g_\zeta)\overline{X} + b(d_p g_\zeta)X \in T_{0,1}(M)_{g_\zeta(p)}.$$

Definition 7.5. Let $\mu : S^1 \to \mathbf{C} \cup \{\infty\}$ be defined by $\mu(\zeta) = b/a$. μ is called a *Beltrami differential* associated with the orbit of p. □

Note that $\mu(\zeta)$ is uniquely determined. Indeed, if $(a', b') \in \mathbf{C}^2_*$ is such that

$$a'(d_p g_\zeta)\overline{X} + b'(d_p g_\zeta)X \in T_{0,1}(M)_{g_\zeta(p)},$$

then (since $T_{0,1}(M)_{g_\zeta(p)}$ has complex dimension 1)

$$a'(d_p g_\zeta)\overline{X} + b'(d_p g_\zeta)X = \lambda[a(d_p g_\zeta)\overline{X} + b(d_p g_\zeta)X]$$

for some $\lambda \in \mathbf{C}$; hence $a' = \lambda a$, $b' = \lambda b$ (and $\lambda \neq 0$). □

Next, let us observe that $|\mu(\zeta)| < 1$, for any $\zeta \in S^1$. Indeed, note first that $\mu(1) = \beta/\alpha$; hence $|\mu(1)| < 1$. Moreover, it may be seen that $|\mu(\zeta)| \neq 1$, by an argument entirely analogous to that leading to the conclusion $|\alpha| \neq |\beta|$. Therefore, it must be that $|\mu(\zeta)| < 1$ for each $\zeta \in S^1$. □

The terminology adopted in Definition 7.5 is justified by the following result:

Proposition 7.3. (L. Lempert [278])
Let $p' = g_\omega(p)$ be a point in the orbit of p ($\omega \in S^1$). Let μ' be the Beltrami differential determined by p' and by a vector $X' \in H(M)_{p'} \otimes \mathbf{C}$ of $(1,0)$-type. Then there is an automorphism $\varphi \in \mathrm{Aut}(\Delta)$ of the unit disk $\Delta \subset \mathbf{C}$ such that $\mu'(\zeta) = \varphi(\mu(\omega\zeta))$, for any $\zeta \in S^1$.

In his paper [278], L. Lempert wrote, "*Let us remark here that 'Beltrami coefficients' on CR manifolds have first been introduced by Korányi and Reimann.*[1] ... *Although our Beltrami differentials are in spirit related to theirs, there is no logical relationship, and they are different kind of objects.*" As it turns out, mathematical reality is for once better than the researcher's expectations. In the remainder of this section we shall demonstrate the actual relationship between the Beltrami differential associated with the orbit of a point and the complex dilatation associated with a contact transformation.

Assume M to be strictly pseudoconvex and let θ be a contact form on M with L_θ positive definite. Since $g_\zeta : M \to M$ is a contact transformation,

$$g_\zeta^* \theta = \lambda_\zeta \, \theta,$$

[1] Cf. A. Korányi and H.M. Reimann [254]–[255].

where, with the notation adopted earlier in this section, $\lambda_\zeta = \lambda_{g_\zeta}(\theta, \theta)$. Assume that $\lambda_\zeta > 0$ everywhere on M and let

$$\mu_\zeta = \mu_{g_\zeta} : T_{1,0}(M) \to T_{1,0}(M)$$

be the complex dilatation associated with g_ζ. Let $\{T_1\}$ be a local frame of $T_{1,0}(M)$, defined on an open set $U \subseteq M$. Then

$$(dg_\zeta)T_1 = f_1^1(\zeta)T_1 + f_1^{\overline{1}}(\zeta)T_{\overline{1}},$$
$$(dg_\zeta)T_{\overline{1}} = f_{\overline{1}}^1(\zeta)T_1 + f_{\overline{1}}^{\overline{1}}(\zeta)T_{\overline{1}},$$

for some smooth functions $f_B^A : U \to \mathbf{C}$ such that $f_1^1(\zeta)$ is $(\mathbf{C} \setminus \{0\})$-valued. Then the complex dilatation μ_ζ of g_ζ is given by

$$\mu_\zeta T_1 = \mu_{\overline{1}}^1(\zeta)T_1, \qquad \mu_{\overline{1}}^1(\zeta) := \frac{f_{\overline{1}}^1(\zeta)}{f_1^1(\zeta)}.$$

Let $p \in M$ and let $X \in H(M)_p \otimes \mathbf{C}$ be a $(1,0)$-like vector, i.e.,

$$A_p(X) \subseteq \{(\alpha, \beta) \in \mathbf{C}^2 : |\alpha| > |\beta|\},$$

where $A_p(X) := \{(\alpha, \beta) \in \mathbf{C}_*^2 : \alpha\overline{X} + \beta X \in T_{0,1}(M)_p\}$. Next, let us set

$$B_p(X; \zeta) := \{(a, b) \in \mathbf{C}_*^2 : a(d_p g_\zeta)\overline{X} + b(d_p g_\zeta)X \in T_{0,1}(M)_{g_\zeta(p)}\}.$$

As previously shown, $B_p(X; \zeta) \subseteq \{(a, b) \in \mathbf{C}^2 : |a| \neq |b|\}$. Let $(a, b) \in B_p(X; \zeta)$. If $X = AT_{1,p} + BT_{\overline{1},p}$ then

$$\begin{aligned}
T_{0,1}(M)_{g_\zeta(p)} &\ni a(d_p g_\zeta)\overline{X} + b(d_p g_\zeta)X \\
&= (a\overline{A} + bB)\{f_{\overline{1}}^1(\zeta)T_1 + f_{\overline{1}}^{\overline{1}}(\zeta)T_{\overline{1}}\}_p + (a\overline{B} + bA)\{f_1^1(\zeta)T_1 + f_1^{\overline{1}}(\zeta)T_{\overline{1}}\}_p \\
&\qquad = \{(a\overline{A} + bB)f_{\overline{1}}^1(\zeta)_p + (a\overline{B} + bA)f_1^1(\zeta)_p\}T_{1,p} \\
&\qquad + \{(aA + bB)f_{\overline{1}}^{1}(\zeta)_p + (a\overline{B} + bA)f_1^{\overline{1}}(\zeta)_p\}T_{\overline{1},p} \\
&\qquad = f_1^1(\zeta)_p\{(a\overline{A} + bB)\mu_{\overline{1}}^1(\zeta)_p + a\overline{B} + bA\}T_{1,p} \\
&\qquad\qquad + f_{\overline{1}}^{\overline{1}}(\zeta)_p\{a\overline{A} + bB + (a\overline{B} + bA)\mu_1^{\overline{1}}(\zeta)_p\}T_{\overline{1},p};
\end{aligned}$$

hence (since $f_1^1(\zeta)_p \neq 0$)

$$(a\overline{A} + bB)\mu_{\overline{1}}^1(\zeta)_p + a\overline{B} + bA = 0.$$

Note that $a\overline{A} + bB \neq 0$. Indeed, if it were $a\overline{A} + bB = 0$ then $a\overline{B} + bA = 0$ as well; hence $a(d_p g_\zeta)\overline{X} + b(d_p g_\zeta)X = 0$, i.e., one would get $a = b = 0$, a contradiction. Consequently (by taking into account that $\mu : S^1 \to \Delta$ is given by $\mu(\zeta) = b/a$)

$$\mu_{\frac{1}{1}}(\zeta)_p = -\frac{A\mu(\zeta) + \overline{B}}{B\mu(\zeta) + \overline{A}}. \tag{7.4}$$

Our claim is proved. An elegant formulation of the result is

$$\mu_{\zeta,p}\left(\overline{\alpha} X + \overline{\beta}\, \overline{X}\right) = -\frac{A\mu(\zeta) + \overline{B}}{B\mu(\zeta) + \overline{A}} \left(\overline{\alpha} X + \overline{\beta}\, \overline{X}\right) \tag{7.5}$$

for any $X = AT_{1,p} + BT_{\overline{1},p}$ and any $(\alpha, \beta) \in A_p(X)$. To prove (7.5) let $(\alpha, \beta) \in A_p(X)$. Then $\alpha \overline{B} + \beta A = 0$. If for instance $A \neq 0$ then $\beta = -(\overline{B}/A)\alpha$ and

$$\mu_{\zeta,p}(\overline{\alpha} X + \overline{\beta}\, \overline{X}) = (\overline{\alpha} A + \overline{\beta}\, \overline{B})\mu_{\frac{1}{1}}(\zeta)_p T_{1,p} = \frac{|A|^2 - |B|^2}{\overline{A}}\, \overline{\alpha}\, \mu_{\frac{1}{1}}(\zeta)_p\, T_{1,p}.$$

Then (7.4) and

$$T_{1,p} = \frac{\overline{A}}{|A|^2 - |B|^2}\, X - \frac{B}{|A|^2 - |B|^2}\, \overline{X}$$

yield (7.5). □

7.2 *K*-quasiconformal maps

The following notion is central to the present chapter.

Definition 7.6. Let $K > 0$ be fixed. A contact transformation $f : M \to M'$ satisfying the CR invariant property (7.3) is called a *smooth K-quasiconformal* map if

$$\frac{\lambda_f}{K} G_\theta(X, X) \le G_{\theta'}((df)X, (df)X) \le \lambda_f K G_\theta(X, X), \tag{7.6}$$

for any $X \in H(M)$, for some contact 1-forms θ, θ' with $L_\theta, L_{\theta'}$ positive definite (and thus for all). □

Note that (7.6), when satisfied, is a CR invariant property. The following reformulation of K-quasiconformality is elementary, yet useful.

Proposition 7.4. *Let $f : M \to M'$ be a contact transformation. Then f is a K-quasiconformal map if and only if*

$$\frac{1}{K} G_\theta(X, X) \le G_f(X, X) \le K G_\theta(X, X),$$

for some contact 1-form θ on M and all $X \in H(M)$.

As to compositions of quasiconformal maps, we have the following theorem:

Theorem 7.1. *Let $f_i : M \to M_i$ be K_i-quasiconformal, $i \in \{1, 2\}$, for some $K_i > 0$. If the maps f_i have the same complex dilatation then $\phi = f_2 \circ f_1^{-1}$ is a K-quasiconformal map, of zero dilatation, with $K = K_1 K_2$.*

The proof is left as an exercise to the reader.

Let $f : M \to M'$ be a contact transformation of complex dilatation μ. For further use, let us define $\|\mu\| : M \to [0, \infty)$ by setting

$$\|\mu\|_x = \sup\{L_{\theta,x}(\mu_x Z, \overline{\mu_x Z})^{1/2} : L_{\theta,x}(Z, \overline{Z}) = 1, Z \in T_{1,0}(M)_x\}$$

for any $x \in M$.

7.3 The tangential Beltrami equations

Let $f : \mathbf{H}_n \to \mathbf{C}^{n+1}$ be a C^2 map of components $f = (f^1, \dots, f^{n+1})$. Then

$$(df)T_\alpha = T_\alpha(f^k)\frac{\partial}{\partial z^k} + T_\alpha(\overline{f^k})\frac{\partial}{\partial \bar{z}^k},$$

where $T_\alpha = \partial/\partial z^\alpha + i\bar{z}^\alpha \partial/\partial t$. Since $T_\alpha \in T_{1,0}(\mathbf{H}_n)$, if f is a contact transformation of complex dilatation μ then

$$T_\alpha - \overline{\mu T_\alpha} \in T_{1,0}(\mathbf{H}_n)_f = \text{Ker}\left(\pi_{0,1} \circ (df)\,\big|_{H(\mathbf{H}_n)\otimes\mathbf{C}}\right)$$

(where $\pi_{0,1} : H(M) \otimes \mathbf{C} \to T_{0,1}(M)$ is the projection and $M = f(\mathbf{H}_n)$ carries the CR structure induced from (the complex structure of) $\mathbf{C}^{n+1}$). Throughout, we assume M to be strictly pseudoconvex. The complex dilatation of f is a bundle morphism $\mu : T_{1,0}(\mathbf{H}_n) \to T_{1,0}(\mathbf{H}_n)$; hence we may write

$$\mu T_\alpha = \mu_{\bar{\alpha}}^{\beta} T_\beta$$

and

$$\begin{aligned} 0 &= \pi_{0,1}\left((df)\left(T_\alpha - \mu_\alpha^{\bar{\beta}} T_{\bar{\beta}}\right)\right) \\ &= \pi_{0,1}\left[\left(T_\alpha(f^k)\frac{\partial}{\partial z^k} + T_\alpha(\overline{f^k})\frac{\partial}{\partial \bar{z}^k}\right)\right. \\ &\qquad \left. - \mu_\alpha^{\bar{\beta}}\left(T_{\bar{\beta}}(f^k)\frac{\partial}{\partial z^k} + T_{\bar{\beta}}(\overline{f^k})\frac{\partial}{\partial \bar{z}^k}\right)\right] \\ &= T_\alpha(\overline{f^k})\frac{\partial}{\partial \bar{z}^k} - \mu_\alpha^{\bar{\beta}} T_{\bar{\beta}}(\overline{f^k})\frac{\partial}{\partial \bar{z}^k} \end{aligned}$$

(where $\mu_\alpha^{\bar{\beta}} = \overline{\mu_{\bar{\alpha}}^{\beta}}$), whence

$$T_{\bar{\alpha}}(f^k) = \mu_{\bar{\alpha}}^{\beta} T_\beta(f^k), \quad 1 \le k \le n+1. \tag{7.7}$$

Definition 7.7. The equations (7.7) are referred to as the *tangential Beltrami equations*. □

The reason for the choice of terminology will become apparent when we discuss the boundary behavior of a symplectomorphism (cf. [256]) of the Siegel domain to itself. A. Korányi and H.M. Reimann refer to (7.7) as the Beltrami equation (cf. [255]) by analogy with the Beltrami equation $\partial_{\bar{z}} f = \mu \partial_z f$ in one complex variable (cf., e.g., [1]). However, as we shall see later on, the trace on $\partial\Omega_{n+1}$ of any solution $f \in C^2(\overline{\Omega}_n)$ of the ordinary Beltrami system (cf., e.g., [419]) satisfies the tangential Beltrami equations on some open subset; hence our terminology seems the most appropriate.

Let $f : M \to M'$ be a contact transformation of complex dilatation μ and let $\{T_\alpha\}$ be a (local) frame of $T_{1,0}(M)$. Then (by an earlier observation)

$$(d\theta)(T_\alpha, \overline{\mu T_\beta}) + (d\theta)(\overline{\mu T_\alpha}, T_\beta) = 0,$$

i.e.,

$$\mu_{\alpha\beta} = \mu_{\beta\alpha},$$

where

$$\mu_{\alpha\beta} = \overline{\mu_{\bar{\alpha}\bar{\beta}}}, \quad \mu_{\bar{\alpha}\bar{\beta}} = h_{\gamma\bar{\alpha}} \mu^{\gamma}_{\bar{\beta}} .$$

Let us set

$$\theta_0 = dt + \frac{i}{2} \sum_{\alpha=1}^{n} \left(z^\alpha d\bar{z}^\alpha - \bar{z}^\alpha dz^\alpha \right).$$

We have shown that the components of a contact transformation $f : \mathbf{H}_n \to M = f(\mathbf{H}_n) \subset \mathbf{C}^{n+1}$ satisfy the tangential Beltrami equations (7.7). As to the converse of this statement, we have the following result:

Theorem 7.2. (A. Korányi and H.M. Reimann [255])
Let $f : \mathbf{H}_n \to M = f(\mathbf{H}_n) \subset \mathbf{C}^{n+1}$ be a diffeomorphism each of whose components f^k satisfies the tangential Beltrami equations $T_{\bar{\alpha}}(f^k) = \mu^{\beta}_{\bar{\alpha}} T_\beta(f^k)$ for some smooth functions $\mu^{\beta}_{\bar{\alpha}} : \mathbf{H}_n \to \mathbf{C}$ such that $\mu_{\alpha\beta} = \mu_{\beta\alpha}$. Assume M to be strictly pseudoconvex. Then

(i) The map f is a contact transformation and there is a contact 1-form θ' on M such that $\lambda_f(\theta_0, \theta') = 1$.
(ii) Let μ be the complex dilatation of f. If $\|\mu\| \leq 1$ then $L_{\theta'}$ is positive definite, where $\|\mu\|$ is computed with respect to θ_0 and θ' (furnished by (i)).

Proof. Let us set

$$V_\alpha = T_\alpha(f^k) \frac{\partial}{\partial z^k}, \quad W_\alpha = T_{\bar{\alpha}}(f^k) \frac{\partial}{\partial z^k} .$$

Then, on the one hand,

$$(df)T_\alpha = V_\alpha + W_{\bar{\alpha}} ,$$

where $W_{\bar{\alpha}} = \overline{W_\alpha}$, and on the other, since f^k satisfies the tangential Beltrami equations,

$$W_\alpha = \mu_{\bar{\alpha}}^{\beta} V_\beta \, .$$

Let us show that

$$(df)\left(T_\alpha - \mu_\alpha^{\bar{\beta}} T_{\bar{\beta}}\right) \in T_{1,0}(M). \tag{7.8}$$

Indeed

$$\begin{aligned}(df)\left(T_\alpha - \mu_\alpha^{\bar{\beta}} T_{\bar{\beta}}\right) &= V_\alpha + W_{\bar{\alpha}} - \mu_\alpha^{\bar{\beta}} \left(\overline{V_\beta} + W_\beta\right) \\ &= V_\alpha + W_{\bar{\alpha}} - \overline{W_\alpha} - \mu_\alpha^{\bar{\beta}} W_\beta \in T^{1,0}(\mathbf{C}^{n+1}) \cap [T(M) \otimes \mathbf{C}] = T_{1,0}(M).\end{aligned}$$

As our next step, let us observe that the complex tangent vectors

$$Z_\alpha = (df)\left(T_\alpha - \mu_\alpha^{\bar{\beta}} T_{\bar{\beta}}\right)$$

are (pointwise) linearly independent and span $T_{1,0}(M)$. At this point, we may prove the existence of a pseudo-Hermitian structure θ' on M such that $f^*\theta' = \theta_0$. Indeed, let θ'' be just any pseudo-Hermitian structure on M and let us set

$$B = \operatorname{span}\left\{T_\alpha - \mu_\alpha^{\bar{\beta}} T_{\bar{\beta}}\right\}_{\alpha=1}^{n} \subset T(\mathbf{H}_n) \otimes \mathbf{C}.$$

Then $\dim_{\mathbf{C}} B_x = n$, $x \in \mathbf{H}_n$. Note that $\operatorname{Re}\{B \oplus \overline{B}\} \subseteq H(\mathbf{H}_n)$. Yet $\operatorname{Re}\{B \oplus \overline{B}\}$ has real rank $2n$; hence we have equality, i.e.,

$$H(\mathbf{H}_n) = \operatorname{Re}\{B \oplus \overline{B}\}. \tag{7.9}$$

Taking into account (7.8), we have

$$\left(f^*\theta''\right) B = \theta''(df)B \subseteq \theta''(T_{1,0}(M)) = 0.$$

Then, by complex conjugation $\left(f^*\theta''\right)\overline{B} = 0$; hence (by (7.9))

$$\left(f^*\theta''\right) H(\mathbf{H}_n) = 0.$$

Consequently

$$f^*\theta'' = \lambda\theta_0 \, ,$$

for some $\lambda \in C^\infty(\mathbf{H}_n)$. Note that $\lambda(x) \neq 0$ for any $x \in \mathbf{H}_n$. Indeed, if $\lambda(x) = 0$ for some $x \in \mathbf{H}_n$ then $\left(f^*\theta''\right)_x = 0$, i.e., $\theta''_{f(x)} \circ (d_x f) = 0$. Yet $d_x f$ is a linear isomorphism; hence $\theta''_{f(x)} = 0$, i.e., $\operatorname{Sing}(\theta'') \neq \emptyset$, a contradiction. Then we set

$$\theta' = \frac{1}{\lambda \circ f^{-1}} \theta'' \, .$$

Therefore

$$\left(f^*\theta'\right)_x = \theta'_{f(x)} \circ (d_x f) = \frac{1}{\lambda(x)}\theta''_{f(x)} \circ (d_x f) = \frac{1}{\lambda(x)}\left(f^*\theta''\right)_x = \theta_{0,x},$$

or

$$f^*\theta' = \theta\,,$$

i.e., f is a contact transformation with $\lambda_f(\theta_0, \theta') = 1$. □

To prove the second statement in the theorem, note that, by the definitions, if $Z \in T_{1,0}(\mathbf{H}_n)$ then $Z - \overline{\mu Z} \in T_{1,0}(\mathbf{H}_n)_f$, i.e., $Z - \overline{\mu Z} \in H(M) \otimes \mathbf{C}$ and $(df)(Z - \overline{\mu Z}) \in T_{1,0}(M)$. Consequently

$$T_{1,0}(M) = \operatorname{Span}\{Z_\alpha\}_{\alpha=1}^n,$$

and to see that $L_{\theta'}$ is positive definite it suffices to estimate $L_{\theta'}(Z_\alpha, Z_{\bar\alpha})$ from below. Indeed

$$\begin{aligned} L_{\theta'}(Z_\alpha, Z_{\bar\alpha}) &= -i(d\theta')(Z_\alpha, Z_{\bar\alpha}) \\ &= -i(df^*\theta')(T_\alpha - \mu_\alpha^{\bar\beta} T_{\bar\beta}, T_{\bar\alpha} - \mu_{\bar\alpha}^{\gamma} T_\gamma) \\ &= -i(d\theta_0)(T_\alpha, T_{\bar\alpha}) - i(d\theta_0)(T_{\bar\beta}, T_\gamma)\mu_\alpha^{\bar\beta}\mu_{\bar\alpha}^{\gamma} \\ &= L_{\theta_0}(T_\alpha, T_{\bar\alpha}) - L_{\theta_0}(\mu T_\alpha, \overline{\mu T_\alpha}) \\ &= L_{\theta_0}(T_\alpha, T_{\bar\alpha})\left\{1 - \frac{L_{\theta_0}(\mu T_\alpha, \overline{\mu T_\alpha})}{L_{\theta_0}(T_\alpha, T_{\bar\alpha})}\right\} \\ &\geq L_{\theta_0}(T_\alpha, T_{\bar\alpha})(1 - \|\mu\|^2) \geq 0. \end{aligned}$$

□

7.3.1 Contact transformations of $\mathbf{H}_n$

Theorem 7.3. (A. Korányi and H.M. Reimann [254])
Let $f : \mathbf{H}_n \to \mathbf{H}_n$ be a contact transformation with $\lambda = \lambda_f(\theta_0, \theta_0) > 0$. The following statements are equivalent:

(i) f is a K-quasiconformal map.
(ii) The complex dilatation μ of f satisfies

$$\|\mu\| \leq \frac{K-1}{K+1}\,.$$

Proof. We adopt the following notation:

$$\begin{aligned} e_\alpha = X_\alpha &= \frac{\partial}{\partial x^\alpha} + 2y^\alpha \frac{\partial}{\partial t} = T_\alpha + T_{\bar\alpha}\,, \\ e_{\alpha+n} = Y_\alpha &= \frac{\partial}{\partial y^\alpha} - 2x^\alpha \frac{\partial}{\partial t} = i(T_\alpha - T_{\bar\alpha}), \\ (df)e_A &= F_A^B e_B\,, \\ g &= \left[\frac{1}{\sqrt{\lambda}} F_A^B\right]_{\substack{1\leq A\leq 2n \\ 1\leq B\leq 2n}} : \mathbf{H}_n \to \mathrm{GL}(2n, \mathbf{R}), \end{aligned}$$

where $\{e_A\} = \{e_\alpha, e_{\alpha+n}\}$. Consider

$$\mathrm{Sp}(2n, \mathbf{R}) = \{g \in \mathrm{GL}(2n, \mathbf{R}) : g^t J_0 g = J_0\},$$

where

$$J_0 = \begin{pmatrix} 0 & -I_n \\ I_n & 0 \end{pmatrix}.$$

Since $f^*\theta_0 = \lambda\theta_0$ and exterior differentiation commutes with the pullback by C^1 maps, it follows that

$$f^*(d\theta_0) = (d\lambda) \wedge \theta_0 + \lambda(d\theta_0);$$

hence

$$(d\theta_0)((df)X, (df)Y) = \lambda(d\theta_0)(X, Y),$$

for any $X, Y \in H(\mathbf{H}_n)$. In particular

$$F_A^C F_B^D (d\theta_0)(e_C, e_D) = (d\theta_0)((df)e_A, (df)e_B) = \lambda(d\theta_0)(e_A, e_B).$$

On the other hand,

$$\begin{aligned}(d\theta_0)(e_A, e_B) &= 2i\Big(\sum_{\alpha=1}^{n} dz^\alpha \wedge d\bar{z}^\alpha\Big)(e_A, e_B)\\ &= i\sum_{\alpha=1}^{n}\big(dz^\alpha(e_A)d\bar{z}^\alpha(e_B) - dz^\alpha(e_B)d\bar{z}^\alpha(e_A)\big)\end{aligned}$$

and by

$$\begin{aligned} dz^\alpha(e_\beta) &= \delta^\alpha_\beta, \quad & dz^\alpha(e_{\beta+n} &= i\delta^\alpha_\beta\,,\\ d\bar{z}^\alpha(e_\beta) &= \delta^\alpha_\beta, \quad & d\bar{z}^\alpha(e_{\beta+n}) &= -i\delta^\alpha_\beta\,,\end{aligned}$$

we get

$$\begin{aligned} (d\theta_0)(e_\beta, e_\gamma) &= 0, \qquad & (d\theta_0)(e_\beta, e_{\gamma+n}) &= 2\delta_{\beta\gamma}\,,\\ (d\theta_0)(e_{\beta+n}, e_\gamma) &= -2\delta_{\beta\gamma}, \qquad & (d\theta_0)(e_{\beta+n}, e_{\gamma+n}) &= 0.\end{aligned}$$

Consequently (for $A = \alpha$ and $B = \beta$)

$$\begin{aligned} 0 &= F_\alpha^C F_\beta^D (d\theta_0)(e_C, e_D)\\ &= F_\alpha^\lambda F_\beta^\mu (d\theta_0)(e_\lambda, e_\mu) + F_\alpha^\lambda F_\beta^{\mu+n}(d\theta_0)(e_\lambda, e_{\mu+n})\\ &\qquad + F_\alpha^{\lambda+n} F_\beta^\mu (d\theta_0)(e_{\lambda+n}, e_\mu) + F_\alpha^{\lambda+n} F_\beta^{\mu+n}(d\theta_0)(e_{\lambda+n}, e_{\mu+n})\\ &= 2F_\alpha^\lambda F_\beta^{\mu+n}\delta_{\lambda\mu} - 2F_\alpha^{\lambda+n} F_\beta^\mu \delta_{\lambda\mu}, \end{aligned}$$

or

$$\sum_{\lambda=1}^{n}\left(F_\alpha^\lambda F_\beta^{\lambda+n} - F_\alpha^{\lambda+n} F_\beta^\lambda\right) = 0, \tag{7.10}$$

for any $1 \le \alpha, \beta \le n$. Similar calculations for $A = \alpha$, $B = \beta + n$, respectively for $A = \alpha + n$, $B = \beta$, and for $A = \alpha + n$, $B = \beta + n$, lead to the following identities:

$$\sum_{\lambda=1}^{n}\left(F_\alpha^\lambda F_{\beta+n}^{\lambda+n} - F_\alpha^{\lambda+n} F_{\beta+n}^\lambda\right) = \lambda\, \delta_{\alpha\beta}\,, \tag{7.11}$$

$$\sum_{\lambda=1}^{n}\left(F_{\alpha+n}^\lambda F_\beta^{\lambda+n} - F_{\alpha+n}^{\lambda+n} F_\beta^\lambda\right) = -\lambda\, \delta_{\alpha\beta}\,, \tag{7.12}$$

$$\sum_{\lambda=1}\left(F_{\alpha+n}^\lambda F_{\beta+n}^{\lambda+n} - F_{\alpha+n}^{\lambda+n} F_{\beta+n}^\lambda\right) = 0. \tag{7.13}$$

Note that

$$a = \begin{pmatrix} a_\alpha^\beta & a_{\alpha+n}^\beta \\ a_\alpha^{\beta+n} & a_{\alpha+n}^{\beta+n} \end{pmatrix} \in \mathrm{Sp}(2n, \mathbf{R})$$

if and only if

$$\sum_\gamma \left(a_\alpha^{\gamma+n} a_\beta^\gamma - a_\alpha^\gamma a_\beta^{\gamma+n}\right) = 0,$$

$$\sum_\gamma \left(a_{\alpha+n}^{\gamma+n} a_\beta^\gamma - a_{\alpha+n}^\gamma a_\beta^{\gamma+n}\right) = \delta_{\alpha\beta}\,,$$

$$\sum_\gamma \left(a_\alpha^{\gamma+n} a_{\beta+n}^\gamma - a_\alpha^\gamma a_{\beta+n}^{\gamma+n}\right) = -\delta_{\alpha\beta}\,,$$

$$\sum_\gamma \left(a_{\alpha+n}^{\gamma+n} a_{\beta+n}^\gamma - a_{\alpha+n}^\gamma a_{\beta+n}^{\gamma+n}\right) = 0.$$

Then, by taking into account (7.10)–(7.13), we have the following result:

Proposition 7.5. $g : \mathbf{H}_n \to \mathrm{GL}(2n, \mathbf{R})$ *is actually* $\mathrm{Sp}(2n, \mathbf{R})$*-valued.*

Next, let $K \subset \mathrm{Sp}(2n, \mathbf{R})$ be given by

$$K = \{k \in \mathrm{Sp}(2n, \mathbf{R}) : k^t = k^{-1}\}$$

and let us set

$$A = \{a = \mathrm{diag}\left(e^{t_1}, \dots, e^{t_n}, e^{-t_1}, \dots, e^{-t_n}\right) : t_j \in \mathbf{R},\ 1 \le j \le n\},$$
$$A^+ = \{a \in A : t_1 \ge t_2 \ge \cdots \ge t_n \ge 0\}.$$

We shall need the so-called *Cartan decomposition* of $\mathrm{Sp}(2n, \mathbf{R})$:

Proposition 7.6. $\mathrm{Sp}(2n, \mathbf{R}) = K\, A^+\, K$.

Let

$$j : \mathrm{GL}(n, \mathbf{C}) \to \mathrm{GL}(2n, \mathbf{R}), \quad j : A + iB \mapsto \begin{pmatrix} A & -B \\ B & A \end{pmatrix}.$$

Note that $K = \mathrm{O}(2n) \cap j(\mathrm{U}(n))$. Using

$$\begin{aligned}
(df)T_\alpha &= f_\alpha^\beta T_\beta + f_\alpha^{\bar\beta} T_{\bar\beta}, \\
(df)T_{\bar\alpha} &= f_{\bar\alpha}^\beta T_\beta + f_{\bar\alpha}^{\bar\beta} T_{\bar\beta}, \\
\mu_{\bar\alpha}^\beta f_\beta^\gamma &= f_{\bar\alpha}^\gamma, \\
T_\alpha &= \tfrac{1}{2}\left(e_\alpha - i e_{\alpha+n}\right), \quad T_{\bar\alpha} = \tfrac{1}{2}\left(e_\alpha + i e_{\alpha+n}\right),
\end{aligned}$$

one has

$$\begin{aligned}
\tfrac{1}{2}\left((df)e_\alpha - i(df)e_{\alpha+n}\right) &= (df)T_\alpha = f_\alpha^\beta T_\beta + f^{\bar\beta} T_{\bar\beta} \\
&= \frac{\sqrt{\lambda}}{2}\left(g_\alpha^\beta e_\beta + g_\alpha^{\beta+n} e_{\beta+n} - i g_{\alpha+n}^\beta e_\beta - i g_{\alpha+n}^{\beta+n} e_{\beta+n}\right),
\end{aligned}$$

where $g_B^A = F_B^A/\sqrt{\lambda}$ are the components of $g : \mathbf{H}_n \to \mathrm{Sp}(2n, \mathbf{R})$. Therefore

$$\begin{aligned}
f_\alpha^\beta + f_\alpha^{\bar\beta} &= \sqrt{\lambda}\left(g_\alpha^\beta - i g_{\alpha+n}^\beta\right), \\
f_\alpha^\beta - f_\alpha^{\bar\beta} &= i\sqrt{\lambda}\left(g_\alpha^{\beta+n} - i g_{\alpha+n}^{\beta+n}\right).
\end{aligned}$$

By summing up (respectively by subtracting) these identities, one gets

$$f_\alpha^\beta = \tfrac{1}{2}\sqrt{\lambda}\left(g_\alpha^\beta + g_{\alpha+n}^{\beta+n}\right) + \frac{i}{2}\sqrt{\lambda}\left(g_\alpha^{\beta+n} - g_{\alpha+n}^\beta\right), \tag{7.14}$$

$$f_\alpha^{\bar\beta} = \tfrac{1}{2}\sqrt{\lambda}\left(g_\alpha^\beta - g_{\alpha+n}^{\beta+n}\right) - \frac{i}{2}\sqrt{\lambda}\left(g_\alpha^{\beta+n} + g_{\alpha+n}^\beta\right). \tag{7.15}$$

Since g is $\mathrm{Sp}(2n, \mathbf{R})$-valued, we may write

$$g = j(k)\, a\, j(k'), \tag{7.16}$$

for some functions $k, k' : \mathbf{H}_n \to \mathrm{U}(n)$ and $a : \mathbf{H}_n \to A^+$. Since the matrices $j(k)$, respectively a, have the form

$$\begin{aligned}
j(k) &= \begin{pmatrix} k_\beta^\alpha & k_{\beta+n}^\alpha \\ -k_{\beta+n}^\alpha & k_\beta^\alpha \end{pmatrix}, \\
a &= \mathrm{diag}(e^{t_1}, \dots, e^{t_n}, e^{-t_1}, \dots, e^{-t_n}),
\end{aligned}$$

we have

$$j(k)aj(k') = \begin{pmatrix} k^\alpha_\gamma k'^\gamma_\beta e^{t_\gamma} - k^\alpha_{\gamma+n} k'^\gamma_{\beta+n} e^{-t_\gamma} & k^\alpha_\gamma k'^\gamma_{\beta+n} e^{t_\gamma} + k^\alpha_{\gamma+n} k'^\gamma_\beta e^{-t_\gamma} \\ -k^\alpha_{\gamma+n} k'^\gamma_\beta e^{t_\gamma} + k^\alpha_\gamma k'^\gamma_{\beta+n} e^{-t_\gamma} & -k^\alpha_{\gamma+n} k'^\gamma_{\beta+n} e^{t_\gamma} + k^\alpha_\gamma k'^\gamma_\beta e^{-t_\gamma} \end{pmatrix};$$

hence (7.16) has the following scalar form:

$$\begin{aligned} g^\alpha_\beta &= k^\alpha_\gamma k'^\gamma_\beta e^{t_\gamma} - k^\alpha_{\gamma+n} k'^\gamma_{\beta+n} e^{-t_\gamma}, \\ g^\alpha_{\beta+n} &= k^\alpha_\gamma k'^\gamma_{\beta+n} e^{t_\gamma} + k^\alpha_{\gamma+n} k'^\gamma_\beta e^{-t_\gamma}, \\ g^{\alpha+n}_\beta &= -k^\alpha_{\gamma+n} k'^\gamma_\beta e^{t_\gamma} - k^\alpha_\gamma k'^\gamma_{\beta+n} e^{-t_\gamma}, \\ g^{\alpha+n}_{\beta+n} &= -k^\alpha_{\gamma+n} k'^\gamma_{\beta+n} e^{t_\gamma} + k^\alpha_\gamma k'^\gamma_\beta e^{-t_\gamma}. \end{aligned}$$

At this point, we substitute into (7.14)–(7.15) to obtain

$$\begin{aligned} f^\alpha_\beta &= \sqrt{\lambda}\left(k^\alpha_\gamma - i k^\alpha_{\gamma+n}\right)\left(k'^\gamma_\beta - i k'^\gamma_{\beta+n}\right)\cosh t_\gamma \\ f^{\bar\alpha}_\beta &= \sqrt{\lambda}\left(k^\alpha_\gamma + i k^\alpha_{\gamma+n}\right)\left(k'^\gamma_\beta - i k'^\gamma_{\beta+n}\right)\sinh t_\gamma \,. \end{aligned}$$

Let us substitute into $\mu^\beta_{\bar\alpha} f^\gamma_\beta = f^\gamma_{\bar\alpha}$. We get

$$\mu^\beta_\alpha h^\gamma_\sigma h'^\sigma_\beta \cosh t_\gamma = h^\gamma_\sigma \overline{h'^\sigma_\alpha} \sinh t_\sigma,$$

where $h^\gamma_\sigma = k^\gamma_\sigma - i k^\gamma_{\sigma+n}$, etc. Let us contract with $h^{\bar\gamma}_{\bar\rho} = \overline{h^\gamma_\rho}$. We get

$$\mu^\beta_{\bar\alpha} h^\gamma_\sigma h'^\sigma_\beta h^{\bar\gamma}_{\bar\beta} \cosh t_\sigma = h^\gamma_\sigma h'^{\bar\sigma}_{\bar\alpha} h^{\bar\gamma}_{\bar\rho} \sinh t_\sigma \,. \tag{7.17}$$

On the other hand, because of

$$\begin{gathered} k = (k^i_j) = \begin{pmatrix} A & -B \\ B & A \end{pmatrix}, \\ AA^t + BB^t = I_n \;, \; BA^t - AB^t = 0, \end{gathered}$$

we may compute $h^\gamma_\sigma h^{\bar\gamma}_{\bar\rho}$ as

$$\begin{aligned} h^\gamma_\sigma h^{\bar\gamma}_{\bar\rho} &= (k^\gamma_\sigma - i k^\gamma_{\sigma+n})(k^\gamma_\rho + i k^\gamma_{\rho+n}) \\ &= (k^\gamma_\sigma - i k^\gamma_{\sigma+n})\overline{(k^\rho_\gamma - i k^{\rho+n}_\gamma)} = \delta_{\sigma\rho}, \end{aligned}$$

and (7.17) becomes

$$\mu^\beta_{\bar\alpha} h'^\rho_\beta \cosh t_\rho = h'^{\bar\rho}_{\bar\alpha} \sinh t_\rho$$

(here ρ is not a summation index). Finally, by contracting with $h'^{\bar\rho}_{\bar\gamma}$ we get (because of $h'^\rho_\beta h'^{\bar\rho}_{\bar\gamma} = \delta_{\beta\gamma}$)

$$\mu^\beta_{\bar\alpha} = \sum_{\sigma=1}^n \left(h'^{-1}\right)^\beta_\sigma h'^{\bar\sigma}_{\bar\alpha} \tanh t_\sigma,$$

or

$$\mu = h'^{-1} \operatorname{diag}(\tanh t_1, \ldots, \tanh t_n)\overline{h'} . \tag{7.18}$$

At this point, we may compute $\|\mu\|$. Taking into account

$$(d\theta_0)(T_\alpha, T_{\bar\beta}) = i\delta_{\alpha\beta} ,$$
$$\|\mu\|^2 = \sup_{Z \in T_{1,0}(\mathbf{H}_n)} \frac{L_{\theta_0}(\mu Z, \overline{\mu Z})}{L_{\theta_0}(Z, \overline{Z})} ,$$
$$Z = a^\alpha T_\alpha \in T_{1,0}(\mathbf{H}_n),$$

we get

$$\|\mu\|^2 = \sup_{\substack{a \in \mathbf{C}^n \\ |a|=1}} \sum_{\alpha,\beta,\gamma} a^\alpha a^{\bar\gamma} \mu^\beta_{\bar\alpha} \mu^{\bar\beta}_\gamma .$$

On the other hand (by (7.18)),

$$\begin{aligned} \sum_\beta \mu^\beta_{\bar\alpha}\mu^{\bar\beta}_\gamma &= \sum_\sigma (\tanh t_\sigma)^2\, h'^{\bar\sigma}_\alpha h'^{\sigma}_{\bar\gamma} \\ &\le \max_{t_\sigma} (\tanh t_\sigma)^2 \sum_\sigma h'^{\bar\sigma}_\alpha h'^{\sigma}_{\bar\gamma} = (\tanh t_1)^2\, \delta_{\alpha\gamma} . \end{aligned}$$

Then

$$\|\mu\|^2 \le (\tanh t_1)^2 \sum_{\alpha,\gamma} a^\alpha a^{\bar\gamma} \delta_{\alpha\gamma} = (\tanh t_1)^2 \sum_\alpha |a_\alpha|^2 = (\tanh t_1)^2 ,$$

i.e.,

$$\|\mu\| \le \tanh t_1 . \tag{7.19}$$

On the other hand, let $Z = a^\alpha T_\alpha \in T_{1,0}(\mathbf{H}_n)$ such that $\sum_\alpha a^\alpha h'^{\bar\sigma}_\alpha = \delta_{1\sigma}$. With this choice of Z we have

$$L_{\theta_0}(\mu Z, \overline{\mu Z}) = (\tanh t_1)^2 .$$

Hence (by (7.19) and the definition of $\|\mu\|$)

$$\|\mu\| = \tanh t_1 .$$

Consequently

$$\frac{1 + \|\mu\|}{1 - \|\mu\|} = \frac{1 + \tanh t_1}{1 - \tanh t_1} .$$

Also

$$\frac{\max_{|x|=1}|g(x)|}{\min_{|x|=1}|g(x)|} = \frac{e^{t_1}}{e^{-t_1}} = \frac{1+\|\mu\|}{1-\|\mu\|}.$$

If f is K-quasiconformal then

$$\frac{1}{K}G_{\theta_0}(X,X) \leq G_{\theta_0}(gX,gX) \leq K\,G_{\theta_0}(X,X),$$

for any $X \in H(\mathbf{H}_n)$. Here gX is the ordinary[2] matrix product. Thus

$$\frac{1}{K}|x|^2 \leq |gx|^2 \leq K|x|^2,$$

for any $x \in \mathbf{R}^{2n}$. It follows that

$$\frac{\max_{|x|=1}|g(x)|}{\min_{|x|=1}|g(x)|} \leq K,$$

or

$$\|\mu\| \leq \frac{K-1}{K+1}.$$

The converse is left as an exercise to the reader.

7.3.2 The tangential Beltrami equation on $\mathbf{H}_1$

An interesting question (posed by A. Korányi and H.M. Reimann (cf. [254], p. 65)) is to decide which antilinear morphisms μ may arise as complex dilatations of K-quasiconformal mappings. Let $f : \mathbf{H}_n \to \mathbf{C}^{n+1}$ be a C^∞ map such that $M := f(\mathbf{H}_n)$ is a real hypersurface in $\mathbf{C}^{n+1}$. Assume from now on that M is a strictly pseudoconvex CR manifold (with the CR structure induced from the complex structure of $\mathbf{C}^{n+1}$) and $f : \mathbf{H}_n \to M$ is a K-quasiconformal map. In particular f is a contact transformation possessing the CR invariant property (7.3), so that we may consider its complex dilatation $\mu = \mu_f$. We set

$$J_{\mu,x} = (d_x f)^{-1} \circ J_{f(x)} \circ (d_x f), \quad x \in N,$$

where $J : H(M) \to H(M)$ is the complex structure (of the Levi distribution of M). Then $J_{\mu,x}$ is a new complex structure in $H(\mathbf{H}_n)_x$. As the notation suggests, J_μ is determined by the complex dilatation μ. To see that this is indeed so let $Z_\alpha = \partial/\partial z^\alpha + i\overline{z}^\alpha\,\partial/\partial t$ and let us write

$$J_\mu Z_\alpha = J_\alpha^\beta(\mu) Z_\beta + J_\alpha^{\overline{\beta}}(\mu) Z_{\overline{\beta}} \tag{7.20}$$

[2] One identifies X to a $\mathbf{R}^{2n}$-valued function $\tilde{X}$ on $\mathbf{H}_n$ (by using the (global) frame $\{e_A\}$), and identifies back $g\tilde{X}$ with a section (denoted by gX) in $H(\mathbf{H}_n)$.

for some C^∞ functions $J_\alpha^\beta(\mu)$, $J_\alpha^{\overline{\beta}}(\mu) : \mathbf{H}_n \to \mathbf{C}$. Given a local frame $\{T_\alpha : 1 \leq \alpha \leq n\}$ of $T_{1,0}(M)$, defined on an open set $U \subseteq M$, the left-hand member of (7.20) may also be written $(df)^{-1} J(f_\alpha^\beta T_\beta + f_\alpha^{\overline{\beta}} T_{\overline{\beta}})$ or $i(df)^{-1}(f_\alpha^\beta T_\alpha - f_\alpha^{\overline{\beta}} T_{\overline{\beta}})$. Hence (7.20) yields

$$J_\alpha^\beta(\mu) f_\beta^\gamma + J_\alpha^{\overline{\beta}}(\mu) f_{\overline{\beta}}^\gamma 7 = i f_\alpha^\gamma \,, \tag{7.21}$$

$$J_\alpha^\beta(\mu) f_\beta^{\overline{\gamma}} + J_\alpha^{\overline{\beta}}(\mu) f_{\overline{\beta}}^{\overline{\gamma}} 7 = -i f_\alpha^{\overline{\gamma}} \,, \tag{7.22}$$

on $f^{-1}(U)$. Let us recall that the complex dilatation μ is locally given by $\mu Z_\alpha = \mu_{\overline{\alpha}}^\beta Z_\beta$, where

$$\mu_{\overline{\alpha}}^\beta = f_{\overline{\alpha}}^\gamma g_\gamma^\beta, \qquad [g_\beta^\alpha] := [f_\beta^\alpha]^{-1}.$$

Let us contract with g_γ^β in (7.21) (respectively with $g_{\overline{\gamma}}^{\overline{\beta}}$ in (7.22)). We obtain

$$J_\alpha^\beta(\mu) + \mu_{\overline{\sigma}}^\beta J_\alpha^{\overline{\sigma}}(\mu) = i \delta_\alpha^\beta \,, \tag{7.23}$$

$$\mu_\sigma^{\overline{\beta}} J_\alpha^\sigma(\mu) + J_\alpha^{\overline{\beta}}(\mu) = -i \mu_\alpha^{\overline{\beta}} \,. \tag{7.24}$$

Next (7.23)–(7.24) may be also written as

$$U_\gamma^\beta J_\alpha^\gamma(\mu) = i(\delta_\alpha^\beta + \mu_{\overline{\sigma}}^\beta \mu_\gamma^{\overline{\sigma}}), \quad U_{\overline{\gamma}}^{\overline{\beta}} J_\alpha^{\overline{\gamma}}(\mu) = -2i \mu_\alpha^{\overline{\beta}} \,, \tag{7.25}$$

where

$$U_\gamma^\beta := \delta_\gamma^\beta - \mu_{\overline{\sigma}}^\beta \mu_\gamma^{\overline{\sigma}} \,.$$

Lemma 7.3. $\det[U_\beta^\alpha(x)] \neq 0$ *for any* $x \in f^{-1}(U)$.

Lemma 7.3 shows that $J_\alpha^\gamma(\mu)$ and $J_\alpha^{\overline{\gamma}}(\mu)$ are completely determined by μ. The proof of Lemma 7.3 is by contradiction. If $\det[U_\beta^\alpha(x_0)] = 0$ for some $x_0 \in f^{-1}(U)$ then there is $\zeta = (\zeta^1, \ldots, \zeta^n) \in \mathbf{C}^n \setminus \{0\}$ such that $U_\beta^\alpha(x_0)\zeta^\beta = 0$, i.e.,

$$\zeta^\alpha = \mu_{\overline{\sigma}}^\alpha(x_0) \mu_\beta^{\overline{\sigma}}(x_0) \zeta^\beta \,. \tag{7.26}$$

Let $\theta = dt + i(z^\alpha d\overline{z}_\alpha - \overline{z}_\alpha dz^\alpha)$ (with $z_\alpha = z^\alpha$). Let us set

$$Z = |\zeta|^{-1} \, \zeta^{\overline{\alpha}} Z_{\alpha, x_0} \in T_{1,0}(\mathbf{H}_n)_{x_0} \,,$$

where $|\zeta|^2 = \delta_{\alpha\beta} \zeta^\alpha \zeta^{\overline{\beta}}$ and $\zeta^{\overline{\alpha}} = \overline{\zeta^\alpha}$. Then $L_{\theta, x_0}(Z, \overline{Z}) = 1$ and

$$L_{\theta, x_0}(\mu_{x_0} Z \,, \, \overline{\mu_{x_0} Z}) = |\zeta|^{-2} \zeta^{\overline{\alpha}} \zeta^\beta \mu_{\overline{\alpha}}^\sigma(x_0) \mu_\beta^{\overline{\rho}}(x_0) \delta_{\sigma\rho} \,.$$

On the other hand, as shown earlier in this chapter, μ is symmetric, i.e., $\mu_{\alpha\beta} = \mu_{\beta\alpha}$, where $\mu_{\alpha\beta} = \mu_\alpha^{\overline{\gamma}} h_{\beta\overline{\gamma}}$ (here of course $h_{\alpha\overline{\beta}} = \delta_{\alpha\beta}$). Then (by (7.26))

$$L_{\theta,x_0}(\mu_{x_0}Z\,,\,\overline{\mu_{x_0}Z}) = |\zeta|^{-2}\zeta^{\overline{\alpha}}\zeta^{\beta}\mu^{\sigma}_{\overline{\rho}}(x_0)\mu^{\overline{\rho}}_{\beta}(x_0)\delta_{\sigma\alpha}$$
$$= |\zeta|^{-2}\zeta^{\overline{\alpha}}\zeta^{\sigma}\delta_{\sigma\alpha} = 1.$$

Finally

$$\|\mu\|_{x_0} = \sup\{L_{\theta,x_0}(\mu_{x_0}W, \overline{\mu_{x_0}W}) : L_{\theta,x_0}(W,\overline{W}) = 1, \quad W \in T_{1,0}(\mathbf{H}_n)_{x_0}\} \geq 1,$$

which contradicts $\|\mu\| \leq (K-1)/(K+1) < 1$. Lemma 7.3 is proved. Hence if $[V^{\alpha}_{\beta}] := [U^{\alpha}_{\beta}]^{-1}$ then

$$J^{\gamma}_{\alpha}(\mu) = i\,V^{\gamma}_{\beta}\,(\delta^{\beta}_{\alpha} + \mu^{\overline{\sigma}}_{\alpha}\mu^{\beta}_{\overline{\sigma}}), \quad J^{\gamma}_{\overline{\alpha}}(\mu) = 2i\,V^{\gamma}_{\beta}\,\mu^{\beta}_{\overline{\alpha}}.$$

Let us set

$$V_{\alpha} = Z_{\alpha} - \mu^{\overline{\beta}}_{\alpha}Z_{\overline{\beta}} \in T_{1,0}(\mathbf{H}_n).$$

Then

$$J_{\mu}V_{\alpha} = (J^{\gamma}_{\alpha}(\mu) - \mu^{\overline{\beta}}_{\alpha}J^{\gamma}_{\overline{\beta}}(\mu))Z_{\gamma} + (J^{\overline{\gamma}}_{\alpha}(\mu) - \mu^{\overline{\beta}}_{\alpha}J^{\overline{\gamma}}_{\overline{\beta}}(\mu))Z_{\overline{\gamma}}$$
$$= i(V^{\gamma}_{\alpha} - \mu^{\overline{\sigma}}_{\alpha}\mu^{\beta}_{\overline{\sigma}}V^{\gamma}_{\beta})Z_{\gamma} - i(\mu^{\overline{\beta}}_{\alpha}V^{\overline{\gamma}}_{\overline{\beta}} - \mu^{\overline{\beta}}_{\alpha}\mu^{\rho}_{\overline{\beta}}\mu^{\overline{\sigma}}_{\rho}V^{\overline{\gamma}}_{\overline{\sigma}})Z_{\overline{\gamma}}$$
$$= i\,V^{\gamma}_{\beta}\,U^{\beta}_{\alpha}\,Z_{\gamma} - i\,V^{\overline{\gamma}}_{\overline{\sigma}}\,\mu^{\overline{\beta}}_{\alpha}U^{\overline{\sigma}}_{\overline{\beta}}\,Z_{\overline{\gamma}},$$

i.e., $J_{\mu}V_{\alpha} = i\,V_{\alpha}$. Let $\mathcal{H}(\mu) \subset H(\mathbf{H}_n) \otimes \mathbf{C}$ be the complex subbundle spanned by $\{V_{\alpha} : 1 \leq \alpha \leq n\}$. Then, as we have just shown, $\mathcal{H}(\mu)$ is the eigenbundle of J_{μ} corresponding to the eigenvalue i. Assume from now on that $n = 1$. Then $\mathcal{H}(\mu)$ is a new CR structure on $\mathbf{H}_1$ such that J_{μ} is the corresponding complex structure in $H(\mathbf{H}_n) \otimes \mathbf{C} = \mathcal{H}(\mu) \oplus \overline{\mathcal{H}(\mu)}$.

Conversely, let us assume that $\mu^{1}_{\overline{1}} : \mathbf{H}_1 \to \mathbf{C}$ is a C^{∞} function such that the antilinear morphism $\mu : T_{1,0}(\mathbf{H}_1) \to T_{1,0}(\mathbf{H}_1)$ given by $\mu Z_1 = \mu^{1}_{\overline{1}}Z_1$ satisfies $\|\mu\| \leq (K-1)/(K+1)$. Then a K-quasiconformal map $f : \mathbf{H}_1 \to M \subset \mathbf{C}^2$ with $\mu_f = \mu$ exists if and only if the CR manifold $(\mathbf{R}^3, \mathcal{H}(\mu))$ is embeddable in $\mathbf{C}^2$. Also the CR manifold $(\mathbf{R}^3, \mathcal{H}(\mu))$ is CR equivalent to $\mathbf{H}_1$ if and only if there is a K-quasiconformal mapping $f : \mathbf{H}_1 \to \mathbf{H}_1$ with the complex dilatation μ. The embedding problem is particularly difficult. H. Jacobowitz and F. Trèves [224], have shown that the complex dilatations μ such that $\mathcal{H}(\mu)$ is not embeddable are dense. Clearly, if f^j, $j \in \{1, 2\}$, are C^1 solutions to the tangential Beltrami equation $\overline{Z}f = \mu\, Zf$ (with $Z = Z_1$) satisfying $df^1 \wedge df^2 \neq 0$ then locally $(f^1, f^2) : \mathbf{H}_1 \to \mathbf{C}^2$ gives an embedding of $(\mathbf{R}^3, \mathcal{H}(\mu))$ in $\mathbf{C}^2$. We are left with the natural question, for which μ does the tangential Beltrami equation $\overline{Z}f = \mu\, Zf$ have nonconstant solutions? In the remainder of this section we report on a result of A. Korányi and H.M. Reimann (cf. [254], p. 73) exhibiting a class of functions μ with the property that the tangential Beltrami equation on the lowest-dimensional Heisenberg group $\mathbf{H}_1$ admits nonconstant solutions. We assume throughout that μ is a measurable function of compact support such that

$$\|\mu\|_\infty = \text{ess sup}_{x\in\mathbf{H}_1} |\mu(x)| < 1.$$

Let $\lambda \in \mathbf{R} \setminus \{0\}$ and denote by $\mathcal{H}_\lambda$ the space of all holomorphic functions $\phi : \mathbf{C} \to \mathbf{C}$ such that

$$\|\phi\|_\lambda = \left(\frac{|\lambda|}{\pi} \int_{\mathbf{C}} \exp\left(-|\lambda|\,|\zeta|^2\right) |\phi(\zeta)|^2 d\zeta \right)^{1/2} < \infty.$$

Definition 7.8. Let $T_\lambda : \mathbf{H}_1 \to \text{End}_{\mathbf{C}}(\mathcal{H}_\lambda)$ be the *Bargmann representation* of $\mathbf{H}_1$, i.e., the unitary representation on $\mathcal{H}_\lambda$ given by

$$[T_\lambda(z,t)\phi](\zeta) = \begin{cases} \exp\left(-\frac{\lambda}{2}(|z|^2 + it) - \lambda\bar{z}\zeta\right)\phi(\zeta + z), & \lambda > 0, \\ \exp\left(\frac{\lambda}{2}(|z|^2 - it) + \lambda\bar{z}\zeta\right)\phi(\zeta + z), & \lambda < 0. \end{cases} \qquad \square$$

Definition 7.9. Given a function f in the Schwartz class $\mathcal{S}(\mathbf{H}_1)$ its *Fourier transform at* λ ($\lambda \in \mathbf{R} \setminus \{0\}$) is the operator $T_\lambda(f)$ on $\mathcal{H}_\lambda$ given by

$$[T_\lambda(f)\phi](\zeta) = \int_{\mathbf{H}_1} f(z,t)\,[T_\lambda(z,t)\phi](\zeta)\,dx\,dy\,dt. \qquad \square$$

Definition 7.10. The *trace norm* of $T_\lambda(f)$ is given by

$$\|T_\lambda(f)\|^2 = \text{trace}\{T_\lambda(f)^* T_\lambda(f)\}. \qquad \square$$

Using the inversion and Plancherel formulas (cf. J. Faraut [136]) one may extend the Fourier transform T_λ to functions in $L^2(\mathbf{H}_1)$. Next, let $L^2_+(\mathbf{H}_1)$ be the orthogonal complement of

$$L^2_-(\mathbf{H}_1) = \{f \in L^2(\mathbf{H}_1) : T_\lambda(f) = 0 \text{ for a.e. } \lambda > 0\}.$$

Under the rotation group the space $L^2(\mathbf{H}_1)$ decomposes into the mutually orthogonal subspaces

$$U^k = \{f \in L^2(\mathbf{H}_1) : f(e^{i\varphi}z, t) = e^{ik\varphi} f(z,t)\,,\ \varphi \in \mathbf{R}\}, \quad k \in \mathbf{Z}.$$

Consider the complete orthogonal sums

$$D_j = \bigoplus_{k \le j} U^k\,.$$

We also need the following definition:

Definition 7.11. Let $W_1^p(\mathbf{H}_1)$ be the Sobolev type spaces

$$W_1^p(\mathbf{H}_1) = \{g \in L^1_{loc}(\mathbf{H}_1) : Z(g),\ \overline{Z}(g) \in L^p(\mathbf{H}_1)\}. \qquad \square$$

Theorem 7.4. (A. Korányi and H.M. Reimann [254])
Assume that h is a CR-holomorphic function ($\overline{Z}(h) = 0$) and $\mu \in L^\infty(\mathbf{H}_1)$. If one of the conditions

(1) $\mu \in L^2_+(\mathbf{H}_1)$, $\mu Z(h) \in L^2_+(\mathbf{H}_1)$ *and* $\|\mu\|_\infty < 1/\sqrt{2}$,
(2) $\mu \in D_{-2}$, $\mu Z(h) \in D_{-1}$ *and* $\|\mu\|_\infty < 1/\sqrt{2}$,
(3) $\mu \in D_{-2} \cap L^2_-(\mathbf{H}_1)$, $\mu Z(h) \in D_{-1} \cap L^2_-(\mathbf{H}_1)$ *and* $\|\mu\|_\infty < 1$,

holds, then the tangential Beltrami equation $\overline{Z}(f) = \mu\, Z(f)$ has a unique solution f such that $f - h \in W_1^2(\mathbf{H}_1)$.

The proof is based on the integral representation for solutions of the equation $\overline{Z}(f) = g$, cf., e.g., P.C. Greiner, J.J. Kohn, and E.M. Stein [188]. □

It is an open question whether one may solve the tangential Beltrami equation $\overline{Z}(f) = \mu\, Z(f)$ with $|\mu(x)| < 1$ a.e. in $\mathbf{H}_1$, but not necessarily $\|\mu\|_\infty < 1$. For the ordinary Beltrami equation, G. David has shown (cf. [112]) that given constants $\alpha > 0$ and $C \geq 0$ and a measurable function $\mu : \mathbf{C} \to \mathbf{C}$ such that

$$\text{meas}\,(\{z \in \mathbf{C} : |\mu(z)| > 1 - \epsilon\}) \leq Ce^{-\alpha/\epsilon}$$

for sufficiently small $\epsilon > 0$, then there is a unique homeomorphism f of $\mathbf{C}$ fixing 0, 1, and ∞ such that f admits (locally integrable) partial derivatives and $\partial f/\partial \bar{z} = \mu \cdot (\partial f/\partial z)$ a.e. in $\mathbf{C}$.

7.4 Symplectomorphisms

Let $\Omega \subset \mathbf{C}^n$ be a strictly pseudoconvex domain and $K(z, \zeta)$ the reproducing (or Bergman) kernel for $L^2 H(\Omega)$, the space of square integrable functions, with respect to the Lebesgue measure in $\mathbf{R}^{2n}$, that are holomorphic in Ω (cf., e.g., S. Bergman [58]). Consider the complex tensor field

$$H = \sum_{1 \leq i, j \leq n} \left(\frac{\partial^2}{\partial z_i \partial \bar{z}_j} \log K(z, z) \right) dz_i \otimes d\bar{z}_j$$

and the corresponding real tangent (0, 2)-tensor field g given by

$$g = Re \left\{ H \,\big|_{\mathcal{X}(\Omega) \times \mathcal{X}(\Omega)} \right\}.$$

Then g is a Kählerian metric on Ω (the *Bergman metric* of Ω, cf., e.g., S. Helgason [196]); hence $\omega = -i\partial\overline{\partial} \log K(z, z)$ is a symplectic structure (the Kähler 2-form of (Ω, g)). One of the problems we take up in the present section may be stated as follows. Let $F : \Omega \to \Omega$ be a symplectomorphism of (Ω, ω) into itself, smooth up to the boundary. Does $F : \partial\Omega \to \partial\Omega$ preserve the contact structure of the boundary? The interest in this question may be motivated as follows. If $F : \Omega \to \Omega$ is a biholomorphism, then by a celebrated result of C. Fefferman (cf. Theorem 1 in [139], p. 2) F is smooth up to the boundary; hence $F : \partial\Omega \to \partial\Omega$ is a CR diffeomorphism, and in particular a contact transformation. Also, biholomorphisms are known to be isometries of the Bergman metric g (cf., e.g., [196], p. 370); hence symplectomorphisms of (Ω, ω). On the other hand, one may weaken the assumption on F by requiring only that F be a C^∞ diffeomorphism and $F^*\omega = \omega$. Then, by a result of A. Korányi and H.M. Reimann [256], if F is smooth up to the boundary then $F : \partial\Omega \to \partial\Omega$ is a contact transformation.

7.4.1 Fefferman's formula and boundary behavior of symplectomorphisms

We wish to report on the result by A. Korányi and H.M. Reimann quoted above. The main ingredient in the proof of this result is that a certain negative power of the Bergman kernel ($\rho(z) = K(z,z)^{-1/(n+1)}$) is a defining function of Ω, allowing one to relate the symplectic structure of Ω to the contact structure of its boundary. In turn, this is a consequence of C. Fefferman's asymptotic expansion of $K(z,\zeta)$ (cf. Theorem 2 in [139], p. 9), which we now proceed to recall.

Let Ω be a smoothly bounded strictly pseudoconvex domain $\Omega = \{z \in \mathbf{C}^n : \varphi(z) < 0\}$, where φ is such that the Levi form L_φ satisfies

$$L_\varphi(w)\xi \geq C_1|\xi|^2 \ , \ \xi \in \mathbf{C}^n,$$

for $\varphi(w) < \delta_0$, $\delta_0 > 0$, and C_1 depending only on Ω. Let us set

$$\Psi(\zeta,z) = (F(\zeta,z) - \varphi(z))\chi(|\zeta - z|) + (1 - \chi(|\zeta - z|))|\zeta - z|^2,$$

where

$$F(\zeta,z) = -\sum_{j=1}^{n} \frac{\partial \varphi}{\partial z_j}(z)(\zeta_j - z_j) - \frac{1}{2}\sum_{j,k=1}^{n} \frac{\partial^2 \varphi}{\partial z_j \partial z_k}(z)(\zeta_j - z_j)(\zeta_k - z_k)$$

and χ is a C^∞ cut-off function of the real variable t, with $\chi(t) = 1$ for $|t| < \epsilon_0/2$ and $\chi(t) = 0$ for $|t| \geq 3\epsilon_0/4$.

Theorem 7.5. (C. Fefferman [139])
Let $K(\zeta,z)$ be the Bergman kernel of Ω. Then

$$K(\zeta,z) = c_\Omega|\nabla\varphi(z)|^2 \cdot \det L_\varphi(z) \cdot \Psi(\zeta,z)^{-(n+1)} + E(\xi,z),$$

where $E \in C^\infty(\overline{\Omega} \times \overline{\Omega} \setminus \Delta)$, Δ is the diagonal of $\partial\Omega \times \partial\Omega$, and E satisfies the estimate

$$|E(\zeta,z)| \leq c'_\Omega|\Psi(\zeta,z)|^{-(n+1)+\frac{1}{2}}|\log|\Psi(\zeta,z)||.$$

From now on it is understood that Ω is a strictly pseudoconvex domain satisfying all hypothesis of Theorem 7.5.

Theorem 7.6. (A. Korányi and H.M. Reimann [256])
Let F be a symplectomorphism of (Ω,ω), i.e., a C^∞ diffeomorphism $F : \Omega \to \Omega$ with $F^*\omega = \omega$. If F is smooth up to the boundary then $F : \partial\Omega \to \partial\Omega$ is a contact transformation.

Proof. By a result in [58] one has the representation

$$K(\zeta,z) = \sum_k \phi_k(\zeta)\overline{\phi_k(z)}$$

for any complete orthonormal system $\{\phi_k\}$ in $L^2H(\Omega)$. Hence $K(z,z) > 0$ (because for any $z \in \Omega$ there is $f \in L^2H(\Omega)$ with $f(z) \neq 0$). Then it makes sense to consider the function

$$\rho(z) = K(z, z)^{-1/(n+1)} .$$

By Theorem 7.5

$$\rho(z) \leq |\varphi(z)| \left\{ \Phi(z) + C|\varphi(z)|^{\frac{1}{2}} | \log |\varphi(z)|| \right\}^{-1/(n+1)} ,$$

for some $\Phi \in C^{\infty}(\overline{\Omega})$ such that $\Phi(z) \neq 0$ near $\partial\Omega$. Hence $\rho(z) \to 0$ as $z \to \partial\Omega$. Also, as a corollary of Theorem 7.5 one has

$$K(z, z) = \Phi(z)|\varphi(z)|^{-(n+1)} + \tilde{\Phi}(z) \log |\varphi(z)| ,$$

for some $\Phi, \tilde{\Phi} \in C^{\infty}(\overline{\Omega})$, $\Phi(z) \neq 0$ near $\partial\Omega$; hence $\rho \in C^{\infty}(\overline{\Omega})$ and $\nabla\rho \neq 0$ on $\partial\Omega$, i.e., ρ can be used as a defining function for Ω (and $\Omega = \{\rho > 0\}$).

We need some notation. Let $\mathcal{F}$ be the foliation of U (a one-sided neighborhood of the boundary of Ω) by level sets of ρ (so that $\rho^{-1}(0) = \partial\Omega$). Each leaf $M_c = \rho^{-1}(c)$ is a strictly pseudoconvex CR manifold with the CR structure $T_{1,0}(M_c)$ induced from the complex structure on U. Let $T_{1,0}(\mathcal{F})$ be the subbundle of $T(U) \otimes \mathbf{C}$ whose portion over M_c is $T_{1,0}(M_c)$. Since Ω is strictly pseudoconvex, there is a uniquely defined complex vector field ξ of type $(1, 0)$ on U that is orthogonal to $T_{1,0}(\mathcal{F})$ with respect to $\partial\overline{\partial}\rho$ and for which $\partial\rho(\xi) = 1$ (cf., e.g., [274], p. 163). Let us define $r : U \to \mathbf{R}$ by setting $r = 2\left(\partial\overline{\partial}\rho\right)(\xi, \overline{\xi})$, so that ξ and r are characterized by

$$\xi \rfloor \partial\overline{\partial}\rho = r\,\overline{\partial}\rho\,, \quad \partial\rho(\xi) = 1. \tag{7.27}$$

Let $\theta = i(\overline{\partial} - \partial)\rho/2$ and $N = 2\mathrm{Re}(\xi)$. Then $(d\rho)N = 2$ and $\theta(N) = 0$. Note that

$$\omega = i(n+1)\Big(\frac{\partial\overline{\partial}\rho}{\rho} - \frac{\partial\rho \wedge \overline{\partial}\rho}{\rho^2}\Big). \tag{7.28}$$

Let us set $H(\mathcal{F}) = Re\{T_{1,0}(\mathcal{F}) \oplus \overline{T_{1,0}(\mathcal{F})}\}$ (so that the portion of $H(\mathcal{F})$ over a leaf M_c is the Levi distribution of M_c). Then (by (7.28))

$$\omega(X, N) = 0,$$

for any $X \in H(\mathcal{F})$. On the other hand, we may write (7.28) as

$$\omega = (n+1)\Big(\frac{d\theta}{\rho} - \frac{d\rho \wedge \theta}{\rho^2}\Big);$$

hence (by $F^*\omega = \omega$)

$$\begin{aligned} 0 &= \omega((dF)X, (dF)N) \\ &= (n+1)\rho^{-1}(d\theta)((dF)X, (dF)N) - (n+1)\rho^{-2}(d\rho \wedge \theta)((dF)X, (dF)N) \end{aligned}$$

for any $X \in H(\mathcal{F})$. Since F is smooth up to the boundary,

$$(d\theta)((dF)X, (dF)N)$$

stays finite near $\partial\Omega$. Hence, in the limit,

$$(d\rho)((dF)X)\theta((dF)N) - (d\rho)((dF)N)\theta((dF)X)$$

vanishes on $\partial\Omega$. If X lies in $H(\partial\Omega)$, the Levi distribution of $\partial\Omega$ as a CR manifold, then $(dF)X \in T(\partial\Omega)$; hence $(d\rho)((dF)X) = 0$. Finally $(d\rho)((dF)N) \neq 0$ (since F is a diffeomorphism and $d\rho \neq 0$ on $\partial\Omega$); hence $\theta((dF)X) = 0$ for any $X \in H(\partial\Omega)$. □

7.4.2 Dilatation of symplectomorphisms and the Beltrami equations

Let $\Omega \subset \mathbf{C}^n$ be a bounded domain. Let F be a symplectomorphism of (Ω, ω) into itself. We have

Lemma 7.4. *For any $z \in \Omega$ and any $Z \in T^{1,0}(\Omega)_z$, $Z \neq 0$, one has $(d_z F)\overline{Z} \notin T^{1,0}(\Omega)_{F(z)}$.*

The proof is imitative of that of Lemma 7.1. Assume that $(d_z F)\overline{Z} \in T^{1,0}(\Omega)_{F(z)}$ for some $Z \in T^{1,0}(\Omega)_z$, $Z \neq 0$, and some $z \in \Omega$. Since F is a diffeomorphism, $(d_z F)\overline{Z} \neq 0$. Hence

$$0 < \|(d_z F)\overline{Z}\|^2 = g_{\gamma, F(z)}((d_z F)\overline{Z}, (d_z F)Z) = -i\omega_{\gamma,z}(\overline{Z}, Z) = -\|Z\|^2,$$

a contradiction. □

Let $T^{1,0}(\Omega)_F$ consist of all $Z \in T(\Omega) \otimes \mathbf{C}$ with $(dF)Z \in T^{1,0}(\Omega)$.

Lemma 7.5. *For any symplectomorphism F of (Ω, ω) there is a $\mathbf{C}$-antilinear bundle map*

$$\text{dil}(F) : T^{1,0}(\Omega) \to T^{1,0}(\Omega)$$

such that

$$T^{1,0}(\Omega)_F = \{Z - \overline{\text{dil}(F)Z} : Z \in T^{1,0}(\Omega)\}.$$

The proof is imitative of that of Lemma 7.2. Let $\pi_{0,1} : T(\Omega) \otimes \mathbf{C} \to T^{0,1}(\Omega)$ be the natural projection. Then

$$T^{1,0}(\Omega)_F = \text{Ker}(\pi_{0,1} \circ (dF)).$$

Let $(z^1, \dots, z^n)$ be the natural complex coordinates on $\mathbf{C}^n$. Let us set

$$F_k^j = \frac{\partial F^j}{\partial z^k}, \quad F_{\bar{k}}^j = \frac{\partial F^j}{\partial \bar{z}^k},$$

etc. Then $\det(F_{\bar{j}}^{\bar{k}}) \neq 0$ everywhere on Ω. Indeed, if $\det(F_{\bar{j}}^{\bar{k}}(z_0)) = 0$ at some $z_0 \in \Omega$ then $\sum_k F_{\bar{k}}^{\bar{j}}(z_0)\overline{\zeta^k} = 0$, $1 \leq j \leq n$, for some $(\zeta^1, \dots, \zeta^n) \in \mathbf{C}^n - \{0\}$. Let us set $Z = \sum_j \zeta^j \left(\partial/\partial z^j\right)_{z_0} \in T^{1,0}(\Omega)_{z_0}$. Then $Z \neq 0$ and

$$(d_{z_0}F)\overline{Z} = \sum_{j,k} \overline{\zeta^k} F^j_{\bar{k}}(z_0) \left(\frac{\partial}{\partial z^j}\right)_{F(z_0)} \in T^{1,0}(\Omega)_{F(z_0)},$$

a contradiction (by Lemma 7.4). Let $\mathrm{dil}(F) : T^{1,0}(\Omega) \to T^{1,0}(\Omega)$ be given by $\mathrm{dil}(F)\left(\partial/\partial z^j\right) = \sum_k \mathrm{dil}(F)^k_{\bar{j}} \partial/\partial z^k$ (followed by **C**-antilinear extension), where

$$F^\ell_{\bar{j}} = \sum_k \mathrm{dil}(F)^k_{\bar{j}} F^\ell_k . \tag{7.29}$$

Finally, note that $\partial/\partial z^j - \overline{\mathrm{dil}(F)\partial/\partial z^j} \in \mathrm{Ker}(\pi_{0,1} \circ (dF))$. □

Definition 7.12. The bundle map $\mathrm{dil}(F)$ is referred to as the *complex dilatation* of the symplectomorphism F. □

Proposition 7.7. *Let F be a symplectomorphism of (Ω, ω) and* $\mathrm{dil}(F)$ *its complex dilatation. Then*

$$\omega(Z, \overline{\mathrm{dil}(F)W}) + \omega(\overline{\mathrm{dil}(F)Z}, W) = 0,$$

for any $Z, W \in T^{1,0}(\Omega)$. Also, $\mathrm{dil}(F) = 0$ *if and only if F is holomorphic.*

Indeed, if $Z \in T^{1,0}(\Omega)$ then $(dF)(Z - \overline{\mathrm{dil}(F)Z}) \in T^{1,0}(\Omega)$. Therefore, since ω vanishes on complex vectors of the same type,

$$\begin{aligned} 0 &= \omega_\gamma((dF)(Z - \overline{\mathrm{dil}(F)Z}), (dF)(W - \overline{\mathrm{dil}(F)W})) \\ &= \omega_\gamma(Z - \overline{\mathrm{dil}(F)Z}, W - \overline{\mathrm{dil}(F)W}) \\ &= -\omega_\gamma(Z, \overline{\mathrm{dil}(F)W}) - \omega_\gamma(\overline{\mathrm{dil}(F)Z}, W), \end{aligned}$$

for any $Z, W \in T^{1,0}(\Omega)$.

By (7.29), each component F^j of the symplectomorphism F satisfies the first-order PDE (with variable coefficients)

$$\frac{\partial f}{\partial \bar{z}^j} = \sum_k d^k_{\bar{j}} \frac{\partial f}{\partial z^k}, \tag{7.30}$$

where $d^k_{\bar{j}} = \mathrm{dil}(F)^k_{\bar{j}}$.

Definition 7.13. We refer to (7.30) as the *Beltrami equations.* □

Cf., e.g., [419].

Theorem 7.7. (A. Korányi and H.M. Reimann [256])
Assume that the symplectomorphism $F : \Omega \to \Omega$ with complex dilatation $\mathrm{dil}(F)$ *extends smoothly to the boundary. Then* $\mathrm{dil}(F)$ *restricted to $H(\mathcal{F})$ converges to the complex dilatation of the boundary contact transformation.*

Proof. The complex dilatation $\mathrm{dil}(F)$ describes the pullback of the complex structure. But on the boundary, the Levi distribution $H(\partial\Omega)$ is invariant under the boundary contact transformation. The pullback of the complex structure within $H(\partial\Omega)$ therefore has to be the limit of the pullback of the complex structure in the interior, after restriction to $H(\mathcal{F})$.

7.4.3 Boundary values of solutions to the Beltrami system

Let Ω_n be the Siegel domain in $\mathbf{C}^n$ and let $d_{\bar{j}}^k$ be smooth functions defined on some neighborhood of $\overline{\Omega}_n$. The complex vector fields $\partial/\partial\overline{\zeta}^j - \sum_k d_{\bar{j}}^k \partial/\partial\zeta^k$ span a rank-n complex vector subbundle $B \subset T(\Omega_n) \otimes \mathbf{C}$. For the Siegel domain Ω_n, the vector field ξ (determined by (7.27)) is given by $\xi = 2i\partial/\partial\zeta^1$. The CR isomorphism $\phi : \mathbf{H}_{n-1} \simeq \partial\Omega_n$ maps the Lewy operators $L_{\overline{\alpha}}$ into $Z_{\overline{\alpha}} = \partial/\partial\overline{\zeta}^{\alpha} + \zeta^{\alpha}\overline{\xi}$, $2 \le \alpha \le n$.

Proposition 7.8. *Let D be an open neighborhood of $\overline{\Omega}_n$ and*

$$\mu : T^{1,0}(D) \to T^{1,0}(D)$$

a fiberwise **C***-antilinear bundle morphism that maps $T_{1,0}(\partial\Omega_n)$ into itself. Let $B_b \subset T(\partial\Omega_n) \otimes \mathbf{C}$ be the rank-$(n-1)$ complex subbundle spanned by $Z_{\bar{\alpha}} - \mu_{\bar{\alpha}}^{\beta} Z_{\beta}$, $2 \le \alpha \le n$, where $\mu_{\bar{\alpha}}^{\beta}$ are given by $\mu(Z_\alpha) = \mu_{\bar{\alpha}}^{\beta} Z_\beta$. Let $d_{\bar{j}}^k$ be given by $\mu(\partial/\partial\zeta^j) = d_{\bar{j}}^k \partial/\partial\zeta^k$ and set $h(\zeta) = 2i\sum_\beta d_{\bar{1}}^{\beta}\overline{\zeta}_\beta - d_{\bar{1}}^1 - 1$. Then*

$$B_b = [T(\partial\Omega_n) \otimes \mathbf{C}] \cap B$$

on $\partial\Omega_n \cap \{\zeta : h(\zeta) \neq 0\}$. In particular, the trace on $\partial\Omega_n$ of any solution $f \in C^\infty(\overline{\Omega}_n)$ of the Beltrami equations (7.30) satisfies the tangential Beltrami equations $Z_{\bar{\alpha}} f = \mu_{\bar{\alpha}}^{\beta} Z_\beta f$ on the open set $\{\zeta \in \partial\Omega_n : h(\zeta) \neq 0\}$.

Indeed, since $\mu(T_{1,0}(\partial\Omega_n)) \subseteq T_{1,0}(\partial\Omega_n)$,

$$\mu_{\bar{\alpha}}^{\beta} = d_{\bar{\alpha}}^{\beta} - 2i\zeta_\alpha d_{\bar{1}}^{\beta}, \quad 2i\mu_{\bar{\alpha}}^{\beta}\overline{\zeta}_\beta = d_{\bar{\alpha}}^1 - 2i\zeta_\alpha d_{\bar{1}}^1,$$

where $\zeta_\alpha = \zeta^\alpha$. Consequently $Z = a^j\left(\partial/\partial\overline{\zeta}^j - d_{\bar{j}}^k\partial/\partial\zeta^k\right)$ is tangent to $\partial\Omega_n \cap \{h \neq 0\}$ if and only if $a^1 = -2i\zeta_\alpha a^\alpha$, i.e., $Z \in \Gamma^\infty(B_b)$. □

7.4.4 A theorem of P. Libermann

Let M be a nondegenerate CR manifold and θ a contact 1-form on M. Let $f_s : M \to M$, $|s| < \epsilon$, be a (local) 1-parameter group of contact transformations. Then $f_s^*\theta = \lambda_s\theta$, for some C^∞ function λ_s on M (depending smoothly on the parameter). Let V be the tangent vector field on M induced by $(f_s)_{|s|<\epsilon}$. Let μ be the C^∞ function on M defined by

$$\mu(x) = \frac{d}{ds}\left(\lambda_s(x)\right)_{s=0}, \quad x \in M.$$

Then $f_s^*\theta = \lambda_s\theta$ yields $\mathcal{L}_V\theta = \mu\theta$ (where $\mathcal{L}$ is the Lie derivative). Next, let us set

$$p = \theta(V) \in C^\infty(M).$$

Then (by Cartan's formula $\mathcal{L}_X = i_X\, d + d\, i_X$)

$$i_V d\theta + dp = \mathcal{L}_V \theta - d i_V \theta + dp = \mu\theta \,.$$

Let T be the characteristic direction of (M, θ). Then

$$V = \theta(V)T + h(V), \tag{7.31}$$

where $h(V) = \pi_H V$ (and $\pi_H : T(M) \to H(M)$ is the projection). The local integration of T furnishes a flow $\mathcal{F}$ on M (the *contact flow* of M). Let $\Omega^k_{\frac{1}{2}B}(\mathcal{F})$ be the $C^\infty(M)$-module of all semibasic k-forms (with respect to the foliation $\mathcal{F}$, i.e., $T(\mathcal{F}) \rfloor \eta = 0$ for any $\eta \in \Omega^k_{\frac{1}{2}B}(\mathcal{F})$). Consider the map

$$\alpha_\theta : \Gamma^\infty(H(M)) \to \Omega^1_{\frac{1}{2}B}(\mathcal{F}), \quad \alpha_\theta : X \mapsto i_X d\theta.$$

α_θ is well defined because of $i_T i_X d\theta = 0$. Also, if $i_X d\theta = 0$, then by the nondegeneracy of $d\theta$ on $H(M)$, it follows that $X = 0$, i.e., α_θ is injective. Finally, given $\omega = \omega_\alpha \theta^\alpha + \omega_{\bar\alpha}\theta^{\bar\alpha} \in \Omega^1_{\frac{1}{2}B}(\mathcal{F})$, $\omega^{\bar\alpha} = \overline{\omega_\alpha}$, we may choose $X = Z^\alpha T_\alpha + \overline{Z^\alpha} T_{\bar\alpha} \in H(M)$ given by $Z^\alpha = -i h^{\alpha\bar\beta}\omega_{\bar\beta}$ and with this choice $i_X d\theta = \omega$; hence α_θ is surjective. Then the equation

$$i_V d\theta + dp = \mu\theta$$

may be written as follows. First, apply i_T to get $\mu = T(p)$ and substitute from (7.31) to get

$$i_{\theta(V)T + h(V)} d\theta + dp = T(p)\theta,$$

or

$$\alpha_\theta(h(V)) = T(p)\theta - dp,$$

and (7.31) becomes

$$V = pT + \alpha_\theta^{-1}(T(p)\theta - dp).$$

We have proved the following theorem:

Theorem 7.8. (P. Libermann [285])
Let M be a nondegenerate CR manifold and θ a contact 1-form on M. Then to any C^∞ function p on M there corresponds an infinitesimal contact transformation V_p (given by $V_p = pT + \alpha_\theta^{-1}(T(p)\theta - dp)$) and conversely, to any infinitesimal contact transformation V on M there corresponds a C^∞ function p (i.e., $p = \theta(V)$) such that $V_p = V$.

Let us apply P. Libermann's theorem to $\mathbf{H}_n$. For the contact 1-form

$$\theta_0 = dt + \frac{i}{2}\sum_{j=1}^{n}\left(z^j d\bar z^j - \bar z^j dz^j\right)$$

one has $T = \partial/\partial t$ and $d\theta_0 = i\sum_{j=1}^n dz^j \wedge d\bar{z}^j$, so we may compute the isomorphism $\alpha = \alpha_{\theta_0} : \Gamma^\infty(H(\mathbf{H}_n)) \to \Omega^1_{\frac{1}{2}B}(\mathcal{F})$ (here $\mathcal{F}$ is the foliation of $\mathbf{H}_n$ by the curves C_x, $x \in \mathbf{H}_n$, where $C_x(s) = (z, s+t)$, $s \in \mathbf{R}$, $x = (z,t)$). We have

$$\alpha(X) = (d\theta_0)(X, \cdot) = \sum_j \left(-i\overline{Z}^j dz^j + iZ^j d\bar{z}^j\right),$$

for any $X = Z^j T_j + \overline{Z}^j T_{\bar{j}} \in H(\mathbf{H}_n)$. Finally, we obtain the following corollary:

Corollary 7.1. (P. Libermann [285])
Any infinitesimal contact transformation V on $\mathbf{H}_n$ has a representation of the form

$$V = i\sum_{j=1}^n \left(T_{\bar{j}}(p)T_j - T_j(p)T_{\bar{j}}\right) + pT \tag{7.32}$$

for some real-valued C^∞ function p on $\mathbf{H}_n$. Conversely, integration of any tangent vector field V of the form (7.32) *yields a local* 1*-parameter group of contact transformations of $\mathbf{H}_n$.*

Cf. also A. Korányi and H.M. Reimann [257], p. 70.

7.4.5 Extensions of contact deformations

Let $\Omega \subset \mathbf{C}^n$ be a smoothly bounded strictly pseudoconvex domain satisfying the assumptions of Theorem 7.5.

Definition 7.14. An infinitesimal contact transformation on $\partial\Omega$ is referred to as a *contact deformation* of $\partial\Omega$. □

Let $f \in C^\infty(\Omega)$ and denote by X_f the Hamiltonian vector field associated with f (determined by $X_f \rfloor \omega = df$).

Theorem 7.9. (A. Korányi and H.M. Reimann [256])
Any smooth contact deformation of $\partial\Omega$ extends to a Hamiltonian vector field on Ω.

Proof. Let V be a contact deformation of $\partial\Omega$. Then, by P. Libermann's theorem, there is $p \in C^\infty(\partial\Omega)$ such that V can be expressed as $V = pT + \alpha_\theta^{-1}(T(p)\theta - dp)$, where $\theta = (i/2)(\bar{\partial} - \partial\rho)$ and $\rho(z) = K(z,z)^{-1/(n+1)}$. Then, let us continue the function p to a smooth function on $\overline{\Omega}$ and let us set

$$f(z) = p(z)K(z,z)^{1/(n+1)}.$$

Then X_f is the Hamiltonian vector field one is looking for. Indeed, we may show that X_f is a continuous extension of $[-2/(n+1)]V$. Near the boundary we set $W_1 = \xi/\|\xi\|$ and complete W_1 to a local orthonormal frame $\{W_1, \dots, W_n\}$ of $T^{1,0}(\Omega)$. Next, let us set

$$T_j = \sqrt{\frac{n+1}{2\rho}}\, W_j\,.$$

A calculation shows that

$$X_f = \frac{2i}{n+1}\sum_{j=2}^{n}\Big[-T_j(p)T_{\bar{j}} + T_{\bar{j}}(p)T_j\Big] + \frac{i}{\rho\|T_1\|^2}\left[-T_1(p)T_{\bar{1}} + T_{\bar{1}}(p)T_1\right] - \frac{p}{2\rho^2\|T_1\|^2}T.$$

Then

$$\lim_{z\to\partial\Omega}\frac{1}{2\rho^2\|T_1\|^2} = \frac{2}{n+1}\,,\quad \lim_{z\to\partial\Omega}\frac{1}{\rho\|T_1\|^2} = 0$$

(because $\rho \to 0$ as $z \to \partial\Omega$). Thus X_f converges to

$$\frac{2}{n+1}i\sum_{j=2}^{n}\Big(-T_j(p)T_{\bar{j}} + T_{\bar{j}}(p)T_j\Big) + \frac{2}{n+1}pT = -\frac{2}{n+1}V. \qquad \square$$

The reader may see also L. Capogna and P. Tang [87], and P. Tang [399].

8

Yang–Mills Fields on CR Manifolds

In this chapter, we build a canonical family $\{D_S\}$ of Hermitian connections in a Hermitian CR-holomorphic vector bundle (E, h) over a nondegenerate CR manifold M, parameterized by the elements $S \in \Gamma^\infty(\mathrm{End}(E))$ with S skew-symmetric. Consequently, we prove an existence and uniqueness result for the solution to the inhomogeneous Yang–Mills equation $d_D^* R^D = f$ on M. As an application we solve for $D \in \mathcal{D}(E, h)$ when E is the trivial line bundle, a locally trivial CR-holomorphic vector bundle over a nondegenerate real hypersurface in a complex manifold, or a canonical bundle over a pseudo-Einsteinian CR manifold.

In an attempt to extend the results of S. Donaldson [117], and K. Uhlenbeck and S.T. Yau [411], to CR manifolds (of hypersurface type), H. Urakawa has solved (cf. [412]) the Yang–Mills equation for Hermitian connections $D \in \mathcal{D}(E, h)$ (in a Hermitian CR-holomorphic vector bundle over a strictly pseudoconvex manifold) with $(1, 1)$-type curvature R^D. The solution turned out to be precisely the canonical connection built by N. Tanaka [398]. In the present chapter, we generalize H. Urakawa's result by dealing with the inhomogeneous Yang–Mills equation

$$d_D^* R^D = f.$$

This is solved by a geometric method resembling both Tanaka's construction and the proof of Theorem 2.3 in [412], p. 551. The main difference is that we look for solutions D whose curvature has prescribed (rather than zero) trace $\Lambda_\theta R^D$. The result is contained in Theorem 8.2 and is applied to solve several concrete inhomogeneous Yang–Mills equations.

8.1 Canonical S-connections

Let $(E, \overline{\partial}_E)$ be a CR-holomorphic vector bundle over a strictly pseudoconvex CR manifold M. Let θ be a contact form such that L_θ is positive definite. Let h be a Hermitian structure in E. By a result of N. Tanaka [398], there is a unique connection D in E (the *Tanaka connection*) such that (i) $D^{0,1} = \overline{\partial}_E$, (ii) $Dh = 0$, and (iii)

$\Lambda_\theta R^D = 0$, where R^D is the curvature tensor field of D and Λ_θ is the trace operator. Here $D^{0,1}s$ is the restriction of Ds to $T_{0,1}(M)$.

Definition 8.1. A connection D in E is *Hermitian* if it satisfies the axioms (i)–(ii) above. □

By slightly generalizing Tanaka's result, one may single out a Hermitian connection D whose curvature R^D has prescribed trace (rather than zero trace) with respect to a fixed contact form θ on M. Also, this can be made to work in the arbitrary-signature case.

Theorem 8.1. *Let $(M, T_{1,0}(M))$ be a nondegenerate CR manifold (of hypersurface type) of CR dimension n and θ a contact form on M. Let $(E, \overline{\partial}_E)$ be a CR-holomorphic vector bundle over M. Let h be a Hermitian structure in E and $S \in \Gamma^\infty(\mathrm{End}(E))$ a global field of skew-symmetric endomorphisms, i.e., $h(Su, v) + h(u, Sv) = 0$ for any $u, v \in \Gamma^\infty(E)$. There is a unique connection $D = D(h, \theta, S)$ in E such that*

(i) $D_{\overline{Z}} u = (\overline{\partial}_E u)\overline{Z}$,
(ii) $V(h(u, v)) = h(D_V u, v) + h(u, D_{\overline{V}} v)$, *for any* $Z \in \Gamma^\infty(T_{1,0}(M))$, $V \in \Gamma^\infty(T(M) \otimes \mathbf{C})$ *and* $u, v \in \Gamma^\infty(E)$, *and*
(iii) $\Lambda_\theta R^D = 2nS$.

Proof. We first establish uniqueness. Let

$$D : \Gamma^\infty(E) \to \Gamma^\infty(T^*(M) \otimes E)$$

be a connection in E satisfying axioms (i)–(iii). Consider a complex vector field $Z \in \Gamma^\infty(T_{1,0}(M))$ and a section $u \in \Gamma^\infty(E)$. Then $D_{\overline{Z}} u$ and $D_Z u$ are expressed by

$$D_{\overline{Z}} u = (\overline{\partial}_E u)\overline{Z}, \tag{8.1}$$

$$h(D_Z u, v) = Z(h(u, v)) - h(u, (\overline{\partial}_E v)\overline{Z}), \tag{8.2}$$

for any $v \in \Gamma^\infty(E)$. Let T be the characteristic direction of (M, θ). Since $T(M) \otimes \mathbf{C} = T_{1,0}(M) \oplus T_{0,1}(M) \oplus \mathbf{C}T$ it remains to compute $D_T u$. Let $D^2 u$ be defined by

$$(D^2 u)(X, Y) = D_X D_Y u - D_{\nabla_X Y} u, \quad X, Y \in \Gamma^\infty(T(M)).$$

Here ∇ is the Tanaka–Webster connection of (M, θ). Next, let B be given by

$$B(X, Y)u = (D^2 u)(X, Y) - (D^2 u)(Y, X).$$

A calculation shows that

$$B(X, Y)u = R^D(X, Y)u - D_{T_\nabla(X,Y)} u. \tag{8.3}$$

Note that $T_\nabla(E_\alpha, E_{\overline{\alpha}}) = 2i\epsilon_\alpha T$, for any local orthonormal frame $\{E_\alpha\}$ of $(T_{1,0}(M), L_\theta)$. Then (by taking traces in (8.3)) we obtain

$$D_T u = S(u) - \frac{1}{2n}(\Lambda_\theta B)\, u. \tag{8.4}$$

Clearly (by (8.1)–(8.2)) the trace $\Lambda_\theta B$ of B is determined (in terms of the data $(\overline{\partial}_E, h)$). By (8.1)–(8.2) and (8.4) we get the uniqueness statement in Theorem 8.1. □

To establish existence, let us first prescribe $D_{\overline{Z}} u$ and $D_Z u$ by (8.1)–(8.2), for any $Z \in \Gamma^\infty(T_{1,0}(M))$, $u \in \Gamma^\infty(E)$. Then $\Lambda_\theta B$ is known. Hence we may define $D_T u$ by (8.4). Then D is a connection in E. For instance, the property

$$D_T(fu) = f D_T u + T(f)u$$

follows from

$$(\Lambda_\theta B)(fu) = f(\Lambda_\theta B)u - 2nT(f)u\,, \quad f \in C^\infty(M).$$

Moreover, (8.1) yields $D^{0,1} = \overline{\partial}_E$. Also $\Lambda_\theta R^D = 2nS$ as a consequence of (8.3)–(8.4). It remains to check the axiom (ii). Note that

$$V(h(u, v)) = h(D_V u, v) + h(u, D_{\overline{V}} v), \tag{8.5}$$

for any $V \in \Gamma^\infty(H(M) \otimes \mathbf{C})$ (by (8.2) and its complex conjugate, as an identity in $C^\infty(M)$). A straightforward calculation (based on (8.5)) leads to

$$T_\nabla(V, W)(h(u, v)) = -h(B(V, W)u, v) - h(u, B(\overline{V}, \overline{W})v), \tag{8.6}$$

for any $V, W \in H(M) \otimes \mathbf{C}$. Next, by taking traces in (8.6), we obtain

$$2nT(h(u, v)) = -h((\Lambda_\theta B)u, v) - h(u, (\Lambda_\theta B)v), \tag{8.7}$$

$$\begin{aligned} h(D_T u, v) + h(u, D_T v) &= h(Su, v) + h(u, Sv) \\ &\quad - \frac{1}{2n}\{h((\Lambda_\theta B)u, v) + h(u, (\Lambda_\theta B)v)\} = T(h(u, v)), \end{aligned}$$

and the proof of Theorem 8.1 is complete. □

Definition 8.2. The connection $D = D(h, \theta, S)$ furnished by Theorem 8.1 will be referred to as the *canonical S-connection* (determined by the data (h, θ, S)) in (E, ∂_E). □

It possesses the following curvature properties:

Corollary 8.1. *Let us consider a Hermitian CR-holomorphic vector bundle $(E, \overline{\partial}_E, h)$ and a bundle endomorphism $S \in \Gamma^\infty(\mathrm{End}(E))$ such that $h(Su, v) + h(u, Sv) = 0$ for any $u, v \in \Gamma^\infty(E)$. The curvature R^D of the canonical S-connection D in E satisfies the identities*

$$R^D(Z, W) = R^D(\overline{Z}, \overline{W}) = 0, \tag{8.8}$$

$$h(R^D(Z, \overline{W})u, v) + h(u, R^D(\overline{Z}, W)v) = 0, \tag{8.9}$$

for any $Z, W \in T_{1,0}(M)$ and $u, v \in \Gamma^\infty(E)$. In particular D has a $(1,1)$-type curvature tensor field, i.e., $R^D \in \mathcal{B}^{1,1}(\mathrm{End}(E))$, if and only if $T \rfloor R^D = 0$.

Proof. The integrability property for $\overline{\partial}_E$ yields $R^D(\overline{Z}, \overline{W}) = 0$. Then (by (8.3)) $R^D(Z, W) = 0$, and (8.8) is proved. To prove (8.9) one performs the following calculation (based on (8.3), (8.6)):

$$\begin{aligned} &h(R^D(Z, \overline{W})u, v) + h(u, R^D(\overline{Z}, W)v) \\ &\quad = h(B(Z, \overline{W})u, v) + h(u, B(\overline{Z}, W)v) + h(D_{T_\nabla(Z,\overline{W})}u, v) + h(u, D_{T_\nabla(\overline{Z},W)}v) \\ &\qquad = -T_\nabla(Z, \overline{W})(h(u, v)) + 2iL_\theta(Z, \overline{W})h(D_T u, v) \\ &\qquad\quad + 2iL_\theta(Z, \overline{W})h(u, D_T v) = -2iL_\theta(Z, \overline{W})(D_T h)(u, v) = 0, \end{aligned}$$

and the proof of (8.9) is complete.

8.2 Inhomogeneous Yang–Mills equations

Let (E, h) be a Hermitian CR-holomorphic vector bundle over a compact strictly pseudoconvex CR manifold M, and let

$$\mathcal{YM}(D) = \tfrac{1}{2} \int_M \|R^D\|^2 \theta \wedge (d\theta)^n \tag{8.10}$$

be the *Yang–Mills functional* on the space of all connections D in E. It is known (cf., e.g., [392], p. 125) that D is a critical point of $\mathcal{YM}$ if and only if it satisfies the *Yang–Mills equation*

$$d_D^* R^D = 0,$$

where d_D^* is the formal adjoint of d_D with respect to the L^2 inner product (8.24). H. Urakawa has shown (cf. Theorem 2.3 in [412], p. 551) that a Hermitian connection of $(1, 1)$-type curvature R^D is a solution of the (homogeneous) Yang–Mills equation if and only if D is the Tanaka connection (i.e., $D = D(h, \theta, 0)$). See also [413]–[414]. As an application of Theorem 8.1 we establish the following result:

Theorem 8.2. (H. Urakawa et al. [127])
Let $(M, T_{1,0}(M))$ be a compact orientable nondegenerate CR manifold, of CR dimension n, and θ a contact form on M. Let $(E, \overline{\partial}_E)$ be a CR-holomorphic vector bundle over M and h a Hermitian metric on E. Let $f = f^{1,0} + f^{0,1} + \theta \otimes u \in \mathcal{A}^1(\mathrm{End}(E))$, with $f^{1,0} \in \mathcal{B}^{1,0}(\mathrm{End}(E))$, $f^{0,1} = \overline{f^{1,0}}$, and u skew-symmetric. Let D_S be the canonical S-connection, determined by the data (h, θ, S), $-4nS = u$, and assume that D_S has a $(1, 1)$-type curvature tensor field. Then the inhomogeneous Yang–Mills equation

$$d_D^* R^D = f \tag{8.11}$$

admits a unique solution D of $(1, 1)$-type curvature, provided that f satisfies the compatibility relation

$$\overline{\partial}_{D_S} u = -i f^{0,1}. \tag{8.12}$$

Moreover, if this is the case, then the solution to (8.11) is precisely D_S.

Proof. For $f = 0$ this is, of course, H. Urakawa's result quoted above ([412]). Let D be a Hermitian connection in (E, h) with curvature

$$R^D \in \mathcal{C}^{1,1}(\mathrm{End}(E)) \subset \mathcal{A}^2(\mathrm{End}(E))$$

(one adopts the notation and conventions in Section 8.4). To compute $d_D^* R^D \in \mathcal{A}^1(\mathrm{End}(E))$ we take into account the decomposition

$$\mathcal{A}^1(\mathrm{End}(E)) = \mathcal{B}^{1,0}(\mathrm{End}(E)) \oplus \mathcal{B}^{0,1}(\mathrm{End}(E)) \oplus \Gamma^\infty((\mathbf{C}T)^* \otimes \mathrm{End}(E)).$$

For any $\varphi \in \mathcal{B}^{1,0}(\mathrm{End}(E))$,

$$\left(d_D^* R^D, \varphi\right)_\theta = \left(R^D, d_D \varphi\right)_\theta = \left(R^D, A(\varphi) + d'_D\varphi + d''_D\varphi\right)_\theta = \left(R^D, d''_D\varphi\right)_\theta,$$

by (8.23) in Section 8.4 and because of

$$(p, q) \neq (p', q') \Longrightarrow \mathcal{C}^{p,q}(\mathrm{End}(E)) \perp \mathcal{C}^{p',q'}(\mathrm{End}(E)).$$

From now on, assume that $R^D \in \mathcal{B}^{1,1}(\mathrm{End}(E))$. Recall that the pseudo-Hermitian torsion τ of the Tanaka–Webster connection possesses the property

$$\tau(T_{0,1}(M)) \subseteq T_{1,0}(M),$$

and let us set $\tau(E_{\bar\beta}) = A^\alpha_{\bar\beta} E_\alpha$. Then

$$d''_D\varphi = 2\bar\partial_D\varphi + 2A^\alpha_{\bar\beta}\, \varphi_\alpha\, \theta \wedge \theta^{\bar\beta},$$

where $\varphi_\alpha = \varphi(E_\alpha) \in \Gamma^\infty(\mathrm{End}(E))$. Then (since R^D is of type $(1, 1)$)

$$\left(d_D^* R^D, \varphi\right)_\theta = \left(R^D, 2\,\bar\partial_D\varphi\right)_\theta = \left(2\,\bar\partial_D^* R^D, \varphi\right)_\theta,$$

i.e., $2\bar\partial_D^* R^D$ is the $\mathcal{B}^{1,0}(\mathrm{End}(E))$-component of $d_D^* R^D$. In a similar manner one shows that the $\mathcal{B}^{0,1}(\mathrm{End}(E))$-component of $d_D^* R^D$ is $2\partial_D^* R^D$. Next, let $\varphi = \theta \otimes u$, with $u \subset \Gamma^\infty(\mathrm{End}(E))$. Then

$$\left(d_D^* R^D, \varphi\right)_\theta = \left(R^D, d_D\varphi\right)_\theta = 2\left(R^D, (d\theta) \otimes u - \theta \wedge Du\right)_\theta.$$

Moreover (again because R^D is of type $(1, 1)$),

$$\begin{aligned}\left(d_D^* R^D, \theta \otimes u\right)_\theta &= \left(R^D, 2(d\theta) \otimes u\right)_\theta = \left(R^D, -2\Omega \otimes u\right)_\theta \\ &= \left(R^D, -2Lu\right)_\theta = \left(\Lambda_\theta R^D, -2u\right)_\theta = \left(-2\theta \otimes \Lambda_\theta R^D, \theta \otimes u\right)_\theta\end{aligned}$$

(where the last equality holds due to $g_\theta^*(\theta, \theta) = 1$). Consequently, the $\Gamma^\infty((\mathbf{C}T)^* \otimes \mathrm{End}(E))$-component of $d_D^* R^D$ is $-2\theta \otimes \Lambda_\theta R^D$. Summing up, we have

$$d_D^* R^D = 2\left\{\overline{\partial}_D^* R^D + \partial_D^* R^D - \theta \otimes \Lambda_\theta R^D\right\}. \tag{8.13}$$

Note that

$$D_T R^D = T \rfloor d_D R^D = 0,$$

by the second Bianchi identity. Hence $d_D' R^D = 0$ and the identity (8.25) for $\varphi = R^D$ yield

$$\partial_D R^D = 0.$$

In a similar way one proves

$$\overline{\partial}_D R^D = 0.$$

Hence, by taking into account the commutation formulas (8.26), we obtain

$$\partial_D^* R^D = -i\overline{\partial}_D(\Lambda_\theta R^D)\,,\ \overline{\partial}_D^* R^D = i\partial_D(\Lambda_\theta R^D),$$

which together with (8.13) gives

$$d_D^* R^D = 2i\left[\partial_D - \overline{\partial}_D + i\theta\right]\Lambda_\theta R^D. \tag{8.14}$$

Then, on the one hand,

$$d_{D_S}^* R^{D_S} = 4ni\left[\partial_{D_S} - \overline{\partial}_{D_S} + i\theta\right] S = f$$

(by the compatibility relation (8.12)), i.e., D_S is a solution of the inhomogeneous Yang–Mills equation (8.11). Conversely, let D be a Hermitian connection, of $(1,1)$-type curvature, satisfying (8.11). Then (by (8.14))

$$f^{1,0} = 2i\partial_D\Lambda_\theta R^D\,,\quad f^{0,1} = -2i\overline{\partial}_D\Lambda_\theta R^D\,,\quad u = -2\Lambda_\theta R^D\,.$$

Finally, the last identity and Theorem 8.1 yield $D = D(h, \theta, -\frac{1}{4n}u)$, and the first two are identically satisfied (by (8.12)). □

8.3 Applications

In this section, we apply Theorem 8.2 to solve several concrete inhomogeneous Yang–Mills equations. The bundles dealt with are *quantum bundles* and therefore the condition that the background S-connection have $(1,1)$-type curvature is always satisfied. For the notions we use the main reference is [259].

Let $\pi : L \to M$ be a complex line bundle. Let us set $\mathbf{C}^* = \mathbf{C} \setminus \{0\}$ and $\tilde{L} = L \setminus \{\text{zero section}\}$. If $U \subseteq M$ is an open set and $s, t \in \Gamma^\infty(U, \tilde{L})$ then the unique $f \in C^\infty(U, \mathbf{C}^*)$ such that $t = fs$ on U is denoted by $f = t/s$.

Let D be a connection in L. With each $s \in \Gamma^\infty(U, \tilde{L})$ one associates a (local) 1-form $\alpha(s) \in \Omega^1(U)$ given by

$$\alpha(s)X = \frac{1}{i}\frac{D_X s}{s},$$

for any $X \in \Gamma^\infty(U, T(M))$. Next, let us consider the complex 1-form $(1/(iz))dz$ on $\mathbf{C}^*$. It is a GL(1, **C**)-invariant 1-form on $\mathbf{C}^*$. Consequently, for any $x \in M$ there is a unique 1-form $\beta_x \in \Omega^1(\tilde{L}_x)$ such that

$$\lambda^* \beta_x = \frac{1}{iz} dz,$$

for any GL(1, **C**)-equivariant map $\lambda : \mathbf{C}^* \to \tilde{L}_x$.

Definition 8.3. A complex 1-form $\alpha \in \Omega^1(\tilde{L})$ is a *connection form* on L if (i) α is GL(1, **C**)-invariant, and (ii) $j_x^* \alpha = \beta_x$ for any $x \in M$, where $j_x : \tilde{L}_x \subset L$. □

Given a complex line bundle with connection (L, D) there is (cf. Proposition 1.5.1 in [259], p. 101) a unique connection form $\alpha \in \Omega^1(\tilde{L})$ such that $s^* \alpha = \alpha(s)$ for any local section $s \in \Gamma^\infty(U, \tilde{L})$.

Definition 8.4. Let α be the connection form of (L, D). The *curvature form* of (L, D) is the unique closed 2-form $\text{curv}(L, D) \in \Omega^2(M)$ on M determined by

$$\tilde{\pi}^* \, \text{curv}(L, D) = d\alpha,$$

where $\tilde{\pi}$ is the restriction of π to $\tilde{L}$. □

The following concept is of central importance for the present section.

Definition 8.5. Let M be a C^∞ manifold and $H \to M$ a symplectic subbundle of $T(M)$, with symplectic form Ω. Let $L \to M$ be a complex line bundle, D a connection in L, and h a *D-invariant* (i.e., $Dh = 0$) Hermitian structure in L. Then (L, D, h) is called a *quantum bundle* over (M, H, Ω) if

$$\text{curv}(L, D) = \Omega$$

on $H \otimes H$. □

This slightly generalizes the notion of B. Kostant [259], p. 133 (cf. also K. Gawedzki [169], p. 14), where $H = T(M)$. We shall deal with symplectic vector bundles arising on CR manifolds (and the corresponding quantum bundles).

Let (E, h) be a Hermitian CR-holomorphic vector bundle over a CR manifold M (of hypersurface type) of CR dimension n and denote by $\mathcal{D}(E, h)$ the set of all Hermitian connections in (E, h). Let θ be a pseudo-Hermitian structure on M and consider the inhomogeneous Yang–Mills equation

$$d_D^* R^D = 4ni\, \theta \otimes I, \tag{8.15}$$

where I is the identical transformation.

8.3.1 Trivial line bundles

We look for solutions $D \in \mathcal{D}(L, h)$ to (8.15), where $L = M \times \mathbf{C}$ is the trivial complex line bundle, over a nondegenerate CR manifold M, with the Hermitian structure

$$h_x((x, z), (x, w)) = z\overline{w}\ , \quad z, w \in \mathbf{C}.$$

Let $\pi : M \times \mathbf{C} \to M$ and $\hat{\pi} : M \times \mathbf{C} \to \mathbf{C}$ be the natural projections. Then $\alpha \in \Omega^1(\tilde{L})$ given by

$$\alpha = \hat{\pi}^* \left(\frac{1}{iz} dz \right) - \pi^* \theta$$

is a connection form on L whose curvature form is $-d\theta$. Also h is α-invariant. Indeed, if $s(x) = (x, f(x))$, $f \in C^\infty(M, \mathbf{C}^*)$, is a (nowhere-vanishing) section in L, then (by $s^*\alpha = \alpha(s)$)

$$\alpha(s) = f^* \left(\frac{1}{iz} dz \right) - \theta, \tag{8.16}$$

so that $Dh = 0$. We have proved the following:

Proposition 8.1. *(L, α, h) is a quantum bundle over $(M, H(M), -d\theta)$.*

Indeed, let us consider the differential operator

$$\overline{\partial}_L : \Gamma^\infty(L) \to \Gamma^\infty(T_{0,1}(M)^* \otimes L),$$

$$\left(\overline{\partial}_L s\right)_x = \left(x, \left(\overline{\partial}_M f\right)_x\right)\ , \quad s(x) = (x, f(x))\ , \quad x \in M.$$

Then $(L, \overline{\partial}_L)$ is a CR-holomorphic line bundle. Note that as another consequence of (8.16), one has $D^{0,1} = \overline{\partial}_L$ and $\left(\Lambda_\theta R^D\right) s = -2nis$. Hence D (given by (8.16)) is precisely the canonical S-connection $D = D(h, \theta, S)$ obtained for $S = -iI$. By Theorem 2, we can deduce the following result:

Proposition 8.2. *D is the unique solution (in $\mathcal{D}(L, h)$) to (8.15).*

8.3.2 Locally trivial line bundles

Let M be a CR manifold. Let us recall (cf. Chapter 5 of this book) that a function $u \in C^\infty(M, \mathbf{R})$ is CR-pluriharmonic if for any $x \in M$ there is an open neighborhood U of x in M and a function $v \in C^\infty(U, \mathbf{R})$ such that $\overline{\partial}_b(u + iv) = 0$. By a result of E. Bedford, under fairly general assumptions (cf. [51], p. 334) the boundary values of pluriharmonic functions are locally real parts of CR functions, i.e., CR-pluriharmonic. By a result of J.M. Lee [270], which was discussed in detail in Chapter 5 of this book, if M is nondegenerate of CR dimension $n \geq 2$ then a function $u \in C^\infty(M, \mathbf{R})$ is CR-pluriharmonic if and only if the covariant $(1, 1)$-Hessian (with respect to the

Tanaka–Webster connection) of u is a scalar multiple of the Levi form at each point of M.

Let M be a nondegenerate real hypersurface of a complex manifold V. Let $\pi : F \to V$ be a holomorphic line bundle over V, with a flat Hermitian structure H. Then $E = \pi^{-1}(M)$ is a locally trivial CR-holomorphic line bundle over M. Let h be the Hermitian structure induced by H on E. We look for solutions $D \in \mathcal{D}(E, h)$ to the inhomogeneous Yang–Mills equation (8.15). Let $D = D(h, \theta, S)$ be the canonical S-connection obtained for $S = -iI$.

Proposition 8.3. *(E, D, h) is a quantum bundle over $(M, H(M), -d\theta)$.*

To prove this statement, let $\sigma : U \to F$ be a holomorphic frame in F, hence $s = \sigma|_{M\cap U}$ is a frame of E (and $\overline{\partial}_E s = 0$). The first Chern form of F is $(-1/2\pi i)\partial\overline{\partial} \log H(\sigma, \sigma)$. Hence, since H is flat, $\log H(\sigma, \sigma)$ is pluriharmonic. Then (cf. [51], pp. 334–335) $\log h(s, s)$ is CR-pluriharmonic. Consequently (by Proposition 3.3 in [270], p. 167) there is $\lambda \in C^\infty(M, \mathbf{C})$ such that

$$\nabla_{\bar\beta}\nabla_\alpha \log h(s, s) = \lambda\, h_{\alpha\bar\beta}. \tag{8.17}$$

Contraction with $h^{\alpha\bar\beta}$ in (8.17) gives

$$\lambda = -\frac{1}{n}\left(\Delta_b \log h(s, s) - inT(\log h(s))\right).$$

Let $\omega = \operatorname{curv}(E, D)$. Then (8.30) becomes (by (8.17))

$$2\pi\, \omega(T_\alpha, T_{\bar\beta}) = \frac{S(s)}{s}\, h_{\alpha\bar\beta};$$

hence (since $S = -I$) $\omega = -d\theta$ on $H(M) \otimes H(M)$. By Theorem 8.2 we have the following:

Proposition 8.4. *D is the unique solution (in $\mathcal{D}(E, h)$) to* (8.15).

Let M be a nondegenerate CR manifold of CR dimension n and θ a contact form on M. Let ∇ be the Tanaka–Webster connection of (M, θ) and R its curvature tensor field. Let $\{T_\alpha\}$ be a local frame of $T_{1,0}(M)$ on $U \subseteq M$ and set

$$R(T_C, T_D)T_A = R_A{}^B{}_{CD} T_B,$$

where $A, B, \ldots \in \{0, 1, \ldots, n, \bar 1, \ldots, \bar n\}$ and $T_0 = T$. Let $R_{\lambda\bar\mu} = R_\alpha{}^\alpha{}_{\lambda\bar\mu}$ and $\rho = h^{\alpha\bar\beta} R_{\alpha\bar\beta}$ be the pseudo-Hermitian Ricci tensor and the pseudo-Hermitian scalar curvature of (M, θ). Given a Hermitian CR-holomorphic vector bundle (E, h) over M, we consider the inhomogeneous Yang–Mills equation

$$d_D^* R^D = 4ni\{d_b^c \rho - \rho\theta\} \otimes I, \tag{8.18}$$

where $d_b^c = i(\partial_b - \overline{\partial}_b)$.

8.3.3 Canonical bundles

Let us set $\Lambda^{p,0}(M) = \Lambda^p \hat{T}(M)^*$. Then $K(M) = \Lambda^{n+1,0}(M)$ is a complex line bundle over M (the *canonical bundle* of $(M, T_{1,0}(M))$). Let $\{\theta^\alpha\}$ be the admissible coframe determined by $\{T_\alpha\}$, i.e., θ^α are the complex 1-forms on U determined by $\theta^\alpha(T_\beta) = \delta^\alpha_\beta$, $\theta^\alpha(T_{\bar\beta}) = \theta^\alpha(T) = 0$. If $s, t \in \Gamma^\infty(K(M))$ we set

$$h(s,t) = H^{-1} f \overline{g}, \qquad H = \left|\det(h_{\alpha\bar\beta})\right|,$$
$$s = f\,\theta \wedge \theta^1 \wedge \cdots \wedge \theta^n, \quad t = g\,\theta \wedge \theta^1 \wedge \cdots \wedge \theta^n.$$

Then h is a globally defined Hermitian structure in $K(M)$. We wish to look for solutions to (8.18) in $\mathcal{D}(K(M), h)$. The Tanaka–Webster connection ∇ of (M, θ) induces a connection D in $K(M)$. Precisely, if $s \in \Gamma^\infty(K(M))$ then Ds is the covariant derivative of s, thought of as a (scalar) $(n+1, 0)$-form on M, with respect to ∇. We consider the local frame $\zeta_0 : U \to K(M)$ given by

$$\zeta_0 = \sqrt{H}\,\theta \wedge \theta^1 \wedge \cdots \wedge \theta^n.$$

Let ω^α_β be the connection 1-forms of the Tanaka–Webster connection, i.e., $\nabla\theta^\alpha = -\omega^\alpha_\beta \otimes \theta^\beta$. Then

$$D\zeta_0 = \left\{d \log \sqrt{H} - \omega^\alpha_\alpha\right\} \otimes \zeta_0 \tag{8.19}$$

and $Dh = 0$ is equivalent to

$$\omega^\alpha_\alpha + \omega^{\bar\alpha}_{\bar\alpha} = d \log H. \tag{8.20}$$

Finally (8.20) is a consequence of $\omega_{\alpha\bar\beta} + \omega_{\bar\beta\alpha} = d\,h_{\alpha\bar\beta}$. Hence h is D-invariant. As another consequence of (8.19), the curvature R^D of D is given by

$$R^D(X, Y)\zeta_0 = -2(d\omega^\alpha_\alpha)(X, Y)\zeta_0, \tag{8.21}$$

for any $X, Y \in T(M)$. Let us set $A_{\alpha\beta} = h_{\alpha\bar\mu} A^{\bar\mu}_\beta$. Then

$$R_\alpha{}^\rho{}_{\lambda\mu} = 2i(A_{\mu\alpha}\delta^\rho_\lambda - A_{\lambda\alpha}\delta^\rho_\mu).$$

Thus $R_\alpha{}^\alpha{}_{\lambda\mu} = 0$. Therefore (8.21) together with the identity

$$h^{\alpha\bar\beta} L_\theta(R(X, Y)T_\alpha, T_{\bar\beta}) = 2(d\omega^\alpha_\alpha)(X, Y)$$

leads to $R^D(Z, W) = 0$ for any $Z, W \in T_{1,0}(M)$. Finally, again as a consequence of (8.21), one gets

$$R^D(T_\alpha, T_{\bar\beta})\zeta_0 = -R_{\alpha\bar\beta}\zeta_0. \tag{8.22}$$

Assume M to be a nondegenerate CR manifold admitting a pseudo-Einsteinian structure θ of nowhere-zero pseudo-Hermitian scalar curvature ρ. Let us set

$$\hat{\theta} = -\frac{\rho}{2n}\theta.$$

Then (by (8.22))

$$R^D(Z, \overline{W})\zeta_0 = 2L_{\hat{\theta}}(Z, \overline{W})\zeta_0 \ , \ \ Z, W \in T_{1,0}(M).$$

This is summarized in the following proposition:

Proposition 8.5. (H. Urakawa et al. [127])
$(K(M), D, h)$ is a quantum bundle over $(M, H(M), -d\hat{\theta})$.

Let us set

$$\overline{\partial}_{K(M)} = D^{0,1}.$$

Then $(K(M), \overline{\partial}_{K(M)})$ is a CR-holomorphic (line) bundle. Clearly

$$\overline{\partial}_{K(M)}(f\zeta_0) = f\,\overline{\partial}_{K(M)}\zeta_0 + (\overline{\partial}_M f) \otimes \zeta_0,$$

for any $f \in C^\infty(M, \mathbf{C})$. Next, a calculation based on

$$\overline{\partial}_{K(M)}\zeta_0 = \left\{\overline{\partial}_M \log \sqrt{H} - \omega_\alpha^\alpha \circ j_{0,1}\right\} \otimes \zeta_0$$

(where $j_{0,1} : T_{0,1}(M) \subset T(M) \otimes \mathbf{C}$) leads to

$$\left[\overline{Z}, \overline{W}\right]\zeta_0 = \overline{Z}\,\overline{W}\,\zeta_0 - \overline{W}\,\overline{Z}\,\zeta_0 + 2(d\omega_\alpha^\alpha)(\overline{Z}, \overline{W})\zeta_0$$

(where $\overline{Z}\zeta_0 = \left(\overline{\partial}_{K(M)}\zeta_0\right)\overline{Z}$) for any $Z, W \in \Gamma^\infty(T_{1,0}(M))$. Then the identity

$$R_\alpha{}^\rho{}_{\bar{\lambda}\bar{\mu}} = 2i\left(h_{\alpha\bar{\lambda}}A^\rho_{\bar{\mu}} - h_{\alpha\bar{\mu}}A^\rho_{\bar{\lambda}}\right)$$

and the self-adjointness of τ (with respect to g_θ) yields

$$2(d\omega_\alpha^\alpha)(T_{\bar{\lambda}}, T_{\bar{\mu}}) = R_\alpha{}^\alpha{}_{\bar{\lambda}\bar{\mu}} = 0,$$

that is, $\overline{\partial}_{K(M)}$ satisfies the requested integrability property. Finally (by (8.22))

$$\left(\Lambda_\theta R^D\right)\zeta_0 = i\rho\zeta_0.$$

We therefore have the following result:

Proposition 8.6. *The connection D in $K(M)$ is precisely the canonical S-connection $D(h, \theta, S)$ obtained for $S = i\rho I$; hence it is the unique solution to* (8.18).

8.4 Various differential operators

We recall several technical facts used through the proof of Theorem 8.2. Proofs are omitted. The main references for this section are [398] and [412].

Let $E \to M$ be a complex vector bundle over the nondegenerate CR manifold M. Let θ be a contact form on M and T the corresponding characteristic direction. Let us set

$$\hat{T}(M) = T_{1,0}(M) \oplus \mathbf{C}T,$$

and consider the following spaces of C^∞ sections:

$$\begin{aligned}
\mathcal{A}^k(E) &= \Gamma^\infty\left((\Lambda^k T^*(M) \otimes \mathbf{C}) \otimes E\right),\\
\mathcal{B}^k(E) &= \Gamma^\infty\left((\Lambda^k H(M)^* \otimes \mathbf{C}) \otimes E\right),\\
\mathcal{B}^{p,q}(E) &= \Gamma^\infty\left(\Lambda^p T_{1,0}(M)^* \wedge \Lambda^q T_{0,1}(M)^* \otimes E\right),\\
\mathcal{C}^{p,q}(E) &= \Gamma^\infty(\Lambda^p \hat{T}(M)^* \wedge \Lambda^q T_{0,1}(M)^* \otimes E).
\end{aligned}$$

Note that

$$\mathcal{B}^{p,q}(E) = \mathcal{B}^{p+q}(E) \cap \mathcal{C}^{p,q}(E).$$

Let D and ∇ be respectively a Hermitian connection in (E, h) and the Tanaka–Webster connection of (M, θ). Let X be a tangent vector field on M. We need the differential operators

$$\begin{aligned}
D_X &: \mathcal{A}^k(E) \to \mathcal{A}^k(E),\\
d_D &: \mathcal{A}^k(E) \to \mathcal{A}^{k+1}(E),\\
A &: \mathcal{C}^{p,q}(E) \to \mathcal{C}^{p+2,q-1}(E),\\
d'_D &: \mathcal{C}^{p,q}(E) \to \mathcal{C}^{p+1,q}(E), \quad d''_D : \mathcal{C}^{p,q}(E) \to \mathcal{C}^{p,q+1}(E),
\end{aligned}$$

given by

$$(D_X\varphi)(X_1, \dots, X_k) = D_X(\varphi(X_1, \dots, X_k)) - \sum_{j=1}^{k} \varphi(X_1, \dots, \nabla_X X_j, \dots, X_k),$$

$$\begin{aligned}
(d_D\varphi)(X_1, \dots, X_{k+1}) &= \sum_{j=1}^{k+1}(-1)^{j+1} D_{X_j}(\varphi(X_1, \dots, \hat{X}_j, \dots, X_{k+1}))\\
&+ \sum_{1\le i<j\le k+1} (-1)^{i+j}\varphi([X_i, X_j], X_1, \dots, \hat{X}_i, \dots, \hat{X}_j, \dots, X_{k+1}),
\end{aligned}$$

for any $X, X_i \in \Gamma^\infty(T(M) \otimes \mathbf{C})$. Here a hat indicates, as usual, the suppression of a term. Moreover,

$$(A\varphi)(Z_1, \dots, Z_{p+2}, \overline{W}_1, \dots, \overline{W}_{q-1})$$
$$= \sum_{1 \le i < j \le p+2} (-1)^{i+j+1} \varphi(T_\nabla(Z_i, Z_j), Z_1, \dots, \hat{Z}_i, \dots, \hat{Z}_j, \dots,$$
$$\dots, Z_{p+2}, \overline{W}_1, \dots, \overline{W}_{q-1})(d'_D\varphi)(Z_1, \dots, Z_{p+1}, \overline{W}_1, \dots, \overline{W}_q)$$
$$= \sum_{j=1}^{p+1} (-1)^{j+1} (D_{Z_j}\varphi)(Z_1, \dots, \hat{Z}_j, \dots, Z_{p+1}, \overline{W}_1, \dots, \overline{W}_q)$$
$$(d''_D\varphi)(Z_1, \dots, Z_p, \overline{W}_1, \dots, \overline{W}_{q+1})$$
$$= \sum_{j=1}^{q+1} (-1)^{p+j+1} (D_{\overline{W}_j}\varphi)(Z_1, \dots, Z_p, \overline{W}_1, \dots, \hat{\overline{W}}_j, \dots, \overline{W}_{q+1})$$
$$+ \sum_{i=1}^{p} \sum_{j=1}^{q+1} (-1)^{p+i+j+1} \varphi(T_\nabla(Z_i, \overline{W}_j), Z_1, \dots,$$
$$\dots, \hat{Z}_i, \dots, Z_p, \overline{W}_1, \dots, \hat{\overline{W}}_j, \dots, \overline{W}_{q+1})$$

for any $Z_i \in \Gamma^\infty(\hat{T}(M))$ and $W_j \in \Gamma^\infty(T_{1,0}(M))$. Then

$$d_D\varphi = A(\varphi) + d'_D\varphi + d''_D\varphi, \quad \varphi \in \mathcal{C}^{p,q}(E). \tag{8.23}$$

Proposition 8.7. *For any Hermitian connection D in (E, h) over the nondegenerate CR manifold M one has $R^D \in \mathcal{C}^{1,1}(\mathrm{End}(E))$ and*

$$A(R^D) = 0, \quad d'_D R^D = 0, \quad d''_D R^D = 0.$$

This is Lemma 1.15 in [412], p. 548, stated for Hermitian bundles over a strictly pseudoconvex CR manifold in [412], yet it holds in the case of an arbitrary nondegenerate CR manifold (of hypersurface type). Cf. [398], p. 105.

Proposition 8.8. *There is a real operator*

$$*_{\mathcal{B}} : \mathcal{B}^k(E) \to \mathcal{B}^{2n-k}(E)$$

such that

$$\langle \varphi, \psi \rangle_\theta (d\theta)^n = n!\, \varphi \wedge \overline{*_{\mathcal{B}} \psi},$$
$$*_{\mathcal{B}} *_{\mathcal{B}} \varphi = (-1)^k \varphi,$$

for any $\varphi, \psi \in \mathcal{B}^k(E)$.

Here $\langle\ ,\ \rangle_\theta$ is the pointwise inner product induced on $(H(M)^* \otimes \mathbf{C}) \otimes E$ by g_θ and h.

Definition 8.6. Let us set $\Omega = -d\theta$ and consider the operator

$$L : \mathcal{B}^k(E) \to \mathcal{B}^{k+2}(E), \ L\varphi = \Omega \wedge \varphi.$$

Also let us consider

$$\Lambda_\theta : \mathcal{B}^{k+2}(E) \to \mathcal{B}^k(E), \ \Lambda_\theta \psi = (-1)^k *_{\mathcal{B}} L *_{\mathcal{B}} \psi. \qquad \square$$

Proposition 8.9. *Λ_θ is the formal adjoint of L with respect to the L^2 inner product*

$$(\varphi, \psi)_\theta = \int_M \langle \varphi, \psi \rangle_\theta \; \theta \wedge (d\theta)^n. \tag{8.24}$$

Note that

$$i\,\Lambda_\theta \psi = \sum_{\alpha=1}^{n} \epsilon_\alpha \psi(E_\alpha, E_{\overline{\alpha}})\,, \quad \psi \in \mathcal{B}^{1,1}(E),$$

where $\{E_\alpha\}$ is a local orthonormal (i.e., $L_\theta(E_\alpha, E_{\bar\beta}) = \epsilon_\alpha \delta_{\alpha\beta}$, where $\epsilon_1 = \cdots = \epsilon_r = -\epsilon_{r+1} = \cdots = -\epsilon_{r+s} = 1$) frame in $T_{1,0}(M)$. Let $D = D(h, \theta, 0)$ be the Tanaka connection of the Hermitian CR-holomorphic vector bundle (E, h) over (M, θ) and $\hat\theta = e^f \theta$. Using the frame $\hat E_\alpha = e^{-f/2} E_\alpha$ one shows easily that $\hat\Lambda \varphi = e^{-f} \Lambda_\theta \varphi$; hence $\hat\Lambda_\theta R^D = 0$. Then (by the uniqueness statement in Theorem 8.1) $D(h, \hat\theta, 0) = D(h, \theta, 0)$, which is equivalent to the following:

Proposition 8.10. *The Tanaka connection is a CR invariant.*

If $\{\theta^\alpha\}$ is the admissible coframe determined by $\{E_\beta\}$ then (by Lemma 1.17 in [412], p. 549)

$$d'_D \varphi = \sum_{\alpha=1}^{n} \theta^\alpha \wedge D_{E_\alpha} \varphi + \theta \wedge D_T \varphi. \tag{8.25}$$

Next, we shall need the differential operators

$$\partial_D : \mathcal{B}^{p,q}(E) \to \mathcal{B}^{p+1,q}(E), \quad \overline{\partial}_D : \mathcal{B}^{p,q}(E) \to \mathcal{B}^{p,q+1}(E),$$

given by

$$\partial_D \varphi = \sum_{\alpha=1}^{n} \theta^\alpha \wedge D_{E_\alpha} \varphi,$$

$$\overline{\partial}_D \varphi = \sum_{\alpha=1}^{n} \theta^{\bar\alpha} \wedge D_{E_{\bar\alpha}} \varphi,$$

and their formal adjoints

$$\partial_D^* : \mathcal{B}^{p+1,q}(E) \to \mathcal{B}^{p,q}(E), \quad \overline{\partial}_D^* : \mathcal{B}^{p,q+1}(E) \to \mathcal{B}^{p,q}(E),$$

with respect to $(\ ,\ )_\theta$. Explicitly

$$\partial_D^* \varphi = - *_{\mathcal{B}}\, \overline{\partial}_D *_{\mathcal{B}} \varphi, \quad \overline{\partial}_D^* \psi = - *_{\mathcal{B}}\, \partial_D *_{\mathcal{B}} \psi.$$

Finally, we recall (cf. Lemma 1.20 in [412], p. 550) the following result:

Proposition 8.11. *The commutation formulas:*

$$[\Lambda_\theta, \partial_D] = i \overline{\partial}_D^*, \quad \left[\Lambda_\theta, \overline{\partial}_D\right] = -i \partial_D^* \tag{8.26}$$

are valid on any nondegenerate CR manifold.

8.5 Curvature of S-connections

We wish to compute the $(1, 1)$-component of the curvature form

$$\omega = \mathrm{curv}(L, D)$$

of the canonical S-connection of a locally trivial CR-holomorphic line bundle $(L, \overline{\partial}_L)$. Given a nowhere-vanishing CR-holomorphic section s on U we have (by (8.1)–(8.2))

$$\alpha(s)\overline{Z} = 0,$$
$$\alpha(s)Z = \frac{1}{i} Z(\log \rho(s)),$$

where $\rho(s) = h(s, s)$. Moreover, given a local frame $\{T_\alpha\}$ of $T_{1,0}(M)$ on U, we have

$$(D^2 s)(T_\alpha, T_{\overline{\beta}}) = 0, \tag{8.27}$$

$$(D^2 s)(T_{\overline{\beta}}, T_\alpha) = \left(\nabla_{\overline{\beta}} \nabla_\alpha \log \rho(s)\right) s \tag{8.28}$$

(covariant derivatives are taken with respect to the Tanaka–Webster connection of (M, θ)). Let $\Box_b$ be the Kohn–Rossi operator on (M, θ). Using (8.27)–(8.28) and the identities

$$\Box_b f = -h^{\lambda\overline{\mu}} \nabla_\lambda \nabla_{\overline{\mu}} f, \quad \Delta_b f = \Box_b f - inT(f)$$

(for any $f \in C^\infty(M)$) one may compute the trace of B as

$$(\Lambda_\theta B)\, s = -\left\{ i\Delta_b \log \rho(s) + nT(\log \rho(s)) \right\} s.$$

Hence (by (8.4))

$$\alpha(s)T = \frac{1}{i} \left\{ \frac{S(s)}{s} + \frac{1}{2n} \left[i\Delta_b \log \rho(s) + nT(\log \rho(s)) \right] \right\}.$$

Finally

$$i\alpha(s) = T_\alpha(\log \rho(s))\theta^\alpha + \left\{ \frac{S(s)}{s} + \frac{1}{2n} \left(i\Delta_b \log \rho(s) + nT(\log \rho(s)) \right) \right\} \theta, \tag{8.29}$$

for any nowhere-vanishing CR-holomorphic section $s \in \Gamma^\infty(U, L)$. The curvature form $\omega = \mathrm{curv}(L, D)$ is given by $\omega = d\alpha(s)$ on U. Let us differentiate in (8.29) and use the identity

$$d\theta^\alpha = \theta^\beta \wedge \omega_\beta^\alpha + \theta \wedge \tau^\alpha$$

to obtain

$$\begin{aligned} \omega \equiv & \left\{ \frac{S(s)}{s} + \frac{1}{2n} \left[i\Delta_b \log \rho(s) + nT(\log \rho(s)) \right] \right\} d\theta \\ & - \left(\nabla_{\overline{\beta}} \nabla_\alpha \log \rho(s) \right) \theta^\alpha \wedge \theta^{\overline{\beta}}, \ \mathrm{mod}\ \theta^\alpha \wedge \theta^\beta, \theta^\alpha \wedge \theta, \theta^{\overline{\alpha}} \wedge \theta. \end{aligned} \tag{8.30}$$

9

Spectral Geometry

We present A. Greenleaf's lower bound (cf. Theorem 9.1 below) on the first nonzero eigenvalue of the sub-Laplacian Δ_b of a compact strictly pseudoconvex CR manifold. The methods employed are L^2 (that is, we establish Bochner-type formulas) and the result is a CR analogue of the well-known theorem of A. Lichnerowicz (cf. [59], pp. 179–186) that the first nonzero eigenvalue λ_1 of the Laplace–Beltrami operator Δ of a compact n-dimensional Riemannian manifold (M, g) of Ricci curvature $\geq C\, g$, for some constant $C > 0$, satisfies

$$\lambda_1 \geq \frac{n}{n-1} C. \tag{9.1}$$

No CR analogue of M. Obata's theorem (cf. [328]) [that under the same hypothesis, equality is achieved in (9.1) if and only if M is isometric to the standard sphere S^n] has been obtained as yet.

By the same techniques, and under the same hypothesis as in Theorem 9.1, we give a lower bound on the first nonzero eigenvalue of the operator $\Delta_b - icT$, $|c| < n$.

9.1 Commutation formulas

Let $(M, T_{1,0}(M))$ be a nondegenerate CR manifold, of hypersurface type, and θ a contact form on M. Let $\{T_\alpha : 1 \leq \alpha \leq n\}$ be a (local) frame of $T_{1,0}(M)$, defined on an open set $U \subseteq M$, and let us set $\{T_A\} = \{T, T_\alpha, T_{\bar{\alpha}}\}$, where $A \in \{0, 1, \ldots, n, \bar{1}, \ldots, \bar{n}\}$ and $T_0 = T$).

Let $f \in C^\infty(M) \otimes \mathbf{C}$. Let $\nabla^2 f$ be the pseudo-Hermitian Hessian of f, with respect to the Tanaka–Webster connection ∇ of (M, θ). For all local calculations, we set

$$f_{AB} = (\nabla^2 f)(T_A, T_B),$$

so that

$$f_{\alpha\beta} = T_\alpha(f_\beta) - \Gamma^\gamma_{\alpha\beta} f_\gamma\,, \tag{9.2}$$

$$f_{\alpha\bar\beta} = T_\alpha(f_{\bar\beta}) - \Gamma^{\bar\gamma}_{\alpha\bar\beta} f_{\bar\gamma}\,, \tag{9.3}$$

$$f_{0\beta} = T(f_\beta) - \Gamma^\gamma_{0\beta} f_\gamma\,, \tag{9.4}$$

$$f_{\alpha 0} = T_\alpha(f_0)\,, \tag{9.5}$$

where $f_\alpha = T_\alpha(f)$, $f_{\bar\alpha} = T_{\bar\alpha}(f)$, and $f_0 = T(f)$. The pseudo-Hermitian Hessian $\nabla^2 f$ of f is not symmetric, but rather one has the following:

Proposition 9.1. *The commutation relations*

$$f_{\alpha\beta} = f_{\beta\alpha}\,, \tag{9.6}$$

$$f_{\alpha\bar\beta} = f_{\bar\beta\alpha} - 2ih_{\alpha\bar\beta} f_0\,, \tag{9.7}$$

$$f_{\alpha 0} = f_{0\alpha} + A^\beta_\alpha f_{\bar\beta}\,, \tag{9.8}$$

hold on any nondegenerate CR manifold.

To prove (9.6)–(9.8) we differentiate the identity

$$df = f_\alpha \theta^\alpha + f_{\bar\alpha}\theta^{\bar\alpha} + f_0\theta$$

to get

$$df = (df_\alpha)\wedge\theta^\alpha + f_\alpha d\theta^\alpha + (df_{\bar\alpha})\wedge\theta^{\bar\alpha} + f_{\bar\alpha} d\theta^{\bar\alpha} + (df_0)\wedge\theta + f_0 d\theta\,.$$

Let us recall (1.62) and (1.64) of Chapter 1,

$$d\theta = 2ih_{\alpha\bar\beta}\theta^\alpha\wedge\theta^{\bar\beta}\,,$$
$$d\theta^\alpha = \theta^\beta\wedge\omega^\alpha_\beta + \theta\wedge\tau^\alpha\,,$$

and substitute in the previous identity. We obtain (by (9.2)–(9.5))

$$f_{\alpha\beta}\theta^\alpha\wedge\theta^\beta + \left\{f_{\alpha\bar\beta} - f_{\bar\beta\alpha} + 2ih_{\alpha\bar\beta} f_0\right\}\theta^\alpha\wedge\theta^{\bar\beta} + f_{\bar\alpha\bar\beta}\theta^{\bar\alpha}\wedge\theta^{\bar\beta}$$
$$+\left\{f_{0\beta} - f_{\beta 0} + f_{\bar\alpha} A^{\bar\alpha}_\beta\right\}\theta\wedge\theta^\beta + \left\{f_{0\bar\beta} - f_{\bar\beta 0} + f_\alpha A^\alpha_{\bar\beta}\right\}\theta\wedge\theta^{\bar\beta} = 0\,,$$

which (by comparing types) yields (9.6)–(9.8). □

Definition 9.1. We define a third-order covariant derivative $\nabla^3 f$ of f by setting

$$(\nabla^3 f)(X, Y, Z) = (\nabla_X \nabla^2 f)(Y, Z)$$
$$= X((\nabla^2 f)(Y, Z)) - (\nabla^2 f)(\nabla_X Y, Z) - (\nabla^2 f)(Y, \nabla_X Z)\,,$$

for any $X, Y, Z \in T(M)$. □

Also, with respect to the local frame $\{T_A\}$, set

$$f_{ABC} = (\nabla^3 f)(T_A, T_B, T_C)\,.$$

We shall need the corresponding commutation formulas. These are referred to as *inner commutation relations* if they involve the indices B, C, respectively as *outer commutation relations* if they involve the indices A, B. We start from the identity

$$df_\alpha = f_{\beta\alpha}\theta^\beta + f_{\bar\beta\alpha}\theta^{\bar\beta} + f_{0\alpha}\theta + f_\gamma\omega_\alpha^\gamma\,.$$

Let us differentiate and use the formulas (1.62), (1.64) of Chapter 1. We obtain

$$\begin{aligned}0 = (df_{\beta\alpha}) \wedge \theta^\beta + (df_{\bar\beta\alpha}) \wedge \theta^{\bar\beta} & \\ + (df_{0\alpha}) \wedge \theta + (df_\gamma) \wedge \omega_\alpha^\gamma + f_{\beta\alpha}\{\theta^\gamma \wedge \omega_\gamma^\beta + \theta \wedge \tau^\beta\} & \\ + f_{\bar\beta\alpha}\{\theta^{\bar\gamma} \wedge \omega_{\bar\gamma}^{\bar\beta} + \theta \wedge \tau^{\bar\beta}\} + 2i f_{0\alpha} h_{\beta\bar\gamma}\theta^\beta \wedge \theta^{\bar\gamma} + f_\gamma d\omega_\alpha^\gamma\,. &\end{aligned}$$

At this point, we recall (1.90) of Chapter 1,

$$\begin{aligned}\Pi_\alpha^\beta &= d\omega_\alpha^\beta - \omega_\alpha^\gamma \wedge \omega_\gamma^\beta \\ \Omega_\alpha^\beta &= \Pi_\alpha^\beta - 2i\theta_\alpha \wedge \tau^\beta + 2i\tau_\alpha \wedge \theta^\beta\,, \\ \Omega_\alpha^\beta &= R_\alpha{}^\beta{}_{\lambda\bar\mu}\theta^\lambda \wedge \theta^{\bar\mu} + W_{\alpha\lambda}^\beta\theta^\lambda \wedge \theta - W_{\alpha\bar\lambda}^\beta\theta^{\bar\lambda} \wedge \theta\,,\end{aligned}$$

and substitute in the previous identity. This procedure yields

$$\begin{aligned}0 = \{T_\lambda(f_{\mu\alpha}) + T_\lambda(f_\gamma)\Gamma_{\mu\alpha}^\gamma + f_{\beta\alpha}\Gamma_{\mu\lambda}^\beta + 2i f_\lambda A_{\alpha\mu} - f_\beta\Gamma_{\mu\alpha}^\gamma\Gamma_{\lambda\gamma}^\beta\}\theta^\lambda \wedge \theta^\mu & \\ + \{-T_{\bar\mu}(f_{\lambda\alpha}) + T_\lambda(f_{\bar\mu\alpha}) + T_\lambda(f_\gamma)\Gamma_{\bar\mu\alpha}^\gamma - T_{\bar\mu}(f_\gamma)\Gamma_{\lambda\alpha}^\gamma + f_{\beta\alpha}\Gamma_{\bar\mu\lambda}^\beta - f_{\bar\beta\alpha}\Gamma_{\lambda\bar\mu}^{\bar\beta} & \\ + 2i f_{0\alpha}h_{\lambda\bar\mu} + f_\beta R_\alpha{}^\beta{}_{\lambda\bar\mu} + f_\beta\Gamma_{\lambda\alpha}^\gamma\Gamma_{\bar\mu\gamma}^\beta - f_\beta\Gamma_{\bar\mu\alpha}^\gamma\Gamma_{\lambda\gamma}^\beta\}\theta^\lambda \wedge \theta^{\bar\mu} & \\ + \{T_{\bar\lambda}(f_{\bar\mu\alpha}) + T_{\bar\lambda}(f_\gamma)\Gamma_{\bar\mu\alpha}^\gamma + f_{\bar\beta\alpha}\Gamma_{\bar\mu\bar\lambda}^{\bar\beta} + 2i f_\beta h_{\alpha\bar\lambda}A_{\bar\mu}^\beta - f_\beta\Gamma_{\bar\mu\alpha}^\gamma\Gamma_{\bar\lambda\gamma}^\beta\}\theta^{\bar\lambda} \wedge \theta^{\bar\mu} & \\ + \{T(f_{\mu\alpha}) - T_\mu(f_{0\alpha}) - T_\mu(f_\gamma)\Gamma_{0\alpha}^\gamma + T(f_\gamma)\Gamma_{\mu\alpha}^\gamma - f_{\beta\alpha}\Gamma_{0\mu}^\beta & \\ + f_{\bar\beta\alpha}A_\mu^{\bar\beta} - f_\beta W_{\alpha\mu}^\beta - f_\beta\Gamma_{\mu\alpha}^\gamma\Gamma_{0\gamma}^\beta + f_\beta\Gamma_{0\alpha}^\gamma\Gamma_{\mu\gamma}^\beta\}\theta \wedge \theta^\mu & \\ + \{T(f_{\bar\mu\alpha}) - T_{\bar\mu}(f_{0\alpha}) - T_{\bar\mu}(f_\gamma)\Gamma_{0\alpha}^\gamma + T(f_\gamma)\Gamma_{\bar\mu\alpha}^\gamma & \\ + f_{\beta\alpha}A_{\bar\mu}^\beta - f_{\bar\beta\alpha}\Gamma_{0\bar\mu}^{\bar\beta} + f_\beta W_{\alpha\bar\mu}^\beta - f_\beta\Gamma_{\bar\mu\alpha}^\gamma\Gamma_{0\gamma}^\beta + f_\beta\Gamma_{0\alpha}^\gamma\Gamma_{\bar\mu\gamma}^\beta\}\theta \wedge \theta^{\bar\mu}\,. &\end{aligned}$$

Using (9.2)–(9.5) and

$$f_{\alpha\beta\gamma} = T_\alpha(f_{\beta\gamma}) - \Gamma^\lambda_{\alpha\beta} f_{\lambda\gamma} - f_{\beta\mu}\Gamma^\mu_{\alpha\gamma} , \tag{9.9}$$

$$f_{\alpha\bar\beta\gamma} = T_\alpha(f_{\bar\beta\gamma}) - \Gamma^{\bar\lambda}_{\alpha\bar\beta} f_{\bar\lambda\gamma} - f_{\bar\beta\mu}\Gamma^\mu_{\alpha\gamma} , \tag{9.10}$$

$$f_{\bar\alpha\beta\gamma} = T_{\bar\alpha}(f_{\beta\gamma}) - \Gamma^\lambda_{\bar\alpha\beta} f_{\lambda\gamma} - f_{\beta\bar\mu}\Gamma^{\bar\mu}_{\bar\alpha\gamma} , \tag{9.11}$$

$$f_{\bar\alpha\bar\beta\gamma} = T_{\bar\alpha}(f_{\bar\beta\gamma}) - \Gamma^{\bar\lambda}_{\bar\alpha\bar\beta} f_{\bar\lambda\gamma} - f_{\bar\beta\mu}\Gamma^\mu_{\bar\alpha\gamma} , \tag{9.12}$$

$$f_{0\beta\gamma} = T(f_{\beta\gamma}) - \Gamma^\lambda_{0\beta} f_{\lambda\gamma} - f_{\beta\mu}\Gamma^\mu_{0\gamma} , \tag{9.13}$$

$$f_{\alpha 0\gamma} = T_\alpha(f_{0\gamma}) - f_{0\mu}\Gamma^\mu_{\alpha\gamma} , \tag{9.14}$$

$$f_{0\bar\beta\gamma} = T(f_{\bar\beta\gamma}) - \Gamma^{\bar\lambda}_{0\bar\beta} f_{\bar\lambda\gamma} - f_{\bar\beta\mu}\Gamma^\mu_{0\gamma} , \tag{9.15}$$

$$f_{\bar\alpha 0\gamma} = T_{\bar\alpha}(f_{0\gamma}) - f_{0\mu}\Gamma^\mu_{\bar\alpha\gamma} , \tag{9.16}$$

we obtain

$$\begin{aligned}
0 = \{f_{\lambda\mu\alpha} + 2i f_\lambda A_{\alpha\mu}\}\theta^\lambda \wedge \theta^\mu \\
+ \{-f_{\bar\mu\lambda\alpha} + f_{\lambda\bar\mu\alpha} + 2i f_{0\alpha} h_{\lambda\bar\mu} + f_\beta R_\alpha{}^\beta{}_{\lambda\bar\mu}\}\theta^\lambda \wedge \theta^{\bar\mu} \\
+ \{f_{\bar\lambda\bar\mu\alpha} + 2i f_\beta h_{\alpha\bar\lambda} A^\beta_{\bar\mu}\}\theta^{\bar\lambda} \wedge \theta^{\bar\mu} \\
+ \{f_{0\mu\alpha} - f_{\mu 0\alpha} + f_{\bar\beta\alpha} A^{\bar\beta}_\mu - f_\beta W^\beta_{\alpha\mu}\}\theta \wedge \theta^\mu \\
+ \{f_{0\bar\mu\alpha} - f_{\bar\mu 0\alpha} + f_{\beta\alpha} A^\beta_{\bar\mu} + f_\beta W^\beta_{\alpha\bar\mu}\}\theta \wedge \theta^{\bar\mu} .
\end{aligned}$$

Therefore, we have proven the following:

Proposition 9.2. *The outer commutation relations*

$$f_{\lambda\mu\alpha} = f_{\mu\lambda\alpha} + 2i(f_\mu A_{\alpha\lambda} - f_\lambda A_{\alpha\mu}) , \tag{9.17}$$

$$f_{\lambda\bar\mu\alpha} = f_{\bar\mu\lambda\alpha} - 2i f_{0\alpha} h_{\lambda\bar\mu} - f_\beta R_\alpha{}^\beta{}_{\lambda\bar\mu} , \tag{9.18}$$

$$f_{\bar\lambda\bar\mu\alpha} = f_{\bar\mu\bar\lambda\alpha} + 2i f_\beta (h_{\alpha\bar\mu} A^\beta_{\bar\lambda} - h_{\alpha\bar\lambda} A^\beta_{\bar\mu}) , \tag{9.19}$$

$$f_{\mu 0\alpha} = f_{0\mu\alpha} + f_{\bar\beta\alpha} A^{\bar\beta}_\mu - f_\beta W^\beta_{\alpha\mu} , \tag{9.20}$$

$$f_{\bar\mu 0\alpha} = f_{0\bar\mu\alpha} + f_{\beta\alpha} A^\beta_{\bar\mu} + f_\beta W^\beta_{\alpha\bar\mu} , \tag{9.21}$$

hold on any nondegenerate CR manifold.

Proposition 9.3. *The inner commutation formulas*

$$f_{A\beta\gamma} = f_{A\gamma\beta} , \tag{9.22}$$

$$f_{A\beta\bar\gamma} = f_{A\bar\gamma\beta} - 2i f_{A0} h_{\beta\bar\gamma} , \tag{9.23}$$

$$f_{A\beta 0} = f_{A0\beta} + A^{\bar\mu}_\beta f_{A\bar\mu} + f_{\bar\gamma} A^{\bar\gamma}_{\beta\,,\,A} , \tag{9.24}$$

hold on any nondegenerate CR manifold, where $A^{\bar\gamma}_{\beta\,,\,A}$ *are given by*

$$(\nabla_{T_A}\tau_{1,0})T_\beta = (A^{\bar\gamma}_{\beta\,,\,A})T_{\bar\gamma} .$$

Here $\tau_{1,0}$ is the restriction of τ to $T_{1,0}(M)$ (a bundle morphism $T_{1,0}(M) \to T_{0,1}(M)$, by Lemma 1.2). The identities (9.22)–(9.24) follow from (9.9)–(9.16) and from the commutation formulas (9.6)–(9.8). Only (9.24) needs some care. We have (by (9.8))

$$\begin{aligned} f_{A\beta 0} - f_{A0\beta} &= T_A(f_{\beta 0} - f_{0\beta}) - \Gamma^{\mu}_{A\beta}(f_{\mu 0} - f_{0\mu}) \\ &= T_A(A^{\bar{\mu}}_{\beta} f_{\bar{\mu}}) - \Gamma^{\mu}_{A\beta} A^{\bar{\gamma}}_{\mu} f_{\bar{\gamma}} \end{aligned}$$

and

$$A^{\bar{\mu}}_{\alpha,B} = T_B(A^{\bar{\mu}}_{\alpha}) + A^{\bar{\beta}}_{\alpha}\Gamma^{\bar{\mu}}_{B\bar{\beta}} - \Gamma^{\beta}_{B\alpha} A^{\bar{\mu}}_{\beta};$$

hence (by replacing ordinary derivatives in terms of covariant derivatives from (9.3)–(9.4)) one gets (9.24). □

9.2 A lower bound for λ_1

Let M be an arbitrary nondegenerate CR manifold and $\{T_\alpha\}$ a (local) frame of $T_{1,0}(M)$. Then the formal adjoints (with respect to the volume form $\Psi = \theta \wedge (d\theta)^n$) of T_α, respectively of $T_{\bar{\alpha}}$, are given by

$$T^*_{\alpha} = -T_{\bar{\alpha}} - \Gamma^{\bar{\beta}}_{\bar{\beta}\bar{\alpha}}, \tag{9.25}$$

$$T^*_{\bar{\alpha}} = -T_{\alpha} - \Gamma^{\beta}_{\beta\alpha}. \tag{9.26}$$

To prove (9.25), we may write

$$(f, T^*_{\alpha} g) = (T_\alpha f, g) = \int_M (T_\alpha f)\overline{g}\Psi$$

for $f, g \in C^\infty(M) \otimes \mathbf{C}$, at least one of compact support. Note that

$$\operatorname{div}(f\overline{g}T_\alpha) - T_\alpha(f\overline{g}) + \Gamma^{\beta}_{\beta\alpha} f\overline{g}$$

and also

$$(T_\alpha f)\overline{g} = \operatorname{div}(f\overline{g}T_\alpha) - \Gamma^{\beta}_{\beta\alpha} f\overline{g} - fT_\alpha(\overline{g}).$$

Thus

$$\int_M (T_\alpha f)\overline{g}\Psi = (f, -T_{\overline{\alpha}}(g) - \Gamma^{\bar{\beta}}_{\bar{\beta}\bar{\alpha}} g),$$

i.e., (9.25) holds. Finally, (9.26) follows from (9.25) by complex conjugation (since $\overline{(f, g)} = (\overline{f}, \overline{g})$). □

On a strictly pseudoconvex CR manifold, we shall employ (9.25)–(9.26) for a frame $\{T_\alpha\}$ with respect to which $h_{\alpha\bar\beta} = \delta_{\alpha\beta}$. We recall that (cf. the first of the identities (1.54))

$$\Gamma^\alpha_{\gamma\beta} = h^{\bar\mu\alpha}\left\{T_\gamma(h_{\beta\bar\mu}) - g_\theta(T_\beta, [T_\gamma, T_{\bar\mu}]\right\};$$

hence (since $h_{\alpha\bar\beta} = \delta_{\alpha\beta}$)

$$\Gamma^\alpha_{\gamma\beta} = -g_\theta(T_\beta, [T_\gamma, T_{\bar\alpha}]) = -g_\theta(T_\beta, \nabla_{T_\gamma} T_{\bar\alpha}) = -\Gamma^{\bar\mu}_{\gamma\bar\alpha} h_{\beta\bar\mu},$$

i.e.,

$$\Gamma^\alpha_{\gamma\beta} = -\Gamma^{\bar\beta}_{\gamma\bar\alpha} \tag{9.27}$$

and the formulas (9.25)–(9.26) may be also written as

$$T^*_\alpha = -T_{\bar\alpha} + \sum_{\beta=1}^n \Gamma^\alpha_{\bar\beta\beta}, \tag{9.28}$$

$$T^*_{\bar\alpha} = -T_\alpha + \sum_{\beta=1}^n \Gamma^{\bar\alpha}_{\beta\bar\beta} \tag{9.29}$$

(compare to (3.12) in [186], p. 200). Hence (by (9.28)–(9.29))

$$\begin{aligned}\Delta_b f &= \sum_{\alpha=1}^n (f_{\alpha\bar\alpha} + f_{\bar\alpha\alpha}) = \sum_{\alpha=1}^n \left(T_\alpha(f_{\bar\alpha}) - \Gamma^{\bar\gamma}_{\alpha\bar\alpha} f_{\bar\gamma}\right) + \sum_{\alpha=1}^n \left(T_{\bar\alpha}(f_\alpha) - \Gamma^{\gamma}_{\bar\alpha\alpha} f_\gamma\right) \\ &= \sum_\alpha \Big\{ - T^*_{\bar\alpha} T_{\bar\alpha} f + \sum_\beta \Gamma^{\bar\alpha}_{\beta\bar\beta} f_{\bar\alpha} - \Gamma^{\bar\gamma}_{\alpha\bar\alpha} f_{\bar\gamma}\Big\} \\ &\quad + \sum_\alpha \Big\{ - T^*_\alpha T_\alpha f + \sum_\beta \Gamma^\alpha_{\bar\beta\beta} f_\alpha - \Gamma^\gamma_{\bar\alpha\alpha} f_\gamma \Big\},\end{aligned}$$

so that (by (9.27))

$$\Delta_b f = -\sum_{\alpha=1}^n \left(T^*_\alpha T_\alpha + T^*_{\bar\alpha} T_{\bar\alpha}\right). \tag{9.30}$$

In particular Δ_b is a self-adjoint operator. Finally, let us recall that Δ_b is subelliptic of order $1/2$. In particular, by a result of A. Menikoff and J. Sjöstrand [300], it follows that Δ_b has a discrete spectrum consisting of nonnegative eigenvalues and tending to $+\infty$:

$$0 \le \lambda_1 \le \lambda_2 \le \cdots \le \lambda_k \le \cdots \to +\infty.$$

The reader may consult the paper by A. Menikoff and J. Sjöstrand (cf. op. cit.) for a study of the spectral asymptotics of a larger class of operators.

Let M be a strictly pseudoconvex CR manifold and θ a fixed contact 1-form on M with respect to which L_θ is positive definite. Given $u \in C^\infty(M)$, let ∇u be its gradient with respect to the Webster metric g_θ. Let $\pi_H : T(M) \to H(M)$ be the natural projection (with respect to the direct sum decomposition (1.20)). Then, there is a unique complex vector field $\nabla^{1,0}u \in \Gamma^\infty(T_{1,0}(M))$ such that

$$\pi_H \nabla u = \nabla^{1,0}u + \nabla^{0,1}u,$$

where $\nabla^{0,1}u = \overline{\nabla^{1,0}u}$. With respect to a local frame $\{T_\alpha\}$ of $T_{1,0}(M)$,

$$\nabla^{1,0}u = h^{\alpha\bar{\beta}} u_{\bar{\beta}} T_\alpha,$$

or, if $h_{\alpha\bar{\beta}} = \delta_{\alpha\beta}$, then

$$\nabla^{1,0}u = \sum_{\alpha=1}^{n} u_{\bar{\alpha}} T_\alpha. \tag{9.31}$$

Our next task is to compute $\Delta_b \left(\|\nabla^{1,0}u\|^2\right)$ in terms of covariant derivatives. Note that (by (9.31))

$$\|\nabla^{1,0}u\|^2 = \sum_{\lambda=1}^{n} |u_\lambda|^2.$$

Then (by (9.2)–(9.3))

$$\begin{aligned} T_\alpha\left(\|\nabla^{1,0}u\|^2\right) &= \sum_\lambda \left\{T_\alpha(u_\lambda)u_{\bar{\lambda}} + u_\lambda T_\alpha(u_{\bar{\lambda}})\right\} \\ &= \sum_\lambda \left\{u_{\alpha\lambda}u_{\bar{\lambda}} + \Gamma^\mu_{\alpha\lambda} u_\mu u_{\bar{\lambda}} + u_\lambda u_{\alpha\bar{\lambda}} + u_\lambda \Gamma^{\bar{\mu}}_{\alpha\bar{\lambda}} u_{\bar{\mu}}\right\}. \end{aligned}$$

Next (by (9.27))

$$\sum_\lambda \left\{\Gamma^\mu_{\alpha\lambda} u_\mu u_{\bar{\lambda}} + \Gamma^{\bar{\mu}}_{\alpha\bar{\lambda}} u_\lambda u_{\bar{\mu}}\right\} = \sum_{\lambda,\mu} \left\{-\Gamma^{\bar{\lambda}}_{\alpha\bar{\mu}} u_\mu u_{\bar{\lambda}} + \Gamma^{\bar{\mu}}_{\alpha\bar{\lambda}} u_\lambda u_{\bar{\mu}}\right\} = 0;$$

hence

$$T_\alpha\left(\|\nabla^{1,0}u\|^2\right) = \sum_\lambda \left(u_\lambda u_{\alpha\bar{\lambda}} + u_{\bar{\lambda}} u_{\alpha\lambda}\right). \tag{9.32}$$

Moreover (by (9.3) and (9.32)),

$$\begin{aligned} \left(\|\nabla^{1,0}u\|^2\right)_{\alpha\bar{\alpha}} &= T_\alpha\left(T_{\bar{\alpha}}\|\nabla^{1,0}u\|^2\right) - \Gamma^{\bar{\gamma}}_{\alpha\bar{\alpha}} T_{\bar{\gamma}}\left(\|\nabla^{1,0}u\|^2\right) \\ &= T_\alpha\Big(\sum_\lambda \left(u_{\bar{\lambda}} u_{\bar{\alpha}\lambda} + u_\lambda u_{\bar{\alpha}\bar{\lambda}}\right)\Big) - \Gamma^{\bar{\gamma}}_{\alpha\bar{\alpha}} \sum_\lambda \left(u_{\bar{\lambda}} u_{\bar{\gamma}\lambda} + u_\lambda u_{\bar{\gamma}\bar{\lambda}}\right) \\ &= \sum_\lambda \Big\{T_\alpha(u_{\bar{\lambda}})u_{\bar{\alpha}\lambda} + T_\alpha(u_\lambda)u_{\bar{\alpha}\bar{\lambda}} \\ &\qquad + u_{\bar{\lambda}}\left[T_\alpha(u_{\bar{\alpha}\lambda}) - \Gamma^{\bar{\gamma}}_{\alpha\bar{\alpha}} u_{\bar{\gamma}\lambda}\right] + u_\lambda\left[T_\alpha(u_{\bar{\alpha}\bar{\lambda}}) - \Gamma^{\bar{\gamma}}_{\alpha\bar{\alpha}} u_{\bar{\gamma}\bar{\lambda}}\right]\Big\}. \end{aligned}$$

At this point, we may replace the ordinary derivatives by covariant derivatives, by (9.2)–(9.3) and (9.10)–(9.11), and observe the cancellation of Christoffel symbols (cf. also (9.27)). We obtain

$$\left(\|\nabla^{1,0}u\|^2\right)_{\alpha\bar{\alpha}} = \sum_{\lambda} \left\{u_{\alpha\bar{\lambda}}u_{\bar{\alpha}\lambda} + u_{\alpha\lambda}u_{\bar{\alpha}\bar{\lambda}} + u_{\bar{\lambda}}u_{\alpha\bar{\alpha}\lambda} + u_{\lambda}u_{\alpha\bar{\alpha}\bar{\lambda}}\right\}. \tag{9.33}$$

Finally (by (2.15) and (9.33)), we have the following result:

Proposition 9.4.

$$\Delta_b\left(\|\nabla^{1,0}u\|^2\right) = 2\sum_{\alpha,\lambda}\left(u_{\alpha\bar{\lambda}}u_{\bar{\alpha}\lambda} + u_{\alpha\lambda}u_{\bar{\alpha}\bar{\lambda}}\right) + \sum_{\alpha,\lambda}\left(u_{\bar{\lambda}}u_{\alpha\bar{\alpha}\lambda} + u_{\lambda}u_{\alpha\bar{\alpha}\bar{\lambda}} + u_{\lambda}u_{\bar{\alpha}\alpha\bar{\lambda}} + u_{\bar{\lambda}}u_{\bar{\alpha}\alpha\lambda}\right). \tag{9.34}$$

9.2.1 A Bochner-type formula

By the (inner) commutation formulas (9.22)–(9.23) we have

$$\begin{aligned}
u_{\alpha\bar{\alpha}\lambda} &= u_{\alpha\lambda\bar{\alpha}} + 2iu_{\alpha 0}\delta_{\lambda\alpha},\\
u_{\alpha\bar{\alpha}\bar{\lambda}} &= u_{\alpha\bar{\lambda}\bar{\alpha}},\\
u_{\bar{\alpha}\alpha\bar{\lambda}} &= u_{\bar{\alpha}\bar{\lambda}\alpha} - 2iu_{\bar{\alpha}0}\delta_{\alpha\lambda},\\
u_{\bar{\alpha}\alpha\lambda} &= u_{\bar{\alpha}\lambda\alpha};
\end{aligned}$$

hence (9.34) becomes

$$\begin{aligned}
\Delta_b\left(\|\nabla^{1,0}u\|^2\right) &= 2\sum_{\alpha,\lambda}\left(u_{\alpha\bar{\lambda}}u_{\bar{\alpha}\lambda} + u_{\alpha\lambda}u_{\bar{\alpha}\bar{\lambda}}\right)\\
&+ \sum_{\alpha,\lambda}\left(u_{\bar{\lambda}}u_{\alpha\lambda\bar{\alpha}} + u_{\lambda}u_{\alpha\bar{\lambda}\bar{\alpha}} + u_{\lambda}u_{\bar{\alpha}\bar{\lambda}\alpha} + u_{\bar{\lambda}}u_{\bar{\alpha}\lambda\alpha}\right) + 2i\sum_{\alpha}\left(u_{\bar{\alpha}}u_{0\alpha} - u_{\alpha}u_{0\bar{\alpha}}\right)\\
&+ 2i\sum_{\alpha,\beta}\left(A_{\alpha\beta}u_{\bar{\alpha}}u_{\bar{\beta}} - A_{\bar{\alpha}\bar{\beta}}u_{\alpha}u_{\beta}\right).
\end{aligned} \tag{9.35}$$

Next, by the (outer) commutation relations (9.18)–(9.19) we have

$$\begin{aligned}
u_{\alpha\lambda\bar{\alpha}} &= u_{\lambda\alpha\bar{\alpha}} - 2iu_{\bar{\beta}}\left(\delta_{\alpha\lambda}A^{\bar{\beta}}_{\alpha} - A^{\bar{\beta}}_{\lambda}\right),\\
u_{\alpha\bar{\lambda}\bar{\alpha}} &= u_{\bar{\lambda}\alpha\bar{\alpha}} - 2iu_{0\bar{\alpha}}\delta_{\lambda\alpha} + u_{\bar{\beta}}R_{\bar{\alpha}}{}^{\bar{\beta}}{}_{\bar{\lambda}\alpha},\\
u_{\bar{\alpha}\bar{\lambda}\alpha} &= u_{\bar{\lambda}\bar{\alpha}\alpha} + 2iu_{\beta}\left(\delta_{\alpha\lambda}A^{\beta}_{\bar{\alpha}} - A^{\beta}_{\bar{\lambda}}\right),\\
u_{\bar{\alpha}\lambda\alpha} &= u_{\lambda\bar{\alpha}\alpha} + 2iu_{0\alpha}\delta_{\lambda\alpha} + u_{\beta}R_{\alpha}{}^{\beta}{}_{\lambda\bar{\alpha}}.
\end{aligned}$$

Also, we may use the identity

$$(\Delta_b u)_{\beta} = \sum_{\alpha}\left(u_{\beta\alpha\bar{\alpha}} + u_{\beta\bar{\alpha}\alpha}\right),$$

which allows us to write (9.35) as follows:

Proposition 9.5.

$$\Delta_b\left(\|\nabla^{1,0}u\|^2\right) = 2\sum_{\alpha,\lambda}\left(u_{\alpha\bar\lambda}u_{\bar\alpha\lambda} + u_{\alpha\lambda}u_{\bar\alpha\bar\lambda}\right) + 4i\sum_{\alpha}\left(u_{\bar\alpha}u_{0\alpha} - u_{\alpha}u_{0\bar\alpha}\right)$$
$$+ 2\sum_{\alpha,\beta} R_{\alpha\bar\beta}u_{\bar\alpha}u_{\beta} + 2in\sum_{\alpha,\beta}\left(A_{\alpha\beta}u_{\bar\alpha}u_{\bar\beta} - A_{\bar\alpha\bar\beta}u_{\alpha}u_{\beta}\right)$$
$$+ \sum_{\alpha}\left(u_{\bar\alpha}\left(\Delta_b u\right)_{\alpha} + u_{\alpha}\left(\Delta_b u\right)_{\bar\alpha}\right). \quad (9.36)$$

Here, to recognize $\sum_\lambda R_{\bar\lambda}{}^{\bar\alpha}{}_{\bar\mu\lambda}$ as the Ricci curvature, one uses the symmetries (1.92)–(1.93) (in chapter 1) so that (by $h_{\alpha\bar\beta} = \delta_{\alpha\beta}$)

$$R_{\alpha\bar\mu} = R_{\alpha}{}^{\lambda}{}_{\lambda\bar\mu} = \sum_{\lambda} R_{\alpha\bar\lambda\lambda\bar\mu} = \sum_{\lambda} R_{\bar\lambda\alpha\bar\mu\lambda} = R_{\bar\lambda}{}^{\bar\alpha}{}_{\bar\mu\lambda}.$$

9.2.2 Two integral identities

Assume from now on that M is compact. We shall need the following lemma:

Lemma 9.1. *For any $u \in C^\infty(M)$,*

$$\int_M i\sum_{\alpha}\left(u_{0\alpha}u_{\bar\alpha} - u_{0\bar\alpha}u_{\alpha}\right)\Psi$$
$$= \frac{1}{n}\int_M\left(\sum_{\alpha,\beta}u_{\alpha\bar\beta}u_{\bar\alpha\beta} - \sum_{\alpha,\beta}u_{\alpha\beta}u_{\bar\alpha\bar\beta} - \sum_{\alpha,\beta}R_{\alpha\bar\beta}u_{\bar\alpha}u_{\beta}\right)\Psi, \quad (9.37)$$

$$\int_M i\sum_{\alpha}\left(u_{0\alpha}u_{\bar\alpha} - u_{0\bar\alpha}u_{\alpha}\right)\Psi$$
$$= \int_M\left(-\frac{2}{n}\left|\sum_{\alpha}u_{\alpha\bar\alpha}\right|^2 - \frac{1}{n}\left(\Delta_b u\right)^2 + i\sum_{\alpha,\beta}\left(A_{\bar\alpha\bar\beta}u_{\alpha}u_{\beta} - A_{\alpha\beta}u_{\bar\alpha}u_{\bar\beta}\right)\right). \quad (9.38)$$

Proof. By (9.28) (giving the expression of the formal adjoint of T_α)

$$\int_M\left(\sum_{\alpha,\beta}u_{\alpha\beta}u_{\bar\alpha\bar\beta}\right)\Psi = \sum_{\alpha,\beta}\left(u_{\alpha\beta}, u_{\alpha\beta}\right)$$
$$= \sum_{\alpha,\beta}\left(u_{\alpha\beta}, T_\alpha(u_\beta) - \Gamma^{\gamma}_{\alpha\beta}u_\gamma\right)$$
$$= \sum_{\alpha,\beta}\left(T^*_\alpha u_{\alpha\beta}, u_\beta\right) - \sum_{\alpha,\beta,\gamma}\left(\Gamma^{\bar\gamma}_{\bar\alpha\bar\beta}u_{\alpha\beta}, u_\gamma\right)$$
$$= -\sum_{\alpha,\beta}\left(T_{\bar\alpha}u_{\alpha\beta}, u_\beta\right) + \sum_{\alpha,\beta,\gamma}\left(\Gamma^{\alpha}_{\bar\gamma\gamma}u_{\alpha\beta} - \Gamma^{\bar\beta}_{\bar\alpha\bar\gamma}u_{\alpha\gamma}, u_\beta\right).$$

At this point, we express $u_{\alpha\beta}$ in terms of ordinary derivatives (cf. (9.2)) and use the identity

$$T_{\bar{\alpha}}T_\alpha = T_\alpha T_{\bar{\alpha}} + \Gamma^\gamma_{\bar{\alpha}\alpha}T_\gamma - \Gamma^{\bar{\gamma}}_{\alpha\bar{\alpha}}T_{\bar{\gamma}} + 2iT$$

to switch $T_{\bar{\alpha}}$ and T_α. We have

$$\begin{aligned}\int_M \Big(\sum_{\alpha,\beta} u_{\alpha\beta}u_{\bar{\alpha}\bar{\beta}}\Big)\Psi &= -\sum_{\alpha,\beta}\Big(T_{\bar{\alpha}}T_\alpha u_\beta - T_{\bar{\alpha}}\Big(\Gamma^\gamma_{\alpha\beta}u_\gamma\Big), u_\beta\Big)\\ &\quad + \sum_{\alpha,\beta,\gamma}\Big(\Gamma^\alpha_{\bar{\gamma}\gamma}u_{\alpha\beta} - \Gamma^{\bar{\beta}}_{\bar{\alpha}\bar{\gamma}}u_{\alpha\gamma}, u_\beta\Big)\\ &= -\sum_{\alpha,\beta}(T_\alpha T_{\bar{\alpha}}u_\beta, u_\beta) - 2in\sum_\beta (T(u_\beta), u_\beta)\\ &\quad + \sum_{\alpha,\beta,\gamma}\Big(\Gamma^{\bar{\gamma}}_{\alpha\bar{\alpha}}T_{\bar{\gamma}}(u_\beta) - \Gamma^\gamma_{\bar{\alpha}\alpha}T_\gamma(u_\beta) + \Gamma^\alpha_{\bar{\gamma}\gamma}u_{\alpha\beta}\\ &\qquad - \Gamma^{\bar{\beta}}_{\bar{\alpha}\bar{\gamma}}u_{\alpha\gamma} + T_{\bar{\alpha}}(\Gamma^\gamma_{\alpha\beta})u_\gamma + \Gamma^\gamma_{\alpha\beta}T_{\bar{\alpha}}(u_\gamma), u_\beta\Big).\end{aligned}$$

Let us replace the ordinary derivatives of u_β (all except for $T(u_\beta)$) in terms of covariant derivatives (cf. (9.2)–(9.3)). Then

$$\begin{aligned}\int_M \Big(\sum_{\alpha,\beta} u_{\alpha\beta}u_{\bar{\alpha}\bar{\beta}}\Big)\Psi &= -\sum_{\alpha,\beta}\Big(T_\alpha(u_{\bar{\alpha}\beta}) + T_\alpha(\Gamma^\gamma_{\bar{\alpha}\beta}u_\gamma), u_\beta\Big) - 2in\sum_\beta (T(u_\beta), u_\beta)\\ &\quad + \sum_{\alpha,\beta,\gamma}\Big(\Gamma^{\bar{\gamma}}_{\alpha\bar{\alpha}}u_{\bar{\gamma}\beta} - \Gamma^\gamma_{\bar{\alpha}\alpha}u_{\gamma\beta} + \Gamma^\alpha_{\bar{\gamma}\gamma}u_{\alpha\beta} - \Gamma^{\bar{\beta}}_{\bar{\alpha}\bar{\gamma}}u_{\alpha\gamma} + \Gamma^\gamma_{\alpha\beta}u_{\bar{\alpha}\gamma}, u_\beta\Big)\\ &\quad + \sum_{\alpha,\beta,\gamma}\Big(T_{\bar{\alpha}}(\Gamma^\gamma_{\alpha\beta})u_\gamma, u_\beta\Big)\\ &\quad + \sum_{\alpha,\beta,\gamma,\delta}\Big(\Gamma^{\bar{\gamma}}_{\alpha\bar{\alpha}}\Gamma^\rho_{\bar{\gamma}\beta}u_\rho - \Gamma^\gamma_{\bar{\alpha}\alpha}\Gamma^\rho_{\gamma\beta}u_\rho + \Gamma^\gamma_{\alpha\beta}\Gamma^\rho_{\bar{\alpha}\gamma}u_\rho, u_\beta\Big)\\ &= -\sum_{\alpha,\beta}(u_{\bar{\alpha}\beta}, T^*_\alpha u_\beta) - 2in\sum_\beta (Tu_\beta, u_\beta)\\ &\quad + \sum_{\alpha,\beta,\gamma}\Big(T_{\bar{\alpha}}(\Gamma^\gamma_{\alpha\beta})u_\gamma - T_\alpha(\Gamma^\gamma_{\bar{\alpha}\beta})u_\gamma, u_\beta\Big)\\ &\quad + \sum_{\alpha,\beta,\gamma,\rho}\Big((\Gamma^{\bar{\gamma}}_{\alpha\bar{\alpha}}\Gamma^\rho_{\bar{\gamma}\beta} - \Gamma^\gamma_{\bar{\alpha}\beta}\Gamma^\rho_{\alpha\gamma} + \Gamma^\gamma_{\alpha\beta}\Gamma^\rho_{\bar{\alpha}\gamma} - \Gamma^\gamma_{\bar{\alpha}\alpha}\Gamma^\rho_{\gamma\beta})u_\rho, u_\beta\Big)\\ &\quad + \sum_{\alpha,\beta,\gamma}\Big(\Gamma^{\bar{\gamma}}_{\alpha\bar{\alpha}}u_{\bar{\gamma}\beta} - \Gamma^\gamma_{\bar{\alpha}\alpha}u_{\gamma\beta} + \Gamma^\alpha_{\bar{\gamma}\gamma}u_{\alpha\beta} + \Gamma^\gamma_{\alpha\beta}u_{\bar{\alpha}\gamma}, u_\beta\Big).\end{aligned}$$

Let us use (9.28) and the (local) expression of the curvature tensor

$$R(T_{\bar{\alpha}}, T_\alpha)T_\beta = R_\beta{}^\rho{}_{\bar{\alpha}\alpha}T_\rho,$$

i.e.,

$$R_\beta{}^\rho{}_{\bar\alpha\alpha} = T_{\bar\alpha}(\Gamma^\rho_{\alpha\beta}) - T_\alpha(\Gamma^\rho_{\bar\alpha\beta}) + \Gamma^\gamma_{\alpha\beta}\Gamma^\rho_{\bar\alpha\gamma} - \Gamma^\gamma_{\bar\alpha\beta}\Gamma^\rho_{\alpha\gamma} + \Gamma^{\bar\gamma}_{\alpha\bar\alpha}\Gamma^\rho_{\bar\gamma\beta} - \Gamma^\gamma_{\bar\alpha\alpha}\Gamma^\rho_{\gamma\beta} - 2i\Gamma^\rho_{0\beta},$$

and observe that (by (9.27))

$$\sum_{\alpha,\beta,\gamma}\Big\{\Big(u_{\bar\alpha\beta}, \Gamma^\gamma_{\bar\alpha\beta}u_\gamma\Big) - \Big(\Gamma^{\bar\alpha}_{\gamma\bar\gamma}u_{\bar\alpha\beta}, u_\beta\Big) + \Big(\Gamma^{\bar\gamma}_{\alpha\bar\alpha}u_{\bar\gamma\beta}, u_\beta\Big) - \big(\Gamma^\gamma_{\bar\alpha\alpha}u_{\gamma\beta}, u_\beta\big) + \Big(\Gamma^\alpha_{\bar\gamma\gamma}u_{\alpha\beta}, u_\beta\Big) + \Big(\Gamma^\gamma_{\alpha\beta}u_{\bar\alpha\gamma}, u_\beta\Big)\Big\} = 0.$$

We obtain

$$\int_M \Big(\sum_{\alpha,\beta} u_{\alpha\beta}u_{\bar\alpha\bar\beta}\Big)\Psi = \sum_{\alpha,\beta}(u_{\bar\alpha\beta}, u_{\bar\alpha\beta}) + \sum_{\alpha,\beta,\rho}\Big(R_\beta{}^\rho{}_{\bar\alpha\alpha}u_\rho + 2i\Gamma^\rho_{0\beta}u_\rho, u_\beta\Big) - 2in\sum_\beta (Tu_\beta, u_\beta);$$

hence (by (9.4))

$$\int_M i\Big(\sum_\alpha u_{0\alpha}u_{\bar\alpha}\Big)\Psi = \frac{1}{2n}\int_M \sum_{\alpha,\beta}\Big(u_{\bar\alpha\beta}u_{\alpha\bar\beta} - u_{\alpha\beta}u_{\bar\alpha\bar\beta} - R_{\alpha\bar\beta}u_{\bar\alpha}u_\beta\Big). \tag{9.39}$$

Again, one recognizes the Ricci curvature by

$$\sum_\alpha R_\beta{}^\rho{}_{\bar\alpha\alpha} = \sum_\alpha R_{\beta\bar\rho\bar\alpha\alpha} = -\sum_\alpha R_{\bar\rho\beta\bar\alpha\alpha} = -\sum_\alpha R_{\bar\alpha\beta\bar\rho\alpha} = -\sum_\alpha R_{\bar\alpha}{}^{\bar\beta}{}_{\bar\rho\alpha} = -R_{\beta\bar\rho}$$

(cf. (1.93) and (1.98) in Chapter 1). Let us take the complex conjugate of (9.39) and sum up the resulting identity and (9.39). This procedure leads to (9.37). □

To prove (9.38) let us perform the calculation (by (2.15))

$$\int_M (\Delta_b u)^2\,\Psi = \int_M \Big(\sum_\alpha (u_{\alpha\bar\alpha} + u_{\bar\alpha\alpha})\Big)^2\Psi$$
$$= \int_M \sum_{\alpha,\beta}\Big(u_{\alpha\bar\alpha} + u_{\beta\bar\beta} + u_{\alpha\bar\alpha}u_{\bar\beta\beta} + u_{\bar\alpha\alpha}u_{\beta\bar\beta} + u_{\bar\alpha\alpha}u_{\bar\beta\beta}\Big)\Psi$$
$$= \int_M\Big\{2\Big|\sum_\alpha u_{\alpha\bar\alpha}\Big|^2 + \sum_{\alpha,\beta}\Big((u_{\bar\alpha\alpha} - 2iu_0)\,u_{\beta\bar\beta} + u_{\bar\alpha\alpha}\big(u_{\beta\bar\beta} + 2iu_0\big)\Big)\Big\}\Psi$$

by the commutation formula (9.7). Finally,

$$\int_M (\Delta_b u)^2\,\Psi = 4\int_M \Big|\sum_\alpha u_{\alpha\bar\alpha}\Big|^2\Psi - 2n\int_M i\sum_\alpha u_0\,(u_{\alpha\bar\alpha} - u_{\bar\alpha\alpha})\,\Psi. \tag{9.40}$$

The last term in (9.40) may be computed as

$$\int_M \sum_\alpha u_0\,(u_{\alpha\bar\alpha} - u_{\bar\alpha\alpha})\,\Psi = \sum_\alpha (u_0, u_{\bar\alpha\alpha} - u_{\alpha\bar\alpha})$$
$$= \sum_\alpha \left(u_0, T_{\bar\alpha}(u_\alpha) - \Gamma^\gamma_{\bar\alpha\alpha} u_\gamma - T_\alpha(u_{\bar\alpha}) + \Gamma^{\bar\gamma}_{\alpha\bar\alpha}\right)$$
$$= \sum_\alpha \{(T^*_{\bar\alpha} u_0, u_\alpha) - (T^*_\alpha u_0, u_{\bar\alpha})\} + \sum_{\alpha,\gamma}\left\{-\left(u_0, \Gamma^\gamma_{\bar\alpha\alpha} u_\gamma\right) + \left(u_0, \Gamma^{\bar\gamma}_{\alpha\bar\alpha} u_{\bar\gamma}\right)\right\}$$
$$= \sum_\alpha \{-(T_\alpha u_0, u_\alpha) + (T_{\bar\alpha} u_0, u_{\bar\alpha})\}$$

(the terms containing Christoffel symbols cancel in pairs). Then, by expressing ordinary derivatives in terms of covariant derivatives and using the commutation formula (9.8), we obtain

$$\int_M \sum_\alpha u_0\,(u_{\alpha\bar\alpha} - u_{\bar\alpha\alpha})\,\Psi$$
$$= \int_M \sum_\alpha (u_{0\bar\alpha} u_\alpha - u_{0\alpha} u_{\bar\alpha})\,\Psi + \sum_{\alpha,\beta}\left\{-\left(A_{\alpha\beta} u_{\bar\beta}, u_\alpha\right) + \left(A_{\bar\alpha\bar\beta} u_\beta, u_{\bar\alpha}\right)\right\}. \tag{9.41}$$

Finally, we may substitute from (9.41) into (9.40) and solve for

$$\int_M i \sum_\alpha (u_{0\alpha} u_{\bar\alpha} - u_{0\bar\alpha} u_\alpha)\Psi.$$

This procedure furnishes (9.38). □

9.2.3 A. Greenleaf's theorem

Let us recall that by definition, $\Delta_b u = \operatorname{div}(\pi_H \nabla u)$. In particular

$$\int_M (\Delta_b u)\Psi = 0,$$

for any $u \in C^\infty_0(M)$. Let us assume from now on that M is compact. Integration of (9.36) over M leads to

$$0 = \int_M \Bigg(\sum_{\alpha,\beta} u_{\alpha\bar\beta} u_{\bar\alpha\beta} + \sum_{\alpha,\beta} u_{\alpha\beta} u_{\bar\alpha\bar\beta} + 2i \sum_\alpha (u_{\bar\alpha} u_{0\alpha} - u_\alpha u_{0\bar\alpha})$$
$$+ \sum_{\alpha,\beta} R_{\alpha\bar\beta} u_{\bar\alpha} u_\beta + in \sum_{\alpha,\beta}\left(A_{\alpha\beta} u_{\bar\alpha} u_{\bar\beta} - A_{\bar\alpha\bar\beta} u_\alpha u_\beta\right)$$
$$+ \tfrac{1}{2} \sum_\alpha \left(u_{\bar\alpha}(\Delta_b u)_\alpha + u_\alpha (\Delta_b u)_{\bar\alpha}\right) \Bigg)\Psi. \tag{9.42}$$

We shall need the following identity:

Proposition 9.6.

$$\int_M \sum_\alpha u_\alpha (\Delta_b u)_{\bar\alpha} \Psi = -\int_M (\Delta_b u) \operatorname{div}\left(\nabla^{1,0} u\right) \Psi. \tag{9.43}$$

Let us prove (9.43). We have

$$\int_M \sum_\alpha u_\alpha (\Delta_b u)_{\bar\alpha} \Psi = \int_M u^{\bar\alpha} T_{\bar\alpha} (\Delta_b u) \Psi = \int_M \left(T_{\bar\alpha}\left(u^{\bar\alpha}\Delta_b u\right) - (\Delta_b u) T_{\bar\alpha}(u^{\bar\alpha})\right) \Psi.$$

Let us set $Z = (\Delta_b u) u^\alpha T_\alpha \in T_{1,0}(M)$. Then

$$\operatorname{div}(\overline{Z}) = \operatorname{trace}\{V \mapsto \nabla_V \overline{Z}\} = \left(\nabla_{T_{\bar\alpha}} \overline{Z}\right)^{\bar\alpha} = T_{\bar\alpha}\left(u^{\bar\alpha}\Delta_b u\right) + u^{\bar\beta} (\Delta_b u) \Gamma^{\bar\alpha}_{\bar\alpha\bar\beta},$$

i.e.,

$$T_{\bar\alpha}\left(u^{\bar\alpha}\Delta_b u\right) = \operatorname{div}(\overline{Z}) - u^{\overline{\beta}} (\Delta_b u) \Gamma^{\bar\alpha}_{\bar\alpha\bar\beta}.$$

Finally, by Green's lemma

$$\begin{aligned}\int_M \sum_\alpha u_\alpha (\Delta_b u) \Psi &= \int_M \{-u^{\bar\beta} (\Delta_b u) \Gamma^{\bar\alpha}_{\bar\alpha\bar\beta} - (\Delta_b u) T_{\bar\alpha}(u^{\bar\alpha})\}\Psi \\ &= -\int_M (\Delta_b u) \operatorname{div}\left(u^{\bar\alpha} T_{\bar\alpha}\right) \Psi. \qquad \square\end{aligned}$$

Now let us take the complex conjugate of (9.43) and sum up the resulting identity and (9.43). We obtain

$$\int_M \sum_\alpha \left(u_\alpha (\Delta_b u)_{\bar\alpha} + u_{\bar\alpha} (\Delta_b u)_\alpha\right) \Psi = 2\int_M (\Delta_b u)^2 \Psi;$$

hence (9.42) becomes

$$\begin{aligned}\int_M \Bigg(\sum_{\alpha,\beta} u_{\alpha\bar\beta} u_{\bar\alpha\beta} + \sum_{\alpha,\beta} u_{\alpha\beta} u_{\bar\alpha\bar\beta} - \tfrac{1}{2} (\Delta_b u)^2 \\ + \sum_{\alpha,\beta} R_{\alpha\bar\beta} u_{\bar\alpha} u_\beta + in \sum_{\alpha,\beta}\left(A_{\alpha\beta} u_{\bar\alpha} u_{\bar\beta} - A_{\bar\alpha\bar\beta} u_\alpha u_\beta\right) \\ + 2i \sum_\alpha (u_{\bar\alpha} u_{0\alpha} - u_\alpha u_{0\bar\alpha}) \Bigg)\Psi = 0. \end{aligned} \tag{9.44}$$

Let $C \in \mathbf{R}$ and write

$$\begin{aligned}&\int_M i \sum_\alpha (u_{0\alpha} u_{\bar\alpha} - u_{0\bar\alpha} u_\alpha) \Psi \\ &= C \int_M i \sum_\alpha (u_{0\alpha} u_{\bar\alpha} - u_{0\bar\alpha} u_\alpha) \Psi + (1 - C) \int_M i \sum_\alpha (u_{0\alpha} u_{\bar\alpha} - u_{0\bar\alpha} u_\alpha) \Psi.\end{aligned}$$

Next, let us calculate the last two integrals using (9.37) and (9.38), respectively. We obtain

$$\begin{aligned}
&\int_M i \sum_{\alpha} (u_{0\alpha} u_{\bar\alpha} - u_{0\bar\alpha} u_{\alpha})\, \Psi \\
&= \int_M \Bigg\{ \frac{C}{n} \sum_{\alpha,\beta} u_{\alpha\bar\beta} u_{\bar\alpha\beta} - \frac{C}{n} \sum_{\alpha,\beta} u_{\alpha\beta} u_{\bar\alpha\bar\beta} - \frac{2(1-C)}{n} \Big| \sum_{\alpha} u_{\alpha\bar\alpha} \Big|^2 + \frac{(1-C)}{2n} (\Delta_b u)^2 \\
&\qquad - \frac{C}{n} \sum_{\alpha,\beta} R_{\alpha\bar\beta} u_{\bar\alpha\beta} u_{\beta} + i(1-C) \sum_{\alpha,\beta} \Big(A_{\bar\alpha\bar\beta} u_{\alpha} u_{\beta} - A_{\alpha\beta} u_{\bar\alpha} u_{\bar\beta} \Big) \Bigg\} .
\end{aligned}$$

Now, we use this identity to replace the term $\int_M i \sum_\alpha (u_{0\alpha} u_{\bar\alpha} - u_{0\bar\alpha} u_{\alpha})\, \Psi$ in (9.44). Then (9.44) becomes

$$\begin{aligned}
&\int_M \Bigg\{ \Big(1 + \frac{2C}{n}\Big) \sum_{\alpha,\beta} u_{\alpha\bar\beta} u_{\bar\alpha\beta} + \Big(1 - \frac{2C}{n}\Big) \sum_{\alpha,\beta} u_{\alpha\beta} u_{\bar\alpha\bar\beta} \\
&- \frac{4(1-C)}{n} \Big| \sum_{\alpha} u_{\alpha\bar\alpha} \Big|^2 + \Big(-\tfrac{1}{2} + \frac{1-C}{n} \Big) (\Delta_b u)^2 + \Big(1 - \frac{2C}{n}\Big) \sum_{\alpha,\beta} R_{\alpha\bar\beta} u_{\bar\alpha} u_{\beta} \\
&\qquad + i(n - 2(1-C)) \sum_{\alpha,\beta} \Big(A_{\alpha\beta} u_{\bar\alpha} u_{\bar\beta} - A_{\bar\alpha\bar\beta} u_{\alpha} u_{\beta} \Big) \Bigg\}\, \Psi = 0. \quad (9.45)
\end{aligned}$$

In particular, for $C = -n/2$ the identity (9.45) furnishes

$$\begin{aligned}
&\int_M \Bigg\{ 2 \sum_{\alpha,\beta} u_{\alpha\beta} u_{\bar\alpha\bar\beta} - \Big(2 + \frac{4}{n}\Big) \Big| \sum_{\alpha} u_{\alpha\bar\alpha} \Big|^2 + \frac{1}{n} (\Delta_b u)^2 \\
&\qquad + 2 \sum_{\alpha,\beta} R_{\alpha\bar\beta} u_{\bar\alpha} u_{\beta} + 2i \sum_{\alpha,\beta} \Big(A_{\bar\alpha\bar\beta} u_{\alpha} u_{\beta} - A_{\alpha\beta} u_{\bar\alpha} u_{\bar\beta} \Big) \Bigg\}\, \Psi = 0. \quad (9.46)
\end{aligned}$$

Also, for $C = 1$, the identity (9.45) furnishes

$$\begin{aligned}
&\int_M \Bigg\{ \Big(1 + \frac{2}{n}\Big) \sum_{\alpha,\beta} u_{\alpha\bar\beta} u_{\bar\alpha\beta} + \Big(1 - \frac{2}{n}\Big) \sum_{\alpha,\beta} u_{\alpha\beta} u_{\bar\alpha\bar\beta} - \tfrac{1}{2} (\Delta_b u)^2 \\
&\qquad + \Big(1 - \frac{2}{n}\Big) \sum_{\alpha,\beta} R_{\alpha\bar\beta} u_{\bar\alpha} u_{\beta} - in \sum_{\alpha,\beta} \Big(A_{\bar\alpha\bar\beta} u_{\alpha} u_{\beta} - A_{\alpha\beta} u_{\bar\alpha} u_{\bar\beta} \Big) \Bigg\}\, \Psi = 0. \quad (9.47)
\end{aligned}$$

Note that

$$\begin{aligned}
\int_m (\Delta_b u)^2\, \Psi &= \int_M \Big(\sum_{\alpha} (u_{\alpha\bar\alpha} + u_{\bar\alpha\alpha}) \Big)^2 \Psi \\
&= \int_M \Bigg\{ \Big(\sum_{\alpha} u_{\alpha\bar\alpha} \Big)^2 + \Big(\sum_{\alpha} u_{\bar\alpha\alpha} \Big)^2 + 2 \Big| \sum_{\alpha} u_{\alpha\bar\alpha} \Big|^2 \Bigg\} \Psi;
\end{aligned}$$

hence

$$\int_M (\Delta_b u)^2 \, \Psi - 2 \int_M \Big| \sum_\alpha u_{\alpha\bar\alpha} \Big|^2 \Psi = \int_M \Big\{ \Big(\sum_\alpha u_{\alpha\bar\alpha} \Big)^2 + \Big(\sum_\alpha u_{\bar\alpha\alpha} \Big)^2 \Big\} \Psi \, . \tag{9.48}$$

Let $u \in C^\infty(M)$ be a real eigenfunction of Δ_b corresponding to the eigenvalue $\lambda > 0$, i.e., $\Delta_b u = \lambda u$. We may distinguish two cases:

$$\text{(I)} \quad \int_M \Big\{ \Big(\sum_\alpha u_{\alpha\bar\alpha} \Big)^2 + \Big(\sum_\alpha u_{\bar\alpha\alpha} \Big)^2 \Big\} \Psi \geq 0$$

and

$$\text{(II)} \quad \int_M \Big\{ \Big(\sum_\alpha u_{\alpha\bar\alpha} \Big)^2 + \Big(\sum_\alpha u_{\bar\alpha\alpha} \Big)^2 \Big\} \Psi < 0.$$

If case I occurs, then (9.48) yields

$$2 \int_M \Big| \sum_\alpha u_{\alpha\bar\alpha} \Big|^2 \Psi \leq \int_M (\Delta_b u)^2 \, \Psi$$

and then (by (9.46))

$$\begin{aligned} 0 = \int_M \Big\{ & 2 \sum_{\alpha,\beta} u_{\alpha\beta} u_{\bar\alpha\bar\beta} - \Big(2 + \frac{4}{n} \Big) \Big| \sum_\alpha u_{\alpha\bar\alpha} \Big|^2 + \frac{1}{n} (\Delta_b u)^2 \\ & + 2 \sum_{\alpha,\beta} \Big[R_{\alpha\bar\beta} u_{\bar\alpha\beta} + i \Big(A_{\bar\alpha\bar\beta} u_\alpha u_\beta - A_{\alpha\beta} u_{\bar\alpha} u_{\bar\beta} \Big) \Big] \Big\} \Psi \\ \geq \int_M \Big\{ & - \Big(1 + \frac{1}{n} \Big) (\Delta_b u)^2 + \sum_{\alpha,\beta} \Big[R_{\alpha\bar\beta} u_{\bar\alpha} u_\beta + i \Big(A_{\bar\alpha\bar\beta} u_\alpha u_\beta - A_{\alpha\beta} u_{\bar\alpha} u_{\bar\beta} \Big) \Big] \Big\} \Psi , \end{aligned}$$

or, if we assume that

$$R_{\alpha\bar\beta} Z^\alpha Z^\beta + i \Big(A_{\bar\alpha\bar\beta} \overline{Z}^\alpha \overline{Z}^\beta - A_{\alpha\beta} Z^\alpha Z^\beta \Big) \geq 2k h_{\alpha\bar\beta} Z^\alpha \overline{Z}^\beta , \tag{9.49}$$

for some $k > 0$ and any $Z = Z^\alpha T_\alpha \in T_{1,0}(M)$, then

$$0 \geq \int_M \Big\{ - \Big(1 + \frac{1}{n} \Big) (\Delta_b u)^2 + 4k \sum_\alpha |u_\alpha|^2 \Big\} \Psi \, . \tag{9.50}$$

We need the following identity:

Proposition 9.7.

$$-2 \int_M \sum_\alpha |u_\alpha|^2 \, \Psi = \int_M u \, (\Delta_b u) \, \Psi \, . \tag{9.51}$$

Indeed (9.51) may be checked as follows

$$\begin{aligned}\int_M \sum_\alpha |u_\alpha|^2 \Psi &= \sum_\alpha (u_\alpha, u_\alpha)\\ &= \sum_\alpha (T_\alpha(u), u_\alpha) = \sum_\alpha \left(u, T_\alpha^* u_\alpha\right)\\ &= -\sum_\alpha (u, T_{\bar\alpha} u_\alpha) + \sum_{\alpha,\beta}\left(u, \Gamma^\alpha_{\bar\beta\beta} u_\alpha\right)\\ &= -\sum_\alpha (u, u_{\bar\alpha\alpha}) - \sum_{\alpha,\beta}\left(u, \Gamma^\beta_{\bar\alpha\alpha} u_\beta\right) + \sum_{\alpha,\beta}\left(u, \Gamma^\alpha_{\bar\beta\beta} u_\alpha\right),\end{aligned}$$

i.e.,

$$\int_M \sum_\alpha |u_\alpha|^2 \Psi = -\sum_\alpha (u, u_{\bar\alpha\alpha}) .$$

Take the complex conjugate and sum up with the resulting identity. We get

$$2\int_M \sum_\alpha |u_\alpha|^2 \Psi = -\sum_\alpha (u, u_{\bar\alpha\alpha} + u_{\alpha\bar\alpha}) ,$$

which is equivalent (by (2.15)) to (9.51). □

We may use (9.51) to rewrite the estimate (9.50) as

$$0 \geq \int_M \left\{-\left(1+\frac{1}{n}\right) + \frac{k}{\lambda}\right\} (\Delta_b u)^2 \Psi,$$

hence

$$0 \geq -\left(1+\frac{1}{n}\right) + \frac{k}{\lambda},$$

or

$$\lambda \geq \frac{kn}{n+1} . \tag{9.52}$$

As to case II, by (9.48) one gets

$$\int_M (\Delta_b u)^2 \Psi \leq 2\int_M \left|\sum_\alpha u_{\alpha\bar\alpha}\right|^2 \Psi . \tag{9.53}$$

On the other hand,

$$\sum_{\alpha,\beta} u_{\alpha\bar\beta} u_{\bar\alpha\beta} = \sum_{\alpha,\beta}\left|u_{\alpha\bar\beta}\right|^2 \geq \sum_\alpha |u_{\alpha\bar\alpha}|^2 \geq \frac{1}{n}\left|\sum_\alpha u_{\alpha\bar\alpha}\right|^2 .$$

Therefore (by (9.53))

$$\int_M \Big(\sum_{\alpha,\beta} u_{\alpha\bar\beta} u_{\bar\alpha\beta}\Big)\Psi \geq \frac{1}{2n}\int_M (\Delta_b u)^2 \Psi ,$$

and the identity (9.47) leads to

$$0 \geq \int_M \left\{ \frac{2}{n}\left(1 + \frac{1}{2n}\right)(\Delta_b u)^2 - \tfrac{1}{2}(\Delta_b u)^2 \right.$$
$$\left. + \left(1 - \frac{2}{n}\right) \sum_{\alpha,\beta} R_{\alpha\bar\beta} u_{\bar\alpha} u_\beta - in \sum_{\alpha,\beta} \left(A_{\bar\alpha\bar\beta} u_\alpha u_\beta - A_{\alpha\beta} u_{\bar\alpha} u_{\bar\beta} \right) \right\} \Psi ,$$

provided that $n \geq 2$. Furthermore, if we assume that $n \geq 3$ and

$$R_{\alpha\bar\beta} Z^\alpha \overline{Z}^\beta - \frac{n^2}{n-2} i \left(A_{\bar\alpha\bar\beta} \overline{Z}^\alpha \overline{Z}^\beta - A_{\alpha\beta} Z^\alpha Z^\beta \right) \geq 2k h_{\alpha\bar\beta} Z^\alpha \overline{Z}^\beta , \tag{9.54}$$

for some $k > 0$ and any $Z = Z^\alpha T_\alpha \in T_{1,0}(M)$, then (by (9.51))

$$0 \geq \int_M \left(\frac{1}{n}\left(1 + \frac{2}{n}\right) - 1 + \frac{(n-2)}{n} \frac{k}{\lambda} \right) \Psi ;$$

hence

$$0 \geq \frac{1}{n}\left(1 + \frac{2}{n}\right) - 1 + \left(1 - \frac{2}{n}\right) \frac{k}{\lambda},$$

or $\lambda \geq nk/(n+1)$, i.e., (9.52) holds. We have proved the following result:

Theorem 9.1. (A. Greenleaf [186])
Let M be a strictly pseudoconvex CR manifold of CR dimension $n \geq 3$. Let θ be a contact 1-form on M with the Levi form L_θ positive definite. Assume that there is a constant $k > 0$ such that

$$R_{\alpha\bar\beta} Z^\alpha \overline{Z}^\beta + i \left(A_{\bar\alpha\bar\beta} \overline{Z}^\alpha \overline{Z}^\beta - A_{\alpha\beta} Z^\alpha Z^\beta \right) \geq 2k h_{\alpha\bar\beta} Z^\alpha \overline{Z}^\beta$$

and

$$R_{\alpha\bar\beta} Z^\alpha \overline{Z}^\beta - \frac{n^2}{n-2} i \left(A_{\bar\alpha\bar\beta} \overline{Z}^\alpha \overline{Z}^\beta - A_{\alpha\beta} Z^\alpha Z^\beta \right) \geq 2k h_{\alpha\bar\beta} Z^\alpha \overline{Z}^\beta$$

for any $Z = Z^\alpha T_\alpha \in T_{1,0}(M)$. Then the first nonzero eigenvalue λ_1 of the sub-Laplacian Δ_b of (M, θ) satisfies the estimate

$$\lambda_1 \geq \frac{kn}{n+1} .$$

9.2.4 A lower bound on the first eigenvalue of a Folland–Stein operator

Let us set $X_\alpha = \frac{1}{2}(T_\alpha + T_{\bar\alpha})$ and $Y_\alpha = \frac{i}{2}(T_{\bar\alpha} - T_\alpha)$. Then (by (9.25)–(9.26))

$$X_\alpha^* = -X_\alpha + \tfrac{1}{2} \sum_\beta \left(\Gamma^\alpha_{\bar\beta\beta} + \Gamma^{\bar\alpha}_{\beta\bar\beta} \right),$$

$$Y_\alpha^* = -Y_\alpha + \frac{i}{2} \sum_\beta \left(\Gamma^\alpha_{\bar\beta\beta} - \Gamma^{\bar\alpha}_{\beta\bar\beta} \right) .$$

Next, we perform the following calculation:

$$
\begin{aligned}
\sum_\alpha (T_\alpha f, T_\alpha f) &= \sum_\alpha ((X_\alpha + iY_\alpha) f, (X_\alpha + iY_\alpha) f) \\
&= \sum_\alpha \left\{ \|X_\alpha f\|^2 + \|Y_\alpha f\|^2 + i\left(X_\alpha^* Y_\alpha f, f\right) - i\left(Y_\alpha^* X_\alpha f, f\right)\right\} \\
&= \sum_\alpha \Bigg\{ \|X_\alpha f\|^2 + \|Y_\alpha f\|^2 - i\,(X_\alpha Y_\alpha f, f) + i\,(Y_\alpha X_\alpha f, f) \\
&\quad + \frac{i}{2} \sum_\beta \left((\Gamma^\alpha_{\bar\beta\beta} + \Gamma^{\bar\alpha}_{\beta\bar\beta}) Y_\alpha f, f \right) + \tfrac{1}{2} \sum_\beta \left((\Gamma^\alpha_{\bar\beta\beta} - \Gamma^{\bar\alpha}_{\beta\bar\beta}) X_\alpha f, f \right) \Bigg\} \\
&= \sum_\alpha \Bigg\{ \|X_\alpha f\|^2 + \|Y_\alpha f\|^2 \\
&\quad - \tfrac{1}{4}\,((T_{\bar\alpha} - T_\alpha)(f_\alpha + f_{\bar\alpha}), f) + \tfrac{1}{4}\,((T_\alpha + T_{\bar\alpha})(f_{\bar\alpha} - f_\alpha), f) \\
&\quad + \tfrac{1}{4} \sum_\beta \left((\Gamma^\alpha_{\bar\beta\beta} - \Gamma^{\bar\alpha}_{\beta\bar\beta})(f_\alpha + f_{\bar\alpha}), f \right) - \tfrac{1}{4} \sum_\beta \left((\Gamma^\alpha_{\bar\beta\beta} + \Gamma^{\bar\alpha}_{\beta\bar\beta})(f_{\bar\alpha} - f_\alpha), f \right) \Bigg\} \\
&= \sum_\alpha \left\{ \|X_\alpha f\|^2 + \|Y_\alpha f\|^2 + \tfrac{1}{2} \left(T_\alpha f_{\bar\alpha} - \Gamma^{\bar\beta}_{\alpha\bar\alpha} f_{\bar\beta} - T_{\bar\alpha} f_\alpha + \Gamma^{\beta}_{\bar\alpha\alpha} f_\beta, f \right) \right\} \\
&= \sum_\alpha \left\{ \|X_\alpha f\|^2 + \|Y_\alpha f\|^2 + \tfrac{1}{2}\,(f_{\alpha\bar\alpha} - f_{\bar\alpha\alpha}, f) \right\};
\end{aligned}
$$

hence (by (9.7))

$$
\sum_\alpha \|T_\alpha f\|^2 = \sum_\alpha \left\{ \|X_\alpha f\|^2 + \|Y_\alpha f\|^2 \right\} - in\,(f_0, f) .
$$

Note that $(iT)^* = iT$; hence $(if_0, f) \in \mathbf{R}$. We obtain

$$
n\,(if_0, f) \le \sum_\alpha \left\{ \|X_\alpha f\|^2 + \|Y_\alpha f\|^2 \right\} .
$$

On the other hand,

$$
\begin{aligned}
\sum_\alpha \left\{ \|X_\alpha f\|^2 + \|Y_\alpha f\|^2 \right\} &= \tfrac{1}{4} \sum_\alpha \{ (f_\alpha + f_{\bar\alpha}, f_\alpha + f_{\bar\alpha}) + (f_{\bar\alpha} - f_\alpha, f_{\bar\alpha} - f_\alpha) \} \\
&= \tfrac{1}{2} \sum_\alpha \{ (f_\alpha, f_\alpha) + (f_{\bar\alpha}, f_{\bar\alpha}) \} = \tfrac{1}{2}\,(\Delta_b f, f)
\end{aligned}
$$

(the last equality follows by integration by parts). Hence

$$
2n\,(if_0, f) \le (\Delta_b f, f) .
$$

Analogously, one may develop $\sum_\alpha \|T_{\bar\alpha} f\|^2$ and prove that

$$
2n\,(if_0, f) \ge -(\Delta_b f, f) .
$$

Therefore

$$
2n\,|(if_0, f)| \le (\Delta_b f, f) , \tag{9.55}
$$

for any $f \in C^\infty(M) \otimes \mathbf{C}$. Consider the operators

$$\mathcal{L}_c = \Delta_b - icT, \quad |c| < n.$$

Then

$$\begin{aligned}(\mathcal{L}_c f, f) &= (\Delta_b f, f) - c\,(i f_0, f) \\ &\geq (\Delta_b f, f) - |c|\,|(i f_0, f)| \geq \left(1 - \frac{|c|}{2n}\right)(\Delta_b f, f)\,.\end{aligned}$$

By a result in [150] each $\mathcal{L}_c$, $|c| < n$, is a subelliptic operator of order $\epsilon = 1/2$; hence $\mathcal{L}_c$ has a discrete spectrum tending to $+\infty$.

Proposition 9.8. (A. Greenleaf [186])
Let λ_1^c be the first eigenvalue of $\mathcal{L}_c$. Assume that there is $\lambda \in \mathrm{Spec}(\Delta_b)$, $\lambda \neq 0$, such that

$$\mathrm{Eigen}\left(\mathcal{L}_c, \lambda_1^c\right) \cap \mathrm{Eigen}\left(\Delta_b, \lambda\right) \neq (0)\,.$$

Then, under the hypothesis of Theorem 9.1, one has

$$\lambda_1^c \geq \left(1 - \frac{|c|}{n}\right) \frac{nk}{n+1}\,.$$

9.2.5 Z. Jiaqing and Y. Hongcang's theorem on CR manifolds

We end this chapter by reporting on a recent result on spectra of CR manifolds, cf. E. Barletta et al. [35]. Let M be a compact strictly pseudoconvex $(2n+1)$-dimensional CR manifold and Δ_b the sub-Laplacian corresponding to a fixed choice of contact 1-form θ on M. Let λ_k be the kth nonzero eigenvalue of Δ_b. Using L^2 methods (i.e., a pseudo-Hermitian analogue of the Bochner formula in Riemannian geometry) A. Greenleaf has shown (see Theorem 9.1 in this chapter) that the first nonzero eigenvalue λ_1 of Δ_b satisfies

$$\lambda_1 \geq \frac{n}{n+1} C_0 \tag{9.56}$$

under a restriction involving the pseudo-Hermitian Ricci tensor $R_{\alpha\bar\beta}$ and the pseudo-Hermitian torsion $A_{\alpha\beta}$. The restriction and proof are imitative of those in A. Lichnerowicz [286], p. 135. A similar result is the following:

Theorem 9.2. (E. Barletta et al. [35])
Let M be a compact strictly pseudoconvex CR manifold (of CR dimension n). Assume that the problem

$$\begin{cases} \Delta_b v = \lambda_k v, & T(v) = 0, \\ \sup v = 1, & \\ \inf v = -C, & 0 < C \leq 1, \end{cases} \tag{9.57}$$

admits some C^∞ solution v. If

$$\mathrm{Ric}(X - iJX, X + iJX) + 2(n-2)A(X, JX) \geq 0, \tag{9.58}$$

for any $X \in H(M)$, then

$$\lambda_k \geq \frac{\pi^2}{d_\theta^2}. \tag{9.59}$$

Here T is the characteristic direction of (M, θ) and d_θ is the diameter of M with respect to the Webster metric g_θ. The proof of Theorem 9.2 is omitted (in contrast to [186], L^∞ methods are employed). If for instance $M = S^{2n+1}$ (the sphere carrying the standard pseudo-Hermitian structure) then both the hypothesis of Theorem 9.1 in this chapter and the assumption (9.58) hold.

Let M be a strictly pseudoconvex CR manifold of vanishing pseudo-Hermitian torsion. Then the assumption (9.58) is weaker than the hypothesis of Theorem 9.1. However, it must be pointed out that while one works under less-restrictive geometric conditions, the proof of (9.59) requires the existence of a solution of (9.57) (rather than a solution of an eigenvalue problem for Δ_b, alone). As a result, one may estimate terms of the form $u^\alpha(\mathcal{L}_2 u_\alpha)$ at a point (where $\mathcal{L}_2$ is a Folland–Stein operator). General existence theorems for the solutions of (9.57) are not known (and this is precisely the limitation of Theorem 9.2). An example in which (9.57) may be solved is indicated below.

If v is a solution of (9.57) then $\Delta_b v = \Delta v$ (where Δ is the Laplace–Beltrami operator of (M, g_θ)) so that actually $\lambda_k \in \mathrm{Spec}(M, g_\theta)$ and the estimate (9.59) follows from work by Z. Jiaqing and Y. Hongcang [231], provided that the metric (here the Webster metric g_θ) has nonnegative Ricci curvature. Nevertheless, this requirement may be seen to be generically stronger than our assumption (9.58). Indeed, let M be a strictly pseudoconvex CR manifold of vanishing pseudo-Hermitian torsion. We have

$$\mathrm{Ric}_\theta(X, Y) = \mathrm{Ric}(X, Y) - \tfrac{1}{2} g_\theta(X, Y) + \frac{n+1}{2}\theta(X)\theta(Y) \tag{9.60}$$

for any $X, Y \in T(M)$. Thus

$$\mathrm{Ric}(X - iJX, X + iJX) = \mathrm{Ric}_\theta(X, X) + \mathrm{Ric}_\theta(JX, JX) + \|X\|^2$$

for any $X \in H(M)$. Consequently, if $\mathrm{Ric}_\theta(X, X) \geq 0$ for any $X \in T(M)$ then $R_{\alpha\overline{\beta}}$ is positive semidefinite (while the converse does not follow from (9.60)).

The problem of the existence of a solution of (9.57) is open. If for instance $M = S^{2n+1}$ then (9.57) has no solution for $k = 1$ (i.e., there is no first-degree harmonic polynomial H on $\mathbf{R}^{2(n+1)}$ satisfying $T(H) = 0$). Next, all solutions of

$$\begin{cases} \Delta_b v = \lambda_2 v, \\ T(v) = 0, \end{cases}$$

are given as $v = H_{|S^{2n+1}}$, where

$$H(x, y) = \sum_{1 \le i < i \le n+1} a_{ij}(x_i x_j + y_i y_j), \ a_{ij} \in \mathbf{R}$$

(and $\lambda_2 = 4(n+1)$). Also, for each $(i, j) \in \{1, \dots, n+1\}^2$, $i < j$, the eigenfunction $v_{ij} = H_{ij\,|S^{2n+1}}$, where $H_{ij} = 2(x_i x_j + y_i y_j)$, has $\sup v_{ij} = 1$ and $\inf v_{ij} = -1$ (i.e., v_{ij} is a solution of (9.57) with $k = 2$ and $C = 1$).

The estimate (9.56) may be thought of as an estimate on λ_k, $k \ge 2$. As such, (9.59) is sharper than (9.56) provided that

$$d_\theta < \pi \sqrt{\frac{n+1}{nC_0}}\,. \tag{9.61}$$

However, among the odd-dimensional spheres only S^3 and S^5 satisfy (9.61) (since $M = S^{2n+1}$ yields $C_0 = n+1$; cf. [422]).

A

A Parametrix for $\Box_b$

For the convenience of the reader we give a few additional details concerning the analysis on the Heisenberg group and, more generally, on a strictly pseudoconvex CR manifold. The source is the fundamental work by G.B. Folland and E.M. Stein [150]. We start by recalling the following lemma:

Lemma A.1. *Let f be a homogeneous function of degree $\lambda \in \mathbf{R}$ that is C^1 away from 0. There is a constant $C > 0$ such that*
(1) $|f(x) - f(y)| \leq C|x - y|\,|x|^{\lambda - 1}$ *for any $x, y \in \mathbf{H}_n$ with $|x - y| \leq \frac{1}{2}|x|$,*
(2) $|f(xy) - f(x)| \leq C|y|\,|x|^{\lambda - 1}$ *for any $x, y \in \mathbf{H}_n$ with $|y| \leq \frac{1}{2}|x|$.*

Proof. To establish (1) one may assume, by homogeneity, that $|x| = 1$ and $|x - y| \leq \frac{1}{2}$. But then y is bounded away from zero, hence (by the mean value theorem and (1.26))

$$|f(x) - f(y)| \leq C\|xy - x\|.$$

For the proof of (2), the same argument leads to

$$|f(xy) - f(x)| \leq C\|xy - x\|.$$

Yet $y \mapsto xy$ is C^∞; hence

$$\|xy - x\| \leq C\|y\| \leq C|y|. \qquad \square$$

There is a right-invariant version of the operator $\mathcal{L}_0$ that we need to recall. Note that

$$\mathcal{L}_0 = -\frac{1}{4}\sum_{\alpha=1}^{n}\left(X_\alpha^2 + Y_\alpha^2\right) = -\sum_{j=1}^{2n} L_j^2 .$$

By analogy, we may consider the right-invariant differential operator

$$\mathcal{R}_0 = -\sum_{j=1}^{2n} R_j^2 .$$

The fact that Φ_0 is a two-sided inverse for $\mathcal{L}_0$ (cf. Proposition 3.2) yields

$$\mathcal{R}_0\Phi_0 = \delta. \tag{A.1}$$

Indeed, if $\mathcal{D}_0 = -\sum_{j=1}^{2n} D_j * D_j$ is the distribution kernel of $\mathcal{L}_0$, then $\mathcal{L}_0\Phi_0 = \Phi_0 * \mathcal{D}_0$ and $\mathcal{R}_0\Phi_0 = \mathcal{D}_0 * \Phi_0$ and both these expressions equal δ. □

For further use one may rewrite (A.1) also as

$$-\sum_{j=1}^{2n} D_j * D_j * \Phi_0 = \delta.$$

One of the deep results of G.B. Folland and E.M. Stein is the following (cf. Theorem 8.13 in [150], p. 450):

Theorem A.1. *Let F be a PV distribution.*[1] *Then there exist homogeneous distributions $F_1, \ldots, F_{2n}$ of degree $-2n-1$ such that*

$$F = \sum_{j=1}^{2n} D_j * F_j \, .$$

Theorem A.1 provides a set of "noncommutative Riesz transforms" with which one may manipulate derivatives. The proof is quite involved and out of the scope of this book. The reader may see [150], pp. 450–454. □

The following result emphasizes the importance of the Folland–Stein spaces S_k^p:

Theorem A.2. *If F is a PV distribution, the map $\varphi \mapsto \varphi * F$, $\varphi \in C_0^\infty(\mathbf{H}_n)$, extends to a bounded operator on S_k^p for $1 < p < \infty$ and $k \in \{0, 1, 2, \ldots\}$.*

Cf. Proposition 9.2 in [150], p. 455, for a proof. □

With these tools, we look at L^p estimates for $\mathcal{L}_\alpha$. Since its fundamental solution Φ_α is regular homogeneous of degree $-2n$, one may establish the following result:

Theorem A.3. (G.B. Folland and E.M. Stein [150])
Let α be admissible.[2] *Then*

(1) The map $\varphi \mapsto K_\alpha\varphi = \varphi * \Phi_\alpha$, $\varphi \in C_0^\infty$, extends to a bounded map from L^p to L^q, where $1/q = 1/p - 1/(n+1)$, provided that $1 < p < q < \infty$, and from L^1 to $L^{(n+1)/n-\epsilon}_{\rm loc}$ for any $\epsilon > 0$.
(2) The maps $\varphi \mapsto L_j K_\alpha\varphi = \varphi * \Phi_\alpha * D_j$, $\varphi \in C_0^\infty$, $1 \le j \le 2n$, extend to bounded maps of L^p to L^r, where $1/r = 1/p - 1/(2n+2)$, provided that $1 < p < r < \infty$ and from L^1 to $L^{(2n+2)/(2n+1)-\epsilon}_{\rm loc}$ for any $\epsilon > 0$.
(3) The maps $\varphi \mapsto L_i L_j K_\alpha\varphi = \varphi * \Phi_\alpha * D_j * D_i$, $\varphi \in C_0^\infty$, $1 \le i, j \le 2n$, extend to bounded operators on S_k^p for $1 < p < \infty$ and $k \in \{0, 1, 2, \ldots\}$.

---

[1] See Definition 3.20 in Chapter 3.
[2] See Definition 3.10 in Chapter 3.

As a corollary one obtains the following estimates for $\mathcal{L}_\alpha$:

Corollary A.1. *If α is admissible, $1 < p < \infty$, and $k \in \{0, 1, 2, \dots\}$, then*

$$\|\varphi\|_{p,k+2} \leq C_{p,k} \left(\|\mathcal{L}_\alpha \varphi\|_{p,k} + \|\varphi\|_p \right),$$

for any $\varphi \in C_0^\infty(\mathbf{H}_n)$.

Indeed, by Proposition 3.2, one has $\varphi = K_\alpha \mathcal{L}_\alpha \varphi$; hence [by part (3) in Theorem A.3 above]

$$\|L_i L_j \varphi\|_{p,k} \leq C_{p,k} \left(\|\mathcal{L}_\alpha \varphi\|_{p,k} + \|\varphi\|_p \right).$$

Yet

$$\|\varphi\|_{p,k+2} \leq \|\varphi\|_p + \sum_{j=1}^{2n} \|L_j \varphi\|_p + \sum_{i,j=1}^{2n} \|L_i L_j \varphi\|_{p,k},$$

so it suffices to prove the estimate

$$\|L_j \varphi\|_p \leq C_j \left(\|L_j^2 \varphi\|_p + \|\varphi\|_p \right).$$

Let $\gamma_j(t)$ be the 1-parameter group of transformations obtained by integrating L_j. Then, by Taylor's formula for the function $\Gamma_j(t) = \varphi(x\gamma_j(t))$ one has

$$\varphi(x\gamma_j(1)) = \varphi(x) + (L_j\varphi)(x) + \int_0^1 (1-t)(L_j^2\varphi)(x\gamma_j(t))dt,$$

or

$$(L_j\varphi)(x) = \varphi(x\gamma_j(1)) - \varphi(x) - \int_0^1 (1-t)(L_j^2\varphi)(x\gamma_j(t))dt.$$

The estimate sought is obtained by using the translation invariance of $\|\cdot\|_p$ and the Minkowski inequality, and by taking the L^p norms of both sides. □

Definition A.1. For $U \subseteq \mathbf{H}_n$ we define the spaces

$$S_k^p(U, \text{loc}) = \{F \in \mathcal{D}'(\mathbf{H}_n) : \varphi F \in S_k^p \ for \ any \ \varphi \in C_0^\infty(U)\}. \qquad \Box$$

We may state the following L^p regularity result:

Theorem A.4. (G.B. Folland and E.M. Stein [150])
Let α be admissible and $F, G \in \mathcal{D}'(\mathbf{H}_n)$ such that $\mathcal{L}_\alpha F = G$ on $U \subseteq \mathbf{H}_n$. Then
(1) *If $G \in S_k^p(U, \text{loc})$ with $1 < p < \infty$ then $F \in S_{k+2}^p(U, \text{loc})$.*
(2) *If $G \in L^p_{\text{loc}}(U)$ and $1/q = 1/p - 1/(n+1) > 0$ then $F \in L^q_{\text{loc}}(U)$ provided that $p > 1$, and $F \in L^{q-\epsilon}_{\text{loc}}(U)$ for any $\epsilon > 0$, provided that $p = 1$.*

Cf. also Theorem 9.5 in [150], p. 457, and the rather involved proof there. □

Let us look now at Lipschitz estimates for $\mathcal{L}_\alpha$. One of the important results by G.B. Folland and E.M. Stein (cf. op. cit.) is the fact that convolution with a PV distribution is a bounded operator on Γ_β for all β.

Theorem A.5. *If K_0 is a PV distribution and $f \in \Gamma_\beta$, $0 < \beta < \infty$, is a function of compact support, then $f * K_0 \in \Gamma_\beta$.*

Cf. also Theorem 10.1 in [150], p. 458. Also, the following result on kernels of higher homogeneity is known to hold (cf. Theorem 10.12 in [150], p. 465):

Theorem A.6. *Let K be a regular homogeneous distribution of degree $k - 2n - 2$, $k \in \{1, 2, 3, \dots\}$ and let f be a function of compact support. Set $g = f * K$. Then*

(1) *If $f \in \Gamma_\beta$, $0 < \beta < \infty$, then $g \in \Gamma_{\beta+k}(\text{loc})$.*

(2) *If $f \in L^p$ and $\beta = k - (2n+2)/p > 0$ then $g \in \Gamma_\beta(\text{loc})$.*

We may now state the Lipschitz regularity result for $\mathcal{L}_\alpha$.

Theorem A.7. (G.B. Folland and E.M. Stein [150])
Let α be admissible and $F, G \in \mathcal{D}'(\mathbf{H}_n)$ satisfy $\mathcal{L}_\alpha F = G$ on $U \subseteq \mathbf{H}_n$. Then

(1) *If $G \in \Gamma_\beta(U, \text{loc})$ with $0 < \beta < \infty$ then $F \in \Gamma_{\beta+2}(U, \text{loc})$.*

(2) *If $G \in L^p_{\text{loc}}(U)$ and $\beta = 2 - (2n+2)/p > 0$ then $F \in \Gamma_\beta(U, \text{loc})$.*

The remainder of Appendix A is devoted to a few results concerning analysis on a CR manifold. Precisely, we review certain classical results, such as (1) the construction of a parametrix for the Kohn–Rossi operator $\Box_b$ of a strictly pseudoconvex CR manifold (inverting $\Box_b$ up to *operators of type* 1; these are smoothing, i.e., are bounded operators of Folland–Stein spaces $S^p_k(M) \to S^p_{k+1}(M)$) and therefore, by general theorems estimating integral operators of type λ, (2) the derivation of estimates for $\Box_b$, e.g.,

$$\|\varphi\|_{p,k+2} \le C\left(\|\Box_b\varphi\|_{p,k} + \|\varphi\|_p\right), \tag{A.2}$$

for any $\varphi \in \Gamma^\infty(\Lambda^{0,q}(M))$ supported in a fixed compact set (here $0 < q < n$, $1 < p < \infty$, $k \ge 0$, and $C = C(p,k) > 0$, cf. Theorem A.14); hence leading to (3) regularity results for $\Box_b$, e.g., (by (A.2)) if φ is a locally integrable $(0,q)$-form, $0 < q < n$, and $\Box_b\varphi \in S^p_k(U, \text{loc})$ for $U \subseteq M$ open, $1 < p < \infty$, $k \in \{0, 1, 2, \dots\}$, then $\varphi \in S^p_{k+2}(U, \text{loc})$.

The main tool are the normal, or Folland–Stein, local coordinates on a strictly pseudoconvex CR manifold, as introduced in Chapter 4. The main feature of Folland–Stein coordinates is to compensate for the lack of a CR analogue of local complex coordinates (of a complex manifold). If M is a complex manifold, and J its almost complex structure (that is, $J_x : T_x(M) \to T_x(M)$, $J_x^2 = -I$, $x \in M$; i.e., J_x is the complex linear structure at the level of the tangent space), then for any point $x \in M$ there is an open neighborhood U and complex coordinates $\varphi = (z^1, \dots, z^n) : U \to \mathbf{C}^n$ such that $T_x(M) \simeq T_{\varphi(x)}(\mathbf{C}^n)$, a *complex linear* isomorphism, i.e., the $\{(\partial/\partial z^j)_x\}$ span $T_x(M)$, and as is well known, this is guaranteed by the Nijenhuis integrability of J. If, in turn, M is a CR manifold, the analogue (1.8)–(1.9) of the integrability property of J doesn't lead in general to the existence of such special coordinates, i.e., we may not infer the existence of, say, local coordinates $(z^1, \dots, z^n, t) : U \to \mathbf{C}^n \times \mathbf{R} \simeq$

$\mathbf{H}_n$ such that $T_x(M)$ is spanned by the Lewy operators $L_{\alpha,x} = (\partial/\partial z^\alpha + i\bar{z}^\alpha \partial/\partial t)_x$. In other words, a choice of local coordinates φ around a point $x \in M$ gives an $\mathbf{R}$-linear isomorphism $T_x(M) \simeq T_{\varphi(x)}(\mathbf{H}_n)$ that will not preserve, in general, the CR structures. G.B. Folland and E.M. Stein [150], constructed a sort of exponential map, which takes into account the splitting $T(M) = H(M) \oplus \mathbf{R}T$, such that the resulting local coordinates (at x) $\Theta_x : V_x \subset M \to \mathbf{R}^{2n+1}$ map a given pseudo-Hermitian frame $T_{\alpha,x}$ not to $L_{\alpha,x}$ as argued above, but rather to $L_{\alpha,x}$ plus "error terms," i.e.,

$$(\Theta_x)_* T_\alpha = L_\alpha + \sum_{\beta=1} \left(O^1 \frac{\partial}{\partial z^\beta} + O^1 \frac{\partial}{\partial \bar{z}^\beta} \right) + O^2 \frac{\partial}{\partial t},$$

where O^k denotes a function of Heisenberg-type order k (cf. Chapter 4). It was then remarkable that $\Theta(x, y) := \Theta_x(y)$ is $x^{-1}y$ when $M = \mathbf{H}_n$, that Θ is C^∞ on $\Omega = \bigcup_{x \in M} \{x\} \times V_x$, and that $\rho(x, y) = |\Theta(x, y)|$ (Heisenberg norm) is a sort of distance function (though not satisfying the triangle inequality, but only (3.22)).

Furthermore, we shall look at estimates on integral operators (with kernel of the form

$$K_\alpha(x, y) = \Phi_\alpha(\Theta(x, y)),$$

for x close to y), which rely on several, nowadays classical, results in functional analysis, and due to A.W. Knapp and E.M. Stein [240], R. Coifman and G. Weiss [105]. The proofs are but sketchy; yet the amount we give invites the reader to take note of material involving deep ideas in other disciplines, such as the theory of singular integrals (cf., e.g., the Hardy–Littlewood–Sobolev fractional integration theorem ([384], p. 119) based itself, as well as its generalization (cf. Lemma A.4) on the Marcinkiewicz interpolation theorem ([384], p. 272)), possessing of course their own history and exegetes (cf., e.g., E.M. Stein [384]).

The main results we seek to state are the existence of a parametrix and the regularity for $\Box_b$. It is worth noting that the same arguments apply to the sub-Laplacian Δ_b with only minor alterations (because Δ_b is modeled on $\mathcal{L}_0$, which possesses the fundamental solution Φ_0), resulting in fundamental estimates and regularity theorems for Δ_b; cf. Theorems 3.16 and 3.17 (which, in the end, yield regularity results for the CR Yamabe equation, cf. Theorem 3.22).

Let us start by discussing estimates for integral operators. Following [150], we are going to build a parametrix for $\Box_b$ out of integral operators whose kernels are of the form $K_\alpha(x, y) = \Phi_\alpha(\Theta(y, x))$ for x close to y. To work with such operators, one needs the following general lemmas:

Lemma A.2. *Let $A^1, A^2, A^3, \ldots$ be a sequence of bounded operators on a Hilbert space such that*

$$\|A^j \left(A^\ell\right)^* \| \le C\epsilon^{|j-\ell|}, \quad \| \left(A^j\right)^* A^\ell \| \le C\epsilon^{|j-\ell|},$$

for some $\epsilon \in (0, 1)$ and some $C > 0$. There is a constant $C_1 > 0$ independent of N such that

$$\Big\| \sum_{j=1}^{N} A^j \Big\| \le C_1 .$$

For a proof the reader may see A.W. Knapp and E.M. Stein [240]. □

Definition A.2. Let (X, μ) be a measure space and f a measurable function on X. The *distribution function* $\beta_f : [0, \infty) \to [0, \infty]$ of f is given by

$$\beta_f(s) = \mu\left(\{x \in X : |f(x)| > s\}\right) . \qquad \square$$

Then, with the notation in Definition A.2,

$$\int |f|^p \, d\mu = p \int_0^{\infty} s^{p-1} \beta_f(s) \, ds, \tag{A.3}$$

for any $0 < p < \infty$.

Definition A.3. We say that f is *weak* L^p if

$$\beta_f(s) \le \left(\frac{C}{s}\right)^p ,$$

for some $C > 0$. An operator A mapping $L^p(X, \mu)$ into measurable functions is of *weak type* (p, q), $q < \infty$, if

$$\beta_{Af}(s) \le \left(\frac{C\|f\|_p}{s}\right)^q$$

for some $C > 0$ and any $f \in L^p(X, \mu)$. □

We recall the following result:

Theorem A.8. (E.M. Stein [384])
If A is defined on $L^{p_0} + L^{p_1}$ and is of weak type (p_0, q_0) and (p_1, q_1) with $1 \le p_j \le q_j < \infty$ and $q_0 \ne q_1$, then A is bounded from L^{p_t} to L^{q_t} for any $0 < t < 1$, where $\frac{1}{p_t} = \frac{1-t}{p_0} + \frac{t}{p_1}$ and $\frac{1}{q_t} = \frac{1-t}{q_0} + \frac{t}{q_1}$.

Theorem A.8 is referred to as the *Marcinkiewicz interpolation theorem*. See [384], p. 272.

Lemma A.3. *Let k be a measurable function on $X \times X$. Assume that there is a constant $C_0 > 0$ such that*

$$\int |k(x, y)| \, d\mu(y) \le C_0, \qquad \int |k(x, y)| \, d\mu(x) \le C_0,$$

for any $x, y \in X$. Then the operator

$$(Af)(x) = \int k(x, y) f(y)\, d\mu(y)$$

satisfies

$$\|Af\|_p \le C_0 \|f\|_p$$

for any $p \in [1, \infty]$.

Proof. For $p = \infty$ the statement is obvious. Let $p \in [1, \infty)$ and q be such that $1/p + 1/q = 1$. Then (by Hölder's inequality)

$$\begin{aligned}
|(Af)(x)| &\le \int |k(x, y)|\, |f(y)| d\mu(y) = \int |k(x, y)|^{1/q} |k(x, y)|^{1/p} |f(y)| d\mu(y) \\
&\le \Big(\int |k(x, y)| d\mu(y) \Big)^{1/q} \Big(\int |k(x, y)|\, |f(y)|^p d\mu(y) \Big)^{1/p} \\
&\le C_0^{1/q} \Big(\int |k(x, y)|\, |f(y)|^p d\mu(y) \Big)^{1/p}.
\end{aligned}$$

Next, by Fubini's theorem

$$\begin{aligned}
\int |(Af)(x)|^p d\mu(x) &= C_0^{p/q} \int d\mu(x) \int |k(x, y)|\, |f(y)|^p d\mu(y) \\
&\le C_0^{1/q+1/p} \|f\|_p = C_0 \|f\|_p .
\end{aligned}$$

□

Lemma A.4. *Let k be a measurable function on $X \times X$. Assume that there is $r > 0$ such that (1) $k(x, \cdot)$ is weak L^r uniformly in x, and (2) $k(\cdot, y)$ is weak L^r uniformly in y. Then the operator*

$$(Af)(x) = \int k(x, y) f(y) d\mu(y)$$

is bounded from L^p to L^q whenever $1/q = 1/p + 1/r - 1$ and $1 < p < q < \infty$, and from L^1 to $L^{r-\epsilon}_{\text{loc}}$ for any $\epsilon > 0$.

This extends (cf. also Lemma 15.3 in [150], p. 478) a result in [384], p. 119, i.e., the theorem of *fractional integration* there. To prove Lemma A.4, note that (by the Marcinkiewicz interpolation theorem with $p_0 = p_1 = p$ and $q_0 = q_1 = q$) when $p > 1$ it suffices to show that A is of weak type (p, q), whenever $1/q = 1/p + 1/r - 1$ and $1 \le p < q < \infty$. Let us set

$$\begin{aligned}
k_1(x, y) &= \begin{cases} k(x, y), & |k(x, y)| \ge \gamma, \\ 0, & \text{otherwise}, \end{cases} \\
k_2(x, y) &= k(x, y) - k_1(x, y),
\end{aligned}$$

where $\gamma > 0$ is to be chosen later. Let A_i be the integral operators with kernels k_i, $i = 1, 2$. Then $A = A_1 + A_2$; hence

$$\beta_{Af}(2s) \le \beta_{A_1 f}(s) + \beta_{A_2 f}(s),$$

and (to produce an estimate of the form $\beta_{Af}(t) \le \big(C\|f\|_p/t\big)^q$, $t > 0$) it suffices to estimate the last two terms. Let $f \in L^p$. Without loss of generality we may assume $\|f\|_p = 1$. Let p' be such that $1/p + 1/p' = 1$. Then (by Hölder's inequality)

$$\begin{aligned} |(A_2 f)(x)| &\le \int |k_2(x, y)|\, |f(y)|\, d\mu(y) \\ &\le \Big(\int |k_2(x, y)|^{p'} d\mu(y)\Big)^{1/p'} \|f\|_p = \Big(\int |k_2(x, y)|^{p'} d\mu(y)\Big)^{1/p'}. \end{aligned}$$

Note that $1/r - 1/p' = 1/q > 0$. Then (by (A.3))

$$\begin{aligned} \int |k_2(x, y)|^{p'} d\mu(y) &= p' \int_0^\infty s^{p'-1} \beta_{k_2(x,\cdot)}(s) ds = p' \int_0^\gamma s^{p'-1} \beta_{k(x,\cdot)}(s) ds \\ &\le p' \int_0^\gamma s^{p'-1} \Big(\frac{C}{s}\Big)^r ds \le \frac{p' C^r}{p' - r} \gamma^{p'-r} \end{aligned}$$

(because $k(x, \cdot)$ is weak L^r uniformly with respect to x). Then

$$|(A_2 f)(x)| \le C_0 \gamma^{1-r/p'} = C_0 \gamma^{r/q},$$

where

$$C_0 = \Big(\frac{p' C^r}{p' - r}\Big)^{1/p'} > 0.$$

At this point we may choose $\gamma = (s/C_0)^{q/r}$ such that $|(A_2 f)(x)| \le s$; hence $\beta_{A_2 f}(s) = 0$. Moreover, since $r > 1$,

$$\begin{aligned} \int |k_1(x, y)|\, d\mu(y) &= \int_0^\infty \beta_{k_1(x,\cdot)}(s)\, ds \\ &= \int_\gamma^\infty \beta_{k(x,\cdot)}(s)\, ds \le \int_\gamma^\infty \Big(\frac{C}{s}\Big)^r ds = \frac{C^r}{1 - r} \gamma^{1-r}. \end{aligned}$$

Likewise

$$\int |k_1(x, y)|\, d\mu(x) \le \frac{C^r}{1 - r} \gamma^{1-\gamma}.$$

Therefore (by Lemma A.3) the operator A_1 satisfies

$$\|A_1 f\|_p \le \frac{C^r}{1 - r} \gamma^{1-r} \|f\|_p = \frac{C^r}{1 - r} \gamma^{1-r},$$

or

$$\int |(A_1 f)(x)|^p d\mu(x) \le \Big(\frac{C^r}{1 - r} \gamma^{1-r}\Big)^p$$

and we have the estimates

$$\left(\frac{C^r}{1-r}\gamma^{1-r}\right)^p \geq \int |(A_1 f)(x)|^p d\mu(x)$$
$$\geq \int_{\{x:(A_1 f)(x)>s\}} |(A_1 f)(x)|^p d\mu(x) > \int_{\{x:(A_1 f)(x)>s\}} s^p d\mu(x) = s^p \beta_{A_1 f}(s),$$

so that

$$\beta_{A_1 f}(s) \leq \left(\frac{C^r}{1-r}\frac{\gamma^{1-r}}{s}\right)^p = C_1 s^{-q} = C_1\left(\frac{\|f\|_p}{s}\right)^q,$$

where C_1 is the constant $C^{rp}/[(1-r)^p C_0^{q/r}]$. This completes the proof in the case $p > 1$. □

The proof above shows that A is of weak type $(1, r)$; hence $f \in L^1$ yields

$$Af \in \{weak\ \ L^r\} \subset L^{r-\epsilon}_{\text{loc}},$$

for any $\epsilon > 0$. □

The reader may well note that Lemma A.4 for $X = \mathbf{H}_n$ yields Theorem 3.10 by merely taking $k(x, y) = F(y^{-1}x)$ because (by (3.41)) F is weak L^r with $r = -\lambda/(2n+2)$. □

Let K be a singular integral kernel. Let us set

$$K^j(x, y) = \begin{cases} K(x, y) & 1/2^j \leq \rho(x, y) \leq 1/2^{j-1}, \\ 0 & \text{otherwise}, \end{cases}$$

for any $j \in \{1, 2, 3, \dots\}$. Then

$$K = \sum_{j=1}^{\infty} K^j.$$

Let A^j be the corresponding integral operator, i.e.,

$$(A^j f)(x) = \int K^j(x, y) f(y)\, dy.$$

Lemma A.5. *The operators A^j are uniformly bounded on L^2. Moreover, $\| \sum_{j=1}^N A^j \|_2 \leq C$, where the constant C is independent of N.*

Proof. Note that

$$|K(x, y)| \leq C\rho(x, y)^{-2n-2}.$$

Then (since $\Theta_x^* dV(\xi) = (1 + O^1)\, dy$)

$$\int |K^j(x, y)|\, dy \leq C \int_{1/2^j \leq |\xi| \leq 1/2^{j-1}} |\xi|^{-2n-2}\, dV(\xi) \leq C \log 2.$$

Similarly

$$\int |K^j(x,y)|\, dx \le C \log 2,$$

so that (by Lemma A.3) the operators A^j are uniformly bounded on L^2. To prove the second statement in Lemma A.5 it suffices to show that, for j sufficiently large, the operators A^j obey to the hypothesis of Lemma A.2. This follows from Lemma A.3 and the estimate

$$\int |G_{j\ell}(x,y)| dy \le C 2^{(\ell - j)/2} \tag{A.4}$$

for ℓ large and $j >> \ell$. The proof of (A.4) is quite technical and would take us too far afield (cf. [150], pp. 481–483, for the details).

Lemma A.6. *Let K be a singular integral kernel and set*

$$K_\epsilon(x,y) = \begin{cases} K(x,y) & \rho(x,y) > \epsilon, \\ 0 & \textit{otherwise.} \end{cases}$$

There exist constants C_1, C_2 such that

$$\int_{\rho(x,z) > C_1 \rho(x,y)} |K_\epsilon(z,x) - K_\epsilon(z,y)| dz \le C_2\,,$$

for any $x, y \in V$ and $\epsilon > 0$.

The proof follows from the estimates (3.21)–(3.22) together with Lemma A.1 (cf. [150], pp. 484–485, for the details).

Lemma A.7. *Let $K'(x,y)$ be a function with support in*

$$\Omega \cap \{(x,y) : \rho(x,y) \le 1\}.$$

Assume that there exist constants $C_1, C_2, C_3 > 0$ such that

$$\int_{\rho(x,z) > C_1 \rho(x,y)} |K'(z,x) - K'(z,y)| dz \le C_2$$

and

$$(A' f)(x) = \int K'(x,y) f(y) dy$$

exists a.e. for any $f \in L^p$, $1 \le p \le 2$, and

$$\|A' f\|_2 \le C_3 \|f\|_2$$

for any $f \in L^2$. Then

$$\|A' f\|_p \le C \|f\|_p$$

for any $f \in L^p$, $1 < p \le 2$, where C depends only on C_1, C_2, C_3, and p.

This is a special case of a result by R. Coifman and G. Weiss [105], which generalizes the classical Calderon–Zygmund theorem; cf., e.g., [384]. Similar generalizations are due to A. Korányi and S. Vági [258], and N.M. Rivière [352]. For a proof the reader should see [105], p. 74. As pointed out by G.B. Folland and E.M. Stein (cf. [150], p. 485) the hypothesis $K' \in L^2(M \times M)$ in [105] should be replaced by the hypothesis that $(A'f)(x)$ exists for a.e. x and all $f \in L^p$, $1 \leq p \leq 2$.

Theorem A.9. *Let A be a singular integral operator with kernel $K(x, y)$ and let $1 < p < \infty$. The operators A_ϵ are uniformly bounded on L^p and converge strongly as $\epsilon \to 0$. Hence A may be defined on L^p by $Af = \lim_{\epsilon \to 0} A_\epsilon f$ and A is bounded on L^p.*

Proof. Let K_ϵ be defined as in Lemma A.6. Then

$$(A_\epsilon f)(x) = \int K_\epsilon(x, y) f(y)\, dy$$

exists for all x and $f \in L^p$ because

$$|(A_\epsilon f)(x)| \leq \left(\int K_\epsilon(x, y)^q dy \right)^{1/q} \|f\|_p < \infty$$

for all x (where $q \in \mathbf{R}$ is such that $1/p + 1/q = 1$) since $K_\epsilon(x, \cdot)$ is a bounded function of compact support. By Lemma 3.9, A_ϵ converges strongly as $\epsilon \to 0$ on a dense subspace of L^p. Therefore, it remains to be shown that A_ϵ is bounded on L^p, $1 < p < \infty$, uniformly in ϵ. For $p = 2$ this follows from Lemma A.5. Next, Lemmas A.6 and A.7 yield the statement for $1 < p \leq 2$. To end the proof, set

$$\tilde{K}(x, y) = \overline{K(y, x)}.$$

Then $\tilde{K}$ is a singular integral kernel; hence the corresponding integral operators

$$(\tilde{A}_\epsilon f)(x) = \int_{\rho(x,y) > \epsilon} \tilde{K}(x, y) f(y)\, dy$$

are, by the first part of this proof, bounded on L^p, $1 < p \leq 2$, uniformly in ϵ. Yet

$$\int (A_\epsilon f)(x) \overline{g(x)}\, dx = \int f(x) \overline{(\tilde{A}_\epsilon g)(x)}\, dx;$$

hence by Hölder's inequality, A_ϵ is uniformly bounded on L^p, $2 \leq p < \infty$. □

At this point we introduce a class of kernels on M that play the role of regular homogeneous distributions on $\mathbf{H}_n$.

Definition A.4. Let $\lambda \geq 0$. A function $K(x, y)$ on $M \times M$ is a *kernel of type* λ if for any $m \in \mathbf{Z}$, $m > 0$, one can write K in the form

$$K(x, y) = \sum_{i=1}^{N} a_i(x) K_i(x, y) b_i(y) + E_m(x, y),$$

where (1) $E_m \in C_0^m(M \times M)$, (2) $a_i, b_i \in C_0^\infty(M)$, $1 \le i \le N$, and (3) K_i is C^∞ away from the diagonal and is supported in $\Omega \cap \{(x, y) : \rho(x, y) \le 1\}$, and $K_i(x, y) = k_i(\Theta(y, x))$ for $\rho(x, y)$ sufficiently small, where k_i is homogeneous of degree $\lambda_i = \lambda - 2n - 2 + \mu_i$ for some $\mu_i \in \mathbf{Z}$, $\mu_i \ge 0$. In addition, if $\lambda = 0$ and $\lambda_i = -2n - 2$ one requires that k_i have mean value zero. □

In other words, a kernel of type λ is one whose "principal part" is regular homogeneous of degree $\lambda - 2n - 2$ near the diagonal.

Definition A.5. An *operator of type* λ, $\lambda > 0$, is an operator A of the form

$$(Af)(x) = \int K(x, y) f(y)\, dy,$$

where K is a kernel of type λ. An *operator of type* 0 is an operator A of the form

$$(Af)(x) = \lim_{\epsilon \to 0} \int_{\rho(x,y) > \epsilon} K(x, y) f(y)\, dy + a(x) f(x),$$

where K is a kernel of type 0 and $a \in C_0^\infty$. □

Theorem A.10. *Let A be an operator of type $\lambda > 0$. Then A is bounded on L^p, $1 \le p \le \infty$. Moreover, if $0 < \lambda < 2n+2$ then A is bounded from L^1 to $L^{(2n+2)/(2n+2-\lambda)-\epsilon}$ for any $\epsilon > 0$ and from L^p to L^q, $1/q = 1/p - \lambda/(2n+2)$, whenever $1 < p < q < \infty$.*

For a proof the reader should see [150], p. 487. It may be also shown that operators of type 0 are bounded on L^p, $1 < p < \infty$. □

All constructions so far rely on the local orthonormal frame $\{X_j\}$. These vector fields (as differential operators) interact with operators of type $\lambda \ge 0$ as follows.

Proposition A.1. *Let A be an operator of type λ. If $\lambda \ge 1$ then $X_j A$ and $A X_j$ are operators of type $\lambda - 1$ for $1 \le j \le 2n$, while if $\lambda \ge 2$ then $X_0 A$ and $A X_0$ are operators of type $\lambda - 2$.*

Cf. Proposition 15.14 in [150], p. 487. Also Theorem 3.10 in Chapter 4 admits the following analogue

Theorem A.11. *If A is an operator of type 0 then there exist operators $A_0, \dots, A_{2n}$ of type 1 such that*

$$A = \sum_{j=1}^{2n} A_j X_j + A_0 .$$

Cf. Theorem 15.15 in [150], p. 490.

Definition A.6. The space $S_k^p(M)_{\text{loc}}$ consists, by definition, of all $f \in L^p_{\text{loc}}$ such that $Df \in L^p_{\text{loc}}$, as D runs over all differential operators that are sums of monomials of order at most k formed with vector fields in $\Gamma^\infty(H(M) \otimes \mathbf{C})$. □

We wish to remark that although the Folland–Stein space $S_k^p(M)$ is defined in terms of the pseudo-Hermitian frame $\{X_j\}$, the space $S_k^p(M)_{\mathrm{loc}}$ is a CR invariant.

Theorem A.12. *Let A be an operator of type m, $m \in \{0, 1, 2, \dots\}$. Then A is bounded from $S_k^p(M)$ to $S_{k+m}^p(M)$ for $k \in \{0, 1, 2, \dots\}$ and $1 < p < \infty$.*

Cf. Theorem 15.19 in [150], p. 491. □

Now we recall the Lipschitz-type spaces on M:

Definition A.7. If $0 < \beta < 1$, let $\Gamma_\beta(M)$ consist of all bounded functions f for which

$$\sup_{x,y} \frac{|f(y) - f(x)|}{\rho(x, y)^\beta} < \infty.$$

Also, let $\Gamma_1(M)$ consist of all bounded functions f for which

$$\sup_{x,y} \frac{|f(y) + f(\tilde{y}) - 2f(x)|}{\rho(x, y)} < \infty,$$

where $\tilde{y} = \Theta_x^{-1}(-\Theta_x(y))$. Finally, let us set

$$\Gamma_{m+\beta'}(M) = \{f : Df \in \Gamma_{\beta'}(M) \text{ for all } D \in \mathcal{A}_m\},$$

for any $m \in \{1, 2, 3, \dots\}$ and $0 < \beta' \leq 1$. □

The spaces $\Gamma_\beta(M)_{\mathrm{loc}}$ are intrinsically defined, i.e., they are independent of the choices made in constructing normal coordinates. It may be shown (cf. Theorem 15.20 in [150], p. 493) that operators of type m satisfy the following Lipschitz regularity result:

Theorem A.13. *Let A be an operator of type m, $m \in \{0, 1, 2, \dots\}$, and let us set $Af = g$. If $f \in \Gamma_\beta(M)$, $0 < \beta < \infty$, then $g \in \Gamma_{\beta+m}(M)$. If $f \in L^p$, $p \geq 1$, and $\beta = m - (2n+2)/p > 0$ then $g \in \Gamma_\beta(M)$.*

Let us discuss now the existence of a parametrix for $\Box_b$. Let φ be a $(0, q)$-form on M. Locally

$$\varphi = \sum_J \varphi_J \theta^{\bar{J}},$$

where $J = (\alpha_1, \dots, \alpha_q)$ is a multi-index and $\theta^{\bar{J}} = \theta^{\bar{\alpha}_1} \wedge \dots \wedge \theta^{\bar{\alpha}_q}$.

Definition A.8. We say that $\varphi \in S_k^p$ (respectively $\varphi \in \Gamma_\beta$) if $\varphi_J \in S_k^p$ (respectively $\varphi_J \in \Gamma_\beta$) for all multi-indices J. □

Clearly this is independent of the choice of pseudo-Hermitian frame.

Definition A.9. An operator A on forms is said to be *of type* λ if

$$A = \sum_{J,K} (A_{JK}\varphi_J)\theta^{\bar{K}},$$

where A_{JK} is an operator (on functions) of type λ, for any multi-indices J, K. □

We adopt the following notation for "error terms":

Definition A.10. $\mathcal{E}(\varphi)$ will denote an expression of the form

$$\sum_{JK} a_{JK}\varphi_J\theta^{\overline{K}}, \quad a_{JK} \in C^\infty.$$

Also $\mathcal{E}(W\varphi)$ will denote an expression of the form

$$\sum_{J,K,\alpha} a_{JK\alpha}(W_\alpha\varphi_J)\theta^{\overline{K}}, \quad a_{JK\alpha} \in C^\infty,$$

with similar conventions for $\mathcal{E}(\overline{W}\varphi)$, $\mathcal{E}(T\varphi)$, etc., where $\{W_1, \dots, W_n\}$ is a given (local) frame of $T_{1,0}(M)$. Also, $\mathcal{E}(A, B)$ is short for $\mathcal{E}(A) + \mathcal{E}(B)$. □

For instance, if M is strictly pseudoconvex and $\{W_\alpha\}$ is chosen such that $\langle W_\alpha, W_\beta\rangle_\theta = \delta_{\alpha\beta}$, then the equation

$$\nabla_{W_\alpha} W_{\overline{\beta}} - \nabla_{W_{\overline{\beta}}} W_\alpha - [W_\alpha, W_{\overline{\beta}}] = 2ih_{\alpha\overline{\beta}}T$$

may be written

$$[W_\alpha, W_{\overline{\beta}}]f = -2i\delta_{\alpha\beta}Tf + \mathcal{E}(Wf, \overline{W}f),$$

for any $f \in C^2(M)$.

We already computed $\overline{\partial}_b$, $\overline{\partial}_b^*$, and $\Box_b$ in terms of the given frame $\{W_\alpha\}$; cf. our identities (1.135), (1.142), and (1.146). A rough form, but at times easier to handle, of these equations, in which use is made of the above error term notation, follows immediately. For instance

$$\begin{aligned}\overline{\partial}_b\varphi &= \sum_{J,\alpha}(W_{\overline{\alpha}}\varphi_J)\theta^{\overline{\alpha}} \wedge \theta^{\overline{J}} + \sum_J \varphi_J \overline{\partial}_b\theta^{\overline{J}} \\ &= \sum_{J,\alpha}(W_{\overline{\alpha}}\varphi_J)\theta^{\overline{\alpha}} \wedge \theta^{\overline{J}} + \mathcal{E}(\varphi).\end{aligned}$$

Similarly (1.146) yields

$$q!\Box_b\varphi = \sum_J \left(\mathcal{L}_{n-2q}\varphi_J\right)\theta^{\overline{J}} + \mathcal{E}(W\varphi, \overline{W}\varphi, \varphi), \tag{A.5}$$

for any $\varphi \in \Omega^{0,q}(M)$. Therefore $\Box_b$ is expressed by the same formula as for the Heisenberg group, modulo lower-order error terms (compare to (3.25)).

Let $\{W_\alpha\}$ be a pseudo-Hermitian frame, defined on the open set $V \subseteq M$. Let $\xi \in V$ and let $\Theta_\xi : V_\xi \to \Theta_\xi(V_\xi) \subseteq \mathbf{H}_n$ be a normal coordinate system at ξ, associated with $\{W_\alpha\}$. We let $\mathcal{L}_\alpha$ act on functions $f : V_\xi \to \mathbf{C}$ by setting

$$\mathcal{L}_\alpha f := \mathcal{L}_\alpha(f \circ \Theta_\xi^{-1}).$$

Let us set $x = (z, t) = \Theta_\xi$. Then (by (A.5))

$$\Box_b \varphi = \sum_J \left(\mathcal{L}_{n-2q}\varphi_J\right)\theta^{\overline{J}} + \mathcal{E}(\varphi, \frac{\partial \varphi}{\partial z}, \frac{\partial \varphi}{\partial \overline{z}})$$
$$+ O^1\mathcal{E}\Big(\frac{\partial \varphi}{\partial t}, \frac{\partial^2 \varphi}{\partial z^2}, \frac{\partial^2 \varphi}{\partial z \partial \overline{z}}, \frac{\partial^2 \varphi}{\partial \overline{z}^2}\Big) + O^2\mathcal{E}\Big(\frac{\partial^2 \varphi}{\partial z \partial t}, \frac{\partial^2 \varphi}{\partial \overline{z} \partial t}\Big) + O^3\mathcal{E}\Big(\frac{\partial^2 \varphi}{\partial t^2}\Big), \quad \text{(A.6)}$$

where $O^k\mathcal{E}(\cdot)$ is an expression of the form $\mathcal{E}(\cdot)$ whose coefficients are of type O^k. In spite of the rather complicated appearance of the error terms in (A.6), its use is quite simple: one should think of φ as a sum of homogeneous terms and note the fact that $\mathcal{L}_{n-2q}$ lowers the homogeneity by 2 degrees, while the error terms lower it by at most one.

At this point, we may build a parametrix for $\Box_b$. By a partition of unity argument, we need only look at forms supported in some compact set W. Let $\psi(\xi, \eta)$ be a real-valued C_0^∞ function supported in $\Omega \cap \{(\xi, \eta) : \rho(\xi, \eta) \le 1\}$ and such that $\psi(\xi, \eta) = \psi(\eta, \xi)$ and $\psi(\xi, \eta) = 1$ in some neighborhood of the diagonal in $W \times W$. Let Φ_α be the fundamental solution for $\mathcal{L}_\alpha$ on $\mathbf{H}_n$. Let $0 < q < n$ and set

$$K_q(\xi, \eta) = \psi(\xi, \eta)\Phi_{n-2q}(\Theta(\eta, \xi)).$$

Since Φ_{n-2q} is homogeneous of degree $-2n$, K_q is a kernel of type 2. Moreover,

$$K_q(\eta, \xi) = \overline{K_q(\xi, \eta)}$$

because of $\Phi_\alpha(-x) = \overline{\Phi_{\overline{\alpha}}(x)}$. Let A_q be the operator (of type 2) on $(0, q)$-forms given by

$$A_q\varphi(\xi) = \sum_J \Big(\int K_q(\xi, \eta)\varphi_J(\eta)d\eta\Big)\theta_\xi^{\overline{J}}.$$

Also, consider the operator

$$B_q\varphi(\xi) = \psi(\xi, \xi)\varphi(\xi) - \Box_b A_q\varphi(\xi).$$

Proposition A.2. *B_q is an operator of type* 1.

Proof. Let φ be a smooth form of compact support. Define A_q^ϵ by

$$A_q^\epsilon\varphi(\xi) = \sum_J \Big(\int K_q^\epsilon(\xi, \eta)\varphi_J(\eta)d\eta\Big)\theta_\xi^{\overline{J}},$$
$$K_q^\epsilon(\xi, \eta) = \psi(\xi, \eta)\Phi_{n-2q,\epsilon}(\Theta(\eta, \xi)),$$
$$\Phi_{\alpha,\epsilon} = \frac{1}{c_\alpha}\varphi_{\alpha,\epsilon},$$
$$\varphi_{\alpha,\epsilon} = \rho_\epsilon^{-\frac{n+\alpha}{2}}\overline{\rho}_\epsilon^{-\frac{n-\alpha}{2}}, \quad \rho_\epsilon(z, t) = |z|^2 + \epsilon^2 - it$$

(we replaced Φ_{n-2q} in the definition of A_q by its regularized version $\Phi_{n-2q,\epsilon}$, as in the proof of Theorem 3.9). Then (by the dominated convergence theorem)

$$\|A_q^\epsilon\varphi - A_q\varphi\|_\infty \leq \|\varphi\|_\infty \int \Big| K_q^\epsilon(\xi,\eta) - K_q(\xi,\eta)\Big| d\eta \to 0, \ \epsilon \to 0;$$

hence $A_q^\epsilon\varphi \to A_q\varphi$ as distributions. Then $\Box_b A_q^\epsilon\varphi \to \Box_b A_q\varphi$. All functions entering $A_q^\epsilon\varphi$ are C^∞, so we may differentiate under the integral sign. We get (by (A.6))

$$\begin{aligned}\Box_b A_q^\epsilon\varphi(\xi) &= \sum_J \Big(\mathcal{L}_{n-2q}(A_q^\epsilon\varphi)_J(\xi)\Big)\theta_\xi^{\overline{J}} + \text{ (error terms) } (\epsilon,\xi)\\ &= \sum_J \Big[\int \mathcal{L}_{n-2q}^\xi(K_q^\epsilon(\xi,\eta))\varphi_J(\eta)d\eta\Big]\theta_\xi^{\overline{J}} + \text{ (error terms) } (\epsilon,\xi). \quad \text{(A.7)}\end{aligned}$$

As $\epsilon \to 0$, the error terms give rise to an integral operator applied to φ whose kernel consists of ξ-derivatives of $\psi(\xi,\eta)\Phi_{n-2q}(\Theta(\eta,\xi))$, which lower the homogeneity by $j \in \{0,1,2,3,4\}$, multiplied by coefficients of type O^{j-1}. In particular, the coefficients kill any part (of these distribution derivatives) supported on the diagonal. Then it may be shown that the error terms form an operator of type 1 applied to φ (cf. also [150], p. 495). As to the first term in the expression of $\Box_b A_q^\epsilon\varphi$, we have

$$\begin{aligned}&\mathcal{L}_{n-2q}^\xi\left[\psi(\xi,\eta)\Phi_{n-2q,\epsilon}(\Theta(\eta,\xi))\right]\\ &\quad = \left(\mathcal{L}_{n-2q}^\xi\psi\right)\Phi_{n-2q,\epsilon}(\Theta(\eta,\xi)) + \psi(\xi,\eta)\mathcal{L}_{n-2q}^\xi\left[\Phi_{n-2q,\epsilon}(\Theta(\eta,\xi))\right]. \quad \text{(A.8)}\end{aligned}$$

Note that $\mathcal{L}_{n-2q}^\xi\psi$ is a first-order operator involving only the derivatives $X_1,\dots,X_{2n}$ (and not involving X_0). Therefore (by Proposition A.1)

$$\left(\mathcal{L}_{n-2q}^\xi\psi\right)\Phi_{n-2q,\epsilon}(\Theta(\eta,\xi))$$

is a kernel of type 1. Let us set $u = \Theta_\eta(\xi)$. Then the second term in (A.8) is $\psi(\xi,\eta)\mathcal{L}_{n-2q}(\Phi_{n-2q,\epsilon}(u))$. Since $d\xi$ and $dV(u)$ for $\xi = \eta$ coincide, it follows (by Theorem 3.10) that

$$\psi(\xi,\eta)\mathcal{L}_{n-2q}(\Phi_{n-2q}(u)) \to \psi(\xi,\eta)\delta(\xi,\eta), \quad \epsilon \to 0,$$

where $\delta(\xi,\eta)$ is the distribution on $M \times M$ given by

$$\int \delta(\xi,\eta)f(\xi)g(\eta)d\xi d\eta = \int f(\xi)g(\xi)d\xi.$$

Thus (by (A.7)–(A.8))

$$\Box_b A_q\varphi(\xi) = \lim_{\epsilon\to 0}\Box_b A_q^\epsilon\varphi(\xi) = \psi(\xi,\xi)\varphi(\xi) + (H_q\varphi)(\xi)$$

for some operator H_q of type 1. Yet $H_q = -B_q$ (by the definitions). □

Proposition A.3. *Let φ be a $(0, q)$-form, $1 \leq q \leq n-1$, with* $\text{supp}(\varphi) \subset W$. *Then*

$$\Box_b A_q \varphi = \varphi - B_q \varphi, \quad A_q \Box_b \varphi = \varphi - B_q^* \varphi,$$

and B_q^* is of type 1.

Proof. The formula $\Box_b A_q \varphi = \varphi - B_q \varphi$ is merely the definition of B_q, since $\psi(\xi, \xi) = 1$ for any $\xi \in W$. To prove the second formula, let λ, μ be $(0, q)$-forms. Then

$$\begin{aligned}(A_q \lambda, \mu) &= \sum_J \int K_q(\xi, \eta) \lambda_J(\eta) \overline{\mu_J(\xi)} d\eta d\xi \\ &= \sum_J \int \lambda_J(\eta) \overline{K_q(\eta, \xi) \mu_J(\xi)} d\eta d\xi = (\lambda, A_q \mu),\end{aligned}$$

i.e., A_q is symmetric. Next, since $\Box_b$ is symmetric on forms with compact support,

$$A_q \Box_b \varphi = \left(\Box_b A_q\right)^* \varphi = (I - B_q^*) \varphi.$$

Finally, the adjoint of an operator of type 1 is of type 1. □

Therefore, A_q is a two-sided parametrix for $\Box_b$ on W, i.e., $I - \Box_b A_q$ and $I - A_q \Box_b$ are smoothing operators, i.e., are bounded operators of S_k^p into S_{k+1}^p, $1 < p < \infty$, $k \in \{0, 1, \ldots\}$ (by Theorem A.12). If better smoothing is desired, one may set

$$A_q^{(m)} = A_q \sum_{j=0}^{m-1} (B_q)^j, \quad A_q^{[m]} = \sum_{j=0}^{m-1} (B_q^*)^j A_q,$$

and then

$$I - \Box_b A_q^{(m)} = I - (I - B_q) \sum_{j=0}^{m-1} (B_q)^j = (B_q)^m,$$

$$I - A_q^{[m]} \Box_b = I - \sum_{j=0}^{m-1} (B_q^*)^j (I - B_q^*) = (B_q^*)^m,$$

and $(B_q)^m$ and $(B_q^*)^m$ are bounded operators from S_k^p to S_{k+m}^p. Again by Theorem A.12, $A_q^{(m)}$ and $A_q^{[m]}$ are bounded operators from S_k^p to S_{k+2}^p. □

Consequently, one has the following estimates for $\Box_b$:

Theorem A.14. *If $0 < q < n$, $1 < p < \infty$, and $k \geq 0$, there is a constant $C = C(p, k) > 0$ such that*

$$\|\varphi\|_{p,k+2} \leq C \left(\|\Box_b \varphi\|_{p,k} + \|\varphi\|_p\right),$$

for any $\varphi \in \Gamma_0^\infty(\Lambda^{0,q})$ supported in a fixed compact set W.

Proof.

$$\begin{aligned}\|\varphi\|_{p,k+2} &\leq \|A^{[k+2]}\Box_b\varphi\|_{p,k+2} + \|(B_q^*)^{k+2}\varphi\|_{p,k+2} \\ &\leq C\left(\|\Box_b\varphi\|_{p,k} + \|\varphi\|_p\right).\end{aligned}$$

□

We are now in a position to state the main regularity theorem for $\Box_b$.

Theorem A.15. *Let φ and ψ be locally integrable $(0, q)$-forms, $0 < q < n$, such that $\Box_b\varphi = \psi$ on an open set $U \subseteq M$. Then*

(a) $\varphi \in L^{(n+1)/n-\epsilon}(U, \text{loc})$ for any $\epsilon > 0$.
(b) If $\psi \in L^p(U, \text{loc})$, $1 < p < n+1$, then $\varphi \in L^r(U, \text{loc})$, where $\frac{1}{r} = \frac{1}{p} - \frac{1}{n+1}$.
(c) If $\psi \in L^p(U, \text{loc})$, $n+1 < p \leq \infty$, then $\varphi \in \Gamma_\beta(U, \text{loc})$, where $\beta = 2 - (2n+2)/p$.
(d) If $\psi \in S_k^p(U, \text{loc})$ with $1 < p < \infty$ and $k \in \{0, 1, 2, \ldots\}$, then $\varphi \in S_{k+2}^2(U, \text{loc})$.
(e) If $\psi \in \Gamma_\beta(U, \text{loc})$, $0 < \beta < \infty$, then $\varphi \in \Gamma_{\beta+2}(U, \text{loc})$.
(f) If $\psi \in C^\infty(U)$ then $\varphi \in C^\infty(U)$.

For a proof the reader should see see [150], pp. 497–498.

References

[1] L.V. Ahlfors, *Lectures on quasiconformal mappings*, D. Van Nostrand Co., Inc., Princeton, New Jersey-Toronto-New York-London, 1966.

[2] T. Akahori, *A new approach to the local embedding theorem of CR structures, the local embedding theorem for $n \geq 4$*, A.M.S. Memoirs, No. 366, 1987.

[3] T. Akahori, *Intrinsic formula for Kuranishi's $\overline{\partial}_b^{\varphi}$*, Publ. RIMS, Kyoto Univ., 14(1978), 615–641.

[4] T. Akahori, *Complex analytic construction of the Kuranishi family on a normal strongly pseudo-convex manifold*, Publ. RIMS, Kyoto Univ., 14(1978), 789–847.

[5] T. Akahori, *The new Neumann operator associated with deformations of strongly pseudo-convex domains and its application to deformation theory*, Invent. Math., 68(1982), 317–352.

[6] T. Akahori, *A criterion for the Neumann type problem over a differential complex on a strongly pseudo convex domain*, Math. Ann., 264(1983), 525–535.

[7] T. Akahori and K. Miyajima, *Complex analytic construction of the Kuranishi family on a normal strongly pseudo-convex manifold*, II, Publ. RIMS, Kyoto Univ., 16(1980), 811–834.

[8] D.V. Alekseevskii, *Riemannian spaces with exceptional holonomy groups*, Funkcional. Anal. i Prilozen., 2(1968), 1–10.

[9] D.V. Alekseevski and A. Spiro, *Compact homogeneous CR manifolds*, preprint, 2000.

[10] D.V. Alekseevski and A. Spiro, *Invariant CR structures on compact homogeneous manifolds*, Hokkaido Math. J., 32(2003), 209–276.

[11] L. Amoroso, *Sopra un problema al contorno*, Rend. Circ. Matem. Palermo, 33(1912), 75–85.

[12] A. Andreotti, *Complexes of partial differential operators*, Yale University Press, New Haven-London, 1975.

[13] A. Andreotti and C.D. Hill, *Complex characteristic coordinates and the tangential Cauchy-Riemann equations*, Ann. Scuola Norm. Sup., Pisa, 26(1972), 299–324.

[14] A. Andreotti and C.D. Hill, *E.E. Levi convexity and the Hans Lewy problem*, I-II, Ann. Sc. Norm. Sup., Pisa, 26(1972), 325–363, 747–806.

[15] A. Andreotti and G.A. Fredricks, *Embeddability of real analytic Cauchy-Riemann manifolds*, Ann. Sc. Norm. Sup. Pisa, (4)6(1979), 285–304.

[16] A. Andreotti, G. Fredricks, and M. Nacinovich, *On the absence of Poincaré lemma in tangential Cauchy-Riemann complexes*, Ann. Sc. Norm. Sup., Pisa, (3)8(1981), 365–404.

[17] A. Andreotti, C.D. Hill, S. Lojasiewicz, and B. Mackichan, *Mayer-Vietoris sequences for complexes of differential operators*, Bull. A.M.S., (3)82(1976), 487–490.
[18] A. Andreotti, C.D. Hill, S. Lojasiewicz, and B. Mackichan, *Complexes of differential operators*, Invent. Math., 35(1976), 43–86.
[19] A. Andreotti and M. Nacinovich, *Noncharacteristic hypersurfaces for complexes of differential operators*, Annali Matem. Pura Appl., (4)CXXV(1980), 13–83.
[20] A. Andreotti and M. Nacinovich, *On analytic and C^∞ Poincaré lemma*, Mathematical Analysis and Applications, Part A, Advances in Math. Supplementary Studies, 7A(1981), 41–93.
[21] A. Andreotti and Y.T. Siu, *Projective imbeddings of pseudo concave spaces*, Ann. Scuola Norm. Sup. Pisa, (3)23(1970), 231–278.
[22] M. Ara, *Geometry of F-harmonic maps*, Kodai Math. J., (2)22(1999), 243–263.
[23] T. Aubin, *Métriques Riemanniennes et courbure*, J. Diff. Geometry, 4(1970), 383–424.
[24] T. Aubin, *Équations differentielles non linéaires et problème de Yamabe concernant la courbure scalaire*, J. Math. Pures Appl., 55(1976), 269–296.
[25] M. Audin, *The topology of torus actions on symplectic manifolds*, Progress in Math., vol. 93, ed. by J. Oesterlé and A. Weinstein, Birkhäuser Verlag, Basel-Boston-Berlin, 1991.
[26] H. Azad, A. Huckleberry, and W. Richthofer, *Homogeneous CR manifolds*, J. Reine Angew. Math., 358(1985), 125–154.
[27] A. Bahri, *Proof of the Yamabe conjecture for locally conformally flat manifolds*, Nonlinear Analysis, Theory, Methods and Applications, 20(1993), 1261–1278.
[28] P. Baird and S. Gudmundson, *p-Harmonic maps and minimal submanifolds*, Math. Ann., 294(1992), 611–624.
[29] Z.M. Balogh and C. Leuenberger, *Transversal group actions on CR manifolds and embedability*, Tohoku Math. J., 55(2003), 507–527.
[30] M.S. Baouendi, L.P. Rothschild, and F. Treves, *CR structures with group action and extendability of CR functions*, Invent. MAth., 82(1985), 359–396.
[31] M.S. Baouendi, P. Ebenfelt, and L.P. Rothschild, *Real submanifolds in complex space and their mappings*, Princeton Mathematical Series, vol. 47, Princeton University Press, Princeton, New Jersey, 1999.
[32] E. Barletta, *Hörmander systems and harmonic morphisms*, Annali della Scuola Normale Superiore di Pisa, Scienze Fisiche e Matematiche, Serie V, (2)2(2003), 379–394.
[33] E. Barletta, *Subelliptic F-harmonic maps*, Riv. Mat. Univ. Parma, (7)2(2003), 33–50.
[34] E. Barletta and S. Dragomir, *On $\mathcal{G}$-Lie foliations with transverse CR structure*, Rendiconti di Matem., Roma, 16(1996), 169–188.
[35] E. Barletta and S. Dragomir, *On the spectrum of a strictly pseudoconvex CR manifold*, Abhandlungen Math. Sem. Univ. Hamburg, 67(1997), 143–153.
[36] E. Barletta and S. Dragomir, *Pseudohermitian immersions, pseudo-Einstein structures, and the Lee class of a CR manifold*, Kodai Math. J., 19(1996), 62–86.
[37] E. Barletta and S. Dragomir, *On the CR structure of the tangent sphere bundle*, Le Matematiche, Catania, (2)L(1995), 237–249.
[38] E. Barletta and S. Dragomir, *New CR invariants and their application to the CR equivalence problem*, Ann. Scuola Norm. Sup. Pisa, (1)XXIV(1997), 193–203.
[39] E. Barletta and S. Dragomir, *Transversally CR foliations*, Rendiconti di Matematica, Roma, 17(1997), 51–85.
[40] E. Barletta and S. Dragomir, *On the Djrbashian kernel of a Siegel domain*, Studia Mathematica, (1)127(1998), 47–63.
[41] E. Barletta and S. Dragomir, *On boundary behaviour of symplectomorphisms*, Kodai Math. J., 21(1998), 285–305.

[42] E. Barletta and S. Dragomir, *Differential equations on contact Riemannian manifolds*, Annali della Scuola Normale Superiore di Pisa, Cl. Sci., Ser. IV, (1)XXX(2001), 63–96.

[42.1] E. Barletta and S. Dragomir, *On the regularity of weak subelliptic F-harmonic maps*, Tsukuba J. Math., (2)28(2004), 417–436.

[43] E. Barletta, S. Dragomir, and H. Urakawa, *Pseudoharmonic maps from a nondegenerate CR manifold into a Riemannian manifold*, Indiana University Mathematics Journal, (2)50(2001), 719–746.

[44] E. Barletta, S. Dragomir, and H. Urakawa, *Yang–Mills fields on CR manifolds*, preprint, 2004.

[45] E. Barletta and C. Parrini, *Bounded solutions for $\overline{\partial}$-problem in pseudo-Siegel domains*, Annali di Matematica pura e applicata, (IV)CLXVIII(1995), 119–132.

[46] M. Barros and F. Urbano, *Submanifolds of complex Euclidean space which admit a holomorphic distribution*, Quart. J. Math. Oxford, (2)34(1983), 141–143.

[47] M. Barros, B.Y. Chen, and F. Urbano, *Quaternion CR-submanifolds of quaternion manifolds*, Kodai Math. J., 4(1981), 399–417.

[48] W. Barthel, *Nichtlineare Zusammenhänge und deren Holonomiegruppen*, J. Reine Angew. Math., 212(1963), 120–149.

[49] H. Baum, *Lorentzian twistor spinors and CR geometry*, Differential Geometry and its Applications, 11(1999), 69–96.

[50] R. Beals, P.C. Greiner, and N.K. Stanton, *The heat equation on a CR manifold*, J. Diff. Geometry, 20(1984), 343–387.

[51] E. Bedford, $(\partial\overline{\partial})_b$ *and the real parts of CR functions*, Indiana Univ. Math. J., (3)29(1980), 333–340.

[52] E. Bedford, *The Dirichlet problem for some overdetermined systems on the unit ball in* $\mathbf{C}^n$, Pacific Journal of Mathem., (1)51(1974), 19–25.

[53] E. Bedford and P. Federbush, *Pluriharmonic boundary values*, Tôhoku Math. J., 26(1974), 505–511.

[54] J.K. Beem and P.E. Ehrlich, *Global Lorentzian geometry*, Marcel Dekker, Inc., New York - Basel, 1981.

[55] A. Bejancu, *CR submanifolds of a Kaehler manifold*, I-II, Proc. A.M.S., 69(1978), 134–142, Trans. A.M.S., 250(1979), 335–345.

[56] A. Bejancu, *Geometry of CR submanifolds*, D. Reidel Publ. Co., Dordrecht-Boston-Lancaster-Tokyo, 1986.

[57] V. Benci, D. Fortunato, and A. Masiello, *On the geodesic connectedness of Lorentzian manifolds*, Math. Z., 217(1994), 73–93.

[58] M. Berger, *Sur les groupes d'holonomie des variétés à connexion affine et des variétés Riemanniennes*, Bull. Soc. Math. France, 83(1955), 279–230.

[59] M. Berger, P. Gauduchon, and E. Mazet, *Le spectre d'une varieté Riemannienne*, Lecture Notes in Math., 194, Springer-Verlag, Berlin-New York, 1971.

[60] A.L. Besse, *Einstein manifolds*, Ergebnisse der Mathematik und ihrer Grenzgebiete, 3. Folge · Band 10, A Series of Modern Surveys in Mathematics, Springer-Verlag, Berlin-Heidelberg-New York-London-Paris-Tokyo, 1987.

[61] M. Biroli and U. Mosco, *Sobolev inequalities on homogeneous spaces*, Pot. Anal., 4(1995), 311–324.

[62] D.E. Blair, *Contact manifolds in Riemannian geometry*, vol. 509, Springer-Verlag, Berlin-Heidelberg-New York, 1976.

[63] D.E. Blair, *On the geometric meaning of the Bochner tensor*, Geometriae Dedicata, 4(1975), 33–38.

[64] D.E. Blair and B.Y. Chen, *On CR submanifolds of Hermitian manifolds*, Israel J. Math., 34(1979), 353–363.

[65] D.E. Blair and D. Perrone, *A variational characterization of contact metric manifolds with vanishing torsion*, Canad. Math. Bull., (4)35(1992), 455–462.

[66] D.E. Blair and D. Perrone, *Second variation of the "total scalar curvature" on contact manifolds*, Canad. Math. Bull., (1)38(1995), 16–22.

[67] D.E. Blair and S. Dragomir, *The Rossi-Vergne theorem and constant coefficient PDEs on the Heisenberg group*, in *Selected Topics in Cauchy-Riemann geometry* (ed. by S. Dragomir), Quaderni di Matematica, Ser. Ed. by Dipartimento di Matematica, Seconda Università di Napoli, Caserta, 9(2001), 89–99.

[68] D.E. Blair and S. Dragomir, *Pseudohermitian geometry on contact Riemannan manifolds*, Rendiconti di Matematica, Roma, 22(2002), 275–341.

[69] E. Boeckx and L. Vanhecke, *Harmonic and minimal vector fields on tangent and unit tangent bundles*, Preprint, 2000.

[70] A. Boggess, *CR manifolds and the tangential Cauchy-Riemann complex*, Studies in Advanced Math., CRC Press, Inc., Boca Raton-Ann Arbor-Boston-London, 1991.

[71] J.M. Bony, *Détermination de axiomatiques de théorie du potentiel dont les fonctions harmoniques sont différentiables*, Ann. Inst. Fourier, Grenoble, (1)17(1967), 353–382.

[72] J.M. Bony, *Sur la régularité des solutions du problème de Dirichlet pour les opérateurs elliptiques dégénérés*, C.R. Acad. Sci. Paris, 267(1968), 691–693.

[73] J.M. Bony, *Principe du maximum, inégalité de Harnak et unicité du problème de Cauchy pour les opérateurs elliptiques dégénérés*, Ann. Inst. Fourier, Grenoble, (1)19(1969), 277–304.

[74] W.M. Boothby, *Some fundamental formulas for Hermitian manifolds with nonvanishing torsion*, American J. Math., 76(1954), 509–534.

[75] W.M. Boothby and H.C. Wang, *On contact manifolds*, Ann. of Math., (3)68(1958), 721–734.

[76] R. Bott and P. Baum, *Singularities of holomorphic foliations*, J. Diff. Geometry, 7(1972), 279–342.

[77] L. Boutet de Monvel, *Intégration des équations de Cauchy-Riemann induites formelles*, Sém. Goulaouic-Lions-Schwartz (1974–75), Centre. Math. Ecole Polytech., Paris (1975).

[78] R.O. Buchweitz and J.J. Millson, *CR geometry and deformation of isolated sigularities*, preprint, 1991.

[79] D. Burns and S. Shnider, *Spherical hypersurfaces in complex manifolds*, Inv. Math., 33(1976), 223–246.

[80] D. Burns and E.L. Stout, *Extending functions from submanifolds of the boundary*, Duke Math. J., 43(1976), 391–404.

[81] D. Burns, K. Diederich, and S. Shnider, *Distinguished curves in pseudoconvex boundaries*, Duke Math. J., 44(1977), 407–431.

[82] E. Brieskorn, *Beispiele zur Differentialtopologie von Singularitäten*, Invent. Math., 2(1966), 1–14.

[83] E. Brieskorn and A. Van de Ven, *Some complex structures on products of homotopy spheres*, Topology, 7(1968), 389–393.

[84] R.L. Bryant, *Holomorphic curves in Lorenzian CR manifolds*, Trans. A.M.S., (1)272(1982), 203–221.

[85] L. Capogna, D. Danielli, and N. Garofalo, *An embedding theorem and the Härnak inequality for nonlinear subelliptic equations*, Commun. in Partial Differential Equations, (9&10)18(1993), 1765–1794.

[86] L. Capogna, D. Danielli, and N. Garofalo, *Subelliptic mollifiers and basic pointwise estimates of Poincaré type*, Math. Z., 226(1997), 147–154.

[87] L. Capogna and P. Tang, *Uniform domains and quasiconformal mappings on the Heisenberg group*, Manuscripta Math., (3)86(1995), 267–281.

[88] J.A. Carlson and C.D. Hill, *On the maximum modulus principle for the tangential Cauchy-Riemann equations*, Math. Ann., 208(1974), 91–97.

[89] H. Cartan, *Théorie élémentaire des fonctions analytiques d'une ou plusieurs variables complexes*, Herman, Paris, 1961.

[90] B.Y. Chen, *CR submanifolds of a Kaehler manifold*, I-II, J. Diff. Geometry, 16(1981), 305–322, 493–509; *Cohomology of CR submanifolds*, An. Fac. Sci. Toulouse, 3(1981), 167–172.

[91] B.Y. Chen, *Geometry of submanifolds*, Marcel Dekker, Inc., New York, 1973.

[92] B.Y. Chen, *Riemannian submanifolds*, Handbook of Differential Geometry, Vol. I, pp. 187–418, North-Holland, Amsterdam, 2000.

[93] B.Y. Chen and H.S. Lue, *On normal connection of Kähler submanifolds*, J. Math. Soc. Japan, 27(1975), 550–556.

[94] Jih-Hsin Cheng, *Chain-preserving diffeomorphisms and CR equivalence*, Proc. A.M.S., (1)103(1988), 75–80.

[95] Jih-Hsin Cheng, *Rigidity of automorphisms and spherical CR structures*, preprint, 2001.

[96] Jih-Hsin Cheng and J.M. Lee, *The Burns-Epstein invariant and deformation of CR structures*, Duke Math. J., (1)60(1990), 221–254.

[97] S.Y. Cheng and S.T. Yau, *On the regularity of the Monge-Ampère equation* $\det(\partial^2 u/\partial x^i \partial x^j) = F(x, u)$, Comm. Pure Appl. Math., 30(1977), 41–68.

[98] S.S. Chern and R.S. Hamilton, *On Riemannian metrics adapted to three-dimensional contact manifolds*, Springer Lecture Notes 1111 Springer-Verlag (1985), 279–308.

[99] S.S. Chern and J.K. Moser, *Real hypersurfaces in complex manifolds*, Acta Math., 133(1974), 219–271.

[100] S.S. Chern and J. Simons, *Characteristic forms and geometric invariants*, Annals of Math., 99(1974), 48–69.

[101] L.F. Cheung and P.F. Leung, *Some results on stable p-harmonic maps*, Glasgow Math. J., 36(1994), 77–80.

[102] E.M. Chirka, *Introduction to the geometry of CR manifolds*, Russian Math. Surveys, (1)46(1991), 95–197.

[103] W.L. Chow, *Über Systeme von Linearen Partiellen Differentialgleichungen erster Ordnung*, Math. Ann., 117(1939), 98–105.

[104] G. Citti, N. Garofalo, and E. Lanconelli, *Harnak's inequality for sum of squares plus potential*, Amer. J. Math., 115(1993), 699–734.

[105] R. Coifman and G. Weiss, *Analyse harmonique non-commutative sur certains espaces homogènes*, Lecture Notes in Math., vol. 242, Springer-Verlag, Berlin-New York, 1971.

[106] L.A. Cordero, M. Fernández, L. Ugarte, and A. Gray, *A general description of the terms in the Frölicher spectral sequence*, Diff. Geometry and its Appl., 7(1997), 75–84.

[107] M. Crampin, *Pseudo-Riemannian and related structures on differentiable manifolds*, Ph.D. thesis, King's College, London, 1967.

[108] J.P. D'Angelo, *Polynomial proper holomorphic mappings between balls, II*, Michigan Math. J., 38(1991), 53–65.

[109] J.P. D'Angelo, *Polynomial proper maps between balls*, Duke Math. J., 57(1988), 211–219.

[110] J.P. D'Angelo, *Proper holomorphic maps between balls of different dimensions*, Michigan Math. J., 35(1988), 83–90.

[111] J.P. D'Angelo, *An explicit computation of the Bergman kernel function*, The Journal of Geometric Analysis, (1)4(1994), 23–34.

[112] G. David, *Solutions de l'equation de Beltrami avec* $\|\mu\|_\infty = 1$, Ann. Acad. Sci. Fennicae, 13(1988), 25–70.

[113] E.T. Davies and K. Yano, *On the tangent bundle of Finsler and Riemannian manifolds*, Rend. Circ. Mat. Palermo, 12(1963), 211–228.

[114] F. De Mari, M.M. Peloso, and F. Ricci, *Analysis of second order differential operators with complex coefficients on the Heisenberg group*, J. Reine Angew. Math., 464(1995), 67–96.

[115] P. De Bartolomeis and G. Tomassini, *Traces of pluriharmonic functions*, Compositio Math., 44(1981), 29–39.

[116] P. Dombrowski, *On the geometry of the tangent bundle*, J. Reine Angew. Math., 210(1962), 73–88.

[117] S. Donaldson, *Anti self-dual Yang–Mills connections over complex algebraic surfaces and stable vector bundles*, Proc. London Math. Soc., 50(1985), 1–26.

[118] S. Donnini and G. Gigante, *Classification of left invariant CR structures on* $GL^+(3, \mathbf{R})$, Riv. Mat. Univ. Parma, (4)16(1990), 343–355.

[119] S. Dragomir, *Generalized Hopf manifolds, locally conformal Kaehler structures, and real hypersurfaces*, Kodai Math. J., 14(1991), 366–391.

[120] S. Dragomir, *Pseudohermitian immersions between strictly pseudoconvex CR manifolds*, American J. Math., (1)117(1995), 169–202.

[121] S. Dragomir, *On a conjecture of J.M. Lee*, Hokkaido Math. J., (1)23(1994), 35–49.

[122] S. Dragomir, *Pseudohermitian geometry and interpolation manifolds*, Complex Variables, 27(1995), 105–115.

[123] S. Dragomir, *A survey of pseudohermitian geometry*, The Proceedings of the Workshop on Differential Geometry and Topology, Palermo (Italy), June 3–9, 1996, in *Supplemento ai Rendiconti del Circolo Matematico di Palermo*, Serie II, 49(1997), 101–112.

[124] S. Dragomir and J.C. Wood, *Sottovarietà minimali ed applicazioni armoniche*, Quaderni dell'Unione Matematica Italiana, vol. 35, Pitagora Editrice, Bologna, 1989.

[125] S. Dragomir and L. Ornea, *Locally conformal Kähler geometry*, Progress in Mathematics, Vol. 155, Birkhäuser, Boston-Basel-Berlin, 1998.

[126] S. Dragomir and P. Nagy, *Complex Finsler structures on CR-holomorphic vector bundles*, Rendiconti di Matematica, Roma, Serie VII, 19(1999), 427–447.

[127] S. Dragomir and H. Urakawa, *On the inhomogeneous Yang–Mills equation* $d_D^* R^D = f$, Interdisciplinary Information Sciences, (1)6(2000), 41–52.

[128] S. Dragomir and J. Masamune, *Cauchy-Riemann orbifolds*, Tsukuba J. Math., (2)26(2002), 351–386.

[129] S. Dragomir and S. Nishikawa, *Foliated CR manifolds*, J. Math. Soc. Japan, (4)56(2004), 1031–1068.

[130] R. Dwilewicz, *Bergman-Szegö type theory for CR structures*, in Complex Analysis III, Proceedings, Univ. of Maryland 1985–86, ed. by A. Berenstein, Lecture Notes in Math., vol. 1277, pp. 15–34, Springer-Verlag, Berlin-Heidelberg-New York-London-Paris-Tokyo, 1987.

[131] J. Eells and M.J. Ferreira, *On representing homotopy classes by harmonic maps*, Bull. London Math. Soc., 23(1991), 160–162.

[132] L.P. Eisenhart, *Riemannian geometry*, Princeton Univ. Press, Princeton, 1966.

[133] N. Ejiri, *Bochner Kähler metrics*, Bull. Sc. Math., 108(1984), 423–436.

[134] L. Ehrenpreis, *Lewy unsolvability and several complex variables*, Michigan Math. J., 38(1991), 417–439.

[135] D. Ellis, C.D. Hill, and C.C. Seabury, *The maximum modulus principle, I. Necessary conditions*, Indiana Univ. Math. J., (7)25(1976), 709–715.

[136] J. Faraut, *Analyse harmonique et fonctions spéciales*, in *Deux cours d'analyse harmonique*, École d'été d'analyse harmonique de Tunis (1984), Birkhäuser, 1987.
[137] F. Farris, *An intrinsic construction of Fefferman's CR metric*, Pacific J. Math., 123(1986), 33–45.
[138] C. Fefferman, *Monge-Ampère equations, the Bergman kernel, and geometry of pseudoconvex domains*, Ann. of Math., (2)103(1976), 395–416, 104(1976), 393–394.
[139] C. Fefferman, *The Bergman kernel and biholomorphic equivalence of pseudoconvex domains*, Invent. Math., 26(1974), 1–65.
[140] C. Fefferman and D.H. Phong, *Subelliptic eigenvalue problems*, Proceedings of the Conference on Harmonic Analysis in honour of A. Zygmund, Wadsworth Math. Series, 1981, 590–606.
[141] G. Fichera, *Su un teorema di L. Amoroso nella teoria delle funzioni analitiche di due variabili complesse*, Rev. Roumaine Math. Pures et Appl., (3)XXVII(1982), 327–333.
[142] G. Fichera, *Valori al contorno delle funzioni pluriarmoniche: estensione allo spazio* $\mathbf{R}^{2n}$ *di un teorema di L. Amoroso*, Lecture delivered at the University of Rome, 7 July 1982.
[143] G. Fichera, *Problemi di valori al contorno per le funzioni pluriarmoniche*, Actes du 6e Congrés du Groupment des Mathématiciens d'Expression Latine, Luxembourg, 139–151, Ed. Gauthier-Villars, 1981.
[144] G. Fichera, *Boundary values of analytic functions of several complex variables*, Complex Analysis and Applications, 167–177, Sofia, 1981.
[145] G. Fichera, *Boundary values of pluriharmonic functions*, Uspechi Matemat., (2)37(1982), 255–256 (in Russian).
[146] G.B. Folland, *A fundamental solution for a subelliptic operator*, Bull. A.M.S., (2)79(1973), 373–376.
[147] G.B. Folland, *Introduction to partial differential equations*, Princeton Univ. Press, Princeton, New Jersey, 1976.
[148] G.B. Folland, *Weyl manifolds*, J. Diff. Geometry, 4(1970), 145–153.
[149] G.B. Folland, *Subelliptic estimates and function spaces on nilpotent Lie groups*, Arkiv för Mat., 13(1975), 161–207.
[150] G.B. Folland and E.M. Stein, *Estimates for the* $\overline{\partial}_b$*-complex and analysis on the Heisenberg group*, Comm. Pure Appl. Math., 27(1974), 429–522.
[151] D. Fortunato, F. Giannoni, and A. Masiello, *A Fermat principle for stationary space-times and applications to light rays*, J. Geometry and Phys., 15(1995), 159–188.
[152] B. Franchi and E. Lanconelli, *Hölder regularity theorem for a class of non uniformly elliptic operators with measurable coefficients*, Ann. Scuola Norm. Sup. Pisa, 10(1983), 523–541.
[153] B. Franchi, *Weighted Sobolev-Poincaré inequalities and pointwise estimates for a class of degenerate ellitptic operators*, Trans. A.M.S., 327(1991), 125–158.
[154] B. Franchi, G. Lu, and R.L. Wheeden, *Representation formulas and weighted Poincaré inequalities for Hörmander vector fields*, Ann. Inst. Fourier, Greenoble, 45(1995), 577–604.
[155] B. Franchi, R. Serapioni, and F. Serra Cassano, *Approximation and imbedding theorems for weighted Sobolev spaces asociated with Lipschitz continuous vector fields*, Bolletino U.M.I., 11-B(1997), 83–117.
[156] G. Francsics and N. Hanges, *The Bergman kernel of complex ovals and multivariable hypergeometric functions*, Journal of Functional Analysis, 142(1996), 494–510.
[157] M. Freeman, *The Levi form and local complex foliations*, Trans. A.M.S., 57(1976), 369–370.
[158] A. Frölicher, *Relations between the cohomology groups of Dolbeault and topological invariants*, Proc. Nat. Acad. Sci. U.S.A., 41(1955), 641–644.

[159] B. Fuglede, *Harmonic morphisms between Riemannian manifolds*, Ann. Inst. Fourier, 28(1978), 107–144.
[160] T. Fuglede, *Harmonic morphisms between semi-Riemannian manifolds*, Ann. Acad. Sci. Fennicae, 21(1996), 31–50.
[161] S. Funbashi, *Totally complex submanifolds of a quaternionic Kaehlerian manifold*, Kodai Math. J., 2(1979), 314–336.
[162] K. Galicki and H.B. Lawson, *Quaternionic reduction and quaternionic orbifolds*, Math. Ann., 282(1988), 1–21.
[163] N. Gamara, *CR Yamabe conjecture, the case* $n = 1$, J. Eur. Math. Soc., 3(2001), 105–137.
[164] N. Gamara and R. Yacoub, *CR Yamabe conjecture - the conformally flat case*, Pacific J. Math., (1)201(2001), 121–175.
[165] N. Garofalo and E. Lanconelli, *Frequency functions on the Heisenberg group, the uncertainty principle and unique continuation*, Ann. de l'Inst. Fourier, (2)40(1990), 313–356.
[166] N. Garofalo and D.M. Nhieu, *Lipschitz continuity, global smooth approximation and extension theorems for Sobolev functions in Carnot-Carathéodory spaces*, Anal. Math., 74(1998), 67–97.
[167] N. Garofalo and D.M. Nhieu, *Isoperimetric and Sobolev inequalities for Carnot-Carathéodory spaces and the existence of minimal surfaces*, Comm. Pure Appl. Math., 49(1996), 1081–1144.
[168] T. Garrity and R. Mizner, *Vector valued forms and CR geometry*, Advanced Studies in Pure Math., 25(1997), 110–121.
[169] K. Gawedzki, *Fourier-like kernels in geometric quantization*, Dissertationes Mathematicae, vol. CXXVI, Warszawa, 1976.
[170] F. Giannoni and A. Masiello, *On a variational theory of light rays on Lorentzian manifolds*, Rend. Mat. Acc. Lincei, 6(1995), 155–159.
[171] F. Giannoni and A. Masiello, *Geodesics on product Lorentzian manifolds*, Ann. Inst. Henri Poincaré, (1)12(1995), 27–60.
[172] F. Giannoni and A. Masiello, *On a Fermat principle in general relativity. A Morse theory for light rays*, General Relativity and Gravitation, (7)28(1996), 855–897.
[173] G. Gigante and G. Tomassini, *Infinitesimal bendings of CR hypersurfaces*, preprint, 21, Sc. Norm. Sup. Pisa, 1988.
[174] G. Gigante, *Sulla coomologia di varietà pseudohermitiane con torsione nulla*, Riv. Mat. Univ. Parma, (4)2(1986), 157–165.
[175] G. Gigante, *Symmetries on compact pseudohermitian manifolds*, Rend. Circ. Matem. Palermo, 36(1987), 148–157.
[176] G. Gigante and G. Tomassini, *CR structures on a real Lie algebra*, Rend. Mat. Accad. Lincei, 2(1991), 203–205, Advances in Math., 94(1992), 67–81.
[177] D. Gilbarg and N.S. Trudinger, *Elliptic partial differential equations of second order*, Springer-Verlag, Berlin-Heidelberg-New York-Tokyo, 1983.
[178] S.I. Goldberg, *Curvature and homology*, Dover. Publ., Inc., New York, 1982.
[179] S.I. Goldberg, D. Perrone, and G. Toth, *Contact three-manifolds with positive generalized Tanaka-Webster scalar curvature*, C.R. Math. Rep. Acad. Sci. Canada, (6)X(1988), 255–260.
[180] S.I. Goldberg and D. Perrone, *Contact* 3*-manifolds with positive scalar curvature*, Contemporary Math., 127(1992), 59–68.
[181] W. Goldman, *Conformally flat manifolds with nilpotent holonomy and the uniformization problem for* 3*-manifolds*, Trans. A.M.S., 238(1983), 573–583.
[182] C.R. Graham, *On Sparling's characterization of Fefferman metrics*, American J. Math., 109(1987), 853–874.

[183] C.R. Graham, *The Dirichlet problem for the Bergman Laplacian*, I–II, Communications in Partial Differential Equations, 8(1983), 433–476, 563–641.

[184] C.R. Graham, *Compatibility operators for degenerate elliptic equations on the ball and Heisenberg group*, Math. Z., 187(1984), 289–304.

[185] C.R. Graham and J.M. Lee, *Smooth solutions of degenerate Laplacians on strictly pseudoconvex domains*, Duke Math. J., (3)57(1988), 697–720.

[186] A. Greenleaf, *The first eigenvalue of a sublaplacian on a pseudohermitian manifold*, Comm. Partial Differential Equations, 10(1985), 191–217.

[187] S. Greenfield, *Cauchy-Riemann equations in several variables*, Ann. Sc. Norm. Sup. Pisa, 22(1968), 275–314.

[188] P.C. Greiner, J.J. Kohn, and E.M. Stein, *Necessary and sufficient conditions for the solvability of the Lewy equation*, Proc. Nat. Acad. Sci. U.S.A., (9)72(1975), 3287–3289.

[189] W. Greub, S. Halperin, and R. Vanstone, *Connections, curvature, and cohomology*, Academic Press, New York-London, 1976.

[190] P. Hajlasz and P. Koskela, *Sobolev meets Poincaré*, C.R. Acad. Sci. Paris, 320(1995), 1211–1215.

[191] P. Hájlasz, *Sobolev spaces on an arbitrary metric space*, Potential Analysis, 5(1996), 403–415.

[192] P. Hájlasz and P. Strzelecki, *Subelliptic p-harmonic maps into spheres and the ghost of Hardy spaces*, Math. Ann., 312(1998), 341–362.

[193] N. Hanges and H. Jacobowitz, *Involutive structures on compact manifolds*, American J. Math., 117(1995), 491–522.

[194] S.G. Harris, *A triangle comparison theorem for Lorentz manifolds*, Indiana Univ. Math. J., (3)31(1982), 289–308.

[195] Y. Hayashi, *Compatibility conditions and the manifold of solutions of some nonlinear Cauchy-Riemann systems in several complex variables*, Math. Nachr., 159(1992), 153–167.

[196] S. Helgason, *Differential geometry, Lie groups, and symmetric spaces*, Academic Press, New York, 1978.

[197] G.M. Henkin, *Tangent Cauchy-Riemann equations and the Yang–Mills, Higgs and Dirac fields*, Proceedings of the International Congress of Mathematicians, August 16–24, 1983, Warszawa, pp. 809–827.

[198] C.D. Hill, *A kontinuitätssatz for $\overline{\partial}_M$ and Lewy extendibility*, Indiana Univ. Math. J., (4)22(1972), 339–353.

[199] C.D. Hill, *A family of exotic CR structures on $S^3 \times S^2$*, Analyse complexe multivariable: récents développements (Guadeloupe, 1988), 105–110, Sem. Conf., 5, EditEl., Rende, 1991. (MR 94 f : 32039)

[200] C.D. Hill, *What is the notion of a complex manifold with smooth boundary?*, Algebraic Analysis, 1(1988), 185–201.

[201] C.D. Hill, *Counterexamples to Newlander-Nirenberg up to the boundary*, Proc. Sympos. in Pure Math., 52(1991), Part 3, 191–197.

[202] C.D. Hill and M. Nacinovich, *Pseudoconcave CR manifolds*, Complex Analysis and Geometry (V. Ancona, E. Ballico, A. Silva, eds.), Marcel Dekker, Inc., New York, 1996, pp. 275–297; *Aneurysms of pseudoconcave CR manifolds*, Math. Z., 220(1995), 347–367; *Duality and distribution cohomology for CR manifolds*, Ann. Scuola Norm. Sup. Pisa, XXII(1995), 315–339; *On the Cauchy problem in complex analysis*, Ann. Mat. Pura e Appl., CLXXI(1996), 159–179; *The topology of Stein CR manifolds and the Lefshetz theorem*, Ann. Inst. Fourier Grenoble, 43(1993), 459–468; *The Jacobian theorem for*

mapping of pseudoconcave CR hypersurfaces, Boll. U.M.I., 7(1995), 149–155; *A necessary condition for global Stein immersion of compact CR manifolds*, Riv. Mat. Univ. Parma, (1)5(1992), 175–182.

[203] C.D. Hill and M. Nacinovich, *A weak pseudoconcavity condition for abstract almost CR manifolds*, Inventiones Math., 142(2000), 251–283.

[204] C.D. Hill and G. Taiani, *The local family of analytic discs attached to a CR submanifold*, Several Complex Variables, 166–179, Proceedings of International Conf., Cortona, Italy, 1976–1977, Sc. Norm. Sup. Pisa, 1978.

[205] C.D. Hill and G. Taiani, *Families of analytic discs in* $\mathbf{C}^n$ *with boundaries on a prescribed CR submanifold*, Ann. Sc. Norm. Sup., Pisa, (2)5(1978), 327–380.

[206] C.D. Hill and G. Taiani, *Real analytic approximation of locally embeddable CR manifolds*, Compositio Math., 44(1981), 113–131.

[207] C.D. Hill and G. Taiani, *On the Hans Lewy extension phenomenon in higher codimension*, Proc. A.M.S., (4)91(1984), 568–572.

[208] G. Hirsch, *Differential topology*, Springer, New York-Heidelberg-Berlin, 1976.

[209] N.J. Hitchin, *Kählerian twistor spaces*, Proc. London Math. Soc., (3)43(1981), 133.

[210] M.C. Hong, *On the conformal equivalence of harmonic maps and exponentially harmonic maps*, Bull. London Math. Soc., 24(1992), 488–492.

[211] H. Hopf, *Zur Topologie der komplexen Mannigfaltigkeiten*, Studies and Essays presented to R. Courant, New York, 1948.

[212] L. Hörmander, L^2*-estimates and existence theorems for the* $\overline{\partial}$*-operator*, Acta Math., 113(1965), 89–152.

[213] L. Hörmander, *Hypoelliptic second-order differential equations*, Acta Math., 119(1967), 147–171.

[214] S. Ianus, *Sulle varietà di Cauchy-Riemann*, Rend. dell'Accad. Sci. Fis. Mat., Napoli, 39(1972), 191–195. 191–195.

[215] S. Ianus and A.M. Pastore, *Harmonic maps on contact metric manifolds*, Ann. Math. Blaise Pascal, 2(1995), 43–53.

[216] S. Ishihara, *Quaternion Kählerian manifolds*, J. Diff. Geometry, 9(1974), 483–500.

[217] T. Ishihara, *A mapping of Riemannian manifolds which preserves harmonic functions*, J. Math. Kyoto Univ., (2)19(1979), 215–229.

[218] H. Iwamoto, *On the structure of Riemannian spaces whose holonomy groups fix a null system*, Tohoku Math. J., 1(1950), 109–135.

[219] H. Jacobowitz, *Mappings between CR manifolds*, Ann. of Math. Studies, 100(1981), 227–241.

[220] H. Jacobowitz, *Chains in CR geometry*, J. Diff. Geometry, 21(1985), 163–191.

[221] H. Jacobowitz, *An introduction to CR structures*, Mathem. Surveys and Monographs, No. 32, Providence, RI, 1990.

[222] H. Jacobowitz, *The canonical bundle and realizable CR hypersurfaces*, preprint, 1985.

[223] H. Jacobowitz, *Induced connections on hypersurfaces in* $\mathbf{C}^{n+1}$, Invent. Math., 43(1977), 109–123.

[224] H. Jacobowitz and F. Trèves, *Nonrealizable CR structures*, Inventiones Math., 66(1982), 231–249.

[225] D. Jerison, *The Poincaré inequality for vector fields satisfying Hörmander's condition*, Duke Math. J., 53(1986), 503–523.

[226] D. Jerison and J.M. Lee, *A subelliptic, nonlinear eigenvalue problem and scalar curvature on CR manifolds*, Contemp. Math., 27(1984), 57–63.

[227] D. Jerison and J.M. Lee, *The Yamabe problem on CR manifolds*, J. Diff. Geometry, 25(1987), 167–197.

[228] D. Jerison and J.M. Lee, *CR normal coordinates and the Yamabe problem*, J. Diff. Geometry, 29(1989), 303–344.

[229] D. Jerison and J.M. Lee, *Extremals for the Sobolev inequality on the Heisenberg group and the CR Yamabe problem*, J. Amer. Math. Soc., (1)1(1988), 1–13.

[230] D. Jerison and A. Sánchez-Calle, *Subelliptic second order differential operators*, Complex Analysis III, Proceedings, Univ. of Maryland, 1985–86, Lecture Notes in Math., Nr. 1277, ed. by C.A. Berenstein, Springer-Verlag, Berlin-Heidelberg-New York-London-Paris-Tokyo, 1987.

[231] Z. Jiaqing and Y. Hongcang, *On the estimate of the first eigenvalue of a compact Riemannian manifold*, Scientia Sinica, (12)27(1984), 1265–1273.

[232] F. John, *Partial differential equations*, Springer-Verlag, New York-Heidelberg-Berlin, 1986 (fourth edition).

[233] J. Jost, *Riemannian geometry and geometric analysis*, Springer, Berlin-Heidelberg, 2002 (third edition).

[234] J. Jost and C-J. Xu, *Subelliptic harmonic maps*, Trans. of A.M.S., (11)350(1998), 4633–4649.

[235] Y. Kamishima, *Uniformization of Kähler manifolds with vanishing Bochner tensor*, Acta Mathematica, 172(1994), 299–308.

[236] Y. Kamishima, *A geometry of vanishing Bochner tensor on complex manifolds with a geometric structure and their transformation groups*, preprint, Kumamoto University, Department of Mathematics, Kumamoto 860, Japan, 1997.

[237] Y. Kamishima and T. Tsuboi, *CR structures on Seifert manifolds*, Invent. Math., 104(1991), 149–163.

[238] S. Kaneyuki, *Homogeneous bounded domains and Siegel domains*, Lecture Notes in Math., vol. 241, Springer-Verlag, Berlin-New York, 1971.

[239] M.L. Kazlow, *CR functions on tube manifolds*, Ph.D. Thesis, State University of New York at Stony Brook, 1976.

[240] A.W. Knapp and E.M. Stein, *Intertwining operators for semisimple groups*, Ann. of Math., 93(1971), 489–578.

[241] S. Kobayashi& K. Nomizu, *Foundations of differential geometry*, Interscience Publishers, New York, vol. I, 1963, vol. II, 1969.

[242] L. Koch, *Chains and Lorentz geometry*, Ph.D. Thesis, SUNY at Stony Brook, 1986, 36 pp.

[243] L. Koch, *Chains on CR manifolds and Lorentz geometry*, Trans. A.M.S., (2)307(1988), 827–841.

[244] L. Koch, *Chains, null-chains, and CR geometry*, preprint, Rutgers University, New Brunswick, NJ, 1990.

[245] S.E. Koh, *A nonexistence theorem for stable exponentially harmonic maps*, Bull. Korean Math. Soc., 32(1995), 211–214.

[246] J.J. Kohn, *Boundaries of complex manifolds*, Proc. Conf. on Complex Analysis, Minneapolis, 1964, Springer-Verlag, New York, 1965, pp. 81–94.

[247] J.J. Kohn, *Harmonic integrals on strongly pseudo-convex manifolds*, I-II, Ann. of Math., (1)78(1963), 112–148, ibidem, (3)79(1964), 450–472.

[248] J.J. Kohn, *The range of the tangential Cauchy-Riemann operator*, Duke Math. J., (2)53(1986), 525–545.

[249] J.J. Kohn, *Pseudo-differential operators and non-elliptic problems*, Pseudo-Differential Operators (C.I.M.E., Stresa, 1968), Edizioni Cremonese, Roma, 1969, p. 157–165.

[250] J.J. Kohn and L. Nirenberg, *Non-coercive boundary value problems*, Comm. Pure Appl. Math., 18(1965), 443–492.

[251] J.J. Kohn and L. Nirenberg, *Degenerate elliptic-parabolic equations of second order*, Comm. Pure Appl. Math., 20(1967), 797–871.

[252] J.J. Kohn and H. Rossi, *On the extension of holomorphic functions from the boundary of a complex manifold*, Ann. of Math., 81(1965), 451–472.

[253] A. Koohara, *Similarity principle of the generalized Cauchy-Riemann equations for several complex variables*, J. Math. Japan, 23(1971), 213–249.

[254] A. Korányi and H.M. Reimann, *Quasiconformal mappings on the Heisenberg group*, Invent. Math., 80(1985), 309–338.

[255] A. Korányi and H.M. Reimann, *Quasiconformal mappings on CR manifolds*, Lecture Notes in Math., vol. 1422, 59–75, ed. by V. Villani, Springer-Verlag, 1988.

[256] A. Korányi and H.M. Reimann, *Contact transformations as limits of symplectomorphisms*, C.R. Acad. Sci. Paris, 318(1994), 1119–1124.

[257] A. Korányi and H.M. Reimann, *Foundations for the theory of quasiconformal mappings on the Heisenberg group*, Advances in Math., 111(1995), 1–87.

[258] A. Korányi and S. Vági, *Singular integrals in homogeneous spaces and some problems of classical analysis*, Ann. Scuola Norm. Sup. Pisa, 25(1971), 575–648.

[259] B. Kostant, *Quantization and unitary representations*, Lecture Notes in Math., 170, Springer-Verlag, Berlin-Heidelberg-New York, 1970, 87–208.

[260] V.Y. Kraines, *Topology of quaternionic manifolds*, Trans. A.M.S., 122(1966), 357–367.

[261] S.G. Krantz, *Function theory of several complex variables*, John Wiley and Sons, Inc., New York-Chichester-Brisbane-Toronto-Singapore, 1982.

[262] A. Krüger, *Homogeneous Cauchy-Riemann structures*, Ann. Sc. Norm. Sup., Pisa, (2)18(1991), 193–212.

[263] M. Kuranishi, *Strongly pseudoconvex CR structures over small balls*, I-III, Ann. of Math., 115(1982), 451–500, ibidem, 116(1982), 1–64, ibidem, 116(1982), 249–330.

[264] M. Kuranishi, *CR geometry and Cartan geometry*, Forum Math., 7(1995), 147–205.

[265] J. Lafontaine, *The theorem of Lelong-Ferrand and Obata*, in "Conformal Geometry," Aspects of Math., ed. by R. Kulkarni and M. Pinkall, Max-Planck Inst. Für Math., Bonn, 1988, pp. 93–103.

[266] E. Lanconelli, *Regolarità Hölderiana delle soluzioni deboli di certe equazioni ellitiche fortemente degeneri*, Seinario di Analisi Matematica, Istituto Matematico dell'Università di Bologna, pp. IX.1-IX.27, Centro Stampa *Lo Scarabeo*, Bologna, 1983.

[267] C. Laurent-Thiébaut and J. Leiterer, *Uniform estimates for the Cauchy-Riemann equation on q-convex wedges*, Ann. Inst. Fourier, Grenoble, (2)43(1993), 383–436.

[268] C.R. LeBrun, *Twistor CR manifolds and three-dimensional conformal geometry*, Trans. A.M.S., (2)284(1984), 601–616.

[269] C.R. LeBrun, *Foliated CR manifolds*, J. Diff. Geometry, 22(1985), 81–96.

[270] J.M. Lee, *Pseudo-Einstein structures on CR manifolds*, American J. Math., 110(1988), 157–178.

[271] J.M. Lee, *The Fefferman metric and pseudohermitian invariants*, Trans. A.M.S., (1)296(1986), 411–429.

[272] J.M. Lee, *Higher asymptotics of the complex Monge-Ampère equation and geometry of CR manifolds*, Ph.D. Thesis, M.I.T., Cambridge, MA, 1982.

[273] J.M. Lee, *CR manifolds with noncompact connected automorphism groups*, J. Geom. Anal., (1)6(1996), 79–90.

[274] J.M. Lee and R. Melrose, *Boundary behaviour of the complex Monge-Ampère equation*, Acta Mathematica, 148(1982), 159–192.

[275] J.M. Lee and T. Parker, *The Yamabe problem*, Bull. Amer. Math. Soc., 17(1987), 37–91.

[276] D. Lehmann, *J-homotopie dans les espaces de connexions et classes exotiques de Chern-Simons*, C. R. Acad. Sci. Paris, 275(1972), 835–838.

[277] R. Lehmann and D. Feldmueller, *Homogeneous CR hypersurface structures on spheres*, Ann. Sc. Norm. Sup., Pisa, (4)14(1987), 513–527.

[278] L. Lempert, *On three-dimensional Cauchy-Riemann manifolds*, J. Amer. Math. Soc., (4)5(1992), 923–969.

[279] L. Lempert, *Complex structures on the tangent bundle of Riemannian manifolds*, Complex Analysis and Geometry, 235–251, ed. by V. Ancona and A. Silva, Plenum Press, New York, 1993.

[280] L. Lempert and R. Szöke, *Global solutions of the homogeneous complex Monge-Ampère equation and complex structures on the tangent bundle of Riemannian manifolds*, Math. Ann., 290(1991), 689–712.

[281] T. Levi-Civita, *Sulle funzioni di due o più variabili complesse*, Rend. Accad. Naz. Lincei, 14(1905), 492–499.

[282] E.E. Levi, *Studi sui punti singolari essenziali delle funzioni analitiche di due o più variabili complesse*, Ann. Mat. Pura e Appl., 17(1910), 61–88.

[283] J. Lewandowski and P. Nurowski, *Cartan's chains and Lorentz geometry*, preprint, 1989.

[284] H. Lewy, *On the local character of the solution of an atypical linear differential equation in three variables and a related theorem for regular functions of two complex variables*, Ann. of Math., 64(1956), 514–522; *An example of a smooth linear partial differential equation without solution*, Ann. of Math., 66(1957), 155–158.

[285] P. Liebermann, *Sur les automorphismes infinitésimaux des structures symplectiques et des structures de contact*, Colloque de géométrie différentielle globale, Bruxelles, 1958.

[286] A. Lichnerowicz, *Géométrie des groupes de transformations*, Dunod, Paris, 1958.

[287] A. Lichnerowicz, *Applications harmoniques et variétés Kählériennes*, Rend. Sem. Mat. Fis. Milano, 39(1969), 1–12.

[288] F. Loose, *A remark on the reduction of Cauchy-Riemann manifolds*, Math. Nachr., 214(2000), 39–51.

[289] G. Lu, *Weighted Poincaré and Sobolev inequalities for vector fields satisfying Hörmander's condition and applications*, Rev. Mat. Iberoamericana, 8(1992), 367–439.

[290] H.S. Luk, *Normalization of differential equations defining a CR structure*, Invent. Math., 48(1978), 33–60.

[291] G. Lupacciolu and G. Tomassini, *An extension theorem for CR functions*, Ann. Mat. Pura Appl., (4)137(1984), 257–263.

[292] R.A. Macías and C. Segovia, *A decomposition into atoms of distributions on spaces of homogeneous type*, Adv. Math., 33(1979), 271–309.

[293] S. Marchiafava, *Su alcune sottovarietà che ha interesse considerare in una varietà Kähleriana quaternionale*, Rendiconti di Matem., 10(1990), 493–529.

[294] A. Martinez, *Totally complex submanifolds of quaternionic projective space*, preprint, Departamento de Geometria y Topologia, Facultad de Ciencias, Universidad de Granada, 18071 Granada, Spain, 1987.

[295] A. Martinez, J.D. Pérez, and F.G. Santos, *Resultados recientes sobre la geometria de las hipersuperficies reales del espacio proyectivo cuaternionico*, Hom. Pablo Bobillo Guerrero, 1992, 183–190.

[296] A. Masiello, *On the existence of a closed geodesic in stationary Lorentz manifolds*, J. Diff. Equations, (1)104(1993), 48–59

[297] A. Masiello, *Convex regions of Lorentzian manifolds*, Annali di Matem. Pura ed Appl., (4)CLXVII(1994), 299–322.

[298] C. Medori, *Levi-Tanaka algebras and CR manifolds*, Ph.D. Thesis, International School for Advanced Studies, Trieste, 1995/96.

[299] C. Medori and M. Nacinovich, *Pluriharmonic functions on abstract CR manifolds*, preprint, Univ. di Pisa, Sezione di Geometria e Algebra, 1.78(727), 1993.

[300] A. Menikoff and J. Sjöstrand, *On the eigenvalues of a class of hypoelliptic operators*, Math. Ann., 235(1978), 55–58.

[301] L.G. Mikhailov and A.V. Abrosimov, *Generalized Cauchy-Riemann system with several independent variables*, Soviet Math. Dokl., 14(1973), 681–685.

[302] J.J. Millson, *CR geometry and deformations of isolated singularities*, Differential geometry: geometry in mathematical physics and related topics (Los Angeles, CA, 1990), 411–433, Proc. Sympos. Pure Math., 54, Part 2, Amer. Math. Soc., Providence, RI, 1993.

[303] R.R. Miner, *Spherical CR manifolds with amenable holonomy*, preprint, Univ. of Maryland, College Park, 1990.

[304] C. Miranda, *Equazioni alle derivate parziali di tipo ellittico*, Springer-Verlag, 1955.

[305] J. Mitchell, *On Carnot-Carathéodory metrics*, J. Diff. Geometry, 21(1985), 35–45.

[306] K. Miyajima, *Deformations of a complex manifold near a strongly pseudo-convex real hypersurface and a realization of Kuranishi family of strongly pseudo-convex CR structures*, Math. Z., 205(1990), 593–602.

[307] K. Miyajima, *Completion of Akahori's construction of the versal family of strongly pseudo-convex CR structures*, Trans. A.M.S., (1)277(1983), 163–172.

[308] K. Miyajima, *Subellipticity at higher degree of a boundary condition associated with construction of the versal family of strongly pseudo-convex domains*, Publ. RIMS, Kyoto Univ., 23(1987), 337–344.

[309] K. Miyajima, *On realizations of families of strongly pseudo-convex CR structures*, Tokyo J. Math., (1)15(1992), 153–170.

[310] K. Miyajima, *On the existence of Kuranishi family for deformations of holomorphic maps*, Sci. Rep. Kagoshima Univ., 27(1978), 43–76.

[311] K. Miyajima, *Subellipticity of a boundary condition associated with construction of a versal family of complex structures on a domain with non-degenerate boundary*, preprint, 26, Kagoshima Univ., 1988.

[312] R.I. Mizner, *Almost CR structures, f-structures, almost product structures, and associated connections*, Rocky Mountain J. Math., (4)23(1993), 1337–1359.

[313] R.I. Mizner, *CR structures of codimension* 2, J. Diff. Geometry, 30(1989), 167–190.

[314] P. Mollino, *Théorie des G-structures: le problème d'equivalence*, Lecture Notes in Math., vol. 588, Springer-Verlag, Berlin-Heidelberg-New York, 1976.

[315] P. Molino, *Riemannian foliations*, Progress in Mathematics, vol. 73, ed. by J. Oesterlé and A. Weinstein, Birkhäuser, Boston-Basel, 1988.

[316] Y. Morimoto and T. Nagano, *On pseudo-conformal transformations of hypersurfaces*, J. Math. Soc. Japan, 15(1963), 289–300.

[317] J. Moser, *On Harnack's theorem for elliptic differential equations*, Comm. Pure Appl. Math., 14(1961), 577–591.

[318] D. Müller, M.M. Peloso, and F. Ricci, *On the solvability of homogeneous left-invariant diffrential operators on the Heisenberg group*, Journal of Functional Analysis, 148(1997), 368–383; *On local solvability for complex coefficient differential operators on the Heisenberg group*, J. reine angew. Math., 513(1999), 181–234.

[319] E. Musso, *Homogeneous pseudo-hermitian Riemannian manifolds of Einstein type*, American J. Math., 113(1990), 219–241.

[320] M. Nacinovich, *On the absence of Poincaré lemma for some systems of partial differential equations*, Compositio Math., 44(1981), 241–303.

[321] M. Nacinovich, *Poincaré lemma for tangential Cauchy-Riemann complexes*, Math. Ann., 268(1984), 449–471.

[322] M. Nacinovich and G. Valli, *Tangential Cauchy-Riemann complexes on distributions*, Annali Matem. Pura Appl., (4)CXLVI(1987), 123–160.

[323] A. Nagel, E.M. Stein, and S. Wainger, Balls and metrics defined by vector fields I: basic properties, Acta Math., 155(1985), 103–147.
[324] A. Nagel and E.M. Stein, *Lectures on pseudo-differential operators*, Princeton University Press, Princeton, N.J., 1979.
[325] M.S. Narasimhan and S. Ramanan, *Existence of universal connections*, American J. Math., 83(1961), 563–572, 85(1963), 223–231.
[326] L. Nirenberg, *On a question of Hans Lewy*, Russian Math. Surveys, 29(1974), 251–262.
[327] A. Noell, *Interpolation from curves in pseudoconvex boundaries*, Michigan Math. J., (2)37(1990), 275–281.
[328] M. Obata, *Conformal transformations of Riemannian manifolds*, J. Diff. Geometry, 4(1970), 311–333.
[329] M. Obata, *Affine connections on manifolds with almost complex, quaternion or Hermitian structure*, Japanesse J. Math., 26(1957), 43–77.
[330] M. Obata, *The conjectures on conformal transformations of Riemannian manifolds*, J. Differential Geometry, 6(1971), 247–258.
[331] M. Obata, *Certain conditions for a Riemannian manifold to be isometric with a sphere*, J. Math. Soc. Japan, 14(1962), 333–340.
[332] M. Okumura, *Certain almost contact hypersurfaces in Euclidean spaces*, Kodai Math. Sem. Reports, 16(1964), 44–54; *Certain almost contact hypersurfaces in Kaehlerian manifolds of constant holomorphic sectional curvatures*, Tôhoku Math. J., 16(1964), 270–284; *Contact hypersurfaces in certain Kaehlerian manifolds*, ibidem, (1)18(1966), 74–102; S.I. Goldberg, *Totally geodesic hypersurfaces of Kähler manifolds*, Pacific J. Math., 27(1968), 275–281.
[333] O.A. Oleinik, *Linear second order equations with non-negative characteristic form*, Mat. Sb., 69(1966), 111–140, English translation in Amer. Math. Soc. Transl., 65(1967), 167–199.
[334] B. O'Neill, *The fundamental equations of a submersion*, Michigan Math. J., 13(1966), 459–469.
[335] B. O'Neill, *Semi-Riemannian geometry*, Academic Press, New York-London-Paris-San Diego-San Francisco-Sao Paulo-Sydney-Tokyo-Toronto, 1983.
[336] R.S. Palais, *A global formulation of the Lie theory of transformation groups*, Memoirs of A.M.S., no. 22, 1957.
[337] R.S. Palais, *Foundations of global nonlinear analysis*, Math. Lecture Notes Ser., W.A. Benjamin, Inc., New York-Amsterdam, 1968.
[338] P. Pansu, *Métriques de Carnot-Carathéodory et quasiisométries des espaces symétriques de rang un*, Ann. of Math., 129(1989), 1–60.
[339] A. Parmeggiani, *Esistenza e regolarità C^∞ al bordo per il problema di Dirichlet per operatori subellittici di tipo Fefferman-Phong*, Sem. Analisi Matem., Dip. Matem. Univ. Bologna, A.A. 1994/95, 55–67.
[340] A. Parmeggiani and C.J. Xu, *The Dirichlet problem for subelliptic second order equations*, to appear in Annali Matem. Pura Appl.
[341] R. Penrose, Gen. Rel. Grav., 7(1976), 31–52 and 171–176.
[342] R. Penrose, *Physical space-time and nonrealizable CR structures*, Bull. A.M.S., N.S., 8(1983), 427–448; *The twistor programme*, Rep. Math. Phys., 12(1977), 65–76.
[343] D. Perrone, 5-*Dimensional contact manifolds with second Betti number* $b_2 = 0$, Tôhoku Math. J., (1)41(1989), 163–170; *A remark on homogeneous contact five-manifolds*, Boll. U.M.I., (7)3-A(1989), 231–235; *Torsion and critical metrics on contact three-manifolds*, Kodai Math. J., (1)13(1990), 88–100; *Torsion tensor and critical metrics on contact* $(2n+1)$*-manifolds*, Monatshefte Math., 114(1992), 245–259; *Contact Riemannian manifolds satisfying* $R(X, \xi) \cdot R = 0$, Yokohama Math. J., 39(1992), 141–149; *Tangent*

sphere bundles satisfying $\nabla_\xi \tau = 0$, Journal of Geometry, 49(1994), 178–188; *Ricci tensor and spectral rigidity of contact Riemannian 3-manifolds*, Bull. Inst. Math. Acad. Sinica, (2)24(1996), 127–138.

[344] D. Perrone, *Homogeneous contact Riemannian three-manifolds*, Illinois Journal of Mathematics, (2)42(1998), 243–256.

[345] D. Perrone and L. Vanhecke, *Five-dimensional homogeneous contact manifolds and related problems*, Tôhoku Math. J., (2)43(1991), 243–248.

[346] R. Petit, *Harmonic maps and strictly pseudoconvex CR manifolds*, Communications in Analysis and Geometry, (3)10(2002), 575–610.

[347] S. Pinchuk, *On proper holomorphic mappings of strictly pseudoconvex domains*, Siberian Math. J., 15(1974), 644–649.

[348] G. Pitis, *Cohomology of anti-quaternion submanifolds of a quaternion Kaehlerian manifold*, Tensor, N.S., 42(1985), 145–150.

[349] H. Poincaré, *Sur les fonctions de deux variables*, Acta Mathem., 2(1883), 97–113.

[350] H. Poincaré, *Les functions analytiques de deux variables et la représentation conforme*, Rend. Circ. Mat. Palermo, 23(1907), 185–220.

[351] E.V. Radkevic, *Hypoelliptic operators with multiple characteristics*, Math. USSR Sb., 8(1969), 181–205.

[352] N.M. Rivière, *On singular integrals*, Bull. Amer. Math. Soc., 75(1969), 843–847.

[353] J.P. Rosay, *Sur une caractérisation de la boule parmi les domaines de* $\mathbf{C}^n$ *par son groupe d'automorphismes*, Ann. Inst Fourier Grenoble, (4)29(1979), 91–97.

[354] H. Rossi, *Homogeneous strongly pseudoconvex hypersurfaces*, Rice Studies, (3)59(1973), 131–145.

[355] H. Rossi, *LeBrun's nonrealizability theorem in higher dimensions*, Duke Math. J., (2)52(1985), 457–474.

[356] H. Rossi and M. Vergne, *Équations de Cauchy-Riemann tangentielles associées a un domaine de Siegel*, Ann. Scient. Éc. Norm. Sup., 9(1976), 31–80; *Tangential Cauchy-Riemann equations associated with a Siegel domain*, Proc. Sympos. Pure Math., 30(1977), 303–308.

[357] M. Rumin, *Un complexe de formes différentielles sur les variétés de contact*, C.R. Acad. Sci. Paris, 310(1990), 401–404.

[358] M. Rumin, *Formes différentielles sur les variétés de contact*, preprint, 1992.

[359] J-S. Ryu and S. Yi, *Derivations on CR manifolds*, Commun. Korean Math. Soc., (1)19(2004), 135–141.

[360] H. Samelson, *A class of complex-analytic manifolds*, Portugalie Math., 12(1953), 129–132.

[361] J.H. Sampson, *Harmonic maps in Kähler geometry*, CIME Conf., 1984, Lecture Notes in Math., Vol. 1161, 1985, 193–205.

[362] A. Sánchez-Calle, *Estimates for kernels associated to some subelliptic operators*, Thesis, Princeton University, 1983.

[363] A. Sánchez-Calle, *Fundamental solutions and geometry of the sum of squares of vector fields*, Invent. Math., 78(1984), 143–160.

[364] S. Sasaki and C.J. Hsu, *On a property of Brieskorn manifolds*, Tôhoku Math. J., 28(1976), 67–78.

[365] R. Schoen, *Conformal deformation of a Riemannian metric to constant scalar curvature*, J. Diff. Geometry, 20(1984), 479–495.

[366] R. Schoen, *On the conformal and CR automorphism groups*, Geom. Funct. Anal., (2)5(1995), 464–481.

[367] L. Schwartz, *Théorie des noyaux*, Proc. Internat. Congress of Mathematicians, Vol. 1, 220–230, American Mathematical Society, Providence, 1952.

[368] B. Segre, *Questioni geometriche legate colla teoria delle funzioni di due variabili complesse*, Rend. Sem. Mat. Roma, 7(1931), 59–107; *Intorno al problema di Poincaré della rappresentazione pseudoconforme*, Rend. Accad. Lincei, 13(1931), 676–683.

[369] F. Severi, *Risoluzione generale del problema di Dirichlet per le funzioni biarmoniche*, Rend. Accad. Naz. Lincei, 13(1931), 795–804.

[370] F. Severi, *Lezioni sulle funzioni analitiche di più variabili complesse*, C.E.D.A.M., Padova, 1958.

[371] Y. Shibuya, *Real submanifolds in a quaternionic projective space*, Kodai Math. J., 1(1978), 421–439.

[372] N. Shimakura, *Partial differential operators of elliptic type*, Translations of Mathematical Monographs, Vol. 99, American Mathematical Society, Providence, Rhode Island, 1992.

[373] H.B. Shutrick, *Complex extensions*, Quart. J. Math. Oxford Ser., (2)9(1958), 181–201.

[374] Y.T. Siu, *The complex analyticity of harmonic maps and the strong rigidity of compact Kähler manifolds*, Ann. of Math., 112(1980), 73–111.

[375] D.M. Snow, *Invariant complex structures on reductive Lie groups*, J. Reine Angew. Math., 371(1986), 191–215.

[376] F. Sommer, *Komplex analytische Blätterung reeler Mannigfaltigkeiten im* $\mathbf{C}^n$, Math. Ann., 136(1958), 111–133.

[377] G. Sparling, *Twistor theory and the characterization of Fefferman's conformal structures*, Preprint, Univ. of Pittsburg, 1975.

[378] C.M. Stanton, *Intrinsic connections for Levi metrics*, preprint, Indiana University at South Bend, 1983.

[379] N.K. Stanton, *A normal form for rigid hypersurfaces in* $\mathbf{C}^2$, American J. Math., 113(1991), 877–910.

[380] N.K. Stanton, *Infinitesimal CR automorphisms of rigid hypersurfaces in* $\mathbf{C}^2$, J. Geom. Anal., (3)1(1991), 231–267.

[381] N.K. Stanton, *Infinitesimal CR automorphisms of real hypersurfaces*, American J. Math., 117(1995), 141–167.

[382] N. Steenrod, *The topology of fibre bundles*, Princeton Math. Series, 14, ed. by L.A. Caffarelli, J.N. Mather, and E.M. Stein, Princeton Univ. Press, Princeton, New Jersey, 1951.

[383] E.M. Stein, *Singular integrals and estimates for the Cauchy-Riemann equations*, Bull. Amer. Math. Soc., 79(1973), 440–445.

[384] E.M. Stein, *Singular integrals and differentiability properties of functions*, Princeton Univ. Press, Princeton, N.J., 1970.

[385] E.M. Stein, *Harmonic analysis: real-variable methods, orthogonality, and oscillatory integrals*, Princeton Mathem. Ser., 43, Princeton Univ. Press., 1993.

[386] M.B. Stenzel, *Orbits of the geodesic flow and chains on the boundary of a Grauert tube*, Math. Ann., 322(2002), 383–399.

[387] M.B. Stenzel and V. Guilemin, *Grauert tubes and the homogeneous Monge-Ampère equation*, J. Diff. Geometry, 34(1991), 561–570.

[388] S. Sternberg, *Lectures on differential geometry*, Prentince Hall, 1962.

[389] W. Stoll, *The characterization of strictly parabolic manifolds*, Ann. Sc. Norm. Sup., Pisa, (1)7(1980), 87–154.

[390] E.L. Stout, *Interpolation manifolds*, Ann. of Math. Studies, 100(1981), 373–391.

[391] R.S. Strichartz, *Sub-Riemannian geometry*, J. Diff. Geometry, 24(1986), 221–263.

[392] M. Struwe, *The Yang–Mills flow in four dimensions*, Calc. Var., 2(1994), 123–150.

[393] R. Szöke, *Complex structures on tangent bundles of Riemannian manifolds*, Math. Ann., 291(1991), 409–428.

[394] G. Taiani, *Cauchy-Riemann (CR) manifolds*, Pace University, New York, 1989.

[395] T. Takahashi, *Minimal immersions of Riemannian manifolds*, J. Math. Soc. Japan, 18(1966), 380–385.

[396] G. Talenti, *Best constant in Sobolev inequality*, Ann. Mat. Pura Appl., 110(1976), 353–372.

[397] N. Tanaka, *On generalized graded Lie algebras and geometric structures*, J. Math. Soc. Japan, 19(1967), 215–254.

[398] N. Tanaka, *A differential geometric study on strongly pseudo-convex manifolds*, Kinokuniya Book Store Co., Ltd., Kyoto, 1975.

[399] P. Tang, *Quasiconformal homeomorphisms on CR 3-manifolds with symmetries*, Math. Z., (1)219(1995), 49–69; *Regularity and extremality of quasiconformal homeomorphisms on CR 3-manifolds*, Ann. Acad. Sci. Fenn. Math., (2)21(1996), 289–308.

[400] S. Tanno, *A theorem on regular vector fields and its applications to almost contact structures*, Tôhoku Math. J., 17(1965), 235–238.

[401] S. Tanno, *Variational problems on contact Riemannian manifolds*, Trans. A.M.S., (1)314(1989), 349–379.

[402] Y. Tashiro, *On contact structures of tangent sphere bundles*, Tohoku Math. J., 21(1969), 117–143.

[403] G. Tomassini, *Tracce delle funzioni olomorfe sulle sottovarietà analitiche Reali d'una varietà complessa*, Ann. Sc. Norm. Sup., Pisa, 20(1966), 31–43.

[404] G. Tomassini, *Gaetano Fichera's contribution to complex analysis*, Homage to Gaetano Fichera, 325–333, Quad. Mat., 7, Dept. Math., Seconda Univ. Napoli, Caserta, 2000.

[405] G. Tomassini, *Pseudoconformal invariants and differential geometry*, Geometry and complex variables (Bologna, 1988/1990), 369–373, Lecture Notes in Pure and Appl. Math., 132, Dekker, New York, 1991.

[406] G. Tomassini, *Extension d'objets CR*, C.R. Acad. Sci. Paris, Sér. I Math., (1)301(1985), 11–13.

[407] G. Tomassini, *Complessificazione di un gruppo di Lie reale*, Ann. Scuola Norm. Sup. Pisa, (3)22(1968), 101–106.

[408] P. Tondeur, *Foliations of Riemannian manifolds*, Springer-Verlag, New York-Heidelberg-London-Paris-Tokyo, 1988.

[409] N. Trudinger, *Remarks concerning the conformal deformation of Riemannian structures on compact manifolds*, Ann. Scuola Norm. Sup. Pisa, (4)22(1968), 265–274.

[410] K. Uhlenbeck, *Regularity for a class of non-linear elliptic systems*, Acta Mathematica, 138(1977), 219–240.

[411] K. Uhlenbeck and S.T. Yau, *On the existence of hermitian Yang–Mills connections in stable vector bundles*, Commun. Pure Appl. Math., 39(1986), S257–S293.

[412] H. Urakawa, *Yang–Mills connections over compact strongly pseudo-convex CR manifolds*, Math. Z., 216(1994), 541–573.

[413] H. Urakawa, *Variational problems over strongly pseudoconvex CR manifolds*, Differential Geometry, Proceedings of the Sympos. in honour of Prof. Su Buchin on his 90th birthday, 233–242, Shanghai, China, September 17–23, 1991, ed. by C.H. Gu, H.S. Hu, and Y.L. Xin, World Scientific Publ. Co. Pte. Ltd., Singapore-New Jersey-London-Hong Kong, 1993.

[414] H. Urakawa, *Yang–Mills connections and deformation theory over compact strongly pseudoconvex CR manifolds*, Geometric Complex Analysis, ed. by Junijiro Noguchi et al., 1996, World Scientific Publ. Co., 635–652.

[415] I. Vaisman, *On locally and globally conformal Kaehler manifolds*, Trans. A.M.S., (2)262(1980), 533–542.

[416] I. Vaisman, *New examples of twisted cohomologies*, Bolletino U.M.I., (7)**7**-B(1993), 355–368.

[417] J. Veloso and J. Verderesi, *Three dimensional Cauchy-Riemann manifolds*, preprint, Univ. de S. Paulo, Brasil, 1978.

[418] I.N. Vekua, *Generalized analytic functions*, Pergamon Press, Oxford-Addison Wesley-Reading, 1962.

[419] L. Wang, *The globally homeomorphic solutions to Beltrami systems in* $\mathbf{C}^n$, Math. Z., 216(1994), 371–377.

[420] Y. Wang, *On some properties of Riemannian spaces of quasi-constant curvature*, Tensor, N.S., 35(1981), 173–176.

[421] F.W. Warner, *Extension of the Rauch comparison theorem to submanifolds*, Trans. A.M.S., 122(1966), 341–356.

[422] S.M. Webster, *Pseudohermitian structures on a real hypersurface*, J. Diff. Geometry, 13(1978), 25–41.

[423] S.M. Webster, *On the transformation group of a real hypersurface*, Trans. A.M.S., (1)231(1977), 179–190.

[424] S.M. Webster, *The rigidity of CR hypersurfaces in a sphere*, Indiana Univ. Math. J., (3)28(1979), 405–416.

[425] S.M. Webster, *On the pseudo-conformal geometry of a Kähler manifold*, Math. Z., 157(1977), 265–270.

[426] S.M. Webster, *On the Moser normal form at a non-umbilic point*, Math. Ann., 233(1978), 97–102.

[427] S.M. Webster, *Real submanifolds of* $\mathbf{C}^n$ *and their complexifications*, preprint, 1987.

[428] S.M. Webster, *On the local solution of the tangential Cauchy-Riemann equations*, Ann. Inst. Henri Poincaré, (3)6(1989), 167–182.

[429] S.M. Webster, *On the proof of Kuranishi's embedding theorem*, Ann. Inst. Henri Poincaré, (3)6(1989), 183–207.

[430] A. Weinstein, *Lectures on symplectic manifolds*, Memoirs of A.M.S., No. 29, Providence, RI, 1977.

[431] R.O. Wells, Jr., *Differential analysis on complex manifolds*, Springer-Verlag, New York-Heidelberg-Berlin, 1980.

[432] R.O. Wells, Jr., *The Cauchy-Riemann equations and differential geometry*, Bull. A.M.S., (2)6(1982), 187–199.

[433] J.C. Wood, *Harmonic morphisms, foliations and Gauss maps*, Contemporary Mathematics, 49(1986), 145–184.

[434] B. Wong, *Characterization of the unit ball in* $\mathbf{C}^n$ *by its autmorphism group*, Invent. Math., 41(1977), 253–257.

[435] Pit-Mann Wong, *On umbilical hypersurfaces and uniformization of circular domains*, Proceedings of Symposia in Pure Math., 41(1984), 225–252.

[436] Pit-Mann Wong, *Geometry of the complex homogeneous Monge-Ampère equation*, Invent. Math., 67(1982), 261–274.

[437] Y.L. Xin, *Some results on stable harmonic maps*, Duke Math. J., 47(1980), 609–613.

[438] C.J. Xu, *Régularité des solutions d'équations aux dérivées partielles non linéaires associées à un système de champs de vecteurs*, Ann. Inst. Fourier, 37(1987), 105–113.

[439] C.J. Xu, *Opérateurs sous-elliptiques et régularité des solutions d'équations aux dérivées partielles non linéaires du second ordre en deux variables*, Comm. P.D.E., 11(1986), 1575–1603.

[440] C.J. Xu, *Subelliptic variational problems*, Bull. Soc. Math. France, 118(1990), 147–159.

[441] C.J. Xu, *Regularity for quasilinear second order subelliptic equations*, Comm. Pure Appl. Math., XLV(1992), 77–96.

[442] C.J. Xu and C. Zuily, *Smoothness up to the boundary for solutions of the nonlinear and nonelliptic Dirichlet problem*, Trans. Amer. Math. Soc., (1)308(1988), 243–257.

[443] C-J. Xu and C. Zuily, *Higher interior regularity for quasilinear subelliptic systems*, Calc. Var., 5(1997), 323–343.

[444] H. Yamabe, *On a deformation of Riemannian structures on compact manifolds*, Osaka Math. J., 12(1960), 21–37.

[445] K. Yano and M. Kon, *Differential geometry of CR submanifolds*, Geometriae Dedicata, 10(1981), 369–391.

[446] K. Yano and M. Kon, *CR submanifolds of Kaehlerian and Sasakian manifolds*, Progress in Math., vol. 30, ed. by J. Coates and S. Helgason, Birkhäuser, Boston-Basel-Stuttgart, 1983.

[447] K. Yano and M. Kon, *Contact CR submanifolds*, Kodai Math. J., 5(1982), 238–252.

[448] K. Yano and M. Obata, *Conformal changes of Riemannian metrics*, J. Diff. Geometry, 4(1970), 53–72.

[449] Z.R. Zhou, *Uniqueness of subelliptic harmonic maps*, Annals of Global Analysis and Geometry, 17(1999), 581–594.

Index

1-parameter variation, 247
F-frame, 333
F-pseudoharmonic map, 246, 256
G-structure, 68
K-quasiconformal map, 384
n-classifying object, 143
(formally) integrable (almost CR structure), 3

admissible, 258
  (local coframe), 79
  (number $\alpha \in \mathbf{C}$), 184
almost complex manifold, 3
almost contact
  metric structure, 10
  structure, 10
almost CR
  manifold, 3
  structure, 3
amenable, 61
anti-invariant distribution, 41
a priori estimates, 116

Bargmann representation, 397
basic connection, 99
Beltrami
  differential, 381
  equations, 403
Bergman
  kernel, 109, 148
  metric, 398
best constant, 158
Bianchi identity, 46
biholomorphic invariant, 150
Bochner type formulas, 431
Bockstein sequence, 156, 320
Boothby–Wang theorem, 318
Bott connection, 73
Brieskorn sphere, 358
bundle of densities, 161

Calderón–Zygmund theorem, 192
  generalized, 455
canonical 1-form, 21
canonical S-connection, 409
canonical bundle, 84, 119, 416
canonical circle bundle, 119
Carnot–Carathéodory
  ball, 258
  distance, 258
Cartan decomposition, 390
Cartan's formula, 403
Cayley transform, 14, 161
chain, 110
chains, 134
characteristic
  class, 143
  coordinates, 15
  direction, 9
  form, 99, 144
  ring, 317
Chern
  classes, 99, 297
  connection, 128
  CR structure bundle, 128
  tensor field, 61
Chern–Weil theory, 143
Christoffel symbols, 32
classical spinning photon, 23

classifying space, 143
cochain complex, 74
Codazzi equation, 224
compatible (Riemannian metric), 10
complex Bott connection, 83
complex dilatation, 380
complexified tangent bundle, 3
conformal structure, 22
congruence relation, 66
connection form, 413
connection-preserving bundle map, 143
constrained variational
  principle, 157
  problem, 160
contact condition, 10
contact deformation, 405
contact form, 43
contact homothetic (pseudo-Hermitian manifolds), 341
contact metric structure, 10
contact Riemannian geometry, 2
contact transformation, 378
convolution product, 189
CR
  automorphism, 42
  codimension, 3
  dimension, 3
  equivalence, 5
  function, 2
  invariant, 110, 142
  invariant Laplacian, 161, 207
  isomorphism, 5, 35
  Lie algebra, 18
  Lie group, 19, 35
  manifold, 4
  map, 4
  normal bundle, 351
  structure, 4
  submanifold, 345
  submanifold (of a Hermitian manifold), 41
  Yamabe equation, 160, 199, 225
  Yamabe problem, 111, 157
CR-holomorphic (function), 81
CR-holomorphic vector bundle, 83
CR-invariant property, 7
CR-pluriharmonic function, 280, 414
curvature form, 413
  (of an S-connection), 421
cut-off function, 261

de Rham cohomology groups, 84
de Rham complex, 73, 84
degenerate elliptic operator, 113, 207
dilation, 12
Dirac distribution, 177
Dirac field, 25
Dirichlet energy functional, 218
Dirichlet problem
  for the complex Monge–Ampére equation, 109, 147
  for the subelliptic harmonic maps equation, 231
discrete group, 313
distribution function, 450
divergence, 111
Dombrowski map, 322
double-dual identification, 21
doubling constant, 259
doubling measure, 259

eigenfunction, 227
Einstein manifold, 134
elliptic space form, 321
embeddable, 14
embedded (CR manifold), 14
embedding problem, 15
energy surface, 21
error terms, 458
Euler–Lagrange equation, 157
Euler–Lagrange equations, 217
exact sequence (of cohomology groups), 95
extremal functions, 158

Fefferman
  bundle, 119
  metric, 109, 128, 216, 232, 239
Fefferman's asymptotic expansion, 399
fiber coordinate, 122
field of vision, 24
filtration, 73, 84
first variation formula, 247
Folland–Stein
  normal coordinates, 165, 348
  operator, 177
  space, 193
  space $S_k^p(M)_{\rm loc}$, 456
formal (Frobenius) integrability, 70
Fourier transform, 183

frame constant, 174
Frölicher spectral sequence, 73
of a CR manifold, 85
Fuglede–Ishihara theorem, 229
fundamental solution, 176
for the Hörmander operator, 260
fundamental tensors (of a Riemannian submersion), 319

Gauss
equation, 155, 224
formula, 353
generalized CR function, 82
generic (energy surface), 21
generic submanifold, 5
globally embeddable, 15
Graham–Lee connection, 135
Greenleaf's
formula, 227, 228, 360
theorem, 434

Hamiltonian
form, 21
vector field, 405
harmonic
map, 211, 217
morphism, 229, 231
polynomial, 228
Harnack inequality, 198
Heisenberg
group, 11, 34
norm, 13, 313
type order, 166
helicity, 23
Hermitian connection, 408
Higgs field, 25
Hodge operator, 141
holomorphic
distribution, 41
extension problem, 81
foliation, 73, 83
map, 211
tangent bundle, 5
homogeneous CR manifold, 343
homogeneous distribution, 191
homogeneous function, 191
homogeneous pseudo-Hermitian manifold, 341
horizontal gradient, 111
hyper contact form, 333
hyper CR manifold, 332
hyperkählerian manifold, 331
hyperquadric, 351
hypersurface type, 5
hypoelliptic, 111, 189
hypoelliptic operator, 114
Hörmander
condition, 233
operator, 113, 225, 231, 233
system, 225, 233

induced connection, 213
infinitesimal CR automorphism, 42
inhomogeneous Yang–Mills equation, 84, 410
inner commutation relations, 425
integrable (G-structure), 70
integral class, 156
interpolation inequality, 199
invariant polynomial, 143
inverse Pontryagin polynomial, 152
isopseudo-Hermitan, 62
immersion, 345
map, 7
isotropy group, 63

Jerison–Lee theorem, 160, 205

kernel of type λ, 455
Kohn–Rossi cohomology, 73, 74, 309
Kohn–Rossi cohomology groups, 85, 88
Kohn–Rossi Laplacian, 73, 100, 104
Kohn Rossi operator, 177
Kähler metric, 41
Kählerian manifold, 41
Küneth formula, 320

Laplace–Beltrami operator, 217
Lee class, 303, 346, 361
Lee conjecture, 304
left-invariant CR structure, 20
left-invariant pseudo-Hermitian structure, 20
Levi
distribution, 4
flat, 70
form, 5, 7
metric, 106
Lewy operator, 12, 231, 403
Libermann's theorem, 404

light-like geodesic, 135
locally embeddable, 14
locally trivial (CR-holomorphic vector bundle), 83
Lorentz
frame, 144
metric, 128

Marcinkiewicz interpolation theorem, 450
maximal complex distribution, 4
mean value, 191
minimal immersion, 228
minimality (of a distribution), 37
minimizing sequence, 159, 200
Minkowski space, 22
momentum (of a massless particle in flat space-time), 23
Monge–Ampère equation, 151

natural lift, 213
Nijenhuis torsion, 10
nondegenerate (CR manifold), 6
nonlinear connection, 321
normal, 10
normal form, 134, 142
normal Tanaka–Webster connection, 346
null geodesic, 110, 134

operator of type λ, 456
outer commutation relations, 425

parabola (tangent to a CR manifold), 359
parabolic geodesic, 229, 358
parametrix (of the Kohn–Rossi operator), 448
Pauli–Lubanski spin vector, 23
pluriharmonic function, 286
Poincaré
inequality, 198
polynomial, 135
Poisson kernel, 294
Pontryagin form, 110, 142, 154
potential theory, 208
projective twistor space, 23
proper (CR submanifold), 41
properly discontinuous group, 152, 313
pseudo-Einstein, 275
pseudo-Einsteinian
manifold, 275
structure, 361

pseudo-Hermitian
frame, 165
geometry, 2
Hessian, 116
holonomy group, 329
immersion, 68, 345, 354
manifold, 7
map, 7
Ricci curvature, 56
Ricci tensor, 50, 275
scalar curvature, 50, 275
space form, 62
structure, 5
torsion, 26, 36
pseudo-Siegel domain, 131, 375
pseudocomplex, 73
pseudoconformal curvature tensor, 61
pseudoharmonic
map, 111, 212, 213
morphism, 229, 252
pseudoharmonicity, 211
pullback bundle, 212
purely tangential first-order differential operators, 1

quadric submanifold, 20
quantum bundle, 412
quasi-constant curvature, 325
quaternionic
contact form, 333
CR submanifold, 332
distribution, 332
Sasakian manifold, 334
submanifold, 332
-Kähler manifold, 331

Radon–Penrose transform, 25
realizable, 14
realized (CR manifold), 14
reflection operator, 190
regular (vector field), 318
regular distribution, 192
regular homogeneous distribution, 190
restricted conformal class, 110, 135
Ricci operator, 56
Riemann–Christoffel tensor, 51
Riemannian
foliation, 96
submersion, 319

Yamabe problem, 157

Sasaki metric, 10, 229, 322
Sasakian space-form, 360
Sasakian structure, 10
Schwartz class, 397
second fundamental form, 155
sheaf (of CR-pluriharmonic functions), 300
Siegel domain, 14, 403
singular integral
kernel, 194, 454
operator, 194
Siu–Sampson formula, 211
slice, 318
Sobolev
inequality, 158
lemma, 159, 176
norm, 114
space, 159
-type space, 261
spectral asymptotics, 428
spectrum (of the sub-Laplacian), 228
spherical (CR manifold), 61
spinor calculus, 111
standard contact vector (of the sphere), 228
strictly pseudoconvex (CR manifold), 6
sub-Laplacian, 111, 212
subelliptic, 111
p-harmonic map, 246
p-Laplacian, 257
harmonic map, 111, 231
harmonic morphism, 231, 232
operator, 114
theory, 111
submanifold of finite type, 228
symmetric homogeneous polynomial, 99
symplectomorphism, 399

Tanaka connection, 407, 420
Tanaka–Webster connection, 26, 241
tangent sphere bundle, 229, 321
tangential Beltrami equations, 385
tangential Cauchy–Riemann
equations, 1, 81
operator, 73
tautologous form, 120
theorem of Fubini, 141
third-order covariant derivative, 424
time-like curve, 135
totally real foliation, 41
totally umbilical, 223
trace (of a pluriharmonic function), 287
trace norm, 397
traceless Ricci tensor, 299
transgression form, 143, 144
transition functions, 83
translation operator, 190
transverse Levi-Civita connection, 96
transverse symmetry, 309
triangle inequalities, 13
tube, 134
twistor space, 23
twistorial construction, 21
type (of an almost CR structure), 3

uniformly elliptic operator, 198
universal Weil homomorphism, 143
unsolvability (of the Lewy operator), 12

vanishing theorem, 97, 99
volume form, 43
volume-normalized, 276

wave operator, 140, 217, 225
weak maximum principle, 207
weak pseudoconvexity locus, 132
weak type operator, 450
Webster metric, 9
Weingarten
equation, 222
formula, 353
operator, 154, 223
Weyl connection, 317

Yamabe
equation, 157
problem, 142
Yamabe–Trudinger–Aubin theorem, 158
Yang–Mills
equation, 410
field, 25
functional, 410